UAVs for Vegetation Monitoring

UAVs for Vegetation Monitoring

Editors

Ana I. de Castro
Yeyin Shi
Joe Maja
Jose M. Peña

MDPI • Basel • Beijing • Wuhan • Barcelona • Belgrade • Manchester • Tokyo • Cluj • Tianjin

Editors
Ana I. de Castro
Plant Protection Department
National Agricultural and Food Research and Technology Institute
CSIC
Madrid
Spain

Yeyin Shi
Department of Biological Systems Engineering
University of Nebraska-Lincoln
Lincoln
USA

Joe Maja
Agricultural Sciences Department
Clemson University
Blackville
USA

Jose M. Peña
tec4AGRO group
Institute of Agricultural Sciences
CSIC
Madrid
Spain

Editorial Office
MDPI
St. Alban-Anlage 66
4052 Basel, Switzerland

This is a reprint of articles from the Special Issue published online in the open access journal *Remote Sensing* (ISSN 2072-4292) (available at: https://www.mdpi.com/journal/remotesensing/special_issues/uav_vegetation).

For citation purposes, cite each article independently as indicated on the article page online and as indicated below:

LastName, A.A.; LastName, B.B.; LastName, C.C. Article Title. *Journal Name* **Year**, *Volume Number*, Page Range.

ISBN 978-3-0365-2192-3 (Hbk)
ISBN 978-3-0365-2191-6 (PDF)

Contents

About the Editors

Ana I. de Castro

Dr. Ana de Castro is a tenured scientist at INIA working in digital agriculture systems and precision agriculture to address aspects of crop protection, e.g., early detection of pests and weeds, monitoring crop variability, and high-throughput crop phenotyping. She has published more than 40 articles in SCI journals (90% Q1). She has participated as PI and research member of 25 national and international research projects and has been invited as Guest Speaker in numerous conferences and workshops. She has a wide network of collaborations both national and international research centers and universities. She is an Editorial Board Member of *Remote Sensing* and leads the Special Issue "UAVs for Vegetation Monitoring". She was associate editor of the *Spanish Journal of Agricultural Research* and guest editor of the *Journal of Chemistry*. She has been awarded the Best Working Weed Science-2013; 2nd Best Oral Communication received in the 9th ECPA; Jose Luis Labrandero-2015; Premio del Articulo Agrario-2017; and Editor's Choice-2018 by *Remote Sensing*.

Yeyin Shi

Dr. Yeyin Shi is an assistant professor of agricultural information systems in the Department of Biological Systems Engineering at University of Nebraska-Lincoln. The goal of her research program is to develop and apply emerging sensing and information technologies, as well as state-of-the-art data science techniques to generate, gather, analyze, and manage data for actionable decision makings in row crop production and pasture management. The data generated by unmanned aircraft systems (UAS or UAV) are a major source of information in her research. In addition, she and her colleagues also go beyond sensing but incorporating sensing into actuation to use UAS for intelligent aerial applications. Dr. Shi is also passionate about college educations and has been part of the thrust in the development of precision and digital agriculture curriculum.

Joe Maja

Joe M. Maja is a Research Scientist at the Edisto Research and Education Center and a faculty of the Department of Agricultural Science. He provided the vision and expertise to develop a distinctive program for an Experiment Station called the Sensor and Automation Laboratory. His main research effort has focused on emerging technologies, small-unmanned aircraft systems, and mobile robotics. These collective efforts attempt to merge unmanned systems with other appropriate technologies to benefit a variety of agricultural disciplines. His research efforts have involved collaboration with colleagues within Clemson and externally both national and international universities or research institutes. He was elected as Chair of the Central Savannah Rivers Section of the IEEE organization and continues to be a member of the executive committee of the same organization. He has secured more than USD 2 million in extramural funding and has been active in securing federal and commodities funding.

Jose M. Peña

Dr. Jose M Peña is a Scientist at the Institute of Agricultural Sciences of CSIC and co-director of the tec4AGRO research group. His research focuses on the development and validation of geospatial technologies and digitization tools for precision agriculture and integrated pest management (IPM), with the aim of proposing new strategies to optimize crop protection and production for achieving more sustainable agricultural systems. He has participated in 20 research projects, which have resulted in the publication of about 100 SCI and outreach articles, 10 book chapters, and 120 conference contributions. He is honored to have received several distinctions, such as the Fulbright scholarship and the Pierre C. Robert award from the International Society of Precision Agriculture.

Preface to "UAVs for Vegetation Monitoring"

Global plant production faces the major challenge of sustainability under the constraint of a rapidly growing world population and the gradual depletion of natural resources. Remote sensing can play a fundamental role in changing the production model through the development and implementation of new technologies for vegetation monitoring (e.g., advanced sensors and remote platforms, powerful algorithms, etc.) that will lead to higher yields and more sustainable and environmentally friendly food and plant products. Unmanned aerial vehicles (UAVs) or drones have been highlighted as one of the most suitable tools for timely tracking and assessment of vegetation status. They can operate at low altitudes, providing an ultra-high spatial resolution image, have great flexibility of flight schedules for data collection at critical and desired moments, and, also, the generation of digital surface models (DSMs) using highly overlapped images and photo-reconstruction techniques or artificial vision. Therefore, it is essential to advance the research for the technical configuration of UAVs, as well as to improve processing and analysis of the UAV imagery of agricultural and forest scenarios in order to strengthen the knowledge of ecosystems and thereby improve farmers' decision-making processes.

This book compiles a set of original and innovative papers included in the Special Issue on UAVs for vegetation monitoring, which proposes new methods and techniques applied to diverse agricultural and forestry scenarios. It is addressed to all members involved in the development and applications of UAVs on vegetation.

The editors of this book would like to thank the authors for selecting this special issue to publish their most recent findings, as well as the editor team members for their help in this process.

Ana I. de Castro, Yeyin Shi, Joe Maja, Jose M. Peña
Editors

Review

UAVs for Vegetation Monitoring: Overview and Recent Scientific Contributions

Ana I. de Castro [1,*], Yeyin Shi [2], Joe Mari Maja [3] and Jose M. Peña [4]

1 National Agricultural and Food Research and Technology Institute-INIA, Spanish National Research Council (CSIC), 28040 Madrid, Spain
2 Department of Biological Systems Engineering, University of Nebraska-Lincoln, 3605 Fair Street, Lincoln, NE 68583, USA; yshi18@unl.edu
3 Department of Plant and Environmental Sciences, Clemson University, Clemson, SC 29634, USA; jmaja@clemson.edu
4 Institute of Agricultural Sciences (ICA), Spanish National Research Council (CSIC), 28006 Madrid, Spain; jmpena@ica.csic.es
* Correspondence: ana.decastro@inia.es

Abstract: This paper reviewed a set of twenty-one original and innovative papers included in a special issue on UAVs for vegetation monitoring, which proposed new methods and techniques applied to diverse agricultural and forestry scenarios. Three general categories were considered: (1) sensors and vegetation indices used, (2) technological goals pursued, and (3) agroforestry applications. Some investigations focused on issues related to UAV flight operations, spatial resolution requirements, and computation and data analytics, while others studied the ability of UAVs for characterizing relevant vegetation features (mainly canopy cover and crop height) or for detecting different plant/crop stressors, such as nutrient content/deficiencies, water needs, weeds, and diseases. The general goal was proposing UAV-based technological solutions for a better use of agricultural and forestry resources and more efficient production with relevant economic and environmental benefits.

Keywords: drone; RGB; multispectral; hyperspectral; thermal; machine learning; water stress; nutrient deficiency; weed detection; disease diagnosis; plant trails

Citation: de Castro, A.I.; Shi, Y.; Maja, J.M.; Peña, J.M. UAVs for Vegetation Monitoring: Overview and Recent Scientific Contributions. *Remote Sens.* **2021**, *13*, 2139. https://doi.org/10.3390/rs13112139

Academic Editor: Thomas Alexandridis

Received: 16 April 2021
Accepted: 25 May 2021
Published: 29 May 2021

1. Introduction

Global plant production faces the major challenge of sustainability under the constraint of a rapidly growing world population and the gradual depletion of natural resources. In this context, remote sensing can play a fundamental role in changing the production model by developing and implementing new technologies for vegetation monitoring (e.g., advanced sensors and remote platforms, powerful algorithms, etc.) that will lead to higher yields, while also obtaining more sustainable and environmentally friendly food and plant products [1]. Among the recent innovations, the unmanned aerial vehicles (UAVs) or drones have demonstrated their suitability for timely tracking and assessment of vegetation status due to several advantages, as follows: (1) they can operate at low altitudes to provide aerial imagery with ultra-high spatial resolution allowing detection of fine details of vegetation, (2) the flights can be scheduled with great flexibility according to critical moments imposed by vegetation progress over time, (3) they can use diverse sensors and perception systems acquiring different ranges of vegetation spectrum (visible, infrared, thermal), (4) this technology can also generate digital surface models (DSMs) with three-dimensional (3D) measurements of vegetation by using highly overlapping images and applying photo-reconstruction procedures with the structure-from-motion (SfM) technique [2]. The UAVs are therefore a cost-effective tool for obtaining high spatial resolution 3D data of plants and trees that are not feasible with airborne or satellite platforms [3], filling the gap between ground-based devices and these other traditional remote sensing systems. In addition, the UAV-based digital images can be an efficient alternative to data collected with manual fieldwork, which is often arduous, involves destructive sampling, and may provide inconsistent and subjective information. As a result of these strengths, the UAVs are becoming very suitable platforms for agroforestry applications [4,5], such as vegetation characterization and mapping [6], land management [7], environmental monitoring [8], and especially to address diverse precision agriculture objectives [9] and plant phenotypic characterization in breeding programs [10,11].

The large amount of detailed data embedded in UAV imagery requires the development and application of powerful analysis procedures capable of extracting information related to the structural and biochemical composition of vegetation that provides a better understanding of relevant plant traits. Traditional statistical analysis, such as linear classifiers, spatial statistics, or extrema estimators, and pixel-based image analysis are commonly used [12]. However, advanced procedures such as the object-based image analysis (OBIA) paradigm, computer vision techniques, and machine learning algorithms further exploit the advantages of the UAV [13,14], enabling rapid and accurate vegetation analysis and, consequently, helping end-users (e.g., farmers, advisors, and public administrations) in the decision-making process of agricultural and forestry management.

Given the increased interest for new investigations and advances in the areas of UAV technology and UAV-based image processing/analysis in agricultural and forest scenarios, this Special Issue (SI) collected a set of twenty-one original and innovative articles that showed several applications of the UAVs for vegetation monitoring in order to deepen our understanding of these ecosystems. The SI includes the recent scientific contributions organized according to three general categories: (1) type of the sensor employed and, additionally, vegetation index (VI) applied, (2) technological goals pursued, in terms of UAV configuration and specifications, and issues related to image spatial resolution, computation, artificial intelligence, and image processing algorithms, and (3) agroforestry applications, which involved novel UAV-based data-driven approaches for monitoring vegetation features in certain critical dates, measuring plant trails over time for crop dynamics studies, and detecting and modeling biotic (weeds, disease) and abiotic (water, nutrition deficiencies) stress factors. These three categories shared the common objective of proposing new methods and techniques to achieve better use of agricultural and forestry resources and more efficient production with relevant economic and environmental benefits.

2. Sensors and Vegetation Indices Used

A wide range of sensors is available for vegetation monitoring, which can be grouped into visible light (i.e., sensitive to the red, green, blue wavebands, also known as RGB cameras), multispectral, hyperspectral, and thermal. The selection of the proper sensor should be made according to specific agricultural or forestry objectives, i.e., for vegetation characterization (e.g., measuring vegetation features and plant traits) or for assessing certain variables associated to plant production (e.g., nutrient content/deficiencies and water stress) and plant protection (e.g., weed detection and disease diagnosis) issues (Figure 1). For example, RGB and multispectral sensors were selected to determine canopy cover (CC), also known as fractional canopy/vegetation cover or vegetation coverage, where the main challenge consisted of separating vegetation from the background by performing simple linear regression models [15–17]. RGB and multispectral sensors were used to monitor crop dynamics, as these sensors allowed building DEMs through a photogrammetric SfM process, then calculating crop height (CH) values, and finally performing morphological procedures [18–20]. These two sensors are cheaper and easier to use than thermal and hyperspectral sensors, and do not require a complex calibration phase, so they proved to be optimal for assessing plant structural traits.

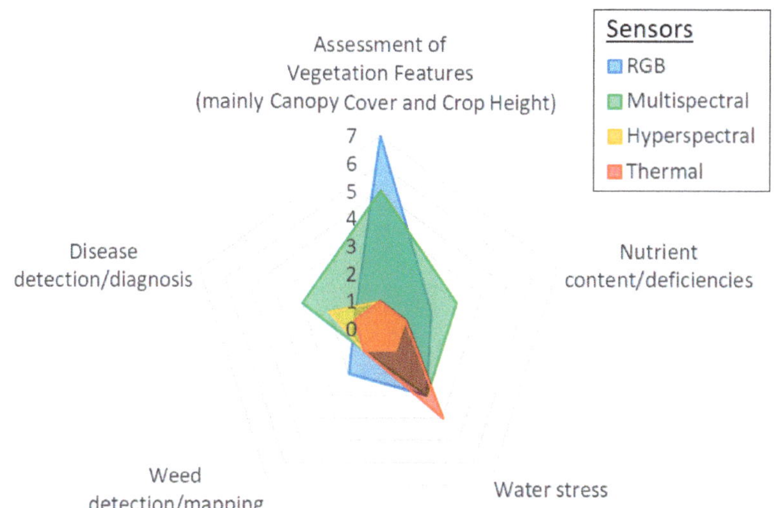

Figure 1. Sensors used in the experiments performed in this SI according to agricultural and forestry objectives.

The RGB and multispectral sensors were also suitable for the target of identifying abiotic stress (e.g., Nitrogen (N) content/deficiencies), although this goal required detecting subtle changes in leaf pigment contents, for which it is necessary to employ more complex procedures, such as machine learning (ML) or partial least squares regressions (PLSR) algorithms. Similar approaches were followed for water deficiencies causing effects on chlorophyll content and organization in the mesophyll and leaf structure. It was possible to address this goal using RGB and thermal-based VIs [21] or multispectral sensors and convolutional neural networks (CNNs) [22]. It is also common to use thermal sensors since water stress causes stomatal closure that reduces transpiration and evaporative cooling and increases temperature [23]. In this way, data from thermal sensors were combined with RGB [21] or multispectral sensors [24,25] to take advantage of the information provided in different areas of the spectrum and the higher spatial resolution of such sensors.

Regarding the assessment of plant biotic stress, the RGB sensors were used to detect weed patches at mid- to late-season using CNN models [26]. However, narrower bandwidth spectral data were required to detect changes in the biochemical and biophysical parameters caused by diseases and, in some cases, texture features (TFs) to improve the diagnosis accuracy. Thus, multispectral and hyperspectral sensors were used for mapping disease occurrence [27–29]. The red-edge information was also relevant for this goal, so regardless of the sensor selected, the images should contain information in this specific spectral region.

It is worth noting that vegetation indices (VIs) were also widely used in many investigations on vegetation monitoring (Table 1), as they enhance the spectral differences of vegetation at specific wavelengths. The applications of the VIs used in this SI are discussed further in Section 4.

Table 1. Vegetation indices (VIs) applied in the papers published in this special issue.

Sensor Type	Vegetation Index	Acronym	Reference
RGB	Excess Green	ExG	[18]
	Red Green Blue VI	RGBVI	[15]
	Hue	H	[21]
	Modified Green Red VI	MGRVI	[21]
	Normalized Difference CIELab Index	NDLab	[21]
Multispectral	Normalized Difference VI	NDVI	[16,17,24,27,29–31]
	Green Normalized Difference VI	GNDVI	[17,30]
	Red-edge Normalized Difference VI	ReNDVI	[16]
	Normalized Difference Red-edge Index	NDRE	[19,27]
	Normalized Ratio VI	NRVI	[17]
	Inverse Ratio VI	IRVI	[17]
	Soil-Adjusted VI	SAVI	[17]
	Modified SAVI	MSAVI	[30]
	Green Chlorophyll Index	CI_{green}	[27]
	Red-edge Chlorophyll Index	CI_{RE}	[27]
	Simplified Canopy Chlorophyll Index	s-CCI	[19]
Hyperspectral	Structural Independent Pigment Index	SIPI	[29]
	Plant Senescence Reflectance Index	PSRI	[29]
	Photochemical Reflectance Index	PRI	[29]
	Modified Simple Ratio	MSR	[29]
Thermal	Crop Water Stress Index	CWSI	[21]

3. Research Focused to Technological Goals

Development and validation of new technologies and computational procedures for vegetation monitoring requires the definition of the optimal flight and sensor configuration, appropriate data format, workflow specifications, and innovative analysis methods that lead to the generation of scientific knowledge on vegetation condition and management. Therefore, this section describes the contributions partially or entirely focused to technological goals, including issues on UAV flight operations, spatial resolution requirements, and computation and data analytics.

3.1. UAV Flight Operations

UAV flight is the first step for image acquisition. Designing a UAV route, i.e., a mission planning or a flight planning, is a critical task requiring several parameters. The quality and viability of acquired imagery and data will largely depend on those operation parameters and on how the flights are performed. The configuration of flight mission and the specifications of the UAV employed must be evaluated in detail and adapted to each specific objective, particularity in the case of agronomical applications [32]. In this regard, Jiang et al. (2020) [31] evaluated the influence of flight altitude, solar zenith angle (SZA), and time of day (TOD) for NDVI calculation at different paddy rice growth levels. The results indicated that both SZA and TOD—which is logical since SZAs depend on the TOD—had the most significant impact on estimated NDVI values following a directly proportional relation. For a flight altitude below 100 m, the correlation with estimated NDVI was inversely proportional for maximum and mean values, while minimum NDVIs increased with higher flight altitude values. In addition, the most accurate reflectivity, and thus the estimated NDVI values, were achieved at the smaller SZA.

The UAVs can also suffer from stability problems, such as shaking or imprecise geographic coordinates (X-Y), which are more pronounced when UAV is piloted manually. They are caused by wind perturbations or improper attitude stabilization, resulting in georeferencing problems with consequent complications for vegetation mapping. Therefore, it is important to ensure a robust stabilization system that provides high-quality image data and, thus, improves the precision of the image registration algorithms. Colorado et al. (2020) [30] developed a novel UAV stabilization control called BS+DAF (backstepping + desired angular acceleration function) to reduce angular wind-induced disturbances. The BS+DAF consisted of a nonlinear trajectory-tracking controller that adds aerodynamics information to control yaw, producing roll and pitch acceleration commands to reduce sudden angular acceleration changes due to external perturbations as those caused by wind. Therefore, precise angular stabilization can be maintained during flight, resulting in a more quality image acquisition and registration. These authors also proved that better correlations in canopy N estimations were obtained when the BS+DAF controller was implemented in the UAV flights.

3.2. *Spatial Resolution Requeriments*

Flight altitude is a critical parameter in remote image acquisition, as it has strong implications in image spatial resolution, flight duration, spatial coverage, and computational cost of the image processing tasks [33,34]. Lower altitudes result in images with finer spatial resolution, but require longer flights and demand higher computational times for image processing [35]. Therefore, evaluation of flight altitude and spatial resolution was a central point in some investigations of this SI [16,24,27,29,36], being a usual objective to resample UAV images to spatial resolutions similar to those of common satellites, e.g., 5 m of Rapideye, 10 m of Sentinel-2, and 30 m of LANDSAT, in order to study the potential of using such satellite platforms instead of UAV for specific agro-forestry goals. Thus, Iizuka et al. (2019) [16] discovered that CC estimation was affected by spatial resolution and coarser resolutions showed stronger correlation with manually delineated data in an RGB orthomosaic, concluding that the satellite images could be feasible for this purpose. These authors explained that variability decreases when finer data is aggregated to a larger unit, thus increasing correlations. However, Nassar et al. (2020) [24] demonstrated that a very high spatial resolution is necessary to use the Two-Source Energy Balance with a dual temperature (TSEB2T) model for successfully estimating evapotranspiration (ET). The authors argued that the selected spatial resolution should be fine enough to discriminate plant canopy and soil temperature, as they evaluated multiple spatial domain scales (3.6, 7.2, 14.4, and 30 m), leading them to conclude that the current satellites were not suitable for this application, especially those with spatial resolution of 30 m or broader, e.g., Landsat.

Image resampling approaches were also carried out in experiments for disease detection, where spatial resolution is a crucial factor as disease-induced biological symptoms are subtle and only detectable at high resolution values. Thus, Guo et al. (2021) [29] and Ye et al. (2020) [27] evaluated several pixel sizes to select the optimal resolution for mapping disease areas, yielding 0.01 m for yellow rust in wheat and a minimum of 0.5 m in the case of *Fusarium wilt* in banana, respectively. In addition, Zhang et al. (2020) [36] explored the influence of image spatial resolution on the segmentation process prior to identify N deficiency in rapeseed fields, concluding that the optimal value by using the U-Net convolutional neural network architecture will depend on the target size and stating an appropriate patch size of 256x256 pixels in the purple rapeseed leaves.

3.3. *Computation and Data Analytics*

Innovations in data analytics are also contributing to overcome diverse challenges in agroforestry scenarios, especially regarding to the implementation of computational advances that speed up the processing times and of new algorithms that provide timely and accurate analysis to assist in complex studies and decision-making processes. One of the disciplines most involved in this progress is artificial intelligence (AI) and, in particular, ML tools such as support vector machine (SVM), random forest (RF), and CNNs algorithms that can solve complex matters using datasets from multiple sources and helping to discover hidden rules and patterns in such large datasets.

Machefer et al. (2020) [37] developed a novel algorithm for counting and sizing potato and lettuce plants based on a deep learning (DL) approach. Their all-in-one algorithm was an adaptation of Mask R-CNN architecture that combined segmentation, plant delineation and counting tasks, avoiding an additional "patching" step and simplifying the computational complexity. This approach was compared to a computer vision baseline showing better accuracy results for CNN-based method in both crops. Another example of

ML techniques was applied by Fu et al. (2021) [38], who estimated winter wheat biomass during several growth stages using multiscale spectral features and TFs and least squares support vector machine (LSSVM) and partial least squares regressions (PLSR) analysis algorithms. The authors determined the optimal TFs for biomass estimation by combining two-dimensional (2D) Gabor filters and gray-level co-occurrence matrix (GLCM) analyses. The author argued that this method performed better using multiscale textures than single-scale textures, and concluded that biomass estimation did not significantly improve using hyperspectral data, showing that LSSVM regression with proper multiscale textural features was an optimal and simpler alternative to hyperspectral-based methods. Bhatnagar et al. (2020) [39] also compared several ML models, but included DL and segmentation algorithms to map raised bog vegetation communities using RGB and TFs. Based on these performances, RF with graph cut as segmentation technique was selected as the best ML classifier. The ResNet50 CNN with the SegNet semantic segmentation architecture was selected for DL. Although slightly better accuracy was reached with ResNet50- SegNet, these authors recommended RF-graph cut for ecotopes mapping, as DL required more dataset training, computation time, and hardware power.

4. Research Focused to Agroforestry Applications

4.1. Assessment of Vegetation Features

Canopy cover (CC), or the proportion of unit area covered by the vertical projection of plant canopy, is an important indicator of plant development, biomass, light interception, photosynthesis activity, and evapotranspiration partitioning [40,41]. Since CC has traditionally been determined using tedious and imprecise manual procedures, the development of an accurate and rapid remote-sensing based tool to measure this vegetation feature has been often a goal in agroforestry. In addition, UAV imagery can generate DSMs and 3D point clouds using photo-reconstruction and artificial vision techniques to provide the height information (or Z-value) at each coordinate (X,Y) as a basis for calculating the CH values of vegetation. Thus, this SI reports on several studies to assess these features and its temporal variation over the growing season for monitoring vegetation or crop dynamics with UAV-based RGB and multispectral images [15–20].

Ashapure et al. (2019) [15] computed CC over the cotton growing season using a model that combined an RGB-based VI with a morphological closing tool, which consisted of a combined dilation and erosion operation to remove the small holes, while keeping the separation boundary intact. The RGBVI was selected to be the most efficient VI for this purpose. These authors showed that the CC estimation model based VI alone was very unstable and failed to identify the canopy when the cotton leaves changed color after ripening. However, applying a morphological closure operation significantly improved the estimation, which led them to conclude that the proposed RGB-based CC model provides an affordable alternative to multispectral-based approaches, which rely on more sophisticated and expensive sensors.

In other study, Iizuka et al. (2019) [16] employed RGB and multispectral data to compute CC of several land uses/land covers in Indonesia peatland forest, including Acacia trees and other types of forests with rapid land transitions. The first step consisted of generating a map of land uses by applying the Multilayer-Perceptron (MLP) classification to the RGB images. Then, the VI that best computed the CC for each land use was selected, resulting in ReNDVI for Acacia and NDVI for grass/shrubs and non-vegetation classes. These authors stated that UAVs' multispectral information can quickly quantify CC in tropical forests, especially those of fast-growing Acacia that can rapidly change land conditions.

Following a similar approach, Lima-Cueto et al. (2019) [17] quantified CC in olive orchards, where cover crops are a critical agro-environmental conservation technique for improving soil fertility and water retention, as well as reducing water erosion. They employed a multispectral sensor to calculate VIs related to ground-truth cover crops at different densities. They showed that the capacity of the evaluated VIs to quantified

CC depends on the amount of cover crops and the range of cover density intervals, i.e., the width of the intervals into which the complete series of densities was divided. This investigation reported the IRVI as the most sensitive VI for a cover interval range of 15–25%, although the IRVI, NRVI, NDVI, GNDVI, and SAVI were valid for differentiating the complete series of densities when the interval rose to 30%.

Borra-Serrano et al. (2020) [18] combined multi-temporal RGB-UAV data with curve fitting to obtain biologically interpretable parameters of a large set of soybean genotypes and derived phenotypic parameters like growth rate, weed suppression, determinacy, and senescence. First, CC and CH were estimated from each flight date based on the ExG and the 90th percentile of the crop height model (CHM) values, respectively. Then, multi-temporal data were adjusted to the sigmoid growth functions Gompertz for CC and Beta for CH. In addition, these authors developed a model to predict seed yield for estimating full maturity, i.e., when each accession reached the R8 stage.

Heidarian Dehkordi et al. (2020) [19] shown another approach to extract multi-temporal CC and CH from several VI and CHM, respectively. These authors also derived crop growth and health maps from RGB and multispectral images by K-means algorithms. These maps were used to monitor crop dynamics and assess century-old biochar's impacts on winter wheat crop performance at the canopy scale over the growing season. In addition, they studied the impact on the predicted multispectral crop yield based on the s-CCI and the NDRE indices (Table 1), which exhibited a significant correlation with crop harvest data. Their results showed a significant positive impact of century-old biochar on the winter wheat CC and CH over time, but no impact on harvested crop yield was found.

Zan et al. (2020) [20] showed another example of monitoring crop dynamic from RGB-UAV imagery with a method to detect and quantify maize tassels in the flowering stage by combining machine learning algorithms and morphological methods. Once the RGB images were taken, a RF algorithm was used to identify and classify tassel pixels, finding tassels of any size and unconnected regions. A refining process based on morphological procedures was performed to remove small objects and to connect pixels producing potential tassel region candidates. Then, VGG16 neural network was used to remove false positives. Their results showed that the tassel can be detected at any stages, although the accuracy was affected by the tasseling development, reporting higher accuracy for late tasseling than for middle and early stages. Finally, these authors also calculated the tassel branch number by an endpoint detection method based on the tassel skeleton.

4.2. Evaluation of Stressor on Vegetation

The biotic and abiotic stressors such as nutrition and water deficiencies and pests (including plagues, diseases, and weeds) can drastically reduce the quality and quantity of vegetation production and crop yield. In fact, the overall impact of the pathogens affecting crop health is estimated to reduce global production by about 40% [42]. The stress factors cause changes in plant physiology and morphology that result in variations in canopy reflectance that can be potentially monitored with UAVs and imaging sensors [43,44]. Therefore, this SI collected papers on the use of UAVs to assess those factors affecting both crop production (e.g., N content/deficiencies and water stress/needs) and crop protection (weed and disease detection/diagnosis) aspects. Many of the cited papers utilized the vegetation features mentioned in the point 4.1., together with various data analytic approaches mentioned in the point 3.3.

4.2.1. Assessment of Nitrogen Content/Deficiencies

Nitrogen is an essential plant nutrient for the growth and yield of crops. Leaf N is directly related to the chlorophyll content, which is a key indicator of photosynthetic capacity and nutrient status. Zhang et al. (2020) [36] developed a method to monitor crop N stress in winter oilseed rape using UAV RGB imagery and ML-based segmentation. Their method consisted in discriminating the rapeseed leaves with purple color, a symptom of N deficiency, using a binary semantic segmentation process and, then, relating the purple leaf area ratio to the N content of the plants. The authors tested five image segmentation approaches at the pixel level for the extraction of the purple rapeseed leaves during the seedling stage. The CNN U-Net model with a patch size of 256 x 256 pixels reported the best results. A negative exponential regression analysis was established between measured N content and estimated purple leaf ratios (leaf pixels over the total pixels) under four N application levels. This method proved to be sensitive to low-medium N application; however, the sensitivity was reduced for N levels over 150 kg/ha.

A ML approach was also employed by Colorado et al. (2020) [30] to monitor canopy N in rice crops with multispectral UAV imagery. The approach was evaluated during the dry season at three rice growth stages: vegetative, reproductive, and ripening. As a preliminary phase, the authors proposed an automated GrabCut image segmentation method, which is an improvement of the regular GrabCut one [45], that adds a guided filtering refinement step to extract the rice canopy pixel data. Next, several ML algorithms accurately correlated ground-truth leaf chlorophyll measurements and canopy N estimates via VIs. The successful ML algorithms were multi-variable linear regressions (MLR), SVM, and artificial neural networks (ANN), which showed that MSAVI, GNDVI, and NDVI have a strong correlation with canopy N at any growth stages. Finally, the authors also proved that these good correlations improve with the implementation of a stabilization control during UAV flight operations.

Similarly, Qiao et al. (2020) [46] described a segmentation-based method to estimate chlorophyll content at different maize growth stages using multispectral imagery. They showed different segmentation results that produced variations in the extraction of maize canopy parameters, with the wavelet method as the best output. Next, a maize canopy chlorophyll content model was developed using VIs from the segmented images. The research demonstrated the importance of image segmentation to remove the soil background and highlight the spectral reflectance of crop canopy for N detection purposes.

4.2.2. Assessment of Water Stress

Fresh water is a finite and scarce resource in many regions of the world. Therefore, determining how much water the plant needs is crucial for efficient crop, water, and irrigation management. Freeman et al. (2019) [22] used multispectral images and cloud-based artificial intelligence for the early detection of water stress in ornamental horticultural plants. Images containing the plants in three types of water stress situations (high, low, and non-stressed) were analyzed by the IBM Watson Visual Recognition platform, a deep CNN hosted in the cloud. The generated models were able to identify the plants with early water stress, i.e., after 48 hours of water deprivation in most of the studied species. In addition, early indicators of stress after 24 hours of water suppression were found for the Buddleia plants. This research opens up the opportunity of identifying water needs in large datasets through cloud-hosted artificial intelligence services.

De Swaef et al. (2021) [21] evaluated the drought tolerance of three common types of forage species (*Festuca arundinacea*, diploid, and tetraploid *Lolium perenne*) with VIs from RGB and thermal imaging. The methodology was tested in five UAV flight dates throughout the summer, which overlapped with visual breeder score-taking. Pearson correlation showed better results between breeder scores and visual-based VIs than with thermal indices, with H from the hue-saturation-value (HSV) color space and NDLab from CIELab color space (Table 1) being the best. These VIs and MGRVI also displayed broad-sense high heritability, i.e., the total variation proportion that may be attributed to genotype variation. This research showed the use of UAV-based RGB-VIs as an efficient alternative to breeder scores for drought tolerance.

In other research conducted by Ellsäßer et al. (2020) [25], RGB and thermal UAV imagery and meteorological data were used to estimate sap flux and leaf stomatal conductance, key inputs for water balance, and plant transpiration assessments. The experiment was conducted in an oil palm plantation, where oil palm (*Elaeis guineensis*) and other several forest tree species (*Peronema canescens, Archidendron pauciflorum, Parkia speciosa, and*

Shorea leprosula) usually appear. These authors tested linear statistical and ML approaches to build the prediction models using in-situ measured transpiration and estimated data. For both parameters, RF model achieved the best accuracy, being less robust for stomatal conductance. R^2 values for sap flux prediction were close to or higher than 0.8 for forest species and 0.58 for palms, however, the model accuracies were moderate (60%) and low (38%) for forest species and palms, respectively. Therefore, image-based variables and meteorological observations used as input to the RF algorithm could be an alternative to sap flux manual measurements only for same of the evaluated species, especially *Parkia speciosa* and *Archidendron pauciflorum*.

In order to model the radiative and convective flux exchanges between soil and canopy for ET estimation, Nassar et al. (2020) [24] tested the TSEB2T model in vineyard using multispectral and thermal images, including a step of mapping canopy and soil temperature. The TSEB2T inputs were calculated from images, as follows: Leaf Area Index (LAI) using a genetic programming model and ground measurements, CH extracted from the DSM and DTM, CC and canopy width (CW) based on NDVI imagery, and canopy and soil/substrate temperature (Tc and Ts, respectively) by averaging the temperature of pixels classified as pure soil and pure canopy based on a linear land surface temperature (LST)-NDVI model. The validation results showed a close agreement of TSEB2T output and ground-truth measurements if spatial image resolution was high enough to discriminate plant canopy and soil temperatures.

4.2.3. Weed Detection and Mapping

Weeds compete with crops for soil resources, thus reducing crop yield and creating a weed seed bank that germinates the following season, increasing the weed-crop competition problem. Weeds are often distributed in patches within fields, which allows designing site-specific weed management (SSWM) strategies to adjust applications according to the weed patches distribution [47]. In this area, Veeranampalayam et al. (2020) [26] evaluated two objects detection-based CNN models and RGB images for mid-to-late season weed identification in soybean fields. Two feature extractors were also evaluated in both NN models, (Faster RCNN and the Single Shot Detector (SSD)), concluding that Inception v2 performs better than Mobilenet v2. Moreover, the result showed that Faster RCNN is the best NN model in terms of weed detection accuracy and inference time compared to SDD and the patch-based CNN, a pre-trained network used for validation. Therefore, due to the rapid and accurate performance for weed detection at mid-late stage, Faster RCNN proves to be a suitable solution for near real-time weed detection. A process improvement could consist of removing manual tasks by using OBIA multiresolution segmentation approach instead, thus automating the process, leading to a real-time processing.

4.2.4. Disease Detection/Diagnosis

Numerous studies on disease detection have been carried out at the leaf scale [48]; however, more research needs to scale up at the canopy level to monitor disease in large areas. Mapping at the field scale would control disease effects through proper management strategies, such as site-specific control tactics. Ye et al. (2020) [27] established a method for mapping *Fusarium wilt* in bananas with multispectral imagery and VIs. Once the more suitable VIs were selected based on an independent t-test analysis, the Binary Logistic Regression (BLR) method was used to relate healthy and infested samples for each selected VIs. The CI_{green}, CI_{RE}, NDVI, and NDRE indices (Table 1) were able to identify healthy and diseased trees, with CI_{RE} exhibiting the best performance. Infestation maps were performed with the BLR equation obtained for CI_{RE}. Moreover, the importance of the red-edge information in the discrimination was highlighted.

Wang et al. (2020) [28] developed a non-supervised method for cotton root rot (CRR) detection with the novelty of the individual identification of CRR-infected plants using multispectral images. This plant-by-plant (PBP) method consists of a segmentation process based on the Superpixel algorithm, which creates multi-pixel pieces (super-pixel), approximately single plant zones, based on shape, color, texture, etc. Then, a classification was performed using k-means algorithms to achieve a two-class super-pixel classification, i.e., CRR and healthy zones. The results highlighted the importance of the segmentation process, as the accuracy increased when it was carried out as a prior step to the classification phase. Compared to conventional supervised methods, the PBP achieved better classifications with the best accuracy parameters in terms of overall accuracy, kappa coefficient and commission errors, and high values in the omission errors. Also, the fungicide treatment based on the PBP-infestation map allowed a high-precision treatment of CRR with a mortality percentage similar to conventional treatment rates.

Finally, Guo et al. (2021) [29] added TFs to VIs extracted from the hyperspectral images to address the wheat yellow rust mapping at different infection stages. To develop the PLSR model for each infection stage, VIs, TFs, and a combination of them were tested and compared to ground truth severity data described by the disease index (DI). Finally, the VI-TF combination, composed of mean (MEA1 and MEA2), variance (VAR2), contrast (CON2), NDVI, SIPI, PRI, PSRI, and MSR, was selected for the PLSR model performance in each infection period, as they achieved the highest accuracy in every stage. The performance accuracy was affected by the infection stage, such that the infection period increased, monitoring accuracy increased, with high values for mid- and late-infection periods, and poor for early ones. Thus, wheat yellow rust infestation maps could be created based on the PLSR model and the VIs and TF selected.

5. Conclusions

The investigations included in this SI proved the wide scope of UAVs in very diverse applications, both in agricultural and forestry scenarios, ranging from the characterization of relevant vegetation features to the detection of plant/crop stressors.

The RGB and multispectral sensors were particularly effective for assessing canopy cover, crop height, growth rate, determinacy, senescence, and maize tassel branch number, among other plant traits. The studies with multispectral sensors also revealed the best vegetation indices for assessing stresses caused by nutrients and water (e.g., NDVI) or by biological agents such as weeds and diseases (e.g., NDRE, CIgreen). The hyperspectral sensors were focused on disease detection, given the ability of these sensors to observe spectral differences at certain wavelengths that are sensitive to biochemical and biophysical changes caused by plant disease. Finally, the research applied the thermal sensors mainly for water stress measures, aiming to develop efficient irrigation systems by optimizing water use at the crop field scale.

A detailed reading of this SI leads us to note some limitations and opportunities of UAVs for vegetation monitoring. UAVs can acquire immense spatial and temporal information of crops and forest areas, although they cannot currently measure all the desired plant traits, most of the measurements being indirect and requiring a rigorous calibration process. Data calibration with downwelling/incident light sensors has already been standardized for multispectral sensors, but there is not a good solution for thermal cameras yet. In addition, there is a problem of flight length and battery duration in UAVs for large areas, so in such cases it could be proposed solutions combined with satellite imagery. The coarser spatial resolution of such images could be an obstacle in certain applications, so new studies combining both UAV and satellite platforms should be pursued.

Regarding data analytics, the machine/deep learning (ML/DL) algorithms are becoming increasingly popular in vegetation characterization, although the large amount of training data they require, as well as their high computational cost, should be carefully considered. Therefore, such powerful methods may not be really necessary in all applications, and in any case, it would be interesting to deepen into generalization issues, i.e., the use of training data from specific locations in other scenes or environments, thus making better use of the data obtained. In this respect, it would be beneficial to adapt the scientific procedures presented here to simpler analytical packages and easy-to-use tools for everyone (not only for remote sensing and data science researchers), thus widening the opportunities for vegetation monitoring and crop/forest management from a practical perspective.

Author Contributions: All authors of this review article have contributed equally to its conceptualization, writing, review, and editing. All authors have read and agreed to the published version of the manuscript.

Funding: This research was funded by the project AGL2017-83325-C4-1R of Agencia Española de Investigación (AEI) and Fondo Europeo de Desarrollo Regional (FEDER). The contribution of Dr. Shi and Dr. Maja were supported by the Nebraska Agricultural Experiment Station through the Hatch Act capacity funding program (Accession Number 1011130) and Project No. SC-1700543 from the USDA National Institute of Food and Agriculture, respectively. This work also represents a contribution to the CSIC Thematic Interdisciplinary Platform PTI TELEDETECT (https://pti-teledetect.csic.es/ (accessed on 27 May 2021)).

Institutional Review Board Statement: Not applicable.

Informed Consent Statement: Not applicable.

Data Availability Statement: The data presented in this study are available in the research articles displayed all along the text or summarized in Table 1.

Conflicts of Interest: The authors declare no conflict of interest.

References

1. Ahmad, A.; Ordoñez, J.; Cartujo, P.; Martos, V. Remotely Piloted Aircraft (RPA) in Agriculture: A Pursuit of Sustainability. *Agronomy* **2021**, *11*, 7. [CrossRef]
2. Dandois, J.P.; Ellis, E.C. High Spatial Resolution Three-Dimensional Mapping of Vegetation Spectral Dynamics Using Computer Vision. *Remote Sens. Environ.* **2013**, *136*, 259–276. [CrossRef]
3. de Castro, A.I.; Peña, J.M.; Torres-Sánchez, J.; Jiménez-Brenes, F.M.; Valencia-Gredilla, F.; Recasens Guinjuan, J.; López-Granados, F. Mapping Cynodon Dactylon Infesting Cover Crops with an Automatic Decision Tree-OBIA Procedure and UAV Imagery for Precision Viticulture. *Remote Sens.* **2020**, *12*, 56. [CrossRef]
4. Dainelli, R.; Toscano, P.; Di Gennaro, S.F.; Matese, A. Recent Advances in Unmanned Aerial Vehicle Forest Remote Sensing—A Systematic Review. Part I: A General Framework. *Forests* **2021**, *12*, 327. [CrossRef]
5. Olson, D.; Anderson, J. Review on Unmanned Aerial Vehicles, Remote Sensors, Imagery Processing, and Their Applications in Agriculture. *Agron. J.* **2021**, 1–22. [CrossRef]
6. Torres-Sánchez, J.; Peña, J.M.; de Castro, A.I.; López-Granados, F. Multi-Temporal Mapping of the Vegetation Fraction in Early-Season Wheat Fields Using Images from UAV. *Comput. Electron. Agric.* **2014**, *103*, 104–113. [CrossRef]
7. Librán-Embid, F.; Klaus, F.; Tscharntke, T.; Grass, I. Unmanned Aerial Vehicles for Biodiversity-Friendly Agricultural Landscapes-A Systematic Review. *Sci. Total Environ.* **2020**, *732*, 139204. [CrossRef]
8. Manfreda, S.; McCabe, M.F.; Miller, P.E.; Lucas, R.; Pajuelo Madrigal, V.; Mallinis, G.; Ben Dor, E.; Helman, D.; Estes, L.; Ciraolo, G.; et al. On the Use of Unmanned Aerial Systems for Environmental Monitoring. *Remote Sens.* **2018**, *10*, 641. [CrossRef]
9. Maes, W.H.; Steppe, K. Perspectives for Remote Sensing with Unmanned Aerial Vehicles in Precision Agriculture. *Trends Plant Sci.* **2019**, *24*, 152–164. [CrossRef]
10. Yang, G.; Liu, J.; Zhao, C.; Li, Z.; Huang, Y.; Yu, H.; Xu, B.; Yang, X.; Zhu, D.; Zhang, X.; et al. Unmanned Aerial Vehicle Remote Sensing for Field-Based Crop Phenotyping: Current Status and Perspectives. *Front. Plant Sci* **2017**, *8*, 1111. [CrossRef] [PubMed]
11. Shi, Y.; Thomasson, J.A.; Murray, S.C.; Pugh, N.A.; Rooney, W.L.; Shafian, S.; Rajan, N.; Rouze, G.; Morgan, C.L.S.; Neely, H.L.; et al. Unmanned Aerial Vehicles for High-Throughput Phenotyping and Agronomic Research. *PLoS ONE* **2016**, *11*, e0159781. [CrossRef] [PubMed]
12. Hassler, S.C.; Baysal-Gurel, F. Unmanned Aircraft System (UAS) Technology and Applications in Agriculture. *Agronomy* **2019**, *9*, 618. [CrossRef]
13. Rehman, T.U.; Mahmud, S.; Chang, Y.K.; Jin, J.; Shin, J. Current and Future Applications of Statistical Machine Learning Algorithms for Agricultural Machine Vision Systems. *Comput. Electron. Agric.* **2019**, *156*, 585–605. [CrossRef]
14. Chen, G.; Weng, Q.; Hay, G.J.; He, Y. Geographic Object-Based Image Analysis (GEOBIA): Emerging Trends and Future Opportunities. *GIScience Remote Sens.* **2018**, *55*, 1–24. [CrossRef]
15. Ashapure, A.; Jung, J.; Chang, A.; Oh, S.; Maeda, M.; Landivar, J. A Comparative Study of RGB and Multispectral Sensor-Based Cotton Canopy Cover Modelling Using Multi-Temporal UAS Data. *Remote Sens.* **2019**, *11*, 2757. [CrossRef]
16. Iizuka, K.; Kato, T.; Silsigia, S.; Soufiningrum, A.Y.; Kozan, O. Estimating and Examining the Sensitivity of Different Vegetation Indices to Fractions of Vegetation Cover at Different Scaling Grids for Early Stage Acacia Plantation Forests Using a Fixed-Wing UAS. *Remote Sens.* **2019**, *11*, 1816. [CrossRef]

17. Lima-Cueto, F.J.; Blanco-Sepúlveda, R.; Gómez-Moreno, M.L.; Galacho-Jiménez, F.B. Using Vegetation Indices and a UAV Imaging Platform to Quantify the Density of Vegetation Ground Cover in Olive Groves (*Olea Europaea* L.) in Southern Spain. *Remote Sens.* **2019**, *11*, 2564. [CrossRef]
18. Borra-Serrano, I.; De Swaef, T.; Quataert, P.; Aper, J.; Saleem, A.; Saeys, W.; Somers, B.; Roldán-Ruiz, I.; Lootens, P. Closing the Phenotyping Gap: High Resolution UAV Time Series for Soybean Growth Analysis Provides Objective Data from Field Trials. *Remote Sens.* **2020**, *12*, 1644. [CrossRef]
19. Heidarian Dehkordi, R.; Burgeon, V.; Fouche, J.; Placencia Gomez, E.; Cornelis, J.-T.; Nguyen, F.; Denis, A.; Meersmans, J. Using UAV Collected RGB and Multispectral Images to Evaluate Winter Wheat Performance Across a Site Characterized by Century-Old Biochar Patches in Belgium. *Remote Sens.* **2020**, *12*, 2504. [CrossRef]
20. Zan, X.; Zhang, X.; Xing, Z.; Liu, W.; Zhang, X.; Su, W.; Liu, Z.; Zhao, Y.; Li, S. Automatic Detection of Maize Tassels from UAV Images by Combining Random Forest Classifier and VGG16. *Remote Sens.* **2020**, *12*, 3049. [CrossRef]
21. De Swaef, T.; Maes, W.H.; Aper, J.; Baert, J.; Cougnon, M.; Reheul, D.; Steppe, K.; Roldán-Ruiz, I.; Lootens, P. Applying RGB- and Thermal-Based Vegetation Indices from UAVs for High-Throughput Field Phenotyping of Drought Tolerance in Forage Grasses. *Remote Sens.* **2021**, *13*, 147. [CrossRef]
22. Freeman, D.; Gupta, S.; Smith, D.H.; Maja, J.M.; Robbins, J.; Owen, J.S.; Peña, J.M.; de Castro, A.I. Watson on the Farm: Using Cloud-Based Artificial Intelligence to Identify Early Indicators of Water Stress. *Remote Sens.* **2019**, *11*, 2645. [CrossRef]
23. Gago, J.; Douthe, C.; Coopman, R.E.; Gallego, P.P.; Ribas-Carbo, M.; Flexas, J.; Escalona, J.; Medrano, H. UAVs Challenge to Assess Water Stress for Sustainable Agriculture. *Agric. Water Manag.* **2015**, *153*, 9–19. [CrossRef]
24. Nassar, A.; Torres-Rua, A.; Kustas, W.; Nieto, H.; McKee, M.; Hipps, L.; Stevens, D.; Alfieri, J.; Prueger, J.; Alsina, M.M.; et al. Influence of Model Grid Size on the Estimation of Surface Fluxes Using the Two Source Energy Balance Model and SUAS Imagery in Vineyards. *Remote Sens.* **2020**, *12*, 342. [CrossRef] [PubMed]
25. Ellsäßer, F.; Röll, A.; Ahongshangbam, J.; Waite, P.-A.; Hendrayanto; Schuldt, B.; Hölscher, D. Predicting Tree Sap Flux and Stomatal Conductance from Drone-Recorded Surface Temperatures in a Mixed Agroforestry System—A Machine Learning Approach. *Remote Sens.* **2020**, *12*, 4070. [CrossRef]
26. Veeranampalayam Sivakumar, A.N.; Li, J.; Scott, S.; Psota, E.; Jhala, A.J.; Luck, J.D.; Shi, Y. Comparison of Object Detection and Patch-Based Classification Deep Learning Models on Mid- to Late-Season Weed Detection in UAV Imagery. *Remote Sens.* **2020**, *12*, 2136. [CrossRef]
27. Ye, H.; Huang, W.; Huang, S.; Cui, B.; Dong, Y.; Guo, A.; Ren, Y.; Jin, Y. Recognition of Banana Fusarium Wilt Based on UAV Remote Sensing. *Remote Sens.* **2020**, *12*, 938. [CrossRef]
28. Wang, T.; Thomasson, J.A.; Isakeit, T.; Yang, C.; Nichols, R.L. A Plant-by-Plant Method to Identify and Treat Cotton Root Rot Based on UAV Remote Sensing. *Remote Sens.* **2020**, *12*, 2453. [CrossRef]
29. Guo, A.; Huang, W.; Dong, Y.; Ye, H.; Ma, H.; Liu, B.; Wu, W.; Ren, Y.; Ruan, C.; Geng, Y. Wheat Yellow Rust Detection Using UAV-Based Hyperspectral Technology. *Remote Sens.* **2021**, *13*, 123. [CrossRef]
30. Colorado, J.D.; Cera-Bornacelli, N.; Caldas, J.S.; Petro, E.; Rebolledo, M.C.; Cuellar, D.; Calderon, F.; Mondragon, I.F.; Jaramillo-Botero, A. Estimation of Nitrogen in Rice Crops from UAV-Captured Images. *Remote Sens.* **2020**, *12*, 3396. [CrossRef]
31. Jiang, R.; Wang, P.; Xu, Y.; Zhou, Z.; Luo, X.; Lan, Y.; Zhao, G.; Sanchez-Azofeifa, A.; Laakso, K. Assessing the Operation Parameters of a Low-Altitude UAV for the Collection of NDVI Values Over a Paddy Rice Field. *Remote Sens.* **2020**, *12*, 1850. [CrossRef]
32. Torres-Sánchez, J.; López-Granados, F.; De Castro, A.I.; Peña-Barragán, J.M. Configuration and Specifications of an Unmanned Aerial Vehicle (UAV) for Early Site Specific Weed Management. *PLoS ONE* **2013**, *8*, e58210. [CrossRef]
33. de Castro, A.I.; Ehsani, R.; Ploetz, R.; Crane, J.H.; Abdulridha, J. Optimum Spectral and Geometric Parameters for Early Detection of Laurel Wilt Disease in Avocado. *Remote Sens. Environ.* **2015**, *171*, 33–44. [CrossRef]
34. Peña, J.M.; Torres-Sánchez, J.; Serrano-Pérez, A.; de Castro, A.I.; López-Granados, F. Quantifying Efficacy and Limits of Unmanned Aerial Vehicle (UAV) Technology for Weed Seedling Detection as Affected by Sensor Resolution. *Sensors* **2015**, *15*, 5609–5626. [CrossRef]
35. Gómez-Candón, D.; De Castro, A.I.; López-Granados, F. Assessing the Accuracy of Mosaics from Unmanned Aerial Vehicle (UAV) Imagery for Precision Agriculture Purposes in Wheat. *Precision Agric.* **2014**, *15*, 44–56. [CrossRef]
36. Zhang, J.; Xie, T.; Yang, C.; Song, H.; Jiang, Z.; Zhou, G.; Zhang, D.; Feng, H.; Xie, J. Segmenting Purple Rapeseed Leaves in the Field from UAV RGB Imagery Using Deep Learning as an Auxiliary Means for Nitrogen Stress Detection. *Remote Sens.* **2020**, *12*, 1403. [CrossRef]
37. Machefer, M.; Lemarchand, F.; Bonnefond, V.; Hitchins, A.; Sidiropoulos, P. Mask R-CNN Refitting Strategy for Plant Counting and Sizing in UAV Imagery. *Remote Sens.* **2020**, *12*, 3015. [CrossRef]
38. Fu, Y.; Yang, G.; Song, X.; Li, Z.; Xu, X.; Feng, H.; Zhao, C. Improved Estimation of Winter Wheat Aboveground Biomass Using Multiscale Textures Extracted from UAV-Based Digital Images and Hyperspectral Feature Analysis. *Remote Sens.* **2021**, *13*, 581. [CrossRef]
39. Bhatnagar, S.; Gill, L.; Ghosh, B. Drone Image Segmentation Using Machine and Deep Learning for Mapping Raised Bog Vegetation Communities. *Remote Sens.* **2020**, *12*, 2602. [CrossRef]
40. Trout, T.J.; Johnson, L.F.; Gartung, J. Remote Sensing of Canopy Cover in Horticultural Crops. *HortScience* **2008**, *43*, 333–337. [CrossRef]

41. Ostos-Garrido, F.J.; de Castro, A.I.; Torres-Sánchez, J.; Pistón, F.; Peña, J.M. High-Throughput Phenotyping of Bioethanol Potential in Cereals Using UAV-Based Multi-Spectral Imagery. *Front. Plant Sci.* **2019**, *10*, 948. [CrossRef] [PubMed]
42. Oerke, E.-C. Crop Losses to Pests. *J. Agric. Sci.* **2006**, *144*, 31–43. [CrossRef]
43. Mahlein, A.-K. Plant Disease Detection by Imaging Sensors–Parallels and Specific Demands for Precision Agriculture and Plant Phenotyping. *Plant Dis.* **2015**, *100*, 241–251. [CrossRef] [PubMed]
44. Messina, G.; Modica, G. Applications of UAV Thermal Imagery in Precision Agriculture: State of the Art and Future Research Outlook. *Remote Sens.* **2020**, *12*, 1491. [CrossRef]
45. Rother, C.; Kolmogorov, V.; Blake, A. "GrabCut": Interactive Foreground Extraction Using Iterated Graph Cuts. *ACM Trans. Graph.* **2004**, *23*, 309–314. [CrossRef]
46. Qiao, L.; Gao, D.; Zhang, J.; Li, M.; Sun, H.; Ma, J. Dynamic Influence Elimination and Chlorophyll Content Diagnosis of Maize Using UAV Spectral Imagery. *Remote Sens.* **2020**, *12*, 2650. [CrossRef]
47. de Castro, A.I.; Jurado-Expósito, M.; Peña-Barragán, J.M.; López-Granados, F. Airborne Multi-Spectral Imagery for Mapping Cruciferous Weeds in Cereal and Legume Crops. *Precis. Agric.* **2012**, *13*, 302–321. [CrossRef]
48. Ferentinos, K.P. Deep Learning Models for Plant Disease Detection and Diagnosis. *Comput. Electron. Agric.* **2018**, *145*, 311–318. [CrossRef]

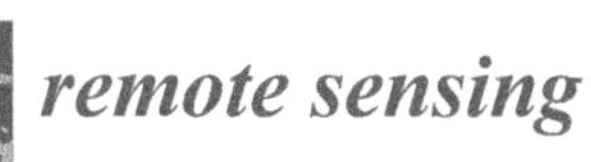 remote sensing

MDPI

Article

Improved Estimation of Winter Wheat Aboveground Biomass Using Multiscale Textures Extracted from UAV-Based Digital Images and Hyperspectral Feature Analysis

Yuanyuan Fu, Guijun Yang [1,2,3], Xiaoyu Song [2,3,*], Zhenhong Li [2,4], Xingang Xu [2,3], Haikuan Feng [2,3] and Chunjiang Zhao [2,3]

1 College of Information and Management Science, Henan Agricultural University, Zhengzhou 450002, China; fuyy@nercita.org.cn
2 Key Laboratory of Quantitative Remote Sensing in Agriculture of Ministry of Agriculture, Beijing Research Center for Information Technology in Agriculture, Beijing 100097, China; songxy@nercita.org.cn (X.S.); Zhenhong.Li@newcastle.ac.uk (Z.L.); xuxg@nercita.org.cn (X.X.); fenghk@nercita.org.cn (H.F.); zhaocj@nercita.org.cn (C.Z.)
3 National Engineering Research Center for Information Technology in Agriculture, Beijing 100097, China
4 School of Engineering, Newcastle University, Newcastle upon Tyne NE1 7RU, UK
* Correspondence: yanggj@nercita.org.cn; Tel.: +86-10-5150-3215

Abstract: Rapid and accurate crop aboveground biomass estimation is beneficial for high-throughput phenotyping and site-specific field management. This study explored the utility of high-definition digital images acquired by a low-flying unmanned aerial vehicle (UAV) and ground-based hyperspectral data for improved estimates of winter wheat biomass. To extract fine textures for characterizing the variations in winter wheat canopy structure during growing seasons, we proposed a multiscale texture extraction method (Multiscale_Gabor_GLCM) that took advantages of multiscale Gabor transformation and gray-level co-occurrency matrix (GLCM) analysis. Narrowband normalized difference vegetation indices (NDVIs) involving all possible two-band combinations and continuum removal of red-edge spectra (SpeCR) were also extracted for biomass estimation. Subsequently, non-parametric linear (i.e., partial least squares regression, PLSR) and nonlinear regression (i.e., least squares support vector machine, LSSVM) analyses were conducted using the extracted spectral features, multiscale textural features and combinations thereof. The visualization technique of LSSVM was utilized to select the multiscale textures that contributed most to the biomass estimation for the first time. Compared with the best-performing NDVI (1193, 1222 nm), the SpeCR yielded higher coefficient of determination (R^2), lower root mean square error (RMSE), and lower mean absolute error (MAE) for winter wheat biomass estimation and significantly alleviated the saturation problem after biomass exceeded 800 g/m^2. The predictive performance of the PLSR and LSSVM regression models based on SpeCR decreased with increasing bandwidths, especially at bandwidths larger than 11 nm. Both the PLSR and LSSVM regression models based on the multiscale textures produced higher accuracies than those based on the single-scale GLCM-based textures. According to the evaluation of variable importance, the texture metrics "Mean" from different scales were determined as the most influential to winter wheat biomass. Using just 10 multiscale textures largely improved predictive performance over using all textures and achieved an accuracy comparable with using SpeCR. The LSSVM regression model based on the combination of the selected multiscale textures, and SpeCR with a bandwidth of 9 nm produced the highest estimation accuracy with $R^2_{val} = 0.87$, $RMSE_{val} = 119.76$ g/m^2, and $MAE_{val} = 91.61$ g/m^2. However, the combination did not significantly improve the estimation accuracy, compared to the use of SpeCR or multiscale textures only. The accuracy of the biomass predicted by the LSSVM regression models was higher than the results of the PLSR models, which demonstrated LSSVM was a potential candidate to characterize winter wheat biomass during multiple growth stages. The study suggests that multiscale textures derived from high-definition UAV-based digital images are competitive with hyperspectral features in predicting winter wheat biomass.

Citation: Fu, Y.; Yang, G.; Song, X.; Li, Z.; Xu, X.; Feng, H.; Zhao, C. Improved Estimation of Winter Wheat Aboveground Biomass Using Multiscale Textures Extracted from UAV-Based Digital Images and Hyperspectral Feature Analysis. *Remote Sens.* **2021**, *13*, 581. https://doi.org/10.3390/rs13040581

Academic Editor: Yeyin Shi
Received: 21 December 2020
Accepted: 5 February 2021
Published: 6 February 2021

Keywords: UAV digital images; winter wheat biomass; multiscale textures; red-edge spectra; least squares support vector machine; variable importance

1. Introduction

As an important grain crop, sustainable wheat production plays a crucial role in ensuring food security in the context of rapidly increasing population worldwide. Aboveground biomass (AGB) usually serves as an important factor in crop growth monitoring and grain yield prediction due to its strong relevance with crop growth status and soil fertility. Moreover, the information on crop biomass is required in calculating critical nitrogen concentration and nitrogen nutrition index, which can well indicate crop nitrogen status and thus facilitate the adjustment of in-season nitrogen management [1,2]. Conventional measurements of crop biomass mainly involve massive manpower investment and destructive sampling in the field [3]. These ground-based methods are time-consuming and not suitable for large-scale crop monitoring, since most of agronomical parameters exhibit high spatial variability within field [4]. Therefore, it is imperative to develop advanced and effective technologies to timely and economically obtain crop biomass information during growing seasons over large areas.

Over the past few decades, remote sensing technology has been gradually recognized as an effective and non-invasive tool for estimating crop biomass and characterizing its spatial and temporal variability. Compared with synthetic aperture radar (SAR) and light detection and ranging (LiDAR), optical remote sensing is still an attractive choice to obtain crop biomass information due to its relatively low cost and the availability of new optical sensors with fine spectral, temporal, and spatial resolutions. The vegetation indices (VIs) derived from multispectral broadband remotely sensed data have long been used as indirect measurements of crop biomass, such as normalized difference vegetation index (NDVI), enhanced vegetation index (EVI), and normalized difference water index (NDWI) [5–7]. However, these vegetation indices have limited value and often lose sensitivity for dense crop canopies when biomass exceeds a certain amount. This VI-AGB insensitivity is known as the saturation problem. Recent studies have asserted that hyperspectral remotely sensed data have great potential to overcome the saturation problem in crop biomass estimation, because they offer approximately continuous spectra and can describe vegetation canopy reflectance in greater detail as compared to multispectral data [3,8–10]. The red-edge region, the transition of red-near infrared, has been shown to have high information content for crop biomass. In the study of Hansen and Schjoerring [3], the newly constructed NDVI (718, 720 nm) yielded high accuracy for the estimation of winter wheat biomass. Fu et al. [9] extracted sensitive features from red-edge spectra using band-depth analysis and combined them with partial least squares regression to estimate winter wheat biomass, achieving higher accuracy than did the optimal NDVI (1097, 980 nm). Marshall and Thenkabail [10] also highlighted the importance of red-edge spectra when comparing the predictive performance of multispectral broadband and hyperspectral narrowband spaceborne remote sensed data for crop biomass estimation. Apart from the wavebands in red edge, several wavebands located in the near-infrared region have been reported to be sensitive to crop biomass. Yao et al. [11] indicated that the continuous wavelet feature located at 1197 nm yielded more accurate estimates of wheat biomass than the existing narrowband hyperspectral VIs. Fu et al. [9] evaluated the predictive performance of all possible two-band combinations in the form of NDVI for winter wheat biomass estimation and found that the optimal NDVI was calculated by the wavebands 1097 and 980 nm. Marshall and Thenkabail [10] indicated that three near-infrared wavebands of Hyperion EO-1 hyperspectral imagery including 914, 1130, and 1320 nm were crucial for crop biomass estimation. In spite of the appealing characteristics of hyperspectral data, many studies associated with hyperspectral remote sensing of crop biomass utilized hand-held devices or ground platforms to acquire hyperspectral data. They are not applicable to crop biomass estimation at large scale. With the accomplishment of upcoming hyperspectral spaceborne missions, the situation will change, and vast hyperspectral data will be available for crop monitoring. However, the weather conditions and specific characteristics of satellite sensors largely limit the availability of high quality hyperspectral images, which satisfy the requirements of precision crop monitoring.

Currently, the rapid development of unmanned aerial vehicles (UAVs) and compact optical sensors opens a new perspective for quantifying crop biomass. The UAV remote sensing systems have shown great potential to fill the gap between ground-based devices and spaceborne and manned airborne remote sensing systems. Compared with ground-based platforms, UAV platforms were no longer confined to a small survey area and can largely improve the efficiency of data collection. The relatively low-cost UAV systems were more accessible for researchers and farmers than spaceborne and manned airborne systems. Moreover, the data acquired by UAV systems usually have higher temporal and spatial resolutions as compared with the existing spaceborne remote sensing systems. A variety of imaging data including red green blue (RGB) digital, multispectral, and hyperspectral images acquired from UAV platforms have been exploited in crop biomass estimation. Bendig et al. [12] indicated that barley biomass was strongly correlated with canopy height, which was extracted from point clouds generated by UAV-based RGB imaging. However, these photogrammetry-derived plant heights were not reliable for dense crop canopies [13]. Moreover, the change in canopy height was not obvious during late growth stages. Therefore, it is not feasible to estimate crop biomass only based on plant height during the entire growth period. Lu et al. [14] reported that the combination of VI and canopy height derived from UAV-based RGB images produced higher estimation accuracy of wheat biomass than the use of VI or canopy height alone. However, the VIs derived from UAV-based RGB images also suffered from the saturation problem. To obtain accurate estimates of potato aboveground biomass, Li et al. [15] established a random forest regression mode, which combined canopy height derived from UAV-based RGB images and multiple hyperspectral indices calculated from UAV-based hyperspectral images. Of these UAV remote sensing systems, the UAV systems equipped with an RGB digital camera have gained great popularity in precision agriculture, owing to its low cost and easy integration with other agricultural management tools. Due to the limitation of UAV-based RGB VIs and canopy height in crop biomass estimation, researchers have begun to mine spatial features from ultrahigh-spatial-resolution RGB images (<10 cm). Texture, a typical spatial feature, is expected to be more effective in imagery with finer spatial resolution, since more discriminative structural details can be shown [16]. Zheng et al. [17] proposed the normalized difference texture index (NDTI), which was calculated by the gray level co-occurrence matrix (GLCM)-based textures. Their experimental results demonstrated the superiority of the NDTI to the evaluated multispectral VIs for the estimation of rice biomass. In the study of Yue et al. [18], the GLCM-based textural features were extracted from UAV-based RGB images with five different spatial resolutions and exhibited better predictive performance in the wheat biomass estimation as compared with traditional hyperspectral VIs and RGB-based VIs. Commonly, crop canopy size and plant density vary with growth stages and field management practices. Therefore, it is difficult to characterize the spatial changes of crop canopy using textural features with a single scale. Better crop estimates may be possible if multiscale textures can be well extracted. Motivated by this idea, a multiscale image texture extraction method is proposed in the study. The proposed texture extraction method takes good advantage of Gabor filters and GLCM analyses. Gabor filters can provide multiscale and multidirectional spatial features that are tolerant to local illumination changes, while GLCM-based textures have the capability of characterizing the local heterogeneity in the tone values of pixels as well as their spatial distribution. To the best of our knowledge, the predictive performance of multiscale textural features has rarely been evaluated in crop biomass estimation.

It is critical to employ appropriate regression methods to correlate the extracted features with crop biomass. Recently, non-parametric regression methods have exhibited superiority over parametric regression methods in crop traits retrieval, especially machine learning based regression techniques. Studies are now saturated with multiple examples of machine learning regression algorithms to estimate a variety of crop traits from remotely sensed data [19–21]. However, it is less informative to treat the established machine learning regression model solely as a black box. Remotely sensed data analyses should always push toward better understanding of the underlying mechanisms and identification of the important spatial and spectral features that contribute most to crop trait of interest. An extensive interest has been shown in applying kernel-based machine learning regession algorithms for crop trait estimation, because they have high generalization performance and good ability to characterize non-linear relationships in a global manner. Owing to the study of Ustun et al. [22], the kernel-based machine learning regression (e.g., support vector machine regression and kernel partial least squares regression) can be interpreted like partial least squares regression (PLSR) rather than a black box technique. The proposed visualization method makes it possible to interpret the driving force behind the established kernel-based regression model. Least squares support vector machine (LSSVM), as a variant of SVM, also applies structural risk minimization and is superior to artifical neural networks (ANNs) in terms of model generalization when limited training samples are available [23]. At present, only limited studies have used LSSVM to retrieve vegetation biophyiscal parameters [24,25]. Moreover, the interpretation of LSSVM model has rarely been conducted. In this study, PLSR and LSSVM were selected as representatives of non-parametric linear and nonlinear regression techniques and used to model wheat biomass. Hyperspectral features describe the average tonal variations in multiple wavebands, whereas textural features contain information concerning the spatial distribution for tonal variations winthin a waveband [26].

The goal of this study was to investigate if the proposed multiscale textures extracted from UAV digital images could compete with hyperspectral information for winter wheat biomass estimation. The specific objectives of this study were to (1) compare the predictive performance of optimal NDVI and continuum removal of red-edge spectra with different bandwidths; (2) evaluate the effectiveness of the proposed multiscale texture extraction method (hereafter referred to as Multiscale_Gabor_GLCM) and determine the optimal textural features for winter wheat biomass estimation; and (3) demonstrate if the combination of multiscale textures and spectral features (i.e., optimal NDVI and continuum removal of red-edge spectra) could further improve estimation accuracy.

2. Materials and Methods

2.1. Experimental Design

Winter wheat field experiments were conducted during three growing seasons (2014–2015, 2017–2018, and 2018–2019) at the Xiaotangshan Experiment Site (40.175°~40.188°N, 116.436°~116.451°E, geographic WGS84), Changping district, Beijing, China. The soil is classified as fine loam with an organic content of 15.3~19.1 mg/kg, a nitrate-nitrogen (NO3-N) content of 8.21~41.71 mg/kg, an ammonium-nitrogen (NH3-N) content of 0.45~8.2 mg/kg, an available potassium content of 60.58~109.61 mg/kg, and an available phosphorus content of 3.14~21.18 mg/kg at top 30 cm soil layer. This experimental station is characterized by a warm temperate and semi-humid continental monsoon climate. The minimum and maximum temperature are −10 °C and 40 °C, respectively. Most of precipitation occurs in summer and the annual amount is about 620 mm.

The trails were undertaken in experimental plots and considered three experimental variables included different cultivars, irrigation rates, and nitrogen (N) application levels (Table 1). Experiment 1 (Exp. 1) followed an orthogonal experimental design and involved three replications of two varieties (Zhongmai 175 and Jing 9843), four N application rates, and three irrigation levels. In this experiment, forty-eight sampling plots were designed with each covering an area of 48 m^2 (6 m × 8 m). Exp. 2 and Exp. 3 were conducted in two consecutive growing seasons (2017–2018 and 2018–2019), and both of them followed a randomized complete block design. There were 64 and 32 sampling plots for the two experiments, respectively. The covering area of each plot was about 135 m^2 (9 × 15 m). The two experiments conducted four repetitions and considered two winter wheat varieties (Lunxuan 167 and Jingdong 18) and four N application rates. The detailed experiment designs can be referred to the studies of Xu et al. [27] and Fu et al. [20]. For all experiments, 50% nitrogen fertilizer was applied prior to seeding, and the left was applied at the jointing stage. Other field management practices followed the local farmlands for winter wheat production.

Table 1. Summary of the treatments used in the three winter wheat field experiments.

Growing Season	Cultivar	Sowing Date	Treatments	Sampling Period
Exp. 1: 2014–2015	Zhongmai175, Jing 9843	7 October	N rate (kg N/ha): 0, 90, 180, 270; Irrigation rate (mm): 0, 192, 384	14 April, jointing (Z.S. 31); 26 April, booting (Z.S. 47); 12 May, anthesis (Z.S. 65); 26 May, filling (Z.S. 75)
Exp. 2: 2017–2018	Lunxuan 167, Jingdong 18	30 September	N rate (kg N/ ha): 0, 90, 180, 270	3 May, booting (Z.S. 47); 16 May, anthesis (Z.S. 65)
Exp. 3: 2018–2019	Lunxuan 167, Jingdong 18	5 October	N rate (kg N/ ha): 0, 90, 180, 270	16 April, jointing (Z.S. 31); 30 April, booting (Z.S. 47)

Note: Z.S. is the abbreviation for Zadoks growth stage.

2.2. Data Collection

2.2.1. UAV Images Acquisition and Pre-Processing

For Exp. 1 and Exp. 2, an eight-rotor UAV equipped with a high-definition digital camera was used to acquire RGB digital images of winter wheat canopy. The camera (Sony DSC-QX100, Sony, Tokyo, Japan; 20.2 megapixel; 64° field of view) was attached to a gimbal, which could compensate the UAV movement during the flight and guaranteed nearly nadir image acquisition. For Exp. 3, a DJI phantom 3 with embedded RGB digital camera (1/2.3-inch CMOS; 12 megapixel) was used to capture images of winter wheat canopy. The UAV systems flew autonomously over the study area according to the predefined flight route programmed by the DJI ground station. The longitudinal and lateral overlaps were set to 80% and 75%, respectively. The flight altitude was about 50 m, corresponding to a ground sampling distance of about 1 cm. The integration time of cameras was adjusted automatically according to the light conditions of each flight to avoid over-saturation and ensure the maximum dynamics. The camera focal length was 10 mm and the aperture was set to f/11 and f/3.2 for Exp. 1 and Exp. 2, respectively. The camera focal length was 9 mm, and the aperture was set to f/5.6 for Exp. 3. All flights were accomplished under cloud-free daylight conditions between 10:00 a.m. and 2:00 p.m. (Beijing local time).

Prior to extracting informative features from UAV-based RGB imagery, image-preprocessing must be conducted, and the main procedures included aligning images, generating dense point cloud, building mesh, constructing texture, and mosaicking image. The generation of ortho-rectified and geo-referred mosaic image depends on the structure from motion algorithm and mosaic blending theory [28,29]. We co-registered the RGB orthomosaics acquired at different growth stages using the image-to-image registration workflow of ENVI (Exelis Visual Information Solution, Boulder, CO, USA) with a geometric accuracy of <0.5 pixel. To maintain the radiometric consistency of multitemporal remotely sensed images, relative radiometric correction was conducted following the algorithm in the literature [30]. The orthomosaic images for the three experiments were all resampled to a ground sampling distance of 1 cm through nearest-neighbor interpolation. To further avoid the influence of illumination variation, the normalization was implemented through dividing each pixel value by the corresponding sum of the pixel value of all bands [31]. For each sampling plot, a region of interest (ROI) was selected from the center of plot area to exclude border effect. Consequently, a total of 384 ROIs were cropped, including 192 ROIs (448 × 553 pixels) for Exp. 1, 128 ROIs (621 × 930 pixels) for Exp. 2, and 64 ROIs (621 × 930 pixels) for the Exp. 3.

2.2.2. Hyperspectral Data Acquisition and Biomass Measurements

After UAV flight missions, the field hyperspectral reflectance measurements of wheat canopy were acquired by a field spectroradiometer (Analytical Spectral Devices, Boulder, CO. USA) with a spectral range of 350~2500 nm under cloud-free conditions around solar noon. The spectral resolution is 1.4 nm between 350 and 1050 nm, while the spectral resolution is 2 nm between 1050 and 2500 nm. Spectral data were usually interpolated to a spectral resolution of 1 nm for further analysis by the corresponding software. The sensor was placed vertically approximately 1.3 m above wheat canopy, with a field of view of 25°. Prior to each measurement, the radiance of a white standard spectral reference panel was recorded to convert the radiance measurements of each plot to reflectance. To ensure the reliability of spectral measurments, ten replicate spectra were taken for each plot, and the average spectrum of them was used to represent the spectrum of the plot. The spectra in the spectral region of 400~1350 nm were used for the subsequent analyses, due to their high signal to noise ratio. To further suppress noise in the spectral measurements, a Savitzky–Golay filter with a frame size of 19 was used.

After the collection of hyperspectral data, ground-based plant sampling was undertaken, and an area of 0.25 m^2 wheat plants was clipped at ground level for each plot. The cut plants were sealed in a plastic bag and sent to laboratory immediately. These samples were firstly green-killed in an oven at 105 °C for about 30 minutes and then dried at 80 °C until a constant weight. The dry aboveground biomass was expressed on the basis of ground area according to the number of plant stems per unit area.

2.3. Multiscale Textures Extraction by Fusing Gabor Filter and GLCM Analyses

To capture the changes in winter wheat canopy during growing season well, multiscale textural features derived from high-definition UAV-based RGB images were required and expected to be more suitable for biomass estimation. To this end, we proposed a new texture extraction method, which took advantages of two-dimensional (2D) Gabor filter and GLCM analyses. For illustrative purposes, Figure 1 shows the schematic of the proposed multiscale texture extraction method (Multiscale_Gabor_GLCM). GLCM is computed for various angular relationships and distances between nearest pixel pairs in the image [26]. The textures derived from GLCM can well represent the spatial distribution of gray tones rather than the solely tones. The human visual system is regarded to have a special function to process information in multiresolution levels, which can be mimicked by changing the scale parameter in 2D Gabor filters [32]. Gabor filters are actually a series of local spatial band-pass filters, which are characterized by good spatial localization, orientation selectivity, and frequency selectivity [32]. In this study, four orientations ([0°, 45°, 90°, 135°]) and five scales ($\nu \in [0, 1, 2, 3, 4]$) were considered to generate a bank of Gabor filters, which approximately uniformly covered the spatial-frequency domain. The settings of other parameters were the same with the literature [20]. According to our previous study [20], for UAV-based RGB images of winter wheat canopy, there was no significant heterogeneity in both Gabor- and GLCM-based textures of different orientations, and the textures of 45° were recommended. Thus, we only considered the orientation of 45° in the subsequent Gabor representation and GLCM analyses. For a ROI of winter wheat, each image band was convoluted with the Gabor filter bank, and a total of 15 Gabor magnitude images were generated. Then GLCM-based textures were extracted from each Gabor magnitude image to form multiscale image textural features. The used GLCM-based textures mainly included mean (Mean), variance (Var), homogeneity (Hom), contrast (Con), dissimilarity (Dis), entropy (Ent), second moment (Sec), and correlation (Cor) [26]. Consequently, a total of 120 textural features were extracted for each ROI image.

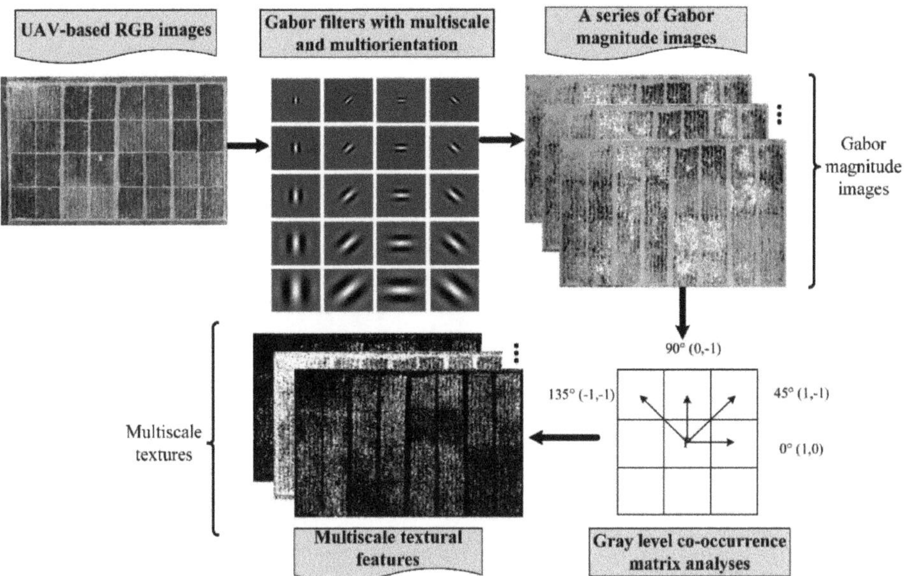

Figure 1. System block diagram of the proposed multiscale texture extraction method (Multiscale_Gabor_GLCM).

The suffixes indicating specific feature scale and image band followed the names of textures for convenience and illustrative purpose. For instance, "Hom-G-S2"represents the texture homogeneity at scale 2 derived from green band, "Sec-R-S1"represents the texture second moment at scale 1 derived from red band, and "Cor-B-S5"represents the texture correlation at scale 5 derived from blue band.

2.4. *Biomass Related Hyperspectral Feature Extraction*

To extract hyperspectral features sensitive to wheat biomass, two methods were conducted including optimal narrowband vegetation index and continuum removal of red-edge spectra.

2.4.1. Narrowband Vegetation Index

Two-band normalized difference vegetation index (NDVI) was used to select biomass specific wavebands. All possible two-pair combinations of 951 wavelengths were used to calculate the narrowband NDVI-like indices. Then, the coefficient of determination (R^2) was obtained in the univariate linear regression of these NDVI-like indices and winter wheat biomass based on the calibration dataset. According to the two-dimension (2D) correlation scalogram, the wavelength combinations resulting in the highest R^2 values formed the optimal NDVI-like indices and were identified as sensitive spectral features to winter wheat biomass.

2.4.2. Continuum Removal of Red-Edge Spectra with Different Bandwidths

Red-edge spectral region has proved to have abundant information for crop biomass. To effectively extract spectral features from red edge, continuum removal was conducted on the spectra between 550 and 750 nm. Comparing with original hyperspectral data, continuum removal enables direct comparison of individual absorption features from a common baseline. Considering the strong intercorrelation among adjacent hyperspectral wavebands, the spectra between 550 and 750 nm under different bandwidths were simulated with full width at half maximum (FWHM) ranging from 1 to 21 nm and a step of 2 nm.

The spectral response function was simulated by Gaussian function (*f*) and calculated with the following Equations (1) and (2).

$$f(b,\sigma) = \exp\left(-\frac{(-b-b_c)^2}{2\sigma^2}\right) \tag{1}$$

$$\sigma = \frac{FWHM}{2\sqrt{2ln2}} \tag{2}$$

where b represents the wavelength in the spectral response range for a certain band width; b_c is the central wavelength of interest; σ is standard variance; and ln is natural logarithm operator.

Apart from the reduction of the hyperspectral dimension, simulating spectra with different bandwidths was conducive to investigating the influence of bandwidth on the predictive performance of red-edge spectral features for winter wheat biomass estimation.

2.5. Non-Parametric Modeling Methods: PLSR and LSSVM

In this study, two different regression techniques were analyzed, including a non-parametric linear regression (i.e., PLSR) and a non-parametric nonlinear regression (i.e., LSSVM), so that improvements in estimation accuracy of features–algorithm pairs could be evaluated. Moreover, the visualization of LSSVM facilitated the identification of influential textural features for winter wheat biomass and improved the understanding of the established predictive models.

2.5.1. Partial Least Squares Regression

As a computationally efficient modeling algorithm, PLSR has been widely used to establish predictive models of crop traits based on a variety of features, due to its ability to cope with multi-collinearity and high-dimensional data. Rather than vegetation indices only using several wavebands, PLSR can incorporate full-spectrum data into modeling analysis through decomposing high-dimensional data into several orthogonal latent factors as well as maximizing the covariance between dependent and independent variables [33]. The leave-one-out cross-validation (LOOCV) was used to select the optimal latent factors, which is crucial to model performance. To make compromises between model predictive ability and simplicity, the criterion of adding an extra factor to the model was that the root mean square error of LOOCV was decreased by no less than 2% [34,35].

2.5.2. Least Squares Support Vector Machine

As a variant of SVM, LSSVM inherits many strengths of standard SVM, including application of structural risk minimization, being capable of processing data with high-dimension and non-normal distribution, and overcoming the problem of small sample [23]. However, the solution of LSSVM is accomplished through solving a set of equality constraints instead of inequality ones in standard SVM [23]. It avoid solving a quadratic programming and largely improves the efficency of model establishement. LSSVM has been widely adopted in characterizing the nonlinear relationships between crop traits and various features. However, the driving force behind the established LSSVM regression models has been rarely interpreted and visualized. In this study, we took advantage of the visualization method for support vector regression (SVR) and investigated the influential textures for winter wheat biomass estimation. In the study of Ustun et al. [22], the interpretation of a SVR model is realized by computing the inner product of the independent variable matrix and the vector of Lagrange multipliers (α-vector). The length of the resulting new vector is the number of independent variables. This new vector can be used to indicate which independent variables are highly correlated to the dependent variables of interest, like the partial least squares (PLS) loadings or regression coefficients. After identifying the most relevant independent variables through the initial LSSVM regression model, the final LSSVM regression was rebuilt with relevant independent variables. During the

LSSVM regression modeling, the regularization parameter and kernel function parameters such as the standard variance of radial basis function (RBF) kernel were determined by LOOCV in the study.

2.6. Modeling Framework and Validation

To construct and validate winter wheat biomass predictive models, the three-year experimental data were split into calibration and validation datasets. The calibration dataset was used to adjust the parameters of models, while the validation dataset was used to confirm the true predictive power of the established models and had no effect on model construction. The validation dataset consisted of the data from the first replicate collected in 2015 and the fourth replicate collected in 2018 and 2019. The left data made up the calibration dataset. Consequently, there were 272 and 112 samples in calibration and validation datasets, respectively. Figure 2 illustrates the modeling framework of this study. To validate the effectiveness of the proposed texture extraction method Multi-scale_Gabor_GLCM, the GLCM-based textures were directly extracted from UAV-based RGB images and used to construct winter wheat biomass predictive model. To illustrate the necessity of the extraction of multiscale textures, the predictive performance of the textures from each scale was evaluated. The spectral-based predictive models included the linear regression model based on the optimal NDVI-like and the models based on the continuum removal of red-edge spectra with different bandwidths. The PLSR and LSSVM analyses based on the combination of multiscale textures and biomass related spectral features were conducted to ascertain whether the combination could improve estimation accuracy. Before modeling analyses, the combined feature matrices were standardized to reduce the influence of different ranges of feature values. The coefficient of determination (R^2), root mean squared error (RMSE), and mean absolute error (MAE) between observations and estimations were used to assess the accuracy of biomass predicted by different models.

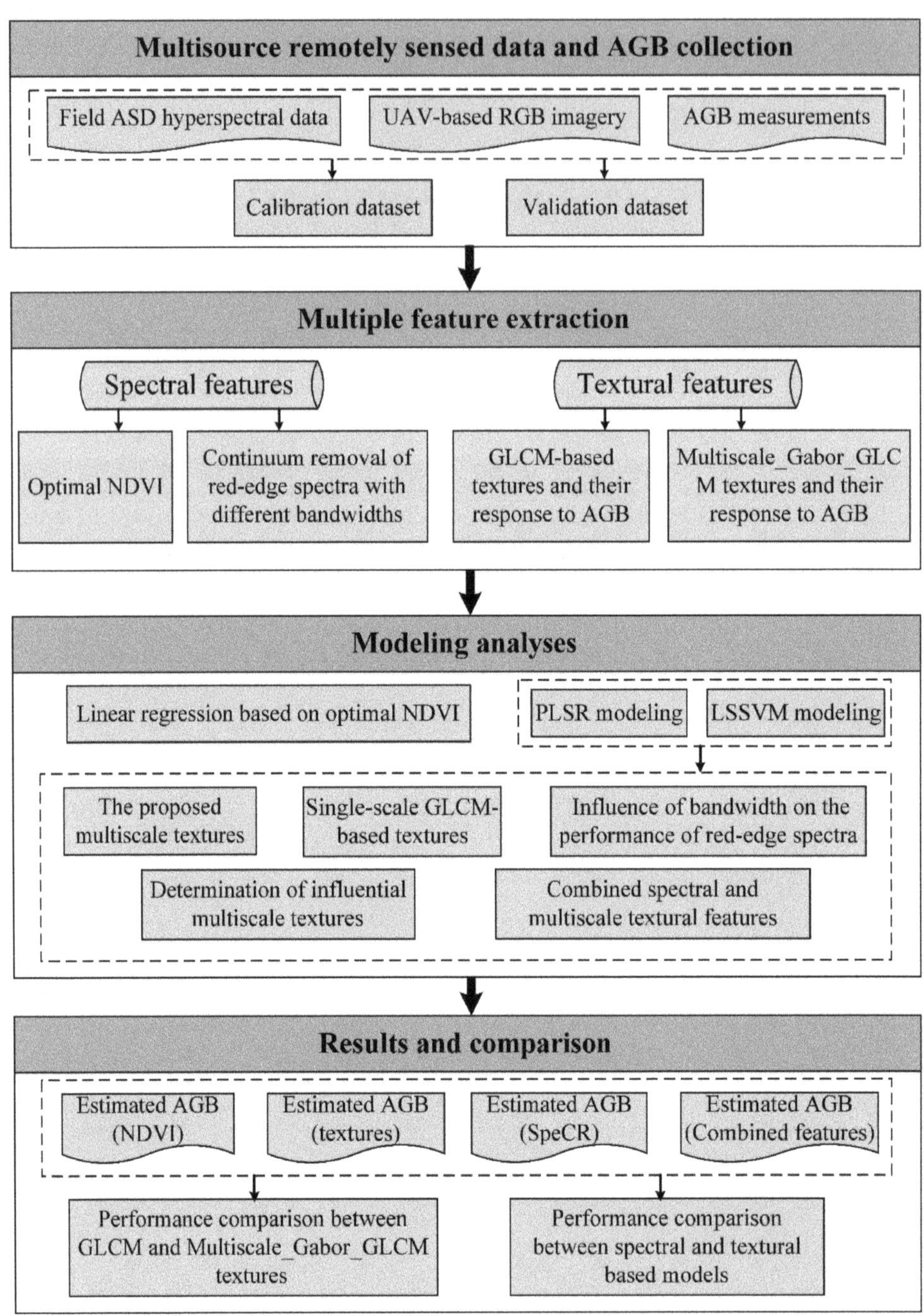

Figure 2. Modeling framework and comparison analyses of the study.

3. Results

3.1. Descriptive Statistics

The differences in winter wheat aboveground biomass were created by multiple treatments including different varieties, different levels of N fertilizer applications, and different irrigation rates. The AGB values under different treatments were compared through the least significant different test (LSD) at the 95% level of significance. The results showed that there existed significant difference in the winter wheat AGB collected from the three experiments ($p < 0.05$). The descriptive statistics of winter wheat AGB exhibited a high degree of variation with a coefficient of variation (CV) of 57.88% for the calibration dataset and a CV of 60.88% for the validation dataset (Table 2). The range of AGB in the calibration dataset is larger than that for the validation dataset. The kurtosis values indicated that the AGB in both calibration and validation datasets approximately followed normal distribution. To help to understand the RMSE and MAE for calibration and validation, the minimum (Min.), mean, maximum (Max.), and standard deviation (Std.) of the observed AGB are also shown in Table 2.

Table 2. Descriptive statistics for winter wheat AGB (g/m^2) of calibration and validation datasets.

Wheat AGB	Min.	Mean	Max.	Std.	CV (%)	Kurtosis
Calibration	50.06	575.85	1759.95	333.30	57.88	3.38
Validation	75.96	526.17	1501.63	320.38	60.88	3.25

3.2. Using Spectral Features to Estimate Winter Wheat AGB

3.2.1. Univariate Linear Regression Based on Narrowband NDVI-Like

Narrowband NDVI-like indices were calculated using all possible two-band combinations. The univariate linear regression analyses of winter wheat biomass against these NDVI-like indices were performed to obtain the R^2 values. As shown in Figure 3a, the formed two-dimension (2D) correlation scalogram exhibits a characteristic pattern with multiple "hot spots" with relatively high R^2 values. The spots were selected by choosing the wavelength combination that had an R^2 larger than 0.55 ($p < 0.01$). It could be observed that most of the wavebands forming these "hot spots" were located in the spectral range of 1168~1276 nm, and several band pairs were located in the spectral range of 675~683 nm. The optimal band combination with the largest R^2 was composed of 1193 and 1222 nm. The univariate linear regression model based on the optimal NDVI-like (1193, 1222 nm) indices was constructed and assessed using the validation dataset. The predictive performance of this model is shown in Figure 3b. The RMSE of this model in the validation dataset reached up to 197.30 g/m^2. The majority error resulted from the underestimation for the biomass samples exceeding 800 g/m^2. It indicated that the optimal NDVI-like (1193, 1222 nm) suffered from saturation problem and consequently resulted in reduced winter wheat AGB estimation accuracy.

3.2.2. Non-Parametric Modeling Based on the Continuum Removal of Red-Edge Spectra

To investigate the influence of bandwidth on the predictive performance of continuum removal of red-edge spectra (SpeCR), both PLSR and LSSVM models were established based on the SpeCR under different bandwidths. In the comparison of Figure 4a,b, it could be observed that the predictive performance of the LSSVM regression models was generally better than that of the PLSR regression models. The RMSE values for the validation dataset generally increased with increasing bandwidth (Figure 4a,b). When the bandwidth exceeded 11 nm, the increasing RMSE values indicated that both LSSVM and PLSR models had an obvious decrease in predictive performance. Interestingly, both PLSR and LSSVM analyses on the SpeCR with a bandwidth of 1 nm did not produce the highest estimation accuracy. The PLSR model based on the SpeCR yielded the highest accuracy when the bandwidth was 5 (R^2_{val} = 0.83, $RMSE_{val}$ = 133.45 g/m^2, and MAE_{val} = 101.65 g/m^2).

The LSSVM regression model based on the SpeCR achieved the highest accuracy when the bandwidth was 9 (R^2_{val} = 0.87, $RMSE_{val}$ = 123.98 g/m^2 and MAE_{val} = 92.08 g/m^2). Compared with the linear model based on the optimal NDVI-like, the optimal models based on the SpeCR significantly improved the estimation accuracy and decreased the RMSE value by 63.85~73.32 g/m^2. As observed in the scatter plots between observed and estimated AGB (Figure 4c,d), to some extent, the models based on the SpeCR alleviated the underestimation problem for the AGB exceeding 800 g/m^2, especially the LSSVM regression model.

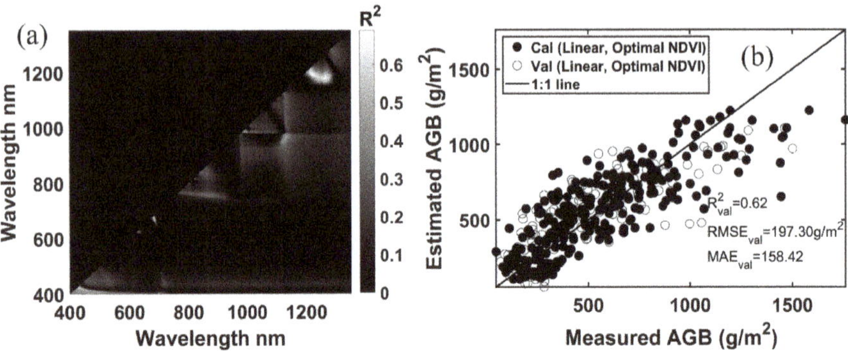

Figure 3. (**a**) Two-dimension scalogram illustrating the coefficient of determination (R^2) between biomass and narrowband NDVI-like indices based on all possible two-band combinations; (**b**) predictive performance of the optimal NDVI-like (1193, 1222 nm) for winter wheat biomass estimation.

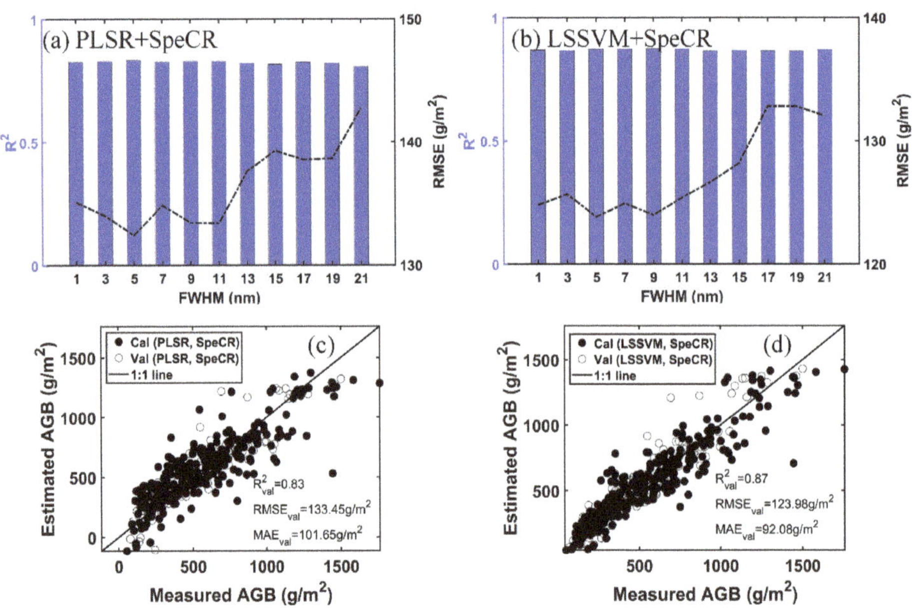

Figure 4. Predictive performance of the continuum removal of red-edge spectra (SpeCR) for winter wheat biomass estimation. Change of the predictive performance of SpeCR using PLSR (**a**) and LSSVM (**b**) under different bandwidths. Predictive performance of the SpeCR using PLSR at the optimal bandwidth (FWHM = 5) (**c**) and using LSSVM at the optimal bandwith (FWHM = 9) (**d**).

3.3. Using Multiscale Textures to Estimate Winter Wheat AGB

The Pearson correlation analyses were conducted between textures and winter wheat AGB during multiple growth stages (Figure 5). For the single-scale GLCM-based textures, the texures from blue band showed higher correlation with AGB than the textures from red and green bands. The texture Var-B produced the highest correlation coefficient (r = 0.41, $p < 0.01$). For the textures extracted by the proposed Multiscale_Gabor_GLCM method, the textures correlating well with AGB could be observed at each scale (Figure 5b). The texture Mea-G-S1 showed the strongest correlation with AGB (r = 0.69, $p < 0.01$). As can be observed from the scatter plots shown in Figure 5c,d, there exist linear relationships between these best-performing textures and AGB for each growth stage. However, the sensitivity of these textures to the variation in AGB decreased when considering multiple growth stages. In the comparison of Figure 5c,d, the best-performing texture Mea-G-S1 exhibited higher correlation with AGB than did the best-performing GLCM-based texture Var-B. For the AGB exceeding 800 g/m^2, the texture Mea-G-S1 still stayed sensitive to AGB, while the texture Var-B approximately approached a saturation level and disorderly response to AGB. Therefore, the models based on the multiscale textures might be more promising than those based on the single-scale GLCM-based textures.

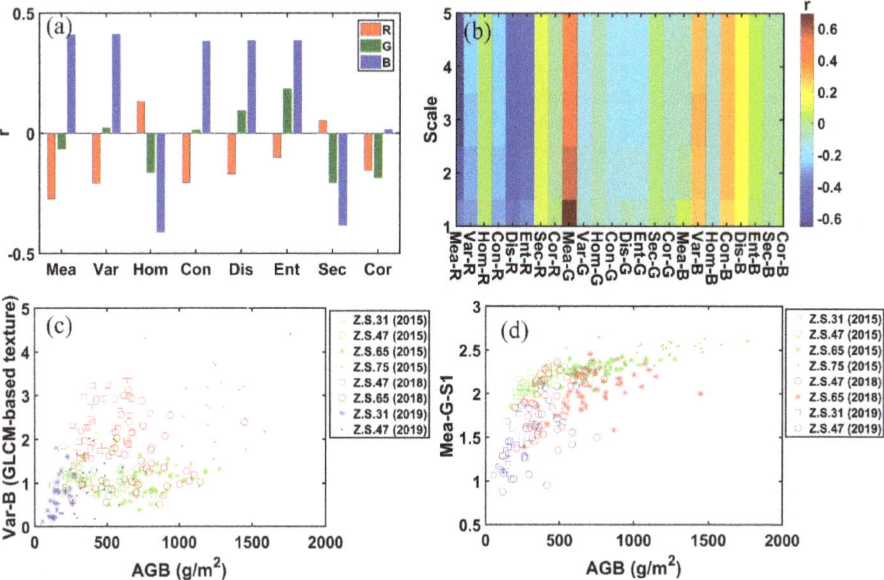

Figure 5. Responses of different textures to winter wheat AGB in the calibration dataset. (**a**) Pearson correlation analyses between GLCM-based textures and AGB; (**b**) Pearson correlation analyses between AGB and Multiscale_Gabor_GLCM textures; (**c**,**d**) Scatter plots between the best-performing textures and AGB for each growth stage.

PLSR analyses were conducted uisng the single-scale GLCM-based textures and multiscale textures, respectively. Figure 6 shows the results of these models for both calibration and validation datasets. For the two PLSR models, the optimal latent factors were 11 and 10, respectively. Compared with the PLSR model based on the single-scale GLCM-based textures, the PLSR model based on the multiscale textures performed better with a R^2 value of 0.78, a RMSE value of 154.54 g/m^2, and a MAE value of 122.86 g/m^2. This model increased the R^2 value by 0.09 and decreased the RMSE value by 24.58 g/m^2.

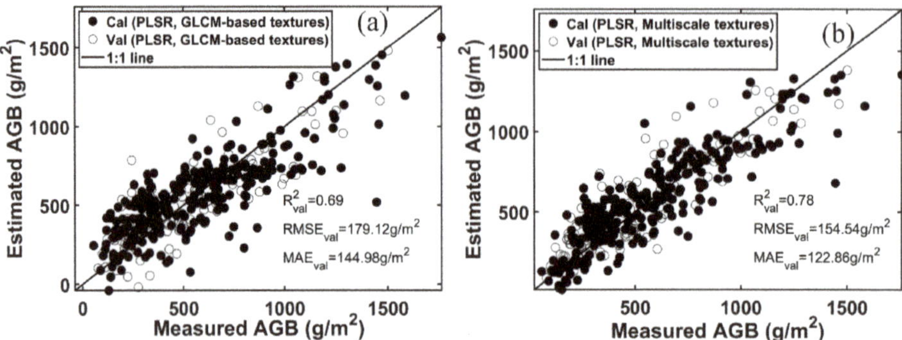

Figure 6. Predictive performance of the PLSR models using (**a**) the single-scale GLCM-based textures and (**b**) multiscale textures for winter wheat AGB estimation.

The influential textures were determined during the construction of LSSVM regression models. According to the importance value of each texture (Figure 7a for the GLCM-based textures; Figure 7b for the multiscale textures), the GLCM-based textures with importance values larger than 5 and the multiscale textures with importance values larger than 200 were selected for the rebuilding of new LSSVM regression models after several trials. A total of seven single-scale GLCM-based textures were selected including Mea-R, Var-R, Con-R, Mea-G, Mea-B, Var-B, and Con-B. The selected multiscale textures were Mea-R-S1, Mea-R-S3, Mea-R-S4, Mea-R-S5, Mea-G-S1, Mea-G-S2, Mea-G-S3, Mea-G-S4, Mea-G-S5, and Mea-B-S1 (10 textures). It can be observed that the multiscale texture metric "Mea" played an important role in AGB estimation. Compared with the initial LSSVM regression models based on all textures, the new LSSVM regression models not only improved the predictive performance, but also reduced the complexity of models (See Figure A1 in Appendix A). The LSSVM regression model based on the selected multiscale textures exhibited higher accuracy with R^2_{val} = 0.85, $RMSE_{val}$ = 129.45 g/m^2, and MAE_{val} = 97.11 g/m^2, compared to the LSSVM regression model using the selected single-scale GLCM-based textures (Figure 7c,d). For both single-scale GLCM-based textures and multiscale textures, the LSSVM regression models performed better than the PLSR models (Figures 6 and 7). For instance, compared with the PLSR model based on the multiscale textures, the LSSVM regression model using the selected multiscale features increased the R^2 value by 0.08 and reduced RMSE value by 25.09 g/m^2. It indicated that LSSVM regression was superior to PLSR in characterizing the nonliner relationship between textures and winter wheat AGB. Compared with the spectral features based models, the LSSVM regression model based on the multiscale textures performed better than the linear model based on the optimal NDVI, but just got comparable estimation accuracy with the LSSVM regression model based on the SpeCR with a bandwidth of 9.

To further demonstrate the necessity of extracting multiscale textures for AGB estimation under multiple growth stages, the LSSVM regression analyses were conducted using the textures at each scale to examine their predictive performance. According to the importance values for the textures at each scale, it could be observed that the texture metric "Mea" still had a higher degree of importance than the other texture metrics. Among the textures from five scales, the textures at scale 1 yielded the best estimation accuracy with R^2_{val} = 0.82, $RMSE_{val}$ = 140.95 g/m^2, and MAE_{val} = 109.10 g/m^2. The textures at scale 2 and scale 3 got the intermediate estimation accuracy. The textures at scale 5 got the worst estimation accuracy and achieved comparable accuracy with the single-scale GLCM-based textures. It could be obviously observed that the LSSVM regression model based on the multiscale textures produced better estimation accuracy than that based on the single-scale textures.

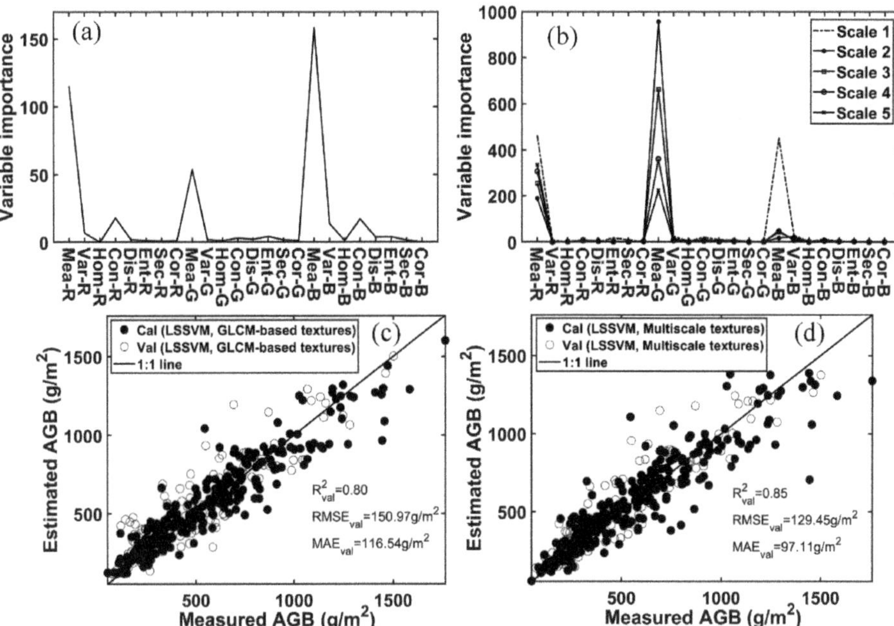

Figure 7. Predictive performance of the LSSVM regression models based on the selected GLCM-based textures (**a**,**c**) and selected multiscale textures (**b**,**d**) for winter wheat AGB estimation.

3.4. The Combined Use of Multiscale Textures and Spectral Features for Winter Wheat AGB Estimation

With combinations of selected multiscale textures, optimal NDVI and SpeCR, the LSSVM regression analyses did not significantly improve estimation accuracy and produced generally better (or similar) results than (or to) those using textures or SpeCR alone. Among these models, the LSSVM regression model based on the combination of selected multiscale textures and SpeCR (FWHM = 9) yielded the highest estimation accuracy of AGB with R^2_{val} = 0.87, $RMSE_{val}$ = 119.76 g/m^2, and MAE_{val} = 91.61 g/m^2 (Table 3). However, the combination of selected mulsticale textures and optimal NDVI just consisted of 11 features and yielded comparable estimation accuracy to the combination of multiscale textures and SpeCR (FWHM = 9) consisting of 29 features (i.e., 10 selected multiscale textures and 19 spectral features).

Table 3. Performance of the LSSVM regression models based on the combination of multiscale textures and biomass-sensitive spectral features for winter wheat biomass estimation.

	Calibration			Validation		
	R^2	RMSE (g/m^2)	MAE (g/m^2)	R^2	RMSE (g/m^2)	MAE (g/m^2)
Multiscale textures + optimal NDVI	0.85	126.98	90.50	0.86	121.28	94.27
Multiscale textures + SpeCR (FWHM = 9)	0.87	121.89	85.85	0.87	119.76	91.61
Multiscale textures + Optimal NDVI + SpeCR (FWHM = 9)	0.87	120.55	84.59	0.87	121.69	91.77

4. Discussion

A three-year field experiment with different cultivars, N fertilizer application, and irrigation rates led to a large number of samples with high variations in winter wheat

AGB. An economical UAV remote sensing system successfully acquired the RGB digital imagery of winter wheat canopy during growing seasons. The spatial resolution of the RGB imagery reached up to 1 cm. It allowed the extraction of fine textural features and the assessment of their predictive performance for winter wheat AGB estimation. The ground-based hyperspectral data and biomass measurements were concurrently collected. A comparison of the predictive performance of hyperspectral features, textural features, and combinations thereof for winter wheat AGB estimation was also possible.

4.1. Performance Comparison of Simple and Complex Spectral Models for Wheat Winter AGB Estimation

Using all the possible two-band combinations for narrowband NDVI-like calculation facilitated the selection of sensitive hyperspectral features for winter wheat AGB. In the linear regression analyses of narrowband NDVI-like indices against winter wheat AGB, it was found that the wavebands forming the "hot spots" in the 2D correlation matrix plot were mainly located between 1168 and 1276 nm (Figure 2). This spectral area is close to 1200 nm, which is affected by the absorption of water, cellulose, starch, and lignin [36]. Yao et al. [11] also indicated that the continuous wavelet features extracted from this spectral region exhibited strong correlation with wheat AGB. Gnyp et al. [37] developed a new hyperspectral vegetation index named GnyLi, which was calculated from wavebands 900, 955, 1050, and 1220 nm. They indicated that the GnyLi vegetation index correlated well with winter AGB during several growth stages (Z.S.30~45) and highlighted the ability of local peak spectra in near-infrared and shortwave infrared regions in alleviating the saturation problem for vegetation AGB estimation. In our study, the best-performing NDVI-like index was composed of wavebands 1193 and 1222 nm. However, in the study of Fu et al. [9], the best narrowband NDVI-like index for winter wheat AGB estimation was calculated from wavebands located at 980 and 1097 nm. The inconsistency in selected wavebands mainly ascribed to the use of different calibration datasets. It is a common problem in band selection for crop AGB estimation based on data-driven methods. In spite of this, the selected wavebands forming the optimal NDVIs can be interpretable in terms of spectral response of crop biomass. Further studies could focus on the improvement in the representativeness of calibration datasets. It can be realized by collecting wheat samples with different cultivars, growth stages, field treatments, and ecological conditions. Based on the comprehensive datasets, the selected sensitive wavebands might be more stable and the predictive model based these wavebands might have better generalization ability. Another way to improve the stability of narrowband vegetation indices is to develop the vegetation indices, which are specific to a certain crop cultivar and environment condition. Obviously, the latter ones are confined to limited application.

Both PLSR and LSSVM analyses using the continuum removal of hyperspectra between 550 and 750 nm produced better estimation accuracy as compared with the linear regression model based on the optimal NDVI (1193 nm, 1222 nm). The results confirm the capability of red-edge spectra in predicting winter wheat AGB, consistent with previous findings reported by Hansen and Schjoerring [3], Fu et al. [9], and Kanke et al. [38]. They have indicated that the red-edge region contains useful information relevant to crop biomass. It is important to note that the LSSVM regression analysis performed better than the PLSR for winter wheat AGB estimation, when combined with the continuum removal of spectra with different bandwidths. It demonstrated that LSSVM regression analysis was advantageous over PLSR in characterizing the relationships between spectral features and winter wheat AGB during multiple growth stages. In the exploration of the influence of bandwidth on the predictive performance of red-edge spectra, both the LSSVM regression and PLSR models generally got decreased estimation accuracy with increasing bandwidths. However, the continuum removal of spectra with a bandwidth of 1 nm did not yield the highest estimation accuracy. It mainly attributes to the high intercorrelation in neighboring hyperspectral wavebands, which has a negative impact on model accuracy and interpretability. However, PLSR and LSSVM have the ability to partly overcome the multi-collinearity problem. Through increasing the bandwidth, the spectral redundancy can be reduced. The PLSR and LSSVM regression models achieved the highest estimation accuracy at the bandwidths of 5 and 9 nm, respectively. The obtained estimation accuracies were comparable to those of the PLSR and LSSVM regression models using all spectral wavebands between 400 and 1350 nm (Figure A2 in Appendix A). After the bandwidth exceeded 11 nm, the predictive performance of both PLSR and LSSVM models had a clear decrease (Figure 4a,b). It means that a fraction of spectral features associated with winter wheat AGB are lost when the hyperspectral data are resampled to multispectral data. It also demonstrates the necessity of exploring hyperspectral features for winter wheat AGB estimation. At the bandwidth of 21 nm, the predictive performance of the LSSVM regression model was still acceptable with a RMSE value of 132.05 g/m^2. This result is conducive to develop hand-held multispectral devices or low-cost digital cameras for proximal crop monitoring.

4.2. Performance Comparison of Single-Scale and Multiscale Textures for Winter Wheat AGB Estimation

With the aim to extract textures that can well characterize the variations of winter wheat canopy during multiple growth stages, a multiscale texture extraction method was proposed (Multiscale_Gabor_GLCM). This method comprehensively utilized the advantages of Gabor filter and GLCM analyses. In the Pearson correlation analyses between winter wheat AGB and textures, the texture metric "Mea-G-S1" extracted by the proposed method exhibited higher correlation with AGB than the best-performing single-scale GLCM-based texture "Var-B" (Figure 5). Yue et al. [18] also reported that the GLCM-based texture metric "Var-B" correlated well with winter wheat AGB. Zheng et al. [17] indicated that the GLCM-based texture metric "Mea" was better than the other seven texture metrics in rice AGB estimation. Through the visualization technique of LSSVM regression, we determined 10 textures (i.e., metric "Mea" from different scales) that contributed most to winter wheat AGB (Figure 7b). In the exploration whether multiscale textures could show superiority to single-scale textures for winter wheat AGB estimation, both PLSR and LSSVM regression models based on the multiscale textures yielded higher R^2 and lower RMSE values for the validation dataset compared to the models based on the single-scale textures including GLCM-based textures and textures at each scale (Figures 6–8). The RGB imagery of winter wheat canopy is usually composed of soils, leaves, stems, ears, and shades resulted from leaf clumping. The proportion of each component changes and wheat canopy structure gradually becomes complicated during growing seasons. At the jointing stage, wheat canopy cannot fully cover soil background, and soils can be seen from imagery. From the heading stage to anthesis stage, wheat ears and flowers emerge, and wheat canopy almost fully covers soil background. From the anthesis stage to early grain filling stage, wheat ears become plump and leaves begin to turn yellow. These wheat canopy dynamic changes reflected in digital images can be captured by fine image textures. Moreover, winter canopies usually exhibit heterogeneity within or among fields due to different cultivars, field treatments, and cropland micro-meteorology, even at the same growth stage. Therefore, using single-scale textures will hardly capture the spatial changes of wheat canopies during growing seasons. This is confirmed by the predictive performance comparison between multiscale and single-scale textures for winter wheat AGB estimation. The proposed multiscale texture extraction method firstly utilized Gabor filters with multiple scales and orientations to generate textural images (i.e., Gabor magnitude images). These Gabor textures are less sensitive to variations of local lighting conditions, which will improve the reliability of subsequent image processing. Then GLCM analyses were conducted on these images to extract finer textures. Our previous study [20] has found that the textures of different orientations did not show significant heterogeneity for winter wheat canopy imagery; thus, this study only focused on the scale properties of textures derived from UAV-based RGB imagery. The optimal scale of textures is highly correlated with crop planting density and canopy size. However, the quantitative relationships among them are still unclear. To obtain abundant discriminative features, we followed the previous research results [20,32] and set five different scales, which determined the spatial frequencies of Gabor filters. In the study of Yue et al. [18], they resampled original UAV-based RGB images to images with spatial resolutions of 2, 5, 10, 15, 20, 25, and 30 cm, respectively. Then they extracted GLCM-based textures from these images for winter wheat AGB estimation. The aim of doing this was also to extract multiscale textures to represent winter wheat canopy variations during multiple growth stages. They got improved estimation accuracy of winter wheat AGB by combining PLSR and the GLCM-based textures derived from RGB images with different spatial resolutions, which was consistent with the findings in our study. However, it seemed a little arbitrary in the determination of spatial resolution of images for analysis. In our proposed multiscale texture extraction method, the scale of textures was controlled by the scale parameter of Gabor filters. Further studies should investigate the quantitative relationships between crop canopy structure and scale parameters. It could help to determine the optimal scale parameter when extracting multiscale textures for crops of interest. In this study, the proposed multiscale texture extraction framework can also involve other multiresolution image analysis methods such as 2D wavelet transformation and 3D Gabor filter.

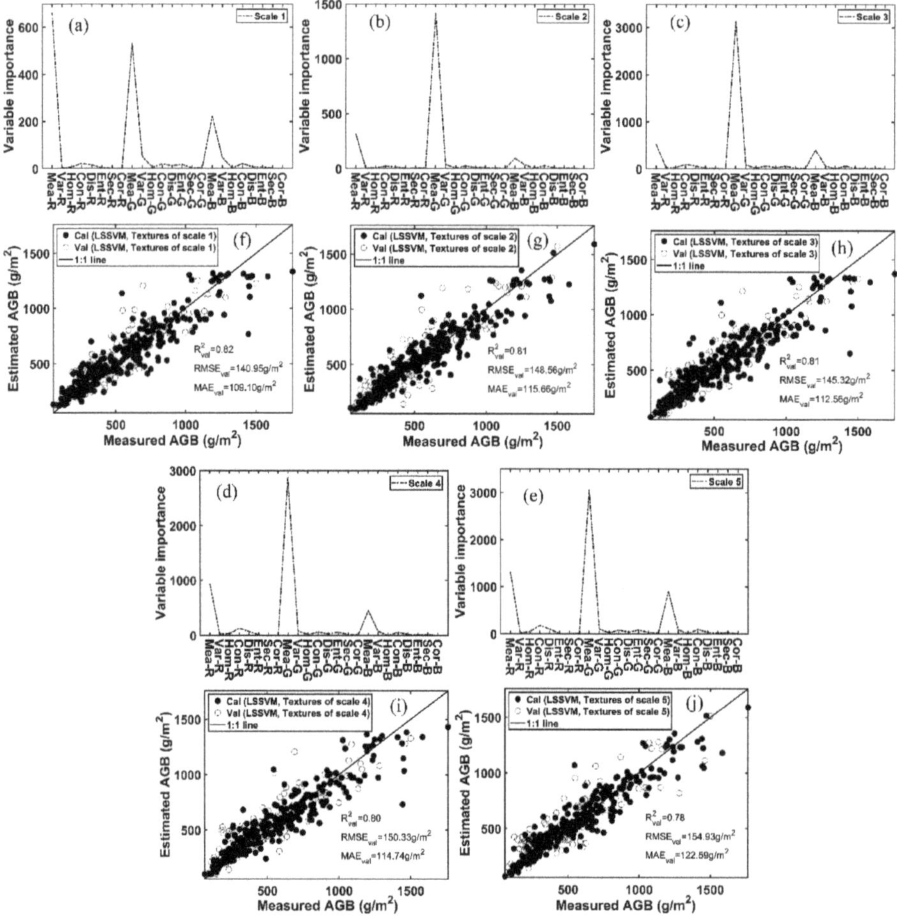

Figure 8. Importance values (**a**–**e**) and predictive performance (**f**–**j**) of the textures at each scale based on LSSVM regession analyses.

4.3. Evaluation of the Combined Use of Hyperspectral and Textural Features for Winter Wheat AGB Estimation

Before the combined use of hyperspectral and textural features, the predictive performances of them were compared for winter wheat AGB estimation. Compared with the models based on the continuum removal of red-edge spectra (SpeCR, FWHM = 9), the models based on the multiscale textures had comparable estimation accuracy. It confirms the utility of textures derived from high-definition RGB images for winter wheat biomass estimation, since UAV-based RGB images are more readily acquired than ground-based, airborne, and spaceborne hyperspectral data. In the exploration of whether LSSVM analyses conducted using the combination of optimal NDVI, SpeCR (FWHM = 9), and multiscale textures could further improve estimation accuracy of winter wheat AGB, the LSSVM regression model based on the combination of the SpeCR (FWHM = 9) and selected multiscale features yielded the higher R^2 and lower RMSE for validation dataset, compared to the other combinations (i.e., multiscale textures + optimal NDVI and multiscale textures + optimal NDVI + SpeCR). However, the combination did not significantly improve the estimation accuracy compared to the use of SpeCR or multiscale textures only. The results are not in agreement with the findings reported by Yue et al. [18] and Zheng et al. [17], who highlighted that the combination of spectral and textural features could improve the estimation accuracy of crop AGB. The primary reason for this is that the multiscale textures in the study perform better than the GLCM-based textures used in [17,18]. The addition of red-edge hyperspectral information could not make an improvement in estimation accuracy. However, the spectral features used in [17,18] were broadband vegetation indices or some common narrowband vegetation indices, such as NDVI (670, 800 nm), optimized soil adjusted vegetation index (OSAVI (670, 800 nm)), and two-band enhanced vegetation index (EVI2), which did not utilize the informative information from red-edge spectra well.

5. Conclusions

This study investigated the utility of multiscale textures extracted from high-definition UAV-based RGB images in predicting winter wheat biomass during multiple growth stages. The proposed multiscale texture extraction method (Multiscale_Gabor_GLCM) took good advantage of Gabor filters and GLCM analyses. The feature selection function embedded in LSSVM regression helped identify 10 influential multiscale textures to winter wheat biomass. Both the PLSR and LSSVM regression models using the multiscale textures produced better estimation accuracies than those using single-scale GLCM-based textures. Compared to the spectral features based models, the multiscale textures got comparable estimation accuracy to the SpeCR with a bandwidth of 9 nm. Although the selected best-performing NDVI (1193, 1222 nm) got intermediate estimation accuracy, it had great value in practical application due to its ease of use and low computation. The LSSVM regression analyses conducted using the combination of multiscale textures and hyperspectral features did not significantly improve the estimation accuracy for winter wheat AGB, compared to the use of multiscale textures or SpeCR alone. The LSSVM regression and its visualization technique showed great potential in establishing a high-accuracy winter wheat estimation model and understanding the driving force behind the model. This study confirmed the reliability of multiscale textures, red-edge spectra, and optimal NDVI calculated by water-sensitive wavebands for winter wheat AGB estimation. Further studies are required to verify the effectiveness of multiscale texture extraction method in predicting crop biomass at different ecological regions, especially for the high-yielding crops.

Author Contributions: Conceptualization: Y.F.; methodology: Y.F.; software: Y.F.; validation: Y.F.; formal analysis: Y.F.; investigation: Y.F.; resources: G.Y. and H.F.; data curation: G.Y., X.X., H.F., and X.S.; writing—original draft preparation: Y.F.; writing—review and editing: Y.F. and G.Y.; visualization: Y.F.; supervision: G.Y., Z.L., and C.Z.; project administration: G.Y. and C.Z.; funding acquisition: Y.F, G.Y., and C.Z. All authors have read and agreed to the published version of the manuscript.

Funding: This study was supported by the Natural Science Foundation of China (41801225), the Key-Area Research and Development Program of Guangdong Province (2019B020214002, 2019B020216001), the National Key Research and Development Program of China (2017YFE0122500) and the Beijing Million Talent Project (No.2019A10).

Institutional Review Board Statement: Not applicable.

Informed Consent Statement: Not applicable.

Data Availability Statement: The data presented in this study are available on request from the corresponding author.

Acknowledgments: The authors extend gratitude to Bo Xu, Weiguo Li, Lin Wang and Hong Chang for their assistance in field data collection. The detailed comments from the editors and reviewers improved our study.

Conflicts of Interest: The authors declare no conflict of interest.

Appendix A

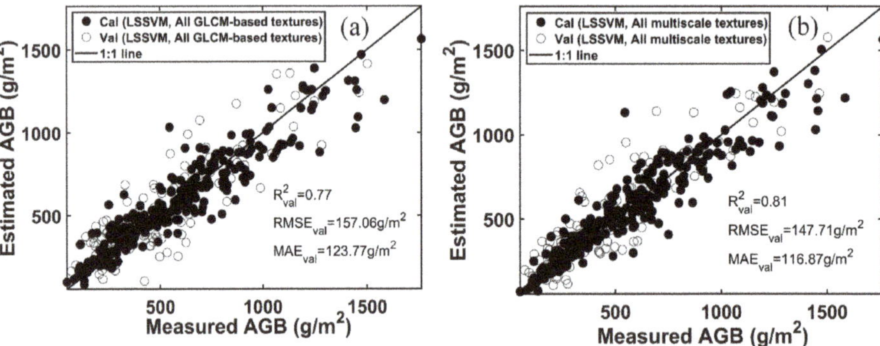

Figure A1. Predictive performance of the LSSVM regression models based on (**a**) all GLCM-based textures and (**b**) all multiscale textures for winter wheat AGB estimation.

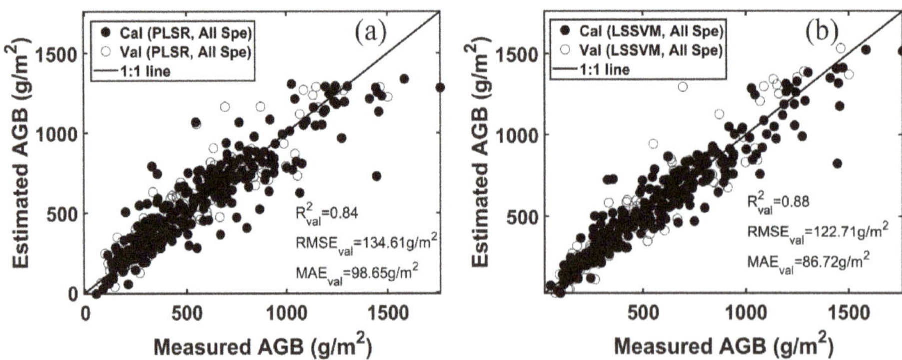

Figure A2. Predictive performance of the PLSR model (**a**) and LSSVM regression model (**b**) using all spectra (400~1350 nm) for winter wheat AGB estimation.

References

1. Mistele, B.; Schmidhalter, U. Estimating the nitrogen nutrition index using spectral canopy reflectance measurements. *Eur. J. Agron.* **2008**, *29*, 184–190. [CrossRef]

2. Ziadi, N.; Brassard, M.; Bélanger, G.; Cambouris, A.N.; Tremblay, N.; Nolin, M.C.; Claessens, A.; Parent, L.-É. Critical Nitrogen Curve and Nitrogen Nutrition Index for Corn in Eastern Canada. *Agron. J.* **2008**, *100*, 271. [CrossRef]
3. Hansen, P.M.; Schjoerring, J.K. Reflectance measurement of canopy biomass and nitrogen status in wheat crops using normalized difference vegetation indices and partial least squares regression. *Remote Sens. Environ.* **2003**, *86*, 542–553. [CrossRef]
4. Jin, X.; Zarco-Tejada, P.; Schmidhalter, U.; Reynolds, M.P.; Hawkesford, M.J.; Varshney, R.K.; Yang, T.; Nie, C.; Li, Z.; Ming, B.; et al. High-throughput estimation of crop traits: A review of ground and aerial phenotyping platforms. *IEEE Geosci. Remote Sens. Mag.* **2020**. [CrossRef]
5. Huang, J.; Chen, D.; Cosh, M.H. Sub-pixel reflectance unmixing in estimating vegetation water content and dry biomass of corn and soybeans cropland using normalized difference water index (NDWI) from satellites. *Int. J. Remote Sens.* **2009**, *30*, 2075–2104. [CrossRef]
6. Gao, S.; Niu, Z.; Huang, N.; Hou, X. Estimating the Leaf Area Index, height and biomass of maize using HJ-1 and RADARSAT-2. *Int. J. Appl. Earth Obs. Geoinf.* **2013**, *24*, 1–8. [CrossRef]
7. Meng, J.; Du, X.; Wu, B. Generation of high spatial and temporal resolution NDVI and its application in crop biomass estimation. *Int. J. Digit. Earth* **2013**, *6*, 203–218. [CrossRef]
8. Mutanga, O.; Skidmore, A.K. Narrow band vegetation indices overcome the saturation problem in biomass estimation. *Int. J. Remote Sens.* **2004**, *25*, 3999–4014. [CrossRef]
9. Fu, Y.; Yang, G.; Wang, J.; Song, X.; Feng, H. Winter wheat biomass estimation based on spectral indices, band depth analysis and partial least squares regression using hyperspectral measurements. *Comput. Electron. Agric.* **2014**, *100*, 51–59. [CrossRef]
10. Marshall, M.; Thenkabail, P. Advantage of hyperspectral EO-1 Hyperion over multispectral IKONOS, GeoEye-1, WorldView-2, Landsat ETM+, and MODIS vegetation indices in crop biomass estimation. *Isprs J. Photogramm. Remote Sens.* **2015**, *108*, 205–218. [CrossRef]
11. Yao, X.; Si, H.; Cheng, T.; Jia, M.; Chen, Q.; Tian, Y.; Zhu, Y.; Cao, W.; Chen, C.; Cai, J.; et al. Hyperspectral Estimation of Canopy Leaf Biomass Phenotype per Ground Area Using a Continuous Wavelet Analysis in Wheat. *Front Plant Sci.* **2018**, *9*, 1360. [CrossRef]
12. Bendig, J.; Bolten, A.; Bennertz, S.; Broscheit, J.; Eichfuss, S.; Bareth, G. Estimating Biomass of Barley Using Crop Surface Models (CSMs) Derived from UAV-Based RGB Imaging. *Remote Sens.* **2014**, *6*, 10395–10412. [CrossRef]
13. Sofonia, J.; Shendryk, Y.; Phinn, S.; Roelfsema, C.; Kendoul, F.; Skocaj, D. Monitoring sugarcane growth response to varying nitrogen application rates: A comparison of UAV SLAM LiDAR and photogrammetry. *Int. J. Appl. Earth Obs. Geoinf.* **2019**, *82*. [CrossRef]
14. Lu, N.; Zhou, J.; Han, Z.; Li, D.; Cao, Q.; Yao, X.; Tian, Y.; Zhu, Y.; Cao, W.; Cheng, T. Improved estimation of aboveground biomass in wheat from RGB imagery and point cloud data acquired with a low-cost unmanned aerial vehicle system. *Plant Methods* **2019**, *15*, 17. [CrossRef] [PubMed]
15. Li, B.; Xu, X.; Zhang, L.; Han, J.; Bian, C.; Li, G.; Liu, J.; Jin, L. Above-ground biomass estimation and yield prediction in potato by using UAV-based RGB and hyperspectral imaging. *ISPRS J. Photogramm. Remote Sens.* **2020**, *162*, 161–172. [CrossRef]
16. Nichol, J.E.; Sarker, M.L.R. Improved Biomass Estimation Using the Texture Parameters of Two High-Resolution Optical Sensors. *IEEE Trans. Geosci. Remote Sens.* **2011**, *49*, 930–948. [CrossRef]
17. Zheng, H.; Cheng, T.; Zhou, M.; Li, D.; Yao, X.; Tian, Y.; Cao, W.; Zhu, Y. Improved estimation of rice aboveground biomass combining textural and spectral analysis of UAV imagery. *Precis. Agric.* **2018**, *20*, 611–629. [CrossRef]
18. Yue, J.; Yang, G.; Tian, Q.; Feng, H.; Xu, K.; Zhou, C. Estimate of winter-wheat above-ground biomass based on UAV ultrahigh-ground-resolution image textures and vegetation indices. *Isprs J. Photogramm. Remote Sens.* **2019**, *150*, 226–244. [CrossRef]
19. Maimaitijiang, M.; Ghulam, A.; Sidike, P.; Hartling, S.; Maimaitiyiming, M.; Peterson, K.; Shavers, E.; Fishman, J.; Peterson, J.; Kadam, S.; et al. Unmanned Aerial System (UAS)-based phenotyping of soybean using multi-sensor data fusion and extreme learning machine. *Isprs J. Photogramm. Remote Sens.* **2017**, *134*, 43–58. [CrossRef]
20. Fu, Y.; Yang, G.; Li, Z.; Song, X.; Li, Z.; Xu, X.; Wang, P.; Zhao, C. Winter Wheat Nitrogen Status Estimation Using UAV-Based RGB Imagery and Gaussian Processes Regression. *Remote Sens.* **2020**, *12*, 3778. [CrossRef]
21. Caicedo, J.P.R.; Verrelst, J.; Munoz-Mari, J.; Moreno, J.; Camps-Valls, G. Toward a Semiautomatic Machine Learning Retrieval of Biophysical Parameters. *IEEE J. Sel. Top. Appl. Earth Obs. Remote Sens.* **2014**, *7*, 1249–1259. [CrossRef]
22. Ustun, B.; Melssen, W.J.; Buydens, L.M. Visualisation and interpretation of Support Vector Regression models. *Anal. Chim. Acta* **2007**, *595*, 299–309. [CrossRef] [PubMed]
23. Suykens, J.A.K.; Vandewalle, J. *Least Squares Support Vector Machine Classifiers*; Kluwer Academic Publishers: Amsterdam, The Netherlands, 1999.
24. Liang, L.; Qin, Z.H.; Zhao, S.H.; Di, L.P.; Zhang, C.; Deng, M.X.; Lin, H.; Zhang, L.P.; Wang, L.J.; Liu, Z.X. Estimating crop chlorophyll content with hyperspectral vegetation indices and the hybrid inversion method. *Int. J. Remote Sens.* **2016**, *37*, 2923–2949. [CrossRef]
25. Wang, F.M.; Huang, J.F.; Wang, Y.; Liu, Z.Y.; Zhang, F.Y. Estimating nitrogen concentration in rape from hyperspectral data at canopy level using support vector machines. *Precis. Agric.* **2013**, *14*, 172–183. [CrossRef]
26. Haralick, R.M.; Shanmugam, K.; Dinstein, I.H. Textural features for image classification. *IEEE Trans. Syst. Man Cybern.* **1973**, 610–621. [CrossRef]

27. Xu, X.; Teng, C.; Zhao, Y.; Du, Y.; Zhao, C.; Yang, G.; Jin, X.; Song, X.; Gu, X.; Casa, R.; et al. Prediction of Wheat Grain Protein by Coupling Multisource Remote Sensing Imagery and ECMWF Data. *Remote Sens.* **2020**, *12*, 1349. [CrossRef]
28. Verhoeven, G. Taking computer vision aloft—archaeological three-dimensional reconstructions from aerial photographs with photoscan. *Archaeol. Prospect.* **2011**, *18*, 67–73. [CrossRef]
29. Brown, M.; Lowe, D.G. Recognising Panoramas. In Proceedings of the International Conference on Computer Vision, Nice, France, 13–16 October 2003; p. 1218.
30. Yue, J.B.; Feng, H.K.; Jin, X.L.; Yuan, H.H.; Li, Z.H.; Zhou, C.Q.; Yang, G.J.; Tian, Q.J. A Comparison of Crop Parameters Estimation Using Images from UAV-Mounted Snapshot Hyperspectral Sensor and High-Definition Digital Camera. *Remote Sens.* **2018**, *10*, 1138. [CrossRef]
31. Cheng, H.D.; Jiang, X.H.; Sun, Y.; Wang, J. Color image segmentation: Advances and prospects. *Pattern Recognit.* **2001**, *34*, 2259–2281. [CrossRef]
32. Tao, D.; Li, X.; Wu, X.; Maybank, S.J. General tensor discriminant analysis and gabor features for gait recognition. *IEEE Trans Pattern Anal Mach Intell* **2007**, *29*, 1700–1715. [CrossRef]
33. Geladi, P.; Kowalski, B.R. Partial Least-Squares Regression: A Tutorial. *Anal. Chim. Acta* **1986**, *185*, 1–17. [CrossRef]
34. Cho, M.A.; Skidmore, A.; Corsi, F.; van Wieren, S.E.; Sobhan, I. Estimation of green grass/herb biomass from airborne hyperspectral imagery using spectral indices and partial least squares regression. *Int. J. Appl. Earth Obs. Geoinf.* **2007**, *9*, 414–424. [CrossRef]
35. Kooistra, L.; Salas, E.A.L.; Clevers, J.G.P.W.; Wehrens, R.; Leuven, R.S.E.W.; Nienhuis, P.H.; Buydens, L.M.C. Exploring field vegetation reflectance as an indicator of soil contamination in river floodplains. *Environ. Pollut.* **2004**, *127*, 281–290. [CrossRef]
36. Curran, P.J. Remote sensing of foliar chemistry. *Remote Sens. Environ.* **1989**, *30*, 271–278. [CrossRef]
37. Gnyp, M.L.; Bareth, G.; Li, F.; Lenz-Wiedemann, V.I.S.; Koppe, W.; Miao, Y.; Hennig, S.D.; Jia, L.; Laudien, R.; Chen, X.; et al. Development and implementation of a multiscale biomass model using hyperspectral vegetation indices for winter wheat in the North China Plain. *Int. J. Appl. Earth Obs. Geoinf.* **2014**, *33*, 232–242. [CrossRef]
38. Kanke, Y.; Tubana, B.; Dalen, M.; Harrell, D. Evaluation of red and red-edge reflectance-based vegetation indices for rice biomass and grain yield prediction models in paddy fields. *Precis. Agric.* **2016**, *17*, 507–530. [CrossRef]

 remote sensing

MDPI

Article

Applying RGB- and Thermal-Based Vegetation Indices from UAVs for High-Throughput Field Phenotyping of Drought Tolerance in Forage Grasses

Tom De Swaef [1,*,†], Wouter H. Maes [2,†], Jonas Aper [1], Joost Baert [1], Mathias Cougnon [3], Dirk Reheul [3], Kathy Steppe [4], Isabel Roldán-Ruiz [1,5] and Peter Lootens [1]

1 Plant Sciences Unit, Research Institute for Agriculture, Fisheries and Food (ILVO), 9090 Melle, Belgium; jonas.aper@ilvo.vlaanderen.be (J.A.); joost.baert@ilvo.vlaanderen.be (J.B.); isabel.roldan-ruiz@ilvo.vlaanderen.be (I.R.-R.); peter.lootens@ilvo.vlaanderen.be (P.L.)
2 UAV Research Centre (URC), Department of Plants and Crops, Ghent University, 9000 Ghent, Belgium; wouter.maes@ugent.be
3 Sustainable Crop Production, Department of Plants and Crops, Ghent University, 9000 Ghent, Belgium; mathias.cougnon@ugent.be (M.C.); dirk.reheul@ugent.be (D.R.)
4 Laboratory of Plant Ecology, Department of Plants and Crops, Ghent University, 9000 Ghent, Belgium; kathy.steppe@ugent.be
5 Department of Plant Biotechnology and Bioinformatics, Ghent University, 9000 Ghent, Belgium
* Correspondence: tom.deswaef@ilvo.vlaanderen.be
† These authors contributed equally to this work.

Citation: De Swaef, T.; Maes, W.H.; Aper, J.; Baert, J.; Cougnon, M.; Reheul, D.; Steppe, K.; Roldán-Ruiz, I.; Lootens, P. Applying RGB- and Thermal-Based Vegetation Indices from UAVs for High-Throughput Field Phenotyping of Drought Tolerance in Forage Grasses. *Remote Sens.* **2021**, *13*, 147. https://doi.org/10.3390/rs13010147

Received: 17 November 2020
Accepted: 29 December 2020
Published: 5 January 2021

Abstract: The persistence and productivity of forage grasses, important sources for feed production, are threatened by climate change-induced drought. Breeding programs are in search of new drought tolerant forage grass varieties, but those programs still rely on time-consuming and less consistent visual scoring by breeders. In this study, we evaluate whether Unmanned Aerial Vehicle (UAV) based remote sensing can complement or replace this visual breeder score. A field experiment was set up to test the drought tolerance of genotypes from three common forage types of two different species: *Festuca arundinacea*, diploid *Lolium perenne* and tetraploid *Lolium perenne*. Drought stress was imposed by using mobile rainout shelters. UAV flights with RGB and thermal sensors were conducted at five time points during the experiment. Visual-based indices from different colour spaces were selected that were closely correlated to the breeder score. Furthermore, several indices, in particular *H* and *NDLab*, from the HSV (Hue Saturation Value) and CIELab (Commission Internationale de l'éclairage) colour space, respectively, displayed a broad-sense heritability that was as high or higher than the visual breeder score, making these indices highly suited for high-throughput field phenotyping applications that can complement or even replace the breeder score. The thermal-based Crop Water Stress Index *CWSI* provided complementary information to visual-based indices, enabling the analysis of differences in ecophysiological mechanisms for coping with reduced water availability between species and ploidy levels. All species/types displayed variation in drought stress tolerance, which confirms that there is sufficient variation for selection within these groups of grasses. Our results confirmed the better drought tolerance potential of *Festuca arundinacea*, but also showed which *Lolium perenne* genotypes are more tolerant.

Keywords: UAV; RGB camera; thermal camera; drought tolerance; forage grass; HSV; CIELab; broad-sense heritability; phenotyping gap; high throughput field phenotyping

1. Introduction

Grasslands cover about 30% of the agricultural area in Europe (EU28) [1] and represent a substantial economic value through feed production. At the same time, they are also crucial for biodiversity and carbon sequestration [2–4]. However, grassland persistence and productivity is very sensitive to drought [5,6]. The increasing frequency and intensity of dry spells in Western Europe as a consequence of climate change substantially impacts the production and carbon sequestration capacity of these grasslands [7–9]. It is therefore important that new varieties of forage grasses possess a good drought tolerance as part of the adaptation strategy of the fodder production system to the changing climate [10].

In temperate regions like Western Europe, perennial ryegrass (*Lolium perenne* L.) and tall fescue (*Festuca arundinacea* Schreb.) are two of the major perennial forage grass species used for fodder production. *Lolium perenne* (Lp) is praised for its excellent feeding quality both for grazing and mowing. Perennial ryegrass occurs in nature as a diploid, but breeders have developed tetraploid varieties after chromosome doubling by e.g., colchicine treatment [11]. Tetraploid ryegrasses have a lower dry matter content, but this is compensated by a higher resistance to crown rust and snow mould and a higher fresh yield. In addition, tetraploid varieties have a more open sward and are less persistent [12]. Although *Festuca arundinacea* (Fa) has a lower feeding quality than Lp, its yield potential is larger, especially under drought conditions, owing to its deeper rooting [13–15].

Attaining a higher level of tolerance to drought in new cultivars of these grass species implies phenotyping for this trait across a wide range of genotypes/accessions in breeding programs. This typically requires thousands of genotypes/plants to be evaluated under water-limited conditions. To this end, drought tolerance testing is preferably done in the field and using rainout shelters, since these block the precipitation, while having a relatively limited impact on temperature rise or light level reduction [16]. In addition, the root growth of the plants is not limited in contrast with pot experiments [17].

In forage grass breeding programs in Western Europe, the early phase of the breeding program aims to select the best genotypes from a large pool of individual plants by visually assigning a score to their appearance. This breeder score is based on an evaluation of the greenness, biomass and health of the plant, and can therefore also be used for scoring drought tolerance. As this is an 'integrative' score, it is not possible to disentangle particular reactions and adaptations of the different genotypes to drought. Hence, these breeder scores cannot infer whether the apparent drought tolerance is the result from e.g., a deep rooting system, or from efficient stomatal behaviour. Nevertheless, this scoring allows to capture a substantial amount of variation that can be attributed to genotypic differences among the plants. The proportion of the total variation that can be attributed to genotype variation is called 'broad-sense' heritability, and is impacted by the experimental conditions and the observation method [18]. Scoring one single plant takes very little time, but since many plants (thousands in one experiment) need to be scored, assigning breeder scores is labour-intensive. Furthermore, factors as fatigue and changing weather conditions can result in 'within-observer' inconsistencies, even for highly skilled observers [19]. Furthermore, visual scores are prone to subjectivity of the observer [20], as individual criteria of observers differ slightly, and thus result in a low 'between-observer' consistency [19]. Because of these limitations, high throughput field phenotyping (HTFP) with close-range remote sensing, including from UAVs (Unmanned Aerial Vehicles), provides an interesting alternative [21]. The initial focus for this kind of screening was on multispectral sensors covering the visual and near infrared wavelengths [21,22], allowing the calculation of standard vegetation indices (VIs) as Normalized Difference Vegetation Index (*NDVI*; [23]) or Soil-Adjusted Vegetation Index (*SAVI*; [24]). UAV-based *NDVI* has already been demonstrated to be well-correlated with visual scoring by breeders for perennial ryegrass [25]. Recently, the use of commercial RGB (Red/Green/Blue) sensors receives increasing attention as low-cost and reliable alternatives [26–29], from which crop height and size or specific VIs can be derived.

Several studies even report that RGB-based remote sensing outperforms conventional multispectral remote sensing, e.g., for assessing crop growth (e.g., [30]), nutrient levels [31,32] or weed detection [26,33]. This seems particularly the case in applications where very high spatial resolution outweighs spectral resolution [26,34]. Vegetation indices that can be calculated directly from the RGB colour space include for instance Excess Green (*ExG*; [35]) and Green-Red Vegetation Index (*GRVI*; [36,37]). In addition, the RGB information can be converted to alternative colour spaces, from which additional indices can be calculated. These other colour spaces, such as CIE (Commission Internationale de l'éclairage) Lab, CIELuv, HSV (Hue-Saturation-Value) or HSL (Hu-Saturation-Lightness), can be calculated from RGB data and were developed to more accurately represent the colours as observed by the human eye [38,39]. They are also less influenced by the specific differences in illumination and shade within and between images [40,41]. The use of different colour spaces is well-established for estimating fractional crop cover and for image segmentation and classification (e.g., [38,42–47], but has also been applied for disease detection [48] and for assessing nitrogen content of leaves and canopies [39,49].

In the context of grass breeding and phenotyping trials, previous research demonstrated the potential of UAV-based RGB data for objectively measuring persistency and biomass [19,50]. In this article, we are investigating whether visual-based indices can also be used to complement, or even replace, visual breeder scores. In this context, we hypothesize that VIs based on advanced colour spaces are better related to the breeder scores, since these are more similar to the colour perception by the human eye. When focusing on tolerance to drought, thermal imagery can be of particular importance in HTFP, as drought stress often leads to the physiological plant response of stomatal closure, and consequently leaf temperature increase [29,51,52]. Unlike RGB-based information, which typically presents an integrative result over time (e.g., growth and vegetation indices), thermal imagery provides an instantaneous readout of the plant's ecophysiological status. UAV-based thermal imagery has been applied in phenotyping for drought stress tolerance in trees [53,54], wheat [29,55], soybean [56] and sugarcane [57].

The overall aim of this study is to evaluate the use of UAV-based remote sensing for complementing or replacing the breeder score and for assessing the drought tolerance of individual genotypes of three forage grass types, belonging to two species: tall fescue (Fa), diploid perennial ryegrass (Lp2) and tetraploid perennial ryegrass (Lp4). More specifically, we address four research questions:

1. Which RGB-based indices correlate best to the breeder score for screening drought stress tolerance in forage grasses?
2. Does the single-trial 'broad-sense' heritability (i.e., the proportion of total variance that can be attributed to genotypic variance) differ between the breeder score and these indices?
3. Is there an added value of thermal imagery for screening drought stress tolerance in forage grasses?
4. Can we infer distinct behaviour in response to drought stress within and between the three grass forage species/types?

2. Materials and Methods

2.1. Field Trial

In the first half of April 2013, 150 genotypes of *Festuca arundinacea* Schreb. (hereafter named Fa), 120 genotypes of diploid *Lolium perenne* L. (hereafter named Lp2) and 120 genotypes of tetraploid *Lolium perenne* L. (hereafter named Lp4) were selected from the Ghent University (Fa) and the ILVO (Research Institute for Agriculture, Fisheries and Food) (Lp2, Lp4) breeding gene pool and planted as spaced plants (planting distance 0.5 m × 0.5 m) in a field (soil texture: sandy loam) equipped with three mobile rainout shelters at ILVO, Melle (50° 59.5′ N, 3° 47.1′ E) (Figure 1). Although we realize that diploid (Lp2) and tetraploid (Lp4) perennial ryegrass are not different species, for ease of interpretation we regarded them as different species in the remainder of the analysis. The experiment was organised as an alpha design with 15 blocks ($s = 15$), 8 to 10 genotypes per block ($k = 8 - 10$) and three replications ($r = 3$, one in each rainout shelter) using the package *agricolae* and the function *design.alpha* in the RStudio software [58,59]. As such, there were 1170 plants $((150 + 120 + 120) \times 3)$ under evaluation (marked by the boxes in Figure 1). Besides the plants under study, 1080 other grass plants were planted (360 per rainout shelter), but these plants were not the subject of the current study. In the middle of each of the three plots, a hole (90 cm deep, 40 cm diameter) was dug, to install soil moisture sensors across a 80 cm soil profile. To avoid impact of this soil disturbance on the Fa plants under study in the middle plot, we planted non-studied plants in the two rows bordering the hole. As a result, the Fa sub-plot in the middle plot was divided into two parts Figure 1.

The three rainout shelters, each 30 m long and 10 m wide, are located next to each other (Figure 1). To reduce edge effects, two additional border rows were planted with individuals of the neighbouring species at the long edges of the three plots. Five rows of border plants were planted at the short edges, as these sides of the rainout shelters were more open, and thus potentially more vulnerable for sideward incoming rain (Figure 1). These five border rows also consisted of individuals of the neighbouring species.

After planting, but before the start of the drought treatment, five plants had died (three of Fa and two of Lp4), thus resulting in a total of 1165 plants under evaluation. Plants were mown on 28 May (DOY 148), 26 June (DOY 177), 30 July (DOY 211) and 24 September (DOY 267). Fertilizer was applied on 29 May (DOY 149, 53 kg N ha^{-1} + 18 kg P_2O_5 ha^{-1} + 77 kg K_2O ha^{-1}), on 29 June (DOY 180, 60 kg N ha^{-1} + 20 kg P_2O_5 ha^{-1} + 88 kg K_2O ha^{-1}), and on 30 July (DOY 211, 40 kg N ha^{-1}).

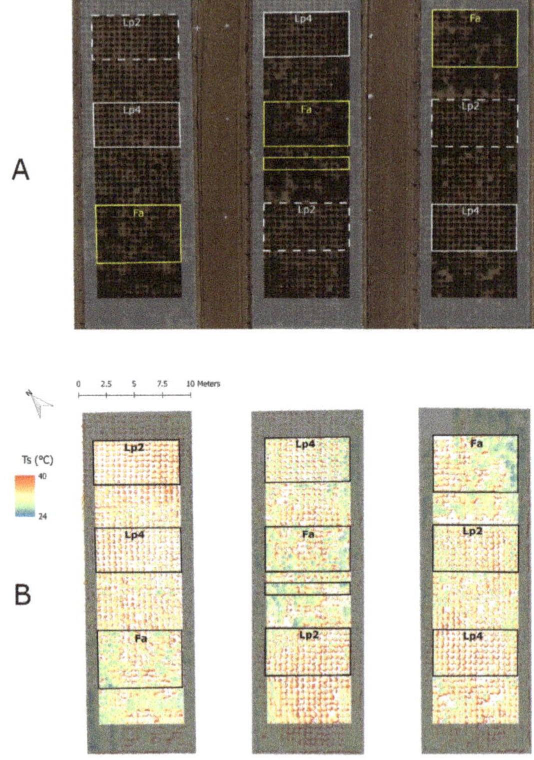

Figure 1. Orthophoto (RGB; **A**) and false colour image of plant surface temperature (T_s; **B**) of the experimental field on 6 September 2013 (DOY 249). The three parts of the field correspond to the three rainout shelters. Grey shaded areas include border plants, the framed areas contain the plants used for this study. In part (**B**) (thermal image) the soil pixels were filtered out to improve visualisation. T_s was calculated according Equation (1).

2.2. Drought Treatment

The rainout shelters were moved over the experimental field during two periods of the 2013 summer season (from DOY 186 until 207 and from DOY 232 until 260, Figure 2). In-between these two periods, plants were mown and fertilized, and were exposed to two rain events to allow the fertilizer to dissolve. Before, during and after the drought periods, meteorological conditions were recorded using a temperature + relative humidity (RH) probe (CS215, Campbell Scientific, Inc., Logan, UT, USA), a precipitation sensor (ARG100, Campbell Scientific, Inc., Logan, UT, USA) and a pyranometer (LP02, Hukseflux, The Netherlands) installed about 15 m from the experimental set-up. Furthermore, the volumetric water content of the soil (VWC in $m^3\ m^{-3}$) was monitored continuously for the 10–40 cm depth at 6 locations per rainout shelter (18 in total) using Time Domain Reflectometry sensors (type CS616, Campbell Scientific Inc., UK) (Figure 2).

CWD was calculated as the accumulation of the difference between daily reference evapotranspiration (ET_0 in mm) and precipitation (P in mm). ET_0 was calculated using the ET.PenmanMonteith function of the R package Evapotranspiration [60,61].

2.3. Breeder Score

During the experiment, plants were scored by one experienced forage grass breeder on 16 July, 29 August and 6 September (T2, T4 an T5 resp. in Figure 2). Plant scores varied from 1 to 9 and integrate both the amount of biomass and the green colour of the plant. Higher values are assigned to plants with high biomass levels (i.e., yield) that are green and healthy-looking.

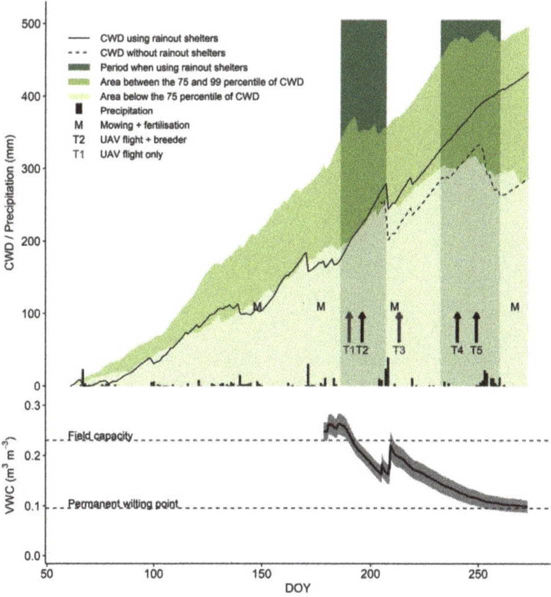

Figure 2. Visual representation of the environmental conditions, management actions and the time points of measurements during the experiment. The upper part displays the evolution of the cumulative water deficit (CWD; (mm)), and precipitation along with the timing of rainout shelter use, mowing and fertilisation, and observations. The lower graph shows the evolution of the volumetric soil moisture content (VWC; 10 to 40 cm profile; ($m^3\ m^{-3}$)) as an average of 18 sensors distributed over the field. The grey shaded area represents the 95% confidence interval on the data.

2.4. *Uav-Based Remote Sensing and Image Analysis*

Aerial imagery was obtained with an AT8 octocopter of AerialTronics (Scheveningen, The Netherlands), from flights carried out on five sunny days between 9 July and 6 September 2013 (Figure 2). The UAV was equipped with both a thermal and an RGB sensor, installed on a 2 axis AV2000 gimbal (PhotoHigher, Wellington, New Zealand), which had a nadir orientation for all flights. Flights were performed following a pre-programmed waypoint route of 9 parallel flight lines, at 4 m from each other, at a flight speed of 1.5 to 2 m s^{-1} and at a flight height of 12–14 m. The thermal sensor, a FLIR SC305 (FLIR Systems, Inc., Wilsonville, OR, USA), had a resolution of 320 × 240 pixels, a thermal accuracy of ±2 °C and a thermal sensitivity of 0.05 °C , and was equipped with a 10 mm lens with a field of view (FOV) of 45° × 34°. At 12 m altitude, each single image covered about 10.0 m × 7.3 m, with a resolution of 2.9 cm. An on-board computer (Olimex) was used to take images every 2 to 2.5 s, resulting in a horizontal and vertical overlap of 60%. The at-sensor radiance was converted into brightness temperature (T_{br}) by using the FLIR ThermaCam Researcher Pro 2.10 software (FLIR Systems, Inc., Wilsonville, OR, USA) for performing the atmospheric correction, with the locally measured air temperature and relative humidity (Table 1), and a flight height of 12 m as input parameters [62]. Surface temperature (T_s) was then calculated as:

$$T_s = \sqrt[4]{\frac{T_{br}^4 - (1-\epsilon)T_{bg}^4}{\epsilon}}, \tag{1}$$

in which ϵ is the emissivity, estimated as 0.985, and T_{bg} is the background temperature, retrieved by measuring the brightness temperature of a reference panel covered with aluminium sheet [62,63].

Table 1. Air temperature (T_{air}), vapour pressure deficit (VPD) and background temperature (T_{bg}) of the reference panels during the UAV flights.

Time Point	Date	DOY	T_{air} (°C)	VPD (kPa)	T_{bg} (°C)
T1	9 July 2013	190	22.5	2.51	9.7
T2	16 July 2013	197	25.4	3.06	5.2
T3	1 August 2013	213	29.9	4.05	10.4
T4	28 August 2013	240	22.0	2.67	5.6
T5	6 September 2013	249	22.8	2.54	7.1

The RGB sensor was a 12MP Canon S110 (Canon, Japan). Its spectral response (Figure S1) is typical of a consumer-grade digital RGB camera, with some overlap between the blue, green and red spectra [64,65]. It was set to the maximal field of view 74° × 53° (5.2 mm focal length, 24 mm equivalent). Image parameters were set manually before each flight and remained the same for the entire flight. The sensor was programmed with CHDK (https://chdk.fandom.com/) to log every second. Individual images cover about 18.0 × 12.0 m with a spatial resolution of 0.4 cm. Horizontal overlap was 75% and vertical overlap was 88%. Georeferenced RGB and thermal orthophotos were assembled with AgiSoft PhotoScan Pro v1.2.6 (AgiSoft LLC, St-Petersburg, Russia), with GPS data extracted from the GPX-logfile of the UAV. The RGB orthophotos of 16 July served as base map to register all other RGB data and all thermal data in PhotoScan, using ground control points visible in both the base map and the other datasets. All further processing then occurred with ArcGIS 10.1 (ESRI, Redlands, CA, USA). On the flight of 1 August (T3, Figure 2), the grass plants had been recently mown, and individual plants did not overlap. The Green Chromatic Coordinate index (*GCC*) and Excess Green index (*ExG*) were calculated from the RGB image of this day, and a vegetation mask was created from these indices based on threshold values. A polygon shapefile was generated automatically from the mask, and this shapefile was checked manually to ensure that each polygon corresponded with a single grass plant. All shapefiles were negatively buffered by 5 cm to avoid edge effects or remaining inconsistencies due to imperfect image co-registration. Row and column numbers were assigned to each polygon and were used to link each plant polygon to the species and block. In total, 1165 polygons were created. For each flight day and for the RGB and surface temperature image separately, the polygons were checked visually to make sure they covered the area of the plant- if not, the polygon location was adjusted for this particular time point and sensor. An example of the polygons for different time points and for both surface temperature and RGB imagery is given in Figure S2. Next, for each time point (T1–T5, Figure 2), the median R, G and B (all in 8-bit digital numbering—so ranging between 0 and 255) and T_s (in °C) per plant were calculated and saved, and these data served as the basis for the calculations of all vegetation indices. On T2 (16 July), because of a problem with the logger system of the thermal sensor, no thermal data were available for a full block of Lp4; for this day, the thermal data for Lp4 were not included in the analyses.

2.5. Calculation of Vegetation Indices

From the RGB data, 35 different VIs were calculated, hereafter referred to as 'visual-based indices', belonging to five different colour spaces (RGB, CIELab, CIELuv, HSL and HSV)—see Table 2 for a full overview. To convert the RGB data to the different colour spaces, scripts were developed in Matlab R2018b (Mathworks Inc., Natic, MA, USA) based on the formulas provided by http://www.easyrgb.com/en/math.php#text2. The HSL and HSV colour spaces are very similar, and break down the data into a colour hue (H), saturation or colour purity (S), and brightness (Value in HSV, Luminosity in HSL) [38]. The CIELab and CIELuv colour spaces are also similar. One dimension represents luminance or lightness (L^*), whereas the other two dimensions represent chrominance and separate the colour spectrum. For instance, negative values of the a^* dimension represent a green colour, positive values a red colour, whereas the b^* dimension expresses a blue (negative) to yellow (positive) spectrum [38,39].

Table 2. Overview of the different RGB-based vegetation indices (VIs) and their colour space used in this study.

VI	Name	Equation	Colour Space	Reference
R	Red		RGB	
G	Green		RGB	
B	Blue		RGB	
RCC	Red Chromatic Coordinate Index	$\frac{R}{R+G+B}$	RGB	[35]
GCC	Green Chromatic Coordinate Index	$\frac{G}{R+G+B}$	RGB	[35,36]
BCC	Blue Chromatic Coordinate Index	$\frac{B}{R+G+B}$	RGB	[35]
ExG	Excess Green Index	$2G-B-R$	RGB	[35]
ExG2	Excess Green Index v2	$\frac{2G-B-R}{R+G+B}$	RGB	[35]
ExR	Excess Red Index	$\frac{1.4R-G}{R+G+B}$	RGB	[66]
ExGR	Excess Green minus Excess Red Index	$ExG2-EXR$	RGB	[67]
GRVI	Green Red Vegetation Index	$\frac{G-R}{G+R}$	RGB	[36,37]

Table 2. *Cont.*

VI	Name	Equation	Colour Space	Reference
GBVI	Green Blue Vegetation Index	$\frac{G-B}{G+B}$	RGB	
BRVI	Blue Red Vegetation Index	$\frac{B-R}{B+R}$	RGB	
G/R	Green-Red Ratio	$\frac{G}{R}$	RGB	[68]
G-R	Green-Red Difference	$G-R$	RGB	
B-G	Blue-Green Difference	$B-G$	RGB	
VDVI	Visible-band Difference Vegetation Index	$\frac{2G-R-B}{2G+R+B}$	RGB	[69]
VARI	Visible Atmospherically Resistant Index	$\frac{G-R}{G+R-B}$	RGB	[37]
MGRVI	Modified Green Red Vegetation Index	$\frac{G^2-R^2}{G^2+R^2}$	RGB	[70]
CIVE	Colour Index Of Vegetation	$0.441R-0.881G+0.385B+18.787$	RGB	[71]
VEG	Vegetative Index	$\frac{G}{R^{0.667}B^{0.334}}$	RGB	[72]
WI	Woebbecke Index	$\frac{G-B}{R-G}$	RGB	[35]
H	Hue		HSV/HSL	[73]
S	Saturation		HSV/HSL	
V	Value		HSV	
I	Intensity		HSL	
L*	Lightness		CIELab	
a*	Green-Red component		CIELab	
b*	Blue-Yellow component		CIELab	
ab		a^*b^*	CIELab	
NDLab	Normalized Difference CIELab Index	$\frac{1-a^*-b^*}{1-a^*+b^*}+1$	CIELab	[39]
u*	Green-Red component		CIELuv	
v*	Blue-Yellow component		CIELuv	
uv		u^*v^*	CIELuv	
NDLuv	Normalized Difference CIELuv Index	$\frac{1-u^*-v^*}{1-u^*+v^*}+1$	CIELuv	[39]

From the thermal data, two indices (further thermal-based indices), ΔT and the Crop Water Stress Index ($CWSI$), were calculated as:

$$\Delta T = T_s - T_{air} \tag{2}$$

$$CWSI = \frac{T_s - T_{pot}}{T_{dry} - T_{pot}} \tag{3}$$

with T_{air} being the air temperature and with T_{pot} and T_{dry} corresponding to the surface temperature of a grass plant transpiring at maximal rate (T_{pot}) and not transpiring at all (T_{dry}) [51,74]. In this case, T_{pot} and T_{dry} were calculated directly per measurement day as the 1st and 99th percentile of the T_s polygon records of that day. This histogram approach is commonly applied [75–77] and assumes that on any given measurement day, (at least)

1% of the plants were not transpiring at all, and (at least) 1% of the plants were transpiring at potential (maximal) level. In this case, given the large number of plants and the large range in drought sensitivity among the species, we believe this is a valid assumption.

2.6. Data Analysis Workflow

On flight times T2, T4 and T5, breeder scores taken from the same or previous day were available. For those days, Pearson correlations were calculated between the different VIs and the breeder scores. Correlations were calculated for each day separately as well as for the data of the three flight days combined. In addition, correlations were calculated per species as well as for all plants together. In all cases, the correlations were calculated using the individual plant data. The adjusted mean value of each VI for each plant and for each time point was calculated taking into account the block effects of the alpha design in the mixed-effects model of the *PBIB.test* function of the *agricolae* package in R [58,59]. Using the variance analysis output of the *PBIB.test* function, the single-trial broad-sense heritability (H^2) for each VI at each time point (T1–T5) and for each species (Fa, Lp2, Lp4) was computed using:

$$H^2 = \frac{Var(G)}{Var(P)} \tag{4}$$

where $Var(G)$ is the genotypic variance, and $Var(P)$ the total phenotypic variance. Next, a Principal Component Analysis (PCA) and a hierarchical clustering analysis was carried out on the scores per plant and per measurement day of *CWSI* and a selection of the visual-based VIs that were promising in terms of correlation with the breeder score and H^2, using the *PCA* and *HCPC* functions of the *FactoMineR* package in R [59,78]. Based on the PCA, we selected one index per colour space and one thermal index for deeper analysis.

3. Results

3.1. Environmental Conditions

Rainout shelters were used for two periods to gradually reduce water availability in field conditions Figure 2. Near the end of the first drought treatment period, the volumetric water content for the 10–40 cm soil profile (VWC) dropped to about 0.16 $m^3\ m^{-3}$, corresponding to 40% of relative extractable water (REW). During the second drought period, VWC dropped to levels close to the wilting point (0% REW). However, at this time (around DOY 270), deeper soil layers (50–80 cm) still had higher VWC levels (Figure S3), and potentially provided water for deeper rooting genotypes. The cumulative water deficit (CWD) resulting from these treatments was situated between the 75th and 99th percentile, based on weather data from 1975 to 2019 for Flanders.

3.2. Breeder Scores

The distributions of the breeder scores showed that for all species (Fa, Lp2 and Lp4), the median values of the breeder score were highest at T2, at the start of the experiment (Figure 3). At T4 and T5, the spread on the scores increased compared to T2. At these time points (T4 and T5), the proportion of both the high scores (9) and the low scores (1) was substantially higher for Fa, compared to Lp2 and Lp4. For Fa, the median score at T5 was higher than at T4, whereas there was almost no difference between these two dates in Lp2. In Lp4 on the other hand, the median value at T5 was lower compared to T4.

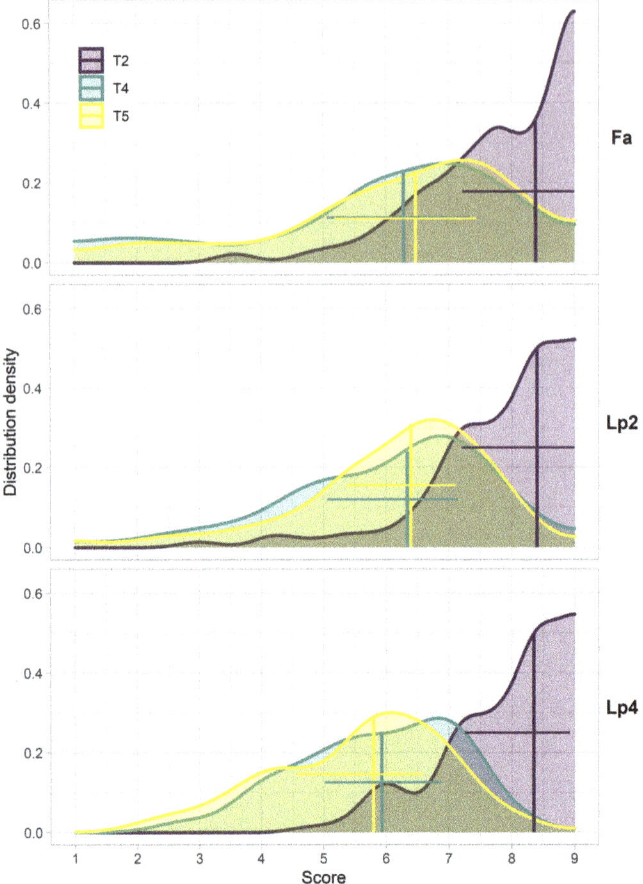

Figure 3. Distributions of the breeder scores at the different time points (T2, T4, T5) for the different species (Fa: *Festuca arundinacea;* Lp2: diploid *Lolium perenne;* Lp4: tetraploid *Lolium perenne*). Data are the adjusted means calculated using the *PBIB.test* function of the *agricolae* package in R [58]. Vertical lines indicate the median, horizontal lines indicate the range between the 25th and 75th percentile.

3.3. Correlating Breeder Scores and Vegetation Indices

The Pearson correlation statistic between each of the 37 indices (35 visual-based and 2 thermal) and the breeder score is given in Figure 4, for each species separately and for all species together as well as for the three time points (T2, T4 and T5) separately and all time points together. In general, correlations between the breeder scores and the VIs were lower at T2 than at T4 and T5. For the visual-based indices, there was no clear species effect on the linear regressions between indices and the breeder score, resulting in a similar correlation when all species were taken together (Panel "All" in Figure 4). However, for some indices (*ExG, G-R, CIVE, a*, ab, u** and *uv*), there was a good correlation with the breeder score for each single time point, even across the species, but this correlation was substantially lower when all time points were pooled. For these overall data, the VIs with the highest correlations with the breeder score were *H* (r = 0.84), *NDLab* (r = 0.82), *MGRVI* (r = 0.79) and *VARI* (r = 0.79). Correlations between the breeder scores and the thermal indices (ΔT and *CWSI*) were considerably lower than for the best-performing visual-based indices, and were stronger for Fa than for Lp2 and Lp4.

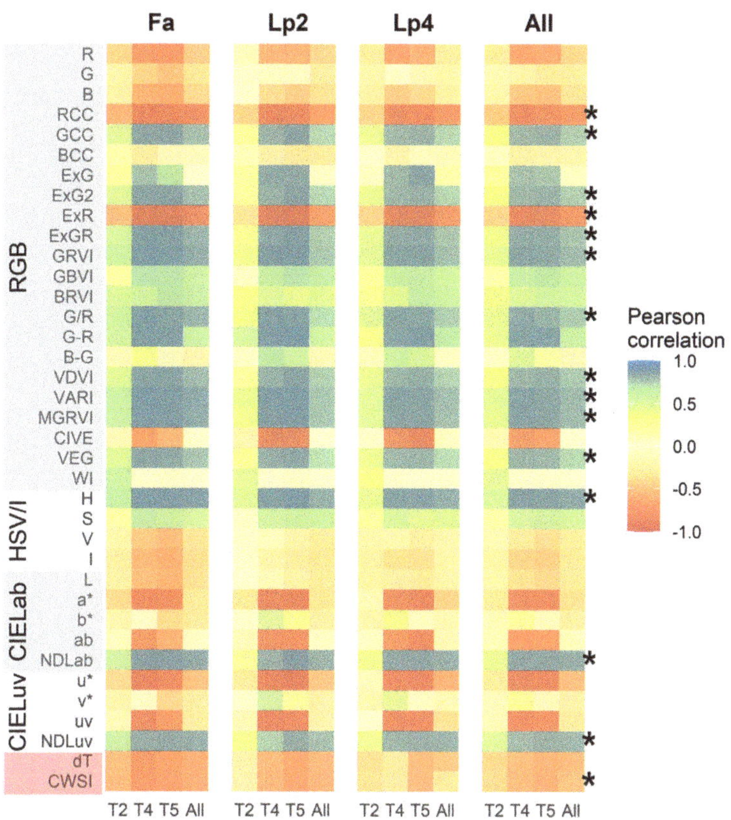

Figure 4. Pearson correlations between 37 different indices and the breeder score per species separately (Fa: *Festuca arundinacea;* Lp2: diploid *Lolium perenne;* Lp4: tetraploid *Lolium perenne*) and all species together (All) and per time point (T2, T4 and T5) and all time points together (All). The different indices are organized by sensor and colour space. Asterisks (*) point out indices of which the overall correlation was above 0.70 or below −0.70 and which were used for further analyses, along with *CWSI*.

We selected the indices with a correlation value above 0.7 or below −0.7 across all time points and species, as well as the thermal based *CWSI*, for further analysis. These selected indices are marked with an asterisk in Figure 4. The relations of the breeder score with *MGRVI*, *H*, *NDLab* and *CWSI* as a function of time are visualised in Figure 5. These density plots indicate that the spread on the data was smallest at T2, whereas T4 and T5 displayed more variation and showed a clearer correlation between UAV-based indices and the breeder score.

Figure 6 shows the broad-sense heritability (H^2), calculated for each species—time point—index combination. Higher values of H^2 for an index or score indicate that more of the variation in the data can be attributed to genotypic variation within each species. This is very relevant in a breeding context, because it allows the breeder to better select suitable plants from a large number of genotypes. The H^2 was substantially higher near the end of the drought treatment (T4 and T5, compared to T1–T3) (Figure 6A). This was the case for all species, but was most striking for Fa. Additionally, for all species, H^2 was higher at T1 compared to T2.

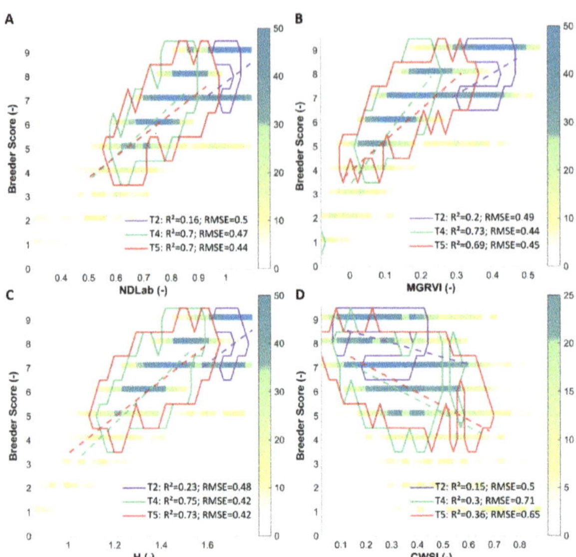

Figure 5. Density plots of *H*, *NDLab*, *MGRVI*, and *CWSI* (panels A–D) versus the breeder score, with linear regression (dashed line) and contour plot (full) per time point (T2, T4 and T5). Contours contain 75% of all data for this time point. R^2 and RMSE (Root Mean Squared Errors—in breeder score units) are given for the regression at each time point.

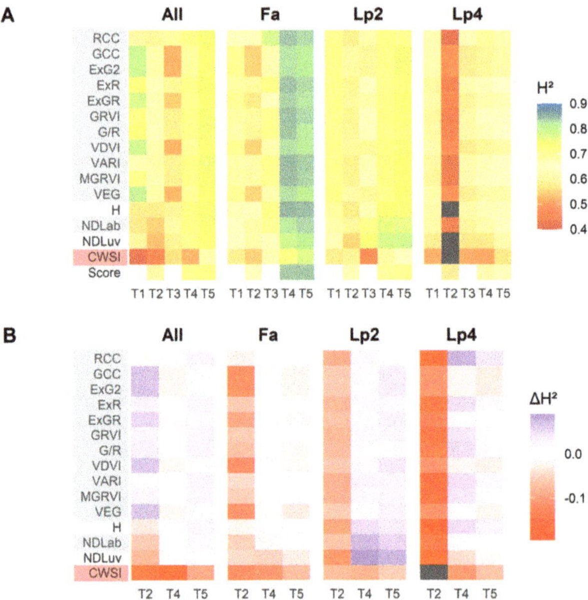

Figure 6. (**A**): Broad-sense heritability (H^2) calculated per species (Fa: *Festuca arundinacea;* Lp2: diploid *Lolium perenne;* Lp4: tetraploid *Lolium perenne*), per time point (T1–T5) and per index/score. (**B**): Difference in broad-sense heritability between the breeder score and each of the indices for each species and time point.

The difference in H^2 (ΔH^2) between the UAV indices and the breeder scores at T2, T4 and T5 (Figure 6B), was in all cases smaller than 0.2. For all species separately, H^2 values of the breeder score were higher than the UAV indices at T2, but this was not the case when the data of all species were analysed together (Figure 6B "All"). The behaviour of the data at T4 and T5 was different for each species. For Lp2, all the visual-based indices tended to have higher H^2 than the breeder score, whereas for Fa the breeder score tended to perform slightly better than the visual-based indices. For Lp4, several indices outperformed the breeder score (e.g., *RCC*, *ExR*, *GRVI*, *G/R*, *VARI*, *MGRVI* and *H*), whereas others had lower (*GCC*, *ExG2*, *VDVI*, *VEG*), or similar H^2 values (*ExGR*, *NDLab*). When data of the three species were pooled, H^2 of the good performing VIs was as good as the breeder score at T4, and slightly better at T5. The thermal index *CWSI* had lower H^2 values than the breeder scores at any time point and any species.

3.4. Contrasting Drought Tolerance Across Species

Adjusted means over the three replications were calculated for each selected index, for each time point and each genotype, by taking the block effects of the alpha design into account. The results of the PCA analysis, based on these adjusted means, is shown in Figure 7. This analysis revealed that the visual-based indices are closely correlated and align on the first principal axis (PC1), either positively or negatively. This axis describes 89.9% of the variation present, and can be interpreted as a descriptor of visual appearance, where high values of PC1 indicate greener plants. The second axis (PC2) describes 5.7% of the variation and is particularly related to *CWSI*. Hence, it can be interpreted as indicating reductions in transpiration. It is noteworthy that all visual indices from the RGB colour space also have a similar score on PC2, whereas the indices from the alternative colour spaces have an opposite score for PC2. These results persist when using only T2, T4 and T5, and including the breeder score in the PCA (Figure S4).

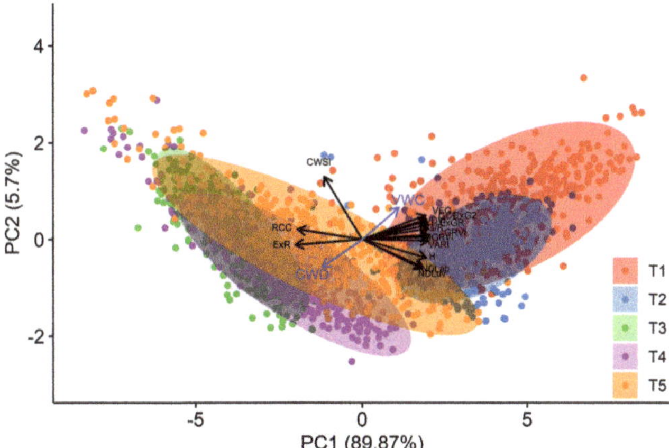

Figure 7. Two first principal component axes of the Principal Component Analysis (PCA) performed on the individual plant scores per day for the indices *RCC*, *ExR*, *GCC*, *ExG2*, *G/R*, *GRVI*, *MGRVI*, *VARI*, *VDVI*, *VEG*, *H*, *NDLab*, and *CWSI*. Dots display the projected mean values per genotype and per time point (T1–T5). The Cumulative Water Deficit (CWD) and Volumetric Water Content (VWC) data are presented as a quantitative supplementary variable, but were not used for the PCA.

At T1 and T2, plants had relatively high values on PC1, but a quite large variation on PC2 across genotypes and species, indicating different behaviour in stomatal closure when soil moisture was not yet limiting. At T3, plants scored very low on PC1 (low greenness), as a result of the mowing two days earlier. Data from T4 and T5 present a large variation on both axes and suggest that there is a link between both dimensions as high values on

PC1 (greener plants) correspond to low values on PC2 (higher stomatal conductance) and vice versa. To further explore the data by species, we clustered the points according to their score on PC1 and PC2 in three groups (Figure 8). Cluster 1 associated with low values for PC1 and high values of PC2, and thus contained data of plants with reduced transpiration and lower greenness. Cluster 2 contained intermediate data points on both axes, whereas Cluster 3 contained data with high values on PC1, but large variations on the PC2 axis. At T1 and T2, the vast majority of genotypes were classified in Cluster 3, particularly for Lp2 and Lp4. At T3, after mowing, Lp2 and Lp4 genotypes were almost exclusively classified in Cluster 1, with genotypes from Fa distributed evenly across Clusters 1 and 2. At T4 and T5, most plants were classified in Cluster 2 for all species, with the remaining genotypes mostly grouped in Cluster 1. The exception here is Fa at T5, where 27% of the genotypes were classified in Cluster 3. On the other hand, a relatively large portion (37%) of genotypes of Lp4 were in Cluster 1. The PCA results indicate a strong correlation among the indices from the RGB-colour space as well as those between *NDLab* and *NDLuv*. For further analysis of species behaviour, we selected the *MGRVI* index from the RGB colour space, because of its high correlation with the breeder score (Figure 4) and high H^2 values for all species at T4 and T5 (Figure 6), along with the *H*, *NDLab*, *CWSI* index and the breeder score.

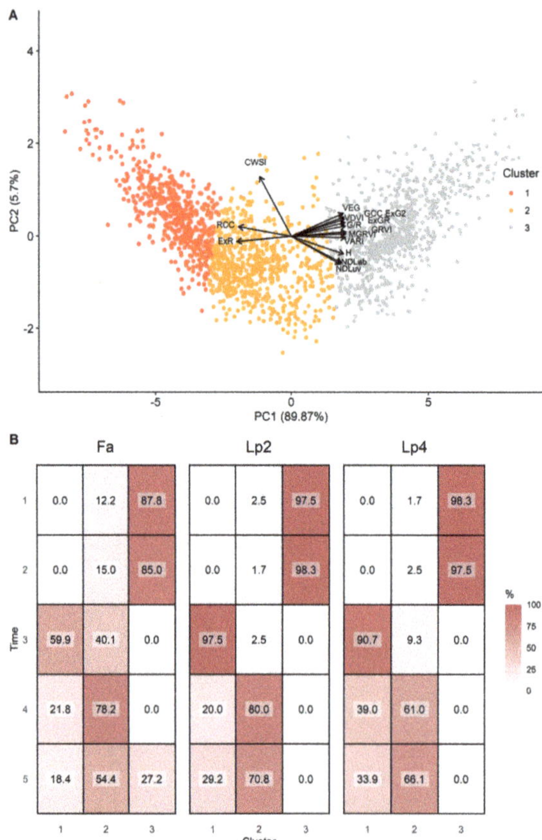

Figure 8. Results of the hierarchical cluster analysis, presented via the two major principal components (panel A). The distribution in % of the genotypes per time point (1–5) and per species (Fa: *Festuca arundinacea*; Lp2: diploid *Lolium perenne*; Lp4: tetraploid *Lolium perenne*) in the three clusters (panel B).

While values of *H* and *NDLab* increased from T1 to T2, values of *MGRVI* index decreased (Figure 9). The reduction in greenness after mowing was clearly visible at T3 for all species and for all visual-based indices. Also for all species, the visual-based indices displayed an overall increase from T3 to T4, which corresponds to growth after mowing, as was also visualised in Figure 9. From T4 to T5, however, the behaviour of the different species is somewhat different and in line with the observations of the PCA. While the median value of all visual-based indices for Fa still increased from T4 to T5, they remained constant or even decreased for Lp2 and Lp4. A similar trend was captured by the breeder score. The median *CWSI* value for Fa remained more or less constant over time. For Lp2 and Lp4, *CWSI* increased from T1 to T2 and to T3, and declined again at T4. The median *CWSI* values of Lp2 at T4 and T5 were higher than those of Lp4. For all species, the median *CWSI* was highest at T3.

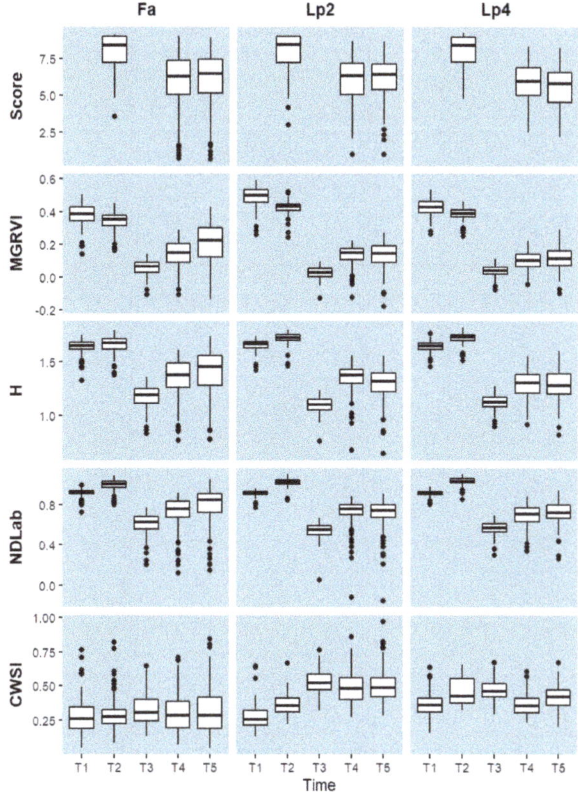

Figure 9. Boxplots represent the distribution per time point (T1–T5) and species (Fa: *Festuca arundinacea*; Lp2: diploid *Lolium perenne*; Lp4: tetraploid *Lolium perenne*) of the breeder score and four selected indices: *MGRVI*, *H*, *NDLab* and *CWSI*. Boxes are delimited by the 25th and 75th percentile, horizontal lines indicate the median values. Data are the mean values per genotype and are grouped per time point.

4. Discussion

4.1. Conceptualisation of the Drought Stress Treatment on Grasses

Forage grasses are typically mown multiple times per growing season. As the soil gradually dries out in the absence of rain, the drying period is very likely to include one of the mowing events, as was the case in the present experiment (Figure 2). Due to the mowing, all visual-based indices indicated a clear change at T3 (Figure 9), caused by the removal of green biomass rather than by sudden drought stress. This complicates the comparison of phenotypes before and after the stress treatment, and therefore makes it more challenging to disentangle a genotype's drought tolerance from other traits that affect regrowth after mowing (e.g., number of growing tips). Regrowth after mowing is substantially impacted by the availability of nutrients, and species and genotypes differ in nutrient use efficiency [79]. Because we wanted to exclude genotypic differences in nutrient use efficiency from biasing the drought response, fertilizer was added after mowing and intermediate rain events were used to ensure that the fertilizer was dissolved and thus available for plant uptake (Figure 2).

4.2. Research Question 1: Visual-Based Indices Are Good Proxies for Breeder Scores

Several VIs correlated very well with the visual breeder scores for the different species (Figure 4). At T2, correlations were lower, but this was related to the lower variation in breeder scores and visual-based indices at that time, when soil moisture was not limiting yet (Figures 3, 5 and 9). Some indices (*ExG*, *G-R*, *CIVE*, a^*, *ab*, u^* and *uv*) displayed high correlations with the breeding score at single time points, but not when data from all time points were pooled. These indices were thus sensitive to the camera settings and to the prevailing environmental conditions at the time of the flight. This sensitivity can possibly be reduced by calibration of the sensor signal and conversion to reflectance values using grey reference panels [80]. However, even without this correction, several other VIs were quite robust to the different sensor settings and conditions, and maintained high correlations with the breeder score when data from all observation days were pooled (Figure 4). The two VIs with the highest overall correlation with the breeder score were *H* ($r = 0.84$), from the HSV colour space, and *NDLab* ($r = 0.82$), from the CIELab colour space. This confirmed our hypothesis that VIs from alternative colour spaces correlate better with the breeder score than RGB-based VIs, as they are more similar to the colour perception by the human eye [38,39] and less influenced by sensor settings and measurement conditions [40,41]. Nevertheless, it was—to our knowledge—the first time that these two indices were related to visual breeder scores. In phenotyping studies, *H* and *NDLab* previously proved useful for predicting grain yield or for assessing disease severity in wheat and maize [39,81–83]. *MGRVI* was the best performing RGB index, although the difference with *VARI* was minimal. Previous studies on barley and rice suggested that *MGRVI* is particularly related to biomass and leaf area index, at least in early growth stages [70,84]. This can explain its good performance in our study.

4.3. Research Question 2: Broad-Sense Heritability

A high overall correlation with the breeder score as such is not sufficient in this context. We additionally tested the VIs for their ability to exploit the genotypic variation present in the experiment, by calculating the broad-sense heritability (H^2). VIs that performed well on H^2 were able to highlight features that explain differences in drought tolerance, while minimizing potential differences that can be attributed to 'random factors'. Several VIs actually showed comparable or even better overall H^2 values than the breeder score itself (Figure 6B), although there was a strong difference across the species and time points. Apparently, the VIs performed worse than the breeder score at T2, when drought was not yet impacting the plants, and the spread on the data was small (Figures 3 and 9). Potentially, this could be attributed to a certain level of saturation of the VIs, as the Leaf Area Index became high. However, this disadvantage at T2 disappeared when taking all species together (Figure 6B). At T4 and T5, several indices performed comparably (especially for Fa), or better (for Lp2 and Lp4) than the breeder score. As a final selection, the VIs with very high correlation to the breeder score, but with comparable or better H^2 values for all species included *H*, *NDLab* and *MGRVI*, and, to a lesser extent, *RCC*, *ExR*, *GRVI*, *G/R* and *VARI* Figures 4, 6 and 7. These are thus the best candidates for replacing or complementing the breeder scores in future studies.

Some of these indices were also highly correlated to each other and were intrinsically reflecting very similar properties. Still, more detailed analyses also revealed differences between on the one hand the RGB-based indices, such as *MGRVI*, and on the other hand *H* and *NDLab* (Figure 9): from T1 to T2, *MGRVI* (as well as other RGB-based indices, data not shown) decreased whereas *H* and *NDLab* increased. As the volumetric water content at T2 was still quite high (83% REW), we expect that plants were still growing between T1 and T2. This would imply that *H* and *NDLab* are better proxies for plant vigour. On the other hand, the *CWSI* index also showed a slight increase, thus indicating a slightly more stressed condition. These higher *CWSI* values, especially for Lp2 and Lp4, could however also be attributed to the higher evaporative demand at T2, compared to T1 (Table 1). The reason for the differences in the trend from T1 to T2 was not entirely clear. The main difference in plant state between T1 and T2 was the size of the plants—which was larger at T2, with grass leaves bending over. This change in leaf orientation could be a possible explanation for the diverging signal – with leaves bending over, more leaves are reflecting very strongly in all bands, causing a drop in *MGRVI* (Table 2). Moreover, the images on T2 were also slightly darker (mean overall $R + G + B$ was 73, versus 206 on T1—See also Figure S2). Colour space transformations as HSV and CIELab are developed to be less sensitive for these illumination issues, caused by camera settings or leaf orientation [41].

4.4. Research Question 3: Added Value of Thermal Indices

Although ΔT and *CWSI* were significantly correlated with the breeder score (Figure 4, Table S1), it was clear that the visual-based VIs were more useful as complement or replacement of the breeder score. Nevertheless, *CWSI* reflects the actual transpiration [51,85], and as the PCA analysis confirmed (Figure 7), *CWSI* can be used as an additional and complementary source of information for selection. The data confirmed that when drought stress was not present or was very moderate, the range in *CWSI* was large, despite a relatively uniform greenness of the canopy. When drought stress became more severe, *CWSI* and RGB-based VIs were more closely correlated (Figure S5). *CWSI* values at T1 and T2 were not highly correlated to *CWSI* at T5 (r = 0.11 and r = 0.28, respectively), suggesting that the genotypes that transpired more vigorously under normal growing conditions did not necessarily maintain high transpiration under drought conditions.

4.5. Research Question 4: Distinct Behaviour within and between the Species

A slight increase in *CWSI* was observed from T1 to T2 for both Lp species, but not for Fa (Figure 9). At T2, soil water availability was not yet limiting (83% REW, Figure 2), but VPD was high (3.06 kPa, versus 2.51 kPa at T1) (Table 1), suggesting that Lp has a more strict stomatal regulation to atmospheric conditions than Fa. This confirmed other studies that reported larger stomatal conductance values of Fa under control conditions [15] and stomatal closure occurring at lower (i.e., more negative) leaf water potential than Lp [86,87]. These differences in stomatal control can be caused by several mechanisms. First, differences in leaf hydraulic conductance can cause variations in water transport limitations [88,89]. Second, hydraulics at the soil-root interface can play a role [90]. In comparison with Lp, Fa has substantially more root biomass in deeper soil layers than Lp grasses [14,91]. Hence, it is plausible that plants with a relatively small rooting system can experience an insufficient hydraulic conductance under conditions of high evaporative demand [90,92]. At T3, all visual-based indices showed a strong reduction in greenness (Figures 3 and 7–9), as a result of mowing, but overall levels of Fa remained higher. The *CWSI* of Fa remained lower than that of Lp2 and Lp4 (Figure 9), probably as a combination of the larger green leaf area and the more limited stomatal control (VPD = 4.0 kPa, Table 1). At T4, the median value of all visual-based indices increased for all species compared to T3 (Figure 9), confirming the regrowth after mowing. In line with this, the *CWSI* of Lp2 and Lp4 decreased compared to T3, although this can also be due to less demanding meteorological conditions (VPD of 2.7 kPa at T4 versus 4.0 kPa at T3). However, the *CWSI* levels for both Lp species remained higher than for Fa (Figure 9). From T4 to T5, with increasing limitation in available water, the species responded differently in terms of visual-based indices *H*, *NDLab* and *MGRVI* (Figure 9). Where the median values of Fa continued to increase, this increment was no longer visible in Lp2 and Lp4 (Figure 9), as was also confirmed by the cluster analysis (Figure 8). The lower *CWSI* of Fa at T5 further confirmed the higher drought tolerance of Fa. *CWSI* values of Lp4 were generally higher than those of Lp2 at T1 and T2 (Figure 9). This confirmed former findings, where modified stomatal density and pore size tend to reduce the transpiration per unit of leaf area of tetraploids compared to diploids for different species [93,94]. In the recovery from T3 to T4, the increase of visual-based indices, and hence the regrowth, was also higher for Lp2 than for Lp4. Yet, at T4 and T5, Lp4 tended to have a lower *CWSI* than Lp2. Potentially, the lower transpiration rates in tetraploid genotypes in non-stressed conditions conserved more water, resulting in a delay of drought stress effects.

4.6. Future Perspectives

In this research, we illustrated that consumer-grade RGB cameras can be used to complement or replace the visual breeder score in forage grasses. This allows breeder score assessments with affordable UAVs by pilots who are not necessarily remote sensing experts. In order to be routinely applied in high throughput phenotyping studies, the data processing should also be straightforward. Nowadays, the computation of orthophotos from RGB imagery using structure-from-motion software is largely automated. In this respect, it is recommended to place permanent ground control points (GCPs) with known coordinates in the experimental field, in a way that they remain visible throughout the entire experiment. The calculation of vegetation indices such as *H*, *NDLab* or *MGRVI* from the orthophotos, and the final estimation of the breeder score, can also be completely automated. Although reference grey panels were not used in this research, they can probably further increase the robustness of vegetation indices.

A further development considers the fate of the visual breeder score itself. With more advanced UAV remote sensing technology, the breeder score's qualitative assessment of greenness and biomass can be translated to more objective and quantitative variables. Plant biomass can be estimated from vegetation indices in the visual and near infrared wavelengths, from the 3D vegetation model, obtained through structure-from-motion software from UAV-based RGB or multispectral data, or from their combination [34].

The plant greenness aspect of the breeder score can relate to two aspects. First, it can relate to the chlorophyll content in the leaves. In breeding studies, this is particularly useful when investigating the stress-sensitivity of the different genotypes [95]. Chlorophyll content can be best quantified with vegetation indices based on spectra in the red edge [34]. Other indices, such as *NDVI*, are sensitive to both leaf area index and chlorophyll content, and have already demonstrated high correlations with breeder scores [25]. On the other hand, plant greenness can also relate to the digestibility of forage grasses. Digestibility can be measured routinely with NIRS (near-infrared spectroscopy) analyses, from which parameters as the D-value (digestibility of organic matter in dry matter), WSC (water soluble carbohydrates) or similar feed quality features can be predicted [96]. However, NIRS uses destructive sampling and is too costly and time-intensive for large scale application in breeding programs [97]. Several studies indicate that hand-held hyperspectral sensors can provide a non-destructive and fast alternative, possibly suited for breeding purposes [97–99]. Hyperspectral sensing from UAVs has been demonstrated to be suitable for estimating forage grass digestibility [96]. However, more research is needed to overcome several remaining challenges, which are particularly related to the poor general applicability of the established empirical regressions across the seasons and years [100,101].

5. Conclusions

Using rainout shelters, a field experiment was set up to evaluate drought stress tolerance on a large number of genotypes of the forage grass species *Festuca arundinacea* (Fa), diploid *Lolium perenne* (Lp2) and *Lolium perenne* (Lp4) in a breeding context using

RGB and thermal UAV imagery. We identified several visual-based indices that showed high correlations with the breeder scores and displayed high 'broad-sense' heritability. These indices can serve as a proxy for breeder scores, for which dedicated pipelines can be set up to automate the processing and potentially increase consistency of selection. In particular, the use of *H*, *NDLab* and *MGRVI* can be considered for this purpose. The thermal index *CWSI* provided complementary information to visual indices, and this enabled the analysis of differences in ecophysiological mechanisms for coping with reduced water availability. The panel of Fa genotypes displayed the largest variation across all indices. Overall, *CWSI* was lowest for Fa, which also showed the best regrowth after mowing under water-limiting conditions, confirming the good drought tolerance of Fa. Nevertheless, substantial variation was found also among the diploid and tetraploid *L. perenne* genotypes, which can be exploited for breeding towards better drought stress tolerance in this species.

Supplementary Materials: The following are available online at https://www.mdpi.com/2072-4292/13/1/147/s1. Figure S1: Spectral response of the RGB camera (Canon S110) used in this study. Figure S2: Detailed view of the Lp2-block (indicated by the thick, dark-red line) of the middle row (see also Figure 1, main text) for the five different time points and for the RGB orthophoto (above) and surface temperature orthophoto (below). The red polygons illustrate the plant polygons in this block used for the data analyses. Figure S3: Volumetric Water Content (VWC) for different soil depths (from 10 to 80 cm). Blue lines are the mean of three sensors (one in each rainout shelter). The grey shaded area represents the 95% confidence interval on the VWC data of the 18 sensors monitoring the 10–40 cm soil profile. Figure S4: Two first principal component axes of the Principal Component Analysis (PCA) performed on the individual genotype values per day for the breeder scores and the indices RCC, ExR, GCC, ExG2, G/R, GRVI, MGRVI, VARI, VDVI, VEG, H, NDLab, and CWSI. Dots display the projected mean values per genotype and per time point (T2, T4 and T5). The CumulativeWater Deficit (CWD) and Volumetric Water Content (VWC) data are presented as a quantitative supplementary variable, but were not used for the PCA. Figure S5: Absolute values of Pearson correlations ($|r|$) between selected vegetation indices for five time points (T1–T5) and for all species (Fa: Festuca arundinacea, Lp2: diploid Lolium perenne, Lp4: tetraploid Lolium perenne. Table S1: Pearson correlations between all 37 indices and the breeder score for the individual species (Fa: Festuca arundinacea; Lp2: diploid Lolium perenne; Lp4: tetraploid Lolium perenne), and all species pooled together (All), and for the individual time points (T2, T4, T5) and all time points pooled together (All).

Author Contributions: Conceptualization, T.D.S. and W.H.M.; methodology, T.D.S., W.H.M., J.A.; formal analysis, T.D.S., W.H.M. and P.L.; investigation, T.D.S., W.H.M., J.A. and P.L.; resources, M.C., D.R., J.A., J.B., W.H.M.; data curation, J.A., P.L., T.D.S. and W.H.M.; writing–original draft preparation, P.L., T.D.S. and W.H.M.; writing–review and editing, P.L., I.R.-R., D.R., M.C., K.S.; visualization, T.D.S. and W.H.M. All authors have read and agreed to the published version of the manuscript.

Funding: At the time of measurement, W.H.M. was funded by the Special Research Fund (BOF) from Ghent University.

Institutional Review Board Statement: Not applicable.

Informed Consent Statement: Not applicable.

Data Availability Statement: Data is publicly available at 10.5281/zenodo.4415643.

Acknowledgments: The authors thank the ILVO field trial team for installing and maintaining the field trial, as well as three anonymous reviewers for their constructive feedback.

Conflicts of Interest: The authors declare no conflict of interest.

Abbreviations

The following abbreviations are used in this manuscript:

CWD	Cumulative Water Deficit
ET_0	Reference evapotranspiration
Fa	diploid *Festuca arundinacea* Schreb.
GCP	Ground Control Point
H^2	broad-sense heritability
HSL	Hue, Saturation, Luminosity
HSV	Hue, Saturation, Value
Lp2	diploid *Lolium perenne* L.
Lp4	tetraploid *Lolium perenne* L.
$NDVI$	Normalized Difference Vegetation Index
NIRS	Near Infra-Red Spectroscopy
P	Precipitation
PCA	Principal Component analysis
r	correlation
REW	Relative Extractable Water
RGB	Red, Green, Blue
T1–T5	Time point 1 to 5
T_{air}	Air temperature
T_{br}	Brightness temperature
T_{dry}	Leaf surface temperature of non-transpiring plant
T_{pot}	Leaf surface temperature of maximal transpiring plant
T_s	Leaf surface temperature
UAV	Unmanned Aerial Vehicle
$Var(G)$	Genotypic variance
$Var(P)$	Phenotypic variance
VPD	Vapour Pressure Deficit
VWC	Volumetric Water Content
WSC	Water Soluble Carbohydrates

References

1. FAOSTAT. Available online: http://www.fao.org/faostat/en/#data/RL (accessed on 27 October 2020).
2. Huyghe, C.; De Vliegher, A.; van Gils, B.; Peeters, A. *Grasslands and Herbivore Production in Europe and Effects of Common Policies*; Editions Quae: Versailles, France, 2014.
3. Soussana, J.F.; Lemaire, G. Coupling carbon and nitrogen cycles for environmentally sustainable intensification of grasslands and crop-livestock systems. *Agric. Ecosyst. Environ.* **2014**, *190*, 9–17. [CrossRef]
4. Kipling, R.P.; Virkajärvi, P.; Breitsameter, L.; Curnel, Y.; De Swaef, T.; Gustavsson, A.M.; Hennart, S.; Höglind, M.; Järvenranta, K.; Minet, J.; et al. Key challenges and priorities for modelling European grasslands under climate change. *Sci. Total. Environ.* **2016**, *566–567*, 851–864. [CrossRef] [PubMed]
5. Blum, A. *Plant Breeding for Water-Limited Environments*; Life Sciences; Springer: New York, NY, USA, 2010.
6. Olesen, J.; Trnka, M.; Kersebaum, K.; Skjelvåg, A.; Seguin, B.; Peltonen-Sainio, P.; Rossi, F.; Kozyra, J.; Micale, F. Impacts and adaptation of European crop production systems to climate change. *Eur. J. Agron.* **2011**, *34*, 96–112. [CrossRef]
7. IPCC. *Climate Change 2013: The Physical Science Basis. Contribution of Working Group I to the Fifth Assessment Report of the Intergovernmental Panel on Climate Change*; Cambridge University Press: Cambridge, UK; New York, NY, USA, 2013. [CrossRef]
8. Chang, J.; Ciais, P.; Viovy, N.; Soussana, J.F.; Klumpp, K.; Sultan, B. Future productivity and phenology changes in European grasslands for different warming levels: Implications for grassland management and carbon balance. *Carbon Balance Manag.* **2017**, *12*, 11. [CrossRef] [PubMed]
9. Dellar, M.; Topp, C.F.E.; Banos, G.; Wall, E. A meta-analysis on the effects of climate change on the yield and quality of European pastures. *Agric. Ecosyst. Environ.* **2018**, *265*, 413–420. [CrossRef]
10. Cyriac, D.; Hofmann, R.W.; Stewart, A.; Sathish, P.; Winefield, C.S.; Moot, D.J. Intraspecific differences in long-term drought tolerance in perennial ryegrass. *PLoS ONE* **2018**, *13*, e0194977. [CrossRef] [PubMed]
11. Hague, L.M.; Jones, R.N. Cytogenetics of *Lolium perenne*. *Theor. Appl. Genet.* **1987**, *74*, 233–241. [CrossRef] [PubMed]
12. Humphreys, M.; Feuerstein, U.; Vandewalle, M.; Baert, J. Ryegrasses. In *Fodder Crops and Amenity Grasses*; Handbook of Plant Breeding; Boller, B., Posselt, U.K., Veronesi, F., Eds.; Springer: New York, NY, USA, 2010; pp. 211–260. [CrossRef]

13. Turner, L.R.; Holloway-Phillips, M.M.; Rawnsley, R.P.; Donaghy, D.J.; Pembleton, K.G. The morphological and physiological responses of perennial ryegrass (*Lolium perenne* L.), cocksfoot (*Dactylis glomerata* L.) and tall fescue (*Festuca arundinacea* Schreb.; syn. *Schedonorus phoenix* Scop.) to variable water availability. *Grass Forage Sci.* **2012**, *67*, 507–518. [CrossRef]
14. Cougnon, M.; De Swaef, T.; Lootens, P.; Baert, J.; De Frenne, P.; Shahidi, R.; Roldán-Ruiz, I.; Reheul, D. In situ quantification of forage grass root biomass, distribution and diameter classes under two N fertilisation rates. *Plant Soil* **2017**, *411*, 409–422. [CrossRef]
15. Fariaszewska, A.; Aper, J.; Van Huylenbroeck, J.; De Swaef, T.; Baert, J.; Pecio, L. Physiological and biochemical besponses of forage grass varieties to mild drought stress under field conditions. *Int. J. Plant Prod.* **2020**, *14*, 335–353. [CrossRef]
16. Parra, A.; Ramírez, D.A.; Resco, V.; Velasco, A.; Moreno, J.M. Modifying rainfall patterns in a Mediterranean shrubland: System design, plant responses, and experimental burning. *Int. J. Biometeorol.* **2012**, *56*, 1033–1043. [CrossRef] [PubMed]
17. Poorter, H.; Bühler, J.; Dusschoten, D.V.; Climent, J.; Postma, J.A. Pot size matters: A meta-analysis of the effects of rooting volume on plant growth. *Funct. Plant Biol.* **2012**, *39*, 839–850. [CrossRef] [PubMed]
18. Downes, S.M.; Matthews, L. Heritability. In *The Stanford Encyclopedia of Philosophy*, 2020th ed.; Zalta, E.N., Ed.; Metaphysics Research Lab, Stanford University: Stanford, CA, USA, 2020.
19. Borra-Serrano, I.; De Swaef, T.; Aper, J.; Ghesquiere, A.; Mertens, K.; Nuyttens, D.; Saeys, W.; Somers, B.; Vangeyte, J.; Roldán-Ruiz, I.; et al. Towards an objective evaluation of persistency of *Lolium perenne* swards using UAV imagery. *Euphytica* **2018**, *214*, 142. [CrossRef]
20. Milberg, P.; Bergstedt, J.; Fridman, J.; Odell, G.; Westerberg, L. Observer bias and random variation in vegetation monitoring data. *J. Veg. Sci.* **2008**, *19*, 633–644. [CrossRef]
21. Araus, J.L.; Cairns, J.E. Field high-throughput phenotyping: The new crop breeding frontier. *Trends Plant Sci.* **2014**, *19*, 52–61. [CrossRef] [PubMed]
22. Shakoor, N.; Lee, S.; Mockler, T.C. High throughput phenotyping to accelerate crop breeding and monitoring of diseases in the field. *Curr. Opin. Plant Biol.* **2017**, *38*, 184–192. [CrossRef] [PubMed]
23. Tucker, C.J. Red and photographic infrared linear combinations for monitoring vegetation. *Remote Sens. Environ.* **1979**, *8*, 127–150. [CrossRef]
24. Huete, A.R. A soil-adjusted vegetation index (SAVI). *Remote Sens. Environ.* **1988**, *25*, 295–309. [CrossRef]
25. Wang, J.; Badenhorst, P.; Phelan, A.; Pembleton, L.; Shi, F.; Cogan, N.; Spangenberg, G.; Smith, K. Using sensors and unmanned aircraft systems for high-throughput phenotyping of biomass in perennial ryegrass breeding trials. *Front. Plant Sci.* **2019**, *10*, 1381. [CrossRef]
26. Araus, J.L.; Kefauver, S.C. Breeding to adapt agriculture to climate change: Affordable phenotyping solutions. *Curr. Opin. Plant Biol.* **2018**, *45*, 237–247. [CrossRef]
27. Rebetzke, G.J.; Jimenez-Berni, J.; Fischer, R.A.; Deery, D.M.; Smith, D.J. Review: High-throughput phenotyping to enhance the use of crop genetic resources. *Plant Sci.* **2019**, *282*, 40–48. [CrossRef] [PubMed]
28. Pieruschka, R.; Schurr, U. Plant phenotyping: Past, present, and future. *Plant Phenomics* **2019**, *2019*, 7507131. [CrossRef] [PubMed]
29. Perich, G.; Hund, A.; Anderegg, J.; Roth, L.; Boer, M.P.; Walter, A.; Liebisch, F.; Aasen, H. Assessment of multi-image unmanned aerial vehicle based high-throughput field phenotyping of canopy temperature. *Front. Plant Sci.* **2020**, *11*, 150. [CrossRef] [PubMed]
30. Travlos, I.; Mikroulis, A.; Anastasiou, E.; Fountas, S.; Bilalis, D.; Tsiropoulos, Z.; Balafoutis, A. The use of RGB cameras in defining crop development in legumes. *Adv. Anim. Biosci.* **2017**, *8*, 224–228. [CrossRef]
31. Vergara-Díaz, O.; Zaman-Allah, M.A.; Masuka, B.; Hornero, A.; Zarco-Tejada, P.; Prasanna, B.M.; Cairns, J.E.; Araus, J.L. A novel remote sensing approach for prediction of maize yield under different conditions of nitrogen fertilization. *Front. Plant Sci.* **2016**, *7*, 666. [CrossRef] [PubMed]
32. Gracia-Romero, A.; Kefauver, S.C.; Vergara-Díaz, O.; Zaman-Allah, M.A.; Prasanna, B.M.; Cairns, J.E.; Araus, J.L. Comparative performance of ground vs. aerially assessed RGB and multispectral indices for early-growth evaluation of maize performance under phosphorus fertilization. *Front. Plant Sci.* **2017**, *8*, 2004. [CrossRef]
33. Jiménez-Brenes, F.M.; López-Granados, F.; Torres-Sánchez, J.; Peña, J.M.; Ramírez, P.; Castillejo-González, I.L.; Castro, A.I.D. Automatic UAV-based detection of *Cynodon dactylon* for site-specific vineyard management. *PLoS ONE* **2019**, *14*, e0218132. [CrossRef]
34. Maes, W.H.; Steppe, K. Perspectives for remote sensing with unmanned aerial vehicles in precision agriculture. *Trends Plant Sci.* **2019**, *24*, 152–164. [CrossRef]
35. Woebbecke, M.D.; Meyer, E.G.; Von Bargen, K.; Mortensen, A.D. Color indices for weed identification under various soil, residue, and lighting conditions. *Trans. ASAE* **1995**, *38*, 259–269. [CrossRef]
36. Hunt, E.R.; Cavigelli, M.; Daughtry, C.S.T.; Mcmurtrey, J.E.; Walthall, C.L. Evaluation of digital photography from model aircraft for remote sensing of crop biomass and nitrogen status. *Precis. Agric.* **2005**, *6*, 359–378. [CrossRef]
37. Gitelson, A.A.; Kaufman, Y.J.; Stark, R.; Rundquist, D. Novel algorithms for remote estimation of vegetation fraction. *Remote Sens. Environ.* **2002**, *80*, 76–87. [CrossRef]
38. Hernández-Hernández, J.L.; García-Mateos, G.; González-Esquiva, J.M.; Escarabajal-Henarejos, D.; Ruiz-Canales, A.; Molina-Martínez, J.M. Optimal color space selection method for plant/soil segmentation in agriculture. *Comput. Electron. Agric.* **2016**, *122*, 124–132. [CrossRef]

39. Buchaillot, M.L.; Gracia-Romero, A.; Vergara-Diaz, O.; Zaman-Allah, M.A.; Tarekegne, A.; Cairns, J.E.; Prasanna, B.M.; Araus, J.L.; Kefauver, S.C. Evaluating maize genotype performance under low nitrogen conditions using RGB UAV phenotyping techniques. *Sensors* **2019**, *19*, 1815. [CrossRef] [PubMed]
40. Philipp, I.; Rath, T. Improving plant discrimination in image processing by use of different colour space transformations. *Comput. Electron. Agric.* **2002**, *35*, 1–15. [CrossRef]
41. Sadeghi-Tehran, P.; Virlet, N.; Sabermanesh, K.; Hawkesford, M.J. Multi-feature machine learning model for automatic segmentation of green fractional vegetation cover for high-throughput field phenotyping. *Plant Methods* **2017**, *13*, 103. [CrossRef] [PubMed]
42. Song, W.; Mu, X.; Yan, G.; Huang, S. Extracting the green fractional vegetation cover from digital images using a shadow-resistant algorithm (SHAR-LABFVC). *Remote Sens.* **2015**, *7*, 10425–10443. [CrossRef]
43. Lootens, P.; Ruttink, T.; Rohde, A.; Combes, D.; Barre, P.; Roldán-Ruiz, I. High-throughput phenotyping of lateral expansion and regrowth of spaced *Lolium perenne* plants using on-field image analysis. *Plant Methods* **2016**, *12*, 32. [CrossRef]
44. Li, L.; Mu, X.; Macfarlane, C.; Song, W.; Chen, J.; Yan, K.; Yan, G. A half-Gaussian fitting method for estimating fractional vegetation cover of corn crops using unmanned aerial vehicle images. *Agric. For. Meteorol.* **2018**, *262*, 379–390. [CrossRef]
45. Suh, H.K.; Hofstee, J.W.; van Henten, E.J. Improved vegetation segmentation with ground shadow removal using an HDR camera. *Precis. Agric.* **2018**, *19*, 218–237. [CrossRef]
46. Rico-Fernández, M.P.; Rios-Cabrera, R.; Castelán, M.; Guerrero-Reyes, H.I.; Juarez-Maldonado, A. A contextualized approach for segmentation of foliage in different crop species. *Comput. Electron. Agric.* **2019**, *156*, 378–386. [CrossRef]
47. Kim, J.; Kang, S.; Seo, B.; Narantsetseg, A.; Han, Y. Estimating fractional green vegetation cover of Mongolian grasslands using digital camera images and MODIS satellite vegetation indices. *Gisci. Remote Sens.* **2020**, *57*, 49–59. [CrossRef]
48. Kerkech, M.; Hafiane, A.; Canals, R. Deep learning approach with colorimetric spaces and vegetation indices for vine diseases detection in UAV images. *Comput. Electron. Agric.* **2018**, *155*, 237–243. [CrossRef]
49. Serret, M.D.; Al-Dakheel, A.J.; Yousfi, S.; Fernández-Gallego, J.A.; Elouafi, I.A.; Araus, J.L. Vegetation indices derived from digital images and stable carbon and nitrogen isotope signatures as indicators of date palm performance under salinity. *Agric. Water Manag.* **2020**, *230*, 105949. [CrossRef]
50. Borra-Serrano, I.; De Swaef, T.; Muylle, H.; Nuyttens, D.; Vangeyte, J.; Mertens, K.; Saeys, W.; Somers, B.; Roldán-Ruiz, I.; Lootens, P. Canopy height measurements and non-destructive biomass estimation of *Lolium perenne* swards using UAV imagery. *Grass Forage Sci.* **2019**, *74*, 356–369. [CrossRef]
51. Maes, W.H.; Steppe, K. Estimating evapotranspiration and drought stress with ground-based thermal remote sensing in agriculture: A review. *J. Exp. Bot.* **2012**, *63*, 4671–4712. [CrossRef] [PubMed]
52. Gerhards, M.; Schlerf, M.; Rascher, U.; Udelhoven, T.; Juszczak, R.; Alberti, G.; Miglietta, F.; Inoue, Y. Analysis of airborne optical and thermal imagery for detection of water stress symptoms. *Remote Sens.* **2018**, *10*, 1139. [CrossRef]
53. Ludovisi, R.; Tauro, F.; Salvati, R.; Khoury, S.; Mugnozza Scarascia, G.; Harfouche, A. UAV-based thermal imaging for high-throughput field phenotyping of black poplar response to drought. *Front. Plant Sci.* **2017**, *8*, 1681. [CrossRef]
54. Dillen, M.; Vanhellemont, M.; Verdonckt, P.; Maes, W.H.; Steppe, K.; Verheyen, K. Productivity, stand dynamics and the selection effect in a mixed willow clone short rotation coppice plantation. *Biomass Bioenergy* **2016**, *87*, 46–54. [CrossRef]
55. Gracia-Romero, A.; Kefauver, S.C.; Fernandez-Gallego, J.A.; Vergara-Díaz, O.; Nieto-Taladriz, M.T.; Araus, J.L. UAV and ground image-based phenotyping: A proof of concept with durum wheat. *Remote Sens.* **2019**, *11*, 1244. [CrossRef]
56. Kaler, A.S.; Ray, J.D.; Schapaugh, W.T.; Asebedo, A.R.; King, C.A.; Gbur, E.E.; Purcell, L.C. Association mapping identifies loci for canopy temperature under drought in diverse soybean genotypes. *Euphytica* **2018**, *214*, 135. [CrossRef]
57. Natarajan, S.; Basnayake, J.; Wei, X.; Lakshmanan, P. High-throughput phenotyping of indirect traits for early-stage selection in sugarcane breeding. *Remote Sens.* **2019**, *11*, 2952. [CrossRef]
58. De Mendiburu, F. Agricolae: Statistical Procedures for Agricultural Research. 2020. Available online: https://CRAN.R-project.org/package=agricolae (accesses on 29 December 2020).
59. R Core Team. *R: A Language and Environment for Statistical Computing;* R Foundation for Statistical Computing: Vienna, Austria, 2020.
60. Allen, R.G.; Pereira, L.S.; Raes, D.; Smith, M. *Crop Evapotranspiration—Guidelines for Computing Crop Water Requirements—FAO Irrigation and Drainage Paper 56;* FAO: Rome, Italy, 1998; p. 15.
61. Guo, D.; Westra, S.; Peterson, T. Evapotranspiration: Modelling Actual, Potential and Reference Crop Evapotranspiration. 2020. Available online: https://CRAN.R-project.org/package=Evapotranspiration (accesses on 29 December 2020).
62. Maes, W.H.; Huete, A.R.; Steppe, K. Optimizing the processing of UAV-based thermal imagery. *Remote Sens.* **2017**, *9*, 476. [CrossRef]
63. Maes, W.H.; Huete, A.R.; Avino, M.; Boer, M.M.; Dehaan, R.; Pendall, E.; Griebel, A.; Steppe, K. Can UAV-based infrared thermography be used to study plant-parasite interactions between mistletoe and eucalypt trees? *Remote Sens.* **2018**, *10*, 2062. [CrossRef]
64. Berra, E.F.; Gaulton, R.; Barr, S. Commercial off-the-shelf digital cameras on unmanned aerial vehicles for multitemporal monitoring of vegetation reflectance and NDVI. *IEEE Trans. Geosci. Remote. Sens.* **2017**, *55*, 4878–4886. [CrossRef]
65. De Kock, M.; Gallacher, D. From Drone Data to Decisions: Turning Images into Ecological Answers. In Proceedings of the Innovation Arabia 9, Dubai, UAE, 7–9 March 2016. [CrossRef]

66. Meyer, G.E.; Hindman, T.W.; Laksmi, K. Machine vision detection parameters for plant species identification. In Proceedings of the Precision Agriculture and Biological Quality, Boston, MA, USA, 3–4 November 1999; Volume 3543, pp. 327–335. [CrossRef]
67. Meyer, G.E.; Camargo Neto, J.; Jones, D.D.; Hindman, T.W. Intensified fuzzy clusters for classifying plant, soil, and residue regions of interest from color images. *Comput. Electron. Agric.* **2004**, *42*, 161–180. [CrossRef]
68. Steele, M.R.; Gitelson, A.A.; Rundquist, D.C.; Merzlyak, M.N. Nondestructive estimation of anthocyanin content in grapevine leaves. *Am. J. Enol. Vitic.* **2009**, *60*, 87–92.
69. Xiaoqin, W.; Miaomiao, W.; Shaoqiang, W.; Yundong, W. Extraction of vegetation information from visible unmanned aerial vehicle images. *Trans. Chin. Soc. Agric. Eng.* **2015**, *31*,152–159.
70. Bendig, J.; Yu, K.; Aasen, H.; Bolten, A.; Bennertz, S.; Broscheit, J.; Gnyp, M.L.; Bareth, G. Combining UAV-based plant height from crop surface models, visible, and near infrared vegetation indices for biomass monitoring in barley. *Int. J. Appl. Earth Obs. Geoinf.* **2015**, *39*, 79–87. [CrossRef]
71. Kataoka, T.; Kaneko, T.; Okamoto, H.; Hata, S. Crop growth estimation system using machine vision. In Proceedings of the 2003 IEEE/ASME International Conference on Advanced Intelligent Mechatronics (AIM 2003), Kobe, Japan, 20–24 July 2003; Volume 2, pp. b1079–b1083. [CrossRef]
72. Hague, T.; Tillett, N.D.; Wheeler, H. Automated crop and weed monitoring in widely spaced cereals. *Precis. Agric.* **2006**, *7*, 21–32. [CrossRef]
73. Cheng, H.D.; Jiang, X.H.; Sun, Y.; Wang, J. Color image segmentation: Advances and prospects. *Pattern Recognit.* **2001**, *34*, 2259–2281. [CrossRef]
74. Idso, S.B.; Reginato, R.J.; Jackson, R.D.; Pinter, P.J. Measuring yield-reducing plant water potential depressions in wheat by infrared thermometry. *Irrig. Sci.* **1981**, *2*, 205–212. [CrossRef]
75. Park, S.; Ryu, D.; Fuentes, S.; Chung, H.; Hernández-Montes, E.; O'Connell, M. Adaptive estimation of crop water stress in nectarine and peach orchards using high-resolution imagery from an unmanned aerial vehicle (UAV). *Remote Sens.* **2017**, *9*, 828. [CrossRef]
76. Bian, J.; Zhang, Z.; Chen, J.; Chen, H.; Cui, C.; Li, X.; Chen, S.; Fu, Q. Simplified evaluation of cotton water stress using high resolution unmanned aerial vehicle thermal imagery. *Remote Sens.* **2019**, *11*, 267. [CrossRef]
77. Maimaitiyiming, M.; Sagan, V.; Sidike, P.; Maimaitijiang, M.; Miller, A.J.; Kwasniewski, M. Leveraging very-high spatial resolution hyperspectral and thermal UAV imageries for characterizing diurnal indicators of grapevine physiology. *Remote Sens.* **2020**, *12*, 3216. [CrossRef]
78. Lê, S.; Josse, J.; Husson, F. FactoMineR: An R package for multivariate analysis. *J. Stat. Softw.* **2008**, *25*, 1–18. [CrossRef]
79. Yang, J.; Udvardi, M. Senescence and nitrogen use efficiency in perennial grasses for forage and biofuel production. *J. Exp. Bot.* **2018**, *69*, 855–865. [CrossRef]
80. Jeong, Y.; Yu, J.; Wang, L.; Shin, H.; Koh, S.M.; Park, G. Cost-effective reflectance calibration method for small UAV images. *Int. J. Remote Sens.* **2018**, *39*, 7225–7250. [CrossRef]
81. Kefauver, S.C.; El-Haddad, G.; Vergara-Diaz, O.; Araus, J.L. RGB picture vegetation indexes for high-throughput phenotyping platforms (HTPPs). In Proceedings of the Remote Sensing for Agriculture, Ecosystems, and Hydrology XVII, Toulouse, France, 22–25 September 2015; Volume 9637, p. 96370J. [CrossRef]
82. Zhou, B.; Elazab, A.; Bort, J.; Vergara, O.; Serret, M.D.; Araus, J.L. Low-cost assessment of wheat resistance to yellow rust through conventional RGB images. *Comput. Electron. Agric.* **2015**, *116*, 20–29. [CrossRef]
83. Rezzouk, F.Z.; Gracia-Romero, A.; Kefauver, S.C.; Gutiérrez, N.A.; Aranjuelo, I.; Serret, M.D.; Araus, J.L. Remote sensing techniques and stable isotopes as phenotyping tools to assess wheat yield performance: Effects of growing temperature and vernalization. *Plant Sci.* **2020**, *295*, 110281. [CrossRef]
84. Qiu, Z.; Xiang, H.; Ma, F.; Du, C. Qualifications of rice growth indicators optimized at different growth stages using unmanned aerial vehicle digital imagery. *Remote Sens.* **2020**, *12*, 3228. [CrossRef]
85. Jones, H.G.; Serraj, R.; Loveys, B.R.; Xiong, L.; Wheaton, A.; Price, A.H. Thermal infrared imaging of crop canopies for the remote diagnosis and quantification of plant responses to water stress in the field. *Funct. Plant Biol.* **2009**, *36*, 978–989. [CrossRef] [PubMed]
86. Durand, J.L.; Gastal, F.; Etchebest, S.; Bonnet, A.C.; Ghesquière, M. Interspecific variability of plant water status and leaf morphogenesis in temperate forage grasses under summer water deficit. In *Developments in Crop Science*; Perspectives for Agronomy; van Ittersum, M.K., van de Geijn, S.C., Eds.; Elsevier: Amsterdam, The Netherlands, 1997; Volume 25, pp. 135–143. [CrossRef]
87. Holloway-Phillips, M.M.; Brodribb, T.J. Contrasting hydraulic regulation in closely related forage grasses: Implications for plant water use. *Funct. Plant Biol.* **2011**, *38*, 594–605. [CrossRef] [PubMed]
88. Thomas, H.; James, A.R.; Humphreys, M.W. Effects of water stress on leaf growth in tall fescue, Italian ryegrass and their hybrid: Rheological properties of expansion zones of leaves, measured on growing and killed tissue. *J. Exp. Bot.* **1999**, *50*, 221–231. [CrossRef]
89. Martre, P.; Cochard, H.; Durand, J.L. Hydraulic architecture and water flow in growing grass tillers (*Festuca arundinacea* Schreb.). *Plant Cell Environ.* **2001**, *24*, 65–76. [CrossRef]
90. Carminati, A.; Javaux, M. Soil rather than xylem vulnerability controls stomatal response to drought. *Trends Plant Sci.* **2020**, *25*, 868–880. [CrossRef] [PubMed]

91. Garwood, E.A.; Sinclair, J. Use of water by six grass species. 2. Root distribution and use of soil water. *J. Agric. Sci.* **1979**, *93*, 25–35. [CrossRef]
92. Lopes, M.S.; Reynolds, M.P. Partitioning of assimilates to deeper roots is associated with cooler canopies and increased yield under drought in wheat. *Funct. Plant Biol.* **2010**, *37*, 147–156. [CrossRef]
93. Chen, S.L.; Tang, P.S. Studies on colchicine-induced autotetraploid barley: III. Physiological studies. *Am. J. Bot.* **1945**, *32*, 177–179. [CrossRef]
94. Levin, D.A. Polyploidy and novelty in flowering plants. *Am. Nat.* **1983**, *122*, 1–25. [CrossRef]
95. Abtahi, M.; Majidi, M.M.; Hoseini, B.; Mirlohi, A.; Araghi, B.; Hughes, N. Genetic variation in an orchardgrass population promises successful direct or indirect selection of superior drought tolerant genotypes. *Plant Breed.* **2018**, *137*, 928–935. [CrossRef]
96. Oliveira, R.A.; Näsi, R.; Niemeläinen, O.; Nyholm, L.; Alhonoja, K.; Kaivosoja, J.; Jauhiainen, L.; Viljanen, N.; Nezami, S.; Markelin, L.; et al. Machine learning estimators for the quantity and quality of grass swards used for silage production using drone-based imaging spectrometry and photogrammetry. *Remote Sens. Environ.* **2020**, *246*, 111830. [CrossRef]
97. Smith, C.; Karunaratne, S.; Badenhorst, P.; Cogan, N.; Spangenberg, G.; Smith, K. Machine learning algorithms to predict forage nutritive value of in situ perennial ryegrass plants using hyperspectral canopy reflectance data. *Remote Sens.* **2020**, *12*, 928. [CrossRef]
98. Bastianelli, D.; Bonnal, L.; Barre, P.; Nabeneza, S.; Salgado, P.; Andueza, D. La spectrométrie dans le proche infrarouge pour la caractérisation des ressources alimentaires. *INRAE Prod. Anim.* **2018**, *31*, 237–254. [CrossRef]
99. Zeng, L.; Chen, C. Using remote sensing to estimate forage biomass and nutrient contents at different growth stages. *Biomass Bioenergy* **2018**, *115*, 74–81. [CrossRef]
100. Pullanagari, R.R.; Yule, I.J.; Tuohy, M.P.; Hedley, M.J.; Dynes, R.A.; King, W.M. Proximal sensing of the seasonal variability of pasture nutritive value using multispectral radiometry. *Grass Forage Sci.* **2013**, *68*, 110–119. [CrossRef]
101. Castro, P.A.; Garbulsky, M.F. Spectral normalized indices related with forage quality in temperate grasses: scaling up from leaves to canopies. *Int. J. Remote Sens.* **2018**, *39*, 3138–3163. [CrossRef]

Article

Wheat Yellow Rust Detection Using UAV-Based Hyperspectral Technology

Anting Guo [1,2], Wenjiang Huang [1,2,3,*], Yingying Dong [1,2], Huichun Ye [1,3], Huiqin Ma [1], Bo Liu [4], Wenbin Wu [5], Yu Ren [1,2], Chao Ruan and Yun Geng [1,2]

1 Key Laboratory of Digital Earth Science, Aerospace Information Research Institute, Chinese Academy of Sciences, Beijing 100094, China; guoat@aircas.ac.cn (A.G.); dongyy@radi.ac.cn (Y.D.); yehc@radi.ac.cn (H.Y.); mahq@aircas.ac.cn (H.M.); renyu@aircas.ac.cn (Y.R.); ruanchao@aircas.ac.cn (C.R.); gengyun17@mails.ucas.ac.cn (Y.G.)
2 University of Chinese Academy of Sciences, Beijing 100190, China
3 Key Laboratory of Earth Observation, Sanya 572029, China
4 State Key Laboratory for Biology of Plant Diseases and Insect Pests, Institute of Plant Protection, Chinese Academy of Agriculture Sciences, No.2 West Yuanmingyuan Road, Beijing 100193, China; bliu@ippcaas.cn
5 Chinese Academy of Agriculture Sciences, Beijing 100081, China; wuwenbin@caas.cn
* Correspondence: huangwj@aircas.ac.cn; Tel.: +86-10-82178169

Abstract: Yellow rust is a worldwide disease that poses a serious threat to the safety of wheat production. Numerous studies on near-surface hyperspectral remote sensing at the leaf scale have achieved good results for disease monitoring. The next step is to monitor the disease at the field scale, which is of great significance for disease control. In our study, an unmanned aerial vehicle (UAV) equipped with a hyperspectral sensor was used to obtain hyperspectral images at the field scale. Vegetation indices (VIs) and texture features (TFs) extracted from the UAV-based hyperspectral images and their combination were used to establish partial least-squares regression (PLSR)-based disease monitoring models in different infection periods. In addition, we resampled the original images with 1.2 cm spatial resolution to images with different spatial resolutions (3 cm, 5 cm, 7 cm, 10 cm, 15 cm, and 20 cm) to evaluate the effect of spatial resolution on disease monitoring accuracy. The findings showed that the VI-based model had the highest monitoring accuracy (R^2 = 0.75) in the mid-infection period. The TF-based model could be used to monitor yellow rust at the field scale and obtained the highest R^2 in the mid- and late-infection periods (0.65 and 0.82, respectively). The VI-TF-based models had the highest accuracy in each infection period and outperformed the VI-based or TF-based models. The spatial resolution had a negligible influence on the VI-based monitoring accuracy, but significantly influenced the TF-based monitoring accuracy. Furthermore, the optimal spatial resolution for monitoring yellow rust using the VI-TF-based model in each infection period was 10 cm. The findings provide a reference for accurate disease monitoring using UAV hyperspectral images.

Keywords: UAV hyperspectral; wheat yellow rust; disease monitoring; vegetation index; texture; spatial resolution

Citation: Guo, A.; Huang, W.; Dong, Y.; Ye, H.; Ma, H.; Liu, B.; Wu, W.; Ren, Y.; Ruan, C.; Geng, Y. Wheat Yellow Rust Detection Using UAV-Based Hyperspectral Technology. *Remote Sens.* **2021**, *13*, 123. https://doi.org/10.3390/rs13010123

Received: 27 October 2020
Accepted: 29 December 2020
Published: 1 January 2021

1. Introduction

As one of the three major grain crops (wheat, rice, and corn) in China, wheat is the key to national food strategic security [1,2]. However, crop pests and diseases (more than 20 common ones) seriously endanger the safety of wheat production [3]. Yellow rust caused by *Puccinia striiformis f.* sp. *tritici* (Pst) causes explosive disease outbreaks and regional epidemics [4]. In epidemic years, yellow rust can reduce wheat yield by more than 40% or total harvest failure [5]. Therefore, the control of wheat yellow rust is crucial for food security. Only by accurately monitoring wheat diseases can we effectively prevent and control them and minimize losses. Traditionally, crop diseases have been assessed using visual monitoring, which has major limitations including small coverage and the inevitable subjectivity of investigators [6]. Therefore, an effective and nondestructive monitoring method for crop diseases is needed to replace the traditional method.

After years of development, remote sensing has become an effective technology for monitoring crop diseases [7–11]. The advantage of remote sensing is that the spatial distribution of crop diseases can be assessed at low cost, allowing for better disease control [12]. Hyperspectral remote sensing is a relatively advanced technology. Compared with multispectral technology, it can provide rich narrow-band information. Previous studies have discussed the advantages of hyperspectral remote sensing in various applications compared with multispectral. For instance, Awad [13] made a comparison between multispectral and hyperspectral images in forest mapping and the results showed that the accuracies based on the hyperspectral image were much higher than that of the multispectral image. Yang et al. [14] compared the performance of airborne multispectral and hyperspectral imagery in mapping cotton root rot. Mariotto et al. [15] compared the performance of hyperspectral and multispectral data in terms of crop productivity modeling and type discrimination. The advantage of hyperspectral technology in crop disease monitoring is that it can capture certain physiological changes (e.g., pigments and water content) caused by diseases. In previous studies, many crop diseases have been monitored by hyperspectral remote sensing including wheat yellow rust [16,17], powdery mildew [18], fusarium head blight [3], peanut leaf spot disease [19], tomato spotted wilt virus [20], and rice bacterial leaf blight [21].

Vegetation indices (VIs) have demonstrated good performance for crop disease monitoring in previous research. For instance, Mahlein et al. [22] successfully identified sugar beet rust and powdery mildew based on some newly constructed spectral disease indices. Zheng et al. [2] used the re-established the photochemical reflectance index (PRI) based on 570 nm, 525 nm, and 705 nm, and anthocyanin reflectance index (ARI) based on 860 nm, 790 nm, and 750 nm to identify yellow rust. The results showed that PRI and ARI had good performance in the early–mid and mid–late growth stages of wheat, respectively. Liu et al. [23] developed a novel red-edge head blight index (REHBI) for monitoring wheat fusarium head blight at a regional scale. The VI is sensitive to the physiological changes inside the leaves caused by diseases, but cannot represent the changes in the surface characteristics of leaves. Texture is a feature that reflects external changes caused by diseases and is proven to have good performance in crop disease monitoring [24]. Previous studies have combined spectral and texture features to improve the ability to characterize diseases. For example, Guo et al. [25] combined texture and spectrum to identify wheat yellow rust on the leaf scales. The results showed that the model based on combined features had better performance than the model based on a single-feature. Al-Saddik et al. [26] successfully identified the yellowness on grapevine leaves using a combined feature (texture and spectral). The accuracy was 99% for both diseases when spectral and texture features were combined in the model. These studies confirm the effectiveness of VIs, texture features (TFs), and their combination for crop disease monitoring. However, previous studies on the fusion of spectral and image features to detect crop diseases have mostly been conducted at the leaf scale, and the feasibility of the method for disease detection at the field scale has not been confirmed.

The technology of unmanned aerial vehicles (UAV) may bring new opportunities for precision agriculture, especially in field-scale crop disease monitoring [27,28]. The advantage of an UAV lies in its flexibility, which can fly at any time under suitable conditions [29]. In general, images with ultra-high spatial and temporal resolution can be acquired by UAV systems in a cost-effective manner [30]. RGB, multispectral, and hyperspectral sensors can all be mounted on UAVs [31]. However, only hyperspectral sensors can simultaneously obtain high spatial and spectral resolution, providing advantages over RGB and multispectral sensors [32]. For instance, Abdulridha et al. [33] extracted VIs from UAV-based hyperspectral images to detect diseases in tomato. The authors concluded that the PRI and the normalized difference vegetation index (NDVI 850) could identify bacterial spot in all stages of disease development. Moreover, Abdulridha et al. [12,34] successfully detected citrus canker disease and squash powdery mildew disease using a UAV-based hyperspectral imaging technique. Deng et al. [35] successfully detected citrus Huanglongbing based on UAV hyperspectral technology. The above studies confirm that UAV technology has good performance for crop disease monitoring; however, few studies have investigated the detection of yellow rust based on UAV hyperspectral images.

Spatial resolution of an image is determined by the instantaneous field of view (IFOV) and altitude of platform [36], which affects the efficiency of image collection and processing [37]. Generally, for the same sensor, the higher the altitude of the platform, the lower the spatial resolution [38]. It is crucial to select optimal spatial resolution according to observation object and specific requirements [39,40]. Few previous studies have focused on the impact of UAV-based image spatial resolution on target detection. Dash et al. [41] used UAV imagery with different resolutions to monitor forest disease outbreaks. They resampled the original images (0.06 m resolution) to new images (0.3, 1, and 3 m) and found that the optimal spatial resolution was 1 m for monitoring forest disease. Zhang et al. [42] assessed the impact of the spatial resolution for monitoring seedling rapeseed. Images with 1.35 cm, 1.69 cm, 2.61 cm, 5.73 cm, and 11.61 cm spatial resolutions were obtained by a UAV at different flying heights of 22 m, 29 m, 44 m, 88 m, and 176 m, respectively. The results showed that 2.61 cm was the optimal spatial resolution for monitoring seedling rapeseed growth. Liu et al. [43] used UAV images with different spatial resolutions to classify vegetation types and concluded that the classification accuracies exhibited slight fluctuations with a decrease in spatial resolution, and obtained the highest value when the spatial resolution was at an intermediate level. Although several studies have explored the effect of spatial resolution on detecting different ground objects, to date, no research has focused on the impact of different spatial resolutions for monitoring crop diseases, especially wheat yellow rust. Due to the limited ground coverage of UAV images, the optimal spatial resolution is an important aspect of the accurate detection of crop diseases.

In this research, a UAV equipped with a hyperspectral sensor was employed to monitor wheat yellow rust in multiple infection stages at different spatial scales. The objectives were to (1) evaluate the performance of monitoring wheat yellow rust at the field scale using VIs, TFs, and their combination obtained from UAV hyperspectral images; (2) determine the optimal image spatial resolution for monitoring wheat yellow rust with UAV images; and (3) use the optimal features and optimal spatial resolution to establish a monitoring model for yellow rust at the field scale in the early infection stage, mid-infection stage, and late infection stage.

2. Materials and Methods

2.1. Study Site

The UAV observation experiments of wheat yellow rust were carried out at the Langfang Experimental Station (39°30′41″N, 116°36′17″E) of the Chinese Academy of Agricultural Sciences; the station belongs to Langfang City, Hebei Province, China. The average annual temperature and precipitation in this area are 12 °C and 550 mm, respectively. The organic matter content, available phosphorus, and available potassium of the soil in the experimental area is about 1.41–1.47%, 20.5–55.8 mg·kg^{-1}, and 117.2–129.3 mg·kg^{-1}, respectively. 'Mingxian 169′, which is a wheat cultivar susceptible to yellow rust, was selected for planting in the experimental area. Three plots were set up in the experimental area including two inoculated plots (A and B in Figure 1c) and one healthy plot (C in Figure 1c), each with an area of 220 m^2. In each plot, we set up eight sample collection regions, each with an area of 1 m^2. Wheat yellow rust fungus (spores) prepared in advance were mixed with water to form a spore suspension with a concentration of 9 mg/100 mL^{-1}. The suspension was sprayed on the wheat leaves with a handheld sprayer. The wheat was then quickly covered with plastic film, which was removed the next morning. The inoculation process followed the Rules for Resistance Evaluation of Wheat to Diseases and Insect Pests—Part 1: Rule for Resistance Evaluation of Wheat to Yellow Rust (NY/T 1443.1-2007). Pesticides were used to spray the wheat in the healthy plot to prevent it from being infected with yellow rust. In addition, the wheat in all plots was managed in the same manner throughout the growth period including the application of the same amount of nitrogen and water. The location of the experimental area and the settings of the three experimental plots are shown in Figure 1. Figure 1c is an image of the entire experimental field obtained on 30 May using a UAV equipped with a hyperspectral sensor.

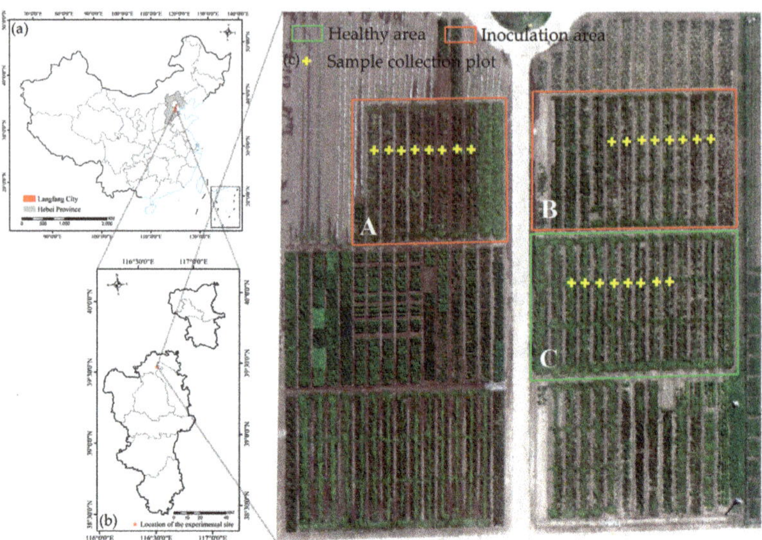

Figure 1. Study area: (**a**) the location of Langfang City; (**b**) the location of the experimental site (red star); (**c**) the image of the experimental field obtained on 30 May using a UAV equipped with a hyperspectral sensor; (**A**,**B**) are the plots inoculated with yellow rust, and (**C**) is the healthy plot; the yellow cross is the location of the sampling point.

2.2. Data Acquisition and Processing

2.2.1. Field Data Acquisition

During the critical growth period of wheat, we conducted six UAV-based hyperspectral observation experiments on 25 April, 4 May, 11 May, 18 May, 24 May, and 30 May, corresponding to 7, 16, 23, 30, 36, and 42 days post-inoculation (DPI), respectively. Figure 2 shows the development of wheat yellow rust at different infection periods. A total of 24 samples were obtained each time including eight samples obtained in the healthy plot and 16 samples obtained in the two inoculated plots. We obtained a total of 144 samples in the six experiments including 57 healthy samples and 87 yellow rust-infected samples. The severity of wheat yellow rust was described by the disease index (DI). Forty wheat plants were randomly selected from each plot with 1 m^2, and the first and second wheat leaves from the top of selected plants were used to assess the disease severity. A total of 80 leaves was selected to calculate the DI of each plot. According to the National Rules for the Investigation and Forecasting of Crop Diseases (GB/T 15795-2011) [6], the disease incidences of leaves were divided into nine categories (0%, 1%, 10%, 20%, 30%, 45%, 60%, 80%, and 100%), where 0% is healthy and 1% is disease level 1, 10% is disease level 2 . . . . . . 100% is disease level 8. The assessment of the incidence level of yellow rust was carried out immediately after the UAV had acquired the images and was performed by the same person under the guidance and supervision of a professional. The equation of the DI calculation is as follows [44]:

$$DI = \frac{\sum xf}{n \sum f} \times 100 \quad (1)$$

where x is the value of the incidence level; n is the value of the highest disease severity gradient; and f is the number of leaves for each degree of disease severity.

Figure 2. Examples of the development of yellow rust during different infection periods.

2.2.2. Unmanned Aerial Vehicle (UAV) Hyperspectral Image Acquisition

A UAV system (S1000) equipped with a hyperspectral imaging sensor (UHD 185) (Figure 3) was employed to acquire the images of wheat. The UAV hyperspectral imaging system consisted of four parts including a six-rotor electric UAV system (DJI Innovations, Shenzhen, China), a UHD 185 hyperspectral data acquisition system (Cubert GmbH, Ulm, Baden-Württemberg, Germany), a three-axis stabilized platform, and a data processing system. The UHD 185 hyperspectral imaging sensor had two charge-coupled device (CCD) detectors with a focal length of 23 mm and a pixel size of 6.45 μm [45]. The available spectral range of the UHD 185 sensor was 450–950 nm, and the spectral resolution was approximately 4 nm. The UHD 185 requires radiometric calibration before the flight and the lens exposure time is automatically matched. The flying height of the UAV was 30 m, the flying speed was 4 m/s, the forward overlap was about 80%, and the lateral overlap was about 60%. At a flying altitude of 30 m, we obtained hyperspectral images with a spatial resolution of 1.2 cm and a spectral resolution of 4 nm. The UAV hyperspectral data acquisition required clear weather with no wind or low wind speed. The image acquisition period in this study was from 11:00 a.m. to 13:00 p.m.

Figure 3. Unmanned aerial vehicle (UAV) hyperspectral imaging system.

2.2.3. UAV Hyperspectral Image Processing

Image fusion and mosaicking were performed after obtaining the UAV hyperspectral images. The Cubert–Pilot software (Cubert GmbH, Ulm, Baden-Württemberg, Germany) was used for image fusion of the hyperspectral data and the corresponding panchromatic image. Agisoft PhotoScan (Agisoft, St. Petersburg, Russia) was employed to image mosaicking. The original hyperspectral images were resampled to coarser resolutions to assess the effect of the spatial resolution on the monitoring accuracy of wheat yellow rust from UAV hyperspectral images. We resampled the original images (1.2 cm resolution) to generate new images (3, 5, 7,

10, 15, 20 cm resolution) using the nearest-neighbor algorithm. The reason for choosing these particular resolutions was that they are representative of most UAV images for monitoring the health status of crops [42,46]. Figure 4 shows an example of the UAV hyperspectral images with different spatial resolutions, which was part of area A in the image obtained on 25 April.

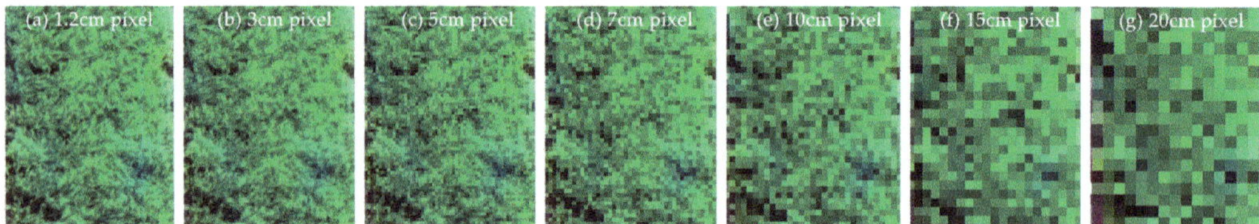

Figure 4. Example of UAV hyperspectral images with different spatial resolution. (**a**) 1.2 cm, (**b**) 3 cm, (**c**) 5 cm, (**d**) 7 cm, (**e**) 10 cm, (**f**) 15 cm, and (**g**) 20 cm.

The ground objects in the UAV images included wheat plants and soil. In this study, the NDVI [47] threshold method was used to extract wheat pixels from the images with a 1.2 cm spatial resolution [48]. The process of this method is as follows. First, set the optimal NDVI threshold. Generally, the NDVI range of vegetation was 0.3–1.0. After many attempts, we found that the NDVI threshold of 0.42 could accurately separate wheat pixels from background pixels. Second, NDVI larger than 0.42 was made into a mask file for extracting wheat pixels. Finally, we used the mask file to crop the original images to obtain the images containing only wheat pixels. For the consistency of the wheat pixels of the images with different spatial resolutions, the mask file made by the NDVI threshold of the original image (1.2 cm) was used to extract the wheat pixels of images at other spatial resolutions. The hyperspectral images containing only wheat plant pixels were used for subsequent processing and analysis.

2.3. Methods

The purpose of the image analysis was to determine which period and spatial scale provided the optimal accuracy for yellow rust identification. The process consisted of three steps: (1) extracting and selecting the VIs and TFs; (2) using partial least square regression (PLSR) to build the monitoring model of yellow rust; and (3) assessing the impact of different spatial resolutions on the monitoring accuracy.

2.3.1. Vegetation Index Extraction

Yellow rust causes abnormal fluctuations in the pigment, water content, and cell structure inside the leaves [49,50]. Therefore, we selected 15 VIs related to crop growth that were used in previous studies to evaluate their ability for monitoring wheat yellow rust. These VIs included the structural independent pigment index (SIPI), PRI, transformed chlorophyll absorption in reflectance index (TCARI), NDVI, normalized pigment chlorophyll index (NPCI), plant senescence reflectance index (PSRI), physiological reflectance index (PhRI), ARI, modified simple ratio (MSR), ratio vegetation structure index (RVSI), modified chlorophyll absorption reflectance index (MCARI), yellow rust index (YRI), greenness index (GI), triangular vegetation index (TVI), and nitrogen reflectance index (NRI). Table 1 shows the details of the VIs.

Table 1. Vegetation indices used for monitoring wheat yellow rust in this study.

VIs.	Equations	Application	Crop	Reference	Publication
SIPI	$(R_{800} - R_{445})/(R_{800} - R_{680})$	Biomass estimation and yield prediction	Potato	[51]	ISPRS Journal of Photogrammetry and Remote Sensing
PRI	$(R_{570}-R_{531})/(R_{570} + R_{531})$	Photosynthetic efficiency	Sunflower	[52]	Remote sensing of Environment
NPCI	$(R_{680} - R_{430})/(R_{680} + R_{430})$	Chlorophyll estimation	Vine	[53]	Remote sensing of Environment
MSR	$(R_{800}/R_{670} - 1)/sqrt(R_{800}/R_{670} + 1)$	Powdery mildew detection	Wheat	[18]	Computers and Electronics in Agriculture
RVSI	$((R_{712} + R_{752})/2) - R_{732}$	Target spot detection	Tomato	[33]	Precision Agriculture
YRI	$(R_{730} - R_{419})/(R_{730} + R_{419}) - 0.5R_{736}$	Yellow rust detection	Wheat	[54]	IEEE J-STARS
GI	R_{554}/R_{677}	Leaf rust detection	Wheat	[49]	Remote sensing
PhRI	$(R_{550}-R_{531})/(R_{550} + R_{531})$	Chlorophyll estimation	Corn	[55]	Remote sensing of Environment
ARI	$(R_{550})^{-1} - (R_{700})^{-1}$	Anthocyanin estimation	Norway maple	[56]	Photochemistry and Photobiology
PSRI	$(R_{680} - R_{500})/R_{750}$	Pigment estimation	Potato	[57]	Physiologia Plantarum
NRI	$(R_{570} - R_{670})/(R_{570} + R_{670})$	Nitrogen status evaluation	Wheat	[58]	Crop science
TCARI	$3((R_{700} - R_{675}) - 0.2(R_{700} - R_{500})/(R_{700}/R_{670}))$	Chlorophyll estimation	Corn	[59]	Remote sensing of Environment
TVI	$0.5 * (120(R_{750} - R_{550}) - 200(R_{670} - R_{550}))$	Laurel wilt detection	Avocado	[36]	Remote sensing of Environment
NDVI	$(R_{830} - R_{675})/(R_{830} + R_{675})$	Diseases detection	Sugar beet	[22]	Remote sensing of Environment
MCARI	$((R_{701} - R_{671}) - 0.2(R_{701} - R_{549}))/(R_{701}/R_{671})$	LAI and chlorophyll estimation	Corn	[60]	European Journal of Agronomy

2.3.2. Texture Feature Extraction

When crops are infected by diseases, changes occur in the pigment, water content, etc. as well as the shape, color, and texture of the leaves [25]. Therefore, texture is an important feature for monitoring wheat diseases. When insufficient spectral features are available to monitor diseases, TFs play an important role in improving the accuracy of disease recognition. Image texture refers to the gray-level distribution of pixels and their neighborhood [61]. Many methods have been used to define TFs including statistical, geometric, signal processing, and structural analysis methods and models [62]. The gray-level co-occurrence matrix (GLCM), a statistical method, is a proven and effective method with strong adaptability and robustness [63–66]. Therefore, the GLCM was used to extract the TFs. Eight frequently used TFs were employed to monitor yellow rust in this study. These TFs included the following features: Mean (MEA), which reflects the average grey level of all pixels in the matrix; Variance (VAR), which describes the dispersion of the values around the mean; Homogeneity (HOM), which indicates the uniformity of the matrix; Contrast (CON), which represents the local variation in the matrix; Dissimilarity (DIS), which reflects the difference in the grayscale; Entropy (ENT), which expresses the level of disorder in the matrix; Second Moment (SEC), which represents the uniformity of the grayscale; and Correlation (COR), which reflects a measurement of image linearity among the pixels [67]. The calculation equations of the TFs based on the GLCM are listed in Table 2.

Table 2. The equation of texture used in this study.

Texture	Equation
Mean, MEA	$MEA = \sum_{i,j=1}^{G} iP(i,j)$
Variance, VAR	$VAR = \sum_{i=1}^{G}\sum_{j=1}^{G}(i-u)^2 P(i,j)$
Homogeneity, HOM	$HOM = \sum_{i=1}^{G}\sum_{j=1}^{G}\frac{P(i,j)}{1+(i-j)^2}$
Contrast, CON	$CON = \sum_{i=1}^{G}\sum_{j=1}^{G}(i-j)^2 P(i,j)$
Dissimilarity, DIS	$DIS = \sum_{i=1}^{G}\sum_{j=1}^{G}P(i,j)\lvert i-j\rvert$
Entropy, ENT	$ENT = -\sum_{i=1}^{G}\sum_{j=1}^{G}P(i,j)\log P(i,j)$
Second moment, SEC	$SEC = \sum_{i=1}^{G}\sum_{j=1}^{G}P^2(i,j)$
Correlation, COR	$COR = \sum_{i=1}^{G}\sum_{j=1}^{G}\frac{(i-MEA_i)(j-MEA_j)P(i,j)}{\sqrt{VAR_i}\sqrt{VAR_j}}$

Note: *i* and *j* are the row and column number of images, respectively; *P* (*i*, *j*) is the relative frequency of two neighboring pixels.

The TFs were extracted from the images transformed by principal component analysis (PCA) [68–70]. Extracting TFs from all bands of the original images will cause a lot of data redundancy, which has adverse effects on model accuracy. Data redundancy can be effectively reduced by converting the original image (125 bands) into several principal component (PC) images containing most of the information through PCA. In this study, PCA was completed using the PCA calculation module in the ENVI 5.1 software. The first three PC images including the first PC image (PC1), the second PC image (PC2), and the third PC image (PC3), which contained more than 96% of the cumulative variance, were used to extract the TFs In addition, the relative distance measured in pixel numbers (d) in the GLCM was set to 1 (d = 1), and the relative orientation (θ) was the average value of four directions (θ = 0°, 45°, 90°, and 135°) [71,72]. The calculation of the TFs was conducted in MATLAB 2016a. The TFs of the samples were extracted by setting the regions of interest (ROI). First, the texture values of the whole image were calculated; then the 50 × 50 pixels regions of interest (ROI) at the sampling point were defined manually; Finally, texture values were extracted within the ROI, and the average value of the pixels in each ROI was calculated to represent the texture feature of one sample. In addition, it should be noted that the 50 × 50 pixel ROI were set in the original image (1.2 cm). For images of different spatial resolutions, we set the ROI corresponding to the size of the ROI in the original image to extract the texture features. The vegetation indices of each sample point were extracted by the same method.

2.3.3. Features Selection

Correlation analysis was used to find features (VIs and TFs) that were sensitive to the wheat yellow rust. For the relationship between DI of yellow rust and TFs, VIs were quantified by correlation coefficient (r). These features may have different sensitivities to yellow rust at different DPI. According to the disease development, we divided the data obtained in the six periods into three categories including the early infection stages (seven and 16 DPI), mid-infection stages (23 and 30 DPI), and late infection stages (36 and 42 DPI). The criteria for the selection of the sensitive features were that the feature and DI were significantly correlated in the three stages, and the average value of the correlation coefficients in the three stages was relatively large. In addition, we used the features extracted from the original images (1.2 cm resolution) to quantify the relationship with the DI of yellow rust.

2.3.4. Severity Estimation Model Based on Partial Least Squares Regression

PLSR was used to build the monitoring model of wheat yellow rust. PLSR has been previously applied for crop growth monitoring and physiological and chemical parameter estimation [18,46]. PLSR is a classic modeling method that includes the characteristics of three methods: PCA, canonical correlation analysis, and multiple linear regression analysis. In PLSR, the latent structure of the variables is determined using the component projection, which projects the predictor variables and the observed variables into a new space [73]. The original variables with high data redundancy are thus transformed into a few variables by selecting the optimal latent variables [51]. PLSR can be expressed as a linear model describing the relationship between the predictor variables and the observed variables. The expression is as follows [74]:

$$Y = \beta_0 + \beta_1 X_1 + \beta_2 X_2 + \beta_3 X_3 + \ldots + \beta_n X_n \quad (2)$$

where β_0 is the intercept; β_n are the regression coefficients; X_n are the independent variables; and n is the number of independent variables. In this study, the independent variables were the VIs, TFs, and their combination; the dependent variable was the DI of wheat yellow rust.

2.3.5. Accuracy Assessment

The coefficient of determination (R^2) and relative root mean square error (RRMSE) were used to assess the performances of the PLSR models based on different features at different spatial scales. The RRMSE is the RMSE divided by the mean of the observations. The equation of the RRMSE is [75]:

$$\text{RRMSE} = \frac{100}{\overline{Q_i}} \sqrt{\frac{1}{n} \sum_{i=1}^{n} (P_i - Q_i)^2} \quad (3)$$

where P_i and Q_i are the estimated and measured values, respectively. $\overline{Q_i}$ is the average of the measured values, and n is the number of samples. In addition, a leave-one-out cross-validation (LOOCV) approach was employed for model training and validation. This approach uses one sample for verification and k−1 samples for model training; k rotation was performed. The average accuracy of obtained for k times was taken as the final accuracy. The advantage of LOOCV is that almost all samples in each round are used to train the model, which can minimize overfitting and allow a more accurate and reliable assessment of model prediction. LOOCV was proven to be effective and is widely used in crop disease monitoring modeling and crop parameter inversion modeling [2,18,23,75–77]. This process was performed in MATLAB 2016 (The Math Works, Inc., Natick, MA, USA).

3. Results

3.1. Spectral Response of Wheat Yellow Rust at Different Inoculation Stages

The spectral reflectance was extracted from the 1.2 cm spatial resolution images to characterize the spectral response of wheat yellow rust at different infection periods. Figure 5 shows the canopy average spectral reflectance curves of healthy and diseased wheat in different inoculation periods (7 DPI, 16 DPI, 23 DPI, 30 DPI, 36 DPI, and 42 DPI). It can be seen from Figure 5a,b that the shape of the canopy spectral reflectance curves of healthy and yellow rust-infected wheat were similar. In the visible light region (450–680 nm), the reflectance of the wheat canopy was affected by various pigments, resulting in relatively low values. The reflectance increased rapidly in the red-edge region (680–740 nm) and the reflectance was highest in the near-infrared region (740–950 nm). In addition, with an increase in the DPI, the canopy reflectance in the near-infrared region first increased and then decreased. The reflectance of both healthy and infected wheat was significantly different in the near-infrared region at different inoculation stages, and the difference was more pronounced in the yellow rust-infected wheat. The canopy reflectance of the yellow rust-infected wheat was lower than that of the healthy wheat in the near-infrared region (740–950 nm) in the six stages. Figure 5c shows the ratio of the spectral reflectance of the healthy and yellow rust-infected wheat in different inoculation periods. Significant differences were observed in the near-infrared region. In the early stages of inoculation (7 DPI and 16 DPI), the reflectance differences between the healthy samples and yellow rust-infected samples were negligible. However, as the DPI increased, the differences became more apparent, and the maximum difference was observed in the late stage of inoculation (42 DPI). As the DPI increased, the extent of damage to the wheat increased. Therefore, the difference between the healthy and diseased wheat became larger. The analysis of the spectral reflectance of the wheat canopy increased our understanding of the changes in disease severity and provided a basis for establishing the monitoring model of yellow rust.

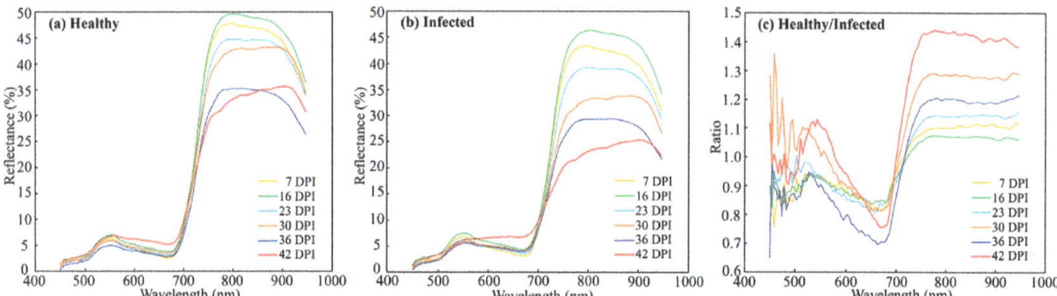

Figure 5. Canopy reflectance curves of (**a**) healthy wheat and (**b**) yellow rust-infected wheat on different inoculation dates; (**c**) the ratio of the spectral reflectance of the healthy and yellow rust-infected wheat on different inoculation dates.

3.2. Features Sensitive to Yellow Rust

Figure 6 shows the correlations between the features (VIs and TFs) and the DI. The different colors represent the different infection periods. Figure 6a shows that most VIs were strongly correlated with the DI, for instance, NDVI, SIPI, PRI, PSRI, MSR, GI, TVI, and NRI were significantly correlated with the DI at the 0.01 level (|R|: 0.390–0.836) in the three infection stages. Several VIs were strongly correlated with the DI in a particular period, for instance, in the late infection stage, PVSI, TCARI, and YRI were significantly correlated with the DI at the 0.01 level, with r values of 0.547, −0.782, and −0.594, respectively. Figure 6b,c shows the correlations between the TFs and the DI in different infection stages. The results showed that most TFs extracted from PC1 were significantly correlated with the DI in the mid- and late infection stages. In the early infection stage, only MEA1, VAR1, and CON1 were significantly correlated with the DI. Except for COR2, the TFs extracted from PC2 were significantly correlated with the DI in all three infection stages (|R|: 0.331–0.761). However, there was a very weak or no correlation between the TFs (except for MEA3 and COR3) extracted from PC3 and the DI in the three infection stages.

Too many variables cause data redundancy and processing difficulties and affect model accuracy. Therefore, we selected features that were sensitive to wheat yellow rust. According to the correlation analysis between the features and the DI, we selected features where the value of r was relatively large and passed the significance test ($p < 0.01$) in each infection period to establish the model. For the VIs, NDVI, SIPI, PRI, PSRI, MSR, GI, TVI, and NRI met the requirements. We then calculated the absolute value of the average correlation coefficient between these VIs and the DI in the three periods and sorted the values by the magnitude. Finally, the five VIs with the largest values (NDVI, SIPI, PRI, PSRI, and MSR) were selected for modeling. Only four TFs (MEA1, MEA2, VAR2, and CON2) met the requirements in the three infection stages. The selected features were input into the PLSR model to establish the monitoring model of yellow rust. The specific values and significance levels of the correlation coefficients between the DI and VIs and TFs in the three infection stages are listed in Tables A1 and A2, respectively.

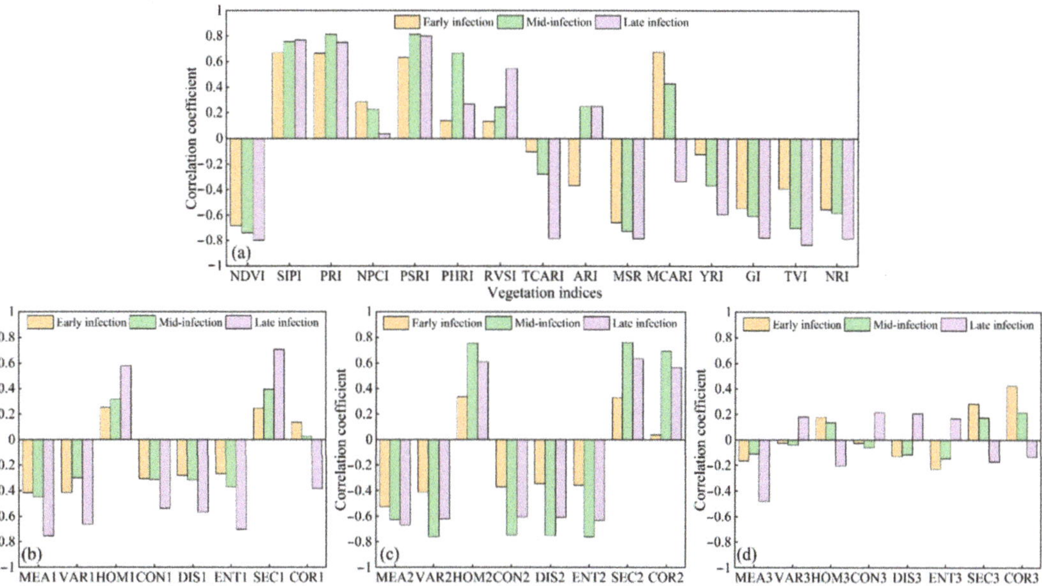

Figure 6. The correlation between the disease index (DI) and different features; (**a**) vegetation indices (VIs) and DI; (**b–d**) texture features (TFs) and DI. Note: MEA1 indicates that the mean (MEA) was extracted from the first principal component (PC) image (PC1), CON2 means that the contrast (CON) was extracted from the second PC image (PC2), and DIS3 means that the dissimilarity (DIS) was extracted from the third PC image (PC3).

3.3. Establishment and Evaluation of the Wheat Yellow Rust Monitoring Models

A suitable image spatial resolution for monitoring wheat yellow rust improves the monitoring accuracy, increases the efficiency of data processing, and reduces costs. In our study, we extracted the features from the images with seven different spatial resolutions (1.2, 3, 5, 7, 10, 15, 20 cm) in each infection period to establish the monitoring model of yellow rust and determine the optimal spatial resolution in different infection periods for monitoring yellow rust based on UAV hyperspectral images. In addition, the VI-based, TF-based, and VI-TF-based models were established to verify the performance of the combination of image features and spectral features extracted from the UAV hyperspectral images for monitoring yellow rust.

Figure 7 shows the comparison of the DI estimated accuracy of PLSR models based on different features at different infection periods and spatial resolutions. The DI estimated accuracy of the VI-based model (NDVI, SIPI, PRI, PSRI, and MSR) is presented in Figure 7a. The results showed that in the early and late infection stages, the accuracy of DI estimation at the seven spatial resolutions had a small fluctuation range (R^2 in early and late infection stages: 0.41–0.52/0.61–0.70). In the early infection stage, the highest accuracy was R^2 of 0.52 at a spatial resolution of 15 cm. In the late infection stage, the highest accuracy was an R^2 of 0.70 at a spatial resolution of 20 cm. In the mid-infection stage, except for the lowest accuracy with R^2 of 0.53 at a resolution of 7 cm, the fluctuation range at other spatial resolutions was very small (R^2: 0.69–0.75), and the highest accuracy was R^2 of 0.75 at a spatial resolution of 15 cm. In addition, we compared the accuracy at the three infection periods. The results showed that the accuracy at the mid-infection stage was the highest, except that at the spatial resolution of 7 cm. The order of accuracy of the three infection periods was the mid-infection stage > late infection stage > early infection stage, indicating that the appropriate stage for monitoring wheat yellow rust using the VI-based model was the mid-infection stage.

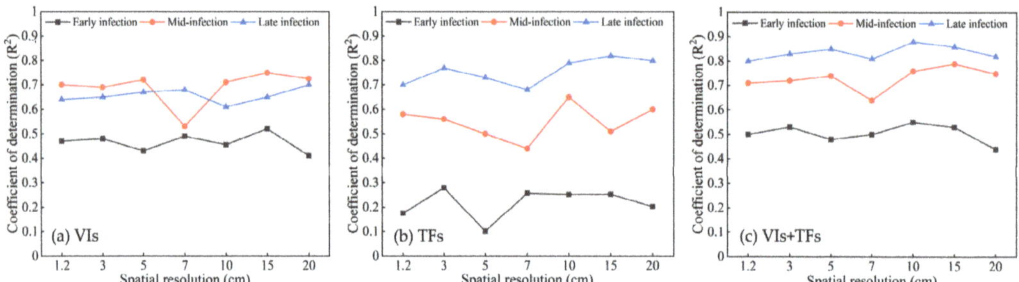

Figure 7. The monitoring accuracy (R^2) of (**a**) VI-based, (**b**) TF-based, and (**c**) VI-TF-based models at different infection stages and spatial resolutions.

For wheat yellow rust monitoring using the TF-based model (Figure 7b), different spatial scales had a greater impact on estimation accuracy. In the early infection stage, the accuracy range was R^2 of 0.1–0.28, and the highest accuracy was only R^2 of 0.28 at the spatial resolution of 3 cm. In the mid-infection stage, the estimation accuracy range was an R^2 of 0.44–0.65, of which the highest was an R^2 of 0.65 at the spatial resolution of 10 cm. In the late infection stage, the accuracy range was an R^2 of 0.68–0.82, of which the highest was an R^2 of 0.82 at the spatial resolution of 15 cm. It can be found that the estimated accuracy of the TF-based model was affected by the spatial resolution in three infection periods. In addition, the accuracy in the early infection stage was very low (the highest R^2 was only 0.28), therefore, it was not suitable to use TFs for monitoring wheat yellow rust in the early infection stage. However, the accuracies were higher in the mid- and late infection stage. In the late infection stage, we obtained the highest accuracy with an R^2 of 0.82 at the spatial resolution of 15 cm, which were the optimal stage and spatial resolution for monitoring yellow rust using the TF-based model. In addition, we compared Figure 7a,b and found that the accuracy based on VIs was higher than that based on TFs at different spatial resolutions in the early and mid-infection stages. In the late infection stage, the accuracy of the TF-based model was better than that of the VI-based model.

Figure 7c shows the monitoring accuracy of the VI-TF-based model at different spatial resolutions. The results showed that the spatial resolution had a negligible impact on the accuracy in the early and late infection stages. In the early infection stages, the range of R^2 was 0.44–0.55, and the highest accuracy was obtained at the spatial resolution of 10 cm (R^2: 0.55). In the mid-infection stage, except for the lowest accuracy at the spatial resolution of 7 cm (R^2: 0.64), the accuracy fluctuation range was relatively small (R^2: 0.64–0.79), and the highest accuracy was an R^2 of 0.79 at the spatial resolution of 15 cm. In the late infection stage, the accuracy range of R^2 was 0.80–0.88, and the highest accuracy was obtained at the spatial resolution of 10 cm (R^2: 0.55). The trend of the monitoring accuracy in different infection periods was the same for the models based on the TFs. As the infection period increased, the monitoring accuracy increased. The highest accuracy of the VI-TF-based model was obtained in the late infection stage at the spatial resolution of 10 cm. In addition, the VI-TF-based model performed better than either the VI-based or TF-based model at each spatial resolution in the three infection periods. For example, at a spatial resolution of 10 cm, the R^2 of the VI-based model in the three infection periods were 0.46, 0.71, and 0.61, those of the TF-based model were 0.25, 0.65, and 0.79, and those of the VI-TF-based model were 0.55, 0.76, and 0.88, respectively. Furthermore, a comparison of the optimal accuracy of the models at the optimal spatial resolution in each infection period (Figure 8) indicated that the VI-TF-based model provided the highest accuracy. In the early infection stage, the monitoring accuracy of the VI-based or TF-based model was very low, especially that of the TF-based model. However, the VI-TF-based model significantly improved the accuracy. The highest R^2 of the TF-based model was only 0.25, that of the VI-based model was 0.52, and that of the VI-TF-based model was 0.55. Therefore, this result provided information to develop an effective monitoring approach for yellow rust in the early stage. The PLSR models for disease monitoring based on different features in the different infection stages at optimal spatial resolutions are shown in Table 3. The specific values of R^2 and RRMSE between the estimated DI and measured DI in the different infection stages and at different spatial resolutions using the VI-based, TF-based, and VI-TF-based models are listed in Table 3.

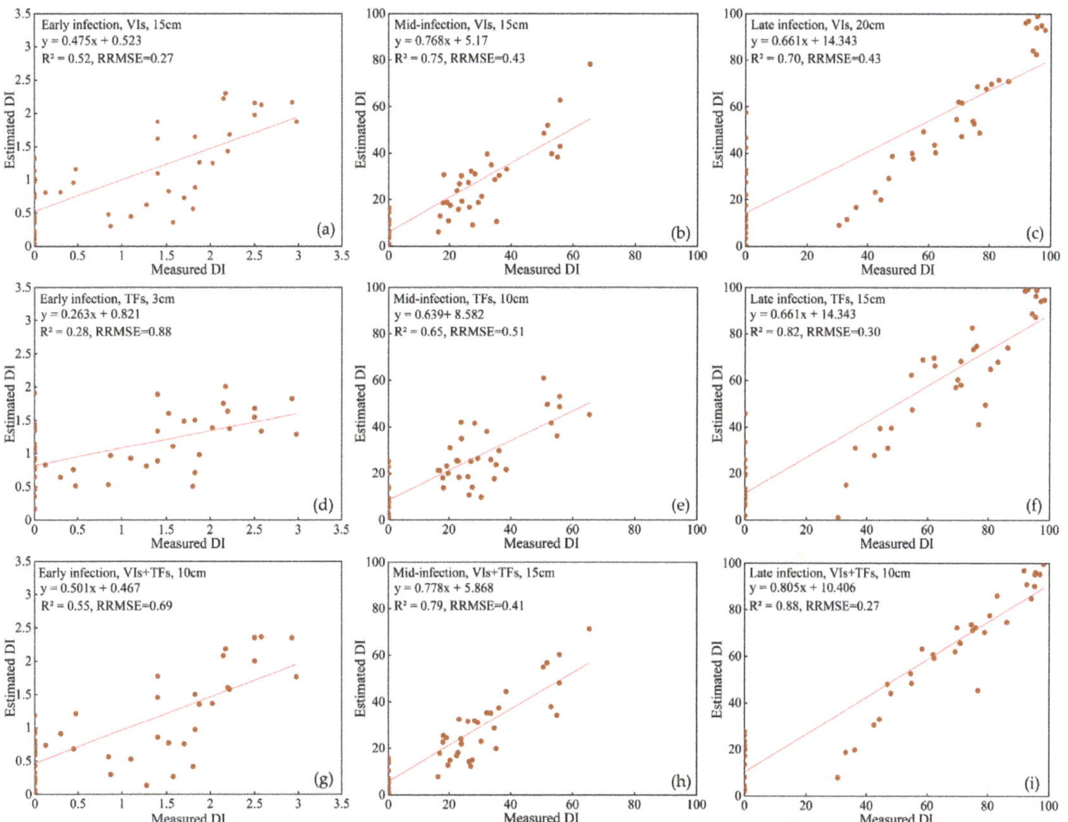

Figure 8. The scatter plots of the measured disease index (DI) versus the estimated DI of (**a–c**) the vegetation index (VI)-based model in the three infection stages at 15 cm, 15 cm, and 20 cm spatial resolution, (**d–f**) the texture feature (TF)-based model in the three infection stages at 3 cm, 10 cm, and 15 cm spatial resolution, and (**g–i**) the VI-TF-based model in the three infection stages at 10 cm, 15 cm, and 10 cm spatial resolution.

Table 3. The partial least squares regression (PLSR) models for disease monitoring based on different features in the different infection stages at optimal spatial resolutions.

Feature	Infection Stages	Spatial Resolution	PLSR-Based Model Equations
VIs	Early infection	15 cm	DI = −1.6143 − 3.9349NDVI + 6.0053SIPI + 7.173PRI + 10.3403PSRI − 0.263MSR
	Mid-infection	15 cm	DI = −391.994 + 374.044NDVI + 59.786SIPI + 338.364PRI + 759.008PSRI − 3.844MSR
	Late infection	20 cm	DI = 26.762 − 51.012NDVI + 39.463SIPI + 178.313PRI + 91.32PSRI − 8.395MSR
TFs	Early infection	3 cm	DI = 16.9718 − 0.1463MEA1 − 0.2437MEA2 − 5.8203VAR2 + 3.1486CON2
	Mid-infection	10 cm	DI = 316.811 + 0.588MEA1 − 5.671MEA2 − 32.4VAR2 + 11.707CON2
	Late infection	15 cm	DI = 479.509 + 0.54MEA1- 8.504MEA2 − 22.926VAR2 + 1.054CON2
Vis + TFs	Early infection	10 cm	DI = −4.355 − 0.008MEA1 − 0.011MEA2 + 0.094VAR2 + 0.175CON2 −4.272NDVI + 8.729SIPI + 14.065PR + 7.918PSRI − 0.189MSR
	Mid-infection	15 cm	DI = 209.235 + 1.339MEA1 − 1.75MEA2 − 109.907VAR2 + 21.766CON2 − 41.262NDVI − 18.082SIPI + 311.905PRI − 10.207PSRI − 13.047MSR
	Late infection	10 cm	DI = −248.697 + 2.211MEA1 − 4.391MEA2 − 14.686VAR2 + 4.341CON2 + 421.597NDVI + 69 SIPI + 288.678PRI + 680.157PSRI − 5.277MSR

The VI-TF-based model with the highest monitoring accuracy in the three infection stages was used to map the yellow rust distribution. For the optimal spatial resolution, in the early and late infection stages, the VI-TF-based model achieved the highest accuracy at a spatial resolution of 10 cm with an R^2 of 0.55 and 0.88, respectively. In the mid-infection stage, the highest accuracy was obtained at a spatial resolution of 15 cm (R^2 = 0.79), and the monitoring accuracy was also high at a spatial resolution of 10 cm (R^2 = 0.76). The accuracy difference was very small between the models at a 10 cm and 15 cm spatial resolution. Therefore, we used the VI-TF-based model to map wheat yellow rust at the 10 cm spatial resolution in the three infection periods. According to the suggestion from the National Plant Protection Department, the DI was classified into four classes including healthy (DI ≤ 5%), slight infection (5 < DI ≤ 20%), moderate infection (20 < DI ≤ 50%), and severe infection (DI > 50%). The spatial distribution of yellow rust in the study area is shown in Figure 9. It can be seen from Figure 9 that the wheat in the inoculation plots (A and B) were seriously infected, and the location of the disease was mainly in the middle of each row in the three infection periods, which was attributed to the location of the inoculation of the pathogen. The infected area and severity increased from the early to the late infection stage, and the infected areas of wheat yellow rust in the early infection stage was very small (Figure 9a), since the wheat had only been inoculated for a short time, the disease was just beginning to develop. The diseases developed rapidly in the mid-infection period (Figure 9b) due to the pathogen accumulation in the early infection stage and suitable environmental factors. In the late infection stage, most wheat in the field was infected with yellow rust (Figure 9c). The wheat yellow rust in the healthy plot (C) was not detected in the early infection stage, and mild diseases were detected in the mid- and late infection stages. The reason may be that we implemented pesticide prevention in the healthy area, resulting in almost no yellow rust on the wheat in this area; the mild disease may have spread from the inoculation area (A and B). In addition, the monitoring model also has certain monitoring errors. The distribution and development of yellow rust in the experiment area were basically in line with the spread of yellow rust and the observations in this experiment.

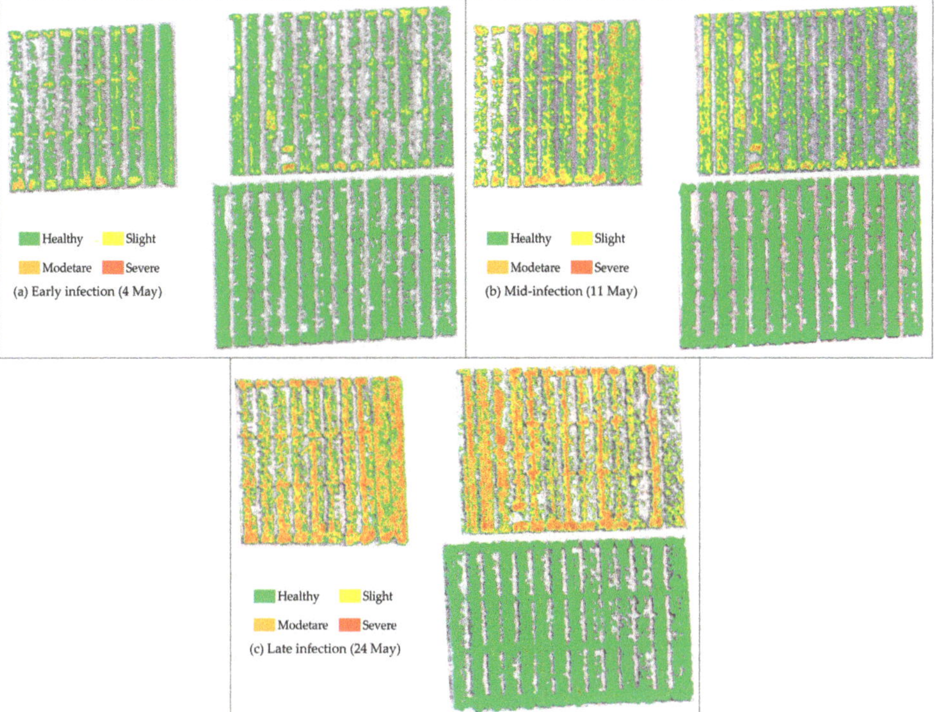

Figure 9. The spatial distribution of wheat yellow rust mapped using the VI-TF-based model in the (**a**) early infection stage (4 May), (**b**) mid-infection stage (11 May), and (**c**) late infection stage (24 May).

4. Discussion

The spectral response of crop diseases is the basis of remote sensing monitoring. Previous studies have confirmed that crops affected by diseases exhibit changes in the pigment, water content, and cell structure of the leaves, and these changes are reflected in the spectrum [78]. Changes in pigment generally cause spectral responses in the visible range, while changes in cell structure cause spectral responses in the near-infrared range. Figure 5 shows that the spectral reflectance of yellow rust-infected wheat was higher than that of healthy wheat in the visible range (450–680 nm), and lower than that of healthy wheat in the near-infrared range (720–950 nm), which may be caused by yellow rust destroying the pigment and altering the cell structure in the leaves. In addition, when wheat is infected by yellow rust, bright yellow spore piles form on the surface of the leaves, which not only affect the spectrum by invading the inside of the leaf and destroying the physiological parameters, but also cause the change in the spectrum through the change in the surface color. These results were consistent with those of Guo et al. [18]. Figure 5 also shows that the spectral reflectance of healthy and yellow rust-infected wheat in the near-infrared range first increased and then decreased with the increase in inoculation time. For healthy wheat, this may be caused by the growth and development of wheat; for diseased wheat, in addition to the effects of wheat growth and development, it may also be caused by the continuous development of diseases over time. The severity of the disease gradually increased over time, and the effects became visible; thus, the spectral reflectance changes were more significant in the later stages. Zheng et al. [2] also observed similar results. In this study, NDVI, SIPI, PRI, PSRI, and MSR were selected to monitor yellow rust. Previous studies have concluded that these indices are closely related to photosynthetic efficiency [52,79], the ratio of carotenoids to chlorophyll, and LAI. Yellow rust destroys pigments and affects the photosynthetic efficiency as well as the LAI at the canopy scale. Therefore, the selected VIs were sensitive to yellow rust.

Based on PLSR, the yellow rust monitoring models were established with different features in different infection periods. The results showed that the VI-TF-based models were better than the VI-based or TF-based models. For example, in the late infection stage the accuracy of the VI-TF-based model was $R^2 = 0.88$ at a spatial resolution of 10 cm, which was 0.09 and 0.27 higher than that of the TF-based and VI -based models, respectively. The VI-based model only reflected the changes in pigment and cell structure inside the leaves caused by yellow rust, while the TF-based model only reflected the changes in color and shape outside the leaves. The advantage of the TF-VI-based model is that it combines VIs (NDVI, SIPI, PRI, PSRI and MSR) and TFs (MEA1, MEA2, VAR2 and CON2), which can reflect both the internal and external changes of leaves caused by yellow rust. These results were consistent with our previous research, which was conducted on the leaf scale [25] This study was the first to demonstrate the effectiveness of monitoring yellow rust using the combination of VIs and TFs at the field scale. In addition, the results showed that the accuracy of the VI-based model was higher than that of the TF-based model in the early and mid-infection stages and lower than that of the TF-based model in the late infection stage. For example, at a spatial resolution of 10 cm, the accuracies of the VI-based models were R^2 of 0.46, 0.71, and 0.61 in the three infection stages, which were 0.21 and 0.06 higher than TF-based models in early and mid-infection stages, respectively, and which was 0.18 lower than the TF-based model in the late infection stage. Disease symptom development may have caused this situation. In the early and mid-infection, yellow rust is asymptomatic or mild, resulting in less information on the external changes being captured by the TFs, while VIs can capture the changes inside the leaves at this time. When the symptoms of yellow rust become obvious in the late infection stage, TFs, which can represent more disease information, have a great advantage over VIs.

In this study, we also evaluated the impact of the spatial resolution on the monitoring accuracy of yellow rust. The results showed that the spatial resolution had a negligible effect on the monitoring accuracy of the VI-based models (Figure 7). The reason may be that the nearest-neighbor algorithm was used to resample the images to change the spatial resolution. Theoretically, the reflectance values of the image pixels will not change after resampling [36]. VIs are converted from spectral bands, so their values will not change. This result was consistent with that of Zhang et al. [42]. The impact of the spatial resolution on the monitoring accuracy of the TF-based models was greater than that of the VI-based models (Figure 7). The texture was calculated based on the size of the pixels and the neighborhood relationship [80]. The pixel size and position may have changed after resampling, resulting in textural changes. Therefore, the accuracy of the TF-based models was significantly affected by spatial resolution. Based on the VI-TF model, the optimal spatial resolution for monitoring wheat yellow rust at the three infection periods was about 10 cm. The reason for obtaining a higher accuracy at the spatial resolution of 10 cm may be related to the size of the wheat plants. Dash et al. [41] obtained similar results when using UAV multispectral images with different spatial resolutions to monitor the occurrence of forest diseases. The optimal spatial resolution for monitoring forest diseases was 1 m, which was similar to the radius of the trees (1.52 m). Therefore, the optimal spatial resolution may not necessarily be the highest resolution, but may depend on the monitoring object.

Crop disease monitoring relies largely on the use of appropriate platforms, sensors, and analysis methods to obtain the optimal spectral, spatial, and temporal features of the diseases. For the monitoring of wheat diseases, near-surface spectral measurement technology is relatively mature including non-imaging sensors (e.g., spectrometers (ASD)) and imaging sensors (e.g., hyperspectral sensors (Headwall Photonics)), which enable remote sensing monitoring of wheat diseases. However, these technologies cannot be applied to the field or larger scales to prevent and monitor diseases. Satellite technology can be used for crop disease monitoring on a large scale, however, due to the limitation of the spectral and time characteristics, current satellite technology has certain limitations for the field-scale monitoring of crop diseases. In addition, satellite sensors are also affected by external factors such as weather and clouds. UAV technology overcomes these limitations with the advantages of low cost and flexibility, providing a new strategy for monitoring crop diseases at the field scale. We successfully acquired hyperspectral images using the DJI UAV equipped with UHD 185 sensors, extracted the optimal features (the combination of VIs and TFs), and selected the optimal spatial resolution for wheat yellow rust monitoring. However, we encountered several limitations. First, we resampled the original images to obtain images with different spatial resolutions to evaluate the impact of different spatial resolutions on the monitoring accuracy. In the future, we will fly the UAV at different altitudes to obtain images at different spatial resolutions rather than simulate the image resolution. Second, in the case of a small sample size, the LOOCV was used to model training and validation. To reduce the impact of spatial autocorrelation of the sample on the modeling accuracy, we will collect more samples, and use samples from different fields to train and verify the model separately. In addition, the proposed method generally did not have high monitoring accuracy in the early stages of disease development, which is the key stage for disease prevention. Therefore, accurate monitoring and effective prevention measures in the early infection stage are essential to minimize the damage caused by the disease. In a follow-up study, we will investigate different methods (e.g., establishing a new spectral index sensitive to yellow rust) to monitor wheat yellow rust in the early infection stage based on UAV images. Furthermore, UAV multispectral images have also been used for disease monitoring and have the advantages of low cost and convenient data acquisition and processing. In the future, we will analyze the use of multispectral images for monitoring yellow rust at the field scale.

5. Conclusions

In this study, UAV hyperspectral images were used to monitor wheat yellow rust at the field scale. We used different features (VIs, TFs, and their combination) to establish PLSR monitoring models of wheat yellow rust in the early, mid-, and late infection stages. We evaluated the impact of different image spatial resolutions (1.2 cm, 3 cm, 5 cm, 7 cm, 10 cm, 15 cm, and 20 cm) on monitoring accuracy. The following conclusions were obtained. (1) The VI-based model provided the highest monitoring accuracy in the mid-infection stage, and the TF-based model yielded the highest monitoring accuracy in the late infection stage. However, the TF-based model was unsuitable to monitor yellow rust in the early infection stage, the highest monitoring accuracy was only an R^2 of 0.28. (2) The VI-TF-based model provided higher accuracy than the VI-based or TF-based model in the three infection periods, and provided the highest accuracy in the late infection stage ($R^2 = 0.88$). In addition, the monitoring accuracy in the early infection stage was significantly improved by using the VI-TF-based model; thus, this model is suitable for the early detection of diseases. (3) The spatial resolution had a negligible influence on the monitoring accuracy of the VI-based models and a greater influence on the TF-based models. The optimal spatial resolution for the VI-TF-based model for monitoring yellow rust was 10 cm. In subsequent studies, we will focus on the precise monitoring of wheat yellow rust in the early infection stage to provide a basis for disease prevention.

Author Contributions: Conceptualization, A.G. and W.H.; Methodology, A.G., H.Y., and Y.D.; Data acquisition and processing, A.G., W.W., B.L., H.M., C.R., Y.R., and Y.G.; Formal analysis, A.G. and C.R.; Writing—original draft preparation, A.G.; Writing—review and editing, A.G., W.H., H.Y., and Y.D.; Funding acquisition, W.H. All authors have read and agreed to the published version of the manuscript.

Funding: This work was supported by the National Key R&D Program of China (2017YFE0122400, 2016YFD0300601); the National Natural Science Foundation of China (41871339, 42071423, 42071320); the National Special Support Program for High-Level Personnel Recruitment (Ten-Thousand Talents Program) (Wenjiang Huang); the Youth Innovation Promotion Association CAS (2017085); the Beijing Nova Program of Science and Technology (Z191100001119089); the Innovation Foundation of Director of Institute of Remote Sensing and Digital Earth, Chinese Academy of Sciences, China, financially supported by Hainan Provincial High Level Talent Program of Basic and Applied Basic Research Plan in 2019 of China (2019RC363).

Institutional Review Board Statement: This study not involving humans.

Informed Consent Statement: This study not involving humans.

Data Availability Statement: Data sharing is not applicable to this article.

Conflicts of Interest: The authors declare no conflict of interest.

Appendix A

Table A1. The correlation coefficients between the DI and VIs in the three infection stages.

VIs	Early Infection	Mid-Infection	Late Infection
NDVI	−0.681 **	−0.739 **	−0.797 **
SIPI	0.669 **	0.757 **	0.768 **
PRI	0.665 **	0.814 **	0.750 **
NPCI	0.284	0.226	0.038
PSRI	0.632 **	0.814 **	0.800 **
PhRI	0.140	0.668 **	0.268
RVSI	0.135	0.243	0.547 **
TCARI	−0.104	−0.277	−0.782 **
ARI	−0.368 *	0.247	0.247
MSR	−0.656 **	−0.728 **	−0.785 **
MCARI	0.675 **	0.428 **	−0.334 *
YRI	−0.123	−0.370 **	−0.594 **
GI	−0.544 **	−0.603 **	−0.777 **
TVI	−0.390 **	−0.702 **	−0.836 **
NRI	−0.554 **	−0.582 **	−0.783 **

Note: ** indicates that the correlation was significant at the 0.01 level; * indicates that the correlation was significant at the 0.05 level. The features highlighted in gray are related to the DI at the 0.01 level in all three infection periods.

Table A2. The correlation coefficients between the DI and TFs in the three infection stages.

TFs	Early Infection	Mid-Infection	Late Infection
MEA1	−0.417 **	−0.446 **	−0.755 **
VAR1	−0.414 **	−0.299 *	−0.664 **
HOM1	0.249	0.318 *	0.577 **
CON1	−0.308 *	−0.312 *	−0.538 **
DIS1	−0.283	−0.316 *	−0.567 **
ENT1	−0.270	−0.369 **	−0.706 **
SEC1	0.247	0.393 **	0.707 **
COR1	0.137	0.029	−0.386 **
MEA2	−0.527 **	−0.627 **	−0.670 **
VAR2	−0.413 **	−0.761 **	−0.623 **
HOM2	0.336 *	0.754 **	0.608 **
CON2	−0.372 **	−0.747 **	−0.605 **
DIS2	−0.346 *	−0.753 **	−0.607 **
ENT2	−0.357 *	−0.764 **	−0.633 **
SEC2	0.331 *	0.760 **	0.634 **
COR2	0.039	0.695 **	0.568 **
MEA3	−0.164	−0.107	−0.483 **
VAR3	−0.023	−0.040	0.182
HOM3	0.177	0.137	−0.202
CON3	−0.024	−0.060	0.216
DIS3	−0.131	−0.115	0.205
ENT3	−0.228	−0.147	0.169
SEC3	0.283	0.175	−0.172
COR3	0.422 **	0.215	−0.132

Note: ** indicates that the correlation was significant at the 0.01 level; * indicates that the correlation was significant at the 0.05 level. The features highlighted in gray are related to the DI at the 0.01 level in all three infection periods.

Table 3. The coefficients of determination (R^2) and RRMSE between the estimated DI and measured DI in the different infection stages at different spatial resolutions for the VI-based, TF-based, and VI-TF-based models.

Feature	Infection Stages	R²/RRMSE						
		1.2 cm	3 cm	5 cm	7 cm	10 cm	15 cm	20 cm
VIs	Early infection	0.47/0.755	0.48/0.757	0.43/0.758	0.49/0.756	0.46/0.765	0.52/0.716	0.41/0.793
	Mid-infection	0.70/0.472	0.69/0.470	0.72/0.469	0.53/0.605	0.71/0.469	0.75/0.425	0.72/0.457
	Late infection	0.64/0.469	0.65/0.468	0.67/0.467	0.68/0.465	0.61/0.467	0.65/0.467	0.70/0.428
TFs	Early infection	0.18/0.942	0.28/0.880	0.10/0.984	0.26/0.893	0.25/0.896	0.25/0.896	0.20/0.926
	Mid-infection	0.58/0.566	0.56/0.566	0.50/0.563	0.44/0.585	0.65/0.513	0.51/0.608	0.60/0.574
	Late infection	0.70/0.388	0.77/0.380	0.73/0.365	0.68/0.367	0.79/0.359	0.82/0.332	0.80/0.352
Vis + TFs	Early infection	0.50/0.694	0.53/0.748	0.48/0.776	0.50/0.784	0.55/0.694	0.53/0.768	0.44/0.812
	Mid-infection	0.71/0.464	0.72/0.460	0.74/0.471	0.64/0.518	0.76/0.415	0.79/0.406	0.75/0.456
	Late infection	0.80/0.332	0.83/0.320	0.85/0.301	0.81/0.285	0.88/0.266	0.86/0.319	0.82/0.311

Note: The numbers before and after "/" are the R^2 and RRMSE, respectively. The values in the gray box are the highest values of each feature in the different infection periods and spatial resolutions.

References

1. Han, J.; Zhang, Z.; Cao, J.; Luo, Y.; Zhang, L.; Li, Z.; Zhang, J. Prediction of Winter Wheat Yield Based on Multi-Source Data and Machine Learning in China. *Remote Sens.* **2020**, *12*, 236. [CrossRef]
2. Zheng, Q.; Huang, W.; Cui, X.; Dong, Y.; Shi, Y.; Ma, H.; Liu, L. Identification of Wheat Yellow Rust Using Optimal Three-Band Spectral Indices in Different Growth Stages. *Sensors* **2018**, *19*, 35. [CrossRef] [PubMed]
3. Ma, H.; Huang, W.; Jing, Y.; Pignatti, S.; Laneve, G.; Dong, Y.; Ye, H.; Liu, L.; Guo, A.; Jiang, J. Identification of Fusarium Head Blight in Winter Wheat Ears Using Continuous Wavelet Analysis. *Sensors* **2020**, *20*, 20. [CrossRef] [PubMed]
4. Moshou, D.; Bravo, C.; West, J.; Wahlen, S.; McCartney, A.; Ramon, H. Automatic detection of 'yellow rust' in wheat using reflectance measurements and neural networks. *Comput. Electron. Agric.* **2004**, *44*, 173–188. [CrossRef]
5. Chen, X. Epidemiology and control of stripe rust [*Puccinia striiformis* f. sp. *tritici*] on wheat. *Can. J. Plant. Pathol.* **2005**, *27*, 314–337. [CrossRef]
6. Zheng, Q.; Huang, W.; Cui, X.; Shi, Y.; Liu, L. New spectral index for detecting wheat yellow rust using sentinel-2 multispectral imagery. *Sensors* **2018**, *18*, 868. [CrossRef]
7. Yu, K.; Anderegg, J.; Mikaberidze, A.; Karisto, P.; Mascher, F.; McDonald, B.A.; Walter, A.; Hund, A. Hyperspectral Canopy Sensing of Wheat Septoria Tritici Blotch Disease. *Front. Plant. Sci.* **2018**, *9*, 1195. [CrossRef]
8. He, L.; Qi, S.-L.; Duan, J.-Z.; Guo, T.-C.; Feng, W.; He, D.-X. Monitoring of Wheat Powdery Mildew Disease Severity Using Multiangle Hyperspectral Remote Sensing. *ITGRS* **2020**. [CrossRef]
9. Hatton, N.M.; Menke, E.; Sharda, A.; van der Merwe, D.; Schapaugh, W., Jr. Assessment of sudden death syndrome in soybean through multispectral broadband remote sensing aboard small unmanned aerial systems. *Comput. Electron. Agric.* **2019**, *167*, 105094. [CrossRef]
10. Chivasa, W.; Onisimo, M.; Biradar, C.M. UAV-Based Multispectral Phenotyping for Disease Resistance to Accelerate Crop Improvement under Changing Climate Conditions. *Remote Sens.* **2020**, *12*, 2445. [CrossRef]
11. Abdulridha, J.; Ehsani, R.; Abd-Elrahman, A.; Ampatzidis, Y. A remote sensing technique for detecting laurel wilt disease in avocado in presence of other biotic and abiotic stresses. *Comput. Electron. Agric.* **2019**, *156*, 549–557. [CrossRef]
12. Abdulridha, J.; Ampatzidis, Y.; Roberts, P.D.; Kakarla, S.C. Detecting powdery mildew disease in squash at different stages using UAV-based hyperspectral imaging and artificial intelligence. *Biosyst. Eng.* **2020**, *197*, 135–148. [CrossRef]
13. Awad, M.M. Forest mapping: A comparison between hyperspectral and multispectral images and technologies. *J. For. Res.* **2017**, *29*, 1395–1405. [CrossRef]
14. Yang, C.; Fernandez, C.J.; Everitt, J.H. Comparison of airborne multispectral and hyperspectral imagery for mapping cotton root rot. *Biosyst. Eng.* **2010**, *107*, 131–139. [CrossRef]
15. Mariotto, I.; Thenkabail, P.S.; Huete, A.R.; Slonecker, E.T.; Platonov, A. Hyperspectral versus multispectral crop-productivity modeling and type discrimination for the HyspIRI mission. *Remote Sens. Environ.* **2013**, *139*, 291–305. [CrossRef]
16. Yao, Z.; Lei, Y.; He, D. Early Visual Detection of Wheat Stripe Rust Using Visible/Near-Infrared Hyperspectral Imaging. *Sensors* **2019**, *19*, 952. [CrossRef] [PubMed]
17. Shi, Y.; Huang, W.; González-Moreno, P.; Luke, B.; Dong, Y.; Zheng, Q.; Ma, H.; Liu, L. Wavelet-Based Rust Spectral Feature Set (WRSFs): A Novel Spectral Feature Set Based on Continuous Wavelet Transformation for Tracking Progressive Host–Pathogen Interaction of Yellow Rust on Wheat. *Remote Sens.* **2018**, *10*, 525. [CrossRef]
18. Zhang, J.-C.; Pu, R.-L.; Wang, J.-H.; Huang, W.-J.; Yuan, L.; Luo, J.-H. Detecting powdery mildew of winter wheat using leaf level hyperspectral measurements. *Comput. Electron. Agric.* **2012**, *85*, 13–23. [CrossRef]
19. Chen, T.; Zhang, J.; Chen, Y.; Wan, S.; Zhang, L. Detection of peanut leaf spots disease using canopy hyperspectral reflectance. *Comput. Electron. Agric.* **2019**, *156*, 677–683. [CrossRef]

20. Gu, Q.; Sheng, L.; Zhang, T.; Lu, Y.; Zhang, Z.; Zheng, K.; Hu, H.; Zhou, H. Early detection of tomato spotted wilt virus infection in tobacco using the hyperspectral imaging technique and machine learning algorithms. *Comput. Electron. Agric.* **2019**, *167* 105066. [CrossRef]
21. Yang, C.-M. Assessment of the severity of bacterial leaf blight in rice using canopy hyperspectral reflectance. *Precis. Agric.* **2010** *11*, 61–81. [CrossRef]
22. Mahlein, A.; Rumpf, T.; Welke, P.; Dehne, H.-W.; Plumer, L.; Steiner, U.; Oerke, E.-C. Development of spectral indices for detecting and identifying plant diseases. *Remote Sens. Environ* **2013**, *128*, 21–30. [CrossRef]
23. Liu, L.; Dong, Y.; Huang, W.; Du, X.; Ren, B.; Huang, L.; Zheng, Q.; Ma, H. A Disease Index for Efficiently Detecting Wheat Fusarium Head Blight Using Sentinel-2 Multispectral Imagery. *IEEE Access* **2020**, *8*, 52181–52191. [CrossRef]
24. Pydipati, R.; Burks, T.; Lee, W.S. Identification of citrus disease using color texture features and discriminant analysis. *Comput Electron. Agric.* **2006**, *52*, 49–59. [CrossRef]
25. Guo, A.; Huang, W.; Ye, H.; Dong, Y.; Ma, H.; Ren, Y.; Ruan, C. Identification of Wheat Yellow Rust Using Spectral and Texture Features of Hyperspectral Images. *Remote Sens.* **2020**, *12*, 1419. [CrossRef]
26. Al-Saddik, H.; Laybros, A.; Billiot, B.; Cointault, F. Using image texture and spectral reflectance analysis to detect Yellowness and Esca in grapevines at leaf-level. *Remote Sens* **2018**, *10*, 618. [CrossRef]
27. Su, J.; Liu, C.; Hu, X.; Xu, X.; Guo, L.; Chen, W.-H. Spatio-temporal monitoring of wheat yellow rust using UAV multispectral imagery. *Comput. Electron. Agric.* **2019**, *167*, 105035. [CrossRef]
28. Su, J.; Liu, C.; Coombes, M.; Hu, X.; Wang, C.; Xu, X.; Li, Q.; Guo, L.; Chen, W.-H. Wheat yellow rust monitoring by learning from multispectral UAV aerial imagery. *Comput. Electron. Agric.* **2018**, *155*, 157–166. [CrossRef]
29. Calou, V.B.C.; Teixeira, A.D.S.; Moreira, L.C.J.; Lima, C.S.; De Oliveira, J.B.; De Oliveira, M.R.R. The use of UAVs in monitoring yellow sigatoka in banana. *Biosyst. Eng.* **2020**, *193*, 115–125. [CrossRef]
30. Ye, H.; Huang, W.; Huang, S.; Cui, B.; Dong, Y.; Guo, A.; Ren, Y.; Jin, Y. Recognition of Banana Fusarium Wilt Based on UAV Remote Sensing. *Remote Sens.* **2020**, *12*, 938. [CrossRef]
31. Franceschini, M.H.D.; Bartholomeus, H.M.; Van Apeldoorn, D.F.; Suomalainen, J.; Kooistra, L. Feasibility of Unmanned Aerial Vehicle Optical Imagery for Early Detection and Severity Assessment of Late Blight in Potato. *Remote Sens.* **2019**, *11*, 224. [CrossRef]
32. Mahlein, A.-K. Plant Disease Detection by Imaging Sensors—Parallels and Specific Demands for Precision Agriculture and Plant Phenotyping. *Plant Dis.* **2016**, *100*, 241–251. [CrossRef] [PubMed]
33. Abdulridha, J.; Ampatzidis, Y.; Kakarla, S.C.; Roberts, P.D. Detection of target spot and bacterial spot diseases in tomato using UAV-based and benchtop-based hyperspectral imaging techniques. *Precis. Agric.* **2019**, *21*, 955–978. [CrossRef]
34. Abdulridha, J.; Batuman, O.; Ampatzidis, Y. UAV-based remote sensing technique to detect citrus canker disease utilizing hyperspectral imaging and machine learning. *Remote Sens.* **2019**, *11*, 1373. [CrossRef]
35. Deng, X.; Zhu, Z.; Yang, J.; Zheng, Z.; Huang, Z.; Yin, X.; Wei, S.; Lan, Y. Detection of Citrus Huanglongbing Based on Multi-Input Neural Network Model of UAV Hyperspectral Remote Sensing. *Remote Sens.* **2020**, *12*, 2678. [CrossRef]
36. De Castro, A.; Ehsani, R.; Ploetz, R.; Crane, J.; Abdulridha, J. Optimum spectral and geometric parameters for early detection of laurel wilt disease in avocado. *Remote Sens. Environ.* **2015**, *171*, 33–44. [CrossRef]
37. Ramírez-Cuesta, J.; Allen, R.; Zarco-Tejada, P.J.; Kilic, A.; Santos, C.; Lorite, I. Impact of the spatial resolution on the energy balance components on an open-canopy olive orchard. *Int. J. Appl. Earth Obs. Geoinf.* **2019**, *74*, 88–102. [CrossRef]
38. Roth, K.L.; Roberts, D.A.; Dennison, P.E.; Peterson, S.H.; Alonzo, M. The impact of spatial resolution on the classification of plant species and functional types within imaging spectrometer data. *Remote Sens. Environ.* **2015**, *171*, 45–57. [CrossRef]
39. Breunig, F.M.; Galvão, L.S.; Dalagnol, R.; Santi, A.L.; Della Flora, D.P.; Chen, S. Assessing the effect of spatial resolution on the delineation of management zones for smallholder farming in southern Brazil. *Remote Sens. Appl. Soc. Environ.* **2020**, *19*, 100325.
40. Zhou, K.; Cheng, T.; Zhu, Y.; Cao, W.; Ustin, S.L.; Zheng, H.; Yao, X.; Tian, Y. Assessing the impact of spatial resolution on the estimation of leaf nitrogen concentration over the full season of paddy rice using near-surface imaging spectroscopy data. *Front Plant Sci.* **2018**, *9*, 964. [CrossRef]
41. Dash, J.P.; Watt, M.S.; Pearse, G.D.; Heaphy, M.; Dungey, H.S. Assessing very high resolution UAV imagery for monitoring forest health during a simulated disease outbreak. *ISPRS J. Photogramm. Remote Sens.* **2017**, *131*, 1–14. [CrossRef]
42. Zhang, J.; Wang, C.; Yang, C.; Xie, T.; Jiang, Z.; Hu, T.; Luo, Z.; Zhou, G.; Xie, J. Assessing the Effect of Real Spatial Resolution of In Situ UAV Multispectral Images on Seedling Rapeseed Growth Monitoring. *Remote Sens.* **2020**, *12*, 1207. [CrossRef]
43. Liu, M.; Yu, T.; Gu, X.; Sun, Z.; Yang, J.; Zhang, Z.; Mi, X.; Cao, W.; Li, J. The Impact of Spatial Resolution on the Classification of Vegetation Types in Highly Fragmented Planting Areas Based on Unmanned Aerial Vehicle Hyperspectral Images. *Remote Sens* **2020**, *12*, 146. [CrossRef]
44. Huang, W.; Lamb, D.W.; Niu, Z.; Zhang, Y.; Liu, L.; Wang, J. Identification of yellow rust in wheat using in-situ spectral reflectance measurements and airborne hyperspectral imaging. *Precis. Agric.* **2007**, *8*, 187–197. [CrossRef]
45. Cubert-GmbH Hyperspectral Firefleye S185 SE. Available online: http://cubert-gmbh.de/ (accessed on 15 October 2020).
46. Yue, J.; Yang, G.; Tian, Q.; Feng, H.; Xu, K.; Zhou, C. Estimate of winter-wheat above-ground biomass based on UAV ultrahigh-ground-resolution image textures and vegetation indices. *ISPRS J. Photogramm. Remote. Sens.* **2019**, *150*, 226–244. [CrossRef]
47. Rouse, J.W., Jr.; Haas, R.H.; Schell, J.A.; Deering, D.W. Monitoring vegetation systems in the great plains with ERTS. In Proceedings of the Third Earth Resources Technology Satellite-1 Symposium, Washington, DC, USA, 10–14 December 1973; pp. 309–317.

48. Zhang, X.; Han, L.; Dong, Y.; Shi, Y.; Huang, W.; Han, L.; González-Moreno, P.; Ma, H.; Ye, H.; Sobeih, T. A Deep Learning-Based Approach for Automated Yellow Rust Disease Detection from High-Resolution Hyperspectral UAV Images. *Remote Sens.* **2019**, *11*, 1554. [CrossRef]
49. Ashourloo, D.; Mobasheri, M.R.; Huete, A. Developing two spectral disease indices for detection of wheat leaf rust (Pucciniatriticina). *Remote Sens.* **2014**, *6*, 4723–4740. [CrossRef]
50. Devadas, R.; Lamb, D.W.; Simpfendorfer, S.; Backhouse, D. Evaluating ten spectral vegetation indices for identifying rust infection in individual wheat leaves. *Precis. Agric.* **2008**, *10*, 459–470. [CrossRef]
51. Li, B.; Li, B.; Zhang, L.; Han, J.; Bian, C.; Li, G.; Xu, X.; Li-Ping, J. Above-ground biomass estimation and yield prediction in potato by using UAV-based RGB and hyperspectral imaging. *ISPRS J. Photogramm. Remote. Sens.* **2020**, *162*, 161–172. [CrossRef]
52. Gamon, J.; Penuelas, J.; Field, C. A narrow-waveband spectral index that tracks diurnal changes in photosynthetic efficiency. *Remote Sens. Environ.* **1992**, *41*, 35–44. [CrossRef]
53. Zarco-Tejada, P.J.; Berjón, A.; López-Lozano, R.; Miller, J.R.; Martín, P.; Cachorro, V.; González, M.; De Frutos, A. Assessing vineyard condition with hyperspectral indices: Leaf and canopy reflectance simulation in a row-structured discontinuous canopy. *Remote Sens. Environ.* **2005**, *99*, 271–287. [CrossRef]
54. Huang, W.; Guan, Q.; Luo, J.; Zhang, J.; Zhao, J.; Liang, D.; Huang, L.; Zhang, D. New Optimized Spectral Indices for Identifying and Monitoring Winter Wheat Diseases. *IEEE J. Sel. Top. Appl. Earth Obs. Remote Sens.* **2014**, *7*, 2516–2524. [CrossRef]
55. Daughtry, C.; Walthall, C.; Kim, M.; De Colstoun, E.B.; McMurtrey Iii, J. Estimating corn leaf chlorophyll concentration from leaf and canopy reflectance. *Remote Sens. Environ.* **2000**, *74*, 229–239. [CrossRef]
56. Gitelson, A.A.; Merzlyak, M.N.; Chivkunova, O.B. Optical properties and nondestructive estimation of anthocyanin content in plant leaves. *Photochem. Photobiol.* **2001**, *74*, 38–45. [CrossRef]
57. Merzlyak, M.N.; Gitelson, A.; Chivkunova, O.B.; Rakitin, V.Y. Non-destructive optical detection of pigment changes during leaf senescence and fruit ripening. *Physiol. Plant.* **1999**, *106*, 135–141. [CrossRef]
58. Filella, I.; Serrano, L.; Serra, J.; Penuelas, J. Evaluating wheat nitrogen status with canopy reflectance indices and discriminant analysis. *Crop. Sci.* **1995**, *35*, 1400–1405. [CrossRef]
59. Haboudane, D.; Miller, J.R.; Tremblay, N.; Zarco-Tejada, P.J.; Dextraze, L. Integrated narrow-band vegetation indices for prediction of crop chlorophyll content for application to precision agriculture. *Remote Sens. Environ.* **2002**, *81*, 416–426. [CrossRef]
60. Zhang, M.; Su, W.; Fu, Y.; Zhu, D.; Xue, J.-H.; Huang, J.; Wang, W.; Wu, J.; Yao, C. Super-resolution enhancement of Sentinel-2 image for retrieving LAI and chlorophyll content of summer corn. *Eur. J. Agron.* **2019**, *111*, 125938. [CrossRef]
61. Zhang, C.; Xie, Z. Combining object-based texture measures with a neural network for vegetation mapping in the Everglades from hyperspectral imagery. *Remote Sens. Environ.* **2012**, *124*, 310–320. [CrossRef]
62. Fu, Y.; Zhao, C.; Wang, J.; Jia, X.; Yang, G.; Song, X.; Feng, H. An Improved Combination of Spectral and Spatial Features for Vegetation Classification in Hyperspectral Images. *Remote Sens.* **2017**, *9*, 261. [CrossRef]
63. Ferreira, M.P.; Wagner, F.H.; Aragão, L.E.; Shimabukuro, Y.E.; de Souza Filho, C.R. Tree species classification in tropical forests using visible to shortwave infrared WorldView-3 images and texture analysis. *ISPRS J. Photogramm. Remote Sens.* **2019**, *149*, 119–131. [CrossRef]
64. Li, S.; Yuan, F.; Ata-Ul-Karim, S.T.; Zheng, H.; Cheng, T.; Liu, X.; Tian, Y.; Zhu, Y.; Cao, W.; Cao, Q. Combining Color Indices and Textures of UAV-Based Digital Imagery for Rice LAI Estimation. *Remote. Sens.* **2019**, *11*, 1763. [CrossRef]
65. Szantoi, Z.; Escobedo, F.; Abd-Elrahman, A.; Smith, S.; Pearlstine, L. Analyzing fine-scale wetland composition using high resolution imagery and texture features. *Int. J. Appl. Earth Obs. Geoinf.* **2013**, *23*, 204–212. [CrossRef]
66. Farwell, L.S.; Gudex-Cross, D.; Anise, I.E.; Bosch, M.J.; Olah, A.M.; Radeloff, V.C.; Razenkova, E.; Rogova, N.; Silveira, E.M.; Smith, M.M.; et al. Satellite image texture captures vegetation heterogeneity and explains patterns of bird richness. *Remote Sens. Environ.* **2021**, *253*, 112175. [CrossRef]
67. Haralick, R.M.; Shanmugam, K.; Dinstein, I. Textural Features for Image Classification. *IEEE Trans. Syst. Man Cybern.* **1973**, *6*, 610–621. [CrossRef]
68. Wang, C.; Wang, S.; He, X.; Wu, L.; Li, Y.; Guo, J. Combination of spectra and texture data of hyperspectral imaging for prediction and visualization of palmitic acid and oleic acid contents in lamb meat. *Meat Sci.* **2020**, *169*, 108194. [CrossRef] [PubMed]
69. Xiong, Z.; Sun, D.-W.; Pu, H.; Zhu, Z.; Luo, M. Combination of spectra and texture data of hyperspectral imaging for differentiating between free-range and broiler chicken meats. *LWT* **2015**, *60*, 649–655. [CrossRef]
70. Liu, D.; Pu, H.; Sun, D.-W.; Wang, L.; Zeng, X.-A. Combination of spectra and texture data of hyperspectral imaging for prediction of pH in salted meat. *Food Chem.* **2014**, *160*, 330–337. [CrossRef]
71. Cen, H.; Lu, R.; Zhu, Q.; Mendoza, F. Nondestructive detection of chilling injury in cucumber fruit using hyperspectral imaging with feature selection and supervised classification. *Postharvest Biol. Technol.* **2016**, *111*, 352–361. [CrossRef]
72. Lu, J.; Zhou, M.; Gao, Y.; Jiang, H. Using hyperspectral imaging to discriminate yellow leaf curl disease in tomato leaves. *Precis. Agric.* **2018**, *19*, 379–394. [CrossRef]
73. Astor, T.; Dayananda, S.; Nidamanuri, R.R.; Nautiyal, S.; Hanumaiah, N.; Gebauer, J.; Wachendorf, M. Estimation of Vegetable Crop Parameter by Multi-temporal UAV-Borne Images. *Remote Sens.* **2018**, *10*, 805. [CrossRef]
74. Li, X.; Zhang, Y.; Luo, J.; Jin, X.; Xu, Y.; Yang, W. Quantification winter wheat LAI with HJ-1CCD image features over multiple growing seasons. *Int. J. Appl. Earth Obs. Geoinf.* **2016**, *44*, 104–112. [CrossRef]

75. Maimaitijiang, M.; Sagan, V.; Sidike, P.; Maimaitiyiming, M.; Hartling, S.; Peterson, K.T.; Maw, M.J.; Shakoor, N.; Mockler, T.; Fritschi, F.B. Vegetation Index Weighted Canopy Volume Model (CVMVI) for soybean biomass estimation from Unmanned Aerial System-based RGB imagery. *ISPRS J. Photogramm. Remote Sens.* **2019**, *151*, 27–41. [CrossRef]
76. Jay, S.; Gorretta, N.; Morel, J.; Maupas, F.; Bendoula, R.; Rabatel, G.; Dutartre, D.; Comar, A.; Baret, F. Estimating leaf chlorophyll content in sugar beet canopies using millimeter- to centimeter-scale reflectance imagery. *Remote. Sens. Environ.* **2017**, *198*, 173–186. [CrossRef]
77. Navrozidis, I.; Alexandridis, T.; Dimitrakos, A.; Lagopodi, A.L.; Moshou, D.; Zalidis, G. Identification of purple spot disease on asparagus crops across spatial and spectral scales. *Comput. Electron. Agric.* **2018**, *148*, 322–329. [CrossRef]
78. Mahlein, A.-K.; Kuska, M.T.; Thomas, S.; Wahabzada, M.; Behmann, J.; Rascher, U.; Kersting, K. Quantitative and qualitative phenotyping of disease resistance of crops by hyperspectral sensors: Seamless interlocking of phytopathology, sensors, and machine learning is needed! *Curr. Opin. Plant. Biol.* **2019**, *50*, 156–162. [CrossRef]
79. Penuelas, J.; Filella, I.; Gamon, J.A. Assessment of photosynthetic radiation-use efficiency with spectral reflectance. *New Phytol.* **1995**, *131*, 291–296. [CrossRef]
80. Rodriguez-Galiano, V.; Chica-Olmo, M.; Abarca-Hernandez, F.; Atkinson, P.M.; Jeganathan, C. Random Forest classification of Mediterranean land cover using multi-seasonal imagery and multi-seasonal texture. *Remote Sens. Environ.* **2012**, *121*, 93–107. [CrossRef]

Article

Predicting Tree Sap Flux and Stomatal Conductance from Drone-Recorded Surface Temperatures in a Mixed Agroforestry System—A Machine Learning Approach

Florian Ellsäßer [1,*], Alexander Röll [1], Joyson Ahongshangbam [1], Pierre-André Waite [2], Hendrayanto [3], Bernhard Schuldt [4] and Dirk Hölscher [1,5]

1 Tropical Silviculture and Forest Ecology, University of Goettingen, Büsgenweg 1, 37077 Göttingen, Germany; aroell@gwdg.de (A.R.); jahongs@gwdg.de (J.A.); dhoelsc@gwdg.de (D.H.)
2 Plant Ecology and Ecosystems Research, University of Goettingen, Untere Karspüle 2, 37073 Göttingen, Germany; pwaite@gwdg.de
3 Forest Management, Kampus IPB Darmaga, Bogor Agricultural University, Bogor 16680, Indonesia; hendrayanto@apps.ipb.ac.id
4 Julius-von-Sachs-Institute for Biological Sciences, Chair of Ecophysiology and Vegetation Ecology, University of Wuerzburg, Julius-von-Sachs-Platz 3, 97082 Wuerzburg, Germany; bernhard.schuldt@uni-wuerzburg.de
5 Centre of Biodiversity and Sustainable Land Use, University of Goettingen, Platz der Göttinger Sieben 5, 37073 Göttingen, Germany
* Correspondence: fellsae@gwdg.de

Received: 25 October 2020; Accepted: 9 December 2020; Published: 12 December 2020

Abstract: Plant transpiration is a key element in the hydrological cycle. Widely used methods for its assessment comprise sap flux techniques for whole-plant transpiration and porometry for leaf stomatal conductance. Recently emerging approaches based on surface temperatures and a wide range of machine learning techniques offer new possibilities to quantify transpiration. The focus of this study was to predict sap flux and leaf stomatal conductance based on drone-recorded and meteorological data and compare these predictions with in-situ measured transpiration. To build the prediction models, we applied classical statistical approaches and machine learning algorithms. The field work was conducted in an oil palm agroforest in lowland Sumatra. Random forest predictions yielded the highest congruence with measured sap flux ($r^2 = 0.87$ for trees and $r^2 = 0.58$ for palms) and confidence intervals for intercept and slope of a Passing-Bablok regression suggest interchangeability of the methods. Differences in model performance are indicated when predicting different tree species. Predictions for stomatal conductance were less congruent for all prediction methods, likely due to spatial and temporal offsets of the measurements. Overall, the applied drone and modelling scheme predicts whole-plant transpiration with high accuracy. We conclude that there is large potential in machine learning approaches for ecological applications such as predicting transpiration.

Keywords: transpiration; method comparison; UAV; oil palm; multiple linear regression; support vector machine; random forest; artificial neural network

1. Introduction

Transpiration is the largest water flux from terrestrial surfaces, accounting for 80%–90% of terrestrial evapotranspiration [1]. Transpiration is strongly affected by changes in land cover and land use [2,3]. In many tropical regions, conversions of forests to agricultural land are ongoing [4,5], resulting in large-scale alterations of the water cycle and transpiration as a key flux. To measure, model and understand the effects of altered transpiration on the hydrological cycle, measurements at the plant and leaf scale with sap flux probes and porometers are frequently applied [3,6,7]. While these hydrometric methods are commonly implemented at the leaf, plant or plot level, measuring transpiration at larger scales remains a challenging task [2,8,9]. Remote sensing techniques are often considered to be more cost-effective and labor-efficient than ground-based approaches, particularly for applications in agricultural and forest landscapes [10]. Remote sensing data at high temporal or spatial resolution, e.g., from satellites, offer opportunities for the extrapolation of point measurements but are associated with large uncertainties, especially for diverse mixed stands and agricultural areas [11]. Recently emerging drone-based remote sensing systems are promising to bridge the gap between leaf and plant scale measurement methods and catchment or landscape scale schemes [12]. Drones can operate close to the surface enabling a delimitation of single plant canopies but can also cover considerable areas with a single flight [13,14]. They can be equipped with a growing variety of light-weight sensors for diverse spectral ranges or structural measurements [15–17]. Image-based remote sensing approaches are limited in that only the top layer of the canopy is recorded and that measurement inaccuracies generally increase with increasing distance to the object of interest. Index approaches based on thermal remote sensing data such as the crop water stress index (CWSI) have been applied in semi-arid and arid areas using drone-recorded thermal data [18,19]. The application of the CWSI is most useful if a scarcity of water causes water stress in the plants [18]. However, if plants are well watered the explanatory power of such an index is reduced. Since leaf surface temperatures are highly variable during the course of the day, depending e.g., on solar irradiance [20] diurnal temperature patterns can be captured and evapotranspiration patterns can be calculated. To some extent, this might also be possible using other non-temperature-based approaches such as the normalized difference vegetation index (NDVI) that can capture the photosynthetic activity during the day [21]. For evapotranspiration, modelled results based on land surface temperature data often follow a linear relationship with ground-based measurements from eddy covariance systems [16,22]. For complex non-linear structures and relationships, e.g., between evapotranspiration and its controlling factors, statistical regression models can be supplemented with machine learning (ML) algorithms [23,24]. ML models are not explicitly programmed to represent biological processes but are data driven models that use a training data set and can later be applied to previously unknown data [10]. Among others, ML algorithms such as support vector machines (SVM), random forests (RF) and artificial neural networks (ANN) have previously been successfully applied to predict evapotranspiration or structural vegetation characteristics in a wide range of ecosystems [23,25–28]. In the SVM algorithm approach a regression plane and two parallel margins are fitted to the data in order to include as many instances as possible between the two margins and thus creating an explanatory model [29,30]. SVM regressors were previously successfully applied to regression problems with spatial data [28,31,32]. RFs are an ensemble learning approach where multiple trained decision trees are combined to provide an improved prediction performance [33,34]. Different designs of RF algorithms have previously been used to successfully predict e.g., crop water stress, evapotranspiration, above ground-biomass and basal area [10,26–28,35]. ANNs consist of multiple neurons that are combined in a set of layers [36] and are frequently used in many different environments [24–26]. Despite successful applications of ML algorithms to predict hydrological fluxes in the biosphere, the reliable quantification of the non-linear processes that govern water fluxes remains a challenging task [23]. Prerequisites for successful quantification include choice of an appropriate algorithm, a set of representative prediction variables and a sufficiently large data set [37].

The objectives of our study were (1) to compare linear statistical and ML approaches to calibrate models to predict sap flux and stomatal conductance measurements from drone remote sensing data and meteorological measurements (2) to identify the most important prediction variables for these models and (3) to compare the direct measurement methods (sap flux, porometry) with the modelling and drone-based methods for bias and interchangeability.

2. Methods

2.1. Study Site

The study was conducted in the lowlands of Sumatra, in Jambi province, Indonesia. Average elevation of the area is 47 m a.s.l., mean annual precipitation is 2235 mm·year^{-1} and average annual temperature 26.7 °C [38]. The sites were situated in an oil palm plantation of the company PT Humusindo in a conventional monocultural oil palm plantation (Figure 1) and oil palm agroforests that were established in the context of a biodiversity enrichment experiment (EFForts-BEE, 103.2536 E, –1.9463 N) [39]. At the experimental agroforest sites and three years prior to this study, 40% of the oil palms were cut and native tree species were planted. In the conventional oil palm monoculture, palms were between 9 and 15 years old and the stem density was approximately 140 palms per hectare. At the time of our study in October 2016, average tree height was 4.7 m and average oil palm meristem height was 6.8 m. The six planted tree species differed in their habitat preferences including early successional species, as well as mid-to late successional species from old-growth (swamp) lowland forests. Therein, the three best performing species in terms of survival and growth (Peronema canescens, Archidendron pauciflorum and Parkia speciosa) are classified as early successional species and the species that experienced higher mortality (Shorea leprosula and Dyera polyphylla) are more closely associated with old-growth forest [40]. Oil palm is regarded as a pioneer species [41] but ecophysiological studies for palm species are generally rare.

Figure 1. The studied monoculture oil palm plantation on the left-hand side and an exemplary oil palm agroforest site on the right side. Canopies of the measured trees and palms are shown with greyscale thermal images. Canopies were marked with 8-bit barcodes (small white dots in the image) to be recognizable from the air. The map shows an exemplary selection of all sampled canopies.

2.2. Data Acquisition

Two groups of data were acquired: the target variables, i.e., point measurements of sap flux and stomatal conductance with in-situ sensor applications, and a set of prediction variables recorded with a drone and a meteorological station.

2.2.1. Sap Flux Measurements

Eight oil palms (*Elaeis guineensis*) and 16 trees of the four most strongly represented native species, *Archidendron pauciflorum*, *Parkia speciosa*, *Peronema canescens* and *Shorea leprosula*, were equipped with sap flux sensors (Table 1). Four palms were located in a conventional monoculture area and four palms and all 16 trees in the agroforest sites. Oil palm sap flux was assessed with thermal dissipation probes [42] installed in the leaf petioles as described in [43]; sap flux density was calculated using calibrated, oil palm specific parameters [43] and then converted to oil-palm water use (mm·h^{-1}). For the dicot trees, we used the heat ratio method [44]; one sensor per tree was installed radially into the xylem at breast height (135 cm above the ground). For oil palm sapwood area was estimated following the methodology described in [43] and since the trees were still very small, the whole stem area was considered as water conducting area. Sap flux was then calculated using Sap Flow Tool version 1.4.1 (ICT International, Australia) and converted to tree water use (mm·h^{-1}). Further details on the applied sap flux methods are provided in [6]. Sap flux measurements were recorded simultaneously for all trees every 10 min over the course of two weeks. We used barcode markers on the sample trees and palms to facilitate their identification in the aerial images.

Table 1. Tree and palm details.

Species	No. of Replicates	DBH * (cm)	Tree/Meristem Height (m)	Crown Projection Area (m^2)
Archidendron pauciflorum, Fabaceae	4	7.6–11.0	6.5–8.8	9.8–17.6
Parkia Speciosa, Fabaceae	4	5.5–9.5	6.0–9.0	4.5–13.0
Peronema canescens, Lamiaceae	4	8.0–11.0	6.0–8.6	7.8–11.0
Shorea leprosula, Dipterocarpaceae	4	4.2–5.8	3.3–5.2	1.1–3.9
Elaeis guineensis, Arecaceae	8	–	5.19–7.11	64.0–103.7

* DBH refers to measurements of the diameter at breast height (135 cm above the ground).

2.2.2. Stomatal Conductance Measurements

Stomatal conductance measurements were conducted using three porometers (AP 4, Delta-T Devices Ltd, Burwell, Cambridge, UK). The sunlit areas of the canopies were reached using scaffoldings. Three palms were assessed using nine fronts per palm as described in [7]. For trees, we successfully measured three *Archidendron pauciflorum*, three *Peronema canescens* and four *Shorea leprosula* (Table 1). The species *Parkia speciosa* is not included in the stomatal conductance dataset because leaflets were too small to fully cover the porometer measurement chamber. Seven to ten sunlit leaves each were measured on three branches per tree and mean values were calculated.

The measurement chamber was placed only on the abaxial epidermis of the leaves, assuming that the majority of stomata of almost all species in tropical forests is located on the abaxial epidermis [45,46]. This approach further supported the use of the built-in light sensor in the porometers sensor head. Each sample palm or tree was measured between three and five times per day, starting from 9 a.m. to 4 p.m. local time over the course of two weeks. This variation in measurement times was mainly caused by differences in relative humidity which affected the time period to complete the measurement cycle of the porometer devices.

2.2.3. Drone-Based Image Acquisition

We used a multicopter drone (MK EASY Okto V3; HiSystems, Moormerland, Germany) equipped with a thermal and an RGB camera mounted in a stereo setup on a gimbal set to nadir perspective. The radiometric thermal camera was a FLIR Tau 2 640 (FLIR Systems Inc., Wilsonville, OR, USA) attached to a TeAx Thermo-capture module (TeAx Technology GmbH, Wilnsdorf, Germany). The sensor covers spectral bands ranging from 7.5 to 13.5 μm with a relative thermal accuracy of 0.04 K and an absolute thermal accuracy of 1–2 K. Since we used a radiometric thermal camera to conduct our canopy surface temperature measurements, the data could directly be exported for analysis from the camera without further need to translate greyscale images to the corresponding temperatures. The RGB camera was a Sony A5000 (Sony Corporation, Tokyo, Japan) with an EPZ 16–55mm F3.5-5.6 OSS lens that was set to 16 mm focal length. During an 11-day campaign in October 2016 a total of 103 flights were conducted to record canopy surface temperatures simultaneously to the sap flux and stomatal conductance measurements. RGB images were merged using Photoscan 1.3.0 (Agisoft LLC, St. Petersburg, Russia) to create a geo-referenced orthomosaic map. All thermal infrared (TIR) images were then referenced to the RGB map, the canopies were delineated and single palm and tree canopies were extracted using QGIS3 version 3.6 Noosa (QGIS Development Team, 2020).

2.2.4. Meteorological Measurements

Meteorological measurements were conducted at a station located in the studied oil palm plantation. The station was equipped with a global radiation sensor (CMP3 Pyranometer, Kipp & Zonen, Delft, The Netherlands), two thermohygrometers (type 1.1025.55.000, Thies Clima, Göttingen, Germany), a net radiometer (NR Lite2, Kipp & Zonen) and a 3-cup anemometer and a wind direction sensor (both Thies Clima). Data were stored on a logger every 10 min (see [47] for details).

2.2.5. Data Pre-Processing

Single canopies were marked with 8-bit barcodes that were visible in the RGB, as well as the thermal spectrum, and canopy surface temperatures were cut from the thermal images for each canopy and flight. We used the QGIS3-plugin QWaterModel [48] to analyze the canopy surface temperatures by extracting the key metrics mean, median, standard deviation, coefficient of variation, kurtosis and FPCS (Fisher-Pearson coefficient of skewness). We then applied the energy balance model DATTUTDUT [49] based on the canopy temperatures, again using the QGIS3-plugin QWaterModel, and extracted the same key metrics for the resulting fluxes. To calculate the fluxes using the DATTUTDUT model we applied both a fully modelled net radiation approach and a short-wave irradiance-based estimation approach [22]. We used the *pandas* library [50] in Python 3.6.9 to merge the datasets according to the individual plant and recording time. We set the maximum time delta to an hour and matched 1710 datasets for sap flux and 877 for stomatal conductance. The average time offsets for sap flux and stomatal conductance with the remote sensing data were 70 and 1481 s, respectively. The final data set contained 96 variables including the two target variables (sap flux and stomatal conductance), micrometeorological variables (short-wave irradiance, barometric pressure, air temperature, wind speed and direction, relative humidity and the vapor pressure deficit (VPD)), duplications in measurement efforts using several measurement devices and a multitude of variables from drone recorded data and modelled products such as land surface temperatures, the leaf based VPD, evaporative fraction and different evapotranspiration estimates (incl. various dispersion metrics such as mean, median, standard deviation, coefficients of variance, kurtosis and the FPCS). Furthermore, the data set contained local time encoded as cyclical feature (with sinus and cosine as a variable), the canopy area, the number of pixels and atmospheric emissivity and transmissivity. All input variables and their abbreviations and units are shown in Table 2.

Table 2. Input variables, abbreviations and units.

Abbreviation	Description	Unit
	Meteorological data	
Short wave irradiance ***	measured short-wave irradiance	($W \cdot m^{-2}$)
Air temperature ***	measured air temperature	(°C)
Barom. pressure	measured barometric pressure	(hPa)
Wind speed	measured wind speed	($m \cdot s^{-1}$)
Wind direction	measured wind direction	(°)
Relative humidity	measured relative humidity	(%)
VPD gen	vapor pressure deficit based on air temperature	(kPa)
	Drone-images/thermal-data	
Canopy area	canopy area derived from aerial image	(m^2)
Number of pixels	sum of canopy area pixels	(-)
LST *	land surface temperatures	(K)
VPD leaf	vapor pressure deficit based on land surface temperatures	(kPa)
	Model results **	
Rn *	net radiation from model output	($W \cdot m^{-2}$)
LE *	latent heat flux from model output	($W \cdot m^{-2}$)
H *	sensible heat flux from model output	($W \cdot m^{-2}$)
G *	ground heat flux from model output	($W \cdot m^{-2}$)
EF *	evaporative fraction from model output	($W \cdot m^{-2}$)
ET *	evapotranspiration from model output	($W \cdot m^{-2}$)
atmos. transmissivity	atmospheric transmissivity from model	(-)
atmos. emissivity	atmospheric emissivity from model	(-)
	Other	
local time sinus	cyclic local time variable	(°)
local time cosinus	cyclic local time variable	(°)

* Data variables used in several metrics: mean, median, std. dev (standard deviation), coef. var. (coefficient of variation), kurtosis, fpcs (Fisher-Pearson coefficient of skewness). ** Model results from two different applications of the DATTUTDUT model [49]: DM is the original version of the model using a fully modelled net radiation, DS is a model version that uses net radiation from a partly measured (measured short-wave irradiance) and otherwise modelled approach. *** Measurements of these variables are duplicated. This is indicated by the numbers of measurement devices (1,2).

2.3. Prediction Models

The data sets for sap flux prediction (n = 1710) and stomatal conductance prediction (n = 877) were each split into a training set and a test set (70% and 30% of the data, respectively) using scikit learn [51] and a pseudo-random seed of 42 to guarantee reproducibility [52]. Further data sets for each species and a joined data set for all trees were created in a similar way. We performed a multicollinearity test based on a variance inflation factor with Pearson's r (cut-off value > 0.9) and backward elimination of variables based on the Akaike information criterion (significance level $p < 0.05$) and removed variables accordingly reducing the number of input variables from 96 to 42. We decided to use a multiple linear regression (MLR) as our baseline method as it represents a well-known standard approach in statistics. We further applied support vector machine regressors (SVM), random forest regressors (RF) and artificial neural network regressors (ANN) to build prediction models.

2.3.1. Multiple Linear Regression

Multiple linear regression (MLR) is an approach to model the relationship between the explanatory input variables and a response variable by fitting a linear equation expressed as a regression plane [36]. In our study, we used a least-squares model that minimizes the sum of squared vertical deviations from each point to this regression plane [53]. MLR was computed using the *LinearRegression* regression class from the *Scikit-learn* package in Python [51]. MLRs were previously successfully used to predict daily transpiration, evapotranspiration and basal area of trees from spatial data [24,27,31].

2.3.2. Support Vector Machine

Support vector machines (SVMs) model the relationship between explanatory variables and the response variable by fitting a regression plane with parallel margins to the data in order to include as many instances as possible between the two margins [29,30]. This regression plane is also referred to as a hyperplane if several explanatory variables are used [54]. SVMs are mostly known as classifier algorithms, but can also be applied for regression problems [55]. We used the SVR method of *Scikit-learn* package and a linear kernel to build SVMs [51].

2.3.3. Random Forest

Decision trees are predictive models that use recursive partitioning for classification or regression tasks [56]. Random forests combine the results of a set of individually trained decision trees [33]. This principle is called ensemble learning and has shown to improve the predictive performance [34,54]. Two widely used methods of ensemble learning are bootstrap aggregation referred to as *bagging* [57] and *boosting* [58,59].

In the bagging method multiple replicates of the original learning data set are created by bootstrapping with replacement [60]. The decision trees are then trained with these different variations of the original data set and the results of the individual decision trees are averaged [60]. Bagging reduces the variance of simple models and helps to avoid overfitting of more complex models [61]. We trained 2000 trees and used the *RandomForestRegressor* method of the *Scikit-learn* package [51] to build our model. The idea behind boosting is to combine a set of weak and moderately inaccurate decision trees and average their predictions to create very accurate predictions [62]. For this approach we trained 4000 trees and used the adaptive boosting algorithm (*AdaBoost*) introduced by [35] from the Scikit-learn package [51].

2.3.4. Artificial Neural Network

Artificial neural networks (ANNs) are inspired by biological brains and consist of multiple neurons that are organized in a set of layers [36]. ANNs are ideal to identify complex non-linear relationships between in- and output data sets and particularly useful in regression problems with processes that are difficult to capture entirely [24,25]. We used the *Sequential*, *Dense* and *KerasRegressor* methods of the *keras* framework [63] to build an ANN with the typical multiple perceptron type (MLP) architecture which was recently used to predict tree metrics from spatial data [31,64]. Similar ANN designs have been used to estimate evapotranspiration and transpiration in a wide range of ecosystems [24–26]. We used rectified linear units (ReLU) for the activation function in the input and the three hidden layers and a linear activation function for the output layer.

2.3.5. Variable Importance

To estimate the importance of each predictor variable for the regression model, we decided against a removal-based approach as described in e.g., [31] and opted for a randomization-based permutation test (also called mean decrease accuracy) using the *PermutationImportance* method from the *eli5* package [33,65]. Hereby, the values of a single input variable of the prediction dataset are randomized (and not left out) and its effect on the prediction accuracy of the regression model is measured [65].

2.3.6. Statistical Analyses of Predicted vs. Measured Values

To evaluate the prediction models, we compared measured and predicted values from the test data set. We calculated model accuracy in % (defined as 100 subtracted by the mean absolute percentage error (MAPE)) as a general prediction performance indicator. We limited the model accuracy to a range of 0%–100 %, setting potential negative values to zero if the prediction error MAPE becomes larger than 100%. The closer this indicator approaches 100% the closer the model predictions are to the actual observations. We further calculated the mean absolute error (MAE) that measures the average magnitude of errors between the prediction and the test data set and is displayed in percentage using the mean of the observation data set as a base. Further the root mean squared error (RMSE) that indicates the square root of the squared errors between prediction and test data set was calculated using the same base for percentage transformation than the MAE. Both the MAE and RMSE are indicators that measure the average prediction error, however the RMSE penalizes more on extreme errors. The MAE focusses on average errors alone and is therefore a more general indicator. Both the MAE and the RMSE are negative indicators, the higher the indicator value the lower the precision of the predictions. We further calculated the coefficient of determination (R^2) to indicate how well the resulting model fits the data. Variable importance of each predictor variable was assessed using a permutation test [65]. Single variable importance was averaged over all data sets for each sap flux and stomatal conductance and might vary for single species. To compare the prediction methods, we used a non-parametric Passing-Bablok regression [66–68]. The python *MethComp* package [69] was used for the computation of the Passing-Bablok regressions. The Passing-Bablok regression outputs a regression line of which the confidence intervals of slope and intercept are especially interesting for a comparison of the different methods. If the confidence intervals of the slope and the intercept include 1 and 0, respectively, there is no statistically significant bias between the methods [53,68]. Linearity of the data is a crucial assumption for Passing-Bablok regressions [68]; and was checked visually. All statistical analyses were computed with Python 3.6.9 using *pandas* [50], *NumPy* [70,71], *SciPy* [72], *statsmodels*, *scikit learn* [51], *keras* and *eli5* packages and libraries. Graphs and figures were created with *Matplotlib* [73] and *seaborn* [74] libraries.

3. Results

Data was acquired on 16 days, between the 1st and 16th of October 2016, from 9 a.m. to 4 p.m. local time. During the field measurements air temperature ranged from 25.2 to 36 °C, incoming short-wave irradiance ranged from 143 to 1124 $W{\cdot}m^{-2}$ and relative humidity ranged from 49.6% to 94.4%. Average wind speed was 1.26 $m{\cdot}s^{-1}$ predominately coming from South-East. Meteorological conditions were therefore highly variable and allow for training the algorithms across multiple weather conditions. Sap flux for both oil palm and trees showed strong diurnal patterns, with near-zero values at night time and often near-noon maxima (following the diurnal patterns of integrated daily radiation Rg and vapor pressure deficit VPD). Daytime maxima for sap flux ranged from 0.831 $mm{\cdot}h^{-1}$ (*Parkia speciosa*) to 2.544 $mm{\cdot}h^{-1}$ (*Archidendron pauciflorum*) daytime maxima for stomatal conductance ranged from 2153.7 $mol{\cdot}m^{-2}{\cdot}h^{-1}$(*Shorea leprosula*) to 11813.4 $mol{\cdot}m^{-2}{\cdot}h^{-1}$(*Peronema canescens*).

3.1. Prediction Performance

Highest model accuracy was achieved by random forest models (both bagging and boosting) predicting sap flux for individual tree species, especially *Archidendron pauciflorum* where a model accuracy of well over 90% was reached (Figure 2). Across all tree species, accuracy was 60% and 52% for oil palm (RF bagging). Prediction of sap flux across all canopies, including dicot trees and palms, was unsuccessful with all applied algorithms. For predictions of stomatal conductance, RF bagging was again the method with the highest model accuracy, consistently yielding prediction accuracies of 60% for the tree species and 48% for oil palm (Figure 2). For stomatal conductance, predictions across a data set comprising all canopies were successful, with 60% accuracy for RF bagging.

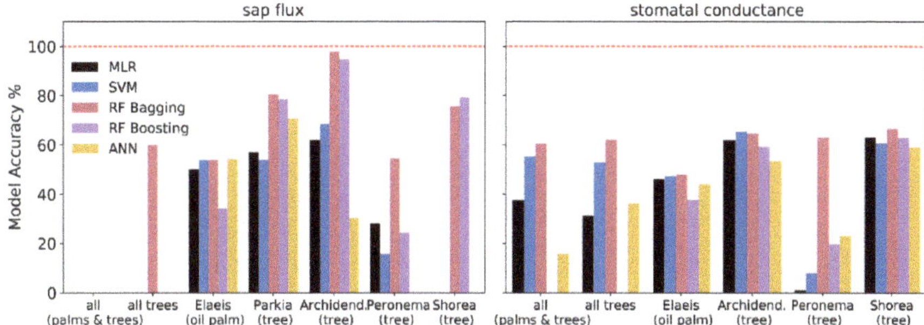

Figure 2. Model accuracy % comparing measured and predicted values of sap flux and stomatal conductance. Abbreviations: MLR—multiple linear regression; SVM—support vector machine; RF—random forest; ANN—artificial neural network.

A comparison of error metrics (MAE and RMSE) showed that errors for the random forest algorithms and particularly RF bagging were comparatively low for both sap flux and stomatal conductance prediction (Figures 3 and 4). The simpler algorithms such as MLR and SVM but also the more complex ANN frequently produced higher errors. Generally, the errors for stomatal conductance predictions are more evenly distributed across algorithms and target groups than for sap flux prediction. For instance, MAE and RMSE for *Shorea leprosula* were relatively small and homogeneous across algorithms for stomatal conductance, but ranged from under 30% (RF bagging) to over 100% (ANN) for sap flux.

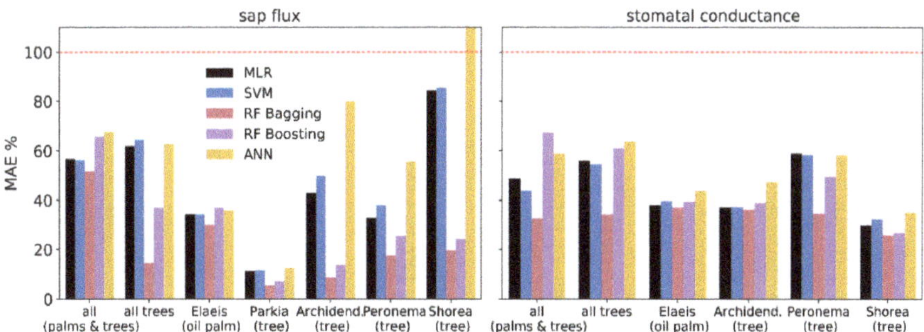

Figure 3. Mean absolute error (MAE) of predicted and measured values for sap flux and stomatal conductance.

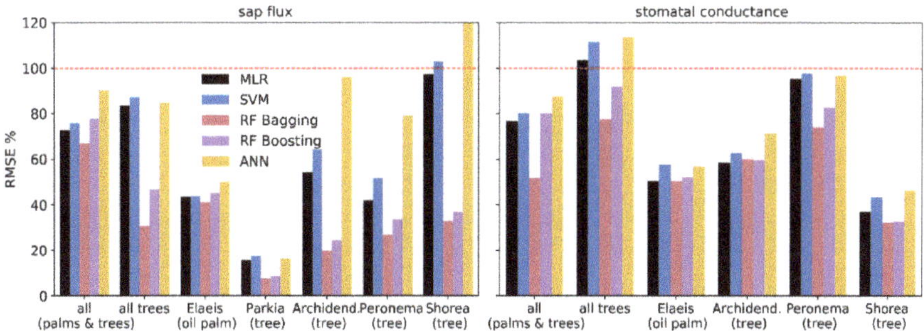

Figure 4. Root mean square error (RMSE) of predicted and measured values for sap flux and stomatal conductance.

Both RF algorithms resulted in predictions highly congruent with the sap flux measurements in terms of R^2 (Figure 5); across all tree species, as well as for each individual tree species, R^2s were close to or higher than 0.8 (RF bagging), while they were close to 0.6 for oil palm. The predictions for stomatal conductance are generally less congruent with the measurements than for sap flux. However, the RF bagging algorithm does achieve R^2s around 0.5 or higher across all canopies and across all tree species, as well as individually for two of the four species.

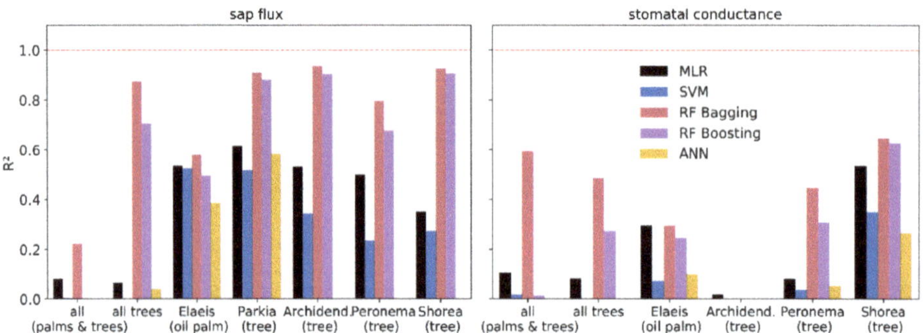

Figure 5. Coefficient of determination (R^2) for predicted and measured values for sap flux and stomatal conductance.

3.2. Method Comparison

Predicted values for sap flux from the RF algorithms almost always showed a linear relationship with their measured counterparts, whereas this can only be observed for the MLR and SVM algorithms for oil palm datasets and was never detected for the ANN algorithm (Figure 6). There was no linear relationship of measured and predicted values for stomatal conductance for all prediction algorithms, the results are therefore not included into a figure. Sap flux predictions for all trees and oil palm (but not all canopies), as well as for *Archidendron pauciflorum* showed no significant continuous or systematic bias from the measurements when the RF bagging algorithm was applied (Figures 6 and 7), which indicates interchangeability of the methods.

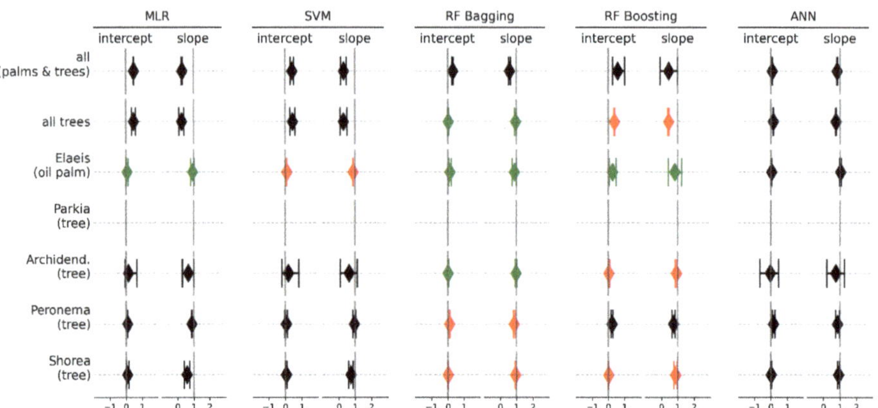

Figure 6. The 99% confidence intervals of slope and intercept from model II Passing-Bablok regression for sap flux. The results in green show no significant bias between measured and predicted values. The results indicated in red fulfilled the assumptions of linearity but a significant difference between measured and predicted values was found. The results displayed in black did not meet the assumption of linearity and are therefore not to be considered.

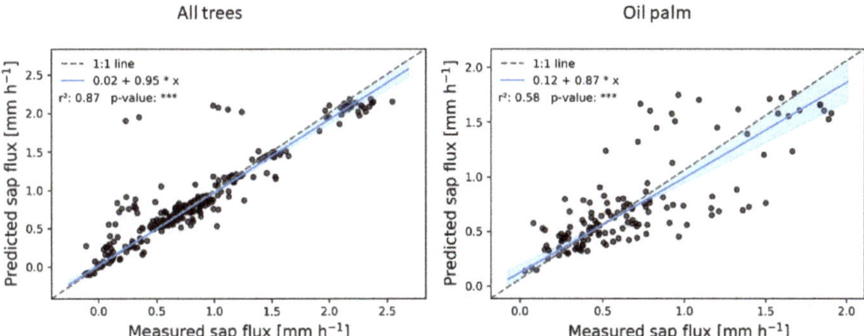

Figure 7. Measured and predicted sap flux for all tree and palm canopies from the RF bagging algorithm. The regression line of the Passing-Bablok regression is displayed in blue. The coefficient of determination is indicated by r^2 and the *** indicate a *p*-value well below 0.001.

Except for some outliers, the measured and RF bagging predicted values for sap flux in trees closely follow the 1:1 line, while showing more variance for oil palm (Figure 7). The same was observed for oil palm canopies where the MLR and RF boosting algorithms predicted sap flux without significant bias or errors (Figure 6). For species specific predictions, linearity was found for both RF algorithms, but except for *Archidendron pauciflorum* a significant systematic and continuous bias was detected. None of the algorithms showed potential in predicting sap flux of *Parkia speciosa*.

3.3. Variable Importance

A total of 96 input variables (or features) were available for our analysis. After applying a multicollinearity test and a backward elimination, the remaining 42 variables were used to train the prediction algorithms. To improve the distribution of measurement efforts and the relevance of input features, we performed a permutation importance analysis. Therein, the numbers of most important variables that explain 95% of the model outcome are highly variable (Figure 8). For the MLR, a very low number of only up to seven input variables (see Figure 9 for details) are required to explain most of the model prediction results for both sap flux and stomatal conductance. For the SVM algorithm, between 10 and 20 variables explain 95% of the model outcome; a smaller number of variables explain the sap flux results, while the number of variables was higher for stomatal conductance predictions. The RF algorithms showed the highest variations ranging from only one variable up to 20 for sap flux and generally using 15 to 20 main variables for stomatal conductance to explain most of the model's outcome. The ANN algorithm uses an intermediate number of around 9 to 17 important variables for both sap flux and stomatal conductance prediction.

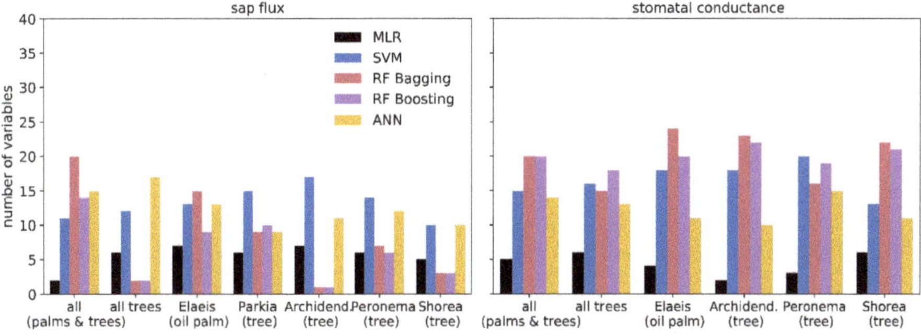

Figure 8. Number of most important input variables that explain 95% of the corresponding model.

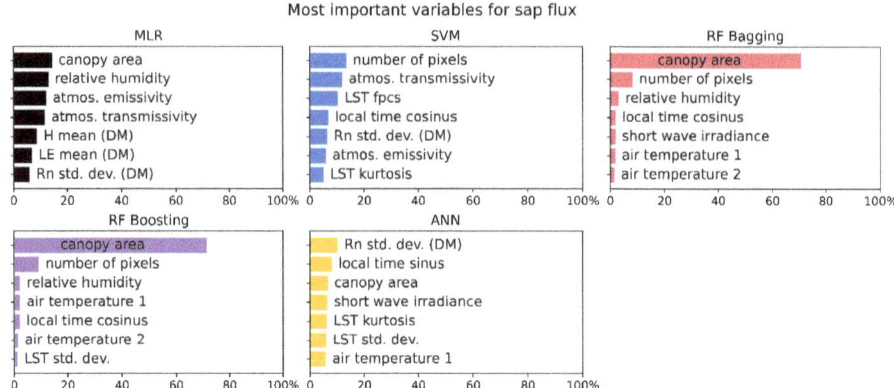

Figure 9. The seven most important input variables averaged over all data sets (different canopies and combinations) for the corresponding models when predicting sap flux.

Variable importance (expressed in %) averaged for sap flux and stomatal conductance over all species shows big differences for the MLR algorithm, where variable importance for sap flux is very homogeneous and the prediction of stomatal conductance is dominated by the latent heat flux as derived from the DATTUTDUT model (Figures 9 and 10). A similar but less pronounced pattern can be observed for the SVM regressor where the main prediction variables for sap flux are very homogeneous and for the prediction of stomatal conductance barometric pressure is the prevalent variable.

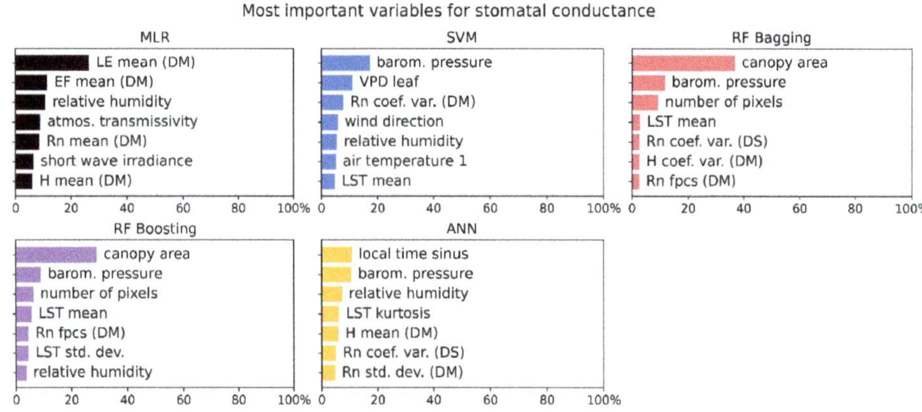

Figure 10. The seven most important input variables averaged over all data sets (different canopies and combinations) for the corresponding models when predicting stomatal conductance.

Both RF algorithms mainly rely on the canopy area as a main prediction variable for both sap flux and stomatal conductance. Another observation is the second variable that is of a much less pronounced importance for the model results: both RF algorithms use the number of pixels for sap flux predictions and the barometric pressure for stomatal conductance predictions (Figures 9 and 10). The ANN algorithm shows a very homogeneous distribution with no prevalent variable for both sap flux and stomatal conductance prediction (Figures 9 and 10).

4. Discussion

This study analyzes the applicability of machine learning approaches for ecological datasets, in our case thermal remote sensing data from drones and micrometeorological data to predict sap flux and stomatal conductance as two key processes related to the water balance of individual plants and ecosystems. Model accuracy was generally higher and errors were lower for sap flux than for the stomatal conductance predictions. Therein, the applied prediction algorithms showed substantial differences regarding input requirements and variable importance. The prediction results from the RF bagging method were found to be interchangeable with ground-based measurements for some canopy types and species.

4.1. Prediction Performance

In our study MLR representing the simplest algorithm produced results of intermediate quality often with lower errors than the more complex SVM algorithm. This stands in contrast to previous studies, where the SVM algorithm produced more congruent estimates, e.g., sap flux from a range of directly measured variables or biomass from remote sensing data [24,28]. Since the creation of a representative model with MLR requires a prevalent linear relationship of at least some input variables with a given target variable, we assume that given a supplementary set of ground-based measurements MLR can still be a simple alternative to more complex machine learning algorithms. In contrast to a previous study where evapotranspiration was successfully predicted from hydro-climatic variables with the SVM algorithm [55], this algorithm did not produce satisfying results in our study. On average, model accuracy was higher for the SVM when predicting stomatal conductance than when predicting sap flux, but coefficients of determination indicate that neither model predicts the target variable accurately. Since we used a linear kernel, the performance of the SVM algorithm highly depends on the linear relationship of at least some input variables with the according target variable.

Overall, both RF algorithms (bagging and boosting) resulted in similar, high quality predictions, with the bagging algorithm performing slightly better. Previous studies predicting water stress from remote sensing data showed similar results, i.e., the bootstrapping-based bagging option slightly outperforming the boosting algorithm but with small overall differences [75]. Compared to the other applied prediction methods, both RF algorithms showed high model accuracy and low errors with a high congruence of measured and predicted values, particularly for sap flux. Likewise, RF algorithms outperformed several other algorithms in studies where above-ground biomass was estimated from remote sensing data [28,31]. A previous study applying similar feed-forward ANN algorithms but using ground-based input data showed satisfactory results for sap flux prediction [76]. While we expected better results from the ANN algorithm due to its complexity, the parameterization of the method is known to be difficult and highly sensitive to variations in input parameters [31,77].

4.2. Method Comparison

Using RF algorithms, sap flux can be predicted from remote sensing data in close congruence to measurements without producing many outliers (Figure 8). While analyzing tree and palm canopies separately, sap flux can well be predicted by the RF bagging algorithm, while for the whole dataset including both, trees and palms, sap flux was not adequately predicted. The reasons are likely underlying differences in physiology between dicot trees and monocot palms; this e.g., includes size and distribution of water-conductive vessels in the stem and crown and leaf architecture. As such, previous ecohydrological assessments in the study region pointed to vast differences in water use between oil palms and trees including rubber trees of similar age, with substantially higher per-tree and stand transpiration rates of oil palms [78,79]. Drone-derived crown metrics of oil palms and adjacent trees further suggested that oil palms transpire two-times more water per unit of crown volume as agroforest trees [6] and about five-times more per unit of crown surface area than rainforest trees [80], which may well be the reason why the joined analysis of trees and palms was unsuccessful.

A further reason for the difficulties in predicting sap flux across tree and oil palm canopies might be of methodological nature, as palm sap flux was assessed in leaf petioles with thermal dissipation probes [42] and tree sap flux was assessed in the lower stems with the heat ratio method [44]. The good prediction performance of RF algorithms compared to e.g., ANN algorithms was previously also described for predicting potential evapotranspiration [26]. RF was further found to be the best of several algorithms for aboveground biomass estimation from remote sensing data [28].

The rather complex ANN algorithm showed no convincing results in our study. Potential reasons include that variables might not have been captured due to a lack of a sufficing number of free weights [24], or that a surplus of free weights might have caused an over-fitting of the model which lacked the generalization to predict reasonable results from our test data set [81]. Even more than for the other algorithms, optimal layout of the ANN highly depends on the input variables [82].

None of the applied models could predict sap flux in *Parkia speciosa,* despite comparatively low errors and a promising model accuracy, as well as high congruence with measurements. Reviewing individual canopies of *Parkia speciosa* showed that, due to its pinnate leaves and open crown structure, canopies are partly barely visible in the thermal images and likely have a high contribution of pixels representing the soil or mixed soil-canopy pixels.

In our study, major challenges remain with the prediction of stomatal conductance. The data set for stomatal conductance differed strongly from the data set for sap flux. Average time offsets of directly and remotely measuring stomatal conductance were much bigger than for sap flux and measured and modelled input variables were not ideally suited for stomatal conductance prediction. The use of an additional set of multispectral images and resulting indices such as the NDVI (normalized difference vegetation index) or the EVI (enhanced vegetation index) could potentially enhance the prediction of stomatal conductance [83] especially from remote sensing data. The use of other TIR based indices such as the crop water stress index (CWSI) might be a useful addition to the training data set. TIR data might not be the ideal source for stomatal conductance predictions, since other studies were able to predict stomatal conductance from hyperspectral reflectance spectroscopy data using RF and ANN algorithms [84]. Further, the sample size for sap flux was much larger than for stomatal conductance due to a limited amount of porometry devices and the non-automated nature of the measurements. Furthermore, the stomatal conductance measurements represent only a small portion (centimeter scale) of a given canopy. The identification and subsequent upscaling from this small leaf area to a whole canopy thus likely introduces substantial errors into the predictions.

Despite the very encouraging results, the genericity of our study is limited as it encompassed only one study region (tropical lowland environment) and only oil palm systems therein and as measurements covered a period of less than one month. On the other hand, the lowland tropics have a large extent, oil palm continues to expand throughout the tropics and the equatorial climate is often quite stable throughout the year. Further studies are necessary to confirm whether the applied schemes can be applied across regions, vegetation types and seasons. A general drawback of machine learning models is that they usually lack causal relations that would enable a biological interpretation [31,85]. Our study is based on approaches that are often described as the traditional statistical (MLR) and traditional machine learning (SVM, RF, ANN) methods [23], while more complex approaches such as the stochastic gradient boosting, which combines the advantages of boosting and bagging [28], were not within the scope of this study. This leaves room for further expanding and refining the here presented approaches.

In summary, RF approaches worked best for predicting sap flux, while a model that can predict stomatal conductance without bias was not found.

4.3. Variable Importance Evaluation

The output quality of prediction algorithms highly depends on how well the input variables represent the ecosystem and the target variables [27]. For sap flux and stomatal conductance, one would generally expect a strong influence of environmental variables such as the vapor pressure deficit (VPD) or solar radiation. The overall number of variables that explained 95% of our MLR model was low, never surpassing seven variables (Figure 8). This is a strong indicator for either no or non-linear relationships between most input variables and the output variables. The seven most important variables averaged over all input data set configurations (mixed and single species) of the MLR model in our study show that canopy area, as well as relative humidity have the greatest impact when predicting sap flux (Figure 8). These results only partly resemble variables from other studies where air temperature and VPD related variables such as relative humidity played a key role in MLR derived models to predict sap flux [76]. The most important variable for stomatal conductance predictions is the mean latent heat flux derived from the DATTUTDUT model (Figure 9). Surprisingly, wind speed was not amongst the most important input variables for the MLR algorithm whereas an influence of wind speed on both target variables was shown in previous studies (e.g., [86]). Asides from meteorological variables, soil moisture is considered to be a classic driver of sap flux and thus transpiration (e.g., [87,88]). While our data set did not comprise soil moisture data, previous assessments in the same study region showed no significant influence of the typically rather small soil moisture fluctuations on (evapo)transpiration of oil palms and trees [3,89,90], with exception of a strong El Niño event [90], which did not occur during our time of study. In contrast, soil moisture can be strongly limiting factor in (semi)arid regions, as was e.g., reported using a MLR approach [24]. We could not identify dominating input variables that would explain the model built by the SVM algorithm for sap flux. The prediction from the SVM algorithm for stomatal conductance was slightly dominated by the variable barometric pressure (Figure 10). The VPD for leaves based on the drone recorded leaf surface temperatures plays a further key role when predicting stomatal conductance with the SVM algorithm. However, none of the variables explained more than 20% of the final model. In contrast, a previous study successfully applied a SVM to predict potential evaporation using solar radiation, relative humidity, air temperature and wind speed as input variables [91]. The number of variables that explains 95% of the model results was very low for both RF algorithms when predicting sap flux. In contrast, the RF algorithms used among the highest numbers of variables when predicting stomatal conductance. For both, canopy area was the (clearly) dominating predictor. With simple linear regression analysis, no relationship between canopy size and sap flux could be found, mainly because the canopy size is a constant variable while sap flux quantity varies during the day. We assume a non-linear relationship between canopy size and sap flux, which requires auxiliary variables to be predicted. Remotely sensed canopy size is an important factor for plant transpiration and was already found to be a suitable predictor in previous studies [6,80]. A general advantage of RF algorithms is their ability to produce accurate predictions even from small samples and with a large set of independent input variables [31]; in our study, they performed better than all other applied algorithms.

The ANN algorithm mainly uses an intermediate number of input variables compared to the small set of MLR based models and the large sets of the SVM models. Compared to the other algorithms, the ANN does not show any clearly dominant input variable. An interesting observation is that the ANN algorithm uses cyclic local time variables among its most important input variables. This seems to be a logical approach as sap flux is known to follow a daily course with near-noon maxima and night time near-zero minima in the study region (e.g., [78,79]). One previous study found that the best results for stand transpiration prediction were achieved by using climate data, soil water content and canopy properties as key input variables [24], which are partly also represented in our ANN derived prediction models.

5. Conclusions

Drone remote sensing and in situ meteorological observations in conjunction with machine learning algorithms can be considered reliable methods for the prediction of sap flux. For tree and palm canopies, random forest regressors predicted interchangeable results without significant bias when compared to direct sap flux measurements. The prediction of stomatal conductance from remotely sensed data was less successful and requires further research. Our study complements the asset of available sap flux approaches by a reliable method for drone-based sap flux prediction.

Author Contributions: The study was conceptualized by D.H. in cooperation with H. (drone and sap flux measurements) and B.S. (stomatal conductance measurements). F.E. led the writing of the paper with help from A.R. and D.H. supervised the work. F.E. collected and processed the drone data, J.A. and A.R. collected the sap flux data and P.-A.W. collected the stomatal conductance data. F.E. conducted data processing, model application, statistical analysis and production of plots in cooperation mainly with D.H. and A.R., F.E., D.H. and A.R. created a first version of the manuscript, which was further improved in a cooperation of all authors. All authors have read and agreed to the published version of the manuscript.

Funding: This study was funded by the Deutsche Forschungsgemeinschaft (DFG, German Research Foundation) project number 192626868—SFB 990 (subprojects A02 and B04) and the Ministry of Research, Technology and Higher Education (Ristekdikti).

Acknowledgments: We thank the Deutsche Forschungsgemeinschaft (DFG, German Research Foundation) and the Ministry of Research, Technology and Higher Education (Ristekdikti) for providing the funds to conduct this study. We thank Ristekdikti for providing the research permit for field work (No. 322/SIP/FRP/E5/Dit.KI/IX/2016). We thank our field assistants Fahrozi, Erwin Pranata, Syahbarudin and Kairul Anwar for great support during the field campaigns. Thanks to all 'EFForTS' colleagues and friends in Indonesia, Germany, and around the world.

Conflicts of Interest: The authors declare no conflict of interest.

References

1. Jasechko, S.; Sharp, Z.D.; Gibson, J.J.; Birks, S.J.; Yi, Y.; Fawcett, P.J. Terrestrial water fluxes dominated by transpiration. *Nat. Cell Biol.* **2013**, *496*, 347–350. [CrossRef] [PubMed]
2. Good, S.P.; Noone, D.; Bowen, G. Hydrologic connectivity constrains partitioning of global terrestrial water fluxes. *Science* **2015**, *349*, 175–177. [CrossRef] [PubMed]
3. Röll, A.; Niu, F.; Meijide, A.; Ahongshangbam, J.; Ehbrecht, M.; Guillaume, T.; Gunawan, D.; Hardanto, A.; Hertel, D.; Kotowska, M.M.; et al. Transpiration on the rebound in lowland Sumatra. *Agric. For. Meteorol.* **2019**, *274*, 160–171. [CrossRef]
4. Hansen, M.C.; Potapov, P.V.; Moore, R.; Hancher, M.; Turubanova, S.A.; Tyukavina, A.; Thau, D.; Stehman, S.V.; Goetz, S.J.; Loveland, T.R.; et al. High-Resolution Global Maps of 21st-Century Forest Cover Change. *Science* **2013**, *342*, 5. [CrossRef] [PubMed]
5. Margono, B.A.; Potapov, P.V.; Turubanova, S.; Stolle, F.; Hansen, M.C. Primary forest cover loss in Indonesia over 2000–2012. *Nat. Clim. Chang.* **2014**, *4*, 730–735. [CrossRef]
6. Ahongshangbam, J.; Khokthong, W.; Ellsäßer, F.; Hendrayanto, H.; Hölscher, D.; Röll, A. Drone-based photogrammetry-derived crown metrics for predicting tree and oil palm water use. *Ecohydrology* **2019**, *12*, e2115. [CrossRef]
7. Waite, P.-A.; Schuldt, B.; Link, R.M.; Breidenbach, N.; Triadiati, T.; Hennings, N.; Saad, A.; Leuschner, C. Soil moisture regime and palm height influence embolism resistance in oil palm. *Tree Physiol.* **2019**, *39*, 1696–1712. [CrossRef]
8. Ford, C.R.; Hubbard, R.M.; Kloeppel, B.D.; Vose, J.M. A comparison of sap flux-based evapotranspiration estimates with catchment-scale water balance. *Agric. For. Meteorol.* **2007**, *145*, 176–185. [CrossRef]
9. Kume, T.; Tsuruta, K.; Komatsu, H.; Kumagai, T.; Higashi, N.; Shinohara, Y.; Otsuki, K. Effects of sample size on sap flux-based stand-scale transpiration estimates. *Tree Physiol.* **2010**, *30*, 129–138. [CrossRef]
10. Virnodkar, S.S.; Pachghare, V.K.; Patil, V.C.; Jha, S.K. Remote sensing and machine learning for crop water stress determination in various crops: A critical review. *Precis. Agric.* **2020**, *21*, 1121–1155. [CrossRef]
11. Wei, Z.; Yoshimura, K.; Wang, L.; MirallesiD, D.G.; Jasechko, S.; Lee, X. Revisiting the contribution of transpiration to global terrestrial evapotranspiration. *Geophys. Res. Lett.* **2017**, *44*, 2792–2801. [CrossRef]
12. Suab, S.A.; Avtar, R. Unmanned Aerial Vehicle System (UAVS) Applications in Forestry and Plantation Operations: Experiences in Sabah and Sarawak, Malaysian Borneo. In *Unmanned Aerial Vehicle: Applications in Agriculture and Environment*; Avtar, R., Watanabe, T., Eds.; Springer International Publishing: Cham, Switzerland, 2019; pp. 101–118.
13. Khokthong, W.; Zemp, D.C.; Irawan, B.; Sundawati, L.; Kreft, H.; Hölscher, D. Drone-Based Assessment of Canopy Cover for Analyzing Tree Mortality in an Oil Palm Agroforest. *Front. For. Glob. Chang.* **2019**, *2*, 10. [CrossRef]

14. Mohan, M.; Silva, C.A.; Klauberg, C.; Jat, P.; Catts, G.; Cardil, A.; Hudak, A.T.; Dia, M. Individual Tree Detection from Unmanned Aerial Vehicle (UAV) Derived Canopy Height Model in an Open Canopy Mixed Conifer Forest. *Forests* **2017**, *8*, 340. [CrossRef]
15. Berni, J.A.J.; Zarco-Tejada, P.J.; Suárez, L.; González-Dugo, V.; Fereres, E. Remote sensing of vegetation from UAV platforms using lightweight multispectral and thermal imaging sensors. *Int. Arch. Photogramm. Remote Sens. Spatial Inform. Sci.* **2009**, *38*, 6.
16. Brenner, C.; Thiem, C.E.; Wizemann, H.-D.; Bernhardt, M.; Schulz, K. Estimating spatially distributed turbulent heat fluxes from high-resolution thermal imagery acquired with a UAV system. *Int. J. Remote Sens.* **2017**, *38*, 3003–3026. [CrossRef] [PubMed]
17. Kellner, J.R.; Armston, J.; Birrer, M.; Cushman, K.C.; Duncanson, L.; Eck, C.; Falleger, C.; Imbach, B.; Král, K.; Krůček, M.; et al. New Opportunities for Forest Remote Sensing Through Ultra-High-Density Drone Lidar. *Surv. Geophys.* **2019**, *40*, 959–977. [CrossRef] [PubMed]
18. Bian, J.; Zhang, Z.; Chen, J.; Chen, H.; Cui, C.; Li, X.; Chen, S.; Fu, Q. Simplified Evaluation of Cotton Water Stress Using High Resolution Unmanned Aerial Vehicle Thermal Imagery. *Remote Sens.* **2019**, *11*, 267. [CrossRef]
19. Matese, A.; Baraldi, R.; Berton, A.; Cesaraccio, C.; Di Gennaro, S.F.; Duce, P.; Facini, O.; Mameli, M.G.; Piga, A.; Zaldei, A. Estimation of Water Stress in Grapevines Using Proximal and Remote Sensing Methods. *Remote Sens.* **2018**, *10*, 114. [CrossRef]
20. Jones, H.G. *Plants and Microclimate: A Quantitative Approach to Environmental Plant Physiology*, 3rd ed.; Cambridge University Press: Cambridge, UK, 2013; ISBN 978-0-511-84572-7.
21. Uudus, B.; Park, K.-A.; Kim, K.-R.; Kim, J.; Ryu, J.-H. Diurnal variation of NDVI from an unprecedented high-resolution geostationary ocean colour satellite. *Remote Sens. Lett.* **2013**, *4*, 639–647. [CrossRef]
22. Ellsäßer, F.; Stiegler, C.; Röll, A.; June, T.; Knohl, A.; Hölscher, D. Predicting evapotranspiration from drone-based thermography—A method comparison in a tropical oil palm plantation. *Biogeosciences* under review. **2020**, 1–37. [CrossRef]
23. Dou, X.; Yang, Y. Evapotranspiration estimation using four different machine learning approaches in different terrestrial ecosystems. *Comput. Electron. Agric.* **2018**, *148*, 95–106. [CrossRef]
24. Fernandes, T.J.; Campo, A.; García-Bartual, R.; González-Sanchís, M. Coupling daily transpiration modelling with forest management in a semiarid pine plantation. *iForest Biogeosci. For.* **2016**, *9*, 38–48. [CrossRef]
25. Antonopoulos, V.Z.; Gianniou, S.K.; Antonopoulos, A.V. Artificial neural networks and empirical equations to estimate daily evaporation: Application to Lake Vegoritis, Greece. *Hydrol. Sci. J.* **2016**, *61*, 2590–2599. [CrossRef]
26. Feng, Y.; Cui, N.; Gong, D.; Zhang, Q.; Zhao, L. Evaluation of random forests and generalized regression neural networks for daily reference evapotranspiration modelling. *Agric. Water Manag.* **2017**, *193*, 163–173. [CrossRef]
27. Pan, S.; Pan, N.; Tian, H.; Friedlingstein, P.; Sitch, S.; Shi, H.; Arora, V.K.; Haverd, V.; Jain, A.K.; Kato, E.; et al. Evaluation of global terrestrial evapotranspiration by state-of-the-art approaches in remote sensing, machine learning, and land surface models. *Glob. Hydrol. Model. Approaches* **2020**, *24*, 1485–1509.
28. Wu, C.; Shen, H.; Shen, A.; Deng, J.; Gan, M.; Zhu, J.; Xu, H.; Wang, K. Comparison of machine-learning methods for above-ground biomass estimation based on Landsat imagery. *J. Appl. Remote Sens.* **2016**, *10*, 35010. [CrossRef]
29. Guo, Y.; Li, Z.; Zhang, X.; Chen, E.-X.; Bai, L.; Tian, X.; He, Q.; Feng, Q.; Li, W. Optimal Support Vector Machines for Forest Above-Ground Biomass Estimation from Multisource Remote Sensing Data. In Proceedings of the 2012 IEEE International Geoscience and Remote Sensing Symposium, Munich, Germany, 22–27 July 2012; pp. 6388–6391.
30. Vapnik, V.N. Introduction: Four Periods in the Research of the Learning Problem. In *The Nature of Statistical Learning Theory*; Springer: New York, NY, USA, 2000; pp. 1–15.
31. Dos Reis, A.A.; Carvalho, M.C.; De Mello, J.M.; Gomide, L.R.; Filho, A.C.F.; Júnior, F.W.A. Spatial prediction of basal area and volume in Eucalyptus stands using Landsat TM data: An assessment of prediction methods. *N. Z. J. For. Sci.* **2018**, *48*, 1. [CrossRef]

32. García-Gutiérrez, J.; Martínez-Álvarez, F.; Troncoso, A.; Riquelme, J.C. A Comparative Study of Machine Learning Regression Methods on LiDAR Data: A Case Study. In *International Joint Conference SOCO'13-CISIS' 13-ICEUTE'13*; Herrero, Á., Baruque, B., Klett, F., Abraham, A., Snášel, V., de Carvalho, A.C.P.L.F., Bringas, P.G., Zelinka, I., Quintián, H., Corchado, E., Eds.; Springer International Publishing: Cham, Switzerland, 2014; Volume 239, pp. 249–258. ISBN 978-3-319-01853-9.
33. Breiman, L. Random Forests. *Mach. Learn.* **2001**, *45*, 5–32. [CrossRef]
34. Opitz, D.W.; Maclin, R. Popular Ensemble Methods: An Empirical Study. *J. Artif. Intell. Res.* **1999**, *11*, 169–198. [CrossRef]
35. Freund, Y.; Schapire, R.E. A Decision-Theoretic Generalization of On-Line Learning and an Application to Boosting. *J. Comput. Syst. Sci.* **1997**, *55*, 119–139. [CrossRef]
36. Nguyen, C.N.; Zeigermann, O. *Machine Learning: Kurz & Gut*; O'Reilly: Berlin/Heidelberg, Germany, 2018; ISBN 978-3-96009-052-6.
37. Granata, F. Evapotranspiration evaluation models based on machine learning algorithms—A comparative study. *Agric. Water Manag.* **2019**, *217*, 303–315. [CrossRef]
38. Drescher, J.; Rembold, K.; Allen, K.; Beckschäfer, P.; Buchori, D.; Clough, Y.; Faust, H.; Fauzi, A.M.; Gunawan, D.; Hertel, D.; et al. Ecological and socio-economic functions across tropical land use systems after rainforest conversion. *Philos. Trans. R. Soc. B Biol. Sci.* **2016**, *371*, 20150275. [CrossRef] [PubMed]
39. Teuscher, M.; Gérard, A.; Brose, U.; Buchori, D.; Clough, Y.; Ehbrecht, M.; Hölscher, D.; Irawan, B.; Sundawati, L.; Wollni, M.; et al. Experimental Biodiversity Enrichment in Oil-Palm-Dominated Landscapes in Indonesia. *Front. Plant Sci.* **2016**, *7*, 1538. [CrossRef] [PubMed]
40. Zemp, D.C.; Gérard, A.; Hölscher, D.; Ammer, C.; Irawan, B.; Sundawati, L.; Teuscher, M.; Kreft, H. Tree performance in a biodiversity enrichment experiment in an oil palm landscape. *J. Appl. Ecol.* **2019**, *56*, 2340–2352. [CrossRef]
41. Maley, J. Elaeis guineensis Jacq. (oil palm) fluctuations in central Africa during the late Holocene: Climate or human driving forces for this pioneering species? *Veg. Hist. Archaeobotany* **2001**, *10*, 117–120. [CrossRef]
42. Granier, A. A new method for sap flow measurement in tree stems (in French). *Ann. Sci. For.* **1985**, *42*, 193–200. [CrossRef]
43. Niu, F.; Röll, A.; Hardanto, A.; Meijide, A.; Köhler, M.; Hölscher, D. Oil palm water use: Calibration of a sap flux method and a field measurement scheme. *Tree Physiol.* **2015**, *35*, 563–573. [CrossRef]
44. Burgess, S.S.O.; Adams, M.A.; Turner, N.C.; Beverly, C.R.; Ong, C.K.; Khan, A.A.H.; Bleby, T.M. An improved heat pulse method to measure low and reverse rates of sap flow in woody plants. *Tree Physiol.* **2001**, *21*, 589–598. [CrossRef]
45. Boeger, M.R.T.; Alves, L.C.; Negrelle, R.R.B. Leaf morphology of 89 tree species from a lowland tropical rain forest (Atlantic forest) in South Brazil. *Braz. Arch. Biol. Technol.* **2004**, *47*, 933–943. [CrossRef]
46. Ichie, T.; Inoue, Y.; Takahashi, N.; Kamiya, K.; Kenzo, T. Ecological distribution of leaf stomata and trichomes among tree species in a Malaysian lowland tropical rain forest. *J. Plant Res.* **2016**, *129*, 625–635. [CrossRef]
47. Meijide, A.; Badu, C.S.; Moyano, F.; Tiralla, N.; Gunawan, D.; Knohl, A. Impact of forest conversion to oil palm and rubber plantations on microclimate and the role of the 2015 ENSO event. *Agric. For. Meteorol.* **2018**, *252*, 208–219. [CrossRef]
48. Ellsäßer, F.; Röll, A.; Stiegler, C.; Hölscher, D. Introducing QWaterModel, a QGIS plugin for predicting evapotranspiration from land surface temperatures. *Environ. Model. Softw.* **2020**, *130*, 104739. [CrossRef]
49. Timmermans, W.; Kustas, W.P.; Andreu, A. Utility of an Automated Thermal-Based Approach for Monitoring Evapotranspiration. *Acta Geophys.* **2015**, *63*, 1571–1608. [CrossRef]
50. McKinney, W. Data Structures for Statistical Computing in Python. In Proceedings of the 9th Python in Science Conference, Austin, TX, USA, 28 June–3 July 2010; Volume 1697900, pp. 56–61.
51. Pedregosa, F.; Varoquaux, G.; Gramfort, A.; Michel, V.; Thirion, B.; Grisel, O.; Blondel, M.; Prettenhofer, P.; Weiss, R.; Dubourg, V.; et al. Scikit-learn: Machine Learning in Python. *J. Mach. Learn. Res.* **2011**, *12*, 6.
52. *Scikit-Learn User Guide*. 2020. Available online: https://scikit-learn.org/stable/user_guide.html (accessed on 11 December 2020).
53. Legendre, P.; Legendre, L. *Numerical Ecology*; Elsevier: Amsterdam, The Netherlands, 2003; ISBN 978-0-444-53868-0.
54. Raschka, S. *Machine Learning mit Python: Das Praxis-Handbuch für Data Science, Predictive Analytics und Deep Learning*; Auflage mitp: Frechen, Germany, 2017; ISBN 978-3-95845-422-4.

55. Shrestha, N.; Shukla, S.K. Support vector machine based modeling of evapotranspiration using hydro-climatic variables in a sub-tropical environment. *Agric. For. Meteorol.* **2015**, *200*, 172–184. [CrossRef]
56. Rokach, L.; Maimon, O. *Data Mining with Decision Trees: Theory and Applications*, 2rd ed.; World Scientific: Hackensack, NJ, USA, 2015; ISBN 978-981-4590-07-5.
57. Breiman, L. Stacked regressions. *Mach. Learn.* **1996**, *24*, 49–64. [CrossRef]
58. Freund, Y.; Schapire, R.E. *Experiments with a New Boosting Algorithm*; Morgan Kaufmann Publishers Inc.: San Francisco, CA, USA, 1996; Volume 96, pp. 148–156.
59. Schapire, R.E. The strength of weak learnability. *Mach. Learn.* **1990**, *5*, 197–227. [CrossRef]
60. Breiman, L. Bagging predictors. *Mach. Learn.* **1996**, *24*, 123–140. [CrossRef]
61. Ghojogh, B.; Crowley, M. The theory behind overfitting, cross validation, regularization, bagging, and boosting: Tutorial. *arXiv* **2019**, arXiv:190512787.
62. Freund, Y.; Schapire, R.E. A Short Introduction to Boosting. *J. Jpn. Soc. Artif. Intell.* **1999**, *14*, 1612.
63. Chollet, F. *Keras*. 2015. Available online: https://github.com/keras-team/keras (accessed on 4 February 2020).
64. Júnior, I.D.S.T.; Da Rocha, J.E.C.; Ebling, Â.A.; Chaves, A.D.S.; Zanuncio, J.C.; Farias, A.A.; Leite, H.G. Artificial Neural Networks and Linear Regression Reduce Sample Intensity to Predict the Commercial Volume of Eucalyptus Clones. *Forests* **2019**, *10*, 268. [CrossRef]
65. Strobl, C.; Boulesteix, A.-L.; Kneib, T.; Augustin, T.; Zeileis, A. Conditional variable importance for random forests. *BMC Bioinform.* **2008**, *9*, 307. [CrossRef]
66. Bilic-Zulle, L. Comparison of methods: Passing and Bablok regression. *Biochem. Med.* **2011**, *21*, 49–52. [CrossRef]
67. Passing, H.; Bablok, W. Comparison of Several Regression Procedures for Method Comparison Studies and Determination of Sample Sizes Application of linear regression procedures for method comparison studies in Clinical Chemistry, Part II. *Clin. Chem. Lab. Med.* **1984**, *22*, 431–445. [CrossRef]
68. Passing, H.; Bablok, W. A New Biometrical Procedure for Testing the Equality of Measurements from Two Different Analytical Methods. Application of linear regression procedures for method comparison studies in Clinical Chemistry, Part I. *Clin. Chem. Lab. Med.* **1983**, *21*, 709–720. [CrossRef]
69. van Doorn, W.P.T.M. Methcomp. 2020. Available online: https://pypi.org/project/methcomp/ (accessed on 2 May 2020).
70. Harris, C.R.; Millman, K.J.; van der Walt, S.J.; Gommers, R.; Virtanen, P.; Cournapeau, D.; Wieser, E.; Taylor, J.; Berg, S.; Smith, N.J.; et al. Array programming with NumPy. *Nature* **2020**, *585*, 357–362. [CrossRef]
71. Van Der Walt, S.; Colbert, S.C.; Varoquaux, G. The NumPy Array: A Structure for Efficient Numerical Computation. *Comput. Sci. Eng.* **2011**, *13*, 22–30. [CrossRef]
72. Virtanen, P.; Gommers, R.; Oliphant, T.E.; Haberland, M.; Reddy, T.; Cournapeau, D.; Burovski, E.; Peterson, P.; Weckesser, W.; Bright, J.; et al. SciPy 1.0: Fundamental algorithms for scientific computing in Python. *Nat. Methods* **2020**, *17*, 261–272. [CrossRef]
73. Hunter, J.D. Matplotlib: A 2D Graphics Environment. *Comput. Sci. Eng.* **2007**, *9*, 90–95. [CrossRef]
74. Waskom, M.; Botvinnik, O.; Ostblom, J. *Mwaskom/Seaborn: V0.10.0*. 2020. Available online: https://seaborn.pydata.org/ (accessed on 4 February 2020).
75. Ismail, R.; Mutanga, O. A comparison of regression tree ensembles: Predicting Sirex noctilio induced water stress in Pinus patula forests of KwaZulu-Natal, South Africa. *Int. J. Appl. Earth Obs. Geoinf.* **2010**, *12*, S45–S51. [CrossRef]
76. Liu, X.; Kang, S.; Li, F. Simulation of artificial neural network model for trunk sap flow of Pyrus pyrifolia and its comparison with multiple-linear regression. *Agric. Water Manag.* **2009**, *96*, 939–945. [CrossRef]
77. Rodriguez-Galiano, V.; Sanchez-Castillo, M.; Chicaolmo, M.; Chica-Rivas, M. Machine learning predictive models for mineral prospectivity: An evaluation of neural networks, random forest, regression trees and support vector machines. *Ore Geol. Rev.* **2015**, *71*, 804–818. [CrossRef]
78. Niu, F.; Röll, A.; Meijide, A.; Hölscher, D. Rubber tree transpiration in the lowlands of Sumatra. *Ecohydrology* **2017**, *10*, e1882. [CrossRef]
79. Röll, A.; Niu, F.; Meijide, A.; Hardanto, A.; Knohl, A.; Holscher, D. Hendrayanto Transpiration in an oil palm landscape: Effects of palm age. *Biogeosciences* **2015**, *12*, 5619–5633. [CrossRef]
80. Ahongshangbam, J.; Röll, A.; Ellsäßer, F.; Hölscher, D. Airborne Tree Crown Detection for Predicting Spatial Heterogeneity of Canopy Transpiration in a Tropical Rainforest. *Remote Sens.* **2020**, *12*, 651. [CrossRef]

81. Kumar, M.; Raghuwanshi, N.S.; Singh, R. Artificial neural networks approach in evapotranspiration modeling: A review. *Irrig. Sci.* **2011**, *29*, 11–25. [CrossRef]
82. Maier, H.R.; Dandy, G.C. Neural networks for the prediction and forecasting of water resources variables: A review of modelling issues and applications. *Environ. Model. Softw.* **2000**, *15*, 101–124. [CrossRef]
83. Panda, S.; Amatya, D.M.; Hoogenboom, G. Stomatal conductance, canopy temperature, and leaf area index estimation using remote sensing and OBIA techniques. *J. Spat. Hydrol.* **2014**, *12*, 25.
84. Vitrack-Tamam, S.; Holtzman, L.; Dagan, R.; Levi, S.; Tadmor, Y.; Azizi, T.; Rabinovitz, O.; Naor, A.; Liran, O. Random Forest Algorithm Improves Detection of Physiological Activity Embedded within Reflectance Spectra Using Stomatal Conductance as a Test Case. *Remote Sens.* **2020**, *12*, 2213. [CrossRef]
85. Özçelik, R.; Diamantopoulou, M.J.; Crecente-Campo, F.; Eler, U. Estimating Crimean juniper tree height using nonlinear regression and artificial neural network models. *For. Ecol. Manag.* **2013**, *306*, 52–60. [CrossRef]
86. Chu, C.-R.; Hsieh, C.-I.; Wu, S.-Y.; Phillips, N.G. Transient response of sap flow to wind speed. *J. Exp. Bot.* **2008**, *60*, 249–255. [CrossRef] [PubMed]
87. Giambelluca, T.W.; Mudd, R.G.; Liu, W.; Ziegler, A.D.; Kobayashi, N.; Kumagai, T.; Miyazawa, Y.; Lim, T.K.; Huang, M.; Fox, J.M.; et al. Evapotranspiration of rubber (Hevea brasiliensis) cultivated at two plantation sites in Southeast Asia: Rubber evapotranspiration in SE Asia. *Water Resour. Res.* **2016**, *52*, 660–679. [CrossRef]
88. Kobayashi, N.; Kumagai, T.; Miyazawa, Y.; Matsumoto, K.; Tateishi, M.; Lim, T.K.; Mudd, R.G.; Ziegler, A.D.; Giambelluca, T.W.; Song, Y. Transpiration characteristics of a rubber plantation in central Cambodia. *Tree Physiol.* **2014**, *34*, 285–301. [CrossRef] [PubMed]
89. Horna, V.; Schuldt, B.; Brix, S.; Leuschner, C. Environment and tree size controlling stem sap flux in a perhumid tropical forest of Central Sulawesi, Indonesia. *Ann. For. Sci.* **2011**, *68*, 1027–1038. [CrossRef]
90. Stiegler, C.; Meijide, A.; Fan, Y.; Ali, A.A.; June, T.; Knohl, A. El Niño–Southern Oscillation (ENSO) event reduces CO2 uptake of an Indonesian oil palm plantation. *Biogeosciences* **2019**, *16*, 2873–2890. [CrossRef]
91. Kişi, O.; Çimen, M. Evapotranspiration modelling using support vector machines/Modélisation de l'évapotranspiration à l'aide de 'support vector machines'. *Hydrol. Sci. J.* **2009**, *54*, 918–928. [CrossRef]

Article

Estimation of Nitrogen in Rice Crops from UAV-Captured Images

Julian D. Colorado [1,*], Natalia Cera-Bornacelli [1], Juan S. Caldas [1], Eliel Petro [2], Maria C. Rebolledo [2,3], David Cuellar [1], Francisco Calderon [1], Ivan F. Mondragon [1] and Andres Jaramillo-Botero [4,5]

1 School of Engineering, Pontificia Universidad Javeriana Bogotá, Carrera 7 No. 40-62, Bogotá 110231, Colombia; ceran@javeriana.edu.co (N.C.-B.); jcaldas@javeriana.edu.co (J.S.C.); david.cuellar@javeriana.edu.co (D.C.); calderonf@javeriana.edu.co (F.C.); imondragon@javeriana.edu.co (I.F.M.)

2 International Center for Tropical Agriculture (CIAT), Km 17 Recta Cali-Palmira, Cali 110231, Colombia; e.e.petro@cgiar.org (E.P.); m.c.rebolledo@cgiar.org (M.C.R.)

3 CIRAD, UMR AGAP, F-34398 | Univ Montpellier, CIRAD, INRA, Montpellier SupAgro, 34000 Montpellier, France

4 Chemistry and Chemical Engineering Division, California Institute of Technology, Pasadena, CA 91125, USA; ajaramil@caltech.edu

5 Electronics and Computer Science, Pontificia Universidad Javeriana Cali, Calle 18 No. 118-250, Cali 760031, Colombia

* Correspondence: coloradoj@javeriana.edu.co

Received: 2 July 2020; Accepted: 22 September; Published: 16 October 2020

Abstract: Leaf nitrogen (N) directly correlates to chlorophyll production, affecting crop growth and yield. Farmers use soil plant analysis development (SPAD) devices to calculate the amount of chlorophyll present in plants. However, monitoring large-scale crops using SPAD is prohibitively time-consuming and demanding. This paper presents an unmanned aerial vehicle (UAV) solution for estimating leaf N content in rice crops, from multispectral imagery. Our contribution is twofold: (i) a novel trajectory control strategy to reduce the angular wind-induced perturbations that affect image sampling accuracy during UAV flight, and (ii) machine learning models to estimate the canopy N via vegetation indices (VIs) obtained from the aerial imagery. This approach integrates an image processing algorithm using the GrabCut segmentation method with a guided filtering refinement process, to calculate the VIs according to the plots of interest. Three machine learning methods based on multivariable linear regressions (MLR), support vector machines (SVM), and neural networks (NN), were applied and compared through the entire phonological cycle of the crop: vegetative (V), reproductive (R), and ripening (Ri). Correlations were obtained by comparing our methods against an assembled ground-truth of SPAD measurements. The higher N correlations were achieved with NN: 0.98 (V), 0.94 (R), and 0.89 (Ri). We claim that the proposed UAV stabilization control algorithm significantly improves on the N-to-SPAD correlations by minimizing wind perturbations in real-time and reducing the need for offline image corrections.

Keywords: UAV; machine learning; plant nitrogen estimation; multispectral imagery; vegetation index; image segmentation

---

1. Introduction

Farmers use soil plant analysis development (SPAD) devices as a field diagnostic tool to estimate nitrogen (N) content in the plant and to predict grain yield [1,2]. For rice crops, N provides critical insight into plant-growth, crop yield, and biomass accumulation [3]. Furthermore, monitoring N allows farmers to properly adapt crop management practices [4,5]. However, using SPAD devices for crop diagnosis is still time consuming. With the advent of low-cost unmanned aerial vehicles (UAVs), several authors have reported faster and accurate remote sensing tools and methods [6,7] to estimate the N canopy from aerial multispectral imagery [8,9]. One of the most common methods used, relies on sensing the canopy reflectance in both visible (VIS) and near-infrared (NIR) wavelengths, using hyperspectral sensors [10–13].

Multiple vegetation indices (VI) have been established to associate specific spectral reflectance wavelengths with the crop variables of interest [14,15]. Although these VIs can be used to estimate the N contents in rice plants, there is no single standard set of VIs that works across all crop stages, rice varieties, soils, and atmospheric conditions. Most of the existing body of work associated with the estimation of N contents in plants, has been conducted using multi-variable regressions [5,16], in which linear functions are defined by combining and weighting each VI according to the regression model. This approach is simple, but sometimes inaccurate since the evolution of the crop tends to exhibit highly nonlinear effects that affect the N content over time. Machine and deep learning methods have recently gained traction, in solving these issues. Machine learning has the potential to drive significant breakthroughs in plant phenotyping [17–20].

For canopy N estimation, training machine learning algorithms requires the precise extraction of VI features from the aerial multispectral imagery. Several authors have applied data fusion methods from different sensors [21–23] for applying image mosaicing techniques [24–27], computing crop surface models based on dense image matching (DIM) techniques [28], or applying photogrammetry methods that are computationally expensive [26,27,29,30]. Other approaches rely on the real-time segmentation and image registration for the extraction of relevant features associated with the leaf/canopy N, including edge detection thresholding, color histograms, and clustering (otsu, K-means, watershed) [31–33].

In this work, we improve on the stabilization control of the UAV for acquiring accurate plot imagery, instead of relying on mosaicing or photogrammetry methods. Commercial UAV quadrotors used for the monitoring of the canopy N usually come with a proportional-integral-derivative (PID) waypoint navigation autopilot. The lack of proper UAV angular stabilization in the presence of large wind perturbations limits the accuracy of image registration algorithms, therefore affecting the estimation of canopy N content through image processing. To address this problem, we demonstrate a nonlinear trajectory-tracking controller that enables precise UAV positioning through a stabilization/attitude control loop. We call this method backstepping+ desired angular acceleration function (BS + DAF). This approach incorporates UAV aerodynamics information within the control law to produce roll and pitch acceleration commands to reduce abrupt angular acceleration changes caused by external wind disturbances. This added compliance enables the UAV to hover over the crop plots precisely, which in turn allows for capturing imagery that can be individually processed in real-time.

Here, we combine an autonomous UAV to acquire multispectral imagery from a crop, with machine learning methods as a means to high-throughput nitrogen content estimation. Three major contributions are involved:

(i) The development and validation of a novel UAV attitude control called BS+DAF. The UAV captures multispectral imagery with relevant above-ground data that must correlate with the SPAD measurements at the ground-level of the crop. The proposed controller is aimed at maintaining precise angular stabilization during flight by properly rejecting wind disturbances, enabling improvements in the estimations of N.
(ii) The application and validation of a segmentation algorithm called GrabCut [34], instead of the traditional edge detection thresholding methods, color histograms, and clustering techniques used in agriculture. The GrabCut approach achieves smooth pixel information with richer detail of the canopy structure, enabling the accurate segmentation of the multispectral imagery and the proper VI-based feature extraction.

(iii) The integration of machine learning methods to process nonlinear N dynamics with the calculations of the VIs during all stages of crop growth. A comprehensive characterization of the crop by designing a ground-truth dataset with several contrasting rice genotypes and accurate direct measurements of leaf nitrogen (training model).

2. Materials and Methods

2.1. System

Figure 1 details the proposed approach. Our UAV is a quadrotor called the AscTec Pelican (manufactured by Intel's Ascending Technologies (https://robots.ros.org/astec-pelican-and-hummingbird/)). This UAV comes with a C++ software development kit (SDK) that enables onboard code development with ease. A high-level Atom Intel processor (HLP) offers enough computing power to run solutions in real-time, whereas a low-level processor (LLP) handles the sensor data fusion and rotor control with an update rate of 1kHz. As shown in Figure 1, we developed a closed-loop attitude controller to drive the low-level rotor's control running in the LLP. This control method depends on the dynamics of the UAV to properly reject wind disturbances and keep the UAV steady. During flight, a dataset of multispectral imagery is precisely collected aiming at the above-ground estimation of nitrogen by using machine learning methods. In previous work [35], our UAV system was tested during two years of in-field experiments with the aim of estimating above ground biomass accumulation. Here, we have extended the capabilities of the system in Figure 1 by developing and integrating novel methods for UAV flight control, multispectral imagery segmentation for VI feature extraction, and their impact on nitrogen estimation using machine learning algorithms.

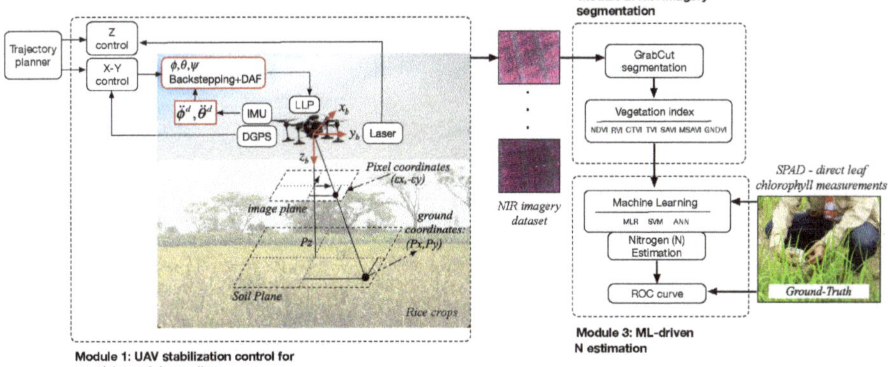

Figure 1. Unmanned aerial vehicle (UAV)-based robotic system for canopy nitrogen estimation in rice crops.

The UAV was equipped with the Parrot Sequoia multispectral sensor (https://www.parrot.com/business-solutions-us/parrot-professional/parrot-sequoia) fabricated with 4 separate bands: green, red, red-edge, and near-infrared. The camera offers a resolution of 1280 × 960 for each independent spectral sensor, yielding a crop-to-image resolution of 1.33 cm/pixel according to the flying altitude of 20 m. In addition, the multispectral camera comes with a radiometric calibration target that enables offline reflectance calibration across the spectrum of light captured by the camera, and an onboard irradiance sensor designed to correct images for illumination differences, all of which enables its outstanding performance in cloudy conditions, as evaluated in [36]. In our application, this calibration procedure was executed prior to any flight mission of the UAV. Additional sensors such as an IMU, magnetometer, and GPS are also embedded within the sunshine sensor. Figure 2a depicts the mounted camera setup, while Table 1 details the general specifications of our system.

Table 1. General specifications.

Specification	Description
UAV	
UAV Processing	High-level: Intel®CoreTM i7 I Low-level: Intel Atom 1.6 GHz
Flight altitude over ground	20 m
Flight linear speed	1.5 ms^{-1}
Autonomy	25 min
NAVIGATION	
DGPS Piksi GNSS Module [1]	HW version: 00108-10
DGPS Processing	dual-core ARM Cortex-A9 at 666 MHz
Real-time kinematic positioning (RTK)	10 Hz: GPS, GLONASS, BeiDou, Galileo
DGPS Position accuracy	35 cm
IMU rate and accuracy	1 KHz I Heading: 0.2° , Pitch/Roll: 0.03°
IMAGERY	
Parrot Sequoia multispectral camera	Spectral bands resolution: 1280 × 960 pixels. Green (530–570 nm), Red (640–680 nm), Red-Edge (730–740 nm), Near-Infrared (770–810 nm). RGB sensor resolution: 4608 × 3456 pixels.
Imagery resolution at 20 m altitude (Ground Sampling Distance—GSD) [2]	RGB: 0.37 cm/pixel. Spectral-bands: 1.33 cm/pixel
Onboard irradiance sensor	real-time illumination image correction
CROPS	
Experiments location	Lat: 4°1′37.85″N, Lon: 73°28′28.65″W International Center for Tropical Agriculture (CIAT), Department of Meta, Colombia [3]
Weather conditions [4]	Dry season (June-September): Average temperatures: L: 22°–H: 31 ° C with average cloudiness of 30%.

[1] Differential Global Positioning System (DGPS) Piksi module: https://www.swiftnav.com/piksi-multi. [2] Ground Sampling Distance (GSD) calculator: https://support.pix4d.com/hc/en-us/articles/202560249-TOOLS-GSD-calculator#gsctab=0. [3] International Center for Tropical Agriculture (CIAT) website: https://ciat.cgiar.org. [4] Weather information: https://weatherspark.com/y/24273/Average-Weather-in-Villavicencio-Colombia-Year-Round.

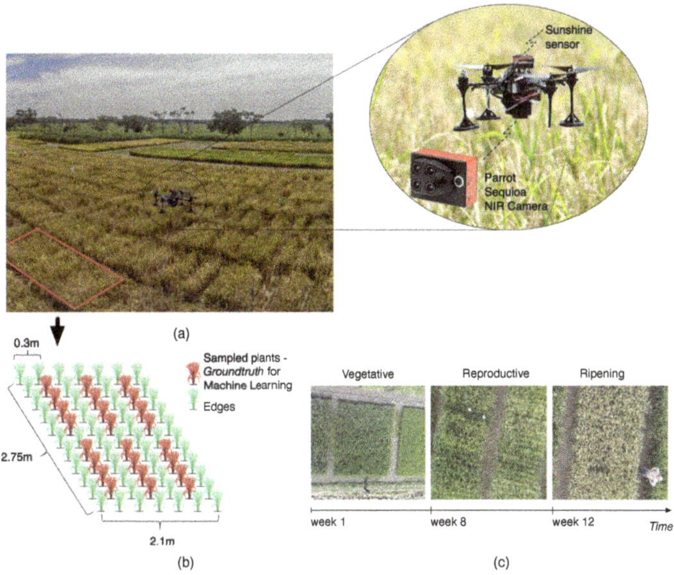

Figure 2. Crop field: (**a**,**b**) each plot was designed with an area of 5.7 m^2 and a rice crop density of 38.4 $kgha^{-1}$. The UAV is programmed to capture multispectral imagery from the plots of interests by using a Sequoia camera manufactured by Parrot [7]. (**c**) The results reported in this work were obtained during three stages of rice growth: vegetative, reproductive, and ripening with an entire phenological cycle ranging between 86–101 days.

2.2. Rice Crops

The crops were designed with 3 spatial repetitions containing 8 contrasting rice genotypes in terms of N concentration, biomass accumulation, and flowering: FED50, MG2, ELWEE, NORUNKAN, IR64, AZUCENA, UPLRI7, and LINE 23. These rice varieties have a phenological cycle ranging between 86-101 days , as detailed in Figure 3b. The experimental design consisted of six months of in-field sampling with three different planting crops. For each experiment, the amount of applied nitrogen and water was modified as follows:

- Experiment 1: 200 $kgha^{-1}$ of nitrogen and limited crop irrigation.
- Experiment 2: 200 $kgha^{-1}$ of nitrogen with permanent crop irrigation.
- Experiment 3: 100 $kgha^{-1}$ of nitrogen with permanent crop irrigation.

Experimental trials were conducted during the dry season. To assess the effects of limited and permanent irrigation on the crops, leaf temperature (MultispeQ (https://www.photosynq.com)), and soil humidity (AquaPro (http://aquaprosensors.com)) were constantly monitored. Fertilizers were applied accordingly in order to maintain the crops through the phenological cycle. Given that, the same amount of fertilizers were applied for the experiments: 60 $kgha^{-1}$ of P_2O_5 (diammonium phosphate) and 130 $kgha^{-1}$ of K_2O (potassium chloride), as detailed in the experimental protocol available in the Supplementary Materials section.

Figure 2 details some characteristics of the crop. As shown in plots (a), (b), the length of a plot was about 2.75 m with a distance between plants of 25 cm and 30 cm between rows. Within each plot, we defined 6 sampled areas with 1 linear meter in length (containing four plants). A ground-truth was defined based on the direct measurements of plant chlorophyll using a SPAD 502 Plus meter (Konica-Minolta) over these sampled areas. Measurements from the crop were obtained during three stages of rice growth: vegetative, reproductive, and ripening. Figure 3b details the specifications of the crop experiments reported here.

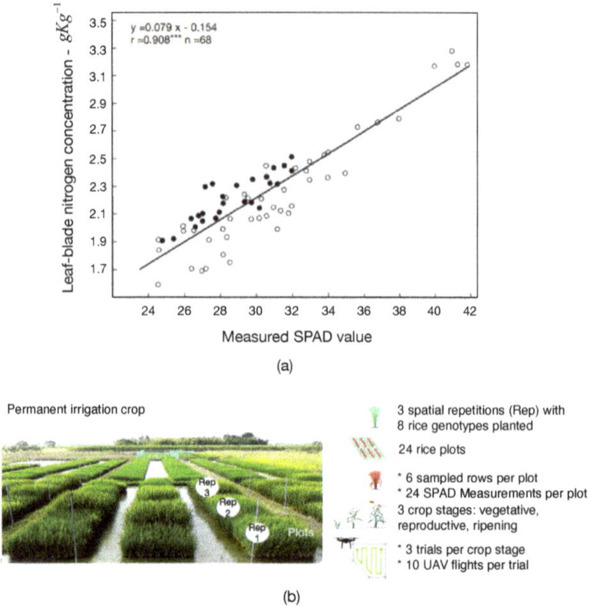

Figure 3. (**a**) Correlation between soil plant analysis development (SPAD) readings and leaf N concentration [37]. (**b**) Experimental specifications.

Table 2 details how the dataset was collected, in which the SPAD value corresponds to the average of 24 measurements conducted over each plot, as depicted in Figure 3b. All ground-truth data are available through the Open Science Framework in the Supplementary Materials section. On the other hand, Figure 3a shows the linear correlation used to relate the measured SPAD value with the leaf-blade N concentration [37].

As mentioned, vegetation indices are widely used to quantify both plant and soil variables by associating certain spectral reflectances that are highly related to variations in canopy chemical components such as nitrogen. From the extensive list of vegetation indices (VIs) available [38–41], we selected 7 VIs with sufficient experimental evidence and quantitative trait loci (QTL)-based characterization regarding their high correlation with rice nitrogen [42]. Table 3 details the selected VIs. These formulas are applied in different wavelengths for taking into account the changes in canopy color, since several factors affect the spectral reflectances of the crop: solar radiation, plant morphology and color, leaf angles, undergrowth, soil characteristics, and water.

In our system, the parrot sequoia camera has a solar radiation sensor that compensates the light variations in the canopy. The change in the rice canopy color is perhaps the most notable variation during the phenological cycle. In the vegetative stage, the green color is predominant whereas in the reproductive stage, panicle formation starts and thus yellow features appear in the images. In ripening, the maturation of the plants occur while the leaves begin to senesce, turning the yellow color predominant. These changes can be observed in Figure 2c. Furthermore, different wavelengths of light have a different level of plant absorption depending on the leaf composition given by several genetic traits. In particular, the relation between the selected VIs in Table 3 with the photosynthetic activity and canopy structural properties has allowed the association of certain spectral reflectances that are highly related to the physico-chemical canopy N variations in plants, especially the green, red, red-edge, and near infrared bands.

Table 2. An example of a ground-truth dataset. The crop field was designed with three spatial repetitions (Rep) containing 8 contrasting rice genotypes.

Plot	Genotype	Rep	SPAD
1	AZUCENA	1	56.55
2	ELWEE	1	47.80
3	LÍNEA 23	1	54.55
4	UPLRI7	1	46.32
5	NORUNKAN	1	43.30
6	IR64	1	32.91
7	FED50	1	47.06
8	MG2	1	43.36
9	AZUCENA	2	49.26
10	IR64	2	42.59
11	LÍNEA 23	2	49.82
12	UPLRI7	2	48.15
13	ELWEE	2	41.29
14	FED50	2	46.29
15	NORUNKAN	2	42.82
16	MG2	2	38.96
17	FED50	3	49.20
18	UPLRI7	3	40.88
19	IR64	3	40.23
20	AZUCENA	3	55.81
21	NORUNKAN	3	43.96
22	ELWEE	3	49.70
23	LÍNEA 23	3	46.45
24	MG2	3	42.67

Table 3. Selected near infrared vegetation indices (extracted features). The ρ_f term denotes the reflectance of the frequency f).

Name	Equation
Normalized Difference Vegetation Index [38]	$\mathrm{NDVI} = \frac{\rho_{780}-\rho_{670}}{\rho_{780}+\rho_{670}}$
Green Normalized Difference Vegetation Index [39]	$\mathrm{GNDVI} = \frac{\rho_{780}-\rho_{500}}{\rho_{780}+\rho_{500}}$
Simple Ratio [38]	$\mathrm{SR} = \frac{\rho_{780}}{\rho_{670}}$
Soil-Adjusted Vegetation Index [39,40]	$\mathrm{SAVI} = (1+\mathrm{L})\left(\frac{\rho_{800}-\rho_{670}}{\rho_{800}+\rho_{670}+\mathrm{L}}\right)$ with $\mathrm{L} = 0.5$
Modified SAVI [40]	$\mathrm{MSAVI} = \frac{1}{2}\left(2\rho_{800}+1-\sqrt{(2\rho_{800}+1)^2-8(\rho_{800}-\rho_{670})}\right)$
Triangular Vegetation Index [39]	$\mathrm{TVI} = \frac{1}{2}\left(120(\rho_{780}-\rho_{500})-200(\rho_{670}-\rho_{500})\right)$
Corrected Transformed Vegetation Index [42]	$\mathrm{CTVI} = \frac{\mathrm{NDVI}+0.5}{\lvert\mathrm{NDVI}+0.5\rvert}\sqrt{\lvert\mathrm{NDVI}+0.5\rvert}$

The selected VIs exhibit a strong dependence on the NIR reflectance due to leaf chlorophyll absorption, providing an accurate approach to determine the health status of the plants and the canopy N. Most of the existing body of research focused on multispectral-based N estimations [14,21,40,42,43], combine the information provided by several vegetation indices in order to avoid saturation issues. For instance, the NDVI, which is one of the most common VIs used, tends to saturate with dense vegetation. In turn, the NDVI alone is not accurate during the reproductive and ripening stages of rice growth. Here, we combine several VIs across the crop stages, to ensure data on wavelengths located in the red-edge and another spectral reflectances that accurately express the healthy status of the leaves (higher NIR and green-band readings).

2.3. UAV Stabilization Control

The estimation of the canopy N requires the precise extraction of Vegetative Index features from the acquired multispectral imagery. Capturing aerial images during UAV hovering that accurately matches with the GPS positions of the SPAD measurements at ground-level is essential.

As previously detailed in Figure 1, three closed-loop controllers are needed to regulate: (i) the X-Y position based on GPS feedback, (ii) the Z altitude based on barometric pressure and laser readings (pointing downwards), and (iii) the ϕ, θ, ψ attitude based on IMU data. The UAV is constantly subjected to wind disturbances that cause unsteady angular motions and therefore imprecise trajectory tracking. Consequently, aerial imagery captured across the crop with the UAV system is affected. To overcome this issue, we replaced both roll and pitch PID-based controllers by a robust nonlinear backstepping (BS) control.

The classical BS method has several advantages. It explicitly takes into account the nonlinearities of the UAV model defined in Equation (A2) (see Appendix A), and most importantly, allows the incorporation of a virtual control law to regulate angular accelerations. For this, we have derived a desired acceleration function (DAF) for roll and pitch. This enhanced controller is called backstepping+DAF (BS+DAF). Our goal is to use the dynamics equations of motion (EoM) defined in Algorithm A1 within the control law in order to compensate for abrupt angular acceleration changes, concretely in roll and pitch. The DAF terms improve the responsiveness of the control law to external perturbative forces. Appendix B details the control law derivation for roll and pitch.

The BS + DAF control supports on the Lyapunov stability concept that guarantee asymptotic stabilization around the equilibrium points. For our application, we require both roll ϕ a pitch θ angles to remain in zero while the UAV is hovering above the crop for capturing multispectral imagery, i.e., $e_\phi = \phi^d - \phi \to 0$ and $e_\theta = \theta^d - \theta \to 0$. Otherwise, the set-point references for both roll and pitch controllers are defined by the X-Y position controller.

As previously mentioned, high wind-speed perturbations affect the UAV during hovering. Nonetheless, our controller law is sensitive to angular acceleration changes caused by external wind disturbances, thanks to both error dynamics (e_2) and the DAF terms ($\ddot{\phi}^d$) in Equation (A18) (Appendix B). Given that, we present how to model the effects of the wind over the UAV aerodynamics, by following the wind effect model (WEM) developed by [44]. Figure 4 details the control architecture and how the WEM module has been incorporated as an external disturbance acting on the UAV body frame.

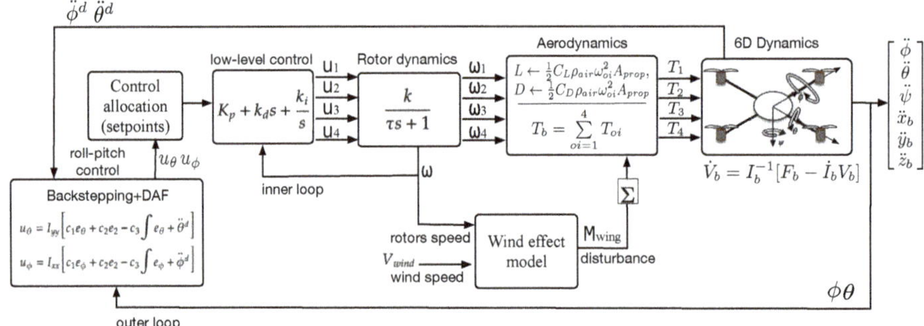

Figure 4. Attitude (roll and pitch) control architecture. The UAV has a low-level PID-based controller to drive each rotor. The proposed backstepping+DAF generates the references to the inner loop according to the dynamics of the system. A wind effect model has been adopted from [44] to add disturbances to our model.

The WEM defines the relationship between the oblique airflow acting on a propeller and the resulting forces and moments applied onto the UAV. In our quadrotor system, the oblique airflow form and angle of 90° with the propeller azimuth angle. In terms of aerodynamics, the incoming airflow induces an increase in vertical thrust (T_{oi}), a side force (F_{wind}), and a pitching moment (M_{wind}). The WEM presented by [44] uses blade element momentum theory to determine the aforementioned effects. Here, we are interested in incorporating the effects caused to the thrust and the pitching moment (M_{wind}) as a function of the magnitude of a given wind speed vector V_{wind} and the rotational speed ω of each propeller driven by the rotor's model. In this regard, the WEM is defined as:

$$M_{wind} = \begin{bmatrix} cos\left(\beta + \frac{\pi}{2}\right) & sin\left(\beta + \frac{\pi}{2}\right) & 0 \\ -sin\left(\beta + \frac{\pi}{2}\right) & cos\left(\beta + \frac{\pi}{2}\right) & 0 \\ 0 & 0 & 1 \end{bmatrix} \begin{bmatrix} M_{prop} \\ 0 \\ 0 \end{bmatrix} + \frac{l}{2} \begin{bmatrix} C_4 T_4 - C_2 T_2 \\ C_1 T_1 - C_3 T_3 \\ 0 \end{bmatrix} \quad (1)$$

The angle β determines the direction of the V_{wind} vector. Depending on the direction of the applied relative wind, certain propellers will be more affected by this external disturbance. Tran et al., in [44,45], propose the use of a weight function that assigns values between 0 to 1, where the maximum value of 1 means the propeller is directly exposed to the wind. In Equation (1), C corresponds to the weighting vector that affects the thrust T generated by each propeller. The parameter l is the distance between non-adjacent propellers and M_{prop} is:

$$M_{prop} = \begin{bmatrix} C_1 & C_2 & C_3 & C_4 \end{bmatrix} \begin{bmatrix} \tau_1 \\ \tau_2 \\ \tau_3 \\ \tau_4 \end{bmatrix} \quad (2)$$

Therefore, the scale of the weighting vector C is defined based on the magnitude of the applied wind speed vector and the corresponding direction (the β angle between the relative wind vector pointing towards the UAV body frame). In addition, the rotor's speeds ω are required to calculate the augmented thrusts T and torques τ (weighted by C). Further details on this model can be found in [44,45].

Sections 3.1 and 3.3 present the results regarding the impact of precise UAV positioning on the accurate estimation of the canopy N through the entire phenological cycle.

2.4. Multispectral Image Segmentation

Multispectral imagery is evaluated by using a multispectral image segmentation method called GrabCut [34]. The original method is widely used for its ease of implementation and for the excellent results in generating a binary classification; however, it suffers from the drawback of being a semi-manual algorithm. The original GrabCut method requires a manual input during the algorithm iteration in order to properly determine both background and foreground pixel values.

This section introduces a modified version of the GrabCut algorithm that is fully automatic, thanks to an added refinement step using a guided filtering [46] to extract the relevant pixel information associated with the plant's canopy. Our approach solves an optimization problem using an energy function that allows the proper labeling of texture in the multispectral image by using a Gaussian mixture model. As mentioned, we use a refinement process based on a guided-filter by taking into account information from all band channels of the multispectral camera: green, red, red-edge, and near-infrared. The resultant multispectral image-mask includes only relevant pixel information that accurately represents the canopy for the estimation of N.

2.4.1. GrabCut Segmentation

The GrabCut method requires an initial model known as a trimap. This model consists of a partition of the input image into three regions: a foreground T_F, a background T_B, and a region with pixels that result from the combination of the foreground and background colors T_U.

The image is modeled as an array of values $\mathbf{z} = (z_1, ..., z_N)$, and the output segmentation can be a channel of values between 0 and 1 or a hard segmentation with a binary assignment (0 or 1). The segmentation is written as $\underline{\alpha} = (\alpha_1, ...\alpha_N)$ with $0 \le \alpha_n \le 1$, or $\alpha_n = \{1, 0\}$.

The GrabCut algorithm also requires a model for the distribution of the foreground/background colors and gray levels. This distribution can be represented as a histogram directly assembled from T_F and T_B, as: $\underline{\theta} = \{h(z; a), a = 0, 1\}$. Under this trimap model, the segmentation algorithm determines the values of α from the image $\mathbf{z}$ and the distribution model $\underline{\theta}$. The α values are calculated from an energy minimization function, as:

$$\mathbf{E}(\underline{\alpha}, \underline{\theta}, \mathbf{z}) = U(\underline{\alpha}, \underline{\theta}, \mathbf{z}) + V(\underline{\alpha}, \mathbf{z}), \tag{3}$$

where the sub-function U evaluates the fitness by assigning a low score if the segmentation (α) is consistent with the image $\mathbf{z}$, defined as follows:

$$U(\alpha, \theta, \mathbf{z}) = \sum_n - \ln h(z_n; \alpha_n) \tag{4}$$

Instead of using the histogram as the estimator of the probability density function, the algorithm uses a Gaussian mixture model in order to take into account the information coming from all channels.

$$V(\underline{\alpha}, \mathbf{z}) = \gamma \sum_{(m,n)\in \mathbf{C}} [\alpha_n \neq \alpha_m] e^{-\beta(z_m - z_n)^2} \tag{5}$$

In Equation (5), the sum set $\mathbf{C} \in 3 \times 3$ refers to the neighbors pixels in a given window, and the the term β can be calculated according to Equation (6).

$$\beta = \frac{1}{\left(2E\,|\langle(z_m - z_n)^2|\rangle\right)} \tag{6}$$

By using the global minimum in 7, the image segmentation is estimated as:

$$\hat{\underline{\alpha}} = \arg\min_{\underline{\alpha}} \mathbf{E}(\underline{\alpha}, \underline{\theta}, \mathbf{z}) \tag{7}$$

Algorithm 1 details the original GrabCut method. In order to eliminate the third manual step of the algorithm, a fixed background image T_B mask is used during the iteration, and a guided-filter refinement process is applied to achieve an automatic segmentation, as detailed in Algorithm 2.

Algorithm 1: Original GrabCut algorithm.

Step 1: Initialization of T_B, T_F, T_U.
Initialize the trimap T_F as the foreground
Initialize the trimap T_B as the background
All remaining pixels are set as a possible foreground pixels T_{UF}
Step 2: Iterative minimization.
Assign and learn GMM
Minimize energy function
Estimate segmentation image α
IF image α have visible errors **GO** to **Step 3**
Step 3: User editing.
Use the segmented image α as the new possible foreground pixels T_{UF}
Input pixel hints for T_B and T_F
GO to **Step 2**

Algorithm 2: Modified GrabCut with guided filter calculation. In the algorithm, $E_r(I)$ denotes a function that calculates the image mean over a radius r, the operations $.*$ and $./$ denotes the matrix element-wise calculation, and q is the image output.

Step 1: Initialization of T_B, T_F and T_U.
Initialize the trimap T_F as the foreground
Initialize the trimap T_B as the background
All remaining pixels are set as a possible foreground pixels T_{UF}
Step 2: Iterative minimization.
Assign and learn GMM
Minimize energy function
Estimate segmentation image α
Step 3:Input image p, input guidance I, radius r, and regularization ϵ.
1: $\mu_I \leftarrow E_r(I), \mu_p \leftarrow E_r(I), Corr_I \leftarrow E_r(I.*I), Corr_{Ip} \leftarrow E_r(I.*p)$.
2: $\sigma_I^2 \leftarrow Corr_I - \mu_I.{*}\mu_I, \sigma_{Ip}^2 \leftarrow Corr_{Ip} - \mu_I.{*}\mu_p$
3: $a \leftarrow \sigma_{Ip}^2./(\sigma_I^2+\epsilon), b \leftarrow \mu_p - a.*\mu_I$
4: $\mu_a \leftarrow E_r(a), \mu_b \leftarrow E_r(b)$
5: $q = \mu_a.{*}I + \mu_b$

2.4.2. Guided Filter Refinement

The guided filter (GF) [46] can be defined as a convolutional Bilateral Filter with a faster response due to its $O(N)$ computational complexity. In this regard, the output of each pixel denoted as q can be expressed as a weighted average over the convolutional window W (i, j denote the pixel coordinates):

$$q_i = \sum_j W_{ij} p_j \tag{8}$$

The GF implies that an image can be filtered using the radiance of another image as guidance. We exploited this concept to filter our segmented plot mask created with the GrabCut algorithm, with the aim of refining the segmented image with the original image as guidance. In this regard, the weight used by the GF is given by:

$$W_{ij}^{GF}(I) = \frac{1}{|\omega|^2} \sum_{k;(i,j)\in\omega_k} \left(1 + \frac{(I_i-\mu_k)(I_j-\mu_k)}{\sigma_k^2+\epsilon}\right), \tag{9}$$

where W_{ij}^{GF} depends entirely of the guidance image I. The parameters μ_k and σ_k^2 are the mean and variance of the guidance image I estimated over a window w_k, ϵ denotes a regularization parameter and $|\omega|$ is the number of pixels in the window w_k.

Section 3.2 presents the results of applying the proposed modified version of GrabCut, by comparing the method against traditional segmentation techniques such as thresholding, K-means, meanshift, but also against the original semi-manual GrabCut method.

2.5. Machine Learning for N Estimations

As detailed in Table 2, the ground-truth for training machine learning (ML) algorithms was defined based on the direct measurements of plant chlorophyll using a SPAD 502 Plus meter (Konica-Minolta) over these sampled areas, as depicted in Figure 1. Datasets contain the measured SPAD value that directly correlates with the leaf-blade N concentrations by following the linear correlation [37] defined in Figure 3a. In this regard, measurements from the crop were obtained during three stages of rice growth: vegetative, reproductive, and ripening, in which 3 trials were conducted per crop stage. These datasets are the result of 10 flights per trial, capturing around 500 images, and yielding a database of 1500 images per stage. Since 3 trials were conducted per crop stage, around 13,500 images were processed in this work. Figure 3b details the experimental specifications.

The ML methods were trained with a set of images accounting for the 60% of the entire database. For the final estimations of leaf nitrogen, we used the remaining 40% of the database (testing phase of the ML methods). The entire imagery dataset and the ground-truth available in the Supplementary Materials section. Here, we report on the use of classical ML methods based on multi-variable linear regressions (MLR), support vector machines (SVM), and artificial neural networks (NN) for the estimation of the canopy N.

For MLR models, the VIs from Table 3 were combined by following the formula: $N = \beta_0 + \sum_{k=1}^{7} \beta_k (VI)_k$, where the parameter β_k changes accordingly to the growth stage of the plant, by weighting each VI independently. In this regard, each crop stage will have an independent MLR model that linearly fits the N content.

With the aim of improving the adaptability of the estimation models by considering several nonlinear effects (dense vegetation, optical properties of the soil, and changes in canopy color), this section presents the estimation results using support vector machines (SVM) and artificial neural networks (NN). SVM models were used with different kernel functions with the aim of determining the proper mathematical function for mapping the input data. Six different kernels based on linear, quadratic, cubic, and Gaussian models will be tested for each crop stage independently.

Contrarily to the MLR and SVM methods, in which N estimation models are determined for each crop stage independently, neural networks enable the combination of the entire dataset (SPAD values and VI extracted features) into a single model that fits for the entire crop phenological stages. Several non-linear training functions are tested with different hidden layers. In addition, we discarded the use of deep-learning methods such as convolutional neural networks (CNN), due to the high computational costs associated to the pooling through lots of hidden layers in order to detect data features. For this application, we use well-known vegetative index features (reported in Table 3) that have been widely used and validated in the specialized literature [14,40,42]. Other image-based features such as color, structure, and morphology do not work well with multispectral imagery of 1.2 mega-pixel in resolution, compared to the 16 mega-pixel in the RGB image. In fact, the main advantage of using VIs (as features for training), relies on having information at different wavelengths, providing key information of the plant health status related to N.

In Section 3.3, we report a comprehensive comparison among multi-variable linear regressions (MLR), support vector machines (SVM), and artificial neural networks (NN) for the estimation of the canopy N. We used three metrics based on the root mean square error ($RMSE$), Pearson's linear correlation (r), and coefficient of determination (R^2), detailed as follows:

$$RMSE = \sqrt{\frac{1}{n}\sum_{i=1}^{n}(\hat{N}_i - N_i)^2}$$

$$r = \frac{\sum_{i=1}^{n}(N_i - \overline{N})}{\sqrt{\sum_{i=1}^{n}(N_i - \overline{N})^2}} \tag{10}$$

$$R^2 = \frac{\sum_{i=1}^{n}(N_i - \hat{N}_i)^2}{\sum_{i=1}^{n}(N_i - \overline{N})^2}$$

where n is the total number of samples, N is the measured value (SPAD scale cf. Figure 3a), $\hat{N}$ is the estimated nitrogen, and $\overline{N}$ is the mean of the N values. In this application, the Pearson's metric indicates the linear relationship between the estimated value of N VS the measured one (ground-truth). On the other hand, the R^2 metric is useful since it penalizes the variance of the estimated value from the measured one.

As mentioned, the ML models require the calculation of the VI formulas presented in Table 3, in order to determine the input feature vector. Every image captured by the UAV is geo-referenced with the DGPS system with a position accuracy of 35 [cm] (see Table 1). Given that, aerial imagery is registered by matching consecutive frames according to a homography computed with the oriented features from accelerated segment test (FAST) and rotated binary robust independent elementary feature (BRIEF) computer vision techniques [47]. The UAV path is planned by ensuring an image overlapping of 60%, which allows for the precise matching with the GPS coordinates of the ground-level markers to ensure a proper comparison between the aerial-estimations and ground-measurements of plant N. Part of this procedure is described in previous work reported in [48]. In turn, the metrics described in Equation (10) report on the performance of the ML models based on the aerial-ground data matching of canopy N. The aforementioned geo-referencing process is conducted by using the affine transformation, a 1st order polynomial function that relates the GPS latitude and longitude values with the pixel coordinates within the image. This procedure is also detailed in previous work reported in [48].

Figure 5 details the procedure required by the machine-learning models. After registration, images are segmented by using the proposed GrabCut method. Although all pixels in the image are evaluated in Algorithm 2, we use pixel clusters of 10×10 to calculate the VI formulas, as shown in Figure 5a. After segmentation, pixels representing the rice canopy are separated from the background, as shown in Figure 5b. All vegetation indices are calculated as the average within each image sub-region.

At canopy-level, several factors affect the spectral reflectances of the crop: solar radiation and weather conditions, plant morphology and color, leaf angles, undergrowth, soil characteristics, and ground water layers. As mentioned in Section 2.1, the multispectral camera comes with an integrated sunshine sensor to compensate light variations in the resultant image. In addition, the GrabCut segmentation method deals with the filtering of undergrowth and other soil noises. Despite that, it remains crucial to validate the accuracy of the selected VIs, since the estimation of N depends on the accuracy and reliability of the extracted features. In this regard, Figure 5c presents several VI calculations in order to analyze the variance of the VI features through the entire phenological cycle of the crop. For this test, we calculated the VI formulas from 360° random images per crop stage under different environmental conditions. We show the most representative VIs to nitrogen variations: simple ratio (SR), normalized difference vegetation index (NDVI), green normalized difference vegetation index (GNDVI), and soil-adjusted vegetation index (MSAVI). As observed, the low variance exhibited through the entire phenological cycle allows us to use the VI features as inputs to the ML models detailed in Figure 5d.

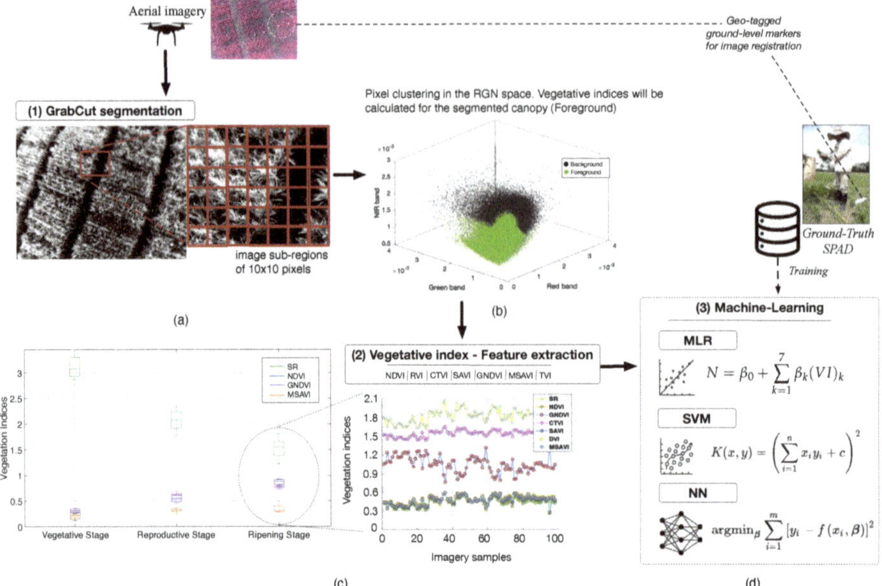

Figure 5. Procedure required by the machine-learning models: (**a**) single images are processed by using the proposed GrabCut segmentation method. Image subregions of 10×10 pixels were analyzed. (**b**) Pixel clustering in the Red+Green+Near-infrared (RGN) space. (**c**) Vegetative index calculation for the foreground cluster. Low index variability was observed at each crop stage. (**d**) Machine-learning methods applied for the estimation of the N canopy.

3. Results

The experiments reported in this paper were carried out during 2018 in the rice farms of the Center of International Agriculture (CIAT). Trials were conducted during the dry season (June-September), as detailed in Table 1. We conducted 3 trials per crop stage (vegetative, reproductive, and ripening).

3.1. UAV Control Results

Figure 6 shows simulation results to evaluate the performance of the proposed controller in terms of wind-disturbance rejection. In plot (a), the desired trajectory (red line) was defined for the UAV to cover the crop while following the trapezoidal velocity profile shown in plot (b). This trajectory profile enables the UAV to hover at certain points (black dots) to capture aerial images. The maximum UAV velocity was set to 1.5 ms^{-1}.

In plot (a), a wind disturbance ($V_{wing} = 9$ ms^{-1}) was added at the starting point of the trajectory, causing the UAV to mismatch the desired path (the V_{wing} vector was applied onto the y + direction). The response of the UAV driven by the BS-DAF attitude controller corresponds to the black line, whereas the PID controller is the green line. As observed, the BS + DAF immediately counteracted the disturbance by generating the corresponding roll command ϕ depicted in plot (d). This response allowed the UAV to follow the path precisely, maintaining the position error along the y axis near to zero, as shown in plot (c). Under this scenario, our system obtained a maximum tracking error of 2%.

A second wind disturbance was added when the UAV reached 10 m in altitude. In this case, V_{wing} was applied onto the z-direction (z points downwards). As observed, an augmented thrust caused a large altitude tracking error that affected the UAV for both control schemes similarly. Unlike for roll and pitch, the BS+DAF control does not drive the altitude loop. In general, the UAV was able to position in the hovering knot-points accurately (black dots in plot a), maintaining a minimum error with the

GPS waypoints. For the rest of the trajectory, the results demonstrate that the backstepping + DAF is accurate and reliable to reject external wind disturbances.

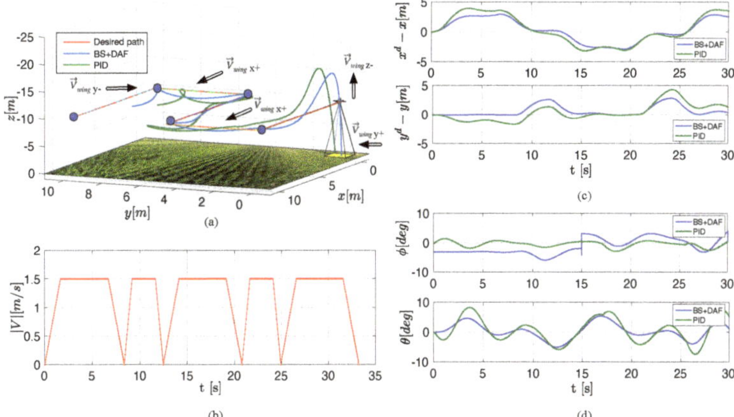

Figure 6. (Simulation) Closed-loop trajectory tracking comparison between the proposed backstepping (BS)-desired acceleration function (DAF) (black lines) and the classical PID control (green lines): (**a**) 3D navigation results. Wind disturbances were added at five instances during the trajectory (black arrows). The black dots indicate the UAV must stop to capture multispectral data. (**b**) Trapezoidal velocity profile for the desired path. (**c**) Position errors. (**d**) Roll and pitch profiles.

3.2. Image Segmentation Results

Algorithms 1 and 2 are applied for each input image in an iterative manner. In Figure 7a, two markers placed as geo-referenced guiding points appear in the RGB image (red dots in the top-right corner). The algorithm provides to the user with the option to manually select these points in order to remove them from the segmentation process. An example of semi-manual GrabCut segmentation with minimal user interaction is shown in Figure 7d. If those points are not manually removed, the corresponding pixels will be classified either in the background of foreground cluster automatically. Subsequently, plots (b) and (c) show how the initial foreground and background are defined. Plot (d) shows an example of applying Algorithm 1 to the input image from (a), whereas plot (e) shows the results when combining the GrabCut with the guided-filter (GF) approach, yielding an automatic segmentation process, as detailed in Algorithm 2. As shown in plot (e), the GF refinement achieves richer detail in the final segmentation. This result can be compared against a traditional Otsu's threshold method shown in plot (f). Although the results from Figure 7e are promising, we implemented a final refinement process based on the so-called GF feathering filtering, in which Algorithm 2 is used with a carefully selected radius r and regularization parameter ϵ. The GF feathering filtering enables a faster implementation of Algorithm 2 by avoiding the $O(N^2)$ computational restriction of the convolution.

The quality of the proposed segmentation method was tested and compared against the binary mask segmentation of three well-known methods: (i) k-means [49], (ii) mean-shift [50], and (iii) manual threshold over the HSV color representation [51]. The acronyms used in the comparison results are listed in Table 4. By applying the fully automatic segmentation method described in Algorithm 2, we used the precision, recall, and the F1-score to measure the performance of the proposed segmentation listed as GCFauto in Table 4. Figure 8 and Table 5 show the results.

For image pre-processing, the automatic GrabCut method in Algorithm 2 with the guided-filter refinement enabled precise plot segmentation of multispectral imagery with richer detail of the canopy structure and proper elimination of the background cluster. In this regard, Figure 9 shows the final segmentation results. In plot (a), the segmented image is achieved by combining the multispectral image shown in plot (b) and the RGB images from plot (c). In plot (d), the values for the radius and ϵ depend on the image size. For this application, those values were experimentally determined as $r = 60$ and $\epsilon = 0.001^2$. Without this segmentation method, significant image corrections would be needed.

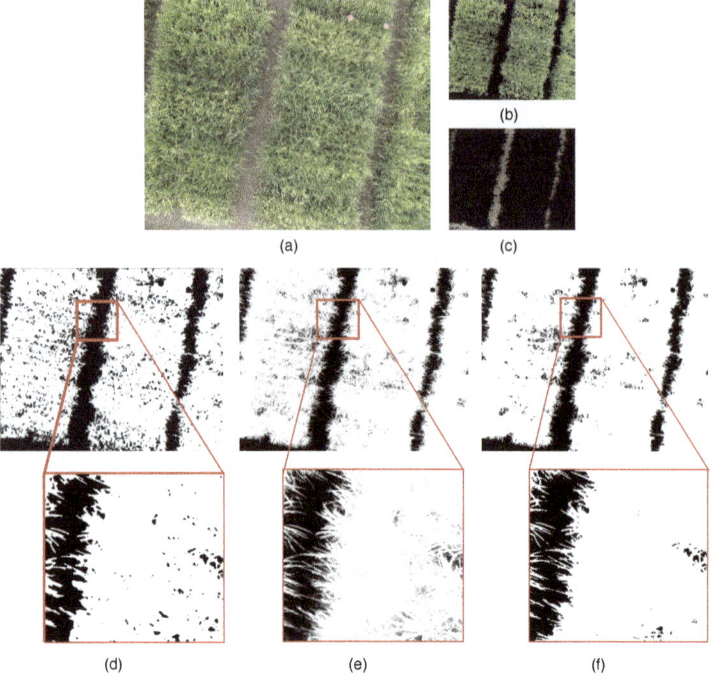

Figure 7. Segmentation results: (**a**) Original image, (**b**) initial T_F, (**c**) initial T_B, (**d**) GrabCut segmentation with manual hints, (**e**) Proposed GrabCut + guided filter (GF) refinement , (**f**) Otsu's threshold + GF refinement. Insets in plots d-f show closeups for each segmentation output.

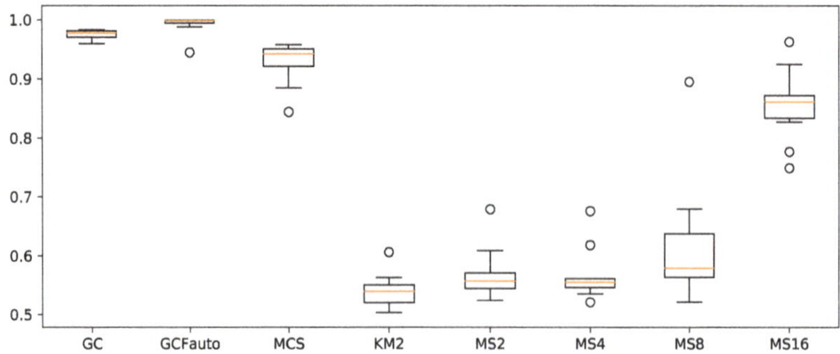

Figure 8. F1 metric of all tested algorithms in Table 4.

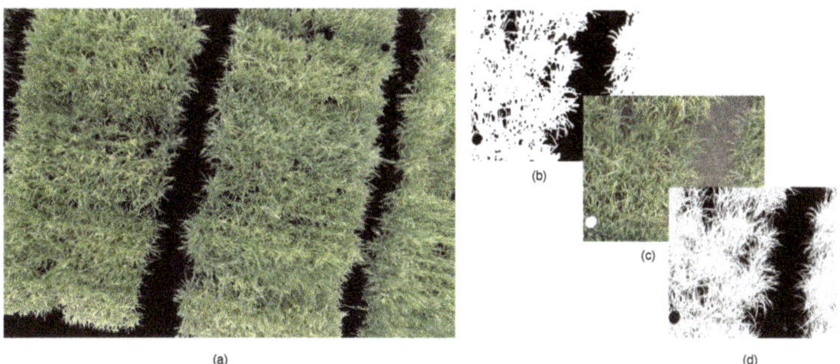

Figure 9. Final GF feathering filtering results for segmentation. (**a**) Example for an output image. (**b**) α image from GrabCut using the complete Algorithm 2, (**c**) original RGB image used as guidance, (**d**) refinement of image in plot (a) using the guidance from plot (b), $r = 60$, and $\epsilon = 0.001^2$.

Table 4. Acronyms for the segmentation methods tested.

Type	Segmentation Algorithm	Acronym
manual	HSV threshold	MCS
manual	Manual GC No refinement	GC
auto	kmeans (K = 2)	KM2
auto	meanshift (BW = 2)	MS2
auto	meanshift (BW = 4)	MS4
auto	meanshift (BW = 8)	MS8
auto	meanshift (BW = 16)	MS16
auto	GrabCut +Fixed T_B + GF	GCFauto

Table 5. Mean Results for precision, recall, and F1-score for 10 images.

Mean	MCS	GC	KM2	MS2	MS4	MS8	MS16	GCFauto
precision	0.959	0.980	0.774	0.802	0.803	0.826	0.773	0.965
Recall	0.964	0.978	0.743	0.811	0.809	0.852	0.777	0.983
F1-Score	0.967	0.981	0.768	0.791	0.791	0.838	0.791	0.978

3.3. Nitrogen Estimations

In the following, we present a comprehensive comparison among multi-variable linear regressions (MLR), support vector machines (SVM), and artificial neural networks (NN) for the estimation of the canopy N. All these models were trained using the ground-truth (cf. Table 2) containing the direct measurements of leaf chlorophyll based on SPAD readings. In this regard, the estimation results are all given in SPAD scale, in which the linear relationship with nitrogen is defined as: $N = 0.079(SPAD) - 0.154$, according to the work reported in [37].

3.3.1. MLR Models

Figure 10 shows the estimation results using the MLR. The samples-axis corresponds to the aerial imagery used for the estimation of nitrogen thought the phenological cycle. As previously mentioned, direct SPAD measurements were conducted for selected crop plots in order to assemble the ground-truth dataset. Given that, our system selects those aerial samples matching with the GPS coordinates of the ground measurements. Overall, the MLR achieved an average N correlations of 0.93 (V), 0.89 (V), and 0.82 (Ri). The data was filtered using a Moving Average Filter to reduce signal fluctuations and smooth the estimation result. Table 6 details the numerical data. Since the coefficients for the linear regressions were found and calibrated for each stage independently, we found strong linear relationships in the vegetative stage between the VIs and the N accumulation. Through the ripening stage, the yellow color becomes predominant and parcels cannot be distinguished with ease (as shown in Figure 2c). This changes in the canopy color and dense biomass accumulation tend to saturate the linear relationships between the VIs and the canopy reflectances.

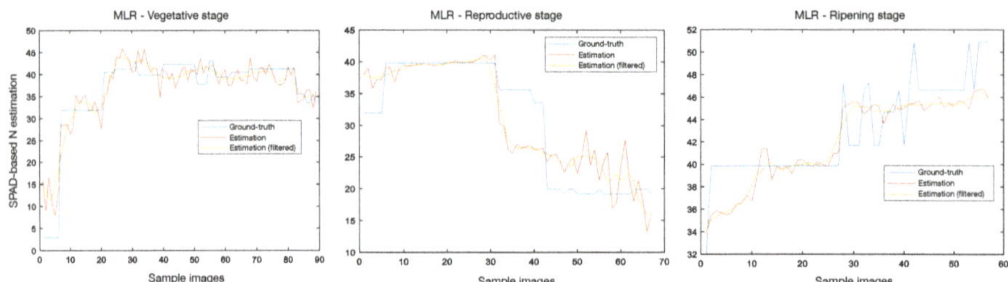

Figure 10. Multi-variable linear regressions (MLR). Numerical data reported in Table 6.

Table 6. Numerical data obtained from the multi-variable linear regressions (MLR).

Crop Stage	Correlation	RMSE
Vegetative	0.929	3.914
Reproductive	0.890	4.356
Ripening	0.822	2.318

3.3.2. SVM Models

We first tried to find a single SVM model for the entire crop phenological cycle; however, the canopy reflectances and the VIs changed drastically from vegetative through ripening. In this regard, an SVM model was defined for each crop stage independently, as shown in Table 7. In addition, Figure 11 details the estimation results for the N dynamics. Based on these results, the following SVM configurations are selected for our estimation system:

- Vegetative: Quadratic, correlation = 0.96, RMSE = 2.34, $R^2 = 0.927$, $\epsilon = 2$.
- Reproductive: Quadratic, correlation = 0.94, RMSE = 2.33, $R^2 = 0.892$, $\epsilon = 2$.
- Ripening: Quadratic, correlation = 0.87, RMSE = 2.04, $R^2 = 0.77$, $\epsilon = 0.5$.

3.3.3. NN Models

Neural networks (NN) with several hidden layers and optimization algorithms were tested. We highlight the results with two and five hidden layers. In Figure 12, the NN with two layers achieved a correlation = 0.964, RSME = 3.315, and $R^2 = 0.94$. Increasing up to five layers, the NN achieved a correlation = 0.986, RSME = 1.703, and $R^2 = 0.97$. Several training functions were tested for each crop stage, where the BFGS Quasi-Newton functions achieved accurate results for most of the vegetative stages, whereas the Levenberg–Marquardt function for both reproductive and ripening. The numerical data are reported in Table 8. Based on the results, the following NN configurations are selected for our estimation system:

- Vegetative: BFGS Quasi-Newton, correlation = 0.986, RMSE = 1.643, $R^2 = 0.972$.
- Reproductive: Levenberg-Marquardt, correlation = 0.944, RMSE = 3, $R^2 = 0.891$.
- Ripening: Levenberg-Marquardt, correlation = 0.890, RMSE = 1.835, $R^2 = 0.792$.

Table 7. Numerical data obtained from support vector machine (SVM) with several kernels.

Kernel	Crop Stage	RMSE	Correlation	R^2
Quadratic	Vegetative	2.34	0.96	0.927
	Reproductive	2.33	0.94	0.892
	Ripening	2.04	0.87	0.770
Cubic	Vegetative	3.75	0.90	0.827
	Reproductive	12.28	0.34	0.119
	Ripening	4.39	0.57	0.326
Linear	Vegetative	4.80	0.86	0.744
	Reproductive	5.88	0.81	0.656
	Ripening	2.38	0.83	0.689
Medium Gaussian	Vegetative	5.42	0.84	0.708
	Reproductive	3.03	0.92	0.887
	Ripening	2.35	0.84	0.713
Coarse Gaussian	Vegetative	5.87	0.79	0.634
	Reproductive	4.60	0.87	0.726
	Ripening	2.97	0.81	0.660
Fine Gaussian	Vegetative	6.07	0.76	0.579
	Reproductive	4.02	0.91	0.882
	Ripening	2.16	0.87	0.761

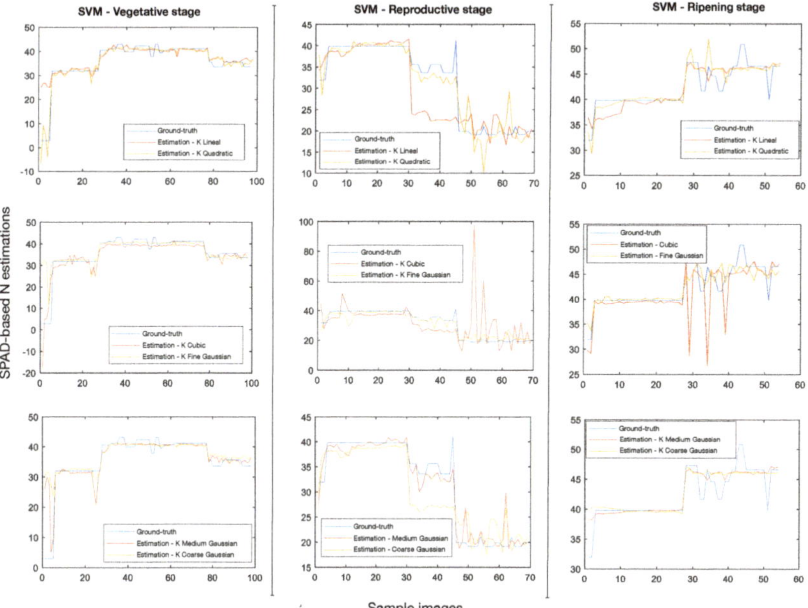

Figure 11. Support vector machine (SVM). On the left, we have several kernels used for the vegetative stage of the crop. Likewise, middle and right plots show the results for reproductive and ripening, respectively. These results were achieved with a margin of tolerance $\epsilon = 2$ for both vegetative and reproductive and $\epsilon = 0.5$ for ripening. Table 7 reports the numerical data.

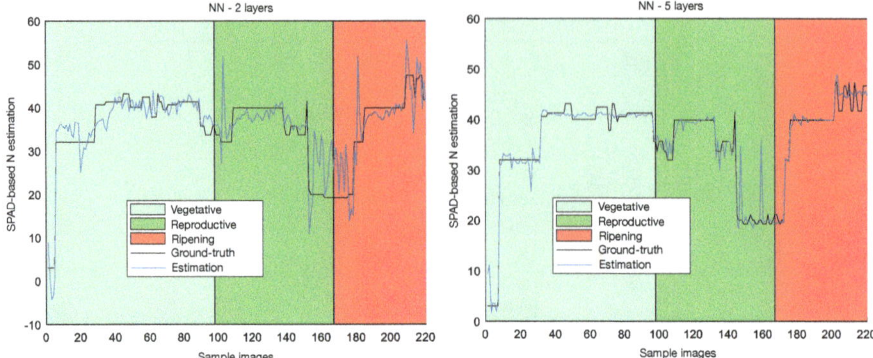

Figure 12. Artificial neural networks (NN). A 9:1 configuration was used for the two layer NN, whereas a 17:12:9:6:1 configuration for the five layer. In addition, several training functions were tested. Table 8 reports the numerical data.

Table 8. Numerical data obtained from the 5-layer NN with several training functions.

NN Training Function	Crop Stage	Correlation	RMSE	R^2
Bayesian regression	Vegetative	0.979	1.986	0.959
	Reproductive	0.921	3.779	0.848
	Ripening	0.67	4.405	0.462
BFGS Quasi-Newton	Vegetative	0.985	1.687	0.971
	Reproductive	0.938	3.203	0.880
	Ripening	0.851	2.149	0.724
Levenberg-Marquardt	Vegetative	0.983	1.855	0.966
	Reproductive	0.9442	3.007	0.891
	Ripening	0.890	1.835	0.792
Scaled Conjugate Gradient	Vegetative	0.952	2.952	0.907
	Reproductive	0.925	3.446	0.856
	Ripening	0.862	2.115	0.742

3.3.4. Machine-Learning Comparative Results

Figure 13 shows overall N estimation results obtained for each machine learning model. In general, we demonstrated strong correlations between the canopy N and the corresponding vegetative indices. In the case of MLR models, the coefficients for the regressions were determined and independently calibrated for each crop stage. We encountered that the N dynamics have strong linear dependencies with MSAVI, GNDVI, and NDVI; concretely, during vegetative and reproductive stages. Table 9 reports the overall numerical data.

From the ROC curve reported in Figure 13d, the accuracy (ACC) was calculated for each ML model. Neural networks achieved an average ACC = 0.85, improving the estimations of canopy N over the other ML models. In fact, by comparing the N-to-SPAD correlations reported on Table 9, the correlation metric was improved over each crop stage. Table 9 compares the mean N estimations achieved by each ML method against the mean SPAD-based N readings measured at ground level. Mean results are presented for each crop stage: vegetative (V), reproductive (R), and ripening (Ri). The last columns in Table 9 report the mean linear correlations between estimations and measurements.

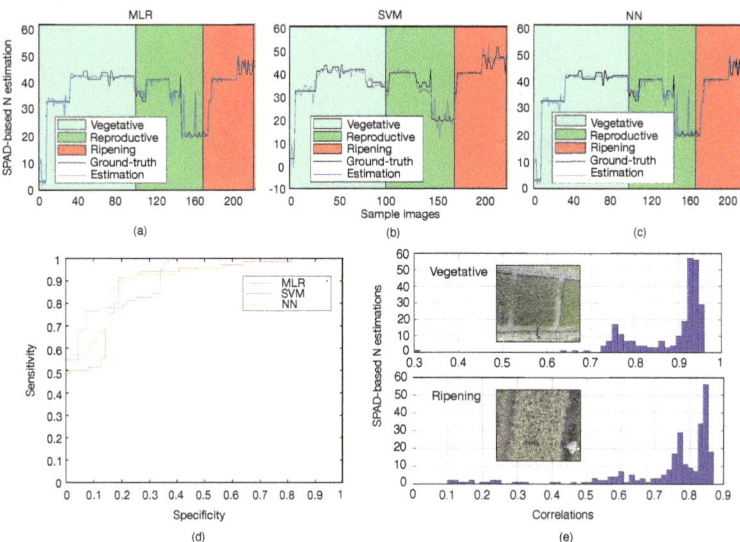

Figure 13. Overall N estimation results. (**a–c**) show average data for the estimated N dynamics from vegetative to ripening stages (between 86–101 days of phenological cycle). We compared the results obtained from MLR (linear regressions), support vector machines (SVM), and neural networks (NN). (**d**) ROC curve obtained for the three ML methods evaluated: ACC = 0.82 for MLR, ACC = 0.78 for SVM, and ACC = 0.85 for NN. (**e**) Histogram of N correlations during crop growth from the initial vegetative stage until ripening. Table 9 reports the numerical data.

Table 9. Overall numerical results for canopy N estimations.

	N Estimations			N SPAD Measurements			Correlation		
Methods	**V**	**R**	**Ri**	**V**	**R**	**Ri**	**V**	**R**	**Ri**
MLR	36.923	30.766	42.669				0.935	0.890	0.82
SVM	35.7472	29.2902	42.5226	35.7906	29.6338	42.4207	0.9699	0.9467	0.8770
NN	36.173	30.495	43.2648				0.986	0.9442	0.890

By comparing the neural networks (NN) against the multi-variable linear regressions (MLR) and support vector machines (SVM), the former estimator achieves higher correlations partly due to the reliability and size of the assembled ground-truth database for training, i.e., reliable N datasets (direct N-leaf measurements) with sufficient variations of the crops during the growth stage (see the experimental protocol in Section 2). In addition, some VIs, such as the simple ratio (SR), tend to saturate as long the the crop grows, but other vegetative indices that behave better with higher biomass and N concentration compensate the effect of the saturated ones. The nonlinear combinations of these VIs enable the NN to achieve accurate estimations.

Furthermore, we tested the effects of the UAV control architecture proposed in Section 2.3, for improving the N estimations by means of precise position tracking during flight. Figure 14 presents the results. Plot (b) shows the X-Y tracking trajectory for each flight control scenario. The UAV was flying at a constant altitude reference of 20 m over the crop at a maximum speed of 1.5 m/s. The black boxes represent the crop plots with ground-level markers where direct N measurements were taken via SPAD. The green dotted circles show the points where the UAV stops and hovers during 4 s to capture canopy imagery.

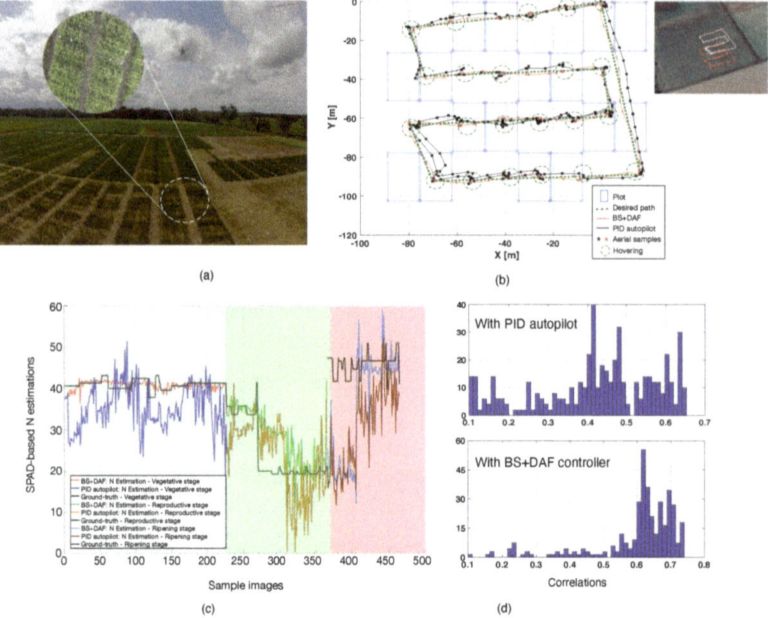

Figure 14. (Experimental) Effects of UAV navigation on N estimations. (**a**) UAV over the crop fields. The white markers placed at ground-level indicate the points for SPAD N measurements. (**b**) UAV X-Y trajectory tracking. Comparison between the proposed BS+DAF attitude controller and PID-based autopilot. The desired path was configured with hovering waypoints to capture canopy imagery. The star-like markers correspond to GPS data of the taken images. (**c**) N estimation results. (**d**) Histograms of N-to-SPAD correlations.

The paths followed by the UAV are shown for both controllers; one driven by the PID that comes standard with the UAV autopilot, and the other driven by the proposed BS+DAF. Likewise, the star markers represent the GPS points of the images taken while hovering (black color for the PID and red for the BS + DAF). As observed, the attitude stabilization of the UAV seems to be more affected with the PID, since the aerial samples within each hovering area (black star markers) are spread over different positions. Instead, the aerial samples obtained with the BS + DAF (red star markers) tend to be more concentrated within the crop plot of interest. The goal is actually capturing most images in the GPS coordinates that accurately matches with the GPS position of the white markers at ground-level, as depicted in Figure 14a.

Finally, Figure 14c presents the N estimation results achieved under both control flight scenarios. Larger fluctuations, noise, and estimation errors are presented when the ML methods are trained and tested with the imagery dataset captured by the PID-driven UAV. This is mainly due to the imprecise X-Y positioning of the aerial vehicle during hovering (caused by improper attitude stabilization), in which several images capture crop areas that do not necessarily match with the ground-level markers, as detailed in Figure 14b. On the other hand, the proposed BS+DAF controller demonstrated being suitable for the proper regulation of the angular motions of the UAV by compensating external disturbances faster and achieving accurate X-Y positioning. As a result, The N estimations correlate higher with the ground-truth. As shown in the histograms in Figure 14d, the N-to-SPAD correlations grouped higher for the experiments driven by the proposed BS+DAF controller.

4. Discussion

The results presented in this work demonstrate a significant improvement in leaf nitrogen prediction in rice crops from images obtained via UAVs, validated with high SPAD correlations. These results are thanks to a combination of technological contributions, namely: (i) a novel UAV stabilization control scheme named BS+DAF that enables precise multispectral aerial image acquisition and registration with ground-truth fiducials, even in the presence of strong wind perturbations, (ii) an automated GrabCut image segmentation method, which leads to finer detail of the plant's canopy in the RGN space, as detailed in Figures 5b and 9d, and (iii) the successful application of machine learning methods trained with the vegetation indices (VI) extracted from the segmented multispectral imagery. Figure 5c confirms the increased reliability and accuracy per VI by calculating the data variance at each crop stage.

In a similar manner, Table 9 summarizes the higher nitrogen-to-SPAD correlations achieved with the implemented neural networks, and confirms the significant nonlinear relationships between leaf spectral reflectances (VI) and chlorophyll-based estimated N concentrations. These N estimation results are comparable to other state-of-the-art results presented in the scientific literature. In [52], Figure 7b shows N estimations in rice using linear regression models, for which the authors report: $RMSE = 0.212$, correlation $r = 0.89$, and $R^2 = 0.803$. Our results (tabulated in Tables 8 and 9) reach mean correlations for the vegetative stage ($r = 0.986$), reproductive ($r = 0.9442$), and ripening stage ($r = 0.89$). In [53], Table 6 and Figure 5 present N-status estimations in rice using vegetation indices calculated with aerial UAV imagery. Authors compared several regression models, with a highest R^2 of 0.74 using the LDM method. In our case, Table 8 lists higher R^2 values from the Levenberg–Marquardt method.

Our approach contributes novel solutions to commercial UAV-based phenotyping technologies, particularly to image-based remote sensing applications that adopt photogrammetric geometric image post-processing methods for image correction, and enables crop/plot analysis in real-time.

One important drawback of our solution was the estimation of canopy N for crops with higher biomass. Counter to the expectation of finding a higher N correlation as a function of crop growth, the correlation results decreased in the ripening stage, as depicted in Figure 13e and from the numerical data presented in Table 9. We attribute this to the disproportionately high number of senescent leaves (yellow color with a bandwidth of 570–590 nm) that do not properly match with the VI dominant wavelengths (see Table 3). Consequently, we suggest introducing plant senescence reflectance as an added variable for the discovery of novel vegetative indices to enhance our system.

5. Conclusions

This paper presents an integrated UAV system with non-invasive image-based methods for the estimation and monitoring of the N dynamics in rice crops. Several challenges were addressed, associated with image segmentation and UAV navigation control, to achieve reliable machine learning training and highly correlated results to SPAD. The proposed BS+DAF attitude controller led to precise UAV way-point tracking, with position errors below 2%. This was key to capture aerial multispectral imagery during hovering that accurately matched with the GPS positions of the SPAD measurements at ground-level. This reduces the need for computationally expensive photogrammetry methods.

Higher N correlations were achieved with neural networks: 0.98 in the vegetative stage, 0.94 in reproductive, and 0.89 in ripening, with an ROC accuracy of 0.85. This is a promising result towards the autonomous estimation of rice canopy nitrogen in real-time. With the advent of small AI-dedicated systems on chip (SoC) and the powerful computing capabilities of our UAV system, we expect upcoming work to achieve real-time computation of the machine learning methods presented in this work.

Several challenges still remain for improving the N estimations, especially when the crop biomass is high. Rice crops commonly have several mixed varieties per small plot areas. Therefore, besides the assessment of the N dynamics as a function of the crop stages, it is also crucial to identify and classify the plants according to their genotype, given that nitrogen is highly sensitive to plant oxygen consumption, root growth, among other variables. To this purpose, the aerial identification of different plant varieties requires sensing spatial resolutions below 30 cm. In addition, high-dimensional data clustering is needed for the extraction of relevant features according to the plant variety.

Supplementary Materials: The following are available online at http://www.mdpi.com/2072-4292/12/20/3396/s1, FILE S1: RAW data supporting image segmentation metrics, multispectral imagery imagery used for machine learning testing, and nitrogen estimation results are available at the Open Science Framework: https://osf.io/cde6h/?view_only=1c4e5e03b9a34d3b96736ad8ab1b2774. FILE S2: Experimental protocol for crop monitoring available at https://www.protocols.io/view/protocol-bjxskpne. VIDEO S3: The video is available in the online version of this article. The video accompanying this paper illustrates the steps performed during the experiments.

Author Contributions: Conceptualization, J.D.C., N.C.-B., F.C., and J.S.C.; methodology, J.D.C., M.C.R., F.C., and A.J.-B.; software, N.C.-B., J.S.C., and D.C.; validation, J.D.C., I.F.M., E.P., D.C., N.C.-B., and J.S.C.; formal analysis and investigation, J.D.C., M.C.R., E.P., F.C., I.F.M., and A.J.-B.; data curation, E.P.; writing—original draft preparation, J.D.C. and F.C.; writing—review and editing, A.J.-B., I.F.M., and M.C.R.; supervision, A.J.-B. and J.D.C. All authors have read and agreed to the published version of the manuscript.

Funding: This work was funded by the OMICAS program: *Optimización Multiescala In-silico de Cultivos Agrícolas Sostenibles (Infraestructura y validación en Arroz y Caña de Azúcar)*, anchored at the Pontificia Universidad Javeriana in Cali and funded within the Colombian Scientific Ecosystem by The World Bank, the Colombian Ministry of Science, Technology and Innovation, the Colombian Ministry of Education and the Colombian Ministry of Industry and Turism, and ICETEX under GRANT ID: FP44842-217-2018.

Acknowledgments: The authors would like to thank all CIAT staff that supported the experiments over the crops located at CIAT headquarters in Palmira, Valle del Cauca, Colombia; in particular to Yolima Ospina and Cecile Grenier for their support in upland and lowland trials. Also, to Carlos Devia from Javeriana University for the insights regarding the structuring of the datasets.

Conflicts of Interest: The authors declare no conflict of interest.

Appendix A. UAV Equations of Motion

Equations of motion (EoM) that describe the inertial and aerodynamics effects are introduced, including a six-dimensional operator describing the spatial accelerations of the body frame as a function of the UAV inertias, moments, and aerodynamics. Second, we introduce a nonlinear MPC-based controller to regulate the stabilization of the UAV, specifically both pitch θ and roll ϕ angles, under external force perturbations.

Our UAV is a Vertical Takeoff and Landing (VTOL) four-rotor drone. As shown in Figure A1a, the body frame $\{b\}$ is a six degree of freedom rigid body. Rotations about the body-frame axes are designated by the Euler angles: roll (ϕ), pitch (θ), and yaw (ψ) following standard aerodynamic conventions. In this sense, the 6D spatial acceleration $\dot{V}_b \in \Re^{6x1}$ of the body frame can be written as:

$$\dot{V}_b = \begin{bmatrix} \dot{\omega}_b \\ \dot{v}_b \end{bmatrix} = \begin{bmatrix} \ddot{\phi} \\ \ddot{\theta} \\ \ddot{\psi} \\ \ddot{x}_b \\ \ddot{y}_b \\ \ddot{z}_b \end{bmatrix} \tag{A1}$$

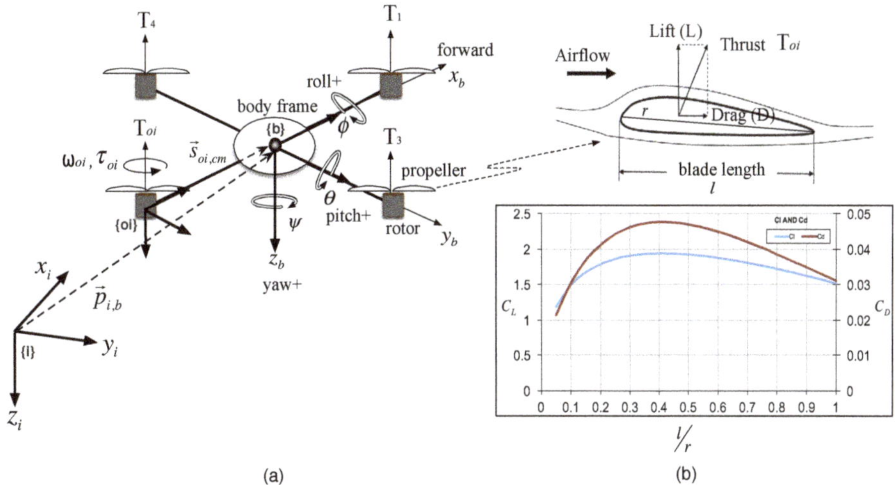

Figure A1. Physical parameters used for UAV modeling and control. (**a**) Equations of motion are derived using three frames of reference: the inertial frame $\{i\}$, the body frame $\{b\}$ located at the center of mass, and the rotors frame $\{oi\}, \forall oi : 1...4$. Each rotor generates a vertical thrust (T_{oi}) that depends on the rotor angular speed (ω_{oi}) and the geometrical properties of the propeller blades. (**b**) Blade properties for deriving aerodynamics equations. We used Blade-Element-Theory computation to calculate both lift and drag coefficients using the FoilSim III simulator provided by NASA [54]. Our UAV has a lift coefficient of $C_L = 1.6$ and a drag coefficient of $C_D = 0.042$, since $l/r = 0.7$.

Both rotational $\dot{\omega}$ and translational $\dot{v}$ accelerations could be derived from the Newton–Euler formulation, as:

$$\dot{V}_b = I_b^{-1}[F_b - \dot{I}_b V_b], \tag{A2}$$

being $I_b \in \Re^{6x6}$ the spatial inertia operator calculated at the Center of Mass (CM) of the body frame $\{b\}$. It can be expressed as:

$$I_b = \begin{bmatrix} J_b & 0 \\ 0 & mU \end{bmatrix} = \begin{bmatrix} I_{xx} & -I_{xy} & -I_{xz} & 0 & 0 & 0 \\ -I_{xy} & I_{yy} & -I_{yz} & 0 & 0 & 0 \\ -I_{xz} & -I_{yz} & I_{zz} & 0 & 0 & 0 \\ 0 & 0 & 0 & m & 0 & 0 \\ 0 & 0 & 0 & 0 & m & 0 \\ 0 & 0 & 0 & 0 & 0 & m \end{bmatrix}, \tag{A3}$$

where $J_b \in \Re^{3x3}$ is the inertial tensor with $diag(I_{xx}, I_{yy}, I_{zz})$ being the moments of inertia, m is the mass of the UAV, and U is a 3×3 identity operator. Likewise, the term $F_b \in \Re^{6x1}$ in Equation (A2) is the 6D spatial force acting on the CM of $\{b\}$. F_b contains the effects caused by both inertial (N_b) and aerodynamics (T_b) forces acting on the body frame:

$$F_b = \begin{bmatrix} N_b \\ f_b \end{bmatrix} = \begin{bmatrix} (N_{b,x}) + (\tau_\phi) \\ (N_{b,y}) + (\tau_\theta) \\ (N_{b,z}) + (\tau_\psi) \\ f_{b,x} \\ f_{b,y} \\ f_{b,z} \end{bmatrix} = \begin{bmatrix} (\dot{\theta}\dot{\psi}\,[I_{yy} - I_{zz}]) + s_{oi,b}\hat{j}(T_4 - T_3) \\ (\dot{\phi}\dot{\psi}\,[I_{zz} - I_{xx}]) + s_{oi,b}\hat{i}(T_1 - T_2) \\ (\dot{\theta}\dot{\phi}\,[I_{xx} - I_{yy}]) + (T_3 + T_4 - T_1 - T_2) \\ (s\psi s\phi + c\psi s\theta c\phi)\,T_b \\ (-c\psi s\phi + s\psi s\theta c\phi)\,T_b \\ mg - (c\psi c\phi)T_b \end{bmatrix} \tag{A4}$$

In Equation (A4), we have determined an expression that incorporates the thrust produced by each independent rotor (T_{oi}) $\forall oi : 1...4$. These aerodynamic terms govern the generation of rolling (τ_ϕ), pitching (τ_θ) and yawing (τ_ψ) torques at the CM of the UAV, where the term $s_{oi,b} = 0.18m$ is the distance between each rotor to the body frame (see Figure A1a). In addition, T_{oi} depends on the lift (L) and drag (D) forces acting on each propeller, as shown in Figure A1b. It can be written as:

$$\begin{aligned} T_{oi} &= L + D \\ &= \tfrac{1}{2}\rho_{air}\omega_{oi}^2 A_{prop}\left(C_L + C_D\right), \end{aligned} \tag{A5}$$

where $\rho_{air} = 1.20$ Kgm^3 is the density of air, ω_{oi}, $\forall oi : 1...4$ is the rotor speed, $A_{prop} = 0.013$ m^2 is the propeller transversal area, C_L is the lift coefficient, and C_D is the drag coefficient. As shown in Figure A1b, we have estimated both values as $C_L = 1.6$ and $C_D = 0.042$, respectively. In this sense, the net vertical thrust (T_b) generated at the CM of the UAV can be calculated as:

$$T_b = \sum_{oi=1}^{4} T_{oi} \tag{A6}$$

As observed in Equation (A4), T_b governs the generation of the linear forces. The expressions $s\psi$, $c\psi$ denote $sin(\psi)$ and $cos(\psi)$, respectively. Finally, the term $m = 0.43$ Kg is the mass of the UAV and $g = 9.81$ ms^{-2} is the gravitational acceleration. In the forthcoming section, we will derive the control strategy to regulate the angular motions precisely. Since our control approach will depend on the UAV model, we introduce the computational steps to calculate the EoM in Algorithm A1.

Algorithm A1: EoM Computation.

Step 1: Aerodynamic forces
Read the rotors speed from encoders: ω_{oi}, $\forall oi : 1...4$
Calculate both lift and drag forces acting on each propeller:
$L \leftarrow \frac{1}{2}C_L\rho_{air}\omega_{oi}^2 A_{prop}$,
$D \leftarrow \frac{1}{2}C_D\rho_{air}\omega_{oi}^2 A_{prop}$
Calculate the thrust produced by each rotor: $T_{oi} = L + D\ \forall oi : 1...4$
Calculate net thrust produced at CM: $T_b = \sum_{oi=1}^{4} T_{oi}$
Rotational forces (rolling, pitching, and yawing) torques onto the body frame:
$\tau_\phi \leftarrow s_{oi,cm}\hat{j}\,(T_4 - T_3)$
$\tau_\theta \leftarrow s_{oi,cm}\hat{i}\,(T_1 - T_2)$
$\tau_\psi \leftarrow \tau_3 + \tau_4 - \tau_1 - \tau_2$
Linear forces acting onto the body frame:
$f_{b,x} \leftarrow (s\psi s\phi + c\psi s\theta c\phi)\,T_b$
$f_{b,y} \leftarrow (-c\psi s\phi + s\psi s\theta c\phi)\,T_b$
$f_{b,z} \leftarrow (-c\psi c\phi)T_b$
6D Aerodynamic Forces: $[\tau_\phi\ \ \tau_\theta\ \ \tau_\psi\ \ f_{b,x}\ \ f_{b,y}\ \ f_{b,z}]^T$
Step 2: Inertial forces
Calculate 6D inertial operator: $I_b = \begin{bmatrix} J_b & 0 \\ 0 & mU \end{bmatrix}$
Calculate inertial terms:
$N_{b,x} \leftarrow \dot{\theta}\dot{\psi}\left[I_{yy} - I_{zz}\right]$
$N_{b,y} \leftarrow \dot{\phi}\dot{\psi}\left[I_{zz} - I_{xx}\right]$
$N_{b,z} \leftarrow \dot{\theta}\dot{\phi}\left[I_{xx} - I_{yy}\right]$
$f_{b,z} \leftarrow mg - cos(\psi)cos(\phi)T_b$
Calculate 6D Forces: $F_b \leftarrow [N_{b,x} + \tau_\phi\ \ N_{b,y} + \tau_\theta\ \ N_{b,z} + \tau_\psi\ \ f_{b,x}\ \ f_{b,y}\ \ f_{b,z}]^T$
Step 3: 6D Equations of Motion (EoM)
$\dot{V}_b \leftarrow I_b^{-1}[F_b - \dot{I}_b V_b]$
Return $\dot{V}_b$

Appendix B. Backstepping + DAF Control Derivation

The proposed control law has to be sensitive to small changes in both angular motions, therefore, an error dynamics could be defined as a function of the angular rates, as:

$$\begin{aligned} \dot{e}_\phi &= \dot{\phi}^d - \omega_x, \\ \dot{e}_\theta &= \dot{\theta}^d - \omega_y \end{aligned} \tag{A7}$$

In Equation (A7), both ω_x and ω_y are measured by the IMU sensor onboard the UAV. The goal is to obtain a desired angular acceleration terms within the control law to account for small angular rate changes. These terms are called desired acceleration function (DAF):

$$\begin{aligned} \ddot{\phi}^d &= f(\phi, \dot{\phi}, F_b) \\ &= \begin{bmatrix} 1 & 0 & 0 & 0 & 0 & 0 \end{bmatrix} \dot{V}_b, \\ \ddot{\theta}^d &= f(\theta, \dot{\theta}, F_b) \\ &= \begin{bmatrix} 0 & 1 & 0 & 0 & 0 & 0 \end{bmatrix} \dot{V}_b \end{aligned} \tag{A8}$$

Both DAF terms $\ddot{\phi}^d$ and $\ddot{\theta}^d$ are extracted from the spatial acceleration $\dot{V}_b \in \Re^{6x1}$ computed in Algorithm A1. To make explicit the DAF terms from Equation (A8) within the backstepping, in the following we focus on deriving the control law for roll (u_ϕ).

From Equation (A6), we introduce a virtual control law that governs the error dynamics, yielding a second tracking error $e_2 = \omega_x^d - \omega_x$ where $\dot{\phi}^d \rightarrow \omega_x^d$. In this sense, a proportional-derivative-integral structure is defined for the virtual control law ω_x^d, as:

$$\omega_x^d = k_p e_\phi + \dot{\phi}^d + k_i \int e_\phi(t)dt \ , \tag{A9}$$

where K_p and K_i must be positive constants, since the BS requires a positive definite Lyapunov function (L) for stabilizing the tracking error: $L\left(e_\phi\right) = \frac{e_\phi^2}{2}$. Now, replacing Equation (A9) into the virtual control error e_2 yields:

$$e_2 = \omega_x^d - \omega_x = k_p e_\phi + \dot{\phi}^d + k_i \int e_\phi(t)dt - \omega_x \tag{A10}$$

By following the same approach from Equation (A7), the error dynamics for $\dot{e}_2$ is determined as:

$$\dot{e}_2 = k_p \dot{e}_\phi + \ddot{\phi}^d + k_i e_\phi - \dot{\omega}_x \tag{A11}$$

In Equation (A11), we have derived $\ddot{\phi}^d$ corresponding to the DAF term for roll (see Equation (A8)). In this regard, the computation of Algorithm A1 is required to close the attitude loop. Now, the control action u_ϕ is determined as:

$$\begin{aligned} u_\phi &\rightarrow \tau_\phi = I_{xx}\ddot{\phi}, \\ \ddot{\phi} &\rightarrow \dot{\omega}_x = I_{xx}^{-1}\tau_\phi \end{aligned} \tag{A12}$$

Replacing $\dot{\omega}_x$ from Equation (A12) in (A11) and isolating the control action, yields:

$$u_\phi = I_{xx}[k_p \dot{e}_\phi + \ddot{\phi}^d + k_i e_\phi - \dot{e}_2] \tag{A13}$$

Since the calculation of $\dot{e}_\phi$ and $\dot{e}_2$ for the real system would introduce accumulative numerical errors, we need to rewrite the control law to be dependent of the tracking errors: e_ϕ and e_2. By isolating $\omega_x = \omega_x^d - e_2$ from Equation (A10) and replacing it into Equation (A7):

$$\dot{e}_\phi = \dot{\phi}^d - (\omega_x^d - e_2) \tag{A14}$$

Replacing Equation (A14) into (A13):

$$u_\phi = I_{xx}[k_p(\dot{\phi}^d - \omega_x^d + e_2) + \ddot{\phi}^d + k_i e_\phi - \dot{e}_2], \tag{A15}$$

and replacing ω_x^d from Equation (A9):

$$\begin{aligned} u_\phi &= I_{xx}[k_p(\dot{\phi}^d - (k_p e_\phi + \dot{\phi}^d + k_i \textstyle\int e_\phi) + e_2) + \ddot{\phi}^d + k_i e_\phi - \dot{e}_2] \\ &= I_{xx}[k_p(e_2 - k_p e_\phi - k_i \textstyle\int e_\phi) + \ddot{\phi}^d + k_i e_\phi - \dot{e}_2] \end{aligned} \tag{A16}$$

Finally, the expression for $\dot{e}_2$ can be rewritten to follow the same form of $\dot{e}_\phi$ in Equation (A14). Likewise, ω_x^d is also replaced by Equation (A9):

$$\begin{aligned} \dot{e}_2 &= \dot{\phi}^d - \omega_x^d + e_2 \\ &= \dot{\phi}^d - (k_p e_\phi + \dot{\phi}^d + k_i \textstyle\int e_\phi) + e_2 \\ &= e_2 - k_p e_\phi - \cancelto{0}{k_i \textstyle\int e_\phi} \end{aligned} \tag{A17}$$

In Equation (A17), the integration of the error can be eliminated since the control law in Equation (A16) already ensures zero steady-state error for e_ϕ. Replacing $\dot{e}_2 = e_2 - k_p e_\phi$ in Equation (A16):

$$\begin{aligned} u_\phi &= I_{xx}[k_p(e_2 - k_p e_\phi - k_i \textstyle\int e_\phi) + \ddot{\phi}^d + k_i e_\phi - e_2 + k_p e_\phi] \\ &= I_{xx}[e_\phi(k_p - k_i - k_p^2) + e_2(k_p - 1) - k_p k_i \textstyle\int e_\phi + \ddot{\phi}^d] \end{aligned} \tag{A18}$$

Equation (A18) presents the control law to regulate ϕ. This controller allows zero steady-state error for roll via $\int e_\phi$, it is sensitive to small variations in roll rate via e_2, and it directly depends on the UAV model via the DAF term $\ddot{\phi}^d$ (Algorithm A1). By following the same structure in Equation (A18), the control law to regulate the pitch angular motion (θ) is:

$$u_\theta = I_{yy}[e_\theta(k_{p,2} - k_{i,2} - k_{p,2}^2) + e_2(k_{p,2} - 1) - k_{p,2}k_{i,2} \textstyle\int e_\theta + \ddot{\theta}^d] \tag{A19}$$

References

1. Zhang, K.; Liu, X.; Tahir Ata-Ul-Karim, S.; Lu, J.; Krienke, B.; Li, S.; Cao, Q.; Zhu, Y.; Cao, W.; Tian, Y. Development of Chlorophyll-Meter-Index-Based Dynamic Models for Evaluation of High-Yield Japonica Rice Production in Yangtze River Reaches. *Agronomy* **2019**, *9*, 106. [CrossRef]
2. Yang, H.; Yang, J.; Lv, Y.; He, J. SPAD Values and Nitrogen Nutrition Index for the Evaluation of Rice Nitrogen Status. *Plant Prod. Sci.* **2014**, *17*, 81–92. [CrossRef]
3. Dong, Z.; Wu, L.; Chai, J.; Zhu, Y.; Chen, Y.; Zhu, Y. Effects of Nitrogen Application Rates on Rice Grain Yield, Nitrogen-Use Efficiency, and Water Quality in Paddy Field. *Commun. Soil Sci. Plant Anal.* **2015**, *46*, 1579–1594. [CrossRef]
4. Xu, G.; Fan, X.; Miller, A.J. Plant nitrogen assimilation and use efficiency. *Annu. Rev. Plant Biol.* **2012**, *63*, 153–182. [CrossRef] [PubMed]
5. Nigon, T.J.; Yang, C.; Dias Paiao, G.; Mulla, D.J.; Knight, J.F.; Fernández, F.G. Prediction of Early Season Nitrogen Uptake in Maize Using High-Resolution Aerial Hyperspectral Imagery. *Remote Sens.* **2020**, *12*, 1234. [CrossRef]
6. Wan L.; Li, Y.; Cen, H.; Zhu, J.; Yin, W.; Wu, W.; Zhu, H.; Sun, D.; Zhou, W.; He, Y. Combining UAV-Based Vegetation Indices and Image Classification to Estimate Flower Number in Oilseed Rape. *Remote Sens.* **2018**, *10*, 1484. doi:10.3390/rs10091484. [CrossRef]
7. Zhang, D.; Zhou, X.; Zhang, J.; Lan, Y.; Xu, C.; Liang, D. Detection of rice sheath blight using an unmanned aerial system with high-resolution color and multispectral imaging. *PLoS ONE* **2018**, *13*. [CrossRef] [PubMed]

8. Liu, Y.; Cheng, T.; Zhu, Y.; Tian, Y.; Cao, W.; Yao, X.; Wang, N. Comparative analysis of vegetation indices, non-parametric and physical retrieval methods for monitoring nitrogen in wheat using UAV-based multispectral imagery, In Proceedings of the 2016 IEEE International Geoscience and Remote Sensing Symposium (IGARSS), Beijing, China, 10–15 July 2016; pp. 7362–7365. [CrossRef]
9. Guan, S.; Fukami, K.; Matsunaka, H.; Okami, M.; Tanaka, R.; Nakano, H.; Sakai, T.; Nakano, K.; Ohdan, H.; Takahashi, K. Assessing Correlation of High-Resolution NDVI with Fertilizer Application Level and Yield of Rice and Wheat Crops Using Small UAVs. *Remote Sens.* **2019**, *11*, 112. [CrossRef]
10. Mishra, P.; Asaari, M.S.M.; Herrero-Langreo, A.; Lohumi, S.; Diezma, B.; Scheunders, P. Close range hyperspectral imaging of plants: A review. *Biosyst. Eng.* **2017**, *164*, 49–67. [CrossRef]
11. Bergstrasser, S.; Fanourakis, D.; Schmittgen, S.; Cendrero-Mateo, M.P.; Jansen, M.; Rascher, H.S.U. HyperART: Non-invasive quantification of leaf traits using hyperspectral absorption-reflectance-transmittance imaging. *Plant Methods* **2015**, *11*, 1. [CrossRef]
12. Pandey, P.; Ge, Y.; Stoerger, V.; Schnable, J.C. High Throughput In vivo Analysis of Plant Leaf Chemical Properties Using Hyperspectral Imaging. *Front. Plant Sci.* **2017**, *8*, 1348. [CrossRef] [PubMed]
13. Sun, D.; Cen, H.; Weng, H.; Wan, L.; Abdalla, A.; El-Manawy, A.I.; Zhu, Y.; Zhao, N.; Fu, H.; Tang, J.; et al. Using hyperspectral analysis as a potential high throughput phenotyping tool in GWAS for protein content of rice quality. *Plant Methods* **2019**, *15*, 54. [CrossRef] [PubMed]
14. Lu, J.; Yang, T.; Su, X.; Qi, H.; Yao, X.; Cheng, T.; Zhu, Y.; Cao, W.; Tian, Y. Monitoring leaf potassium content using hyperspectral vegetation indices in rice leaves. *Precis. Agric.* **2020**, *21*, 324–348. [CrossRef]
15. Wang, W.; Yao, X.; Tian, Y.-C.; Liu, X.-J.; Ni, J.; Cao, W.-X.; Zhu, Y. Common Spectral Bands and Optimum Vegetation Indices for Monitoring Leaf Nitrogen Accumulation in Rice and Wheat. *J. Integr. Agric.* **2012**, *11*, 2001–2012. [CrossRef]
16. Sun, J.; Yang, J.; Shi, S.; Chen, B.; Du, L.; Gong, W.; Song, S. Estimating Rice Leaf Nitrogen Concentration: Influence of Regression Algorithms Based on Passive and Active Leaf Reflectance. *Remote Sens.* **2017**, *9*, 951. [CrossRef]
17. Singh, A.K.; Ganapathysubramanian, B.; Sarkar, S.; Singh, A. Deep Learning for Plant Stress Phenotyping: Trends and Future Perspectives. *Trends Plant Sci.* **2018**, *23*, 883–898. [CrossRef]
18. Xiong, X.; Duan, L.; Liu, L.; Tu, H.; Yang, P.; Wu, D.; Chen, G.; Xiong, L.; Yang, W.; Liu, Q. Panicle-SEG: A robust image segmentation method for rice panicles in the field based on deep learning and superpixel optimization. *Plant Methods* **2017**, *13*, 104. [CrossRef]
19. Ubbens, J.R.; Stavness, I. Deep Plant Phenomics: A Deep Learning Platform for Complex Plant Phenotyping Tasks. *Front. Plant Sci.* **2017**, *8*, 1190. [CrossRef]
20. Ghosal, S.; Blystone, D.; Singh, A.K.; Ganapathysubramanian, B.; Singh, A.; Sarkar, S. An explainable deep machine vision framework for plant stress phenotyping. *Proc. Natl. Acad. Sci. USA* **2018**, *115*, 4613–4618. [CrossRef]
21. Tilly, N.; Aasen, H.; Bareth, G. Fusion of Plant Height and Vegetation Indices for the Estimation of Barley Biomass. *Remote Sens.* **2015**, *7*, 11449–11480. [CrossRef]
22. Wang, C.; Nie, S.; Xi, X.; Luo, S.; Sun, X. Estimating the Biomass of Maize with Hyperspectral and LiDAR Data. *Remote Sens.* **2016**, *9*, 11. [CrossRef]
23. Wang, Y.; Zhang, Z.; Feng, L.; Du, Q.; Runge, T. Combining Multi-Source Data and Machine Learning Approaches to Predict Winter Wheat Yield in the Conterminous United States. *Remote Sens.* **2020**, *12*, 1232. [CrossRef]
24. Yang, Z.; Shen, D.; Yap, P.T. Image mosaicking using SURF features of line segments. *PLoS ONE* **2017**, *12*. [CrossRef] [PubMed]
25. Caruso, G.; Zarco-Tejada, P.J.; González-Dugo, V.; Moriondo, M.; Tozzini, L.; Palai, G.; Rallo, G.; Hornero, A.; Primicerio, J.; Gucci, R. High-resolution imagery acquired from an unmanned platform to estimate biophysical and geometrical parameters of olive trees under different irrigation regimes. *PLoS ONE* **2019**, *14*, e0210804. [CrossRef] [PubMed]
26. Bendig, J.; Bolten, A.; Bareth, G. UAV-based Imaging for Multi-Temporal, very high Resolution Crop Surface Models to monitor Crop Growth Variability. *Photogramm. Fernerkund. Geoinf.* **2013**, *2013*, 551–562. [CrossRef]
27. Zhou, X.; Zheng, H.B.; Xu, X.Q.; He, J.Y.; Ge, X.K.; Yao, X.; Cheng, T.; Zhu, Y.; Cao, W.X.; Tian, Y.C. Predicting grain yield in rice using multi-temporal vegetation indices from UAV-based multispectral and digital imagery. *ISPRS J. Photogramm. Remote Sens.* **2017**, *130*, 246–355. [CrossRef]

28. Han, Y.; Qin, R.; Huang, X. Assessment of dense image matchers for digital surface model generation using airborne and spaceborne images—An update. *Photogram. Rec.* **2020**, *35*, 58–80. [CrossRef]
29. Comba, L.; Biglia, A.; Aimonino, D.R.; Gay, P. Unsupervised detection of vineyards by 3D point-cloud UAV photogrammetry for precision agriculture. *Comput. Electron. Agric.* **2018**, *155*, 84–95. [CrossRef]
30. Maimaitijiang, M.; Sagan, V.; Sidike, P.; Maimaitiyiming, M.; Hartling, S.; Peterson, K.T.; Maw, M.J.W.; Shakoor, N.; Mockler, T.; Fritschid, F.B. Vegetation Index Weighted Canopy Volume Model (CVM VI) for soybean biomass estimation from Unmanned Aerial System-based RGB imagery. *ISPRS J. Photogramm. Remote Sens.* **2019**, *151*, 27–41. [CrossRef]
31. Sun, S.; Li, C.; Chee, P.W.; Paterson, A.H.; Jiang, Y.; Xu, R.; Robertson, J.S.; Adhikari, J.; Shehzad, T. Three-dimensional photogrammetric mapping of cotton bolls in situ based on point cloud segmentation and clustering. *ISPRS J. Photogramm. Remote Sens.* **2020**, *160*, 195–207. [CrossRef]
32. Schirrmann, M.; Hamdorf, A.; Garz, A.; Ustyuzhanin, A.; Dammer, K.H. Estimating wheat biomass by combining image clustering with crop height. *Comput. Electron. Agric.* **2016**, *121*, 374–384. [CrossRef]
33. Kandwal, R.; Kumar, A.; Bhargava, S. Existing Image Segmentation Techniques. *Int. J. Adv. Res. Comput. Sci. Softw. Eng.* **2014**, *4*, 2277–2285.
34. Rother, C.; Kolmogorov, V.; Blake, A. "GrabCut": Interactive foreground extraction using iterated graph cuts. *ACM Trans. Graph.* **2004**, *23*, 309–314. [CrossRef]
35. Devia, C.A.; Rojas, J.P.; Petro, E.; Martinez, C.; Patino, D.; Rebolledo, M.C.; Colorado, J. High-Throughput Biomass Estimation in Rice Crops Using UAV Multispectral Imagery. *J. Intell. Robot. Syst.* **2019**, *96*, 573. [CrossRef]
36. Assmann, J.J.; Kerby, J.T.; Cunliffe, A.M.; Myers-Smith, I.H. Vegetation monitoring using multispectral sensors—Best practices and lessons learned from high latitudes. *J. Unmanned Veh. Syst.* **2018**, 334730. [CrossRef]
37. Thaiparnit, S.; Ketcham, M. A Prediction Algorithm for Paddy Leaf Chlorophyll Using Colour Model Incorporate Multiple Linear Regression. *Eng. J.* **2017**, *21*, 269–280. [CrossRef]
38. Kanke, Y.; Tubaña, B.; Dalen, M.; Harrell, D. Evaluation of red and red edge reflectance-based vegetation indices for rice biomass and grain yield prediction models in paddy fields. *Precis. Agric.* **2016**, *17*, 507–530. [CrossRef]
39. Prabhakara, K.; Hively, W.; McCarty, G. Evaluating the relationship between biomass, percent groundcover and remote sensing indices across six winter cover crop fields in maryland, united states. *Int. J. Appl. Earth Obs. Geoinf.* **2015**, *39*, 88–102. [CrossRef]
40. Gnyp, L.; Miao, Y.; Yuan, F.; Ustin, S.; Yu, K.; Yao, Y.; Huang, S.; Bareth, G. Hyperspectral canopy sensing of paddy rice aboveground biomass at different growth stages. *Field Crops Res.* **2014**, *155*, 42–55. [CrossRef]
41. Cammarano, D.; Fitzgerald, G.J.; Casa, R.; Basso, B. Assessing the Robustness of Vegetation Indices to Estimate Wheat N in Mediterranean Environments. *Remote Sens.* **2014**, *6*, 2827–2844. [CrossRef]
42. Naito, H.; Ogawa, S.; Valencia, M.; Mohri, H.; Urano, Y.; Hosoi, F.; Shimizu, Y.; Chavez, A.; Ishitani, M.; Selvaraj, M.; et al. Estimating rice yield related traits and quantitative trait loci analysis under different nitrogen treatments using a simple tower-based field phenotyping system with modified single-lens reflex cameras. *J. Photogramm. Remote Sens.* **2017**, *135*, 50–62. [CrossRef]
43. Arai, K.; Sakashita, M.; Shigetomi, O.; Miura, Y. Estimation of Protein Content in Rice Crop and Nitrogen Content in Rice Leaves Through Regression Analysis with NDVI Derived from Camera Mounted Radio-Control Helicopter. *Int. J. Adv. Res. Artif. Intell.* **2014**, *3*. [CrossRef]
44. Tran, N.K.; Bulka, E.; Nahon, M. Quadrotor control in a wind field. In Proceedings of the IEEE International Conference on Unmanned Aircraft Systems (ICUAS), Denver, CO, USA, 9–12 June 2015; pp. 320–328. [CrossRef]
45. Tran, N.K. Modeling and Control of a Quadrotor in a Wind Field. Master's Thesis. McGill University, Montreal, CA, USA, 2015. Available online: http://digitool.library.mcgill.ca/webclient/StreamGate?folder_id=0&dvs=1571919888065~753 (accessed on 3 February 2020)
46. He, K.; Sun, J.; Tang, X. Guided Image Filtering. In *Computer Vision—ECCV 2010. ECCV 2010. Lecture Notes in Computer Science*; Daniilidis, K., Maragos, P., Paragios, N., Eds.; Springer: Berlin/Heidelberg, Germany, 2010; Volume 6311.

47. Rublee, E.; Rabaud, V.; Konolige, K.; Bradski, G. ORB: An efficient alternative to SIFT or SURF. In Proceedings of the 2011 International Conference on Computer Vision, Barcelona, Spain, 6–13 November 2011; pp. 3564–3571. [CrossRef]
48. Rojas, J.P.; Devia, C.A.; Petro, E.; Martinez, C.; Mondragon, I.F.; Patino, D.; Rebolledo, M.C.; Colorado, J. Aerial mapping of rice crops using mosaicing techniques for vegetative index monitoring. In Proceedings of the 2018 International Conference on Unmanned Aircraft Systems (ICUAS), Dallas, TX, USA, 12–15 June 2018; pp. 846–855. [CrossRef]
49. Siddiqui, F.U.; Isa, N.A.M. Enhanced moving K-means (EMKM) algorithm for image segmentation. *IEEE Trans. Consum. Electron.* **2011**, *57*, 833–841. [CrossRef]
50. Li, J.; Chen, H.; Li, G.; He, B.; Zhang, Y.; Tao, X. Salient object detection based on meanshift filtering and fusion of colour information. *IET Image Process.* **2015**, *9*, 977–985. [CrossRef]
51. Ganesan, P.; Rajini, V. Assessment of satellite image segmentation in RGB and HSV color space using image quality measures. In Proceedings of the 2014 International Conference on Advances in Electrical Engineering (ICAEE), Vellore, India, 9–11 January 2014; pp. 1–5. [CrossRef]
52. Mahajan, G.R.; Pandey, R.N.; Sahoo, R.N.; Gupta, V.K.; Datta, S.C.; Kumar, D. Monitoring nitrogen, phosphorus and sulphur in hybrid rice (*Oryza sativa* L.) using hyperspectral remote sensing. *Precis. Agric.* **2017**, *18*, 736–761. [CrossRef]
53. Li, S.; Ding, X.; Kuang, Q.; Ata-UI-Karim, S.T.; Cheng, T.; Liu, X.; Tian, Y.; Zhu, Y.; Cao, W.; Cao, Q. Potential of UAV-Based Active Sensing for Monitoring Rice Leaf Nitrogen Status. *Front. Plant Sci.* **2018**, *9*, 1834. [CrossRef]
54. FOILSIM III NASA Aerodynamics Simulator. Glenn Research Center. Available online: https://www.grc.nasa.gov/WWW/K-12/airplane/foil3.html (accessed on 16 July 2019).

Article

Automatic Detection of Maize Tassels from UAV Images by Combining Random Forest Classifier and VGG16

Xuli Zan [1], Xinlu Zhang [1], Ziyao Xing [1], Wei Liu [1], Xiaodong Zhang [1,2], Wei Su [1,2], Zhe Liu [1,2,*], Yuanyuan Zhao [1,2] and Shaoming Li [1,2]

1 College of Land Science and Technology, China Agricultural University, Beijing 100083, China; zanxuli@cau.edu.cn (X.Z.); S20183081345@cau.edu.cn (X.Z.); xingziyao@cau.edu.cn (Z.X.); devilweil@cau.edu.cn (W.L.); zhangxd@cau.edu.cn (X.Z.); suwei@cau.edu.cn (W.S.); zhaoyuanyuan@cau.edu.cn (Y.Z.); lishaoming@cau.edu.cn (S.L.)

2 Key Laboratory of Remote Sensing for Agri-Hazards, Ministry of Agriculture and Rural Affairs, Beijing 100083, China

* Correspondence: liuz@cau.edu.cn; Tel.: +86-1381-072-0768

Received: 17 July 2020; Accepted: 17 September 2020; Published: 18 September 2020

Abstract: The tassel development status and its branch number in maize flowering stage are the key phenotypic traits to determine the growth process, pollen quantity of different maize varieties, and detasseling arrangement for seed maize production fields. Rapid and accurate detection of tassels is of great significance for maize breeding and seed production. However, due to the complex planting environment in the field, such as unsynchronized growth stage and tassels vary in size and shape caused by varieties, the detection of maize tassel remains challenging problem, and the existing methods also cannot distinguish the early tassels. In this study, based on the time series unmanned aerial vehicle (UAV) RGB images with maize flowering stage, we proposed an algorithm for automatic detection of maize tassels which is suitable for complex scenes by using random forest (RF) and VGG16. First, the RF was used to segment UAV images into tassel regions and non-tassel regions, and then extracted the potential tassel region proposals by morphological method; afterwards, false positives were removed through VGG16 network with the ratio of training set to validation set was 7:3. To demonstrate the performance of the proposed method, 50 plots were selected from UAV images randomly. The precision, recall rate and F1-score were 0.904, 0.979 and 0.94 respectively; 50 plots were divided into early, middle and late tasseling stages according to the proportion of tasseling plants and the morphology of tassels. The result of tassels detection was late tasseling stage > middle tasseling stage > early tasseling stage, and the corresponding F1-score were 0.962, 0.914 and 0.863, respectively. It was found that the model error mainly comes from the recognition of leaves vein and reflective leaves as tassels. Finally, to show the morphological characteristics of tassel directly, we proposed an endpoint detection method based on the tassel skeleton, and further extracted the tassel branch number. The method proposed in this paper can well detect tassels of different development stages, and support large scale tassels detection and branch number extraction.

Keywords: maize tassel; tassel branch number; unmanned aerial vehicle; convolution neural network; VGG16; random forest

1. Introduction

Maize (*Zea mays* L.) is a monoecious crop, and the tassel is a branched structure atop the plan. The tassels' size and shape (branch number, compactness, etc.) have influence on the yield and quality of maize through affect the amount of pollen produced, the plant nutrition supply and the light interception of lower leaves [1,2], and such influence will be more significant with the increase of planting density [3,4]. In addition, in order to ensure the purity and quality of maize hybrid seed, cross-pollination should be ensured in the tasseling stage of seed maize production field, so tassels of female plants have to be removed through artificial or mechanical process to prevent self-pollination [5–7]. Therefore, tassel development has always been one of the important phenotypic traits in maize breeding and seed production. Rapid and accurate detection of tassel development status and branch number during maize flowering stage is of great significant for maize production management, and it is also of practical value for the arrangement of detasseling.

The complex planting environment in the field, such as uneven illumination conditions, leaves covered tassels severely and obvious difference between varieties, has brought great challenges to the automatic detection of maize tassels. Traditional crop phenotype acquisition is dependent on artificial field measurement, which is labor-intensive and time-consuming [8,9], and moreover, there is no uniform standard for phenotypical data collection, so the data is subjective. With the improvement of computer performance and the development of image processing, it is possible to extract crop phenotypic traits quickly and automatically. According to the different data acquisition platforms, the tassels detection research in field environment is mainly based on fixed observation tower and unmanned aerial vehicles (UAVs) high-throughput phenotype platform [10–14].

Compared with platform based on ground vehicles [15], the acquisition of images by cameras installed on fixed observation tower or cable-suspend platform is not limited by crop types and soil conditions (for example, the irrigated soil will hinder the mobile vehicle to collect data), which is suitable for real-time monitoring in the field [16,17]. Lu Hao et al. obtained sequence images of maize tassels by using a camera installed on an observation tower, and each sequence covered the tasseling stage to the flowering stage. On this basis, they carried out studies on tassel identification [11], and maize tassels counting which was based on the local counts regression [18]. In terms of distinguishing features, Hue component image in Hue-Saturation-Intensity (HSI) color space has been proved to be able to distinguish tassels from leaves [12,19]. Mao Zhengchong et al. [12] successfully separated the approximate region of tassels through binarization processing and median filtering of Hue component image, and then used LVQ neural network to eliminate the false detection regions, with the final detection accuracy reaching 96.9%. This kind of method can achieve high accuracy, but it is difficult to apply to large scale breeding field because its data acquisition and processing methods are low throughput, and equipment is expensive.

In recent years, UAVs equipped with different sensors have attracted extensive attention from researchers and breeders due to their advantages such as flexibility, high spatial and temporal resolution, and relatively low cost [17,20]. UAVs high-throughput phenotype platform can realize phenotypic analysis on many breeding plots in real time and dynamically, and is regarded as an indispensable tool for plant breeding [21–23]. Some scholars have studied sorghum heads detection [24,25], digital counts of crop plants at early stage [26–28], tree canopy identification through UAV images [29,30], and crop disease monitoring [31,32]. There are two studies on maize tassels detection based on UAV platform, Zhang Qi [13] divided UAV image into several small objects by object oriented classification method, and then identified tassels with vegetation index and spectrum information, with the overall accuracy reaching 85.91%. Yunling Liu et al. [14] used faster region-based convolutional neural network (Faster R-CNN) with ResNet and VGGNet to detect maize tassels, founded that the ResNet as the feature extraction network, was better than the VGGNet, and the detection accuracy can reach 89.96%. Compared with fixed observation tower, UAVs high-throughput phenotype platform is flexible, low-cost and has the potential to be applied in large scale. However, the above methods did not identify tassels in different tasseling stages, which causes the detection scene is single and difficult to meet the demand of dynamic monitoring for tassel development. Moreover, these methods terminate in the detection stage, without more detailed analysis (such as the morphological characteristics of tassels etc.) that made it impossible to show the tassels' development to breeders or production managers intuitively.

To know tassels' development status in maize breeding fields and seed maize production fields in time, and also provide decision support for varieties selection and the detasseling arrangement, a novel tassels detection method suitable for complex scenes is urgently needed. Therefore, the objective of this study is to propose an accurate method for tassels detection that suitable for different maize varieties and tasseling stages based on time series UAV images (from tasseling begins to ends of all breeding plots). Considering that maize tassels vary greatly in shape and size as plant grow over time, and unsynchronized growth stage between the plots, etc. in this study, we first divided images into tassel regions and non-tassel regions by using random forest (RF), and then extracted the potential tassel region proposals by morphological method; afterwards, false positives were removed through VGG16 network, and the influence of different tasseling stages on the model accuracy was analyzed; finally, we proposed an endpoint detection method to explore how to apply detection results to the extraction of tassel branch number.

2. Materials and Methods

2.1. Data Acquisition

The field trial was in Juqiao Town, Hebi City, Henan Province, China. The trial area was approximately 34.5 m from north to south and 72 m from east to west, including 170 breeding plots with a size of 2.4 m × 5 m. The plant density of these plots was 67500 plants/ha, and the row spacing was 0.6 m. In this trial, 167 maize varieties of different genetic backgrounds were sown on 17 June 2019, among which Zhengdan958 was repeated 4 times as the check variety, while the others were not repeated (Figure 1).

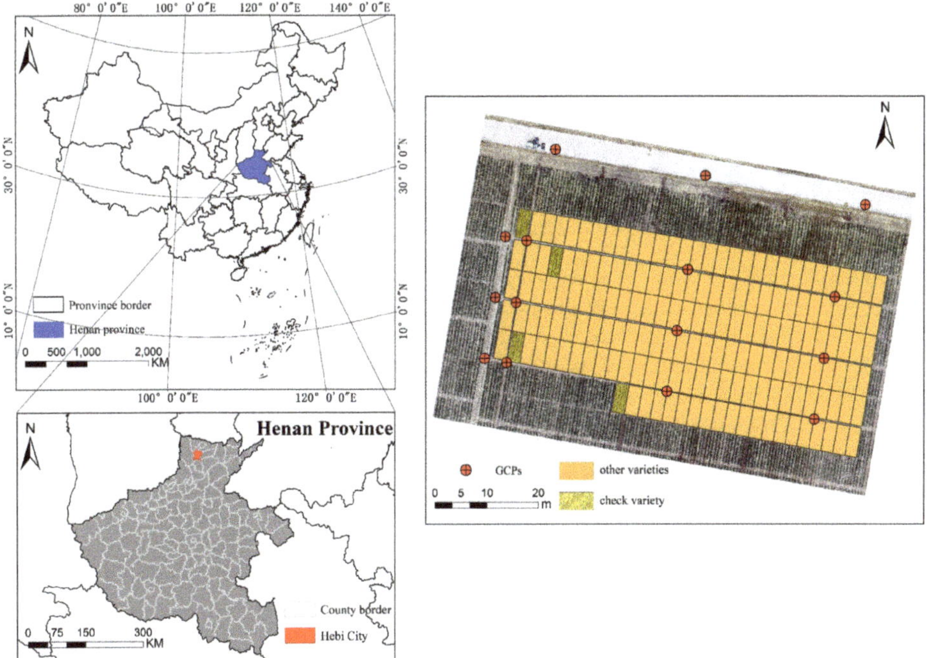

Figure 1. The layout of experimental field.

To improve the image accuracy of UAV in this study, 15 square panels with the size of 50 cm × 50 cm were deployed in the field (Figure 1) to be used as ground control points (GCPs) and measured by Real-time kinematic (RTK) technology (i80, Shanghai Huace Navigation Technology Ltd., Shanghai, China) with centimeter-level accuracy. A DJI ZENMUSE X4s camera (resolution: 5472 × 3648, DJI-Innovations, Inc., Shenzhen, China) was installed on the DJI Inspire 2 drone (DJI-Innovations, Inc.) to record RGB images. The UAV's flight path was set by DJI GSPro software in advance, with 85% forward overlap and 85% lateral overlap. To ensure that the model proposed in this study can realize the tassels detection of different varieties and different tasseling stages, we collected 5 groups of data for 4 days from 7 August 2019 to 12 August 2019 (no image data was collected on 10 and 11 August, due to the strong wind and light rain respectively, and 2 groups of image data were collected at 9:00 and 14:00 on 8 August). The flight altitude was set to 20 m, and about 500 digital images were collected for each flight, so a total of about 2500 images were collected.

2.2. Schematic Diagram of Method

Compared to the controllable greenhouse environment, there are many challenges of monitoring maize tassels in field: (1) complex background (Figure 2a); (2) tassels vary in size, color and shape caused by light conditions, varieties and unsynchronized growth stage (Figure 2b); (3) other factors, such as tassel pose variations caused by wind during shooting (Figure 2c); (4) the newly grown tassels do not have special shape, and the color is similar to the reflective leaves, which further increases the difficulty of tassels detection.

Figure 2. Challenges of maize tassels detection in field. (**a**) Complex background: (1) soil background, (2) green leaf, (3) reflective leaf, (4) shadow. (**b**) Tassels vary in shape and size: (1) (2) tassels vary in shape, and (3) size; (**c**) Tassel pose variations caused by wind.

In view of the problems to be solved and the challenges faced, we proposed an automatic tassels detection method combining RF and VGG16 network. Figure 3 shows the main process of this method. First, RF classifier was used to conduct pixel-based supervised classification of UAV images. The advantage of this process is that we can find tassels of any size, but it would also cause problems such as interference pixels and unconnected tassel regions; therefore, the morphological dilation method was used on unconnected regions belongs to tassels and noises, which are called the potential tassel region proposals; In the process of RF classification, pixels of other categories may be misidentified as tassels, so the potential region proposals have some wrong connected regions. To reduce the false positives, we used VGG16 network to re-classify the potential tassel region proposals and obtain accurate detection results; finally, we explored how to extract branch number of detected tassels.

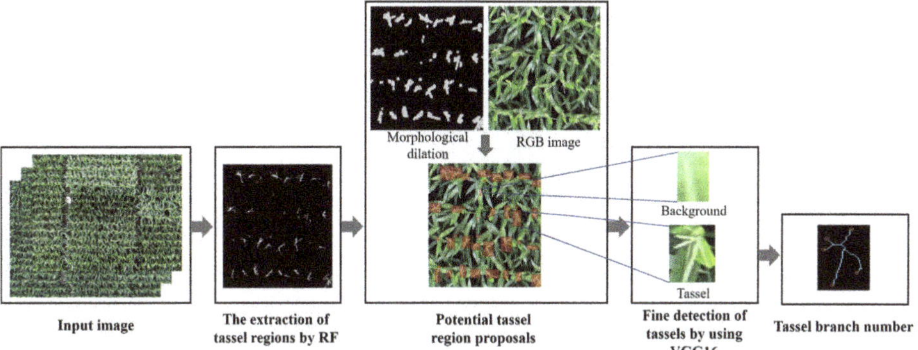

Figure 3. Schematic diagram of method.

2.3. Potential Tassel Region Proposals by RF and Morphological Method

Both the dynamic monitor of tassels development in breeding field, and the detasseling arrangement of the seed maize all require the detection of tassels to have high timeliness. Selective search strategy is a common method in the field of object detection and becomes an essential element for fast detection. It extracts potential bounding boxes based on image segmentation and sub-region merging [33], which can effectively solve the problems of high computational complexity and multiple redundant bounding boxes in exhausted search [11]. Referring to the idea of selective search, we proposed a method for the extraction of potential tassel region proposals that is suitable for our study. This method consists of two stages, including the tassel regions extraction by RF, and the potential tassel region proposals through morphological methods.

2.3.1. The Extraction of Tassel Regions by RF

In this study, we not only need to accurately identify the position of maize tassels, but also to extract the morphological characteristics. We used the RF to separate tassels from the background environment first. RF is an effective integrated learning technology proposed by Breiman [34] in 2001, it can reduce the correlation between decision trees through random selection of features and samples. By combining multiple weak classifiers, the model has high precision and good generalization ability [35], and has been successfully applied to biomedical science, agricultural informatization and other fields [36–39].

Considering the diversity of light conditions, and tassels vary in size and shape, 4 images were randomly selected from 5 groups of UAV data, with a total of 20 images were selected to label samples. To minimize the influence of camera lens distortion, all these images were cropped to 25% of original size (Figure 4b). A total of 3835 sample points were labeled through generating random points (Figure 4c, in this way, samples can represent the actual distribution of field environment type), sample points were divided into 5 categories: leaves, tassels, field path, road (there were hardened roads in some images), and background shadow. Among them, tassel samples accounted for 6%. To compare the influence of different tassel sample proportions on the classification results, some tassel samples were added in the way of artificial labeling.

For each sample, a series of color features were calculated, including R, G, B, H (Hue), S (Saturation), V (Value), ExG (Equation (1)), ExR (Equation (2)) [40,41]. The ratio of training set to test set is 2:1, and the proportion of tassel samples (4%, 6%, 10%, 15%, 20%, and 25%) in training set was adjusted for multiple experiments. From the results of multiple experiments (Table 1), it can be found that the recall of tassel increased gradually while the precision decreased, as the proportion of tassel samples increases. This indicates that the classification model tends to categories with a large number of samples, and the classification results are obviously affected by the sample proportion.

In this study, after the proportion of tassel samples reaches 10%, the identification performance of model for tassel tends to be stable, and based on the overall accuracy (OA), the proportion with 15% is considered to be the best.

$$ExG = 2g - r - b \tag{1}$$

$$ExR = 2r - g - b \tag{2}$$

The corresponding calculation equation of r, g, and b channel characteristics is as follows:

$$\begin{cases} r = \frac{R}{R+G+B} \\ g = \frac{G}{R+G+B} \\ b = \frac{B}{R+G+B} \end{cases} \tag{3}$$

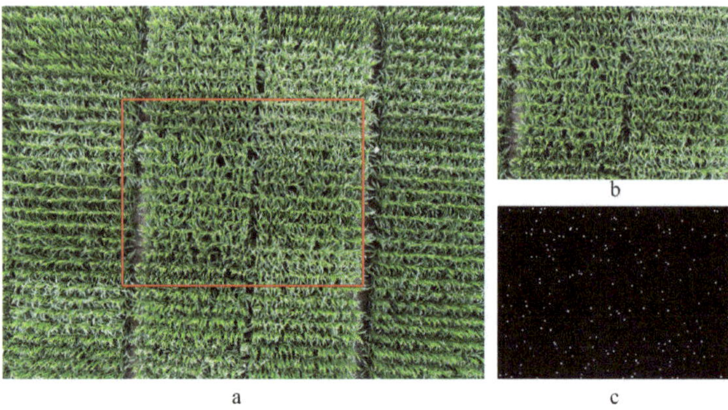

Figure 4. Process of sample acquisition. (**a**) Original image. (**b**) Cropped image. (**c**) Random sample points (white points).

Table 1. Classification results under different proportions of tassel samples.

Proportions of Tassel Samples	Precision	Recall	F1-Score	OA
4%	0.94	0.70	0.80	0.877
6%	0.93	0.72	0.81	0.881
10%	0.89	0.80	0.84	0.886
15%	0.89	0.80	0.85	0.891
20%	0.86	0.84	0.85	0.888
25%	0.85	0.84	0.84	0.880

According to the characteristics of the RF classifier, the importance of features could be evaluated. We applied the permutation importance method based on test set, which was evaluated by permuting the column values of a single feature, rerunning the trained model, and then calculating accuracy change as importance score [34,42]. This method is more reliable than the Gini importance [43]. The rfpimp package with python was used to complete this, and the feature importance obtained is shown in Figure 5. It can be found that ExR has the highest importance score, followed by G (green band), while the B (blue band) has the lowest importance score. Figure 6b shows the classification results of the UAV image, and tassels' morphological characteristics are well displayed. After this, other categories except tassel were selected and integrated together to obtain binary image of tassels (Figure 6c), called tassel regions.

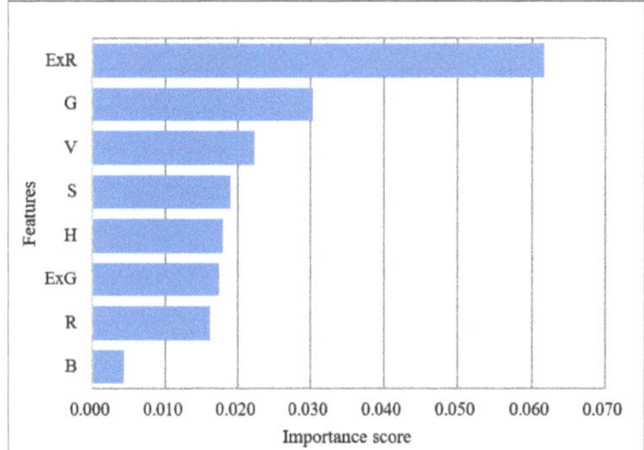

Figure 5. Importance derived by permuting each feature and computing change in accuracy.

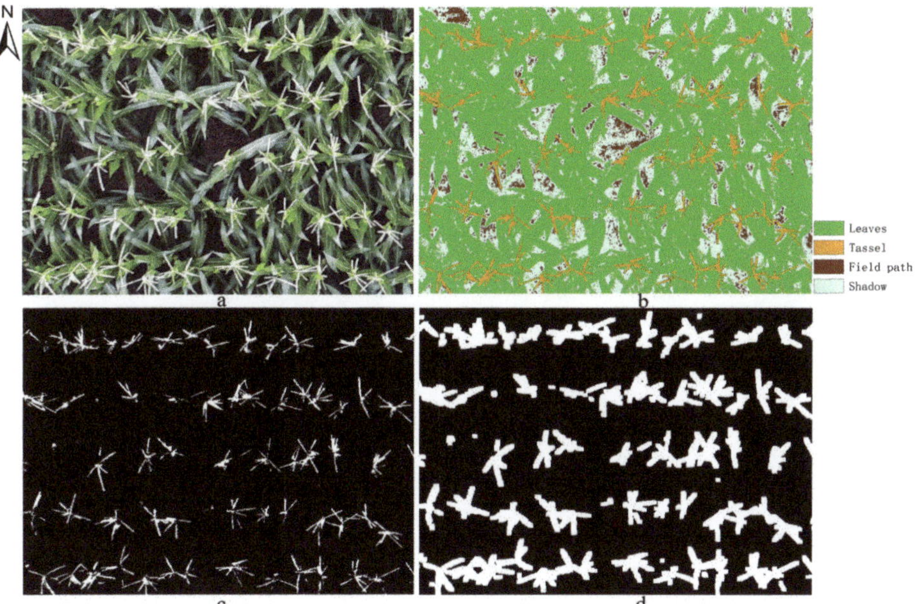

Figure 6. The extraction of potential tassel region proposals. (**a**) Original image. (**b**) Classification result by RF. (**c**) Binary image of tassels (white is tassel regions, black is non-tassel regions). (**d**) Result of morphological dilation (white is tassel regions, black is non-tassel regions).

2.3.2. Potential Tassel Region Proposals Based on Morphological Processing

In the complex field environment, the mutual occlusion between top leaves and tassels is serious, leading to the fact that tassels in the binary image obtained in Section 2.3.1 were not connected regions (Figure 6c), and there is also the problem that other categories' pixels were misidentified as tassels.

To remove some of noises, we applied morphological remove small objects on binary image, and then, used morphological dilation on unconnected pixel regions belongs to tassels and noises in order to obtain potential tassel region proposals (Figure 6d).

2.4. Fine Detection of Tassels by Using VGG16

To reduce the false positives, the potential tassel region proposals obtained after morphological processing were labeled, with 0 representing non-tassel connected region and 1 representing tassel connected region. A total of 2745 samples were labeled, including 1230 positive samples (tassel connected region) and 1515 negative samples (non-tassel connected region).

We found that many labeled samples contained less surrounding information due to the small area of the envelope rectangles, making it difficult for human eyes to recognize these samples. As shown in Figure 7, all the samples were labeled with human recognition attribute (called recognition), and analyzed the distribution of envelope rectangles' pixel number. Almost all of (96%) the envelope rectangles out of unidentifiable samples (recognition = 0) are smaller than 600 pixels, so the envelope rectangles were expanded to 600 pixels according to the aspect ratio before put into VGG network, so that more information can be included. Although nearly half of the envelope rectangles out of identifiable samples (recognition = 1) are less than 600 pixels, after verification, almost all of these samples are non-tassel regions, which means they could be recognized because of their special color.

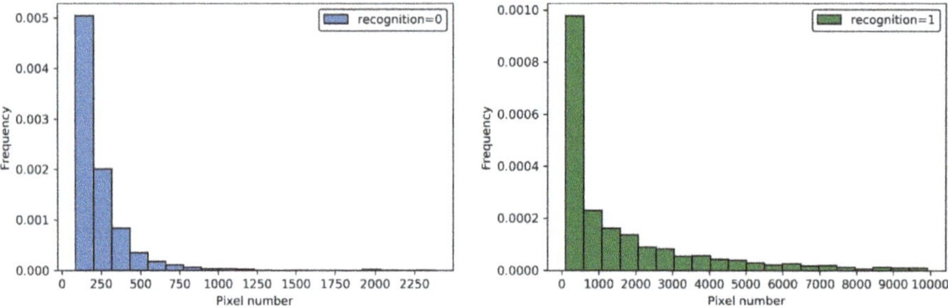

Figure 7. The pixel number distribution of envelope rectangles.

The VGG network was proposed by the Visual Geometry Group of Oxford University, and participated in the 2014 ImageNet Large Scale Visual Recognition Challenge (ILSVRC), won the first and second places in localization and classification tasks. VGG network adopts the superposition of multiple 3×3 convolution filters to replace a large convolution filter, which can increase the network depth and reduce the number of total parameters at the same time [44]. Therefore, the idea of 3×3 convolution filters has become the basis of various subsequent classification models. Two common structures in VGG networks are VGG16 and VGG19, they are differed in the number of convolutional layers that VGG16 has 13 convolutional layers while VGG19 has 16.

VGG16 network structure was selected in this paper, and we used the weight parameters from ImageNet's pre-training model, to achieve better training results and reduce the running time. This study deals with binary classification, so replace the activation function of the output layer with the sigmoid (Figure 8). The ratio of training set to validation set was 7:3. To prevent overfitting due to limited samples and improve the model's generalization, data augmentations through small random transformations with rotate and flip were used. The model's iteration and batch size were 50 and 64, respectively, initial learning rate was 0.001 and adopted the learning rate decay strategy. We fine-tuned the obtained optimal model by releasing the deep network parameters, to make the model parameters more consistent with the data in this study.

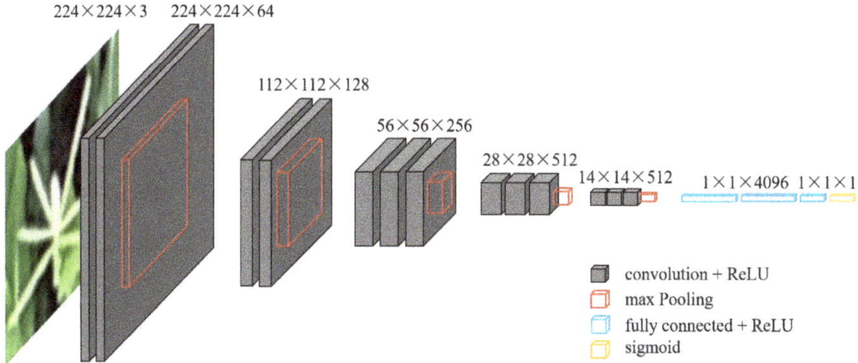

Figure 8. Architecture of VGG16 network.

2.5. Extraction of Tassel Branch Number

As an important factor determining tassel size and pollen quantity, the tassel branch number is one of the most important indicators in maize breeding. Some researchers placed tassels collected in the field in a portable photo boxes for taking pictures. Branch number was estimated by a series of circular arcs from the lowest branch node, the number of intersections between each circle and binary object was calculated, and the greatest value was taken as branch number [4]. However, the data acquisition of this method is low throughput, which is not suitable for large scale application. Moreover, this calculation method of branch number is also not suitable for the UAV images.

In this paper, the tassel branch number was extracted from the result of morphological dilatation (which can be extracted from Section 2.3.2). Skeleton extraction was performed first, and then an endpoint detection method suitable for tassels' shape were proposed based on this (Figure 9). This method abstracts tassel skeleton into a matrix composed of 0 and 1 (0 is the background pixel and 1 represents the skeleton pixel). Based on the principle of most background pixels around the tassel endpoint, the number of background pixels in the window (3 × 3) with skeleton point as center was calculated. The tassels' endpoints are those skeleton points that have the most background pixels.

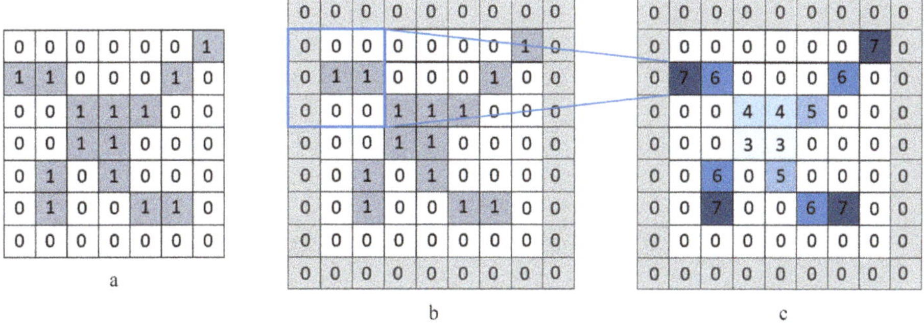

Figure 9. Endpoint detection of tassels. (**a**) 0, 1 matrix of tassel skeleton. (**b**) Matrix after applied padding, the blue box is a 3 × 3 window. (**c**) Pixel with the deepest color are the endpoints of tassel.

2.6. Model Evaluation

To evaluate performance of the proposed method, accuracy, recall rate and F1-score were selected [45]. Precision refers to the proportion of correctly detected in all positive results returned by our model, with a ratio of 1.0 being perfect; recall (1.0 is perfect) indicates for all relevant samples, how many are correctly detected, and F1-score indicates the harmonic average of the precision and recall. The metrics were calculated as follows:

$$Precision = \frac{TP}{TP + FP} \quad (4)$$

$$recall = \frac{TP}{TP + FN} \quad (5)$$

$$F1 - score = \frac{2 \times Precision \times recall}{Precision + recall} \quad (6)$$

where TP, FP, and FN are the number of true positives, false positives and false negatives, respectively.

3. Results

3.1. Influence of the Envelope Rectangles' Size on Model Accuracy

According to the description in Section 2.4, we have trained both the enlarged of envelope rectangles (for samples with envelope rectangle less than 600 pixels, we expanded them to 600 pixels according to the aspect ratio) and the direct use of original envelope rectangles. The iterations of model was set as 50. Figure 10 shows the loss and accuracy curve during model training after the expansion of envelope

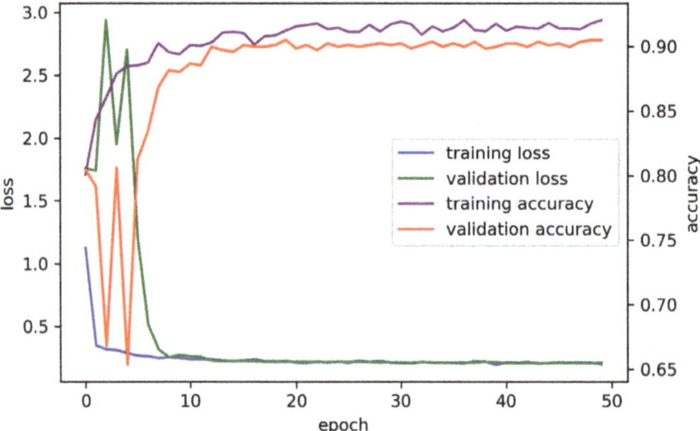

Figure 10. Loss and accuracy curve.

We fixed the shallow network parameters of the optimal model obtained during iteration, and released different number of deep networks for fine-tuning (Table 2). It can be found that no matter how many several layers of deep network were released, the recognition performance of the enlarged of envelope rectangles is better than original envelope rectangles. This is because the more surrounding information was included in samples, the more comprehensive model can learn (especially for the samples whose real category is small part of leaf, after the enlarged of envelope rectangles, it will be found that the surrounding scenes are different from the real tassels). Moreover, when the envelope rectangles were enlarged, the model obtained by releasing networks that are behind the eighth convolution layer has the highest validation accuracy, which is 0.954.

Table 2. Validation accuracy of model under fine-tuning.

Envelope Rectangles' Size	Validation Accuracy_6	Validation Accuracy_8	Validation Accuracy_10
original envelope rectangles	0.944	0.935	0.935
600 pixels	0.952	0.954	0.940

Validation accuracy_6 represents the validation accuracy of model obtained by releasing deep networks that are after the sixth convolution layer.

3.2. Influence of Different Tasseling Stages on Detection Accuracy

To demonstrate the performance of the proposed method, 50 plots were selected from the series UAV images randomly. The detection results of this method is shown in Figure 11. The red boxes, blue boxes, and yellow boxes in the figure represent automatically detection, absent detection, and incorrect detection by proposed method, respectively. It can be found that different shapes, sizes and even tassels that can just be observed by human eyes could be well detected (Figure 12). The precision, recall rate and F1-score were 0.904, 0.979 and 0.94, respectively (Table 3). It should be noted that some of tassels were covered by leaves severely, leading to the failure to connect these tassels through dilation processing of Section 2.3.2. Therefore, a small part of tassels' branches were also marked in the final detection results, which were regarded as false positives by us. In addition, the model error mainly comes from the recognition of leaves vein and reflective leaves as tassels (Figure 12).

It was found that the detection effect was related to the tasseling stage of breeding plots, when analyzed the detection accuracy. Therefore, the tasseling stage of breeding plots were divided into early, middle and late tasseling stages according to the proportion of tasseling plants and whether the tassels have complete morphological characteristics. The definition of tasseling stages are as follows: tasseling proportion is less than half is early tasseling stage (Figure 11a,b); tasseling proportion is more than half, but most of them do not have complete morphological characteristics is middle tasseling stages (Figure 11c,d); tasseling proportion is more than half and tassels have complete morphological characteristics is late tasseling stage (Figure 11e,f).

According to Table 3, the detection effects of three different tasseling stages are as follows: late tasseling stage > middle tasseling stage > early tasseling stage, and the corresponding F1-score are 0.962, 0.914 and 0.863 respectively. The effect of early and middle tasseling stage was worse than that in the late stage, mainly reflected in the low precision. This is because the top leaves of maize plants in these two stages were not fully unfolded, and the high flexibility leads to the large inclination angle. Under the same light condition, the reflection phenomenon of such leaves would be more significant, which is easy to be misidentified as tassels. Moreover, no matter at which tasseling stage, the recall rate was always higher than precision, which indicates that there were fewer tassels missed by the proposed method, but have the phenomenon of other objects (mainly leaves vein and the reflective leaves, Figure 12) were identified as tassels.

Table 3. Evaluation of tassels detection results.

Tasseling Stage	TP	FP	FN	Precision	Recall	F1-Score
Early tasseling stage (include 11 breeding plots)	58	45	5	0.778	0.969	0.863
Middle tasseling stage (include 13 breeding plots)	585	86	24	0.872	0.961	0.914
Late tasseling stage (include 26 breeding plots)	1347	90	15	0.937	0.989	0.962
Overall (include 50 breeding plots)	2090	221	44	0.904	0.979	0.940

Figure 11. Detection results of tassels. The red boxes, blue boxes, and yellow boxes in the figure automatically represent detection, absent detection, and incorrect detection by proposed method respectively.

Figure 12. Detailed diagram of detection results. Red boxes represent automatically detection by proposed method, (1) the newly grown tassels; (2) reflective leaf was misidentified as tassel; (3) and (4) leaves vein were misidentified as tassels.

3.3. The Calculation of Tassel Branch Number

Before extracting the tassel skeleton, we found that the envelope rectangles of detected tassels may contain partial branches of other tassels caused by the high planting density in the field, leading to some interference connected regions (the red boxes at the Figure 13b) in the result of morphological dilatation (which can be extracted from Section 2.3.2). Therefore, before skeleton extraction and endpoint detection, interference connected regions should be removed first.

The tassel branch number was calculated as follows (Figure 13c,d). The uppermost branch of tassel cannot be detected because our image is taken from orthographic angle. Therefore, the final branch number needs to add 1 to the endpoint detection result. In this example, the tassel branch is 7 endpoints plus the uppermost branch, making a total of 8 branches.

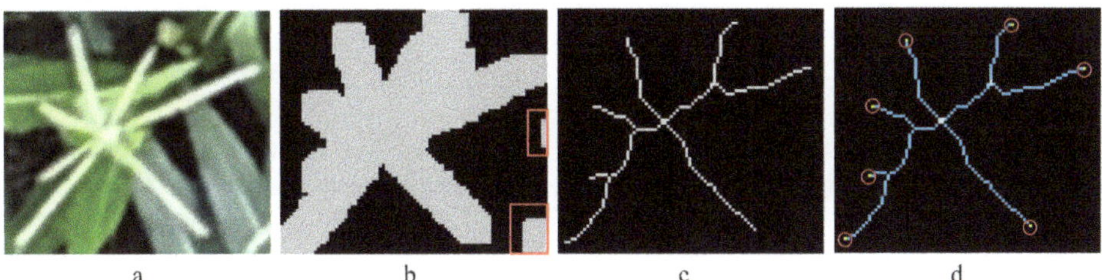

Figure 13. The calculation of tassel branch number. (**a**) Original image. (**b**) Binary image after morphological dilatation. (**c**) Tassel skeleton. (**d**) Endpoint detection.

4. Discussion

4.1. Comparison of the Generation Method of Detection Boxes

The detection of tassels carried out in this paper can be regarded as a common object detection problem in the field of computer vision. Yunling Liu et al. [14] have realized the detection of tassels by using the Faster R-CNN. Faster R-CNN is a two-stage object detection algorithm proposed by Ren Shaoqing et al. [46], which changed the generation method of detection boxes based on Fast R-CNN and proposed the Region Proposal Networks (RPN) strategy. RPN has become a mainstream method, it form a series anchor boxes as initial region proposals by setting different anchor ratios and scales for each pixel on the feature map, which was obtained by the convolution neural network. RPN could improve the generation speed of region proposals, but the smallest anchor box (128^2 pixels) in RPN is also bigger than most of tassels, which will affect the detection accuracy of model [47]. Yunling Liu et al. [14] also payed attention to this problem, so the anchor size was adjusted from [128^2, 256^2, 512^2] to [85^2, 128^2, 256^2], and the final prediction accuracy of tassels increased from 87.27% to 89.96%.

In summary, the size setting of anchor boxes is particularly important in RPN, especially for the tassels detection problem in this paper that including different tasseling stages (tassel size changes significantly) and different shapes. The ratio of the width to height of the tassel samples was used to represent the tassel morphology, and Figure 14 shows the distribution of tassels morphology and size in this paper (tassel samples labeled in Section 2.4). We found that the width to height ratio (Figure 14b) is concentrated in (0.2, 2.0), and the default anchor ratios in RPN are 0.5, 1, 2, which can satisfy the requirements of different shape of tassels; the distributions of pixel width and height of tassel samples (Figure 14a) are mostly within the range of (11, 100), so the default size [128^2, 256^2, 512^2] in anchor parameter cannot cover tassels in our dataset. To better identify tassels, we can adjust the size of anchor and add additional anchor boxes while keeping the default anchor ratios. However, there is just a simple discussion, the optimal parameter setting needs to go through a series of experimental analysis.

Actually, in our proposed method, we divided the images into tassel regions and non-tassel regions by using random forest, and then extracted the potential tassel region proposals through morphological method is also an operation of forming detection boxes. This method can find the detection boxes of tassels at different tasseling stages, even including newly grown tassels (Figure 12), and there is no problem like RPN applied to small size objects. However, it spent more time in labeling samples compared with the object detection method, because we also need a large number of samples when training the random forest classifier.

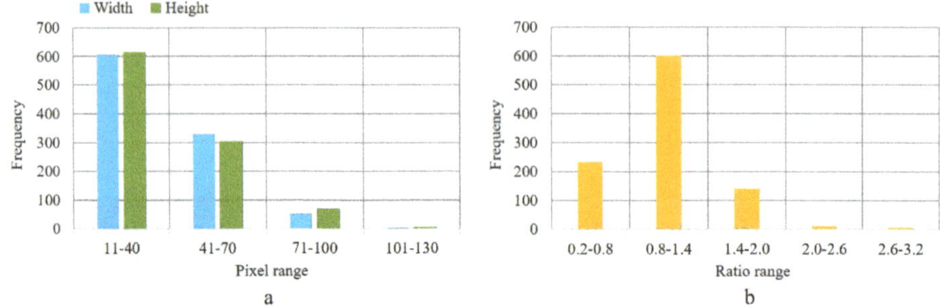

Figure 14. (**a**) The distributions of pixel width and height of tassels. (**b**) The distribution of tassels morphology.

4.2. Comparison of VGG16 and RF in Fine Detection

In the classification of the potential tassel region proposals, we compared the effect of RF and VGG16. The potential tassel region proposals extracted in Section 2.3 was selected, and RF classifier was used. The number of samples, training set and validation set were consistent with the input of VGG16 network, and the Histogram of Oriented Gradient (HOG) method was applied to extract features of each sample.

The results are shown in Table 4, the OA of RF classifier is 0.796, while VGG16 network is 0.954 (Table 2, validation accuracy was calculated in the same way as OA in RF), indicating that VGG16 performs significantly better than RF. The recall rate of tassel obtained by RF is very low, indicating that there are many tassels were missed, which is not suitable for our application scene. This may be caused by the small number of training sets; however, under the same sample size, VGG16 network performs better because we used the ImageNet's pre-training model. This also reflects the advantage of deep learning to transfer existing model on the problem to be solved; moreover, we all know that it does not require feature engineering construction and feature optimization.

Table 4. The result of RF in fine detection of tassels.

Categories	OA	Precision	Recall	F1-Score
tassel	0.796	0.823	0.694	0.753
non-tassel		0.780	0.879	0.826

5. Conclusions

The extraction of tassels development in maize breeding fields and seed maize production fields rapidly and accurately can provide decision support for varieties selection and detasseling arrangement. However, due to the complex planting environment in the field, such as unsynchronized growth stage and tassels vary in size and shape caused by varieties, the detection of maize tassels remains challenging problem. In this paper, based on the time series UAV images of maize flowering stage, we proposed a detection method of maize tassels in complex scenes (different varieties, different tasseling stages) by combining RF and VGG16 network. The main conclusions are as follows:

(1) The potential tassel region proposals with different sizes and shapes could be found by using RF and morphological methods, which is the key point to realize tassels detection in complex scenes. In addition, also the VGG16 network after fine-tuning can distinguish the false positives in the potential tassel region proposals well.
(2) We divided breeding plots into early, middle and late tasseling stages according to the proportion of tasseling plants and whether the tassels have complete morphological characteristics,

and found that the detection effect of tassels was highly correlated with the tasseling stages and the detection effect in late tasseling stage was better than that in middle and early stages.
(3) According to the special morphological characteristics of maize tassels, the endpoint detection method based on tassel skeleton can extract the tassel branch number well.
(4) Moreover, the model performance can be effectively improved after enlarged envelope rectangles of potential tassel region proposals (the threshold value determination needs to consider the actual situation of samples).

The detection method of maize tassels proposed in this paper has the advantages of high precision and fast data acquisition speed, which can be applied to large area of maize breeding fields and seed maize production fields. In the future research, we will carry out experiments on UAV images that collected at different periods of the day, flight heights and other sensors (like multispectral and hyperspectral sensors), and reduce the number of samples to propose a more efficient detection method for maize tassels. As for the problem of unsuccessful photogrammetric processing caused by high similarity of acquired images, we will also look for a solution to this problem, and then replace single images with orthophoto mosaics in the later work.

Author Contributions: Conceptualization, X.Z. (Xiaodong Zhang), S.L. and Z.L.; methodology, Z.L., W.S. and Y.Z.; formal analysis, X.Z. (Xuli Zan), and Z.L.; investigation, X.Z. (Xuli Zan), X.Z. (Xinlu Zhang) and W.L.; data curation, X.Z. (Xinlu Zhang), Z.X. and W.L.; writing—original draft preparation, X.Z. (Xuli Zan); writing—review and editing, Z.L. and X.Z. (Xuli Zan). All authors have read and agreed to the published version of the manuscript.

Funding: This research was funded by National Key Research and Development Plan of China, grant number 2018YFD0100803.

Conflicts of Interest: The authors declare no conflict of interest.

Abbreviations

The following abbreviations are used in this manuscript:

UAV	unmanned aerial vehicle
RF	random forest
GCPs	ground control points

References

1. Lambert, R.; Johnson, R. Leaf angle, tassel morphology, and the performance of maize hybrids 1. *Crop Sci.* **1978**, *18*, 499–502. [CrossRef]
2. Yulan, Y.; Min, Z.; Lei, Y.; Chunguang, L. Research Progress on the Impact of Maize Tassel on Yield. *J. Maize Sci.* **2010**, *018*, 150–152.
3. Hunter, R.; Daynard, T.; Hume, D.; Tanner, J.; Curtis, J.; Kannenberg, L. Effect of tassel removal on grain yield of corn (*Zea mays* L.) 1. *Crop Sci.* **1969**, *9*, 405–406. [CrossRef]
4. Gage, J.L.; Miller, N.D.; Spalding, E.P.; Kaeppler, S.M.; de Leon, N. TIPS: A system for automated image-based phenotyping of maize tassels. *Plant Methods* **2017**, *13*, 21. [CrossRef] [PubMed]
5. Zhang, C.; Jin, H.; Liu, Z.; Li, Z.; Ning, M.; Sun, H. Seed maize identification based on texture analysis of GF remote sensing data. *Trans. Chin. Soc. Agric. Eng.* **2016**, *32*, 183–188.
6. Kurtulmuş, F.; Kavdir, I. Detecting corn tassels using computer vision and support vector machines. *Expert Syst. Appl.* **2014**, *41*, 7390–7397. [CrossRef]
7. Ren, T.; Liu, Z.; Zhang, L.; Liu, D.; Xi, X.; Kang, Y.; Zhao, Y.; Zhang, C.; Li, S.; Zhang, X. Early Identification of Seed Maize and Common Maize Production Fields Using Sentinel-2 Images. *Remote Sens.* **2020**, *12*, 2140. [CrossRef]
8. Ribera, J.; He, F.; Chen, Y.; Habib, A.F.; Delp, E.J. Estimating phenotypic traits from UAV based RGB imagery. *arXiv* **2018**, arXiv:1807.00498.
9. Han, L.; Yang, G.; Dai, H.; Xu, B.; Yang, H.; Feng, H.; Li, Z.; Yang, X. Modeling maize above-ground biomass based on machine learning approaches using UAV remote-sensing data. *Plant Methods* **2019**, *15*, 10. [CrossRef]

10. Madec, S.; Jin, X.; Lu, H.; De Solan, B.; Liu, S.; Duyme, F.; Heritier, E.; Baret, F. Ear density estimation from high resolution RGB imagery using deep learning technique. *Agric. For. Meteorol.* **2019**, *264*, 225–234. [CrossRef]
11. Lu, H.; Cao, Z.; Xiao, Y.; Fang, Z.; Zhu, Y.; Xian, K. Fine-grained maize tassel trait characterization with multi-view representations. *Comput. Electron. Agric.* **2015**, *118*, 143–158. [CrossRef]
12. Zhengchong, M.; Yahui, S. Algorithm of male tassel recognition based on HSI space. *Transducer Microsyst. Technol.* **2018**, *37*, 117–119.
13. Qi, Z. The Research on Extraction of Maize Phenotypic Information Based on Unmanned Aerial Vehicle. Ph.D. Thesis, Northeast Agricultural University, Harbin, China, 2017.
14. Liu, Y.; Cen, C.; Che, Y.; Ke, R.; Ma, Y.; Ma, Y. Detection of maize tassels from UAV RGB imagery with faster R-CNN. *Remote Sens.* **2020**, *12*, 338. [CrossRef]
15. White, J.W.; Andrade-Sanchez, P.; Gore, M.A.; Bronson, K.F.; Coffelt, T.A.; Conley, M.M.; Feldmann, K.A.; French, A.N.; Heun, J.T.; Hunsaker, D.J.; et al. Field-based phenomics for plant genetics research. *Field Crops Res.* **2012**, *133*, 101–112. [CrossRef]
16. Kirchgessner, N.; Liebisch, F.; Yu, K.; Pfeifer, J.; Friedli, M.; Hund, A.; Walter, A. The ETH field phenotyping platform FIP: A cable-suspended multi-sensor system. *Funct. Plant Biol.* **2017**, *44*, 154–168. [CrossRef]
17. Yang, G.; Liu, J.; Zhao, C.; Li, Z.; Huang, Y.; Yu, H.; Xu, B.; Yang, X.; Zhu, D.; Zhang, X.; et al. Unmanned aerial vehicle remote sensing for field-based crop phenotyping: Current status and perspectives. *Front. Plant Sci.* **2017**, *8*, 1111. [CrossRef]
18. Lu, H.; Cao, Z.; Xiao, Y.; Zhuang, B.; Shen, C. TasselNet: Counting maize tassels in the wild via local counts regression network. *Plant Methods* **2017**, *13*, 79. [CrossRef]
19. Tang, W.; Zhang, Y.; Zhang, D.; Yang, W.; Li, M. Corn tassel detection based on image processing. In Proceedings of the 2012 International Workshop on Image Processing and Optical Engineering, Harbin, China, 9–10 June 2012; International Society for Optics and Photonics: The Hague, The Netherlands, 2012; Volume 8335.
20. Shi, Y.; Thomasson, J.A.; Murray, S.C.; Pugh, N.A.; Rooney, W.L.; Shafian, S.; Rajan, N.; Rouze, G.; Morgan, C.L.; Neely, H.L.; et al. Unmanned aerial vehicles for high-throughput phenotyping and agronomic research. *PLoS ONE* **2016**, *11*, e0159781. [CrossRef]
21. Watanabe, K.; Guo, W.; Arai, K.; Takanashi, H.; Kajiya-Kanegae, H.; Kobayashi, M.; Yano, K.; Tokunaga, T.; Fujiwara, T.; Tsutsumi, N.; et al. High-throughput phenotyping of sorghum plant height using an unmanned aerial vehicle and its application to genomic prediction modeling. *Front. Plant Sci.* **2017**, *8*, 421. [CrossRef]
22. Zaman-Allah, M.; Vergara, O.; Araus, J.; Tarekegne, A.; Magorokosho, C.; Zarco-Tejada, P.; Hornero, A.; Albà, A.H.; Das, B.; Craufurd, P.; et al. Unmanned aerial platform-based multi-spectral imaging for field phenotyping of maize. *Plant Methods* **2015**, *11*, 35. [CrossRef]
23. Jang, G.; Kim, J.; Yu, J.K.; Kim, H.J.; Kim, Y.; Kim, D.W.; Kim, K.H.; Lee, C.W.; Chung, Y.S. Review: Cost-Effective Unmanned Aerial Vehicle (UAV) Platform for Field Plant Breeding Application. *Remote Sens.* **2020**, *12*, 998. [CrossRef]
24. Guo, W.; Potgieter, A.; Jordan, D.; Armstrong, R.; Lawn, K.; Kakeru, W.; Duan, T.; Zheng, B.; Iwata, H.; Chapman, S.; et al. Automatic detecting and counting of sorghum heads in breeding field using RGB imagery from UAV. In Proceedings of the CIGR-AgEng Conference—CIGR 2016, Aarhus, Denmark, 26–29 June 2016; pp. 1–5.
25. Guo, W.; Zheng, B.; Potgieter, A.B.; Diot, J.; Watanabe, K.; Noshita, K.; Jordan, D.R.; Wang, X.; Watson, J.; Ninomiya, S.; et al. Aerial imagery analysis–quantifying appearance and number of sorghum heads for applications in breeding and agronomy. *Front. Plant Sci.* **2018**, *9*, 1544. [CrossRef] [PubMed]
26. Gnädinger, F.; Schmidhalter, U. Digital counts of maize plants by unmanned aerial vehicles (UAVs). *Remote Sens.* **2017**, *9*, 544. [CrossRef]
27. Reza, M.N.; Na, I.S.; Lee, K.H. Automatic Counting of Rice Plant Numbers after Transplanting Using Low Altitude UAV Images. *Int. J. Contents* **2017**, *13*, 1–8.
28. Shuaibing, L.; Guijun, Y.; Chenquan, Z.; Haitao, J.; Haikuan, F.; Bo, X.; Hao, Y. Extraction of maize seedling number information based on UAV imagery. *Trans. Chin. Soc. Agric. Eng. (Trans. CSAE)* **2018**, *34*, 69–77.
29. Wang, Y.; Zhu, X.; Wu, B. Automatic detection of individual oil palm trees from UAV images using HOG features and an SVM classifier. *Int. J. Remote Sens.* **2019**, *40*, 7356–7370. [CrossRef]

30. Nevalainen, O.; Honkavaara, E.; Tuominen, S.; Viljanen, N.; Hakala, T.; Yu, X.; Hyyppä, J.; Saari, H.; Pölönen, I.; Imai, N.N.; et al. Individual tree detection and classification with UAV-based photogrammetric point clouds and hyperspectral imaging. *Remote Sens.* **2017**, *9*, 185. [CrossRef]
31. Su, J.; Liu, C.; Coombes, M.; Hu, X.; Wang, C.; Xu, X.; Li, Q.; Guo, L.; Chen, W.H. Wheat yellow rust monitoring by learning from multispectral UAV aerial imagery. *Comput. Electron. Agric.* **2018**, *155*, 157–166. [CrossRef]
32. Zhang, X.; Han, L.; Dong, Y.; Shi, Y.; Huang, W.; Han, L.; González-Moreno, P.; Ma, H.; Ye, H.; Sobeih, T. A deep learning-based approach for automated yellow rust disease detection from high-resolution hyperspectral uav images. *Remote Sens.* **2019**, *11*, 1554. [CrossRef]
33. Uijlings, J.R.; Van De Sande, K.E.; Gevers, T.; Smeulders, A.W. Selective search for object recognition. *Int. J. Comput. Vis.* **2013**, *104*, 154–171. [CrossRef]
34. Breiman, L. Random forests. *Mach. Learn.* **2001**, *45*, 5–32. [CrossRef]
35. Liaw, A.; Wiener, M. Classification and regression by randomForest. *R News* **2002**, *2*, 18–22.
36. Zhang, L.; Liu, Z.; Ren, T.; Liu, D.; Ma, Z.; Tong, L.; Zhang, C.; Zhou, T.; Zhang, X.; Li, S. Identification of Seed Maize Fields With High Spatial Resolution and Multiple Spectral Remote Sensing Using Random Forest Classifier. *Remote Sens.* **2020**, *12*, 362. [CrossRef]
37. Waldner, F.; Lambert, M.J.; Li, W.; Weiss, M.; Demarez, V.; Morin, D.; Marais-Sicre, C.; Hagolle, O.; Baret, F.; Defourny, P. Land cover and crop type classification along the season based on biophysical variables retrieved from multi-sensor high-resolution time series. *Remote Sens.* **2015**, *7*, 10400–10424. [CrossRef]
38. De Castro, A.I.; Torres-Sánchez, J.; Peña, J.M.; Jiménez-Brenes, F.M.; Csillik, O.; López-Granados, F. An automatic random forest-OBIA algorithm for early weed mapping between and within crop rows using UAV imagery. *Remote Sens.* **2018**, *10*, 285. [CrossRef]
39. Li, M.; Ma, L.; Blaschke, T.; Cheng, L.; Tiede, D. A systematic comparison of different object-based classification techniques using high spatial resolution imagery in agricultural environments. *Int. J. Appl. Earth Observ. Geoinf.* **2016**, *49*, 87–98. [CrossRef]
40. Meyer, G.E.; Neto, J.C. Verification of color vegetation indices for automated crop imaging applications. *Comput. Electron. Agric.* **2008**, *63*, 282–293. [CrossRef]
41. Liu, T.; Li, R.; Zhong, X.; Jiang, M.; Jin, X.; Zhou, P.; Liu, S.; Sun, C.; Guo, W. Estimates of rice lodging using indices derived from UAV visible and thermal infrared images. *Agric. For. Meteorol.* **2018**, *252*, 144–154. [CrossRef]
42. Strobl, C.; Boulesteix, A.L.; Kneib, T.; Augustin, T.; Zeileis, A. Conditional variable importance for random forests. *BMC Bioinform.* **2008**, *9*, 307. [CrossRef]
43. Strobl, C.; Boulesteix, A.L.; Zeileis, A.; Hothorn, T. Bias in random forest variable importance measures: Illustrations, sources and a solution. *BMC Bioinform.* **2007**, *8*, 25. [CrossRef]
44. Simonyan, K.; Zisserman, A. Very Deep Convolutional Networks for Large-Scale Image Recognition. *arXiv* **2014**, arXiv:1409.1556.
45. Davis, J.; Goadrich, M. The relationship between Precision-Recall and ROC curves. In Proceedings of the 23rd International Conference on Machine Learning, Pittsburgh, PA, USA, 25–29 June 2006; pp. 233–240.
46. Ren, S.; He, K.; Girshick, R.B.; Sun, J. Faster R-CNN: Towards Real-Time Object Detection with Region Proposal Networks. *IEEE Trans. Pattern Anal. Mach. Intell.* **2015**, *39*, 1137–1149. [CrossRef] [PubMed]
47. Ren, Y.; Zhu, C.; Xiao, S. Small object detection in optical remote sensing images via modified faster R-CNN. *Appl. Sci.* **2018**, *8*, 813. [CrossRef]

Article

Mask R-CNN Refitting Strategy for Plant Counting and Sizing in UAV Imagery

Mélissande Machefer [1,2,*], François Lemarchand [1], Virginie Bonnefond [1], Alasdair Hitchins [1] and Panagiotis Sidiropoulos [1,3]

1 Hummingbird Technologies, Aviation House, 125 Kingsway, Holborn, London WC2B 6NH, UK; francois@hummingbirdtech.com (F.L.); virginie@hummingbirdtech.com (V.B.); alasdair@hummingbirdtech.com (A.H.); panos@hummingbirdtech.com (P.S.)
2 Lobelia by isardSAT, Technology Park, 8-14 Marie Curie Street, 08042 Barcelona, Spain
3 Mullard Space Science Laboratory, University College London, London WC1E 6BT, UK
* Correspondence: melissande@lobelia.earth

Received: 30 June 2020; Accepted: 11 September 2020; Published: 16 September 2020

Abstract: This work introduces a method that combines remote sensing and deep learning into a framework that is tailored for accurate, reliable and efficient counting and sizing of plants in aerial images. The investigated task focuses on two low-density crops, potato and lettuce. This double objective of counting and sizing is achieved through the detection and segmentation of individual plants by fine-tuning an existing deep learning architecture called Mask R-CNN. This paper includes a thorough discussion on the optimal parametrisation to adapt the Mask R-CNN architecture to this novel task. As we examine the correlation of the Mask R-CNN performance to the annotation volume and granularity (coarse or refined) of remotely sensed images of plants, we conclude that transfer learning can be effectively used to reduce the required amount of labelled data. Indeed, a previously trained Mask R-CNN on a low-density crop can improve performances after training on new crops. Once trained for a given crop, the Mask R-CNN solution is shown to outperform a manually-tuned computer vision algorithm. Model performances are assessed using intuitive metrics such as Mean Average Precision (mAP) from Intersection over Union (IoU) of the masks for individual plant segmentation and Multiple Object Tracking Accuracy (MOTA) for detection. The presented model reaches an mAP of 0.418 for potato plants and 0.660 for lettuces for the individual plant segmentation task. In detection, we obtain a MOTA of 0.781 for potato plants and 0.918 for lettuces.

Keywords: UAV; crop mapping; image analysis; precision agriculture; deep learning; individual plant segmentation; plant detection; transfer learning

1. Introduction

Despite the widely accepted importance of agriculture as one of the main human endeavours related to sustainability, environment and food supply, it is only recently that many data science use cases to agricultural lands have been unlocked by engineering innovations (e.g., variable rate sprayers) [1,2]. Two research domains are heavily contributing to this agriculture paradigm shift: remote sensing and artificial intelligence. Remote sensing allows the agricultural community to inspect large land parcels using elaborate instruments such as high-resolution visible cameras, multi-spectral and hyper-spectral cameras, thermal instruments, or LiDAR. On the other hand, artificial intelligence induces informed management decisions by extracting appropriate farm analytics in a fine-grained scale.

In the research frontline of the remote sensing/artificial intelligence intersection lies the accurate, reliable and computationally efficient extraction of plant-level analytics, i.e., analytics that are estimated for each and every individual plant of a field [3]. Individual plant management instead of generalised decisions could lead to major cost savings and reduced environmental impact for low density crops, such as potatoes, lettuces, or sugar beets, in a similar manner as localised herbicide spraying has been demonstrated to be beneficial [4,5]. For example, by identifying each and every potato, farm managers could estimate the emergence rate (the percentage of seeded potatoes that emerged), target the watering strategy to the crop and predict the yield, while the counting and sizing of lettuces determine the harvest and optimise the logistics.

The resolution required for individual plant detection and segmentation (on the order of 1–2 cm per pixel) imposes a strict limitation in the operations. Fields may be several hundreds of hectares, which implies that the acquired image would be of a particularly large size. In this work, an image with a 2 cm resolution per pixel would generate around 0.2 gigabytes per hectare. Therefore, UAV imagery for a single field can reach several tens of thousands of pixel of width and height. Any pixel-level algorithms, such as the one required for plant counting and sizing, would need to be (not only accurate but also) computationally efficient. Additionally, the adopted algorithm should be able to exhibit a near-to-optimal performance without requiring a large volume of labelled data. This constraint becomes particularly important since (a) segmentation annotations, like the ones required for individual plant identification, are typically time-consuming and costly and (b) this type of data exhibits a large variance in appearance, both due to inherent reasons (e.g., different varieties and soil types) and due to acquisition parameters (e.g., illumination, shadows). As a matter of fact, opposite to large-volume natural scene images, which are abundant and rather easy to find (e.g., [6]), UAV imagery datasets with segmentation groundtruth typically contain imagery from a unique location covering only a small area. More specifically, as far as we are aware, no large-volume plant segmentation dataset is currently available.

1.1. Recent Progress on Instance Segmentation and Instantiation

The recent proliferation of deep learning architectures [7–9], exhibiting unprecedented accuracy in a large spectrum of computer vision applications, has triggered the development of a number of variations on a basic theme, most of them focusing on specific challenging problems, which were beyond the reach of the state-of-the-art for decades. The plasticity of deep learning allowed the emergence of new solutions by linking multiple networks in more complex architectures, or even by adding or removing a few layers from a well-known model. The latter was the case in Fully Convolutional Networks (FCN) [10], which derive from models developed for classification purposes, by simply removing the last layer (used for classification), thus causing the network to learn feature maps instead of classes. This paradigm has been extensively used for binary segmentation applications, in which the model learns pixel-level masks (e.g., in our application, 0 corresponding to "not a plant" class, and "1" to "plant" class). Since a FCN can theoretically be produced from almost any Convolutional Neural Network (CNN) [7], many of the architectures typically used for classification has found a variant used for segmentation, such as AlexNet [7], VGG-16 [8], and GoogLeNet [9], or the most recent Inception-v4 [11] and ResNeXt [12].

A main issue with such simple solutions is that, since the spatial size of the final layer is generally reduced compared to the initial input size due to pooling operations within the network, the learned mask is a super-pixel one, i.e., a mask in which several pixels have been aggregated into one value. In order to recover the initial input spatial size, it has been suggested to combine early high-resolution feature maps with skip connections and upsampling [13]. In another line of research, Yu et al. [14] and Chen et al. [15] both propose to use a FCN supported by dilations (or dilated convolutions) to execute the necessary increase of receptive field. FCN used as an encoder and a symmetrical network used as a decoder characterize Segnet [16] and Unet [17] architectures, which achieve state-of-the-art performances in semantic segmentation.

In theory, a network achieving perfect accurate semantic segmentation could be straightforwardly used for instantiation (i.e., counting of blobs of one of the classes), which is merely needed is to count the connected components. However, in practice, this is a sub-optimal approach, since a model trained for segmentation would not discriminate between a false positive that erroneously joints two blobs (or a false negative that erroneously splits one blob) and a false positive (or negative) with no effect on the instantiation result. Hence, a number of deep learning techniques has been suggested to overcome this obstacle, by achieving simultaneously both semantic segmentation and instantiation. Two of the most well-known are YOLO [18–20] and Mask R-CNN [21]. The main difference between the two is that, while YOLO only estimates a bounding box of each instance (blob), Mask R-CNN goes further, by predicting an exact mask within the bounding box.

Mask R-CNN is the latest "product" of a series of deep learning algorithms developed to achieve object instantiation in images. The first version is R-CNN [22], which was introduced in 2014 as a four-step object instantiation algorithm. The first step of this algorithm consists of a selective search which finds a fixed number of candidate regions (around 2000 per image) possibly containing objects of interest. Next, the AlexNet model [7] is applied to assess if the region is valid, before using a SVM to classify the objects to the set of valid classes. A linear regression concludes the framework to obtain tighter boxes coordinates, wrapping up the bounding box closer onto the object. R-CNN achieved a good accuracy, but the elaborate architecture caused a series of issues, mainly, a very high computational cost. Fast R-CNN [23], followed by Faster R-CNN [24], gradually overcame this limitation by turning it into an end-to-end fully trainable architecture, combining models and replacing the slow candidate selection step. Despite the progress, these R-CNN variations were limited on estimating object bounding boxes, i.e., they did not perform pixel-wise segmentation. Mask R-CNN [21] accomplishes this task by adding a FCN branch to the Faster R-CNN architecture, which predicts a segmentation mask of each object. As a result, the classification and the segmentation parts of the algorithm are independently executed; hence, the competition between classes does not influence the mask retrieval stage. Another important contribution of Mask R-CNN is the improvement of pixel accuracy by refining the necessary pooling operations obtained by bilinear interpolation (instead of rounding operation) with the so-called ROIAlign algorithm.

1.2. Plant Counting, Detection and Sizing

Most of the existing "plant counting" (i.e., object instantiation in this particular application) techniques of the literature are following a regression rationale. Typically, the full UAV image is split into small tiles, before a model estimates the number of plants in the tile. Finally, the tiles are stitched and the total number of plants, as well as some localisation information, is returned.

For example, in Ribera et al. [25], the authors train an Inception-V3 CNN model on frames of perfectly row-planted maize plants to perform a direct estimation of the number of plants. On the other hand, both Aich et al. [26] and Li et al. [27] suggest a two-step regression approach (with randomly-drilled crops, respectively wheat and potato). In both of these works, initially, a segmentation mask of the plants is derived, either by training a Segnet model [26] or by simply (Otsu) thresholding a relevant vegetation index [27]. Next, a regression model is run, a CNN in [26] and a Random Forest in [27].

This two step-approach is also used in other works, e.g., [28] for Sorghum heads, which have the advantage of presenting well separated red-orange rounded shapes, [29] for denser crop stages and [30] for tomato fruits, the two last ones, though, using in field, and not remote sensing imagery. One of the weaknesses of such algorithms is that they do not consider the possibility of a plant split between two successive tiles, leading to being falsely counted twice. An interesting solution to overcome this issue was proposed in [3] and in [31] (improving [29]), who employ a network which additionally predicts a density map of the center of the plants for the entire input image.

In general, methods exploiting regression additionally provide at least some localisation information. In order to improve the accuracy, as well as to achieve an easier to be visually evaluated instantiation, several authors have suggested the estimation of the center of the plants. e.g., Pidhirniak et al. [32] propose a solution to palm tree counting with a U-Net [17] predicting a density map of the center of the plants completed by a blob detector to extract the geo-centers. A similar method is considered in [33].

In this work, a segmentation by thresholding on a vegetation index is carried out, before implementing a Watershed algorithm to extract vegetation at a sub-pixel level. Subsequently, a CNN is trained to predict pixels corresponding to the center of these plants and a post-processing step concludes the pipeline to remove outliers. A similar approach has also been developed to detect corn plants in [34] and in [35]. Kitano et al. [34] firstly use Unet [17] for segmentation and then morphological operations for detection. Kitano et al. [34] benefit from significantly higher resolution than in [35], for reference, which may lead to scalability issues when having to fly a UAV over hundreds of hectares of land. García-Martínez et al. [35] use normalized cross-correlation of samples of thresholded vegetation index map, which may introduce false positives in the presence of weeds. Malambo et al. [36] demonstrate similar work on sorghum heads detection and [37] with cotton buddings, two small rounded shaped plants. Firstly, for the segmentation task, [36] train a Segnet [16] while [37] use a Support Vector Machine model coupled with morphological operations for detection, following the approach of Kitano et al. [34].

Finally, in [38], in [39] and in [40], three methods are introduced which are closer to the context of our paper. In [38], a YOLO model is used to perform palm tree counting and bounding-box instantiation. In [39], a RetinaNet [41] is used in a weakly supervised learning model aiming at annotating a small sample of sorghum heads imagery. In [40], the authors customize a U-Net model to generate two outputs, a field of vectors pointing to the nearest object centroid and a segmentation map. A voting process estimates the plants, while also producing a binary mask, similarly to our output. Oh et al. [42] achieve the task of individual sorghum heads segmentation. Detection models for plant counting have also flourished in late 2019, with the democratization of Faster R-CNN [24]. The precursor of Mask R-CNN has been adopted in works such as in [43] for banana plant detection with three different sets of imagery at different resolutions, of which one at a similar resolution to the dataset proposed in this paper. Collecting UAV data at three different flight heights enable them to triple their dataset capacity by reusing annotations. Liu et al. [44] also benefit from Faster R-CNN [24] using high resolution maize tassels imagery from only one field and one date of acquisition, which limits the variability of encountered growth stages. In the works of [45,46], they rely on Faster R-CNN for detecting plants and [47] use Mask R-CNN for individual segmentation of oranges.

1.3. Aim of the Study

Based on the above rationale, the backbone of the selected method for achieving a fully practical and reliable individual plant segmentation and detection is an algorithm which exhibits state-of-the-art accuracy, is computationally fast during inference, and exhibits a good performance when applying transfer learning from a pre-trained model on natural scene images to a rather small domain dataset. A proposed method, which combines all of the desired characteristics, is the so-called Mask R-CNN. This architecture is suggested to be used for this task, after of course adjusting it to the ad-hoc characteristics of this particularly challenging setup. Apart from being one of the first end-to-end remotely sensed individual plant segmentation and detection methods in the literature, the main contributions of this work are (a) the adjustment of Mask R-CNN to individual plant segmentation and detection for plant sizing (b) the thorough analysis and evaluation of transfer learning in this setup, (c) the experimental evaluation of this method on datasets from multiple crops and multiple geographies, which generate statistically significant results, and (d) the comparison of the data-driven models derived with a computer vision baseline for plant detection.

The rest of this paper is organized as follows. In Section 2 Material and Methods, at first, the datasets created are detailed. Then, the Mask R-CNN architecture is discussed with a great level of detail in conjunction with an open source implementation to adapt the complex parametrization of this framework to the plant sizing problem. Subsequently, the transfer learning strategy adopted for training this network on this specific task is presented before explaining the hyperparameters' selection strategy. To follow, a computer vision baseline algorithm for plant detection is presented.

At last, all the metrics used for the experimental evaluation of this approach are described. Section 3 presents the obtained results followed by discussion in Section 4. Section 5 concludes this work.

2. Materials and Methods

2.1. Datasets

2.1.1. Study Area

UAV RGB imagery has been acquired over three different potato UK fields (P1, P2, P3), one Australian field (P4), and two different UK lettuce fields (L1, L2). Table 1 summarizes the information related to the acquisition of the imagery fore each field. The acquisition was planned in order to cover the full span of growth stages from emergence (plant size of more than 8 cm) to the point where the canopy closes. This translates into a time window for the UAV flights from 4 to 7 weeks after emergence. Having imagery over several dates allows for corregistering our imagery and reuse annotations for all images, making the data annotation process more efficient while capturing data at different growth stages. The imagery for lettuce plants only includes two acquisition dates for two different fields. However, the lettuce fields present a wider range of growth stages, compensating for the reduced number of flight dates and increasing the variety in growth stages in the dataset. Imagery acquisition is carried out on diffuse light days two hours before or after solar noon to avoid shadows and hot-spots. UAV photographs are then stitched up together into a unique orthomosaic image using a Structure-from-Motion algorithm. The annotation process will be considerably facilitated by the use of orthomosaic images. From these images, image tiles are generated with a size of 256 × 256 pixels to be fed into the model. RGB UAV images of potato plants and lettuce plants have been acquired with a resolution between 1.7 cm and 2.0 cm per pixel. At such resolutions, the canopy size and shape can be correctly identified with enough discrete pixels from the soil and vegetation classes. The fields are composed of sandy or silty soils which look similar in UAV RGB imagery. Figure 1 shows a 10 × 10 m sample from each orthomosaic image included in our datasets.

2.1.2. Datasets' Specifications

The COCO dataset features a large-scale object instance segmentation dataset with 1.5 million objects over 200 K images across 80 object categories. Imagery covers natural scenes including many different objects. Therefore, not all of the objects are labeled and the groundtruth segmentation can be coarse.

In-house task-related datasets have been created thanks to the imagery acquired in the study area, and Table 2 provides details with the following naming conventions: **C** for "Coarse", **R** for "Refined", **Tr** for "Training", and **Te** for "Test". POT_CTr, is a coarsely annotated dataset of 340 images of potato plants. Annotation work was completed by copying bounding boxes of the same size over the most merged surveys and these boxes were manually, but coarsely, adjusted. The mask was obtained by applying an Otsu threshold [27] to separate soil and vegetation within each bounding box. Growth stages selected enabled the human annotator to visually separate plants with bounding box delimitation and allowing for a small overlap for the coarse dataset. Consequently, this semi-automatic labelling displays imprecision in the groundtruthing as shown in Figure 2d. POT_CTr was used as a proof of concept that fine-tuning Mask-R-CNN with small and coarse datasets could lead to satisfactory results. However, metric results could not be trusted considering the quality of the annotation. As opposed to, POT_RTr is a training set carefully annotated at a pixel level covering sampled patches from the same fields from the UK and containing an extra field from Australia. It brings variability in the size of plants, variety and type of soils. POT_RTe presents the same characteristics and it was kept exclusively as a test set to compare all models presented. Finally, LET_RTr is also a training set accurately annotated with lettuce plants in imagery acquired in the UK. LET_RTe is the corresponding test set. Figure 2 displays samples from all these datasets. As specified in Section 1.2, [47] achieved state-of-the art accuracy training Mask R-CNN on RGB in-field images of oranges with a dataset size of 150 images of 256 × 256 × 3 and a total of 60 instances which is smaller than what our datasets look like.

Table 1. Characteristics of flight acquisition of imagery in the study area.

Field Name	Region	Crop	Date	UAV	Camera RGB Sensor	Relative Flight Height
P1	Fife, Scotland, UK	potato	13 June 2018	DJI S900	Panasonic GH4	50
P1	Fife, Scotland, UK	potato	22 June 2018	DJI S900	Panasonic GH4	50
P1	Fife, Scotland, UK	potato	28 June 2018	DJI S900	Panasonic GH4	50
P1	Fife, Scotland, UK	potato	6 July 2018	DJI S900	Panasonic GH4	50
P2	Fife, Scotland, UK	potato	22 June 2018	DJI S900	Panasonic GH4	50
P2	Fife, Scotland, UK	potato	28 June 2018	DJI S900	Panasonic GH4	50
P3	Cambridgeshire, England, UK	potato	4 June 2018	DJI S900	Panasonic GH4	50
P4	South Australia, Australia	potato	13 November 2018	DJI Inspire 2	DJI Zenmuse X4S	50
L1	Cambridgeshire, England, UK	lettuce	29 July 2018	DJI S900	Panasonic GH4	50
L2	West Sussex, England, UK	lettuce	31 May 2018	DJI S900	Panasonic GH4	50

Table 2. Summary of the charateristics of the datasets. Naming convention relies on **C** for "Coarse", **R** for "Refined", **Tr** for "Training" and **Te** for "Test".

Name	Type of Set	Number of Images	Type of Labeling	Fields	Objects
COCO	Train	200 K	Coarse		Natural Scene
POT_CTr	Train	347	Coarse	P1, P2	Potato
POT_RTr	Train	124	Refined	P1, P2, P3, P4	Potato
POT_RTe	Test	31	Refined	P1, P2, P3, P4	Potato
LET_RTr	Train	153	Refined	L1, L2	Lettuce
LET_RTe	Test	39	Refined	L1, L2	Lettuce

Figure 1. Samples of 10 × 10 m from every single orthomosaic image from our datasets: (**a–d**) P1; (**e**,**f**) P2; (**g**) P3; (**h**) P4; (**i**) L2; (**j**) L1.

Figure 2. Three examples of remote sensing imagery of low density crops (**a**–**c**), as well as their corresponding counting and sizing annotations (**d**–**f**). The (**a**,**d**), (**b**,**e**) are potato fields, respectively, from POT_CTr and POT_RTr while (**c**,**f**) are from a lettuce field from LET_RTr. Please note that the colour coding is random, apart from the fact that each single-colour blob represents a single plant, while the green area not marked (especially in the top image) signifies irrelevant to the crop vegetation (weeds).

2.2. Mask R-CNN Refitting Strategy

The deep learning model introduced to tackle the plant counting and sizing tasks is based on the Mask R-CNN architecture, which is adjusted for this problem. As a matter of fact, if the original model (i.e., the one trained on a large number of natural scene images with coarsely-annotated natural scenes objects) is trained or applied for inference without modifications of the default parameters on UAV images of plants, the results are particularly poor. The main reason behind this failure is the large number of free parameters (around 40) in the Mask R-CNN algorithm. This section reviews thoroughly the Mask R-CNN's architecture and evaluates how each parameter may affect performances. By following the Matterport's implementation [48] terminology and architecture, the goal is to unfold the parameters' tuning process followed in this paper to ensure a fast and accurate individual plant segmentation and detection output while facilitating reproducibility. This approach sets up a bridge of knowledge between the theoretical description of this network and the operational implementation on a specific task. Subsequently, strategies of transfer learning which obtain state-of-the-art results with the Mask R-CNN based architecture on the panel of datasets explored are described. To conclude this section, a description of a computer vision baseline targeting the plant detection task is presented.

2.2.1. Architectural Design and Parameters

The use of Mask R-CNN in the examined setup presents several challenges, related to the special characteristics of high-resolution remote sensing images of agriculture fields. Firstly, most of the fields have a single crop, the classification branch of the pipeline is a binary classification algorithm, a parameter which affects the employed loss function. Secondly, the remotely sensed images of plants impose the target objects (i.e., plants) to not present the same features, scales, and spatial distribution as natural scene objects (e.g., humans, cars) included in the COCO datasets used for the original Mask R-CNN model training. Thirdly, the main challenges of this setup are different than many natural scene ones. For example, false positives due to cluttered background (a main source of concern in multiple computer vision detection algorithms) is expected to be rather rare in plant counting/sizing setup, while object shadows affect the accuracy than in several computer vision applications much more. Due to these differences, Mask R-CNN parameters need to be carefully fine-tuned to achieve an optimal performance. In the following sub-sections, we analyse this fine-tuning, which, in some cases, is directly linked to the ad-hoc nature of the setup and in others is the result of a meticulous trial-and-error process. Figure 3 describes the different blocks composing the Mask R-CNN model and Table A1, in Appendix A, associates acronyms representing the examined model free parameters with the variable names used in Matterport's implementation [48] for reproducibility.

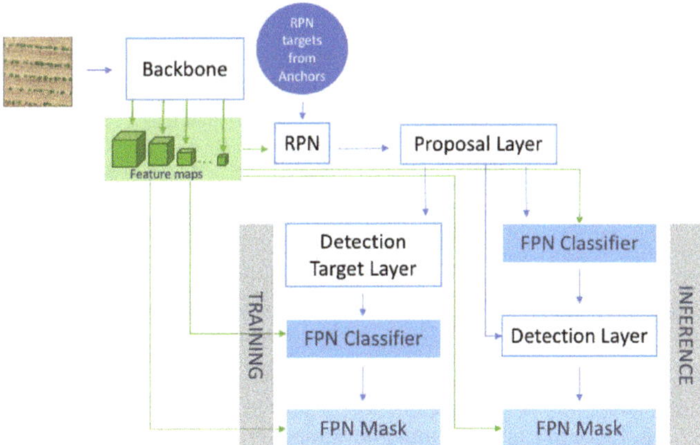

Figure 3. Mask R-CNN architecture.

2.2.2. Backbone and RPN Frameworks

The input RGB image is fed into the Backbone network in charge of visual feature extraction at different scales. This Backbone is a pre-trained ResNet-101 [49] (in our study). Each block in the ResNet architecture outputs a feature map and the resulting collection is served as an input to different blocks of the architecture: the Region Proposal Network (RPN) but also Feature Pyramid Network (FPN). By setting the backbone strides Bs_t, we can choose the sizes of the feature maps Bs_s which feed into the RPN, as the stride controls downscaling between layers within the backbone. The importance of this parameter lies in the role of the RPN. For example, $Bs_t = [4, 8, 16, 32, 64]$ induces $Bs_s = [64, 32, 16, 8, 4]$ (all units in pixels) if the input image size is squared of width 256 pixels. The RPN targets generated from a collection of anchor boxes form an extra input for the RPN. These predetermined boxes are designed for each feature map with base scales $RPNas_s$ linked to the feature map shapes Bs_s, and a collection of ratio $RPNar$ is applied to these $RPNas_s$. Finally, anchors are generated at each pixel of each feature map with a stride of $RPNas_t$. Figure 4 explains the generation of anchors for one feature map. In total, with R_l the number of RPN anchors ratios introduced, nb_a, the total number of anchors generated is defined as

$$nb_a = \sum_i int\left(\frac{Bs_s[i]}{RPNas_t}\right)^2 \times R_l \quad (1)$$

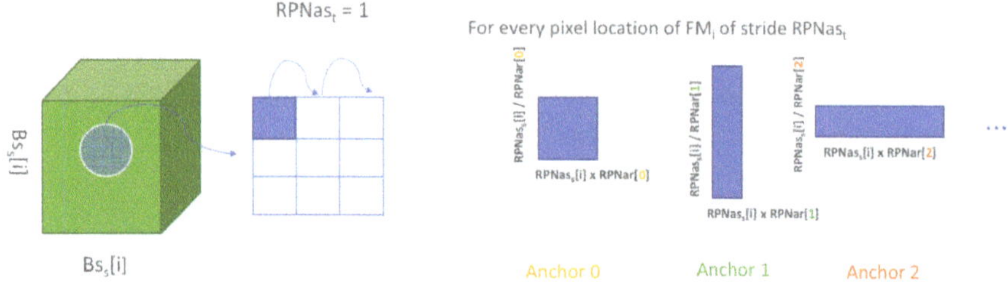

Figure 4. Anchors generation for the ith feature map FM_i feeding the RPN of shape $Bs_s[i]$. FM_i is related to anchors of corresponding anchor scale $RPNas_s[i]$ on which the collection of ratios $RPNar$ is applied to to generate R_l (number of ratios) anchors at each pixel location obtained from the FM_i with stride $RPNas_s$.

All the coordinates are computed in the original input image pixel coordinates system. All of the nb_a anchors do not contain an object of interest, implying that anchors matching the most with the groundtruth bounding boxes filters will be selected. This matching process is carried out by computing the Intersection Over Union (IoU) between anchor boxes and groundtruth bounding boxes locations. If $IoU > 0.7$, then the anchor is classified as positive, if $0.3 < IoU \leq 0.7$ as neutral and finally if $IoU \geq 0.3$ as negative. Then, the collection is resampled to ensure that the number of positive and negative anchors is greater than half of *RPNtapi*, which is a share of the total nb_a anchors kept to train the RPN. Eventually, the RPN targets have two components for each image: a vector which states if each of the nb_a anchors is positive, neutral or negative, and the second component is represented by delta coordinates between groundtruth boxes and positive anchors among the *RPNtapi* selected anchors to train the RPN. It is essential to note that only *mGTi* groundtruth instances are kept per image to avoid training on images with too many objects to detect. This parameter is important for training on natural scene images composing the COCO dataset as they might contain an overwhelming number of overlapping objects. Dimension of the targets for one image are $[nb_a]$ and $[RPNtapi, (dy, dx, \log(dh), \log(dw))]$, where dy and dx are the normalised distance of the coordinates centers between groundtruth and anchor boxes, whereas $log(dh)$ and $log(dw)$ respectively deal with the logarithm delta between height and width. Finally, the RPN is a FCN aiming at predicting these targets.

2.2.3. Proposal Layer

The Proposal Layer is plugged onto the RPN and does not consist of a network but of a filtering block which only keeps relevant suggestions from the RPN. As stated in the previous section, the RPN produces scores for each of the nb_a anchors with the probability to be characterised as positive, neutral or negative and the Proposal Layer begins by keeping the highest scores to select the best *pNMSl* anchors. Predicted delta coordinates from the RPN are coupled to the selected *pNMSl* anchors. Then, the Non-Maximum Suppression (PNMS) algorithm [50] is carried out to prune away predicted RPN boxes overlapping with each other. If two boxes among the *pNMSl* have more than *RPN_NMSt* overlap, the box with the lowest score is discarded. Finally, the top $pNMSr_{tr}$ for training phase and $pNMSr_{inf}$ for inference phase are kept based on their RPN score.

At this stage, the training and inference paths begin to be separated despite the inference path relyinng on blocks previously trained.

2.2.4. Detection Target Layer

As seen in Figure 3, the training path after the Proposal Layer begins with the Detection Target Layer. This layer is not a network but yet another filtering step of the $pNMSr_{tr}$ Regions of Interest (ROIs) outputted by the Proposal Layer. However, this layer uses the *mGTi* groundtruth boxes to compute the overlap with the ROIs and set them as a boolean based on the condition $IoU > 0.5$. Finally, these $pNMSr_{tr}$ ROIs are subsampled to a collection of *trROIpi* ROIs but randomly resampled to ensure a ratio *ROIpr* of the total *trROIpi* as positive ROIs. As the link with groundtruth boxes is established in this block and the notion of anchors is dropped, the output of this Detection Target Layer is composed of *trROIpi* ROIs with corresponding groundtruth masks, instance classes, and delta with groundtruth boxes for positive ROIs. The groundtruth boxes are padded with 0 values for the elements corresponding to negative ROIs. These generated groundtruth features corresponding to the introduced ROIs features will serve as groundtruth to train the FPN.

2.2.5. Feature Pyramid Network

The Feature Pyramid Network (FPN) is composed of the Classifier and by the Mask Graph. The input to these layers will be referred to as Area of Interest AOIs, as they are essentially a collection of regions with their corresponding pixel coordinates. The nature of these AOI can vary during the training and inference phase as shown in Figure 3. Both of the extensions of the FPN (Classifier and Mask) present the same succession of blocks, composed of a sequence of ROIAlign and convolution layers with varying goals. The ROIAlign algorithm, as stated in the beginning of this section, has to pool all feature maps from FPN lying within the AOIs by discretising them into a set of fixed square pooled size bins without generating a pixel misalignment unlike traditional pooling operations. After ROIAlign is applied, the input of the convolution layers is a collection of same size squared feature maps, and the size of the batch is the number of AOIs. In the case of the Classifier, the output of these deep layers is composed of a classifier head with logits and probabilities for each item of the collection to be an object and belong to a certain class as well as refined box coordinates which should be as close as possible from the groundtruth boxes used at this step. In the case of the Mask graph, the output of this step is a collection of masks of fixed squared size which will later on be re-sized to the shape in pixels of the corresponding bounding box extracted in the classifier.

In the previous Section 2.2.4, it has been explained that, for the training phase, the Detection Layer Target outputs *trROIpi* AOIs and computes corresponding groundtruth boxes, class instances, and masks which are used for the training of a sequence of FPN Classifier and Mask Graph as shown in Figure 3. For the inference path, the trained FPN is used in prediction, but the FPN Classifier is firstly applied to the $pNMSr_{inf}$ AOIs extracted from the Proposal Layer, followed by a Detection Layer, as detailed in the following Section 2.2.6. It ensures the optimal choice of AOIs to only keep *Dmi* AOIs. Finally, these masks are extracted by the final FPN Mask Graph in prediction mode.

2.2.6. Detection Layer

This block is dedicated to filtering the $pNMSr_{inf}$ proposals coming out from the Proposal Layer based on probability scores per image per class extracted from the FPN Classifier Graph in inference mode. AOIs with low probability scores under *Dmc* are discarded, and NMS is applied to remove lowest score AOI overlapping more than *DNMSt* with a higher score. Finally, only the best *Dmi* AOIs are selected to extract their masks with the following block FPN Mask Graph.

Each of these blocks is trained with a respective loss in regard to the nature of its function. Boxes' coordinates prediction is associated with a smooth L1-loss, binary mask segmentation with binary cross-entropy loss, and instance classification with categorical cross-entropy loss. The Adam optimiser [51] is used to minimize these loss functions.

2.2.7. Transfer Learning Strategy

Transfer learning describes the process of using a model pretrained for a specific task and retraining this model for a new task [52]. The benefits of this process are to improve training time and performance due to previous domain knowledge demonstrated by the chosen model. Transfer learning can only be successful if the features learnt from the first model are general enough to envelope the targeted domain of the new task [53].

Deep supervised learning methods require a large amount of annotated data to produce the best results possible, especially with large architecture with several millions of parameters to learn such as Mask R-CNN. Therefore, the necessary dataset size is the main factor for a successful model achieving state-of-the-art instance segmentation. Zlateski et al. [54] advocate for 10 K images per class *accurately labeled* to obtain satisfying results on instance segmentation tasks with natural scene images. At a resolution between 1.5 cm and 2 cm, digitising 150 masks of a low density crop takes around an hour for a trained annotator. As a result, precise pixel labeling of 10 K images implies a significant amount of person-hour. However, Zlateski et al. [54] prove that pre-training CNN on a large coarsely-annotated dataset, such as the COCO dataset (200 K labeled images) [6] for natural scene images, and fine-tuning on a smaller refined dataset of the same nature leads to better performances than training directly on a larger finely-annotated dataset. Consequently, transferring the learning from the large and coarse COCO dataset to coarse and refined smaller plant datasets makes the complex Mask R-CNN architecture portable to the new plant population task.

2.2.8. Hyper Parameters Selection

The parametrization setup of the Mask R-CNN model is key to facilitating the training of the model and maximizing the number of correctly detected plants. It was observed that the default parameters in the original implementation of [48] would lead to poor results due to an excessive amount of false negatives. Therefore, an extensive manual search guided by the the understanding the complex parameterization process of Mask R-CNN detailed in Section 2.2.1. Mask R-CNN is originally trained on the COCO dataset which is composed of objects of highly varying scales that can sometimes fully overlap between each other (leading to the presence of the so-called crowd boxes). In consideration of the datasets presented in our study, most of the objects of interest have a smaller range of scales, and two plants cannot have fully overlapping bounding boxes. Regarding the scale of both potato and lettuce crops, an individual plant goes from 4 to 64 pixels at our selected resolutions. Based on these observations, optimising the selection of the regions of interest through the RPN and the Proposal Layer can be solved by tuning the size of the feature maps Bs_s and the scale of the anchors $RPNas_s$. Bs_t cannot be modified due to the Backbone being pre-trained on COCO weights and the corresponding layers frozen. Then, following the explanation in Section 2.2.2, the pixel size of 256×256 was chosen to include sizes of feature maps corresponding to the range of scales of the plant "objects". In addition, taking into account the imagery resolution (see Section 2.1.2) and an estimation of the range of the plant drilling distance, $mGTi$ can easily be inferred. We estimated that not more than $mGTi = 128$ potatoes and $mGTi = 300$ lettuces could be found in an image of 256×256 pixels. This observation also allows for setting the number of anchors per image *RPNtapi* to train the RPN and the number of *trROIpi* in the Detection Target layer to be set to the same value. Starting from this known estimation of the maximum number of expected plants, this bottom-up view of the architecture is key to finding a more accurate number of AOIs to keep at each block and phase for each of the crops investigated. The parameters involved are $pNMSr_{tr}$ and $pNMSr_{inf}$. Manual variations of these parameters by step of 100 have been attempted. Thresholds used for IoU in the NMS (*RPN_NMSt*, *DNMSt*) and confidence scores (*Dmc*) can also be tuned but the default values were kept due to our tuning attempts being inconclusive.

2.3. Computer Vision Baseline

A computer vision baseline for plant detection is considered to compare predicted plant centers by the Mask R-CNN model. As a starting point of this method, the RGB imagery acquired by UAV has to be transformed to highlight the location of the center of the plants. Computing vegetation indexes to highlight image properties is a mastered strategy in the field of remote sensing. Each vegetation index presents their own strengths and weaknesses as detailed in Hamuda et al. [55]. The Colour Index of Vegetation Extraction (CIVE) [56] is presented as a robust vegetation index dedicated to green plant extraction in RGB imagery and defined as

$$CIVE = 0.441R - 0.811G + 0.385B + 18.78745 \tag{2}$$

where R, G and B respectively stand for red, green and blue channels defined between 0 and 255. In order to have a normalised index, all channels are individually divided by 255. Segregating between vegetation and soil pixels in an automated manner is possible using an adaptive threshold such as the Otsu method [27]. In this classification approach, foreground and background pixels are extracted by finding the threshold value which minimizes the intra-class intensity variance. By applying Otsu thresholding on CIVE vegetation index, a binary soil vegetation map is obtained, unaltered by highly varying conditions of illumination and shadowing. The Laplacian filtered CIVE map with masked soil pixels (LfC) highlights regions of rapid intensity change. Potato plants and lettuce plants pixels of this map are meant to have minimal change close to their center due to homogeneous and isotropic properties of their visual aspect. Therefore, finding the local minima of the LfC map should output the geolocation of these plant centers. This is why all LfC pixels are tested as geo-centers of a fixed size disk window and are only considered as the center of a plant if their value is the minimum of all pixels located within the window area. This framework is summarized in Figure 5.

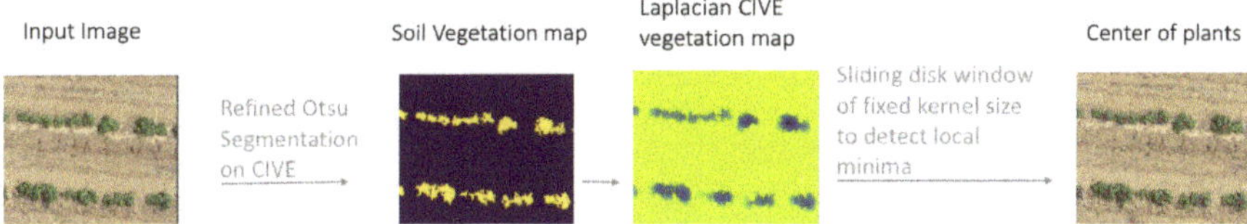

Figure 5. Computer vision based plant detection framework. A CIVE vegetation index is computed from a UAV RGB image before applying Otsu thresholding to discriminate pixels belonging to vegetation and soil. Laplacian filtering is applied to a soil masked CIVE map and, finally, a disk window is slid over this latest to extract local minima, corresponding to the center of the plants.

2.4. Evaluation Metrics

Metrics dedicated to the evaluation of the performance of algorithms in computer vision are strongly task and data dependent. In this paper, our first interest is to compare different strategies of transfer learning using various datasets adopting a data hungry Mask R-CNN model for the instance segmentation task. A plant is considered as a true positive if the intersection over union of the true mask $\mathbf{M_t}$ and a predicted one $\mathbf{M_p}$ is over a certain threshold $\mathbf{T}$ as $\frac{M_t \cap M_p}{M_t \cap M_p} > T$. Varying this threshold from 0.5 to 0.95 with a step scale of 0.05 with averaging allows the widening of the definition of a detected plant. It defines the mean Average Precision (mAP) such as the average precision for a set of different thresholds.

Detection evaluation differs from the individual segmentation metric as the output of the algorithm is a collection of points. The Multiple Object Tracking Benchmark [57] comes with a measure which covers three error sources: false positives, missed plants, and identity switches. A distance map between predicted and true locations of the plants is computed to determine the possible pairs and derive **m** the number of misses, $\mathbf{f_p}$ number of false positives, **mm** the number of mismatches, and **g** the number of objects which allows for computing the multiple object tracking accuracy (MOTA)

$$\text{MOTA} = 1 - \frac{m + f_p + mm}{g} \tag{3}$$

MOTA metric presents the advantage to synthesize error sources monitoring while the conventional Precision and Recall metrics allow for more straightforward interpretability. With $\mathbf{t_p}$ being the number of true positives, Precision and Recall are defined as follows:

$$\text{Precision} = \frac{t_p}{t_p + f_p} \tag{4}$$

$$\text{Recall} = \frac{t_p}{g} \tag{5}$$

3. Results

All models have been trained using an Nvidia GPU TITAN X 12 GB. A GPU with 12 GB of memory that typically allows a maximum batch size of eight images of 256 × 256 × 3 dimensions. We did not use batch normalization due to performance losses demonstrated on small batches [58].

Live data augmentation was performed on the training images with vertical and horizontal flips to artificially increase the training set size. Training a model without data augmentation can lead to overfitting and trigger an automated early stopping of the training phase.

Datasets used for training phases were split into 80% for training and 20% for validation to evaluate the model at the end of each training epoch.

Hyper parameters for each of the following models have been set in accordance with the method stated in Section 2.2.8. Table 3 states the different settings used for training on each dataset described in Table 2.

Table 3. Derived parameters for each Mask R-CNN model thanks to the architecture examination and hyper parameters selection (see Sections 2.2.1 and 2.2.8).

Parameter	COCO	Model M1_POT	Models M2_POT and M3_POT	Models M1_LET and M2_LET and for M3_POT in Inference on LET_RTe Dataset
Bs_t	[4, 8, 16, 32, 64]	[4, 8, 16, 32, 64]	[4, 8, 16, 32, 64]	[4, 8, 16, 32, 64]
$RPNas_s$	(32, 64, 128, 256, 512)	(8, 16, 24, 32, 48)	(8, 16, 24, 32, 48)	(8, 16, 24, 32, 48)
$RPNas_t$	2	2	2	2
$RPNar$	[0.5, 1, 2]	[0.5, 1, 2]	[0.5, 1, 2]	[0.5, 1, 2]
$RPNtapi$	256	128	128	300
$pNMSl$	6000	6000	6000	6000
RPN_NMSt	0.7	0.7	0.7	0.7
$pNMSr_{tr}$	2000	1500	1500	1800
$pNMSr_{inf}$	1000	800	800	1100
$mGTi$	100	128	128	300
$trROIpi$	128	128	128	300
$ROIpr$	0.33	0.33	0.33	0.33
Dmi	100	80	120	300
Dmc	0.5	0.9	0.7	0.7
$DNMSt$	0.3	0.3	0.5	0.5

3.1. Individual Plant Segmentation with Transfer Learning

First, we performed transfer learning from the COCO weights (obtained by training the Mask R-CNN with parameters found in Table 3) on the POT_CTr dataset (model M1_POT). Mean Average Precision (mAP) over POT_RTe is 0.084 which is low, as initially expected, due to the POT_CTr training set being coarsely annotated. We then performed transfer learning from the COCO weights to the POT_RTr dataset and obtained a higher mAP of 0.406 on the POT_RTe dataset due to the refined annotations of the POT_RTr dataset (model M2_POT). Finally, retraining on POT_RTr dataset from the POT_CTr weights (model M3_POT) led to the best mAP of 0.418 demonstrating that pre-training a network on a coarsely labeled dataset before fine-tuning on a polished dataset can definitely boost the accuracy of a model. Figure 6a–c respectively show the inference of M1_POT with mAP of 0.011, M2_POT with mAP of 0.284 and M3_POT with mAP of 0.316 on an image from the POT_RTe dataset. The resulting segmentation and the number of potato plants detected improve significantly as the mAP values are increased. Moreover, the boundaries of the predicted masks drawn on the images fit the contours of the plants better.

Figure 6. Predictions of the Mask R-CNN and Computer Vision (CV) methods. Yellow boundaries depict the groundtruth masks, red boundaries represent the predicted masks of the Mask R-CNN model, and red crosses surrounded by blue disks are the predictions of the computer vision method for plant centers. Samples shown are a subset of the entire image of spatial size 256 × 256 for visualization purpose but metrics computed stand for the whole image. In first row, one cropped sample image from POT_RTe is shown: (**a**) M1_POT; (**b**) M2_POT; (**c**) M3_POT; (**d**) CV. In the second row, one cropped sample image from LET_RTe is shown: (**e**) M3_POT; (**f**) M1_LET; (**g**) M2_LET; (**h**) CV.

In order to assess whether M3_POT, which has been trained on potato plants, presents the same performance on another low density green-colored crop, M3_POT was used in inference on LET_RTe, composed of aerial images containing lettuce plants in fields. A mAP of 0.042 was calculated and by dissecting the parameters used in Mask R-CNN, this low accuracy score was explained by an excessively small $RPNtapi = 128$ limiting the number of possible lettuce plants which could be found in one image. By raising $RPNtapi$ to 300 as well as the number of ROIs kept after using the NMS in training and inference phases (parameters $pNMSr_{tr}$ and $pNMSr_{inf}$), the likelihood of detecting all lettuce plants is significantly increased. As the number of AOIs to detect is higher, *trROIpi* was also increased to obtain more ROIs training the FPN. Consequently, the number of groundtruth objects to consider *mGTi* and the number of predicted objects *Dmi* were also inflated. Following these adjustments, a mAP of 0.095 for M3_POT used in inference on LET_RTe still proved to be low, implying that the updated training parameters would not make any difference in prediction mode. Therefore, we conclude that a model trained on potato plants transfers poorly to lettuce plants detection and segmentation. Two models based on the COCO weights and M3_POT weights were then trained separately on the LET_RTr dataset. The models (M1_LET and M2_LET) respectively obtained a mAP of 0.641 and 0.660 on the LET_RTe dataset. A similar conclusion as for models trained on potato plants can be drawn since pre-training on another crop seems to improve the final model. Figure 6e–g respectively show the inference outputs of M3_POT with mAP of 0.178, M1_LET with mAP of 0.309, and M2_LET with mAP of 0.595 on an image from the LET_RTe dataset. The resulting segmentation and the number of potato plants detected greatly improve as the mAP values demonstrate, as well as the boundaries of the predicted masks. Higher values of mAP are observed for lettuce plants than for potato plants. This could imply that the individual plant segmentation task is easier to achieve on the lettuce crop considering the conditions of acquisition of the imagery in both of the datasets, detailed in Section 2.1.

3.2. A Comparison of Computer Vision Baseline and Mask R-CNN Model for Plant Detection

The presented Mask R-CNN model for individual plant segmentation encompasses the fulfillment of the task of detection. The models with highest mAP, respectively M3_POT for potato plants and M2_LET for lettuce plants, have been compared to the Computer Vision based model detailed in Section 2.3. This latest model outputs coordinates corresponding to the center of the plants and postprocessing was also conducted on the masks predicted by the Mask R-CNN based model to calculate the plants' centroids for comparison. Table 4 displays results for both of the methods applied on the POT_RTe dataset. MOTA, precision, and recall scores are all higher for the Mask R-CNN based model than for the traditional Computer Vision (CV) method. Figure 6c and Figure 6d respectively show the predictions of M3_POT and CV on a sample from POT_RTe. It should be noted that some weeds are wrongly classified as potato plants by the CV method and a higher number plants were missed than with the Mask R-CNN based model. These false negatives can be explained for the CV method by the distance between two local minima of the LfC map being smaller than the size of the chosen window which occurs when plants are very close to each other. It is suggested that a small and isolated plant can be missed due to a possible inaccuracy of the Otsu thresholding or an excessively strong smoothing effect of the Laplacian. However, it shall be noted that the imagery within POT_RTe represents fine farming practices and data collection timing, implying that there is close to no weed in the field and the potato plants' canopy is not closed. It is thought that M3_POT could have encountered difficulties if any of these two edge cases would have appeared in the data.

Table 4. Results of the comparison of Computer Vision (CV) and Mask R-CNN base algorithm for plant detection on POT_RTe.

	CV	M3_POT
MOTA	0.766886	0.781426
Precision	0.983295	0.997166
Recall	0.800657	0.825047

The same experiment is repeated to compare the CV method and M2_LET on LET_RTe. Once again, better MOTA values are observed for the Mask R-CNN based model as seen in Table 5. Lettuce plants display the strong advantage of being of a round shape, compact and well-separated from each other. Lettuce is also a high-value crop, meaning that the amount of care per hectare of crop is high and may result in more organised fields with a reduced presence of weeds. These characteristics translate into a more systematic uniqueness of the local minima of the

LfC. Consequently, the corresponding center of a lettuce plant is easier to detect than for a potato plant. Indeed, potato plants can present an irregular canopy and, often, merge with other plants. The CV method reaches the high scores of MOTA (0.858), Precision (0.997), and Recall (0.882). The deep learning model *M2_LET* also benefits from the advantageous visual features displayed by lettuce plants in comparison with potato plants. A perfect precision score of 100% is reached for the Mask R-CNN refitted model, meaning that no element is wrongly identified as a plant for the entire LET_RTe test set. The model also outperforms the CV method with a MOTA of 0.918 and a Recall of 0.954. A sample extracted from the LET_RTe dataset is processed by M2_LET and illustrated in Figure 6g. It can be observed that only lettuce plants on the edges of the image are missed due to the convolutional structure of the network. In comparison to the CV method, with its results illustrated in Figure 6h, is small lettuces that are missed.

Table 5. Results of the comparison of Computer Vision (CV) and Mask R-CNN base algorithm for plant detection on LET_RTe.

	CV	M2_LET
MOTA	0.857887	0.918403
Precision	0.997476	1.0
Recall	0.882192	0.954365

To emphasise the real-world limits of our models, we have converted every pixel mask predicted by *M2_LET* and *M3_POT* into a square centimetre area using the resolution per pixel. We can observe that the smallest and largest lettuce plants are respectively 122 and 889 cm^2, with a mean leaf area of 405 cm^2. The smallest and largest potato plants are respectively 40 and 391 cm^2, with a mean at 126 cm^2. These values inform us that the *M3_POT* can detect potato plants as small as 4–5 pixels in diameter in the imagery. Moreover, smaller plants detected by *M3_POT* can be interpreted by the fact that potato fields need to be flown early in the season before the plants' canopy merges. Figure 7 shows a sample image from the L1 field with the predicted masks by *M2_LET*. The corresponding size in square centimeter is superimposed over each plant.

Figure 7. Sizing of each lettuce plant in cm^2 for a sample image patch of L1 field. Each lettuce is overlaid with its predicted mask. The masks' green colour becomes increasingly darker with the plant size.

4. Discussion

In the existing literature, detailed in Section 1.2, the methodology applied to tackle the plant detection and counting problems often involves a two-step procedure which, firstly, performs segmentation of the vegetation and, secondly, a detection of the plants' centroids, using either computer vision or deep learning methods [25–27,32–37]. By relying on the Mask R-CNN architecture, our work presents an all-in-one model allowing to output individual plant masks, aiming at perfecting the complex instance segmentation task and intrinsically leading to outstanding plant detection and counting performances.

In this paper, the focus is both to accurately detect plants to estimate a count per area and to delineate every single plant's boundaries at pixel-level. In contrast with our work, previous works have adapted regression-based models to predict a count for a given area without any plant location or plant segmentation [25–27]. Moreover, we avoid the use of vegetation indices [27,33,35] because (a) the most renowned ones require a multi-spectral camera, which typically has a coarser spatial resolution and is more expensive than an RGB camera and (b) other indices, such as CIVE (Color Index of Vegetation Extraction), are linear combinations of the RGB channels, i.e., learnable from the neural network without explicitly estimating them in a pre-processing stage. Finally, we use an all-in-one model which has already demonstrated its potential in agricultural scenarios using in-field imagery taken from the ground [45–47] instead of UAV photographic imagery. By using this unique Mask R-CNN model, we avoid an additional "patching" step seen in a number of techniques explored in the literature which leads to more complex processing pipelines and performance comparability. [25–27,32–37].

Previous studies have also developed algorithms aiming at solving individual plant segmentation such as [38] on palm trees and [39] on sorghum heads. However, the imagery resolutions and plant types studied are not comparable with our presented work. Moreover, [38,39,42] solely assess their solutions on the detection task and could not evaluate them on a sizing task due to a lack of accurately-digitized groundtruth mask. Dijkstra et al. [40] propose a method carrying out individual remotely sensed potato plant bounding box detection. This work presents a notable difference with ours as the objective was not to generate accurate masks but exclusively to predict the plants' centroids within a targeted area associated with each plant. Hence, none of the existing methods enumerated above is directly comparable with our work due to differences in task complexity and metrics being too great.

Bringing together remote sensing and deep learning for plant instance segmentation with the Mask R-CNN embodies a direct automatic cutting-edge approach. Moreover, it outperforms parametrically and a multistep conventional computer vision baseline when used for plant detection. These results were based on the mean Average Precision as a metric for instance segmentation. This metric not only takes into account the matching between the groundtruth masks and their corresponding predictions, but also gradually penalizes the correctness of the resulting mask, at a pixel level, from 0.5 to 0.95 of Intersection Over Union. The images collected for our datasets have been manually digitized with a high precision. This allows for assessing the quality of the models' predictions by comparing with the annotations for each plant. This specific characteristic of our datasets enabled our study to be the first one, to the best of our knowledge, to use mAP using the masks representing plants in UAV images to quantify a model's performance on a plant sizing task. Ganesh et al. [47] also used Mask R-CNN for in-field orange fruit detection in the trees. However, they don't evaluate their algorithm for sizing such as we do thanks to the mean Average Precision metric. The MOTA metric, derived from The Multiple Object Tracking Benchmark [57], has recently been introduced in the remote sensing field and represents with fidelity the performance of algorithms developed for object detection. The best model obtained for the potato crop reaches a mAP of 0.418, and 0.660 for the lettuce crop. For plant detection only, these same models respectively obtain a MOTA score of 0.781 for potato plants and 0.918 for lettuce plants. In comparison, the traditional computer vision baseline solution tested in this paper only obtains 0.767 and 0.858 for the same crops, respectively. Ganesh et al. [47], studying in-field orange fruit detection, obtain a Precision of 0.895 and a Recall of 0.867 with their Mask R-CNN model. With our refitted Mask R-CNN models on remotely sensed images, we reach a Precision of 0.997 (potato plants) and 1.0 (lettuce plants) and a Recall of 0.825 (potato plants) and 0.954 (lettuce plants).

Annotated data in the remote sensing field is a common limitation to projects leveraging the power of deep learning as supervised techniques are data-greedy. The sequential training phases on large natural scenes datasets such as COCO [6], on the coarsely-annotated potato plants dataset and, finally, on the smaller and refined labeled potato plants dataset, allowed to build the best model and highlights a strategy to tackle data scarcity. The resulting model is crop specific despite the fact that potato and lettuce crops are two low-density green-coloured crops. Nonetheless, using the same weights as a basis for a new model for lettuce plant instance segmentation, turns to be a successful strategy. This demonstrates transfer learning capabilities across datasets containing imagery of different plant types. Models yielded poor results when trained on a first specific dataset and used in inference on a second one with other types of objects. This justifies the necessity of understanding the complex parameterization process of Mask R-CNN and this study is the first one, to the best of our knowledge, which disseminates in detail the nodes and effects of this complex model. It also facilitates reproducibility by using the notations of variable names used in an open-source implementation [48].

Our Mask R-CNN based model is robust as it is able to accurately detect and segment plants affected by shadowing effects, occluded by foliage or some degree of overlap between plants. Limitations occur once plants reach a greater merging state but humans annotators also encounter difficulties when digitising plant masks from UAV imagery. The computer vision baseline method for detection has shown to be more sensitive to merged plants. Our Mask R-CNN based model is also capable of rightly ignoring weeds which could be mistaken for potato plants. In contrast, the computer vision baseline algorithm frequently generates false positives when encountering weeds or human-made objects.

To apply this deep learning model presented in this paper to a real-world problem, it is important to note that it has only been trained on frames of 256×256 pixels. Moreover, the individual plant segmentation outputted is over one frame instead of an image sometimes covering hundreds of hectares. We suggest that, in order to process an entire potato or lettuce field, a preprocessing gridding step should be added so that each frame is fed into the model individually. Then, a postprocessing mosaicking step should be used to reconstitute the whole image and display all predictions at once. Due to the model having issues with plants located right on the image border, a sliding window with overlap between two frames could be developed for the preprocessing module. In regard to the postprocessing module, it should include the cropping of the corresponding overlap whilst plants segmented on two frames should be merged as a unique one. Such preprocessing and postprocessing steps would help to validate the model, but also to potentially create systems that automatically count an entire field and possibly assess the maturity of the crops based on their size. The development of such tool could lead to novel field management practices such as variable fertiliser spraying per plant depending on their size or plant growth stage estimation.

5. Conclusions

Combining remote sensing and deep learning for plant counting and sizing using the Mask R-CNN architecture embodies a direct and automatic cutting-edge approach. Moreover, it outperforms the manually-parametrized computer vision baseline requiring multiple processing steps when used for plant detection. This study is the first one, to the best of our knowledge, to evaluate an algorithm of individual instance segmentation on remotely sensed images of plants. The datasets are composed of high quality digitized masks of potato and lettuces. The success of this approach is conditioned by transfer learning strategies and by the correct tuning of the numerous parameters of the Mask R-CNN training process, both detailed in this study. This justifies the necessity of understanding the complex parameterization process of Mask R-CNN and disseminating the implications and effects of this complex model's parameters for remote sensing applications. It also facilitates reproducibility by using the notations of variable names of the open-source Matterport implementation [48]. The experiment design has been established to account for practical constraints of the remote sensing field for precision agriculture: commercial farming practices represented in the variability of images of the dataset, scalability for operational use of the model, and scarcity of annotated images at a pixel level.

Author Contributions: Conceptualization, M.M. and F.L.; Data curation, F.L. and V.B.; Investigation, M.M.; Methodology, M.M.; Supervision, F.L.; Validation, M.M.; Visualization, M.M.; Writing—original draft, M.M.; Writing—Review and editing, M.M., F.L., A.H., and P.S. All authors have read and agreed to the published version of the manuscript.

Funding: This research received no external funding.

Conflicts of Interest: The authors declare no conflict of interest.

Appendix A

Table A1. Acronyms correspondence with Matterport implementation [48].

Acronym	Variable Name in Matterport Implementation [48]	Description
Bs_t	*BACKBONE_STRIDES*	List of strides in pixels of each convolution layer which generates the feature maps used from the Backbone.
Bs_s	*BACKBONE_SHAPE*	List of width in pixels of each squared feature map used to feed the RPN obtained from Bs_t
$RPNas_s$	*RPN_ANCHOR_SCALE*	List of base width in pixels of each anchor used for each feature map.
$RPNas_t$	*RPN_ANCHOR_STRIDE*	Stride in pixels of the locations of the anchors generated for each feature map.
$RPNar$	*RPN_ANCHOR_RATIO*	List of ratio to apply on each element of $RPNas_s$ aiming at generating non squared anchors (see Figure 4).
RPNtapi	*RPN_TRAIN_ANCHORS_PER_IMAGE*	Number of anchors per image selected to train the RPN.
pNMSl	*PRE_NMS_LIMIT*	Number of kept proposals outputted by the RPN based on their RPN scores.
RPN_NMSt	*RPN_NMS_THRESHOLD*	IoU threshold for stating overlapping RPN predicted boxes.
$pNMSr_{tr}$	*POST_NMS_ROIS_TRAINING*	Number of ROIs to keep after NMS in the Proposal Layer for the training phase.
$pNMSr_{inf}$	*POST_NMS_ROIS_INFERENCE*	Number of ROIs to keep after NMS in the Proposal Layer for the inference phase.
mGTi	*MAX_GROUNDTRUTH_INSTANCES*	Number of groundtruth instances per image kept to train the network.
trROIpi	*TRAIN_ROI_PER_IMAGE*	Number of ROIs to keep after the Detection Target Layer.
ROIpr	*ROI_POSITVE_RATIO*	Ratio of positive ROIs among the *trROIpi* .
Dmi	*DETECTION_MAX_INSTANCES*	Maximal number of instances that the Mask R-CNN is allowed to output per image in the prediction mode. It also corresponds to the number of AOIs outputted by the Detection Layer.
Dmc	*DETECTION_MIN_CONFIDENCE*	Minimum FPN Classifier probability for an AOI to be considered as containing an object of interest.
DNMSt	*DETECTION_NMS_THRESHOLD*	NMS threshold used in the Detection Layer.

References

1. Kamilaris, A.; Prenafeta-Boldú, F.X. Deep learning in agriculture: A survey. *Comput. Electron. Agric.* **2018**, *147*, 70–90. [CrossRef]
2. Han, S.; Hendrickson, L.; Ni, B. A variable rate application system for sprayers. In Proceedings of the 5th International Conference on Precision Agriculture, Bloomington, MN, USA, 16–19 July 2000; pp. 1–9.
3. Wu, J.; Yang, G.; Yang, X.; Xu, B.; Han, L.; Zhu, Y. Automatic Counting of in situ Rice Seedlings from UAV Images Based on a Deep Fully Convolutional Neural Network. *Remote Sens.* **2019**, *11*, 691. [CrossRef]
4. Melland, A.R.; Silburn, D.M.; McHugh, A.D.; Fillols, E.; Rojas-Ponce, S.; Baillie, C.; Lewis, S. Spot Spraying Reduces Herbicide Concentrations in Runoff. *J. Agric. Food Chem.* **2015**, *64*, 4009–4020. [CrossRef]

5. Rees, S.; McCarthy, C.; Baillie, C.; Burgos-Artizzu, X.; Dunn, M. Development and evaluation of a prototype precision spot spray system using image analysis to target Guinea Grass in sugarcane. *Aust. J. Multi-Discip. Eng.* **2011**, *8*, 97–106. [CrossRef]
6. Lin, T.; Maire, M.; Belongie, S.J.; Bourdev, L.D.; Girshick, R.B.; Hays, J.; Perona, P.; Ramanan, D.; Dollár, P.; Zitnick, C.L. Microsoft COCO: Common Objects in Context. In Proceedings of the European Conference on Computer Vision, Zurich, Switzerland, 6–12 September 2014; Springer: Cham, Switzerland, 2014; pp. 740–755.
7. Krizhevsky, A.; Sutskever, I.; Hinton, G.E. ImageNet Classification with Deep Convolutional Neural Networks. In Proceedings of the Advances in Neural Information Processing Systems 25, Lake Tahoe, NV, USA, 3–6 December 2012; pp. 1097–1105.
8. Simonyan, K.; Zisserman, A. Very Deep Convolutional Networks for Large-Scale Image Recognition. *arXiv* **2014**, arXiv:1409.1556.
9. Szegedy, C.; Liu, W.; Jia, Y.; Sermanet, P.; Reed, S.E.; Anguelov, D.; Erhan, D.; Vanhoucke, V.; Rabinovich, A. Going Deeper with Convolutions. In Proceedings of the IEEE Conference on Computer Vision and Pattern Recognition (CVPR), Boston, MA, USA, 7–12 June 2015.
10. Kemker, R.; Salvaggio, C.; Kanan, C. Algorithms for semantic segmentation of multispectral remote sensing imagery using deep learning. *ISPRS J. Photogramm. Remote Sens.* **2018**, *145*, 60–77. J.ISPRSJPRS.2018.04.014. [CrossRef]
11. Szegedy, C.; Ioffe, S.; Vanhoucke, V. Inception-v4, Inception-ResNet and the Impact of Residual Connections on Learning. *arXiv* **2016**, arXiv:1602.07261.
12. Xie, S.; Girshick, R.B.; Dollár, P.; Tu, Z.; He, K. Aggregated Residual Transformations for Deep Neural Networks. In Proceedings of the IEEE Conference on Computer Vision and Pattern Recognition, Honolulu, HI, USA, 21–26 July 2017.
13. Shelhamer, E.; Long, J.; Darrell, T. Fully Convolutional Networks for Semantic Segmentation. *IEEE Trans. Pattern Anal. Mach. Learn.* **2017**, *39*, 640–651. [CrossRef]
14. Yu, F.; Koltun, V. Multi-Scale Context Aggregation by Dilated Convolutions. *arXiv* **2015**, arXiv:1511.07122.
15. Chen, L.; Papandreou, G.; Kokkinos, I.; Murphy, K.; Yuille, A.L. DeepLab: Semantic Image Segmentation with Deep Convolutional Nets, Atrous Convolution, and Fully Connected CRFs. *IEEE Trans. Pattern Anal. Mach. Intell.* **2017**, *40*, 834–848. [CrossRef]
16. Badrinarayanan, V.; Kendall, A.; Cipolla, R. SegNet: A Deep Convolutional Encoder-Decoder Architecture for Image Segmentation. *IEEE Trans. Pattern Anal. Mach. Intell.* **2017**, *39*, 2481–2495. [CrossRef]
17. Ronneberger, O.; Fischer, P.; Brox, T. U-Net: Convolutional Networks for Biomedical Image Segmentation. In Proceedings of the International Conference on Medical Image Computing and Computer-Assisted Intervention, Munich, Germany, 5–9 October 2015; pp. 234–241.
18. Redmon, J.; Divvala, S.K.; Girshick, R.B.; Farhadi, A. You Only Look Once: Unified, Real-Time Object Detection. In Proceedings of the IEEE Conference on Computer Vision and Pattern Recognition, Las Vegas, NV, USA, 27–30 June 2016; pp. 779–788.
19. Redmon, J.; Farhadi, A. YOLO9000: Better, Faster, Stronger. In Proceedings of the IEEE Conference on Computer Vision and Pattern Recognition, Las Vegas, NV, USA, 27–30 June 2016; pp. 7263–7271.
20. Redmon, J.; Farhadi, A. YOLOv3: An Incremental Improvement. *arXiv* **2018**, arXiv:1804.02767.
21. He, K.; Gkioxari, G.; Dollár, P.; Girshick, R.B. Mask R-CNN. In Proceedings of the International Conference on Computer Vision, Venice, Italy, 22–29 October 2017; pp. 2961–2969.
22. Girshick, R.B.; Donahue, J.; Darrell, T.; Malik, J. Rich feature hierarchies for accurate object detection and semantic segmentation. In Proceedings of the IEEE Conference on Computer Vision and Pattern Recognition, Columbus, OH, USA, 23–28 June 2014; pp. 580–587.
23. Girshick, R.B. Fast R-CNN. In Proceedings of the IEEE International Conference on Computer Vision, Santiago, Chile, 11–18 December 2015.
24. Ren, S.; He, K.; Girshick, R.B.; Sun, J. Faster R-CNN: Towards Real-Time Object Detection with Region Proposal Networks. In *Advances in Neural Information Processing Systems;* IEEE: New York, NY, USA, 2015; pp. 91–99.
25. Ribera, J.; Chen, Y.; Boomsma, C.; Delp, E.J. Counting plants using deep learning. In Proceedings of the IEEE Global Conference on Signal and Information Processing (GlobalSIP), Montreal, QC, Canada, 14–16 November 2017; pp. 1344–1348.

26. Aich, S.; Ahmed, I.; Ovsyannikov, I.; Stavness, I.; Josuttes, A.; Strueby, K.; Duddu, H.S.; Pozniak, C.; Shirtliffe, S. DeepWheat: Estimating Phenotypic Traits From Images of Crops Using Deep Learning. In Proceedings of the IEEE Winter Conference on Applications of Computer Vision (WACV), Lake Tahoe, NV, USA, 12–15 March 2018; pp. 323–332.
27. Li, B.; Xu, X.; Han, J.; Zhang, L.; Bian, C.; Jin, L.; Liu, J. The estimation of crop emergence in potatoes by UAV RGB imagery. *Plant Methods* **2019**, *15*, 15. [CrossRef]
28. Guo, W.; Zheng, B.; Potgieter, A.B.; Diot, J.; Watanabe, K.; Noshita, K.; Jordan, D.R.; Wang, X.; Watson, J.; Ninomiya, S.; et al. Aerial Imagery Analysis—Quantifying Appearance and Number of Sorghum Heads for Applications in Breeding and Agronomy. *Front. Plant Sci.* **2018**, *9*, 1544. [CrossRef]
29. Lu, H.; Cao, Z.; Xiao, Y.; Zhuang, B.; Shen, C. TasselNet: Counting maize tassels in the wild via local counts regression network. *Plant Methods* **2017**, *13*, 79. [CrossRef]
30. Rahnemoonfar, M.; Sheppard, C. Deep Count: Fruit Counting Based on Deep Simulated Learning. *Sensors* **2017**, *17*, 905. [CrossRef]
31. Xiong, H.; Cao, Z.; Lu, H.; Madec, S.; Liu, L.; Shen, C. TasselNetv2: In-field counting of wheat spikes with context-augmented local regression networks. *Plant Methods* **2019**, *15*, 150. [CrossRef]
32. Pidhirniak, O. Automatic Plant Counting Using Deep Neural Networks. Master's Thesis, Department of Computer Sciences, Ukrainian Catholic University, Lviv, Ukraine, 2019.
33. Fan, Z.; Lu, J.; Gong, M.; Xie, H.; Goodman, E.D. Automatic Tobacco Plant Detection in UAV Images via Deep Neural Networks. *IEEE J. Sel. Top. Appl. Earth Obs. Remote Sens.* **2018**, *11*, 876–887. JSTARS.2018.2793849. [CrossRef]
34. Kitano, B.T.; Mendes, C.C.T.; Geus, A.R.; Oliveira, H.C.; Souza, J.R. Corn Plant Counting Using Deep Learning and UAV Images. *IEEE Geosci. Remote Sens. Lett.* **2019**. [CrossRef]
35. García-Martínez, H.; Flores-Magdaleno, H.; Khalil-Gardezi, A.; Ascencio-Hernández, R.; Tijerina-Chávez, L.; Vázquez-Peña, M.A.; Mancilla-Villa, O.R. Digital Count of Corn Plants Using Images Taken by Unmanned Aerial Vehicles and Cross Correlation of Templates. *Agronomy* **2020**, *10*, 469.
36. Malambo, L.; Popescu, S.; Rooney, W.; Zhou, T. A Deep Learning Semantic Segmentation-Based Approach for Field-Level Sorghum Panicle Counting. *Remote Sens.* **2019**, *11*, 939. [CrossRef]
37. Xia, L.; Zhang, R.; Chen, L.; Huang, Y.; Xu, G.; Wen, Y.; Yi, T. Monitor Cotton Budding Using SVM and UAV Images. *Appl. Sci.* **2019**, *9*, 4312. [CrossRef]
38. Epperson, M.; Rotenberg, J.; Lo, E.; Afshari, S.; Kim, B. *Deep Learning for Accurate Population Counting in Aerial Imagery*; Technical Report; Kastner Research Group: La Jolla, CA, USA, 2014.
39. Ghosal, S.; Zheng, B.; Chapman, S.C.; Potgieter, A.B.; Jordan, D.; Wang, X.; Singh, A.K.; Singh, A.; Hirafuji, M.; Ninomiya, S.; et al. A weakly supervised deep learning framework for sorghum head detection and counting. *Plant Phenomics* **2019**, *2019*, 1525874. [CrossRef]
40. Dijkstra, K.; van de Loosdrecht, J.; Schomaker, L.R.B.; Wiering, M.A. *CentroidNet: A Deep Neural Network for Joint Object Localization and Counting*; Springer: Cham, Switzerland, 2019; pp. 585–601. [CrossRef]
41. Lin, T.; Goyal, P.; Girshick, R.B.; He, K.; Dollár, P. Focal Loss for Dense Object Detection. In Proceedings of the IEEE International Conference on Computer Vision, Venice, Italy, 22–29 October 2017; pp. 2980–2988.
42. Oh, M.H.; Olsen, P.A.; Ramamurthy, K.N. Counting and Segmenting Sorghum Heads. *arXiv* **2019**, arXiv:1905.13291.
43. Neupane, B.; Horanont, T.; Hung, N.D. Deep learning based banana plant detection and counting using high-resolution red-green-blue (RGB) images collected from unmanned aerial vehicle (UAV). *PLoS ONE* **2019**, *14*, e0223906. [CrossRef] [PubMed]
44. Liu, Y.; Cen, C.; Ke, Y.C.R.; Ma, Y. Detection of Maize Tassels from UAV RGB Imagery with Faster R-CNN. *Remote Sens.* **2020**, *12*, 338. [CrossRef]
45. Jiang, Y.; Li, C.; Paterson, A.H.; Robertson, J.S. DeepSeedling: Deep convolutional network and Kalman filter for plant seedling detection and counting in the field. *Plant Methods* **2019**, *15*, 141. [CrossRef]
46. Song, Z.; Fu, L.; Wu, J.; Liu, Z.; Li, R.; Cui, Y. Kiwifruit detection in field images using Faster R-CNN with VGG16. *IFAC Pap.* **2019**, *52*, 76–81. [CrossRef]
47. Ganesh, P.; Volle, K.; Burks, T.; Mehta, S. Deep Orange: Mask R-CNN based Orange Detection and Segmentation. *IFAC Pap.* **2019**, *52*, 70–75. [CrossRef]
48. Abdulla, W. Mask R-CNN for Object Detection and Instance Segmentation on Keras and TensorFlow. 2017. Available online: https://github.com/matterport/Mask_RCNN (accessed on 19 September 2019).

49. He, K.; Zhang, X.; Ren, S.; Sun, J. Deep Residual Learning for Image Recognition. In Proceedings of the IEEE Conference on Computer Vision and Pattern Recognition, Las Vegas, NV, USA, 27–30 June 2016; pp. 770–778.
50. Chen, X.; Gupta, A. An Implementation of Faster RCNN with Study for Region Sampling. *arXiv* **2017**, arXiv:1702.02138.
51. Kingma, D.P.; Ba, J. Adam: A Method for Stochastic Optimization. In Proceedings of the 3rd International Conference for Learning Representations, San Diego, CA, USA, 7–9 May 2015.
52. Goodfellow, I.; Bengio, Y.; Courville, A. *Deep Learning*; MIT Press: Cambridge, MA, USA, 2016; p. 526. Available online: http://www.deeplearningbook.org (accessed on 19 September 2019).
53. Yosinski, J.; Clune, J.; Bengio, Y.; Lipson, H. *How Transferable Are Features in Deep Neural Networks?* NIPS: Montréal, QC, Canada, 2014.
54. Zlateski, A.; Jaroensri, R.; Sharma, P.; Durand, F. On the Importance of Label Quality for Semantic Segmentation. In Proceedings of the IEEE/CVF Conference on Computer Vision and Pattern Recognition, Salt Lake City, UT, USA, 18–22 June 2018; pp. 1479–1487. [CrossRef]
55. Hamuda, E.; Glavin, M.; Jones, E. A survey of image processing techniques for plant extraction and segmentation in the field. *Comput. Electron. Agric.* **2016**, *125*, 184–199. [CrossRef]
56. Kataoka, T.; Kaneko, T.; Okamoto, H.; Hata, S. Crop growth estimation system using machine vision. In Proceedings of the IEEE/ASME International Conference on Advanced Intelligent Mechatronics, Kobe, Japan, 20–24 July 2003; pp. 1079–1083.
57. Bernardin, K.; Stiefelhagen, R. Evaluating Multiple Object Tracking Performance: The CLEAR MOT Metrics. *EURASIP J. Image Video Process.* **2008**, *2008*, 1–10. [CrossRef]
58. Ioffe, S.; Szegedy, C. Batch Normalization: Accelerating Deep Network Training by Reducing Internal Covariate Shift. *arXiv* **2015**, arXiv:1502.03167.

Article

Dynamic Influence Elimination and Chlorophyll Content Diagnosis of Maize Using UAV Spectral Imagery

Lang Qiao [1], Dehua Gao [1], Junyi Zhang [2], Minzan Li [1], Hong Sun [1,*] and Junyong Ma [3]

1 Key Laboratory of Modern Precision Agriculture System Integration Research, Ministry of Education, China Agricultural University, Beijing 100083, China; b20193080667@cau.edu.cn (L.Q.); dehua_gao@cau.edu.cn (D.G.); limz@cau.edu.cn (M.L.)
2 Henan University of Animal Husbandry and Economy, Zhengzhou 450044, China; junyizh@cau.edu.cn
3 Dry-Land Farming Institute of Hebei Academy of Agricultural and Forestry Sciences, Hengshui 053000, China; mjydfi@126.com
* Correspondence: sunhong@cau.edu.cn; Tel.: +86-010-62737838

Received: 28 June 2020; Accepted: 13 August 2020; Published: 17 August 2020

Abstract: In order to improve the diagnosis accuracy of chlorophyll content in maize canopy, the remote sensing image of maize canopy with multiple growth stages was acquired by using an unmanned aerial vehicle (UAV) equipped with a spectral camera. The dynamic influencing factors of the canopy multispectral images of maize were removed by using different image segmentation methods. The chlorophyll content of maize in the field was diagnosed. The crop canopy spectral reflectance, coverage, and texture information are combined to discuss the different segmentation methods. A full-grown maize canopy chlorophyll content diagnostic model was created on the basis of the different segmentation methods. Results showed that different segmentation methods have variations in the extraction of maize canopy parameters. The wavelet segmentation method demonstrated better advantages than threshold and ExG index segmentation methods. This method segments the soil background, reduces the texture complexity of the image, and achieves satisfactory results. The maize canopy multispectral band reflectance and vegetation index were extracted on the basis of the different segmentation methods. A partial least square regression algorithm was used to construct a full-grown maize canopy chlorophyll content diagnostic model. The result showed that the model accuracy was low when the image background was not removed (Rc^2 (the determination coefficient of calibration set) = 0.5431, RMSEF (the root mean squared error of forecast) = 4.2184, MAE (the mean absolute error) = 3.24; Rv^2 (the determination coefficient of validation set) = 0.5894, RMSEP (the root mean squared error of prediction) = 4.6947, and MAE = 3.36). The diagnostic accuracy of the chlorophyll content could be improved by extracting the maize canopy through the segmentation method, which was based on the wavelet segmentation method. The maize canopy chlorophyll content diagnostic model had the highest accuracy (Rc^2 = 0.6638, RMSEF = 3.6211, MAE = 2.89; Rv^2 = 0.6923, RMSEP = 3.9067, and MAE = 3.19). The research can provide a feasible method for crop growth and nutrition monitoring on the basis of the UAV platform and has a guiding significance for crop cultivation management.

Keywords: UAV; crop canopy; multispectral image; chlorophyll content; remote sensing technique

1. Introduction

Chlorophyll content is one of the important indicators that reflect the photosynthetic ability and nutrient status of maize plants [1–3]. The traditional crop chlorophyll diagnosis is mainly carried out by chemical analysis, which requires destructive sampling, takes a long time, and is costly.

These conditions might not satisfy the requirements of rapid chlorophyll monitoring on field crops for making management decision. According to the principle of light absorption and reflectance, technologies of spectral analysis, imaging spectroscopy, and other nondestructive methods have been widely used in crop monitoring [4–8]. Combined with the development of airborne or unmanned aerial vehicle (UAV) platforms [9], imaging spectroscopy obtained with high spatial and temporal resolution has become a preferred method and research topic in farmland estimation owing to its advantages of high efficiency and non-invasion [10–12]. Thus, this article aims to use the multispectral sensor carried by the UAV to collect maize canopy spectral data in the field and conduct a rapid diagnosis of the chlorophyll content to estimate the growth status and guide the field management.

Most current studies on spectral image focus on the diagnosis of chlorophyll content [11–13]. The three directions of these studies include the analysis of spectral response [14–16], quantification and selection of sensitive parameters [17,18], and optimization of models [19–22] on the basis of the visible and near-infrared images. Yu et al. [23] found that the ratio of reflection difference indexes (RRDIs) can eliminate the influence of the crop canopy structure on the spectral reflection characteristics to improve the estimation of chlorophyll content. Gauray et al. [24] collected spectral data of maize canopy by multispectral camera and constructed a diagnostic model of chlorophyll content by machine learning. The optimal model determination coefficient (R^2) was 0.904. The results showed that the combination of an airborne multispectral sensor and machine learning could effectively improve the detection accuracy of chlorophyll content. The above-mentioned studies outline the capability of chlorophyll content estimating by vegetation indices, especially collected on the basis of UAV in the visible, red edge, and near infrared bands, which has sensitive responses to the physiology and biochemistry of vegetation [14].

Many researches have attempted to improve vegetation indices for crop monitoring. Soil adjusted vegetation index (SAVI) used to reduce the influences from the soil back ground shows capability to improve the estimation accuracy [25]. Wu et al. [26] found that the integrated indexes, such as transformed chlorophyll absorption reflectance index (TCARI), modified the chlorophyll absorption ratio index (MCARI) and optimized the soil-adjusted vegetation index (OSAVI), are more suitable for chlorophyll estimation than the traditional vegetation indexes because many interference factors, such as shadow, soil background, and nonphotosynthetic materials, are considered. At the same time, Liu et al. [3] found that with the advance of maize growth period, crop spectrum had characteristic migration in different growth periods, but the vegetation index based on fixed wavelength calculation could not effectively reflect this dynamic migration phenomenon, which reduced its applicability in different growth periods. However, there is no unified mathematical expression that defines all vegetation indices due to the complexity of light spectra combinations, instruments and so on. For UAV applications, Xue et al. [27] investigated more than 100 vegetation indices and classified their specific applicability into basic vegetation indices, adjusted-soil vegetation indices, vegetation indices sensing in spectra regions, and so on. The survey showed that the amounts of vegetation indices, combining visible and near-infrared bands, had significantly improved the sensitivity of the detection of green vegetation.

The influencing factors and reducing methods for crop estimation via spectroscopy have been explored in many studies. During the processing of the canopy spectral image of field crops, Qian et al. [22] found that the factors of canopy structure, soil background, and weeds in the spectral image would interfere with the spectral reflection characteristics of crop canopy; thus, the methods for chlorophyll content estimation were proposed via precision image segmentation and multispectral calibration. Fei et al. [28] used the canopy spectrum to detect the nitrogen nutrition of maize. They found that the ability of the NDRE index to diagnose maize nitrogen gradually increased with the decrease of soil coverage. This finding indicated that the soil background would interfere with the diagnostic ability of the NDRE vegetation index. Roosjen et al. [29] found that the crop canopy structure and other external information had an effect on the crop spectral reflectance. The diagnostic ability of the leaf area index and chlorophyll content were improved when the PROSAIL model was used to retrieve multiple angle reflectance data of crop canopy and reduce external information interference. Several factors, such as soil coverage, vegetation canopy geometry, and weeds, will interfere with the spectral characteristics of plant canopy. Thus, the influencing factors on crop diagnosis are difficult to reduce on the basis of the spectral imaging in the field.

We address the primary concern with image segmentation in eliminating the influence from soil background to solve these problems that greatly limit the reliability of the spectral imaging technology for crop nutrition diagnosis. The several alternative methods of spectral image segmentation include spectral segmentation methods using vegetation indices [21,22], threshold segmentation methods based on spatial grey value distribution [30,31], and learning-based segmentation methods [32]. The method based on the vegetation index mainly distinguishes green plants from the background according to the reflectance differences between the crop and the background special. This method is greatly affected by outdoor ambient light; a strong or weak ambient light will reduce the image segmentation effect [17]. The threshold-based segmentation method classifies crops and background according to the grayscale characteristics of the image, and the calculation is simple; however, this method is sensitive to noise [33]. The segmentation methods based on learning mainly include supervised and unsupervised machine learning classification methods. These methods have high accuracy and strong adaptability to the changes of light environment; however, the calculation is complex, and the results rely on a large amount of data in the training stage [34,35].

The dynamic crop monitoring during different stages by UAV-based spectral images is more complex than the static data collected by manual or farm machinery devices. The growth developments of crop from the soil background are difficult to separate or estimate because of the changes of crops and environment, such as light spots and shadows. The noise removal methods, such as median filtering [36], Gaussian filtering [37], and wavelet transform filtering [38], are also involved to eliminate such influences caused by the light environment. Wavelets have good capability in UAV imaging processing due to their multiresolution and multiscale analytical property. Fang et al. [39] used wavelet transform to process a UAV image and decompose the image signals of different scales to eliminate noise. The research proved that wavelet transform could help eliminate the noise generated during image acquisition and transmission [39]. The wavelet segmentation algorithm has been recently proposed combined with wavelet transform and threshold segmentation methods. The mechanism is also combined with the advantages of the noise filtering and object classification [40].

These studies highlight the critical need for dynamic influence elimination during crop growth stages on the basis of UAV spectral imaging. Several methods could be selected to improve the UAV-based imaging spectrum. However, much uncertainty still exists about the effects of these methods and their influence on the results of the chlorophyll content estimation of maize canopy during growth stages. We aim to clarify several aspects of chlorophyll content diagnosis of maize at different stages by using UAV spectral imagery. It also aims to eliminate the dynamic influences and explore its effects on the results. Moreover, the proposed methods could improve the accuracy of chlorophyll content diagnosis of field maize.

In order to better remove the dynamic influencing factors of the canopy multispectral images of maize obtained by UAV, and to diagnose the chlorophyll content of maize in the field our specific objectives are as follows: (1) to study the effect of soil background noise on the spectral reflectance of maize canopy; (2) to evaluate the performance of different segmentation methods to eliminate the noise in a UAV image; and (3) to establish the diagnostic model of maize canopy chlorophyll content on the basis of the different segmentation methods. These tasks are conducted to improve the diagnostic accuracy of the maize canopy chlorophyll content and provide the basis for crop growth dynamic monitoring on the basis of UAV multispectral images.

2. Materials and Methods

2.1. Spectral Data Collection

The experiment was conducted from June to August 2019, at the Dry Farming and Water-saving Agricultural Experimental Station in Hengshui, Hebei Academy of Agricultural Sciences, China. Figure 1 shows the location of the experimental station. The experimental station of Dry Farming and Water-saving Agriculture is located in Central China (latitude 37.9035011244, east longitude 115.7083898640), 22 m above sea level, and covers an area of 853 acres. The region belongs to the continental monsoon climate zone and is warm and semi-arid. The climate is characterized by four distinct seasons with large differences between warm, dry, and wet. The region has an average annual rainfall of 642.1 mm and dry winters.

Figure 1 shows the following six experimental gradients in this study: A1—(nitrogen 0 kg; phosphorus 0 kg)/mu, A2—(nitrogen 6 kg; phosphorus 4 kg)/mu, A3—(nitrogen 12 kg; phosphorus 8 kg)/mu, A4—(nitrogen 24 kg; phosphorus 16 kg)/mu, A5—(nitrogen 36 kg; phosphorus 24 kg)/mu, and A6—(nitrogen 48 kg; phosphorus 32 kg)/mu. Each gradient includes four straw levels: B1—no straw, B2—150 kg/mu, B3—300 kg/mu, and B4—600 kg/mu. The nitrogen test was combined with the straw test. There were 72 experimental areas in total, and 24 experimental areas were repeated for one time, 3 times in total. The UAV remote sensing data acquisition experiment was simultaneously carried out with the field data acquisition and sampling. The experiment was divided into seedling (July 15, 2019), jointing (July 25, 2019), and ear stages (August 1, 2019), according to the growing season of maize.

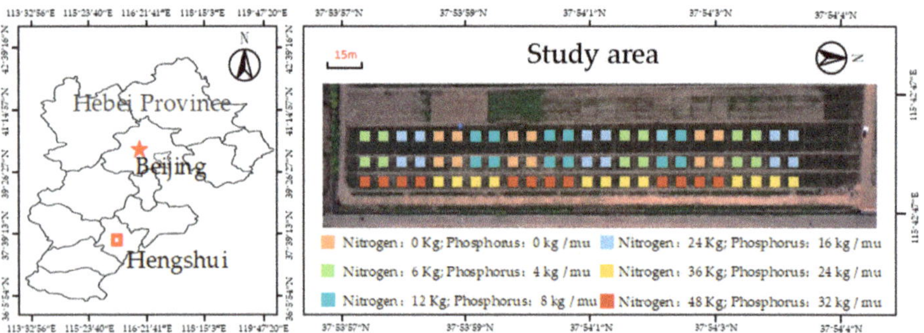

Figure 1. Locations and treatments of the experiments in this study.

2.2. Data Collection

2.2.1. UAV Image Acquisition

The DJI M600 Pro UAV is used as the loading platform (10 kg UAV weight, 5 kg maximum load weight, 18 min flight time, and 30 m flight height) and equipped with a Red Edge MX multispectral camera for remote sensing data acquisition. The quality of the Red Edge MX multispectral camera is 232 g, and the resolution is 1280 × 960 pixels (the sensor size is 8.7 cm × 5.9 cm × 4.54 cm). This camera can collect spectral images of five bands, namely, blue(B), green(G), red(R), red edge (REG)e, and near infrared (NIR). The central wavelength in each band is at 475, 560, 668, 717, and 840 nm, with the bandwidth 32, 27, 14, 12, and 57 nm, respectively. The specific parameters are shown in Table 1. The time period from 12 noon to 13 noon is selected to stabilize the intensity of solar radiation and clear the sky without clouds for UAV data collection. Such initiative is carried out obtain a low remote sensing effect of noise. The flying height of the UAV is 30 m, the flying speed is 4 m/s, and the image overlap is set to 80%.

Table 1. Multispectral camera parameters.

Parameter	Parameter Value	Parameter	Parameter Value
Model	Red Edge MX	Perspective	47.2°
Weight	232 g	Trigger method	Overlap model
Size	8.7 cm × 5.9 cm × 4.54 cm	Bands/nm	475, 560, 668, 717, and 840
Shutter	Global	Resolution	1280 × 960

2.2.2. Ground Data Acquisition

After the UAV data collection was completed, the ground data collection, which mainly included the sampling point selection and GPS position information recording, was performed. A maize plant was selected as a sample point at the center of each sample area, and a part of the leaves was placed into a sealed bag and refrigerated, then taken back to the laboratory to determine the chlorophyll content via stoichiometry and with a spectrophotometer. First, a 4 cm × 4 cm (excluding veins) leaf tissue was cut out from the middle of each leaf. Second, the chopped leaves were submerged in 25 mL of a mixture of acetone and ethanol and then soak in the dark for 24 h. After chlorophyll extraction, absorbance was measured with a 752 UV spectrophotometer. The spectrophotometer requires internal thermal equilibrium of the instrument. The device needs to be preheated for 30 min. The sample solution was poured into three cuvettes. The absorbance was measured at the wavelengths of 645 and 663 nm [2]. The formulas for calculating the total chlorophyll content are as follows:

$$C_a = 12.72A_{663} - 2.59A_{645}, \quad (1)$$

$$C_b = 22.88A_{645} - 4.67A_{663}, \quad (2)$$

$$C_t = C_a + C_b, \quad (3)$$

where A_{645} and A_{663} are the absorbance of 645 and 663 nm, respectively; C_a is chlorophyll a content (mg/L); C_b is chlorophyll b content (mg/L); and C_t is the total chlorophyll content (mg/L).

2.3. Data Processing

2.3.1. Data Processing Flow

After the UAV flight was over, the acquired remote sensing and ground data was processed. The processing flow is shown in Figure 2. The preprocessing mainly includes image stitching and geometric correction, which are completed by Pix4dmapper software. The Red Edge MX sensor combines GPS data and light intensity values in the spectral image. The Pix4dmapper software can import the spectral image to complete the stitching. During stitching, a vector file of the ground control points determined in advance is imported to geometrically correct the stitched image to ensure that the geographic coordinates of the stitched image are consistent with the real coordinates. The stitched multiple band images are fused, and the regions of interest are extracted using ENVI 5.2 software.

Radiation calibration of spectral images is conducted by using black and white calibration boards. Radiation calibration refers to converting the gray values presented light intensity information of the maize multispectral image to reflectivity information. Radiation calibration boards are placed in the study area where the ground is flat and free of shadows. The UAV performs data collection. Equation (4) is used to complete the radiation calibration of the spectral image.

$$\frac{\rho - \rho_1}{\rho_2 - \rho_1} = \frac{DN - DN_1}{DN_2 - DN_1} \quad (4)$$

where DN, DN_1, and DN_2 are the digital numbers to show the light intensity values of the crop, black, and white calibration boards, respectively; ρ, ρ_1, and ρ_2 are the calculated reflectance values of the crop, black, and white calibration boards, respectively.

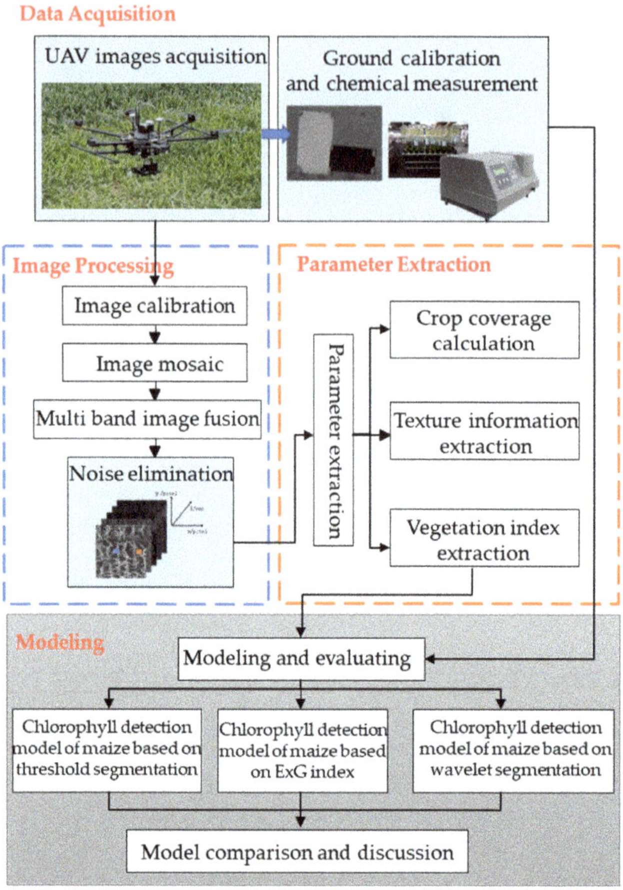

Figure 2. Data processing flowchart.

2.3.2. Extraction of Multispectral Vegetation Index

The multispectral image given by the UAV can provide spectral images in five bands with wavelengths of 475, 560, 668, 717, and 840 nm. After radiation calibration, the original image with the DN value was converted into a reflectance image, and the corresponding vegetation index image was calculated. Twenty types of vegetation indices were selected for construction of a chlorophyll content diagnostic model on the basis of the existing vegetation indices and combined with the characteristics of multispectral images. These selected vegetation indices are shown in Table 2.

2.3.3. Texture Information Extraction

Texture is an expression of image features. In a multispectral image of a crop canopy, texture can represent the structural characteristics of the maize canopy. In this study, the gray distribution statistical method was used to extract texture features from the UAV multispectral image. The standard deviation (σ), smoothness, and entropy were selected to characterize the texture features. They were calculated for input image following Equations (5)–(7), respectively. The standard deviation is a measure of the average contrast of the image. The smaller the value, the more uniform the value of adjacent pixels in the image. Smoothness is a relative smoothness measure of the brightness of the image. The value ranges from zero to one. The closer the value is to 0, the smoother the image is.

The entropy is a measure of the randomness of the image. The smaller the value is, the lower the randomness of pixels in the image is, and the more uniform the image is.

$$\sigma = \sqrt{\mu_2(Z)} \tag{5}$$

$$Smoothness = \frac{\sigma^2}{(1+\sigma^2)} \tag{6}$$

$$Entropy = -\sum_{i=0}^{L-1} p(Z_i)\log_2 P(Z_i) \tag{7}$$

Table 2. Selected multispectral vegetation index.

Vegetation Index	Formula	Reference
Ratio vegetation index (RVI)	RVI = NIR/R	Jordan et al. [41]
Green ratio vegetation index (RVIGRE)	RVIGRE = NIR/G	Jordan et al. [41]
Normalized difference vegetation index (NDVI)	NDVI = (NIR − R)/(NIR + R)	Rouse et al. [27]
Normalized difference red edge (NDRE)	NDRE = (NIR − REG)/(NIR + REG)	Barnes et al. [42]
Difference vegetation index (DVI)	DVI = NIR − R	Jordan et al. [41]
Green difference vegetation index (DVIGRE)	DVIGRE = NIR − G	Jordan et al. [41]
Red edge difference vegetation index (DVIRED)	DVIRED = NIR − REG	Jordan et al. [41]
Chlorophyll index with red edge (CIredege)	CIredege = (NIR/REG) − 1	Gitelson et al. [43]
Chlorophyll index with green (CIgreen)	CIgreen = (NIR/G) − 1	Dash et al. [44]
Green normalized difference vegetation index (GNDVI)	GNDVI = (NIR − G)/(NIR + GRE)	Kataoka et al. [45]
Soil-adjusted vegetation index (SAVI)	SAVI = 1.5(NIR − R)/(NIR + R + 0.5)	Huete et al. [46]
Soil-adjusted vegetation index with green (SAVIGRE)	SAVIGRE = 1.5(NIR − G)/(NIR + G + 0.5)	Verrelst et al. [47]
Optimized soil-adjusted vegetation index (OSAVI)	OSAVI = (1 + 0.16)(NIR − R)(NIR + R + 0.16)	Rondeaux et al. [48]
Optimized soil-adjusted vegetation index with green (OSAVIGRE)	OSAVIGRE = (1 + 0.16)(NIR − G)(NIR + G + 0.16)	Rondeaux et al. [48]
Optimized soil-adjusted vegetation index with red edge (OSAVIREG)	OSAVIREG = (1 + 0.16)(NIR − REG)(NIR + REG + 0.16)	Rondeaux et al. [48]
Red difference vegetation index (RDVI)	RDVI = (NIR − R)/(NIR + R)^0.5	Roujean et al. [49]
Red difference vegetation index with red edge (RDVIREG)	RDVIREG = (NIR − REG)/(NIR + REG)^0.5	Roujean et al. [49]
Modified simple ratio (MSR)	MSR = (NIR/R − 1)/(NIR/R + 1)^0.5	Chen et al. [50]
Modified simple ratio with red edge (MSRREG)	MSRREG = (NIR/REG − 1)/(NIR/REG + 1)^0.5	Chen et al. [50]
Meris terrestrial chlorophyll index (MTCI)	MTCI = (NIR − REG)/(NIR − R)	Dash et al. [51]

2.3.4. Coverage Extraction

The soil background in the crop canopy multispectral image will interfere with the spectral characteristics of the crop canopy. Segmentation of the crop canopy with different segmentation methods will yield different results. Coverage indicates the proportion of the crop canopy in the multispectral image. The coverage can be used to characterize the segmentation results of different segmentation methods. The calculation method is shown in Equation (9), where V_c is the maize canopy coverage, S_c is the number of pixels in the maize canopy vector file area, and S_a is the total of the number of pixels in the multispectral image. For the convenience of calculation, we normalize the coverage value, and its value range is 0 to 1.

$$V_c = \frac{S_c}{S_a} \tag{8}$$

2.4. Segmentation Method

2.4.1. Threshold Segmentation

The premise of the threshold segmentation method is to obtain the segmentation interval of the grayscale image. In the multispectral image, the reflectance difference between the maize plant and the soil background in the near-infrared band reaches the maximum. The difference characteristic of the grayscale interval is significant; the near-infrared image is regarded as a segmented image. The segmentation threshold extracted by the maximum interclass method difference (Otsu) is used to binarize the near-infrared image to finally obtain a maize canopy vector file.

2.4.2. ExG Index Segmentation

Green plants have obvious spectral reflection characteristics in the visible light band. The green (g) band has a high reflectance, the chlorophyll has a strong absorption characteristic in the red (r) band, and the blue (b) band is sensitive to the chlorophyll concentration response. However, the soil does not have such spectral reflection characteristics. The plant can be effectively separated from the soil background by using this spectral difference characteristic. Ex-green vegetation index (ExG), as a typical visible light vegetation index, has been widely used in research.

$$ExG = 2 * g - r - b \quad (9)$$

2.4.3. Wavelet Segmentation

During the acquisition of multispectral images by UAV, the acquired images will have a certain degree of noise due to the influence of the external environment and equipment differences. Wavelet transform, as a signal processing tool, can multiply and scale the image space domain signals to obtain low-frequency and high-frequency wavelets through scaling and translation operations. Accordingly, noise signals are eliminated, and useful ones are retained. In this study, the near-infrared image is denoised by wavelet denoising on the basis of the threshold segmentation method. The noise is eliminated to achieve maize canopy image segmentation.

2.5. Model Establishment and Accuracy Evaluation

Partial least squares regression (PLS) is one of the multivariate statistical data analytical methods that integrates the advantages of principal component analysis, canonical correlation analysis, and linear regression analysis. This method has the advantage of handling multiple correlations between independent variables of small samples. In this study, the experimental data of PLS on July 15, 2019, July 25, 2019, and August 1, 2019 are used to establish a model of the relationship between maize canopy chlorophyll content and multispectral remote sensing images. Fifty-four validation set data are obtained.

The root mean square error (RMSE), determination coefficient (R^2), and the mean absolute error (MAE) of the actual and detected values are used for evaluating the diagnostic capability of the model. Among the parameters, RMSE is used to measure the degree of dispersion of the experimental results, and the model effect is good when the value is small; the R^2 represents the degree of fitting of the model, and the model diagnostic accuracy is high when the value is close to one; MAE can better reflect the actual situation of predicted value error, and the model effect is good when the value is small. In order to distinguish the precision of modeling set and validation set, Rc^2 (the determination coefficient of calibration set) and RMSEF (the root mean squared error of forecast) are used to represent the precision of modeling set, and Rv^2 (the determination coefficient of validation set) and RMSEP (the root mean squared error of prediction) are used to represent the precision of validation set.

3. Results

3.1. Ground Statistics

A total of three experiments were performed in this study. Seventy-two maize leaves were collected at the seedling, jointing, and ear stages for chlorophyll content extraction. A total of 216 samples were obtained during the three growth stages. The box plots of chlorophyll content in maize leaves at each growth stage are shown in Figure 3. The KS algorithm was used to divide the entire growth period samples according to different segmentation methods. The results are shown in Table 3. Table 3 illustrates that the training and validation sets contain a large number of chlorophyll content values for the total sample data.

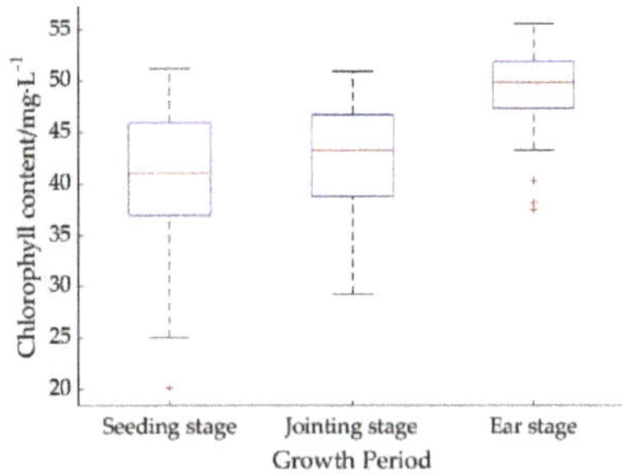

Figure 3. Box plots of chlorophyll content in maize leaves at three growth stages.

Table 3. Chlorophyll content statistics.

Segmentation Method		Number of Samples	Maximum Value ($mg \cdot L^{-1}$)	Minimum Value ($mg \cdot L^{-1}$)	Average Value ($mg \cdot L^{-1}$)	Standard Deviation ($mg \cdot L^{-1}$)
With background	Total sample	216	55.579	20.108	44.189	6.491
	Training set	162	55.579	20.108	44.411	6.241
	Validation set	54	53.809	24.990	43.527	7.150
Threshold segmentation	Total sample	216	55.579	20.108	44.189	6.491
	Training set	162	55.579	20.108	44.407	6.152
	Validation set	54	54.572	24.990	43.539	7.377
ExG index segmentation	Total sample	216	55.579	20.108	44.189	6.491
	Training set	162	55.579	20.108	43.218	6.589
	Validation set	54	54.721	34.878	47.104	5.192
Wavelet segmentation	Total sample	216	55.579	20.108	44.616	6.491
	Training set	162	55.579	20.108	49.030	6.245
	Validation set	54	53.809	24.989	42.911	7.025

3.2. Results of Spectral Image Segmentation by Different Segmentation Methods

ENVI 5.2 software was used to draw the regions of interest of the maize canopy and the soil background in the multispectral image for exploring the effect of soil background on the reflectance of maize canopy in the multispectral image. The average gray value of the region of interest was extracted. The results are shown in Figure 4. The average gray values of the maize canopy and soil are different in five bands. The maize canopy gray values at the blue and red bands are lower than that of soil. The gray value of crop canopy in the green, red edge, and near-infrared bands is higher than that of soil, and the difference between the two reaches the maximum at near-infrared band. Therefore, accurate maize canopy spectral information can be obtained by removing the soil background from multispectral images.

The ExG index segmentation, threshold segmentation method, and wavelet segmentation method were used to remove the soil background in the multispectral image to obtain an accurate maize canopy spectrum. The segmentation results are shown in Figure 5. The maize canopy image shows that the ExG index segmentation will retain several noise points when removing the soil background, and the segmentation effect is poor. The reflectance of maize canopy and soil in the near infrared band has a large difference. On this basis, threshold segmentation method was used because it can efficiently remove the soil background. However, the background close to the maize leaves is retained. The wavelet segmentation method first removes the edges and noise points of the near-infrared image and then uses the maize canopy. The spectral difference characteristics of the soil are segmented; thus, better results can be obtained than the ExG index segmentation and threshold segmentation method.

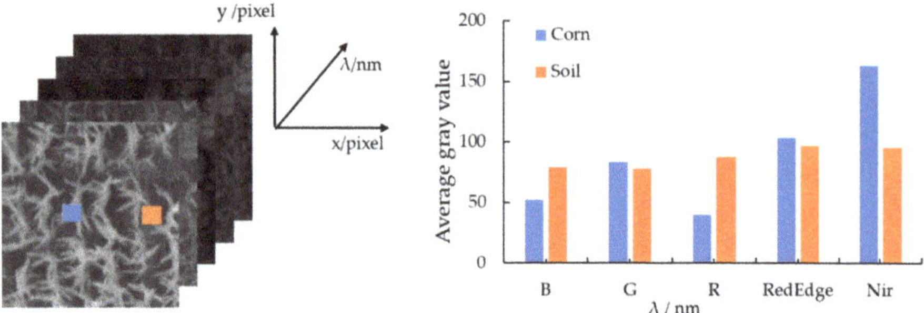

Figure 4. Grayscale distribution of crops and soil in multispectral images.

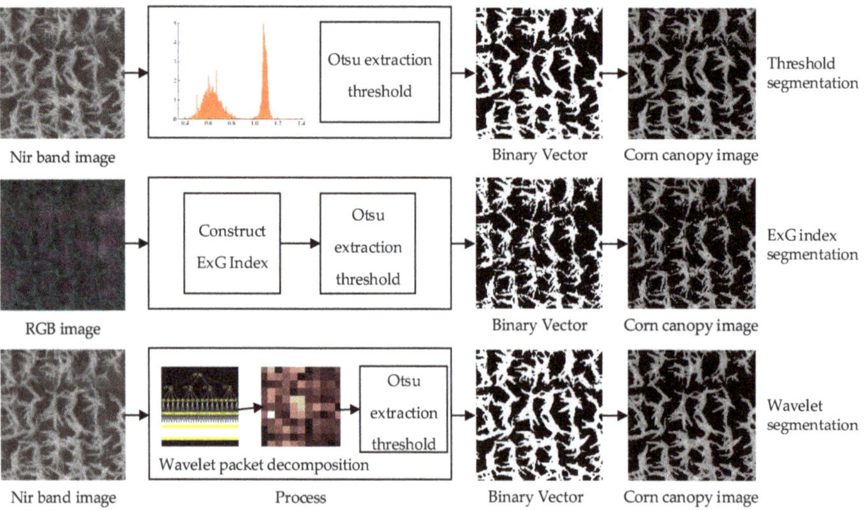

Figure 5. Maize canopy multispectral image segmentation.

3.3. *Effects of Different Segmentation Methods on Maize Growth Dynamics*

3.3.1. Spectral Reflectance Response

Figure 6 lists the maize canopy spectral reflectance trends extracted from the original spectral images and the three segmentation methods obtained during the three growth periods. In general, the spectral reflectance of the maize canopy decreases in the blue and red bands and increases in the green, red edge, and near-infrared bands after the multispectral image is segmented. This phenomenon presents the typical spectral feature of green vegetation with light absorption by chlorophyll and carotene pigment in blue and red bands, and relative higher light reflectance in green and NIR ranges. The effects of the three segmentation methods in different growth periods in Figure 6 vary. As shown in Figure 6a, the maize canopy coverage at the seedling stage is low, the soil background is highly visible, and the segmentation is easy; thus, the results obtained by the three segmentation methods are similar. The maize grows to the jointing and ear stages, and the maize canopy structure is complex, which is not conducive to background culling. Figure 6b,c show that the reflectance of ExG index segmented image is lower than image segmented by the wavelet or threshold methods. However, the performances of wavelet and threshold segmentation are very close in all three stages.

The segmentation results are difficult to evaluate by ideal images, which are without influence factors in the field. We therefore compared the segmented image with the original image to explore the differences between three methods. Table 4 shows results of RMSE operated by three segmentation methods. At the seeding stage, the RMSE values at NIR bands are higher than other band, and decrease following ExG, threshold and wavelet segmentation in blue, green, red, and red edge bands; the value of wavelet segmentation method is lowest in each band. At jointing and ear stage, RMSE values of threshold and wavelet segmentation are close and higher than ExG segmentation results, especially for green and NIR bands. Meanwhile, RMSE value of the wavelet method is higher than that of the threshold method in each band. Higher values show that changes and quality improvement between segmented image and original image in each band are more significant. Thus, the wavelet segmentation method shows a better performance for field difference enhancement and image quality improvement.

Table 4. Comparison of root mean square error (RMSE) between different segmentation and original image.

Segmentation Method	Growth Period	RMSE				
		B	G	R	Red Edge	NIR
Threshold segmentation	Seeding stage	0.021561	0.026965	0.047821	0.057941	0.191166
	Jointing stage	0.012054	0.02807	0.006986	0.039441	0.06589
	Ear stage	0.01235	0.031656	0.008716	0.044059	0.092574
ExG index segmentation	Seeding stage	0.017366	0.022469	0.043826	0.056476	0.197419
	Jointing stage	0.023383	0.069644	0.001567	0.106638	0.206049
	Ear stage	0.024629	0.059465	0.015756	0.085482	0.179996
Wavelet segmentation	Seeding stage	0.014016	0.021435	0.03936	0.053233	0.181855
	Jointing stage	0.024369	0.068886	0.002774	0.104372	0.197246
	Ear stage	0.028779	0.066664	0.019238	0.094825	0.191218

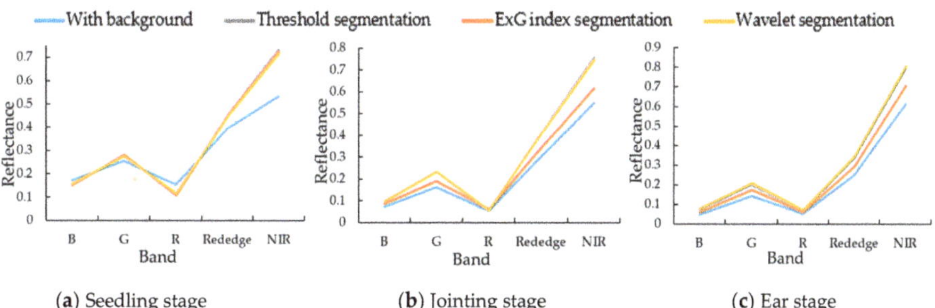

(**a**) Seedling stage (**b**) Jointing stage (**c**) Ear stage

Figure 6. Calculation of maize canopy spectral reflectance by different segmentation methods.

3.3.2. Texture Response

The maize canopy image in the multispectral image was extracted by background culling to generate a maize canopy vector file. We wanted to further discuss texture responses in processed images. The texture change trend of the maize canopy in different growth periods was extracted by various segmentation methods. Figure 7 demonstrates that the standard deviation, smoothness, and entropy of the near-infrared image of the maize canopy gradually increase with the growth of the maize when the maize canopy is not segmented. This phenomenon indicates that the canopy structure of the plant canopy became more and more complicated with the growth of the maize.

Three segmentation methods were used to eliminate influences form the soil background, all of which reduce the standard deviation, smoothness, and entropy of the image to a certain extent. Figure 7 shows that compared with ExG index segmentation method, threshold segmentation method and wavelet segmentation method can better reduce the standard deviation, smoothness, and entropy of the image. Firstly, the results of threshold segmentation method were higher than others. Even the standard deviation and entropy valued of the threshold segmented image are close to the image segmented by wavelet and ExG, respectively, its performances are not the best. Secondly, the standard deviation of ExG processed image is lowest at the seedling stage due to the soil background removal; the results of standard deviation and smoothing are close between ExG and wavelet segmentation methods at the jointing and the ear stage. When compared to the wavelet method, although the ExG method has lower values of standard deviation and smoothness, the entropy value of ExG segmented image is close to threshold segmentation and higher than wavelet segmentation at the jointing and the ear stage. It means that the randomness performances processed by threshold and ExG segmentation are higher than wavelet segmentation because of the limited elimination of canopy structure influences during these stages. Thus, the wavelet segmentation method has achieved stable results in the three growth stages.

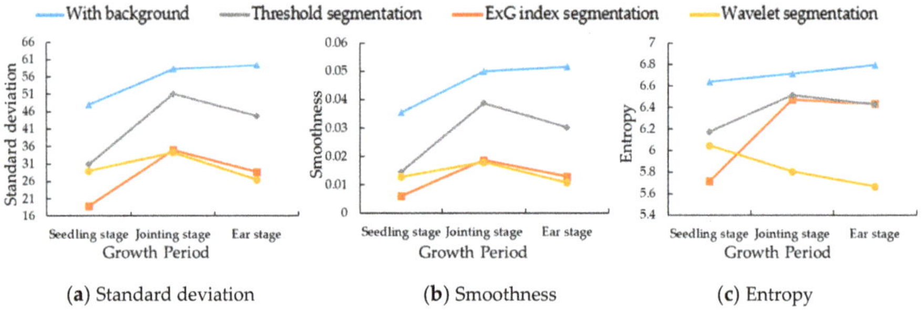

Figure 7. Calculation of the maize canopy texture features using different segmentation methods.

3.3.3. Coverage Response

Figure 8 shows the change trend of maize plant coverage calculated by extracting the canopy area by using three segmentation methods during the above-mentioned three growth stages. The maize canopy coverage was calculated by ratio of segmented canopy pixels and full image pixels and presented between 0 and 1. In general, the values of canopy coverage acquired by all of segmentation methods increased with the growth stages. Some differences in the maize canopy coverage calculated by different segmentation methods can be observed in the same growth period. Firstly, canopy coverage values calculated based on three segmented results at the seedling stage were relatively small and lower than 0.45. This phenomenon is attributed to the simple structure of the maize plant at the seedling stage. The visible part of the soil is large, and the division is simple when the leaf volume is low. Secondly, those at the jointing and ear stages were different, especially ExG segmentation results that were significantly higher than others above 0.75; the coverage values of threshold and wavelet segmented images were similar and higher than 0.45; the result of the threshold segmentation was a little bit higher than wavelet segmentation method at the earing stage. The significantly differences between three segmentation methods at the jointing and ear stages also show that the segmentation was more complicated than the seedling stage due to a large number of leaves crossing each other and a complex canopy structure of maize plant.

As mentioned above, in the three segmentation methods used in this study, the maize canopy coverage calculated by the ExG vegetation index method was high. Meanwhile, the maize canopy coverage calculated by the threshold and wavelet segmentation methods was low. This study takes the ear period as an example. The maize coverage calculated by the ExG vegetation index method, threshold segmentation method, and wavelet segmentation method was 0.764, 0.562, and 0.520, respectively. This result is attributed to the complicated canopy structure at the ear stage of maize and the presence of many background effects, such as soil and shadow. The results are difficult to prove by actual coverage values because of the limitation of ideal image acquisition, so that we resort to the diagnosis modeling of chlorophyll content to further estimate the capability and validity of three segmentation methods.

However, the phenomena presented in Figure 8 could be discussed based on the combination of the spectral response in the field and the principal of each segmentation method. The detail is in discussion part.

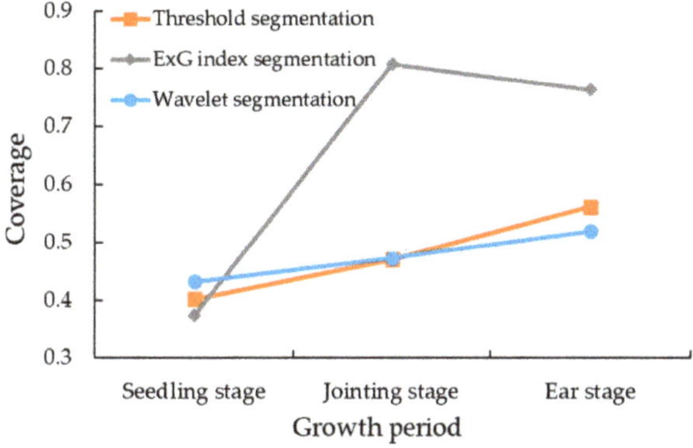

Figure 8. Variation trend of maize canopy coverage by different segmentation methods.

3.4. Modeling and Analysis of Maize Canopy Chlorophyll Content

On the basis of the above analysis, the maize canopy multispectral image is segmented, and the background interference in the image is eliminated. The multispectral band reflectance, multiple vegetation indices, and maize canopy chlorophyll content were established to construct a PLS model. Correlation analysis was performed using the maize canopy chlorophyll content calculated by the model and the real value measured in the experiment. The maize canopy chlorophyll based on different segmentation methods was established by comparing the R^2 and RMSE of the two groups of variables. The test model was analyzed, and the results are shown in Table 5. Figure 9 shows scatter plots of the predicted and true values obtained under different segmentation methods to intuitively display their relationship.

Table 5 shows the model accuracy of the reflectivity modeling obtained using the original reflectivity modeling and the three segmentation methods. The model built using the original reflectivity has low accuracy and large errors. Compared with the original model, the model constructed based on the reflectance obtained by the three segmentation methods has improved performance to a certain extent in terms of accuracy and error. This indicates that removing the soil background from the original image can improve the diagnostic accuracy of chlorophyll content. Compared with the ExG index segmentation method and the threshold segmentation method, the model based on the wavelet segmentation method shows the best performance and the lowest error (Rc^2 = 0.6638, RMSEF = 3.6211, MAE = 2.89; Rv^2 = 0.6923, RMSEP = 3.9067, and MAE = 3.19).

Table 5. Diagnostic results of maize canopy chlorophyll with different segmentation methods.

Segmentation Method	Modeling			Verification		
	Rc^2	RMSEF	MAE	Rv^2	RMSEP	MAE
With background	0.5431	4.2184	3.24	0.5894	4.6947	3.63
Threshold segmentation	0.6330	3.7270	2.93	0.6302	4.5248	3.49
ExG index segmentation	0.6584	3.8512	3.07	0.5601	4.6992	2.96
Wavelet segmentation	0.6638	3.6211	2.89	0.6923	3.9067	3.19

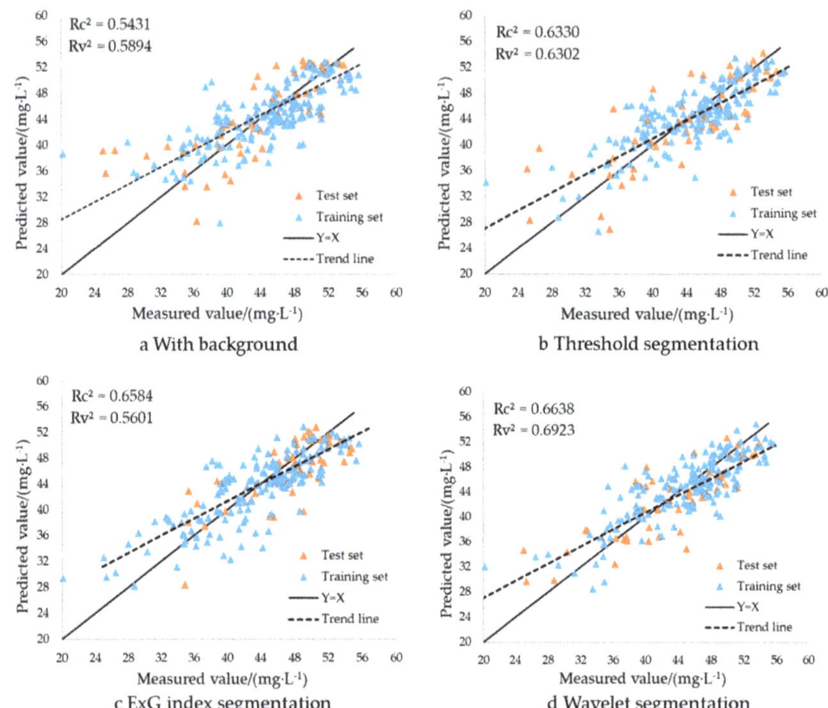

Figure 9. Maize canopy chlorophyll test results.

4. Discussion

Spectral analysis and imaging spectroscopy, a fast and nondestructive diagnosis method, can comprehensively reflect the changes of physiological and biochemical indexes and canopy structure of crops by analyzing the spectral reflectance characteristics of crops; accordingly, a rapid nondestructive diagnosis of crop nutrition is achieved [6,7]. The reflectance spectral data can be used to diagnose the chlorophyll content of maize in the field. The image acquired by a UAV-based spectral platform contains rich information, including not only spectral information of ground objects but also texture and coverage data.

4.1. Elimination of the Impact of Background and Noise

This study showed that the spectral reflectance of maize canopy greatly changed after the soil background was removed, which was consistent with the results of Wu et al. [22]. Compared with the original image with the soil background, the spectral reflectance values reduce in blue and red bands, and increase in green, red edge, and NIR bands. It is similar to that of the maize canopy in Figure 4. The gray value of the maize canopy is lower than that of the soil in the blue and red bands and is higher than that of the soil in the green, red, and near-infrared bands. It shows that the canopy segmentation could help to enhance the spectral characteristic of crop and reduce the influences from the soil background.

Figure 7 shows that the texture features (standard deviation, smoothness, and entropy) of maize canopy multispectral images become more complex with the extension of growth period. The main reason is that with the growth of maize, the area and number of leaves increase and cross each other, which increases the complexity of canopy structure. Three segmentation methods are used to segment the soil background in the multispectral image, and only the corn part of the image is preserved, which reduces the type of elements in the image and reduces the complexity of the image. Therefore, the standard deviation, smoothness, and entropy in Figure 7 are reduced to some extent by three segmentation methods. Compared with threshold segmentation method and wavelet segmentation method, ExG exponential segmentation method has lower performance. The possible reason is that the illumination change during UAV image acquisition interferes with the calculation of vegetation index, which is easily saturated under high biomass conditions. Previous studies have shown that wavelet transform can be used to process UAV images and decompose different proportions of image signals to eliminate noise [38]. The wavelet segmentation method combines the advantages of wavelet transform and threshold segmentation, and can achieve good results in noise elimination and object classification and classification [39]. This is also confirmed by the results of this study. The wavelet segmentation method is used to remove the soil background, reduce the complexity of crop canopy, and achieve stable performance in three growth periods. Figure 8 shows the results of the three segmentation methods to calculate the coverage of maize at three growth stages. Although the ExG segmentation method is widely used in the field crop detection, it is operated based on different spectral bands. According to the spectral characteristics of crop and soil background, the ExG index shown in Equation (9) is established by grey values of blue, green, and red bands to amplify the spectral difference between crop and soil background. As shown in Figure 4, this way pays attention to the light absorption by chlorophyll and carotene pigment in visible bands related to blue and red, it reduces the soil background which has close grey values among blue, green, and red band image. While threshold and wavelet segmentation methods are applied in NIR image in which the grey values are significant different between corn and soil to show the biological vitality. Compared with the NIR image, grey values are easily affected by environment in visible bands. For example, a strong light radiation might reduce grey contrast and lead miss pixels of leaves, the canopy coverage might be lower than those of threshold and wavelet segmentation methods at the seedling stage. At the jointing and ear stage, the coverage values calculated by ExG show significant changes, because the soil background is misclassified as the crop due to more complex structure of plants and the expanded shading elements from the crossing leaves. The coverage value at ear stage is lower than those at jointing stage because of closed crop lines by luxuriant leaves and reduced misclassification of soil back ground.

Regards to the segmented results of the threshold and wavelet method, the texture analysis helps to evaluate the segmented results and explain how it conducts to improve the quality from image contrast, brightness, and randomness. It is discussed based on texture analysis results shown in Figure 7a,b; standard deviation and smoothness values of wavelet segmented images are higher than that of threshold segmented image at seedling stage, lower at ear stage, and similar at jointing stage, respectively. The higher value of standard deviation indicates the higher image contrast, and it means the edge pixels of leaves are more clearly to classify. The higher smoothness value indicates the lower influence from noise or background, and it shows better performances of image enhancement. As a result, the coverage result of the wavelet segmentation method is higher than that of the threshold segmented image at seedling stage, lower at ear stage, and similar at jointing stage, respectively.

In addition, the wavelet analysis is a method usually used in signal denoising and enhancement. Figure 7c shows that the randomness indicated by entropy is improved by wavelet method during jointing and ear stages. It is contributed by the image denoising to reduce the effects of canopy structure and leaves crossing. The wavelet method also helps to enhance useful edge or detail information in the NIR image so that the randomness locates between ExG and the threshold method at seedling stage. From this view, the wavelet segmentation method not only eliminates the soil background in the image, but also segments the edge and noise points in the image [38], so the calculated coverage by wavelet method is more reliable than the ExG index and threshold segmentation method.

4.2. Diagnosis of Chlorophyll Content in Maize Based on Different Segmentation Methods

The vegetation index has been widely used to diagnose crop nutritional parameters, and its performance varies with wavelength configuration [23,52]. Duan et al. [53] found saturation in estimating LAI (Leaf Area Index) by using NDVI alone. Cheng et al. [16] found that the four NIR-based vegetable indices outperformed the two-color indices in AGB estimation across all growth stages. Therefore, a single band or vegetation index was not easy to show good model accuracy in all growth periods. In this study, 20 commonly used vegetation indices were selected to construct a chlorophyll content diagnostic model for the entire growth period of maize.

The spectral reflectance of maize canopy was obtained on the basis of the three segmentation methods to remove soil background, and the vegetation index was calculated. A chlorophyll content diagnostic model of maize canopy based on different segmentation methods was constructed using the PLS method. The results showed that the accuracy of the diagnostic model constructed using the original canopy spectral data was low (Rc^2 = 0.5431, RMSEF = 4.2184, MAE = 3.24; Rv^2 = 0.5894, RMSEP = 4.6947, and MAE = 3.36), and the accuracy of the diagnosis of chlorophyll content in maize canopy could be improved to varying degrees by using three segmentation methods to remove background noise in maize multispectral images. However, among the three segmentation methods selected in this study, the accuracy of the training set based on the ExG index segmentation method was 0.6584, but the accuracy of the verification set was only 0.5601, which indicates that the model is not robust enough. This result is consistent with the previous discussion, and the color index is easily affected by ambient light. In the future, spectral radiometer can be used to correct the reflectivity of UAV images, so as to further optimize the performance of the model. The diagnostic model of maize canopy chlorophyll content based on wavelet segmentation had the highest quasi determination (Rc^2 = 0.6638, RMSEF = 3.6211, MAE = 2.89; Rv^2 = 0.6923, RMSEP = 3.9067, and MAE = 3.19). This also confirmed that the wavelet segmentation method can effectively remove the soil background and noise points in the UAV multispectral image and restore the spectral reflection characteristics of crops in the canopy image. The good performance of wavelet segmentation also shows that image enhancement has great potential in UAV image noise elimination.

5. Conclusions

In the present study, we studied the background noise in the multispectral images of maize canopy acquired by UAV and discussed the performance of three noise rejection methods. The results show that the soil background removal from multispectral images can reduce the complexity of image texture and improve the spectral reflectance characteristics of crop canopy. The wavelet segmentation method achieved better noise rejection effect in the three growth stages of maize compared with the threshold and ExG index segmentation methods. We extracted 20 common vegetation indices on the basis of the original spectral images and the spectral images after background removal and constructed a diagnostic model of maize canopy chlorophyll content using the partial least squares regression method. The results showed that the removal of the soil background was helpful in improving the diagnostic accuracy of the maize canopy chlorophyll content in the field. The diagnostic model of the chlorophyll content in maize canopy based on wavelet segmentation achieved the optimal results (Rc^2 = 0.6638, RMSEF = 3.6211; Rv^2 = 0.6923, and RMSEP = 3.9067). With the popularity of the UAV remote sensing platform in recent years, the results of this study can provide a valuable reference for noise removal during UAV image crop nutrition monitoring. However, this study was based on small-area plot experiments, and other noises besides soil background still exist in field experiments. Such issue was not discussed in depth in this study. Future work will be conducted in this direction.

Author Contributions: Conceptualization, L.Q., D.G., J.Z., M.L., H.S. and J.M.; Data curation, L.Q., D.G., J.Z., M.L., H.S. and J.M.; Formal analysis, L.Q., D.G., J.Z., M.L., H.S. and J.M.; Funding acquisition, L.Q., J.Z., M.L., H.S. and J.M.; Investigation, L.Q., D.G., J.Z., M.L., H.S. and J.M.; Methodology, L.Q., D.G., J.Z., M.L., H.S. and J.M.; Project administration, M.L., H.S. and J.M.; Resources, L.Q., D.G., J.Z., M.L., H.S. and J.M.; Software, L.Q., D.G., J.Z., M.L. and H.S.; Supervision, L.Q., D.G., J.Z., M.L., H.S. and J.M.; Validation, L.Q., D.G. and M.L.; Visualization, L.Q., J.Z. and H.S.; Writing—original draft, L.Q.; Writing—review & editing, L.Q., M.L., H.S. and J.M. All authors have read and agreed to the published version of the manuscript.

Funding: The project was supported by the National Key Research and Development Program (2018YFD0300505-1), the National Natural Science Fund (Grant No. 31971785 and 31501219), the Fundamental Research Funds for the Central Universities (Grant No. 2020TC036), and the Graduate Training Project of China Agricultural University (JG2019004 and YW2020007).

Acknowledgments: We would like to thank Zizheng Xing, Zhiyong Zhang, Xuying Ma, Di Song, Ruomei Zhao, and Song Li for their help with field data collection.

Conflicts of Interest: The authors declare no conflict of interest.

References

1. Gitelson, A.A.; Gritz, Y.; Merzlyak, M.N. Relationships between leaf chlorophyll content and spectral reflectance and algorithms for non–destructive chlorophyll assessment in higher plant leaves. *J. Plant Physiol.* **2003**, *160*, 271–282. [CrossRef] [PubMed]
2. Sun, H.; Li, M.Z.; Zhang, Y.E.; Zhao, Y.; Wang, H.H. Detection of chlorophyll content in maize growth stage. *Spectrosc. Spectr. Anal.* **2010**, *30*, 2488–2492.
3. Liu, H.J.; Li, M.Z.; Zang, J.Y.; Gao, D.H.; Sun, H.; Wu, J.Z. A Modified Vegetation index for Spectral Migration During Crop Growth. *Spectrosc. Spectr. Anal.* **2019**, *39*, 3040–3046.
4. Yuan, Z.; Ata–Ul–Karim, S.T.; Cao, Q.; Lu, Z.; Cao, W.; Zhu, Y.; Liu, X.J. Indicators for diagnosing nitrogen status of rice based on chlorophyll meter readings. *Field Crop. Res.* **2016**, *185*, 12–20. [CrossRef]
5. Gong, Y.; Duan, B.; Fang, S.; Zhu, R.; Wu, X.; Ma, Y.; Peng, Y. Remote estimation of rapeseed yield with unmanned aerial vehicle (UAV) imaging and spectral mixture analysis. *Plant Methods* **2018**, *14*, 70. [CrossRef] [PubMed]
6. Du, M.M.; Noguchi, N. Monitoring of Wheat Growth Status and Mapping of Wheat Yield's within–Field Spatial V ariations Using Color Images Acquired from UAV–camera System. *Remote Sens.* **2017**, *9*, 289. [CrossRef]
7. Battude, M.; Bitar, A.A.; Morin, D.; Cros, J.; Huc, M.; Sicre, C.M.; Dantec, V.D. Estimating maize biomass and yield over large areas using high spatial and temporal resolution Sentinel–2 like remote sensing data. *Remote Sens. Environ.* **2016**, *184*, 668–681. [CrossRef]
8. Kross, A.; Mcnairn, H.; Lapen, D.; Sunohara, M.; Champagne, C. Assessment of RapidEye vegetation indices for estimation of leaf area index and biomass in corn and soybean crops. *Int. J. Appl. Earth Obs. Geoinf.* **2015**, *34*, 235–248. [CrossRef]
9. Matese, A.; Toscano, P.; Di Gennaro, S.F.; Genesio, L.; Vaccari, F.P.; Primicerio, J.; Gioli, B. Intercomparison of UAV, aircraft and satellite remote sensing platforms for precision viticulture. *Remote Sens.* **2015**, *7*, 2971–2990. [CrossRef]
10. Zhang, Y.; Su, Z.; Shen, W.; Jia, R.; Luan, J. Remote monitoring of heading rice growing and nitrogen content based on UAV Images. *Int. J. Smart Home* **2016**, *10*, 103–114. [CrossRef]
11. Jiang, J.; Zheng, H.; Ji, X.; Cheng, T.; Tian, Y.C.; Zhu, Y.; Cao, W.X.; Reza, E.; Yao, X. Analysis and evaluation of the image preprocessing process of a six–band multispectral camera mounted on an unmanned aerial vehicle for winter wheat monitoring. *Sensors* **2019**, *19*, 747. [CrossRef] [PubMed]
12. Colomina, I.; Molina, P. Unmanned aerial systems for photogrammetry and remote sensing: A review. *ISPRS J. Photogramm. Remote Sens.* **2014**, *92*, 79–97. [CrossRef]
13. Mao, Z.H.; Deng, L.; Sun, J.; Zhang, A.W.; Chen, X.Y.; Zhao, Y. Application of multispectral remote sensing of UAV in corn canopy chlorophyll prediction. *Spectrosc. Spectr. Anal.* **2018**, *38*, 2923–2931. (In Chinese)
14. Jin, S.-Y.; Su, Z.-B.; Xu, Z.-N.; Jia, Y.-J.; Yan, Y.-G.; Jiang, T. Chlorophyll Content Retrieval of Rice Canopy with Multi–spectral Inversion Based on LS–SVR Algorithm. *J. Northeast Agric. Univ.* **2019**, *26*, 53–63.

15. Torressánchez, J.; Lópezgranados, F.; Serrano, N.; Arquero, Q.; Jose, M.S.P. High–Throughput 3–D monitoring of agricultural–tree plantations with unmanned aerial vehicle (UAV) technology. *PLoS ONE* **2015**, *10*, e0130479.
16. Zheng, H.; Cheng, T.; Zhou, M.; Li, D.; Yao, X.; Tian, Y.C.; Cao, W.X.; Zhu, Y. Improved estimation of rice aboveground biomass combining textural and spectral analysis of UAV imagery. *Precis. Agric.* **2018**, *20*, 611–619. [CrossRef]
17. Cheng, T.; Song, R.; Li, D.; Zhou, K.; Zheng, H.; Yao, X.; Tian, Y.C.; Cao, W.X.; Zhu, Y. Spectroscopic estimation of biomass in canopy components of paddy rice using dry matter and chlorophyll indices. *Remote Sens.* **2017**, *9*, 319. [CrossRef]
18. Cinat, P.; Gennaro, S.F.D.; Berton, A.; Matese, A. Comparison of Unsupervised Algorithms for Vineyard Canopy Segmentation from UAV Multispectral Images. *Remote Sens.* **2019**, *11*, 1023. [CrossRef]
19. Ahmed, I.; Eramian, M.; Ovsyannikov, I.; van der Kamp, W.; Nielsen, K.; Duddu, H.S.; Rumali, A.; Shirtliffe, S.; Bett, K. Automatic Detection and Segmentation of Lentil Crop Breeding Plots From Multi–Spectral Images Captured by UAV–Mounted Camera. In Proceedings of the 2019 IEEE Winter Conference on Applications of Computer Vision (WACV), Waikoloa Village, HI, USA, 7–11 January 2019; 2019; pp. 1673–1681. [CrossRef]
20. Zhou, X.; Zheng, H.B.; Xu, X.Q.; He, J.Y.; Ge, X.K.; Yao, X.; Cheng, T.; Zhu, Y.; Cao, W.X.; Tian, Y.C. Predicting grain yield in rice using multi–temporal vegetation indices from UA V–based multispectral and digital imagery. *ISPRS J. Photogramm. Remote Sens.* **2017**, *130*, 246–255. [CrossRef]
21. Sun, H.; Xing, Z.Z.; Qiao, L.; Long, Y.W.; Gao, D.H.; Li, M.Z.; Qin, Z. Spectral Imaging Detection of Crop Chlorophyll Distribution Based on Optical Saturation Effect Correction. *Spectrosc. Spectr. Anal.* **2019**, *39*, 3897–3903.
22. Wu, Q.; Sun, H.; Li, M.Z.; Song, Y.Y.; Zhang, Y. Research on Maize Multispectral Image Accurate Segmentation and Chlorophyll Index Estimation. *Spectrosc. Spectr. Anal.* **2015**, *35*, 178–183.
23. Yu, K.; Lenz–Wiedemann, V.; Chen, X.; Bareth, G. Estimating leaf chlorophyll of barley at different growth stages using spectral indices to reduce soil background and canopy structure effects. *ISPRS J. Photogramm. Remote Sens.* **2014**, *97*, 58–77. [CrossRef]
24. Gaurav, S.; Babankumar, B.; Mathew, L.N.; Goswami, J.; Choudhury, B.U.; Raju, P.L.N. Chlorophyll estimation using multi–spectral unmanned aerial system based on machine learning techniques. *Remote Sens. Appl. Soc. Environ.* **2019**, *15*, 100235.
25. Haboudane, D.; Tremblay, N.; Miller, J.R.; Vigneault, P. Remote Estimation of Crop Chlorophyll Content Using Spectral Indices Derived From Hyperspectral Data. *IEEE Trans. Geosci. Remote Sens.* **2018**, *46*, 423–437. [CrossRef]
26. Wu, C.; Niu, Z.; Tang, Q.; Huang, W. Estimating chlorophyll content from hyperspectral vegetation indices: Modeling and validation. *Agric. For. Meteorol.* **2008**, *148*, 1230–1241. [CrossRef]
27. Li, F.; Miao, Y.X.; Feng, G.H.; Yuan, F.; Yue, S.C.; Gao, X.W.; Liu, Y.Q.; Liu, B.; Ustin, S.L.; Chen, X.P. Improving estimation of summer maize nitrogen status with red edge–based spectral vegetation indices. *Field Crop. Res.* **2014**, *157*, 111–123. [CrossRef]
28. Roosjen, P.P.; Brede, B.; Suomalainen, J.; Bartholomeus, H.M.; Kooistra, L.; Clevers, J.G. Improved estimation of leaf area index and leaf chlorophyll content of a potato crop using multi–angle spectral data—Potential of unmanned aerial vehicle imagery. *Int. J. Appl. Earth Obs. Geoinf.* **2018**, *66*, 14–26.
29. Torres–Sánchez, J.; López–Granados, F.; Peña, J.M. An automatic object–based method for optimal thresholding in UAV images: Application for vegetation detection in herbaceous crops. *Comput. Electron. Agric.* **2015**, *114*, 43–52. [CrossRef]
30. Jeon, H.Y.; Tian, L.F.; Zhu, H.P. Robust crop and weed segmentation under uncontrolled outdoor illumination. *Sensors* **2011**, *11*, 6270–6283. [CrossRef]
31. Hayat, M.A.; Wu, J.X.; Cao, Y.L. Unsupervised Bayesian learning for rice panicle segmentation with UAV images. *Plant Methods* **2020**, *16*, 1.
32. Hamuda, E.; Glavin, M.; Jones, E. A survey of image processing techniques for plant extraction and segmentation in the field. *Comput. Electron. Agric.* **2016**, *125*, 184–199. [CrossRef]
33. Sang, H.L.; Hyung, I.K.; Nam, I.C. Image segmentation algorithms based on the machine learning of features. *Pattern Recognit. Lett.* **2010**, *31*, 2325–2336.
34. Guo, W.; Rage, U.K.; Ninomiya, S. Illumination invariant segmentation of vegetation for time series wheat images based on decision tree model. *Comput. Electron. Agric.* **2013**, *96*, 58–66. [CrossRef]

35. Zhu, Y.; Huang, C. An Improved Median Filtering Algorithm for Image Noise Reduction. *Phys. Procedia* **2012**, *25*, 609–616. [CrossRef]
36. Verrelst, J.; Rivera, J.P.; Moreno, J.; Campsvalls, G. Gaussian processes uncertainty estimates in experimental Sentinel–2 LAI and leaf chlorophyll content retrieval. *ISPRS J. Photogramm. Remote Sens.* **2013**, *86*, 157–167. [CrossRef]
37. Lv, J.; Yan, Z. Retrieval of chlorophyll content in maize from leaf reflectance spectra using wavelet analysis. International Symposium on Optoelectronic Technology and Application 2014: Imaging Spectroscopy and Telescopes and Large Optics. *Int. Soc. Opt. Photonics* **2014**, *29*, 7107–7127.
38. Fang, L.; Yang, B. UAV Image denoising using adaptive dual–tree discrete wavelet packets based on estimate the distributing of the noise. In Proceedings of the 10th World Congress on Intelligent Control and Automation, Beijing, China, 6–8 July 2012; pp. 4649–4654.
39. Gao, J.Q.; Wang, B.B.; Wang, Z.Y.; Wang, Y.F.; Kong, F.Z. A wavelet transform–based image segmentation method. *Optik* **2020**, *208*, 164123.
40. Jordan, C.F. Derivation of leaf-area index from quality of light on the forest floor. *Ecology* **1969**, *50*, 663–666. [CrossRef]
41. Rouse, J.W.J.; Haas, R.H.; Schell, J.A.; Deering, D.W. Monitoring Vegetation Systems in the Great Plains with ERTS. *NASA Spec. Publ.* **1973**, *351*, 309–317.
42. Barnes, E.M.; Clarke, T.R.; Richards, S.E.; Colaizzi, P.D.; Haberland, J.; Kostrzewski, M.; Waller, P.M.; Choi, C.Y.; Riley, E.; Thompson, T.L.; et al. Coincident detection of crop water stress, nitrogen status and canopy density using ground–based multispectral data. In Proceedings of the International Conference on Precision Agriculture and Other Resource Management, Bloomington, MN, USA, 16–19 July 2000.
43. Gitelson, A.A.; Andrés, V.; Verónica, C.; Rundquist, D.C.; Arkebauer, T.J. Remote estima–tion of canopy chlorophyll content in crops. *Geophys. Res. Lett.* **2005**, *32*, 93–114. [CrossRef]
44. Dash, J.; Jeganathan, C.; Atkinson, P.M. The use of MERIS terrestrial chlorophyll index to study spatio–temporal variation in vegetation phenology over India. *Remote Sens. Environ.* **2010**, *114*, 1388–1402. [CrossRef]
45. Kataoka, T.; Kaneko, T.; Okamoto, H.; Hata, S. Crop growth estimation system using machine vision. In Proceedings of the 2003 IEEE/ASME International Conference on Advanced Intelligent Mechatronics (AIM 2003), Kobe, Japan, 20–24 July 2003; Volume 2, pp. 1079–1083.
46. Huete, A.R. A soil-adjusted vegetation index (SAVI). *Remote Sens. Environ.* **1988**, *25*, 295–309. [CrossRef]
47. Verrelst, J.; Schaepman, M.E.; Koetz, B.; Kneubühler, M. Angular sensitivity analysis of vegetation indices derived from CHRIS/PROBA data. *Remote Sens. Environ.* **2008**, *112*, 2341–2353. [CrossRef]
48. Rondeaux, G.; Steven, M.; Baret, F. Optimization of soil–adjusted vegetation indices. *Remote Sens. Environ.* **1996**, *55*, 95–107.
49. Roujean, J.L.; Breon, F.M. Estimating PAR absorbed by vegetation from bidirectional reflectance measurements. *Remote Sens. Environ.* **1995**, *51*, 375–384. [CrossRef]
50. Chen, J.M. Evaluation of vegetation indices and a modified simple ratio for boreal applications. *Can. J. Remote Sens.* **1996**, *22*, 229–242. [CrossRef]
51. Dash, J.; Curran, P.J. The MERIS terrestrial chlorophyll index. *Int. J. Remote Sens.* **2004**, *25*, 5403–5413.
52. Cao, Q.; Miao, Y.; Feng, G.; Gao, X.; Li, F.; Liu, B.; Yue, S.C.; Cheng, S.S.; Ustin, S.L.; Khosla, R. Active canopy sensing of winter wheat nitrogen status: An evaluation of two sensor systems. *Comput. Electron. Agric.* **2015**, *112*, 54–67.
53. Duan, B.; Liu, Y.; Gong, Y.; Peng, Y.; Wu, X.T.; Zhu, R.S.; Fang, S.H. Remote estimation of rice LAI based on Fourier spectrum texture from UAV image. *Plant Methods* **2019**, *15*, 1–12.

remote sensing

Article

Drone Image Segmentation Using Machine and Deep Learning for Mapping Raised Bog Vegetation Communities

Saheba Bhatnagar *, Laurence Gill and Bidisha Ghosh

Department of Civil, Structural and Environmental Engineering, Trinity College, University of Dublin, D02 PN40 Dublin, Ireland; laurence.gill@tcd.ie (L.G.); bghosh@tcd.ie (B.G.)
* Correspondence: sbhatnag@tcd.ie

Received: 12 June 2020; Accepted: 11 August 2020; Published: 12 August 2020

Abstract: The application of drones has recently revolutionised the mapping of wetlands due to their high spatial resolution and the flexibility in capturing images. In this study, the drone imagery was used to map key vegetation communities in an Irish wetland, Clara Bog, for the spring season. The mapping, carried out through image segmentation or semantic segmentation, was performed using machine learning (ML) and deep learning (DL) algorithms. With the aim of identifying the most appropriate, cost-efficient, and accurate segmentation method, multiple ML classifiers and DL models were compared. Random forest (RF) was identified as the best pixel-based ML classifier, which provided good accuracy (≈85%) when used in conjunction graph cut algorithm for image segmentation. Amongst the DL networks, a convolutional neural network (CNN) architecture in a transfer learning framework was utilised. A combination of ResNet50 and SegNet architecture gave the best semantic segmentation results (≈90%). The high accuracy of DL networks was accompanied with significantly larger labelled training dataset, computation time and hardware requirements compared to ML classifiers with slightly lower accuracy. For specific applications such as wetland mapping where networks are required to be trained for each different site, topography, season, and other atmospheric conditions, ML classifiers proved to be a more pragmatic choice.

Keywords: semantic segmentation; machine learning; random forest; deep learning; CNN

---

1. Introduction

The use of drones for different types of vegetation classification has increased many folds over the last decade. This is due to the technological development of affordable and lightweight drones. With drones, a very high and flexible spatial resolution can be achieved, which is not possible with satellite imagery due to their fixed orbits. Satellites are both open-source and commercial. Some of the most popular open-source satellites include the Sentinel and Landsat series. These satellites provide global information but lack high spatial resolution with the best resolution possible of 10 m using Sentinel-2 (S2). The S2 imagery has been widely used for classifications; for example, a study carried out by [1] used S2 imagery for temporal mapping of wetland vegetation communities. However, one conclusion from the study was that accuracy decreases for smaller wetlands. In many cases in Ireland, at least, the area of wetlands can be relatively small, whereby satellite-based classification is not sensitive enough and can produce large errors. One of the significant problems is pixel-mixing; when the size of the pixel is 10 m, for example, each pixel can have a combination of species present in it. This affects the overall reflectance value of the pixel, and hence, a good boundary or extent of the species cannot be achieved. There are several ways to reduce the error in satellite images, but most of them require extensive hyperspectral bands. However, another method to get detailed monitoring of small areas is to use unmanned aerial vehicles (UAVs), more commonly known as drones.

Most drones typically carry optical cameras (RGB) and occasionally can support a thermal sensor, but some drones can also support more expensive hyperspectral cameras. The presence of a thermal/hyperspectral sensor allows more details to be gathered and improves spectral resolution. However, the dilemma about spectral versus spatial remains unanswered. Missions like Airborne Visible InfraRed Imaging Spectrometer (AVIRS) and hyperspectral satellite Hyperion provides high spectral resolution. A study [2] has used AVIRIS hyperspectral data with 224 bands and 20 m spatial resolution to detect invasive plant species (*Colubrina asiatica* (Brongniart, Adolphe Théodore) in Florida. Another study [3] used Hyperion (30 m) hyperspectral data to detect *Phragmites australis* (Steudel, Ernst Gottlieb) in coastal wetlands and states that, due to low spatial resolution, the analysis was affected by pixel mixing. Therefore, apart from high spectral resolution, a proper spatial resolution is also required for monitoring vegetation communities closely. Drone images have much higher spatial resolution when compared to satellite images. Drones have been explicitly used for species detection [4–7]. A study by [8] states many advantages of using a drone over satellite imagery for identification of land cover communities such as water, land *Avicennia alba* (Blume, Carl (Karl) Ludwig), *Nypa fruticans* (Verh. Batav. Genootsch. Kunst.), *Rhizophora apiculata* (Blume, Carl (Karl) Ludwig), and *Casuarina equisetifolia* (Linnaeus, Carl). Drone imagery has also been applied for specific applications such as analysing vegetation under shallow water, tracking waterbirds, and their habitats [9,10]. A study by [11] concluded that a thermal (infrared) sensor on its own performs comparable to an RGB sensor, but a multispectral sensor (with multiple spectral bands and indices) is required for the best analysis of nitrogen on rice fields. Multiple spectral sensors, however, although useful, are costly [12]. A study by [13] supports the hypothesis that proper spatial resolution with an RGB sensor is sufficient for the analysis of wetland delineation, classification, and health assessment. Therefore, taking all the points into consideration, as an alternative to an expensive camera, an RGB camera was used in this study.

For the analysis of drone data, many techniques are available. The state-of-the-art techniques in drone image analysis consist of both machine learning (ML) and deep learning (DL). A study by [14] demonstrates the application of ML techniques to classify drone images into roads, vineyards, asphalt, and roofs. The study uses ensemble decision trees with an object-oriented approach. The study [15] has used object-based multi-resolution segmentation (eCognition software) of UAV imagery for the segmentation of the agricultural field. Other than object-based, there are multiple pixel-based studies, such as [16,17] use support vector machine (SVM) classifier for agricultural mapping and reef monitoring using drone imagery. The study [18] applies multiple ML algorithms including random forest (RF), SVM, and gradient boosting decision tree (GBDT) to classify trees, grasses, bare gravel/sand bed, and water surface. The study achieved a high accuracy of up to 98% using RF classifier on UAV images. ML algorithms have also been used for vegetation segmentation. A study by [19] has used simple linear iterative clustering (SLIC) for mangrove segmentation. Another study by [1] has used graph cut for the segmentation of vegetation communities in wetlands. Hence, the segmentation of drone images can help in the identification of subtle changes in vegetation communities. Apart from ML, advanced deep learning techniques are also now being widely applied for drone image segmentation. A recent state-of-the-art review [20] shows the surge of applying DL in the field of RS. It also gives details about various convolutional neural network (CNN) models and suggests that ≈20% of all studies since 2012, uses DL with UAV imagery. The study [21] has used DL to segment concrete-cracks in drone images. The segmentation using a CNN is known as semantic segmentation. It has been applied for various applications like urban land classification [22,23], forest cover classification [24], and wetland type classification [25,26]. A study by [27] uses ResNet50 and UNet for classification of forest tree-species, and [28] has used transfer learning to get the best semantic segmentation of the aerial images AeroScapes dataset. Both [27,28] suggest that the usage of transfer learning enhances the analysis. A study by [29] has utilised both ML (linear regression) and DL (neural network) for predicting water and chlorophyll content in citrus leaves. The study suggested that both ML and DL give comparable results for predictions using UAV images. From the literature, it is apparent that both ML and DL can be applied for drone image segmentation. However, it is not clear which technique, the traditional state-of-the-art machine learning or the advanced deep learning is better for the identification of the communities. Therefore, in this study, we applied both ML and DL techniques for vegetation classification of different vegetation communities on a raised bog wetland. Our study also demonstrates the pros and cons of both methods. It also gives a clear insight into both the techniques and their applicability for future studies on vegetation identification.

2. Study Area and Materials

The area of study is one of the largest intact raised bogs present in Ireland, covering approximately 460 ha area located in the midlands called Clara bog. The two sides of the bog are divided by a road: East Clara is a restored bog (after years of drainage and peat cutting), whereas, the West Clara remains a natural active raised bog. This study concentrates on a small part of the bog located in West Clara bog (as shown in Figure 1). The different vegetation species have been grouped into different communities on the basis of similar habitats, which are termed 'ecotopes' [30].

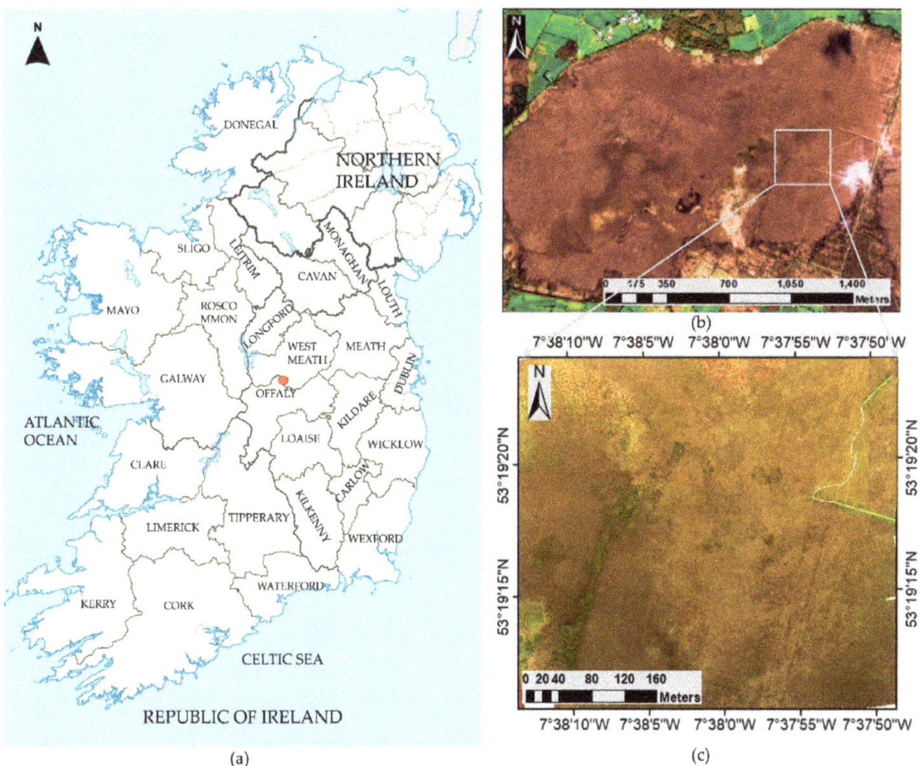

Figure 1. Study area: (**a**) map of Ireland (with the highlighted area: Clara Bog). (**b**) West of Clara Bog, County Offaly (with the highlighted area covered by drone). (**c**) Area covered by DJI Inspire 1™ drone.

The major ecotopes present in Clara bog are Central (C), Subcentral (SC), Submarginal (SM), Marginal (M), and Active flush (or flush) (AF). Other ecotopes like Inactive flush (IAF), Facebank (FB) are also present in this bog but have not been considered in this study due to their low ecological impact. Out of all these ecotopes, the main focus is on the conservation of the active peat-forming areas [1,30,31], which are considered to be C, SC, and AF ecotopes. These areas have high sphagnum moss coverage, with hummocks, hollows, lawns, and many pools. The SM ecotope that appears at the boundaries of the SC ecotope can appear to be almost homogenous, which makes it hard to distinguish between them. The SM and M ecotopes are located on drier areas with vegetation reflective of such conditions.

For capturing high-resolution images, a DJI Inspire 1™ drone was used. The camera used with the drone was Zenmuse X3. It is an optical camera with 100–1600 ISO range (for photo) and 94° field of view (FOV). The lens is anti-distortion and autofocus (20mm of 35mm format equivalent). The aspect ratio, while clicking the images, was kept at 4:3. The images were captured on 21st April 2019 at around noon time. The highest temperature on the day was recorded at 19 °C. The height of the flight was ≈100m, and the spatial resolution of the images captured was 1.8 cm. The drone mission was pre-loaded using Google maps in Pix4DCapture application to capture ≈8 ha of the area using an iOS-12 device. The images were captured individually with 70% frontal and 80% sideways overlap at an average speed of 3 m/s. Figure 1c provides the drone imagery of the study area. For georeferencing, the drone imagery had geo-tags (lat-long locations) present in it. For better orientation, imagery was overlayed on high-resolution DigitalGlobe World Imagery (spatial resolution = 30cm) available as a base map in ArcMap v.10.6.1 [32,33]. Using 'georeferencing' toolbox present in [32], 3–4 ground control points (GCPs) were identified for every image, and projection was rectified to Geographic Coordinate System—World Geodetic System 84 (GCS WGS 84). In this study, C, SC, SM, M, and AF ecotopes were all captured using high-resolution drone imagery (Figure 2).

Submarginal	Subcentral	Marginal	Central	Flush

Figure 2. Ecotopes in Clara bog. Drone images, April 2019.

The SM and SC ecotopes are highly homogenous and appear to be mixed throughout the bog [1]. These communities were therefore merged for the rest of the study. In total, around ≈75 images of dimension 3000 × 4000 were captured. Out of these images, 15 images were discarded due to differences in light intensity, motion blur, and camera tilt. The usable 60 images were divided into 70% training and 30% testing randomly, which is around 40 images for training and 20 images for testing. In order to have a correct idea of mapping accuracy, all the images were labelled for the four vegetation communities (M, SMSC, C, AF). For ML only a part of the labelled training data was required, whereas for DL fully labelled images were used. This is discussed further in Sections 3 and 4. For the creation of a training dataset, it is essential for all the images to have a similar intensity range. Depending on the lighting situation when the picture was taken, the colour properties may be changed, even though the textural properties remain unchanged. In a temperate climate like Ireland, this change in sunlight while capturing drone images is unavoidable. Therefore, going forward in future studies, the usage of colour correction techniques for drone mages is recommended such that all the captured images can be used.

3. Segmentation Using Machine Learning

The segmentation of the images using machine learning techniques utilises combinations of intensity, colour, texture, and motion attributes to come up with hierarchical segments [34]. The drone images used for this study have intensity and colour information. Although textural information is not present in the original image, textural features were subsequently calculated using the parameters mentioned in Table 1, [35]. This was done by converting the RGB image into a grayscale image. The textural information presented in Table 1 was added as features along with the RGB layers.

The entire computation of machine learning techniques and the steps described below was performed using MATLAB v.2019b using image processing toolbox [36].

Table 1. Textural properties calculated using drone imagery.

Property	Description
Contrast	Intensity difference between pixels compared to its neighbour for the whole image [37].
Correlation	Correlation of a pixel and its neighbour for the whole image [38].
Energy	Sum of squared elements in gray level co-occurrence matrix (GLCM) [39].
Homogeneity	Closeness of the distribution of pixels in the GLCM to its diagonal [40].
Mean	Mean of the area across the window
Variance	Variance of the area across the window
Entropy (e)	Statistical measure of randomness $e = -\sum (h \times \log_2 h)$; where h contains the normalised histogram counts
Range	Range of the area across the window [41].
Skewness (S)	Asymmetry of the data over the mean value [42]. $S = E(p - \mu)^3/\sigma^3$, where μ is the mean of the pixel p, σ is the standard deviation of p, and E represents the expected value.
Kurtosis (K)	Distribution to be prone to outliers [42]; $K = E(p - \mu)^4/\sigma^4$

The segmentation technique used in this study, called graph cut, is based on max-flow min-cut [43]. This is done using posterior probabilities associated with every pixel for every class. In order to calculate the posterior probabilities, an initial classification of the drone images was carried out. Based on the texture and colour intensity, a total of 13 bands are used for further classification of the drone images. The type and choice of classifier used are discussed in the following subsection.

3.1. Choice of the ML Classifier

For efficient classification, the choice of the classifier is the most crucial decision that has to be made. Multiple studies have applied hyperspace based SVM [44,45] for image classification. Other studies, like [46], have used decision trees. Studies [47,48] suggest that there is an advantage of using ensemble classifiers over other state-of-the-art classifiers. The most commonly used ensemble classifier consists of a tree model. The tree models are easy to understand and could be used for both classification and regression. There is no need for variable selection (since it is automatic) or variable transformation. They are robust to outliers and missing data, and particularly useful for large datasets.

In this study, in order to provide proper comparative analyses, the drone images captured on 21st April 2019, were classified using multiple classifiers. The training dataset ($\approx$12k pixels from 40 images) was the input for all the classifiers. The classifiers were tested on model accuracy, misclassification cost (i.e., the total number of incorrectly identified pixels per 10,000 pixels), and training time (time taken by the classifier for training). The model accuracy for each ML model was calculated using 5-fold cross-validation for the entire 70% training dataset. This accuracy indicates the capability of the model to label the pixels correctly. The results (Table 2) describes all the classifiers and the corresponding accuracy metric. All the calculations were performed using MATLAB v.2019b [36].

The preliminary comparison was made using six classifiers, namely, decision trees [49], naïve Bayes [50], discriminant analysis [51], SVM [52], k-nearest neighbour (KNN) [53], and random forest (RF) [54]. Based on the misclassification rate, model accuracy, and training time (see Table 2), RF was found to be best classifier. Random forest or bagging is a general-purpose procedure for reducing the variance of a predictive model. When applied on trees, the number of trees (t) is bootstrapped, each having a variance σ^2. In RF each tree can split on only a random subset of the samples (hence, the name). RF requires an attribute (sample) selection and a pruning method. Information gain ratio criterion [55] and Gini Index [56] are the most common attributes selection methodology. For this study, the Gini index criterion was used to decide the attributes. The Gini index (G) is given in Equation (1). Based on the value of G, the attribute was decided automatically.

$$G = \sum_{n} \sum_{i=1}^{N} (p_i \times (1 - p_i))_n \quad (1)$$

where p_i is the proportion of the pixel (i = 1 to N) belonging to a particular class n, i.e., it is the prior probability. A minimum of 10% of the entire ground truth image should be given as training and rest could be used for testing [1]. The samples were divided into 100 random subsets (with repetition), and for each tree, and the attributes (splitting criteria: which of the RGB bands) were decided using Equation (1). The final class selection for every pixel was made using majority voting. The workflow of the RF classifier is given in Figure 3.

Table 2. Comparison of ML classification techniques.

Name	Parameter	Model Accuracy	Misclassification Cost	Training Time (s)
Decision trees	Max. no. of splits = 20; split criterion = Gini's diversity index	87.4	736	7.3
Discriminant analysis	Kernel = quadratic	89.4	618	8.6
Naïve Bayes	Kernel = Gaussian	78.3	1271	19.5
Support vector machine	Kernel = radial basis function (rbf) = 0.25	91.9	472	112.5
K nearest neighbour	No. of neighbours = 2; distance = Euclidean	91.0	528	378.8
Random forest	No. of trees (t) = 100 (1000 samples with repetition); total no. of splits = 5853	92.9	454	59.2

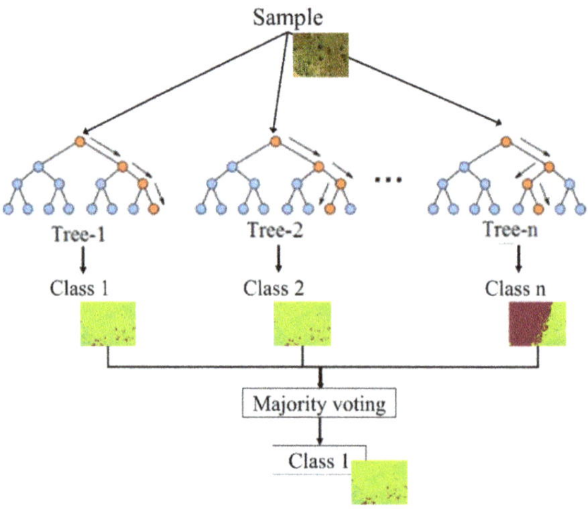

Figure 3. Workflow—random forest classifier.

3.2. Segmentation

Once the drone images were classified, they were segmented using the maximum-a-posterior (energy minimisation) technique. The technique uses contextual (area) information to form proper segments from pixels. The pixels, therefore, are no longer treated as a single entity but part of a more significant segment. It can be considered as a post-classification smoothing process based on spatial similarities. The formation of segments was done using a max-flow min-cut algorithm, commonly known as graph cut. This algorithm uses data cost and smoothness costs [57]. The graph cut segmentation was performed in MATLAB v.2019b [40] using MATLAB wrapper mex file function that enables the user to call C/C++ files [58]. The steps for the segmentation include calculation of the data cost, smoothness cost, and energy using posterior probability from the pixel-based classification map. Based on the maximum probability of the pixels, the segments were formed, and the pixels were joined.

The data cost (D_p) is based on individual labels of pixels and their likelihood function. The data cost D_p measures the cost of assigning the class n to the pixel p for a given set of features U_N in the vectorised image having N pixels. In image processing D_p can be typically expressed as [59], given by Equation (2).

$$D_p = \left\| U_p(n) - I(p) \right\|^2 \quad (2)$$

where, $I(p)$ was the observed reflectance of the pixel p.

The smoothness cost ($V_{p,q}$) on the other hand, was used to promote groups. It was assumed that the neighbouring pixels should belong to the same class, and hence, this cost was given based on the likelihood of pixels p, q belonging to same class n. n_p , n_q are labels of pixels p, q respectively. It was defined using described in Equation (3).

$$Vp,q\left(n_p\ ,\ n_q\right) = c \times exp\ (-\Delta(p,\ q)/\sigma) \times T\left(n_p\ \neq\ n_q\right) \quad (3)$$

where $\Delta(p,\ q) = \left|I(p) - I(q)\right|$ denotes how different the reflectance values of p and q are, $c > 0$ is a smoothness factor, standard deviation $\sigma > 0$ is used to control the contribution of $\Delta(p,\ q)$ to the penalty, and $T = 1$ if $n_p \neq n_q$ and 0 otherwise.

As described in [1], the steps followed for drone images were the same as for the satellite image segmentation. The main difference comes in the choice of the smoothness factor. Since a drone image was much more detailed for forming distinct segments compared to a satellite image, a high smoothness factor was required. After an iterative parametrisation optimisation exercise, a smoothness factor (c) of $c > 5$ was chosen for the drone images. This can be compared to the optimum value of $c < 1$ when processing satellite images [1]. Therefore, it was seen that for a high resolution (1.8 cm), a higher value of c was required, whereas, when working with the 10 m spatial resolution from satellite images, a small value of c suffices.

The pioneering work done by [59] explains energy (E) minimisation can be interpreted directly as posterior maximising. Using probability functions from previous steps, we get the energy function as described in Equation (4).

$$E\ (U_N, n) = \sum_{p \in N} Dp + \sum_{p,q \in N} Vp,q \quad (4)$$

Therefore, $E(U_N,\ n)$, i.e., energy for the image vector with total N pixels (U_N) for all n classes is minimised, leading to the formation of smooth segments. The pixels with least E are joined together to form the segments depending on their initial labels as obtained from the pixel-based RF classification. The results of the segmentation are further discussed in Section 5.1.

4. Segmentation Using Deep Learning

4.1. Parameters in Convolutional Neural Network

Convolution neural networks (CNNs or Covnets) have caused a step-change in pattern recognition progress. Here each neuron is connected to a local region of the input only, making the network faster and less prone to overfitting for a large dataset. Therefore, CNNs, when compared to traditional NNs, can have fewer parameters. In addition, the same parameters are used in more than one place on CNN, making the model both statistically and computationally efficient. The initial layers of the CNN identify lines, corners, edges, textures, and then the deeper the network goes, the more precisely it can learn from the features, as shown in Figure 4, which gives the architecture of CNN. The different layers used in CNN are described in detail in the following subsections.

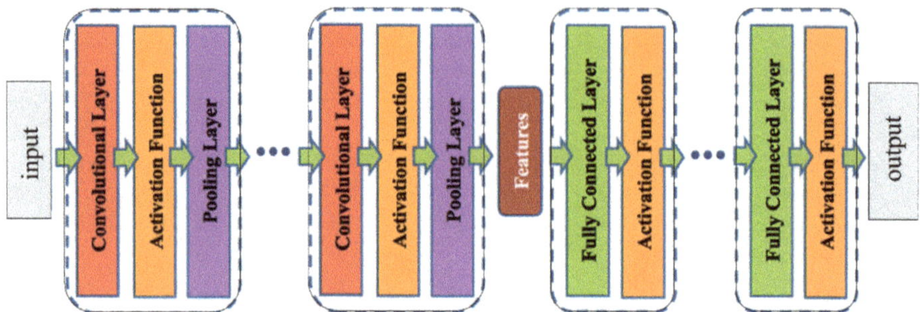

Figure 4. The general architecture of convolutional neural network (CNN).

4.1.1. Convolutional Layer

Convolution in CNN is the mathematical operation that combines signals a and b $[a * b]$, i.e., filtering input a with kernel b. It is a process to overlay 'b' on 'a', multiply the numbers and sum the products and move. In CNN the convolutional layer is used instead of only fully connected layers. For visualising, convolution may look like a sliding window operation, but it is implemented as matrix multiplication. The input is divided into arrays as well as the kernels and rearranged into columns.

4.1.2. Pooling Layer

The pooling layer downsamples the input by locally summarising the data in it. The two types of pooling are shown in Figure 5.

1. Max Pooling: where the local maxima of the filtered region are carried forward.
2. Average pooling: where the local average of the filtered region is carried forward.

Of the two methods, max-pooling was used for this study, as it is a more efficient pooling technique [60]. A feature existing in the input layer is fed forward regardless of its initial position (as the local maxima will still make it to the next layer). The advantages of pooling include decreasing the size of the activation layer that is fed forward to the next layer and increasing the receptive field of the subsequent units.

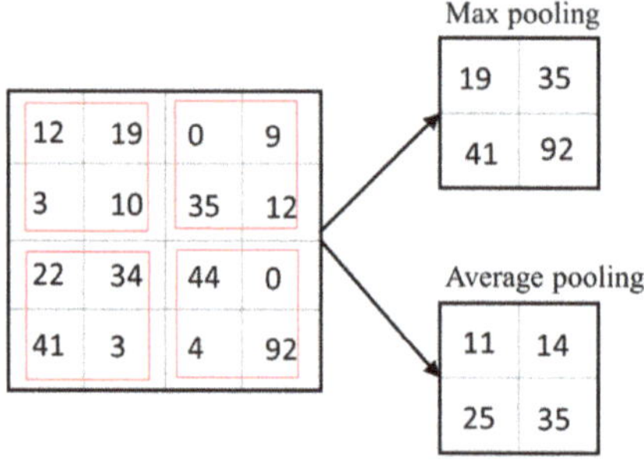

Figure 5. Types of pooling used in CNN.

4.1.3. Kernel Size

Kernels or the filters are used in order to down-sample the layers in CNN. It is preferable to use smaller kernels stacked on top of one another than using a large kernel [61]. Using smaller kernels decreases the number of parameters and also increases the nonlinearity (see Section 4.1.6). For example, a stack of two 3×3 kernels and one 5×5 kernel will have the same receptive field. However, 3×3 will have fewer parameters (as the same kernel is used twice) and more nonlinearity. Therefore, in this study, the kernel of size 3 × 3 was used.

4.1.4. Stride

Stride defines by how much the kernel will move in the convolution layer. The stride can be used to increase the receptive field. Example, stride = 2. Using stride > 1 provides a down-sampling effect and can be used as an alternative to the pooling layer.

4.1.5. Padding

Padding is required to maintain the spatial resolution of the input image. Padding can be of two types, valid and same. In valid padding, the spatial dimension of the output shrinks by one pixel less than the kernel spatial dimension. Whereas, in same padding, the input is surrounded with zeros such that the spatial dimension of output is the same as the input layer. Therefore, the same padding was used in this study in order to maintain the same dimension between input and output.

4.1.6. Activation Function

The activation function ($f(x)$) defines the output for a given input. It also imparts nonlinearity to the input.

Why do we need nonlinearity?

Combining linear functions yields a linear function; however, in order to compute more in-depth features, nonlinearity is required. With just linear functions, the model is no more expressive than a logistic regression model without any hidden layer. Hence, without any nonlinearity, the entire network behaves as a single linear function.

The study [62] describes the types of activation functions. Some of the most commonly used and well-known activation functions are identity (when linear relation is required), binary step (nonlinear, good for binary classification), sigmoid (nonlinear function, ranging from 0 to 1), tangent hyperbolic (tanH) (same as sigmoid, but ranges from −1 to +1), rectified linear unit (ReLu) (nonlinear function, removes all the –ve part of the input). Sigmoid, tanH, and ReLu also has other variants, see [63]. Other studies like [64,65] compare the various activation functions. A study by [66] presents a comparison between 11 activation functions and suggests ReLu to be the best. Additionally, the ReLu function is much more computationally effective, and therefore, for this study, the ReLu activation function was used. Equation (5) describes the ReLu function.

$$\begin{aligned} f(x) &= 0\,; x < 0 \\ f(x) &= x\,; x \geq 0 \end{aligned} \tag{5}$$

4.1.7. Softmax Classifier

Softmax Classifier is an activation function, typically used as the top layer (after a fully connected layer). It imparts probabilities of each input belonging to each output when there are more than two outputs. For the n number of classes, the Softmax activation (σ) can be defined by Equation (6).

$$\sigma(x)_j = \frac{e^{x_j}}{\sum_{k=1}^{n} e^{x_k}}\,; j = 1, \ldots, n \tag{6}$$

4.1.8. Batch Normalisation

It is apparent that each layer is dependent on its previous layer; therefore, even the smallest error in one layer can be magnified in further layers, causing much more significant errors in the final output. To avoid this, a batch normalisation layer is used. This layer normalises the hidden nodes before they are fed into an activation function.

4.1.9. Additional Parameters in CNN

An essential parameter in CNN is optimisation. Training a network can be considered to be an optimisation problem where the goal is to minimise the loss function. There are various optimisation algorithms that can be used to minimise the loss function, such as online learning [67], batch learning [68], and stochastic gradient descent (SGD) [69]. As described in [70] for faster and efficient processing, a subset of the data is taken one at a time, and therefore a stochastic gradient descent was used for optimisation in this study. The subset of data is called a mini-batch, and the number of samples in a mini-batch is called batch size.

Another important parameter in CNN is regularisation. Regularisation of the model can be carried out to make the model simple but effective. This reduces overfitting and adds additional information. This ensures that augmenting the input will not change the quality of the output. Regularisation can be done by adding a weight penalty term to the loss function (Equation (7)).

$$Loss = Loss + weight\ penalty\ (w) \tag{7}$$

L2 or ridge regularisation leads to the formation of small weights [71]. Additionally, L2 regularisation never causes a degradation in performance, even with the addition of kernels [72]. Therefore, L2 regularisation was used in CNNs for this study. For a given input x and its corresponding output $\hat{x}$ the regularisation function is given in Equation (8).

$$Loss = \sum_i (x_i - \hat{x}_i)^2 + \alpha \sum_i {w_i}^2 \tag{8}$$

A third, important parameter for CNN architecture is the learning rate (LR). The LR is defined as the rate at which the weights are updated during the training of the network. The study [73] suggests to start with a bigger learning rate and gradually decrease the gradient when getting closer to the local minima of the loss function. Since adaptive momentum estimation (ADAM) is fast and requires low memory for computation [74], it was selected as the optimisation parameter for the network used in this study. ADAM is a method that learns the LR on a parameter basis and is a combination of both adaptive gradient (AdaGrad) and root mean square (RMSProp).

4.1.10. Popular CNN Models

CNN models are formed using the combinations of parameters mentioned in the above subsections. The combinations of layers and the type of parameters used are often application-based and applied to solve a bigger problem. In this study, VGG16 [75] and ResNet50 [76] were applied based on the work done by [77,78], the models with their salient features are briefly discussed as follows.

VGGNet

- Stands for Visual Geometry Group
- Consists of 13 convolutional layers with three fully connected layers, hence the name VGG16.
- Each convolutional layer has kernel size = 3 with stride = 1 and padding = same.
- Each max-pooling layer has kernel = 2 and stride =2.

ResNet50

- Stands for Residual Network.
- A deep network, having 50 layers.
- It popularised batch normalisation.
- It uses skip connection to add information on output from a previous layer to the next layer.

4.2. CNN for Semantic Segmentation

Semantic segmentation is a process of assigning a label to each pixel in an image such that pixels with the same label are connected via some visual or semantic property [79]. In order to carry out semantic segmentation, the spatial information needs to be retained. Hence no fully connected layers are used, which is why they are called fully convolutional networks.

4.2.1. Moving from a Fully Connected to a Fully Convolution Network

This is where all fully connected layers are converted into 1×1 convolutional layers. In the case of labelling, the output is a 1D vector giving probabilities of the input belonging to n classes. In the case of segmentation, an output layer is a group of 2D probability-maps of each pixel belonging to each class. These are known as score maps. The score maps are coarse as throughout the network; the information (image) has been down-sampled to absorb minute details. Therefore, to make the output compatible with the input in size, up-sampling is required.

Up-sampling can be done using either bilinear interpolation or cubic interpolation (or similar techniques). One way of up-sampling is via skip-connections or shortcut connections. In skip-connection, the feature maps obtained as the output from the max-pooling layers are up-sampled using bilinear interpolation and added to the output score maps. The method works well but requires some amount of learning to up-sample the score maps and feature maps to match it to the size of the input image. In order to minimise the amount of learning, another method encoder-decoder is widely used. Here, the layers which down-sample the input are the part of the encoder and the layers which up-sample are part of the decoder. Three key fully connected models, SegNet [80], UNet [81], and Pyramid Scene Parsing Network (PSPNet) [82] are used in this study. A brief description of the models is given in the following subsections.

4.2.2. SegNet Model

SegNet works with encoder-decoder architecture, followed by a pixel-wise classification layer for multiple classes. Encoders extract the most relevant features from the given input. The decoder uses the information from encoder to up-sample the output (Figure 6).

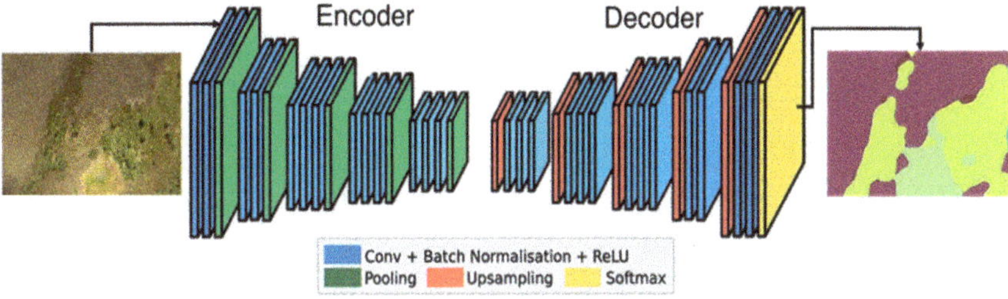

Figure 6. SegNet architecture for semantic segmentation for bog-ecotope semantic segmentation.

The up-sampling technique used by the decoder is known as max-unpooling. Max-unpooling eliminates the need for learning to up-sample (as was required in skip-connections) as shown in Figure 7. Based on the location of the maximum value, the max-pooled values are placed. The remainder of the matrix is loaded with zeros. Convolution is done using any CNN models (as discussed in Section 4.1) using this layer.

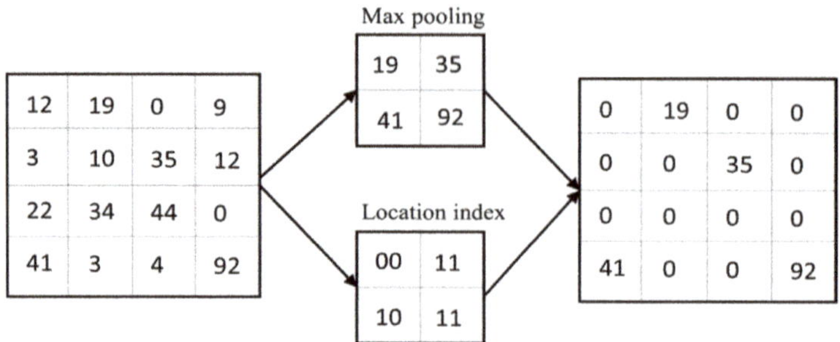

Figure 7. Pooling and unpooling for semantic segmentation.

4.2.3. UNet Model

UNet network carries out the transpose convolution (encoder-decoder) and also uses skip connections (Figure 8).

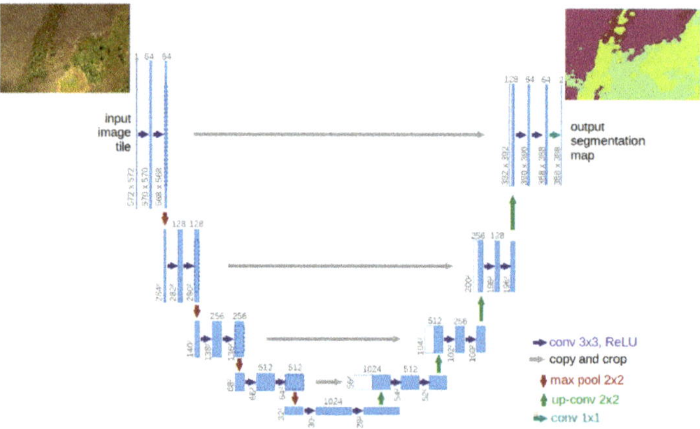

Figure 8. UNet architecture for semantic segmentation for bog-ecotope semantic segmentation.

At every layer in the decoder side, the network finds a corresponding feature map (of the same size) from the encoder and adds (1 × 1 convolution) to the score map. This way, the size of the feature map is always in sync. Due to its architecture and depth, UNet is most widely used in biomedical image analysis.

4.2.4. PSPNet Model

PSPNet stands for Pyramid Scene Parsing Network. This network incorporates the scene and global features for scene parsing and semantic segmentation as shown in Figure 9.

The pyramid pooling module in PSPNet fuses the features in four scales: coarse (1 × 1), 2 × 2, 3 × 3, and 6 × 6. The up-sampling done is a bilinear interpolation, and all the features are concatenated as the final pyramid pooling global feature [82]. The spatial pyramid pooling technique eliminates the need for using the input image of a specific size, which is used in SPPNet [83].

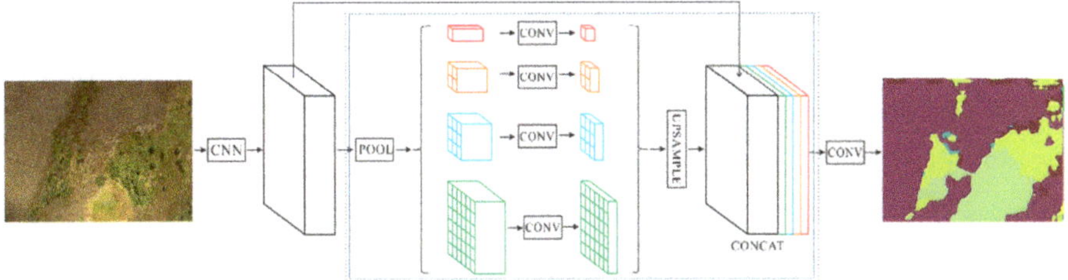

Figure 9. Pyramid Scene Parsing Network (PSPNet) architecture for semantic segmentation for bog-ecotope semantic segmentation.

4.3. Methodology for the Comparison between CNN Models for the Case Study on Raised Bog Drone Images

Using the drone images captured on 21st April 2019, semantic segmentation using various CNN architectures was applied to identify and label the ecotopes present on Clara Bog. The entire computation was performed in python v3.7 [84] using GPU (NVIDIA Tesla K40C 12GB CUDA), accessed remotely from trinity college high performing computer (TCHPC), and partly on google virtual machine (Tesla K40C 12GB). The study uses the repository in [85].

4.3.1. Training Data Preparation

In order to smoothly run the semantic segmentation, the preparation of training data was done as follows

1. Forty drone images were manually labelled using MATLAB-Image Labeler app [36].
2. The labels (in .mat format) were converted into JPG.
3. The images and labels were resized in order to use the GPU memory efficiently and to speed up the process. For resizing, the images were shrunk in the order of 2^n such that the classes were clearly distinguishable. The resizing of the images was done using a bilinear interpolation technique.
4. The images were resized from 3000 × 4000 to 512 × 1024 ($2^9 \times 2^{10}$) for further use. The size of the image is kept rectangular in order to maintain the aspect ratio of the original drone imagery. The ratio can be decided with respect to the application. For this study, to have a fair comparison between ML and DL methods, the size of the imagery was not reduced to smaller patches. Alternatively, patches of the same size ($2^9 \times 2^{10}$) can be extracted with overlapping. For this study, the small patches did not cover all the ecotopes. In a single patch, at maximum, only two ecotope classes were covered. This is due to the large size of the raised bog in the application. Therefore, to incorporate the maximum number of ecotope classes in a single image and to avoid any information loss, resizing of the images was done (instead of extracting the patches).
5. After reshaping, the images were renamed such that the images and their corresponding labels can be identified.

Steps (2–5) were repeated for all 40 images having all four ecotope classes mentioned in step 1. The final training data consisted of 40 images (both RGB and labelled) of the size $2^9 \times 2^{10,}$ which was fed to the CNN models described in the next subsection. The testing was carried out on the rest of the 20 images.

4.3.2. Models Used for Semantic Segmentation

The models were created using a base network (tested on ImageNet) along with a segmentation architecture. Since CNN takes a considerable amount of time to train, only the most frequently used and tested models (in the literature) were compared. The optimisation algorithm used was SGD Adam with initial LR = 0.05 and L2 regularisation. Initially, a high LR was used, as it is reduced throughout the epochs by a factor of 10. The max number of epochs = 100, and images were shuffled at every epoch and a mini-batch size of 64. The loss between the labels given by the model and the actual (training) label at every epoch was calculated using cross-entropy loss described in Equation (9). The cross-entropy loss is commonly applied for classification applications, whereas loss like half mean square error is more common for regression tasks. Therefore, a cross-entropy loss was used here.

$$cross\ entropy = -\frac{1}{N}\sum_{i=1}^{N}\sum_{j=1}^{n}\left(\hat{x}_i \log x_{i,j}\right) + \left(1-\hat{x}_{i,j}\right)\log\left(1-x_{i,j}\right) \tag{9}$$

where N is the total number of pixels, n is the total number of classes, x is the training label (input), and $\hat{x}$ is the output label as predicted by the models. Instead of training the network from scratch, one of the most common techniques is to use a pre-trained network. The idea is to transfer the information learned by the network and then fine-tune and train the classification layer of the model for our specific task. In this manner, given that the weights are already pre-trained for a large dataset, even with a small dataset, the performance is much more improved. Pre-trained weights also speed up the convergence process (to reach local minima, i.e., to overall minimise the loss). It is also considered better than random initialisation. For the four models listed below, 'ImageNet' dataset [86] was used to initialise the weights. Other details are mentioned in detail in [85]. The architecture for these models is shown in Figure 10.

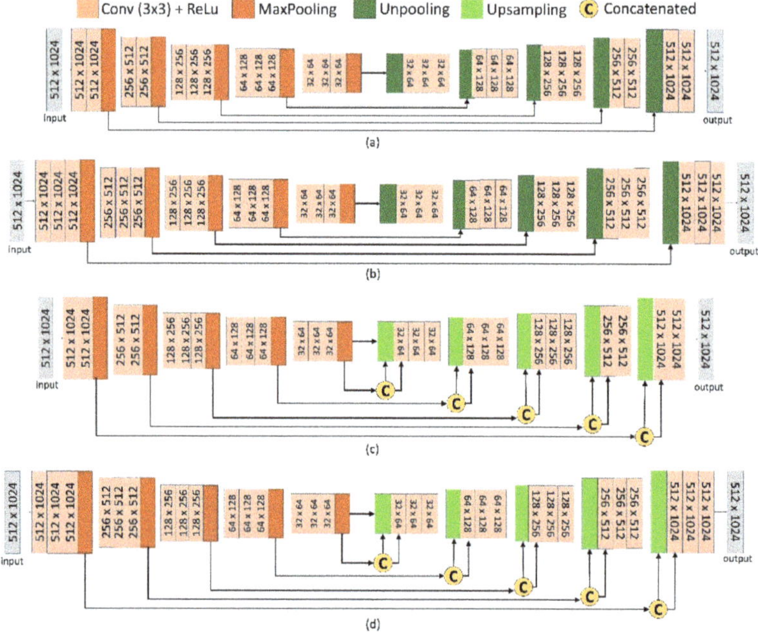

Figure 10. CNN Network architecture. (**a**) VGG16 + SegNet, (**b**) ResNet50 + SegNet, (**c**) VGG16 + UNet, (**d**) ResNet50 + UNet.

1. VGG16 base model with SegNet architecture.
2. ResNet50 base model with SegNet.
3. VGG16 with UNet.
4. ResNet50 with UNet

Figure 10a,b represents the SegNet architecture with VGG16 and ResNet50 as the base model, respectively. The left-hand side is the encoder, which has five blocks, and the layers are from the original VGG16 and ResNet models. The Max pooling operation is depicted by the red arrows. This operation reduces the image dimensions by 2 × 2. The Unpooling is depicted by the green arrows on the right-hand side of the figure(s). The operation ensures the size of the image was restored to the original size, and the output image has the same spatial-dimension as the input image. Figure 10c,d represents the UNet architecture with VGG16 and ResNet50 models, respectively. The network uses the original layers from the VGG16, ResNet50, with the UNet architecture. A clear U-connection can be seen in the figure. The skip connections were used, and upsampling was performed to restore the image dimensions. A concatenated operation was applied to implement the skip connections to combine them with the corresponding feature map (image). The unpooling, skip connections, and upsampling functions were used to ensure that the size of the output image is the same as the input image mentioned in Section 4.3.1.

For a specific task of semantic segmentation, dedicated segmentation based dataset was also used for initialising the weights. For the PspNet, the pre-training was done using ADE20K data [87], and Cityscapes dataset [88]. The ADE20K dataset has 21,200 images of various day to day scenes. The Cityscapes data contains images taken from video frames (≈20,000 coarse images) from 50 cities taken in spring, summer, and fall seasons. The models used are listed below, and the layers and architecture are described in Figure 9.

5. PspNet trained on ADE 20K dataset.
6. PspNet trained on Cityscapes dataset.

5. Results

Figure 11 depicts the segmentation results from both ML and DL techniques for a drone image (sized 512 × 1024) taken of Clara bog. The segmentation was carried out for four ecotope classes present in the drone image captured in the spring season. The accuracy and results are further discussed in this section.

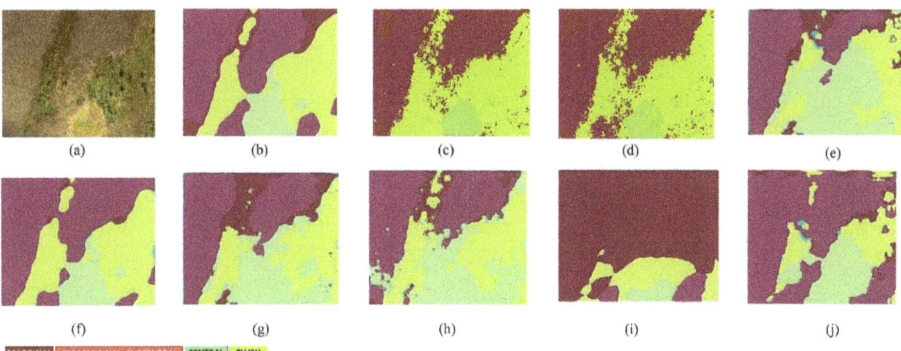

Figure 11. Segmentation results. (**a**) Drone image, (**b**) ground truth labelled image, (**c**) machine learning (ML) (random forest (RF) + Graph cut) segmentation using RGB features, (**d**) ML (RF + Graph cut) segmentation using RGB and textural features, (**e**) deep learning (DL) semantic segmentation using SegNet and VGG 16 model, (**f**) DL semantic segmentation using SegNet and ResNet50 model, (**g**) DL semantic segmentation using UNet with VGG16 model, (**h**) DL semantic segmentation using UNet with ResNet50 model, (**i**) DL semantic segmentation using PSPNet (Cityscapes), (**j**) DL semantic segmentation using PSPNet (ADE 20k).

5.1. Machine Learning

As discussed in Section 3, the ML classifiers were tested for model accuracy (5-fold validation), misclassification cost, and training time. Table 2 depicts the metric calculated over the entire 70% training data (as discussed in Section 3).

RF was chosen to be the best performing classifier, and further segmentation using Graphcut algorithm was performed using the results from RF. The segmentation is a post-classification area based smoothing process. The final segmented image was checked against a fully manually labelled image to give overall accuracy (OA). The OA is the ratio of true positives (TP) with a total number of pixels (Equation (10)).

$$OA = \frac{TP}{TP + FP + FN + TN} \tag{10}$$

where, TP = true positives, FP = false positives, FN = false negatives, and TN = true negatives. This was done for visible bands (RGB) and RGB + textural features. For proper comparison between ML and DL, the image was resampled from its original size (3000 × 4000) to a smaller scale (512 × 1024). The resampling of the image was done using bilinear interpolation [89]. Table 3 depicts the accuracies obtained by using a random forest classifier along with graph cut segmentation for both the sizes of the image. Since, the image used in segmentation using DL techniques is also resized, for an accurate comparison, the resized image (512 × 1024) was used in the further analysis.

Table 3. Segmentation accuracies using ML.

	RGB Features	RGB + Textural Features
Original size (3000 × 4000)	83.3	85.1
Resized (512 × 1024)	82.9	84.8

As can be seen from Figure 11b,c, there is not much difference in the segmentation using RGB with or without textural features. However, the textural features do add extra information and are known to be highly useful when there is a terrain variation in the scene. However, in this application, where the ecotopes under consideration are low-lying, homogenous communities, the addition of textural features did not improve accuracy very significantly—the OA only increasing by approximately 2%.

5.2. Deep Learning

The semantic segmentation using CNNs was performed for 100 epochs. The LR was decreased by a factor of 10 each time a model's accuracy was saturated. The overall accuracy performed on testing data (OA is also calculated for testing data) of all the models is shown in Figure 12.

There is a jump of an average ≈32% in OA from the first to last epoch, with the PspNet model and ResNet50+SegNet showing the maximum increase in OA (≈30%, 25% respectively). The cross-entropy loss decreased by an average of ≈28% for the CNN models under consideration. This decrease happens by reducing the LR. Although accurate, a detailed analysis of per-class accuracy is required to make an informed decision about the best CNN architecture for the segmentation for this particular application in the identification of raised bog vegetation ecotopes. The per-class analysis is done to make sure there is no overfitting. As seen from Figure 11i, a model can lead to overfitting, giving sufficient accuracy but incorrect classification.

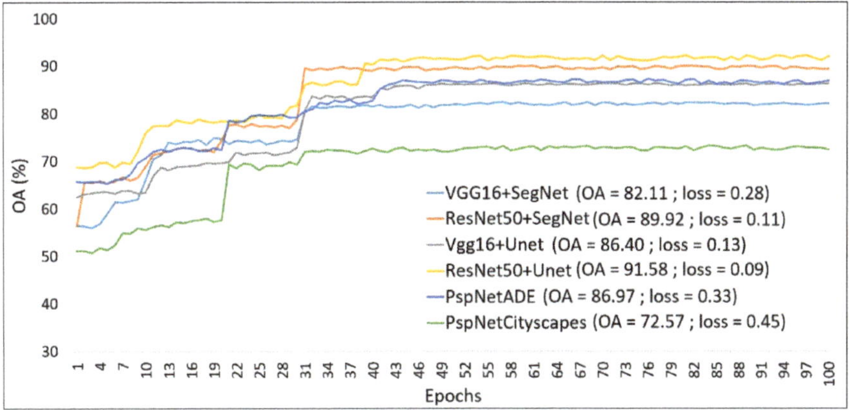

Figure 12. Overall accuracy over 100 epochs used for all CNN architectures for semantic segmentation of ecotopes in Clara Bog.

Table 4 describes the confusion matrix for every community, and both ML and DL algorithms, which is discussed further in Section 6. Other accuracy checking parameters like Precision, Recall, and F1-score were also calculated for every community (ecotope) under consideration. Equations (11)–(13) give the formula to calculate the above-stated accuracy parameters.

$$Precision = \frac{TP}{TP + FP} \tag{11}$$

$$Recall = \frac{TP}{TP + FN} \tag{12}$$

$$F1\ Score = 2 \times \frac{Precision \times Recall}{Precision + Recall} \tag{13}$$

Table 4. Confusion matrix per community per segmentation model (ML and DL).

	RF (RGB)				RF (RGB + TEXTURAL)				SEGNET + VGG16				SEGNET + RESNET50			
	M	SMSC	C	AF	M	SMSC	C	AF	M	SMSC	C	AF	M	SMSC	C	AF
M	58,405	1012	988	36,321	59,781	1598	1002	38,241	43,872	8854	2500	71,005	62,870	1631	835	16,360
SMSC	734	155,583	4979	2033	600	188,296	4608	587	7952	122,544	15,691	9514	3000	162,651	3005	2639
C	328	3862	77,939	44,321	256	4150	83,930	34,658	2831	18,529	77,369	73,108	967	4895	98,584	28,330
AF	38,010	3211	43,509	142,707	39,023	3079	32,584	112,589	65,896	21,251	64,211	99,833	18,470	14,110	10,358	148,383
Precision	0.59	0.94	0.61	0.62	0.59	0.96	0.68	0.66	0.35	0.79	0.45	0.40	0.77	0.95	0.74	0.78
Recall	0.60	0.95	0.61	0.63	0.60	0.95	0.69	0.66	0.36	0.72	0.48	0.39	0.74	0.89	0.87	0.76
F1 score	0.60	0.94	0.61	0.62	0.60	0.95	0.68	0.66	0.36	0.75	0.47	0.40	0.75	0.92	0.80	0.77
	UNET + VGG16				UNET + RESENET50				PSPNET ADE20K				PSPNET CITYSCAPES			
	M	SMSC	C	AF	M	SMSC	C	AF	M	SMSC	C	AF	M	SMSC	C	AF
M	82,589	9510	3258	36,951	73,897	1008	258	3371	36,351	9822	631	5311	128,890	78,353	36,118	63,001
SMSC	10,254	146,933	9800	19,759	4096	152,363	3690	15,892	15,200	210,052	6323	28,200	73,570	107,781	2988	4820
C	4523	12,967	96,582	35,489	982	5183	90,258	28,105	987	7921	96,587	3715	38,562	5815	32,510	12,377
AF	32,563	19,638	34,822	83,417	5101	14,852	22,110	155,446	3074	21,520	5300	127,296	58,360	7826	13,524	57,450
Precision	0.62	0.79	0.65	0.49	0.94	0.87	0.72	0.79	0.70	0.81	0.88	0.81	0.42	0.57	0.36	0.42
Recall	0.64	0.78	0.65	0.47	0.88	0.88	0.78	0.77	0.65	0.84	0.89	0.77	0.43	0.54	0.38	0.42
F1 score	0.63	0.78	0.66	0.48	0.91	0.87	0.75	0.78	0.67	0.83	0.89	0.79	0.43	0.55	0.37	0.42

6. Discussion

The study describes methods to map vegetation communities in a raised bog 'Clara Bog' located in Ireland using drone images from DJI Inspire 1™ drone captured during the spring season. The size of the images were 3000×4000, and 40 images were used as training. Furthermore, both ML, DL algorithms were tested for the rest of the 20 images. The study shows that high-resolution (1.8 cm) RGB images are adequate for mapping vegetation communities. However, a key challenge associated with RGB images is the change in intensity due to sunlight conditions, particularly in a temperate climate like Ireland, where sunlight levels are rarely constant for long. Therefore, in this study, all the images with significantly different light conditions were removed. The use of a colour correction technique could be a possible solution to this problem, which is a domain yet to be explored. Similarly, the addition of textural properties does create the challenge of increasing the computations (time and complexity). The segmentation is done using 13 features instead of three, thereby being more computationally expensive.

Initially, a comparative analysis of the state-of-the-art classifiers was performed (Table 2). It was seen that the RF ensemble classifier outperformed the other classifiers. The RF classifier uses bootstrapping for forming multiple trees leading to reduced possibilities of overfitting of the data. The SVM classifier with RBF kernel had similar accuracy and misclassification cost as RF, but with twice the training time. Hence, RF was deemed to be the best choice for drone image classification with model accuracy of 92%. As pixel-based segmentation often fails to take the contextual (area-based) information into account; therefore, to form segments based on area, graph cut segmentation was subsequently applied. Out of the 40 training images, only a part of labelled pixels ($\approx$12k) was input to the ML model. The entire processing time of ML segmentation was $\approx$30 min.

This was done for both RGB and RGB + textural images. The images were resampled to $2^9 \times 2^{10}$ for proper comparison with deep learning algorithms (discussed later). It has to be noted that the aspect ratio of the imagery was maintained while resampling it. This was done mainly to keep the textural properties intact. The authors of [37] explain that in order to capture textural properties, the size of the image/sliding window should be carefully chosen. Therefore, a decrease in the size of the image (or change in aspect ratio) can lead to a change in textural properties. Table 3 shows that the resampling using bilinear interpolation did not make a big difference in the OA. The resampled image with textural properties performs comparably to the original image. The OA with textural properties is also comparable to OA with just RGB for this application with a low-lying homogenous area of interest. Overall, the textural properties perform the best segmentation with both the original-sized image and the rescaled image.

From Figure 11c,d, it can be seen that using textural properties, the ecotopes like SMSC and AF are differentiated better (see Table 4). Likewise, from Table 4, it can be seen TP for the C ecotope increases with the addition of textural properties but decreases for the AF ecotope. The decrease in misclassified pixels (FP, FN) between SMSC and AF has led to an increase in precision and recall for the SMSC ecotope. There is a definite increase in accuracy for the C and AF ecotopes by using textural features, whereas, the SMSC and M ecotopes are identified with similar precision, recall, and f1 score values for both the images.

The deep learning technique used for segmentation is semantic segmentation using CNN models. In this study, six different models were tested for the semantic segmentation to identify the different bog-ecotopes. The training data for the CNN models consisted of 40 images containing all the ecotopes in different orientations and lighting (brightness). The size of the training data is a notable factor in this study, as for many applications, 100s of images are more usually required for training such CNN models. This study demonstrates the usage of minimum labelled training images for attaining the segmentation, given that 40 images seemed to be sufficient for this application as the weights were initialised using ImageNet dataset having 1000 different classes. This reduces the dependence on the extensive training dataset and also is faster [90]. All these 40 images were resized to $2^9 \times 2^{10}$ for efficiently performing semantic segmentation. For an application involving a prominent area such as this, the classes are also sparsely located. Therefore, cropping or extracting patches from the images was leading to a reduction in classes (ecotopes) covered in an image. In order to make sure that the model identifies all the ecotopes, the images were resized. Nevertheless, for an application where the classes are located close enough (spatially), cropping/extracting patches can be a viable option.

The algorithms were run for 100 epochs, after which the accuracy was becoming saturated. The computation time was ≈700 min per model for 100 epochs. It was decided not to increase the number of epochs as it may lead to overfitting of the model [91]. The LR was decreased with epochs when OA saturated. This decrease leads to faster convergence and an increase in accuracy. Without decreasing the LR, if the same LR is continued, one may still get high accuracy, but it would require a massive number of epochs; therefore, it is not recommended. There is an apparent increase in accuracy using DL methods when compared to ML methods. At the end of the epochs, it is clear that SegNet and UNet architecture with ResNet50 yield the best results for the semantic segmentation of bog-ecotopes. In comparison, the VGG16 base model has led to the over-classification of ecotopes such as M, AF. The VGG model has been shown to be effective when there is noise in the data but does not perform well when the brightness of the images changed [92]. This explains the low accuracy of the model, as the images had different lighting due to variable weather conditions. Figure 11e–j depicts the DL segmentation results. It can be seen that the segmentation using SegNet and UNet is similar for ecotopes like SMSC and C, but is different for AF and M ecotopes.

The study also demonstrates the use of transfer learning by using a segmentation specific pre-trained PspNet model. This model was pre-trained using ADE20K and cityscapes image set instead of widely applicable ImageNet. In our application, the usage of these segmentation datasets was not successful as the weights were calculated for a specific task of segmenting areas of traffic, cars, houses, pavements, etc. Additionally, due to the uniqueness of these communities, the weights transferred from the pre-trained models were not accurate. In order to use transfer learning, the model selected should be pre-trained using similar categories as the application.

For making the final decision of the best CNN architecture, the accuracy parameters for every ecotope were considered. Table 4 shows that the SMSC ecotope is identified quite well using all the CNN models, with the exception of the PspNet model pre-trained with cityscapes images. Using the base model ResNet50, the ecologically important, peat-forming communities (the SMSC, and C ecotopes) are better identified using SegNet than UNet. Using PspNet (ADE20K), the C ecotope was identified the best, although the OA of the model is low. Therefore, taking into consideration the OA, precision, recall, and f1-score of all the communities, SegNet architecture with the ResNet50 base model appears to be the best choice for drone image segmentation in relation to identifying raised bog vegetation types.

The best OA recorded from ML was 85%, and from DL was 91%. However, the most appropriate technique for this study was not decided just on the basis of OA. For applying the technique to new applications, other parameters cannot be ignored. For example, a lot more training data was required for using DL as compared to ML. Similarly, time and hardware also play a significant role in deciding the best technique. Table 5 summarises the essential pros and cons of the two techniques.

It is clear that there are many pros and cons of both techniques, as described in this study. The main idea behind using remote sensing techniques is to reduce the amount of manual fieldwork that is required for monitoring the wetlands. This includes minimising the training data given as input to the classifiers. Additionally, the idea is to automate the process in the simplest way possible, given that the availability of high performing computers or GPUs cannot always be guaranteed, in order to optimise the speed. Keeping in mind the above requisites, the ML technique is the clear choice for our application. Whilst DL techniques can be used once there is enough labelled data created from all the wetlands such that all the species are covered, in the case of a new wetland, which contains new species to be mapped, a clear indication of the species (with full coverage) is required. Therefore, DL is more advantageous to use for more global or applied applications, whereas for a more specific application such as this where not enough training data is available, ML can produce accuracies almost comparable to the DL.

Table 5. Pros and cons of ML vs. DL for mapping ecotopes, Clara bog.

Pros	Cons
MACHINE LEARNING ALGORITHMS	
1. Very fast training (≈30 min per model) 2. Needs less training data (≈12k labelled pixels in 40 images) 3. No need of size alteration; accepts the input of any size. 4. No need for HPCs or GPUs; works well with CPU. 5. Provides good accuracy (OA = 85%)	1. Cannot be applied globally, i.e., does not work on augmented data. 2. In order to re-use the same model, input has to be consistent with original images or to be modified. 3. Can require parameterisation, manual handling. 4. Textural and additional features have to be input separately; does not learn by itself.
DEEP LEARNING ALGORITHMS	
1. Can be applied globally for multiple applications and works well with augmented data. 2. Works well with low resolution imagery. 3. Learns textural features on its own, require no additional inputs. 4. Provides good results (OA = 91%)	1. Slower training process (≈700 min per model). 2. Needs a considerable amount of training data or requires a pre-trained model (≈48 × 106 labelled pixels in 40 images). 3. Requires alteration of images with respect to size for faster computation and analysis. 4. It is recommended the use of GPUs with good RAM for running CNN efficiently.

Finally, the drone images are very high resolution and can be captured at any given time. However, although practical, drone image capturing does have certain limitations. The first limitation of using drones is the length of the battery life. For example, on average, the drone (such as DJI Inspire 1) the battery will only last approximately 15 min and so to cover a large area, many batteries are required. Therefore, in the future, in order to make the process cost-effective, drone images can be used in conjunction with the freely available satellite images. Satellite images give excellent coverage and proper temporal resolution meaning that the usage of drones and satellites together should be advantageous.

7. Conclusions

This study aimed at providing mapping of vegetation in wetlands using image segmentation. For this, ML and DL algorithms were compared by applying them to a set of drone images of Clara Bog, a raised bog located in the middle of Ireland. The images were captured using DJI Inspire 1™ drone (RGB sensor), with the open-source and freely available Pix4DCapture application. Using ML, a total of six different state-of-the-art pixel-based classifiers were compared, out of which, the best ML algorithm for the given dataset was shown to be the RF classifier (model accuracy = 92.9%). For ML image segmentation, RF classifier was used with maximum a-posteriori graph cut segmentation. It was seen that accuracy is improved by ≈2% after addition of textural features (OA = 85.1%) when compared to the original RGB image (83.3%), and ecologically important communities such as central ecotope were mapped better.

Apart from ML, the study also describes image segmentation or 'semantic segmentation' using DL methods. A detailed account on the selection of variables for the DL segmentation was presented. The study was done on a combination of six architectures and base models. For the given dataset, the ResNet50 base model with both UNet (OA = 91.5%) and SegNet (OA = 89.9%) architecture performs very well. ResNet50+SegNet model was deemed best, as it was able to identify complex vegetation communities, such as SMSC ecotope better. All the models were run for 100 epochs, and it was seen the accuracy saturated after ≈35 epochs. It was seen that for mapping ecotopes in a raised bog, transfer of initial weights from wide-ranged ImageNet outperforms the segmentation-specific datasets like ADE20K or Cityscapes.

Overall, the accuracy of the DL was ≈4% higher than the ML methods. Additionally, the DL method does not require any colour correction or the addition of extra textural features. However, DL requires a large amount of initial labelled training data (≈48×10^6 pixels). On the other hand, the ML algorithm requires much less training data (≈12,000 labelled pixels) and is much faster (≈30 times) when compared to CNNs. Therefore, in retrospect, for such a specific application as the wetland mapping application, it was considered that the ML approach is more suitable. This would be particularly useful for any un-surveyed wetland, where the minimum amount of information of the vegetation communities is required to produce accurate maps.

Author Contributions: Conceptualization, S.B. and B.G.; methodology, S.B.; software, S.B.; validation, S.B., B.G. and L.G.; formal analysis, S.B.; investigation, S.B., B.G.; resources, S.B., L.G.; data curation, S.B.; writing—original draft preparation, S.B.; writing—review and editing, S.B., L.G., B.G.; visualisation, B.G.; supervision, B.G.; project administration, L.G.; funding acquisition, L.G. All authors have read and agreed to the published version of the manuscript.

Funding: This study is funded by the Environmental Protection Agency of Ireland (EPAIE) (Grant No.: 2016-W-LS-13).

Acknowledgments: The authors would like to thank Jonathan Munro (University of Bristol) and Yunis Ahmad Lone for their advice, and Trinity College High Power Computing (TCHPC) department for providing the necessary hardware and support.

Conflicts of Interest: The authors declare no conflict of interest.

References

1. Bhatnagar, S.; Gill, L.; Regan, S.; Naughton, O.; Johnston, P.; Waldren, S.; Ghosh, B. Mapping Vegetation Communities Inside Wetlands Using Sentinel-2 Imagery in Ireland. *Int. J. Appl. Earth Obs. Geoinf.* **2020**, *88*, 102083. [CrossRef]
2. Hirano, A.; Madden, M.; Welch, R. Hyperspectral image data for mapping wetland vegetation. *Wetlands* **2003**, *23*, 436–448. [CrossRef]
3. Pengra, B.W.; Johnston, C.A.; Loveland, T.R. Mapping an invasive plant, Phragmites australis, in coastal wetlands using the EO-1 Hyperion hyperspectral sensor. *Remote. Sens. Environ.* **2007**, *108*, 74–81. [CrossRef]
4. Álvarez-Taboada, F.; Araújo-Paredes, C.; Julián-Pelaz, J. Mapping of the Invasive Species Hakea sericea Using Unmanned Aerial Vehicle (UAV) and WorldView-2 Imagery and an Object-Oriented Approach. *Remote. Sens.* **2017**, *9*, 913. [CrossRef]
5. Baena, S.; Moat, J.; Whaley, O.; Boyd, D.S. Identifying species from the air: UAVs and the very high resolution challenge for plant conservation. *PLoS ONE* **2017**, *12*, e0188714. [CrossRef]
6. Dvořák, P.; Müllerová, J.; Bartaloš, T.; Brůna, J. Unmanned Aerial Vehicles for Alien Plant Species Detection and Monitoring. *ISPRS Int. Arch. Photogramm. Remote. Sens. Spat. Inf. Sci.* **2015**, *40*, 83–90. [CrossRef]
7. Hill, D.; Tarasoff, C.; Whitworth, G.E.; Baron, J.; Bradshaw, J.; Church, J.S. Utility of unmanned aerial vehicles for mapping invasive plant species: A case study on yellow flag iris (Iris pseudacorus L.). *Int. J. Remote. Sens.* **2016**, *38*, 2083–2105. [CrossRef]
8. Ruwaimana, M.; Satyanarayana, B.; Otero, V.; Muslim, A.M.; Muhammad, A.M.; Ibrahim, S.; Raymaekers, D.; Koedam, N.; Dahdouh-Guebas, F. The advantages of using drones over space-borne imagery in the mapping of mangrove forests. *PLoS ONE* **2018**, *13*, e0200288. [CrossRef]
9. Chabot, D.; Dillon, C.; Shemrock, A.; Weissflog, N.; Sager, E.P.S. An Object-Based Image Analysis Workflow for Monitoring Shallow-Water Aquatic Vegetation in Multispectral Drone Imagery. *ISPRS Int. J. Geo Inf.* **2018**, *7*, 294. [CrossRef]
10. Han, Y.-G.; Yoo, S.H.; Kwon, O. Possibility of applying unmanned aerial vehicle (UAV) and mapping software for the monitoring of waterbirds and their habitats. *J. Ecol. Environ.* **2017**, *41*, 21. [CrossRef]
11. Zheng, H.; Cheng, T.; Li, D.; Zhou, X.; Yao, X.; Tian, Y.; Cao, W.; Zhu, Y. Evaluation of RGB, Color-Infrared and Multispectral Images Acquired from Unmanned Aerial Systems for the Estimation of Nitrogen Accumulation in Rice. *Remote. Sens.* **2018**, *10*, 824. [CrossRef]
12. Govender, M.; Chetty, K.; Bulcock, H. A review of hyperspectral remote sensing and its application in vegetation and water resource studies. *Water SA* **2009**, *33*. [CrossRef]
13. Boon, M.A.; Greenfield, R.; Tesfamichael, S. Wetland assessment using unmanned aerial vehicle (UAV) photogrammetry. *Remote. Sens. Spat. Inf. Sci.* **2016**, *XLI-B1*, 781–788.
14. Treboux, J.; Genoud, D. Improved Machine Learning Methodology for High Precision Agriculture. In Proceedings of the 2018 Global Internet of Things Summit (GIoTS), Bilbao, Spain, 4–7 June 2018; Institute of Electrical and Electronics Engineers (IEEE): Piscataway, NJ, USA, 2018; pp. 1–6.

15. Pap, M.; Király, S.; Moljak, S. Investigating the usability of UAV obtained multispectral imagery in tree species segmentation. *ISPRS Int. Arch. Photogramm. Remote. Sens. Spat. Inf. Sci.* **2019**, *XLII-2/W18*, 159–165. [CrossRef]
16. Zuo, Z. Remote Sensing Image Extraction of Drones for Agricultural Applications. *Rev. Fac. Agron. Univ. Zulia* **2019**, *36*, 1202–1212.
17. Parsons, M.; Bratanov, D.; Gaston, K.J.; Gonzalez, F. UAVs, hyperspectral remote sensing, and machine learning revolutionising reef monitoring. *Sensors* **2018**, *18*, 2026. [CrossRef]
18. Miyamoto, H.; Momose, A.; Iwami, S. UAV image classification of a riverine landscape by using machine learning techniques. *EGU Gen. Assem. Conf. Abstr.* **2018**, *20*, 5919.
19. Zimudzi, E.; Sanders, I.; Rollings, N.; Omlin, C. Segmenting mangrove ecosystems drone images using SLIC superpixels. *Geocarto Int.* **2018**, *34*, 1648–1662. [CrossRef]
20. Höser, T.; Kuenzer, C. Object Detection and Image Segmentation with Deep Learning on Earth Observation Data: A Review-Part I: Evolution and Recent Trends. *Remote. Sens.* **2020**, *12*, 1667. [CrossRef]
21. Lee, D.; Kim, J.; Lee, D.-W. Robust Concrete Crack Detection Using Deep Learning-Based Semantic Segmentation. *Int. J. Aeronaut. Space Sci.* **2019**, *20*, 287–299. [CrossRef]
22. Zhang, C.; Wang, L.; Yang, R. Semantic segmentation of urban scenes using dense depth maps. In Proceedings of the European Conference on Computer Vision, Crete, Greece, 5–10 September 2010; Springer: Berlin/Heidelberg, Germany, 2010; pp. 708–721.
23. Montoya-Zegarra, J.A.; Wegner, J.D.; Ladický, L.; Schindler, K. Semantic segmentation of aerial images in urban areas with class-specific higher-order cliques. *ISPRS Ann. Photogramm. Remote. Sens. Spat. Inf. Sci.* **2015**, *2*, 127–133. [CrossRef]
24. Dechesne, C.; Mallet, C.; Le Bris, A.; Gouet-Brunet, V. Semantic segmentation of forest stands of pure species combining airborne lidar data and very high resolution multispectral imagery. *ISPRS J. Photogramm. Remote. Sens.* **2017**, *126*, 129–145. [CrossRef]
25. Cui, B.; Zhang, Y.; Li, X.; Wu, J.; Lu, Y. WetlandNet: Semantic Segmentation for Remote Sensing Images of Coastal Wetlands via Improved UNet with Deconvolution. In Proceedings of the International Conference on Genetic and Evolutionary Computing, Qingdao, China, 1–3 November 2019; Springer: Singapore, 2019; pp. 281–292.
26. Jiang, J.; Feng, X.; Liu, F.; Xu, Y.; Huang, H. Multi-Spectral RGB-NIR Image Classification Using Double-Channel CNN. *IEEE Access* **2019**, *7*, 20607–20613. [CrossRef]
27. Kentsch, S.; Caceres, M.L.L.; Serrano, D.; Roure, F.; Donoso, Y.D. Computer Vision and Deep Learning Techniques for the Analysis of Drone-Acquired Forest Images, a Transfer Learning Study. *Remote Sens.* **2020**, *12*, 1287. [CrossRef]
28. Nigam, I.; Huang, C.; Ramanan, D. Ensemble knowledge transfer for semantic segmentation. In Proceedings of the 2018 IEEE Winter Conference on Applications of Computer Vision (WACV), Lake Tahoe, NV, USA, 12–15 March 2018; Institute of Electrical and Electronics Engineers (IEEE): Piscataway, NJ, USA, 2018; pp. 1499–1508.
29. Do, D.; Pham, F.; Raheja, A.; Bhandari, S. Machine learning techniques for the assessment of citrus plant health using UAV-based digital images. In Proceedings of the Autonomous Air and Ground Sensing Systems for Agricultural Optimization and Phenotyping III, Baltimore, MD, USA, 15–16 April 2019; International Society for Optics and Photonics: Bellingham, WA, USA, 2019; Volume 10664, p. 1066400.
30. Bhatnagar, S.; Ghosh, B.; Regan, S.; Naughton, O.; Johnston, P.; Gill, L. Monitoring environmental supporting conditions of a raised bog using remote sensing techniques. *Proc. Int. Assoc. Hydrol. Sci.* **2018**, *380*, 9–15. [CrossRef]
31. Bhatnagar, S.; Ghosh, B.; Regan, S.; Naughton, O.; Johnston, P.; Gill, L. Remote Sensing Based Ecotope Mapping and Transfer of Knowledge in Raised Bogs. *Geophys. Res. Abstr.* **2019**, *21*, 1.
32. ESRI. *ArcMap Desktop*; (Version 10.6.1); Esri Inc.: Redlands, CA, USA, 2019.
33. ESRI "World Imagery" [High Resolution 30 cm Imagery]. Scale ~1:280 (0.03 m). Available online: http://www.arcgis.com/home/item.html?id=10df2279f9684e4a9f6a7f08febac2a9 (accessed on 25 November 2019).
34. Shi, J.; Malik, J. Normalised cuts and image segmentation. *IEEE Trans. Pattern Anal. Mach. Intell.* **2000**, *22*, 888–905.
35. Feng, Q.; Liu, J.; Gong, J. UAV remote sensing for urban vegetation mapping using random forest and texture analysis. *Remote Sens.* **2015**, *7*, 1074–1094. [CrossRef]

36. *MATLAB*; Version R2019b; The MathWorks Inc.: Natick, MA, USA, 2019.
37. Tavares, J.; Jorge, R.N. Computational Vision and Medical Image Processing V. In Proceedings of the 5th Eccomas Thematic Conference on Computational Vision and Medical Image Processing, VipIMAGE 2015, Tenerife, Spain, 19–21 October 2015; CRC Press: Boca Raton, FL, USA, 2015.
38. Schwenker, F.; Abbas, H.M.; El Gayar, N.; Trentin, E. Artificial Neural Networks in Pattern Recognition. In Proceedings of the 7th IAPR TC3 Workshop, ANNPR 2016, Ulm, Germany, 28–30 September 2016; Springer: New York, NY, USA, 2018.
39. Chai, H.Y.; Wee, L.K.; Swee, T.T.; Hussain, S. Gray-level co-occurrence matrix bone fracture detection. *WSEAS Trans. Syst.* **2011**, *10*, 7–16. [CrossRef]
40. Salem, Y.B.; Nasri, S. Texture classification of woven fabric based on a GLCM method and using multiclass support vector machine. In Proceedings of the 2009 6th International Multi-Conference on Systems, Signals and Devices, Jerba, Tunisia, 23–26 March 2009; Institute of Electrical and Electronics Engineers (IEEE): Piscataway, NJ, USA, 2009; pp. 1–8.
41. Wu, Y.; Zhou, Y.; Saveriades, G.; Agaian, S.; Noonan, J.P.; Natarajan, P. Local Shannon entropy measure with statistical tests for image randomness. *Inf. Sci.* **2013**, *222*, 323–342. [CrossRef]
42. Mardia, K.V. Measures of multivariate skewness and kurtosis with applications. *Biometrika* **1970**, *57*, 519–530. [CrossRef]
43. Stoer, M.; Wagner, F. A simple min-cut algorithm. *J. ACM* **1997**, *44*, 585–591. [CrossRef]
44. Ishida, T.; Kurihara, J.; Viray, F.A.; Namuco, S.B.; Paringit, E.C.; Perez, G.J.; Marciano, J.J.J. A novel approach for vegetation classification using UAV-based hyperspectral imaging. *Comput. Electron. Agric.* **2018**, *144*, 80–85. [CrossRef]
45. Braun, A.C.; Weidner, U.; Hinz, S. Support vector machines for vegetation classification–A revision. *Photogramm. Fernerkund. Geoinf.* **2010**, *2010*, 273–281. [CrossRef]
46. Laliberte, A.S.; Rango, A. Texture and scale in object-based analysis of subdecimeter resolution unmanned aerial vehicle (UAV) imagery. *IEEE Trans. Geosci. Remote Sens.* **2009**, *47*, 761–770. [CrossRef]
47. Özlem, A. Mapping land use with using Rotation Forest algorithm from UAV images. *Eur. J. Remote Sens.* **2017**, *50*, 269–279. [CrossRef]
48. Meng, X.; Shang, N.; Zhang, X.; Li, C.; Zhao, K.; Qiu, X.; Weeks, E. Photogrammetric UAV Mapping of Terrain under Dense Coastal Vegetation: An Object-Oriented Classification Ensemble Algorithm for Classification and Terrain Correction. *Remote Sens.* **2017**, *9*, 1187. [CrossRef]
49. Friedl, M.A.; Brodley, C.E. Decision tree classification of land cover from remotely sensed data. *Remote Sens. Environ.* **1997**, *61*, 399–409. [CrossRef]
50. Cheng, J.; Greiner, R. Comparing Bayesian network classifiers. In Proceedings of the Fifteenth conference on Uncertainty in artificial intelligence, Stockholm, Sweden, 30 July–1 August 1999; Morgan Kaufmann Publishers Inc.: Burlington, MA, USA, 1999; pp. 101–108.
51. Balakrishnama, S.; Ganapathiraju, A. Linear discriminant analysis-a brief tutorial. *Inst. Signal Inf. Process.* **1998**, *18*, 1–8.
52. Cortes, C.; Vapnik, V. Support vector machine. *Mach. Learn.* **1995**, *20*, 273–297. [CrossRef]
53. Laaksonen, J.; Oja, E. Classification with learning k-nearest neighbors. In Proceedings of the International Conference on Neural Networks (ICNN'96), Washington, DC, USA, 3–6 June 1996; Institute of Electrical and Electronics Engineers (IEEE): Piscataway, NJ, USA, 1996; Volume 3, pp. 1480–1483.
54. Liaw, A.; Wiener, M. Classification and regression by randomForest. *News* **2020**, *2*, 18–22.
55. Ross, Q.J. *C4. 5: Programs for Machine Learning*; Morgan Kaufmann Publishers: San Mateo, CA, USA, 1993.
56. Breiman, L.; Friedman, J.; Stone, C.J.; Olshen, R.A. *Classification and Regression Trees*; CRC Press: Boca Raton, FL, USA, 1984.
57. Boykov, Y.; Kolmogorov, V. An experimental comparison of min-cut/max-flow algorithms for energy minimisation in vision. *IEEE Trans. Pattern Anal. Mach. Intell.* **2004**, *26*, 1124–1137. [CrossRef] [PubMed]
58. MATLAB Wrapper for Graph Cut. Shai Bagon. Available online: https://github.com/shaibagon/GCMex (accessed on 12 December 2019).
59. Boykov, Y.; Veksler, O.; Zabih, R. Fast approximate energy minimisation via graph cuts. *IEEE Trans. Pattern Anal. Mach. Intell.* **2001**, *23*, 1222–1239. [CrossRef]

60. Masci, J.; Meier, U.; Ciresan, D.; Schmidhuber, J.; Fricout, G. Steel defect classification with max-pooling convolutional neural networks. In Proceedings of the 2012 International Joint Conference on Neural Networks (IJCNN), Brisbane, Australia, 10–15 June 2012; Institute of Electrical and Electronics Engineers (IEEE): Piscataway, NJ, USA, 2012; pp. 1–6.
61. Li, Q.; Cai, W.; Wang, X.; Zhou, Y.; Feng, D.D.; Chen, M. Medical image classification with convolutional neural network. In Proceedings of the 2014 13th International Conference on Control Automation Robotics & Vision (ICARCV), Singapore, 10–12 December 2014; Institute of Electrical and Electronics Engineers (IEEE): Piscataway, NJ, USA, 2014; pp. 844–848.
62. Sharma, S. Activation functions in neural networks. *Towards Data Sci.* **2017**, *6*, 310–316.
63. Nwankpa, C.; Ijomah, W.; Gachagan, A.; Marshall, S. Activation functions: Comparison of trends in practice and research for deep learning. *arXiv* **2018**, arXiv:1811.03378.
64. Özkan, C.; Erbek, F.S. The comparison of activation functions for multispectral Landsat TM image classification. *Photogramm. Eng. Remote Sens.* **2003**, *69*, 1225–1234. [CrossRef]
65. Karlik, B.; Olgac, A.V. Performance analysis of various activation functions in generalised MLP architectures of neural networks. *Int. J. Artif. Intell. Expert Syst.* **2011**, *1*, 111–122.
66. Bircanoğlu, C.; Arıca, N. A comparison of activation functions in artificial neural networks. In Proceedings of the 2018 26th Signal Processing and Communications Applications Conference (SIU), Izimir, Turkey, 2–5 May 2018; Institute of Electrical and Electronics Engineers (IEEE): Piscataway, NJ, USA, 2018; pp. 1–4.
67. Bottou, L. Online learning and stochastic approximations. *On-line Learn. Neural Netw.* **1998**, *17*, 142.
68. Fukumizu, K. Effect of batch learning in multilayer neural networks. *Gen* **1998**, *1*, 1E–03E.
69. Bottou, L. Stochastic gradient learning in neural networks. *Proc. Neuro Nımes* **1991**, *91*, 12.
70. Paine, T.; Jin, H.; Yang, J.; Lin, Z.; Huang, T. Gpu asynchronous stochastic gradient descent to speed up neural network training. *arXiv* **2013**, arXiv:1312.6186.
71. Han, S.; Pool, J.; Tran, J.; Dally, W. Learning both weights and connections for efficient neural network. In Proceedings of the Advances in Neural Information Processing Systems, Montreal, QC, CA, 7–12 December 2015; Neural Information Processing Systems Foundation, Inc. (NIPS): San Diego, CA, USA, 2015; pp. 1135–1143.
72. Van Den Doel, K.; Ascher, U.; Haber, E. *The Lost Honour of l2-Based Regularization*; (Radon Series in Computational and Applied Math); De Gruyter: Berlin, Germany, 2013.
73. Atienza, R. *Advanced Deep Learning with Keras: Apply Deep Learning Techniques, Autoencoders, GANs, Variational Autoencoders, Deep Reinforcement Learning, Policy Gradients, and More*; Packt Publishing Ltd.: Birmingham, UK, 2018.
74. Kingma, D.P.; Ba, J. Adam: A method for stochastic optimisation. *arXiv* **2014**, arXiv:1412.6980.
75. Simonyan, K.; Zisserman, A. Very deep convolutional networks for large-scale image recognition. *arXiv* **2014**, arXiv:1409.1556.
76. He, K.; Zhang, X.; Ren, S.; Sun, J. Deep residual learning for image recognition. In Proceedings of the IEEE Conference on Computer Vision and Pattern Recognition, Las Vegas, NV, USA, 27–30 June 2016; Institute of Electrical and Electronics Engineers (IEEE): Piscataway, NJ, USA, 2016; pp. 770–778.
77. Cheng, K.; Cheng, X.; Wang, Y.; Bi, H.; Benfield, M.C. Enhanced convolutional neural network for plankton identification and enumeration. *PLoS ONE* **2019**, *14*, e0219570. [CrossRef] [PubMed]
78. Qassim, H.; Verma, A.; Feinzimer, D. Compressed residual-VGG16 CNN model for big data places image recognition. In Proceedings of the 2018 IEEE 8th Annual Computing and Communication Workshop and Conference (CCWC), Las Vegas, NV, USA, 8–10 January 2018; Institute of Electrical and Electronics Engineers (IEEE): Piscataway, NJ, USA, 2018; pp. 169–175.
79. Ghosh, S.; Das, N.; Das, I.; Maulik, U. Understanding deep learning techniques for image segmentation. *ACM Comput. Surv. (CSUR)* **2019**, *52*, 73. [CrossRef]
80. Badrinarayanan, V.; Kendall, A.; Cipolla, R. Segnet: A deep convolutional encoder-decoder architecture for image segmentation. *IEEE Trans. Pattern Anal. Mach. Intell.* **2017**, *39*, 2481–2495. [CrossRef]
81. Ronneberger, O.; Fischer, P.; Brox, T. U-net: Convolutional networks for biomedical image segmentation. In Proceedings of the International Conference on Medical Image Computing and Computer-Assisted Intervention, Munich, Germany, 5–9 October 2015; Springer: Cham, Switzerland, 2015; pp. 234–241.
82. Zhao, H.; Shi, J.; Qi, X.; Wang, X.; Jia, J. Pyramid scene parsing network. In Proceedings of the IEEE Conference on Computer Vision and Pattern Recognition, Honolulu, HI, USA, 21–26 July 2017; Institute of Electrical and Electronics Engineers (IEEE): Piscataway, NJ, USA, 2017; pp. 2881–2890.

83. He, K.; Zhang, X.; Ren, S.; Sun, J. Spatial pyramid pooling in deep convolutional networks for visual recognition. *IEEE Trans. Pattern Anal. Mach. Intell.* **2015**, *37*, 1904–1916. [CrossRef]
84. Python Software Foundation. Python Language Reference, Version 3.7. Available online: http://www.python.org (accessed on 20 June 2020).
85. Divamgupta. Image-Segmentation-Keras. 2019. Available online: https://github.com/divamgupta/image-segmentation-keras.git (accessed on 20 June 2020).
86. Russakovsky, O.; Deng, J.; Su, H.; Krause, J.; Satheesh, S.; Ma, S.; Berg, A.C. Imagenet large scale visual recognition challenge. *Int. J. Comput. Vis.* **2015**, *115*, 211–252. [CrossRef]
87. Zhou, B.; Zhao, H.; Puig, X.; Fidler, S.; Barriuso, A.; Torralba, A. Scene parsing through ade20k dataset. In Proceedings of the IEEE Conference on Computer Vision and Pattern Recognition, Honolulu, HI, USA, 21–26 July 2017; Institute of Electrical and Electronics Engineers (IEEE): Piscataway, NJ, USA, 2017; pp. 633–641.
88. Cordts, M.; Omran, M.; Ramos, S.; Rehfeld, T.; Enzweiler, M.; Benenson, R.; Schiele, B. The cityscapes dataset for semantic urban scene understanding. In Proceedings of the IEEE Conference on Computer Vision and Pattern Recognition, Las Vegas, NV, USA, 27–30 June 2016; Institute of Electrical and Electronics Engineers (IEEE): Piscataway, NJ, USA, 2016; pp. 3213–3223.
89. Andrews, H.C.; Patterson, C.L. Digital interpolation of discrete images. *IEEE Trans. Comput.* **1976**, *100*, 196–202. [CrossRef]
90. Liu, Y.; Starzyk, J.A.; Zhu, Z. Optimised approximation algorithm in neural networks without overfitting. *IEEE Trans. Neural Netw.* **2008**, *19*, 983–995.
91. Grm, K.; Štruc, V.; Artiges, A.; Caron, M.; Ekenel, H.K. Strengths and weaknesses of deep learning models for face recognition against image degradations. *IET Biom.* **2017**, *7*, 81–89. [CrossRef]
92. Yim, J.; Joo, D.; Bae, J.; Kim, J. A gift from knowledge distillation: Fast optimisation, network minimisation and transfer learning. In Proceedings of the IEEE Conference on Computer Vision and Pattern Recognition, Honolulu, HI, USA, 21–26 July 2017; Institute of Electrical and Electronics Engineers (IEEE): Piscataway, NJ, USA, 2017; pp. 4133–4141.

Article

Using UAV Collected RGB and Multispectral Images to Evaluate Winter Wheat Performance across a Site Characterized by Century-Old Biochar Patches in Belgium

Ramin Heidarian Dehkordi [1,*], Victor Burgeon [1], Julien Fouche [2], Edmundo Placencia Gomez [3], Jean-Thomas Cornelis [1], Frederic Nguyen [3], Antoine Denis [4] and Jeroen Meersmans [1]

1 TERRA Teaching and Research Centre, Gembloux Agro-Bio Tech, University of Liège, 5030 Gembloux, Belgium; victor.burgeon@uliege.be (V.B.); jtcornelis@uliege.be (J.-T.C.); jeroen.meersmans@uliege.be (J.M.)

2 LISAH, Univ Montpellier, INRAE, IRD, Institut Agro, 34060 Montpellier, France; julien.fouche@supagro.fr

3 Urban and Environmental Engineering, University of Liège, 4000 Liège, Belgium; eplacencia@uliege.be (E.P.G.); f.nguyen@uliege.be (F.N.)

4 Department of Environmental Sciences and Management, University of Liège, 6700 Arlon, Belgium; antoine.denis@uliege.be

* Correspondence: ramin.heidariandehkordi@uliege.be; Tel.: +32-81-622-189

Received: 7 June 2020; Accepted: 3 August 2020; Published: 4 August 2020

Abstract: Remote sensing data play a crucial role in monitoring crop dynamics in the context of precision agriculture by characterizing the spatial and temporal variability of crop traits. At present there is special interest in assessing the long-term impacts of biochar in agro-ecosystems. Despite the growing body of literature on monitoring the potential biochar effects on harvested crop yield and aboveground productivity, studies focusing on the detailed crop performance as a consequence of long-term biochar enrichment are still lacking. The primary objective of this research was to evaluate crop performance based on high-resolution unmanned aerial vehicle (UAV) imagery considering both crop growth and health through RGB and multispectral analysis, respectively. More specifically, this approach allowed monitoring of century-old biochar impacts on winter wheat crop performance. Seven Red-Green-Blue (RGB) and six multispectral flights were executed over 11 century-old biochar patches of a cultivated field. UAV-based RGB imagery exhibited a significant positive impact of century-old biochar on the evolution of winter wheat canopy cover (p-value = 0.00007). Multispectral optimized soil adjusted vegetation index indicated a better crop development over the century-old biochar plots at the beginning of the season (p-values < 0.01), while there was no impact towards the end of the season. Plant height, derived from the RGB imagery, was slightly higher for century-old biochar plots. Crop health maps were computed based on principal component analysis and k-means clustering. To our knowledge, this is the first attempt to quantify century-old biochar effects on crop performance during the entire growing period using remotely sensed data. Ground-based measurements illustrated a significant positive impact of century-old biochar on crop growth stages (p-value of 0.01265), whereas the harvested crop yield was not affected. Multispectral simplified canopy chlorophyll content index and normalized difference red edge index were found to be good linear estimators of harvested crop yield (p-value$_{(Kendall)}$ of 0.001 and 0.0008, respectively). The present research highlights that other factors (e.g., inherent pedological variations) are of higher importance than the presence of century-old biochar in determining crop health and yield variability.

Keywords: UAV remote sensing; crop monitoring; RGB imagery; multispectral imagery; century-old biochar

1. Introduction

Biochar is a carbon-rich material obtained through the pyrolysis of feedstocks to intentionally amend soil [1,2]. In the context of climate-smart agriculture, special attention has been given to biochar application owing to its supposed ability to improve soil fertility and long-term organic carbon storage [3–6]. To date, short-term biochar effects have been well addressed by studying the soil nutrient availability [7,8], soil water holding capacity [9–11], and plant available water content [12,13]. The aforementioned improvements of soil quality could consequently lead to an impact on the crop productivity [14]. Moreover, biochar was found to increase nutrient uptake and crop yield while being incorporated into soil matrix together with fertilizers [15]. The short-term improvement of crop productivity can be attributed to the addition of nutrients related to the biochar application [7]. However, long-term biochar effects on soil and agro-ecosystems properties are still poorly understood. Several authors have reported a higher nutrient availability in the century-old biochar sites originating from earth kiln sites [16,17]. The higher nutrient availability in long-term biochar-enriched soils can be explained by cation exchange capacity that increases over time [18]. Moreover, century-old biochar significantly affects physico-chemical properties of soils in relic hearths of agricultural fields [16].

Across the globe, many efforts have been devoted to unravel biochar effects on aboveground biomass productivity. In many of these studies, a short-term effect of biochar on increasing crop productivity was reported [7,19,20], however, Güereña et al. [21] and van Zwieten et al. [22] showed a reductive influence. Similarly, Hernandez-Soriano et al. [23] reported a long-term biochar impact on increasing aboveground productivity, whereas Mastrolonardo et al. [24] observed a decreased tree ring width in relic charcoal hearths of Belgium. Oak growth depression was also found in aged charcoal hearth patches in the United States [25].

The studies listed above all explored the biochar effects on crop productivity. However, it is highly likely that biochar could affect crop development only during the green-up phase, while the harvested crop yield could remain unaffected [26]. More specifically, long-term biochar effects on crop performance and its development over the entire growing season have received much less attention than biochar effects on total crop productivity.

In recent years, remote sensing has been promoted as a means in precision agriculture for detailed spatio-temporal monitoring of crop dynamics [27,28]. A particular focus at present is the use of satellite imagery for precision agricultural practices due to the images' fine spatial, spectral, and temporal resolution [28]. Moreover, high-resolution unmanned aerial vehicle (UAV)-based imagery can be used to address the data gap in satellite products suffering from cloud coverage [29]. Recent advances in Red-Green-Blue (RGB) and multispectral UAV remote sensing have improved the ability to accurately monitor crop performance and vegetation characteristics at the canopy level, such as canopy cover, leaf area index, and chlorophyll content, allowing for site-specific decision making in the context of precision agriculture [29–32]. In addition, due to remote UAV flights, the experimental surface is not damaged [33], thus enabling a complete crop monitoring over the entire season. Therefore, the spatio-temporal information provided by the UAV images can reveal variability in crop performance due to the presence of century-old biochar.

Every year, agricultural products suffer significant economic losses caused by multiple pests and diseases, such as yellow rust or aphid, which may lead to a shortage of food production globally [34,35] and particularly in Europe [36]. Cameras onboard UAVs are able to illustrate changes in crop dynamics as a function of a particular disease at early onset [37]. The pioneering work of Franke et al. [38] reported that pathogens reduces plant chlorophyll content in the visible and red-edge spectral bands due to necrotic and chlogotic lesions. Su et al. [37] highlighted that the red band of the spectrum has strong potential in discriminating healthy from diseased winter wheat plants. Su et al. [37] also showed that the multispectral optimized soil adjusted vegetation index (OSAVI) has a strong discriminating capability between healthy and diseased winter wheat plants. In addition, UAV imagery is able to detect changes in crop traits linked to disease incidence [39]. Therefore, high-resolution UAV remote sensing can be successfully applied for spatial explicit assessment of crop health as a consequence of the presence of century-old biochar within agricultural soils.

The primary objective of this paper is to evaluate crop performance by studying crop growth and health through the usage of high-resolution RGB and multispectral information, respectively, derived from two UAV platforms. This approach also allowed us to investigate, as a secondary objective, the impacts of century-old biochar on winter wheat crop performance. The available imagery sources were RGB and multispectral data collected using two UAV platforms. UAV remote sensing data in combination with ground-based measurements of crop traits were used to determine the influence of century-old biochar from kiln sites of an agricultural field on winter wheat crop growth. The present study evaluated, for the first time, the impact of century-old biochar on crop performance at the canopy level using UAV imagery, while paving the way to quantify the long-term effects of biochar on winter wheat crop health.

2. Materials and Methods

2.1. Site Description and Data Acquisition

The data were collected in a farm of approximately 13 hectares cultivated with winter wheat located in Isnes, Belgium (Figure 1). Winter wheat (*Triticum aestivum* L.) seeds were sown on 6 December 2018. The region is of temperate climate with dry and hot summers. The soil type is a Luvisol characterized by a silt loam texture [40]. Homogenous agricultural treatments were applied throughout the field. The detailed description is available in Heidarian Dehkordi et al. [26], in which the same experimental configuration was followed by eleven 10 × 10 m century-old biochar plots (red squares in Figure 1) together with the corresponding reference plots (blue squares in Figure 1). Ground control point (GCP) targets, with known centroid coordinates through a real-time kinematic GPS, were deployed throughout the farm (Figure 1) to help with accurate geo-referencing of the data (i.e., 3 cm horizontal accuracy).

Two remote sensing datasets (i.e., RGB and multispectral) were acquired during the 2019 growing season (Table 1) using high-resolution sensors onboard two UAV platforms to monitor the crop status over time (Figure 2). RGB images were collected using a DJI Phantom 4 Pro and the multispectral images were obtained using a MicaSense RedEdge-M and downwelling light sensor, onboard a DJI Matrice 100 platform specifically designed by Heidarian Dehkordi et al. [26], capturing blue (475 nm central wavelength), green (560 nm), red (668 nm), red-edge (717 nm), and near-infrared (840 nm) spectral channels with bandwidths of 20, 20, 10, 10, and 40 nm, respectively. The UAV platforms were programmed to fly at 65 and 60 m above ground altitude (AGL) with an airspeed of 5 and 7 ms^{-1} for RGB and multispectral sensors, respectively. The forward and sideway image overlaps were 85% and 75% for RGB, and 75% and 75% for multispectral acquisitions, reaching a ground sampling distance (GSD) of 1.8 and 3.7 cm for RGB and multispectral images, respectively.

Table 1. An overview of the data acquisition using the unmanned aerial vehicles within the 2019 growing season. Flight times (start time–end time) are expressed in local time.

Date	UAV Imagery	
	RGB	Multispectral
02/22/2019	10:18–11:41	-
03/20/2019	11:02–12:20	12:46–13:40
03/28/2019	11:33–12:59	13:37–14:32
04/16/2019	12:01–13:31	13:49–14:44
04/29/2019	11:00–12:26	12:44–13:54
05/13/2019	11:03–12:36	13:05–14:18
06/24/2019	11:04–12:26	13:16–14:36

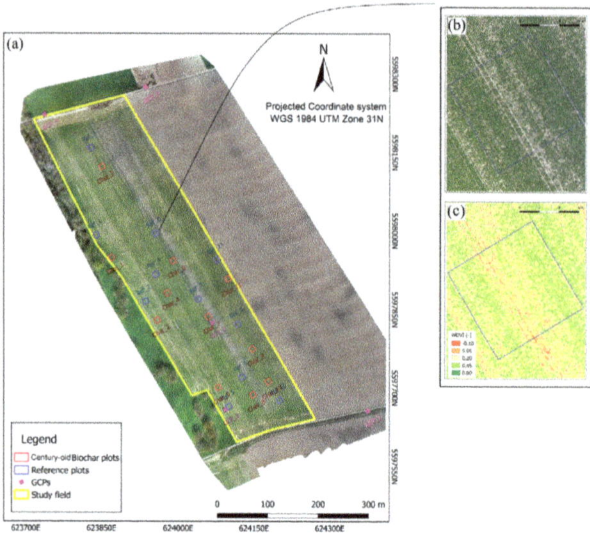

Figure 1. (**a**) A schematic layout of the experimental pairs (reference versus century-old biochar plots) in the winter wheat field and distribution of the ground control points. Background image corresponds to the Red-Green-Blue (RGB) orthomosaic captured by the unmanned aerial vehicle on the 16 April 2019. The right magnified windows present an example of a plot (reference plot 5) for visual RGB (**b**) and multispectral weighted difference vegetation index (**c**) monitored on 16 April 2019.

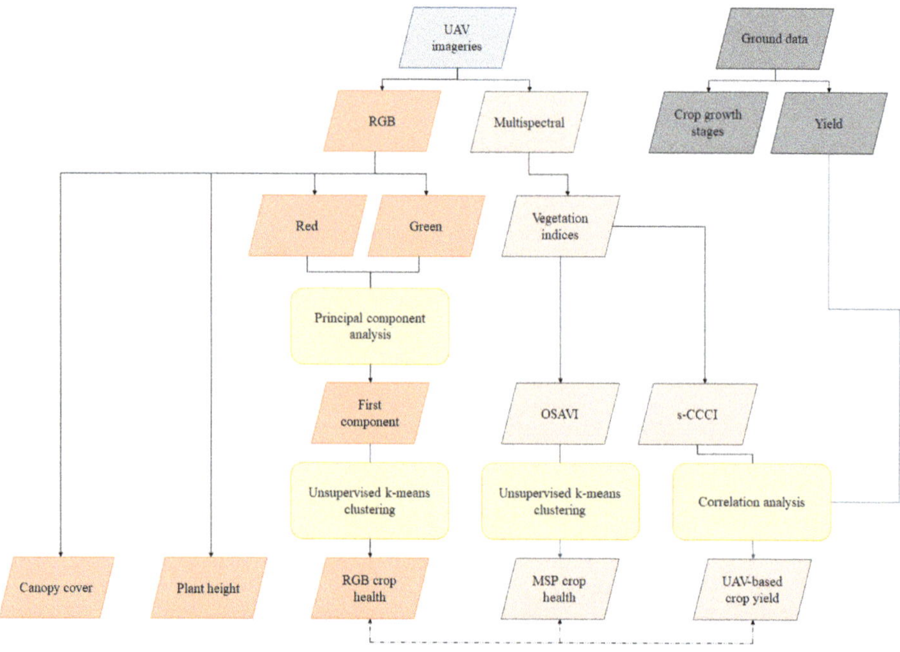

Figure 2. Methodological flowchart for crop traits retrieval from the unmanned aerial vehicle (UAV) imageries in combination with ground-based data.

The photogrammetric processing of the acquired UAV images was conducted in Pix4Dmapper photogrammetric software (Pix4D S.A., Lausanne, Switzerland). After the aerial triangulation of the UAV images, the geometric processing was optimized in Pix4Dmapper using the coordinates of the GCPs, and dense point clouds and final orthomosaic images were then created. For the multispectral data, the radiometric calibration was applied using MicaSense calibrated reference panels with known reflectance values (MicaSense, Seattle, WA, USA) taken immediately before and after each flight to convert the raw image digital number (DN) to the calibrated reflectance values. In addition, sensor's sensitivity to light, expressed as ISO, values recorded by the downwelling light sensor were used to aid the radiometric calibration. Both RGB and multispectral datasets were converted to GeoTIFF format with the projected geographic coordinates system WGS 1984 UTM Zone 31N.

In addition to the aforementioned remotely-sensed datasets, ground based measurements of crop traits were performed (Figure 2). Crop growth stages were determined on 28 May 2019 according to the stabiologische bundesanstalt, bundessortenamt and chemical industry (BBCH) scale for winter wheat [41]. For each plot, a regular non-destructive sampling method of 5 plants, i.e., 4 in the corners and 1 in the middle to capture the intra-plot variability, was followed. The BBCH code of each plant was visually determined and the median value of the five plants represented the plot growth stage.

Yield measurements were conducted on 20 July 2019. For each experimental plot, an area of approximately 20 m^2 was harvested to determine the yield.

2.2. UAV-Based Crop Monitoring

Canopy cover was estimated from the acquired high-resolution RGB images (Figure 2) based on the classification method by Heidarian Dehkordi et al. [26]. This algorithm classifies the pixels in the orthomosaic to vegetation and non-vegetation classes according to the spectral differences between the green band and the blue and red bands as below:

$$[(\text{Green} - \text{Blue}) = \alpha] \text{ and } [(\text{Green} - \text{Red}) = \beta] \quad (1)$$

where Blue, Green, and Red show the digital number (DN) of the corresponding spectral channel of the visible spectrum, α and β are the classification threshold values with α representing the spectral difference between the green and red bands and β revealing the spectral difference between the green and red spectral channels. Similar to the finding of Heidarian Dehkordi et al. [26], a threshold value of greater than 20 for both α and β was also reached in the present study through the visual inspection for the best vegetation classification. Canopy cover estimates were used as a potential discriminative parameter between the reference and century-old biochar plots in each acquisition date [26].

Plant height, as an important phenotyping crop trait, has received no attention in the previous studies exploring the impacts of century-old biochar on crop performance. In this study, plant height was estimated from the RGB images (Figure 3) as below:

$$\text{Height (t)} = \text{DSM (t)} - \text{DTM} \quad (2)$$

where Height is the computed wheat height, DSM and DTM are the UAV-based digital surface model and digital terrain model, respectively, with index t representing the acquisition date. DSMs were generated from the RGB flights for each acquisition date. The RGB flight on 22 February 2019 (Table 1) was executed when the field was completely bare soil and, hence, the UAV-based DSM, revealing the pure terrain surface topography, was used as the DTM for this study. GCP_2 (Figure 1) was used as the reference levelling point to calibrate the generated UAV-based DSMs at each acquisition date, bringing the elevation values onto the same scale. The created RGB-based vegetation mask (Equation (1)), with GSD of 1.8 cm, was used to retrieve the DSM from the true canopy pixels for the height computation.

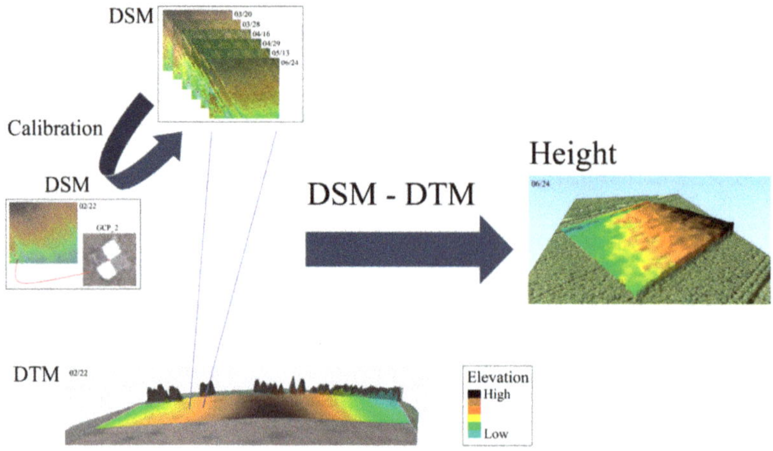

Figure 3. Methodological workflow to derive plants height from RGB images collected by the unmanned aerial vehicle. DTM is the digital terrain model acquired on 22 February when the field was completely bare soil. DSM represents the digital surface model at each acquisition date. Ground control point 2 was used as a reference levelling point to calibrate the DSM images.

In addition to the RGB derivatives listed above, narrow-band vegetation indices given in Table 2 were computed from the multispectral dataset (Figure 2) as a proxy to crop traits allowing for demonstrating spectral changes related to crop structure [39]. Furthermore, the multispectral information derived from the vegetation indices were analyzed to study the impacts of century-old biochar on winter wheat crop spectral information. The normalized difference vegetation index (NDVI) is a well-known proxy of photosynthetic activity and vegetation greenness. Canopy cover and leaf area index (LAI) were accurately predicted based on the weighted difference vegetation index (WDVI) in several studies [42–44]. The normalized difference red edge index (NDRE) was shown as a good proxy of wheat nitrogen concentration [45]. The optimized soil adjusted vegetation index (OSAVI) was found to be a robust indicator of leaf chlorophyll content [46], and the same conclusion was reached for the chlorophyll vegetation index (CVI). The enhanced vegetation index (EVI) was shown to be sensitive to canopy structural parameters such as LAI while adjusting soil background effects and atmospheric resistance using the blue band of the spectrum [47]. The chlorophyll index red (CI-red) and the simplified canopy chlorophyll content index (s-CCCI) are known alternatives of structural indices which are often used for estimating canopy chlorophyll and nitrogen contents [48,49].

The aforementioned vegetation indices were used for identifying the crop spectral differences between reference and century-old biochar plots throughout the growing season. First, WDVI was used to generate a multispectral-based vegetation mask. A threshold of WDVI higher than 0.30 was used to retrieve the so-called true canopy pixels because this threshold yielded the best results through the visual inspection. This method was based on the approach adopted by Heidarian Dehkordi et al. [26]. Finally, the created WDVI-based mask was applied to the vegetation indices (Table 2) to remove the soil pixels from the analysis.

Unsupervised k-means clustering [50] was applied to the multispectral OSAVI imagery in order to generate a MSP crop health map with "good", "moderate", and "poor" classes (Figure 2). The k-means algorithm identifies the k number of cluster centroids (i.e., 3 in the present study) within the image and allocates each pixel to the nearest cluster by minimizing its Euclidean distance from the cluster centroid; this iterative algorithm applies until a steady state with the optimized centroid positions is reached [50]. Moreover, principal component analysis [51] was performed on the red and green spectral channels retrieved from the RGB imagery (Figure 2). The first component (PC1) of the principal component analysis (PCA), which accounts for the maximum proportion of variance [52] from the red and green bands of the RGB orthomosaic (i.e. 99.06% in the present study), was used in the unsupervised k-means clustering to create a RGB crop health map (Figure 2). More precisely, as recommended by Heidarian Dehkordi et al. [26], the need to explore the impact of century-old biochar on crop health during the senescence phase was addressed in this study.

Moreover, the use of multispectral vegetation indices for the remote sensing-based estimation of crop yield at the field-scale, as suggested by [53], was evaluated in this study (Figure 2). In addition, the impact of century-old biochar on the predicted multispectral crop yield was studied.

Table 2. List of the multispectral vegetation indices we have examined in this study.

Index Name	Index Acronym	Formula	Reference
normalized difference vegetation index	NDVI	$(R_{NIR} - R_{red})/(R_{NIR} + R_{red})$	[54]
weighted difference vegetation index	WDVI	WDVI = $R_{NIR} - a.R_{red}$ with a = (R_{NIR}/R_{red}) of the soil	[55]
normalized difference red edge index	NDRE	$(R_{NIR} - R_{rededge})/(R_{NIR} + R_{rededge})$	[45]
optimized soil adjusted vegetation index	OSAVI	$1.16\ (R_{NIR} - R_{red})/(R_{NIR} + R_{red} + 0.16)$	[56]
chlorophyll vegetation index	CVI	$(R_{NIR}/R_{green}) \times (R_{red}/R_{green})$	[57]
enhanced vegetation index	EVI	$2.5\ (R_{NIR} - R_{red})/(R_{NIR} + 6\ R_{red} - 7.5\ R_{blue} + 1)$	[58]
chlorophyll index red	CI-red	$(R_{NIR}/R_{red}) - 1$	Similar to [44]
simplified canopy chlorophyll content index	s-CCCI	NDRE/NDVI	Similar to [49]

2.3. *Statistical Analysis*

Statistical paired and global t-tests [59] were used to determine whether the contrasts between reference and century-old biochar plots were statistically significant. Several levels of significance were considered to investigate the observed differences. Throughout the manuscript, asterisks *, **, ***, and **** indicate the significance levels of p-values less than 0.05, 0.01, 0.001, and 0.0001, respectively.

Relationships between UAV-based derivatives and ground-based measurements were evaluated using the statistical F-test. Subsequently, the significance of the obtained correlations was evaluated through the non-parametric Kendall's rank correlation tau (τ) test which is less sensitive to the small statistical sample size, i.e., 11 experimental plots in the present study [60].

3. Results and Discussion

3.1. *UAV-Based Crop Assessment*

3.1.1. Canopy Cover

The temporal pattern of canopy cover shows vegetation development from 20 March to 29 April (Figure 4). At the beginning of May the canopy cover was stable around the maximum until mid-May. From this time onwards, canopy cover slightly decreased towards the end of the season since leaves lost their greenness and began yellowing. The canopy cover clearly showed larger values in biochar plots than the reference plots for most of the season (Figure 4). The Area under the curve (AUC) of the canopy cover was significantly higher for the century-old biochar plots than the reference plots over the season (p-value of 0.00007 and 0.00002 for paired and global t-tests, respectively). This finding is in accordance with the positive impact of century-old biochar patches on the evolution of chicory canopy cover in the same experimental farm [26]. Another study by Kerré et al. [6] also identified this positive impact of century-old biochar from relict charcoal hearths on maize crop growth in Belgium.

The greater canopy cover development over the century-old biochar plots could be explained by improved soil physical properties [6] such as water holding capacity (WHC), and soil porosity and infiltration rates, leading to higher water content available to plants that, in turn, increased the vegetation development in the biochar plots. Moreover, higher soil chemical properties (i.e., organic carbon and nitrogen content) were observed over the same century-old biochar plots [26] characterized by a positive relationship with the evolution of canopy cover in the 2018 growing season.

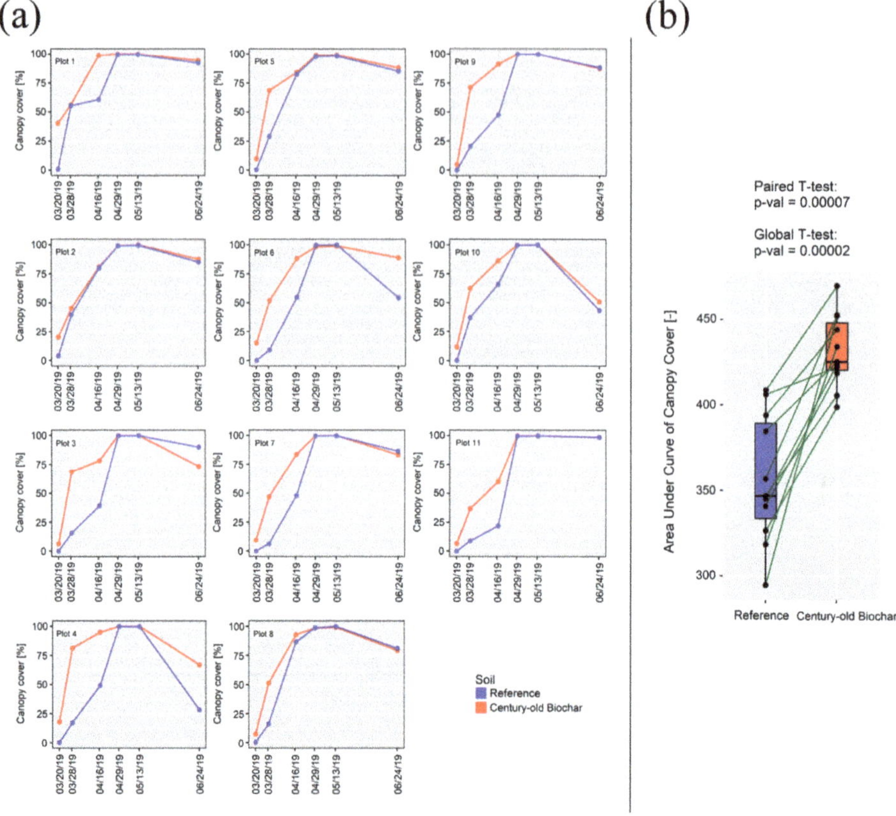

Figure 4. (**a**) Temporal profiles of the canopy cover, derived from the RGB imagery, comparing the 11 century-old biochar plots with the corresponding reference plots throughout the season. (**b**) Box plot visualization of the area under the curve of canopy cover between the century-old biochar and reference plots (i.e., obtained from graphs displayed in sub-plot a for each site). The horizontal black line displays the median value, surrounded by box edges representing the 25th and 75th percentiles. The black circles show all of the experimental plots and the grey dark green line indicates the corresponding pairs of a reference and century-old biochar plot.

In addition, the earlier germination of winter wheat plants over the century-old biochar plots on the first acquisition (Figure 4) could be attributed to the dark background soils increasing the soil temperature and subsequently also plant temperature, leading to a faster crop phenological development. Hence, future research using high-resolution thermal imagery should investigate the impact of century-old biochar on soil–plant water exchanges and their relationship with crop development.

3.1.2. Plant Height

Figure 5 illustrates the outcome of the height analysis. Winter wheat height increased from 20 March to 29 April (Figure 5) which is in accordance with our earlier finding of canopy cover evolution (Figure 4). The height estimates were stable in May and June around the maximum. Wheat height was rather accurately derived with an accuracy (SD/mean) of 4.92 cm counting the entire dataset (Figure 5), while the previous work by ten Harkel et al. [61] reached a higher accuracy of 3.4 cm for wheat height estimates using UAV-based light detection and ranging (LiDAR) data. The usage of UAV-based LiDAR enables a proper selection of the top of the canopy pixels to be used in height computations [61] increasing the modeling accuracy, which was not possible in the present study. The highest variation in height estimates (Figure 5) was observed on 28 March (SD of 10.73 and 11.25 cm for the century-old biochar and reference plots, respectively), which could be attributed to the erectophile and open structure of winter wheat at the beginning of the season, lowering the number of true canopy pixels used for height estimates. However, the observed variations were much lower during the maturity phase (with a maximum SD of 5.45 cm on 13 May) because wheat canopies are rather dense, increasing the number of true canopy pixels in height computations.

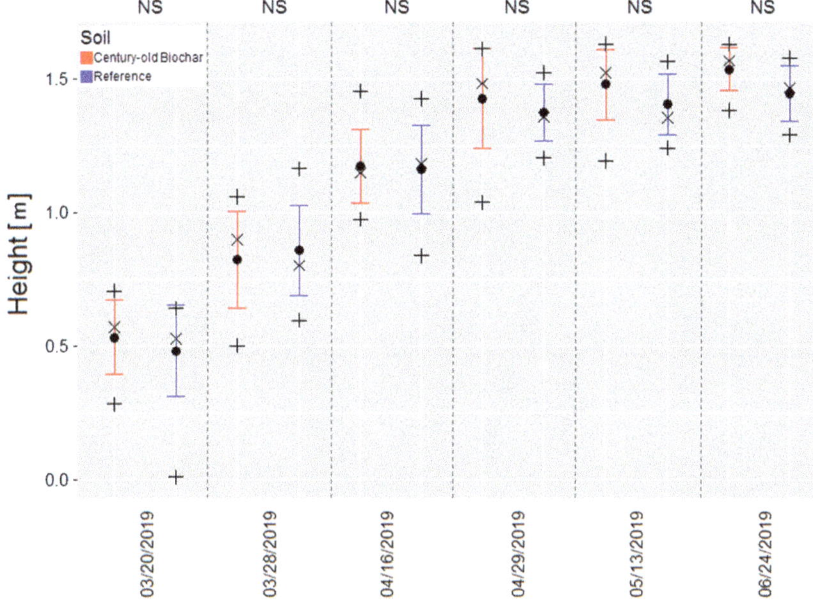

Figure 5. Comparison of the plant height of all the 11 century-old biochar plots (red) versus the 11 reference plots for each acquisition date. The black circle and cross represent the mean and median height, respectively. The error-bar presents the corresponding standard deviation. The bottom and top black pluses (+) indicate the minimum and maximum height, respectively. Outside of the plot, asterisks *, **, ***, and **** reveal the statistical levels of significance; the acronym NS stands for non-significant.

The impact of century-old biochar on winter wheat height considering all of the 11 biochar plots versus the reference plots is shown in Figure 5 for each acquisition date. The mean height (mean value of the 11 plots) was higher in the century-old biochar plots than in the reference plots for each acquisition date except 28 March (Figure 5). It is worth noting that the observed mean height differences between reference and century-old biochar plots were statistically non-significant for all the monitoring dates (Figure 5). This slight promotion of wheat height as a consequence of biochar presence was previously reported for lettuce–cabbage–lettuce [62] and oat [63] fields. The greater plant height over the century-old biochar patches could be attributed to the higher soil organic carbon and nitrogen contents as previously found by Schulz et al. [63].

Figure 6 shows the outcome of the height analysis for each experimental plot. Most of the experimental plots showed greater height values in century-old biochar plots compared to their corresponding reference plots throughout the season, except for plots 8 and 11 where an inverse trend was observed (Figure 6). This can be explained by the location of the experimental plots 8 and 11 which were situated in a part of the field with a high soil moisture content (images not shown). The observed difference of plant height between the century-old biochar plots and the corresponding reference plots at each acquisition date was significant for all of the experimental plots (Figure 6) except for plots 5, 8, and 10 on 28 March and plot 10 on 16 April. The observed height differences between reference and century-old biochar plots were significant for most of the experimental pairs separately (as shown in Figure 6), whereas the differences were not of particular concern when counting the entire dataset (Figure 5). The latter could be explained by the within-field spatial variability of plant height that accounts for the spatial variation in field topographic derivatives such as elevation, groundwater flow, or surface wetness. Validation with the ground-based measurements of plant height should be performed in future studies to explore the robustness of the proposed plant height derivation method in this research.

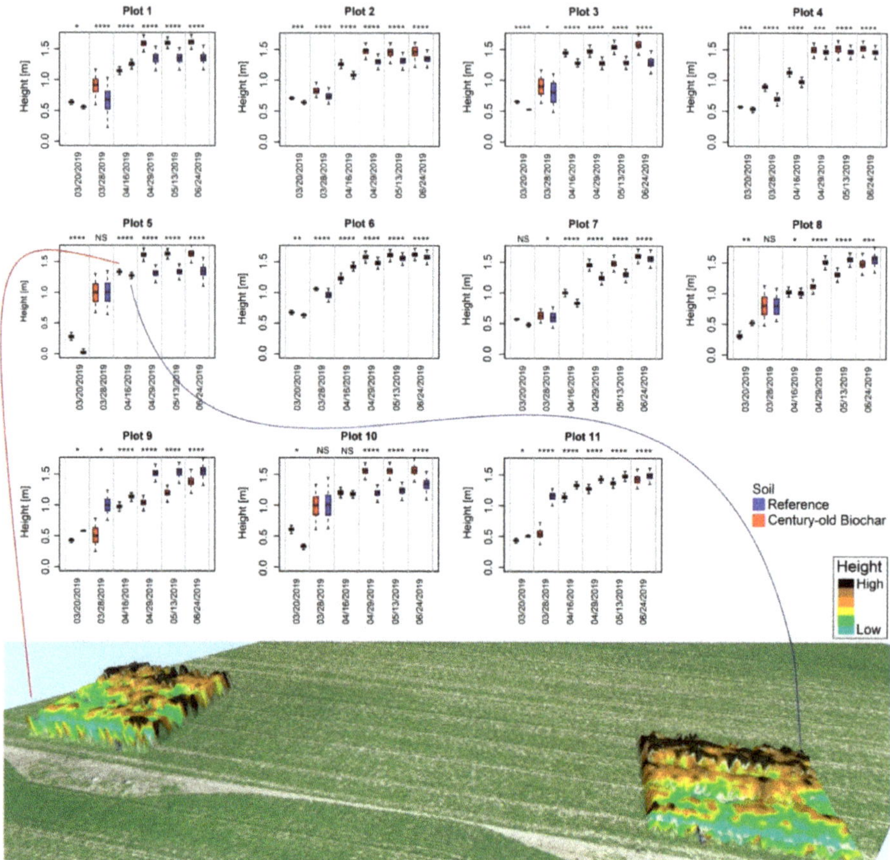

Figure 6. Box plot visualization of winter wheat height development for the 11 experimental pairs of century-old biochar versus reference plots. The horizontal black line displays the median value, surrounded by box edges representing the 25th and 75th percentiles. Outside of the plot, asterisks *, **, ***, and **** reveal the statistical levels of significance and the acronym NS stands for non-significant. The bottom magnified window displays an example of a height comparison between century-old biochar and reference plot 5 (i.e., on 16 April 2019).

3.1.3. Vegetation Indices

Figure 7 illustrates the temporal profiles for OSAVI. There is a slight vegetation development from 20 March until the end of that month. The OSAVI then increased gradually until the end of April and remained constant at the maximum towards mid-May. From this time onwards, the OSAVI started to decrease as the leaves started to lose their greenness. The overall OSAVI pattern (Figure 7) is completely in line with the trend observed from the canopy cover evolution in Figure 4.

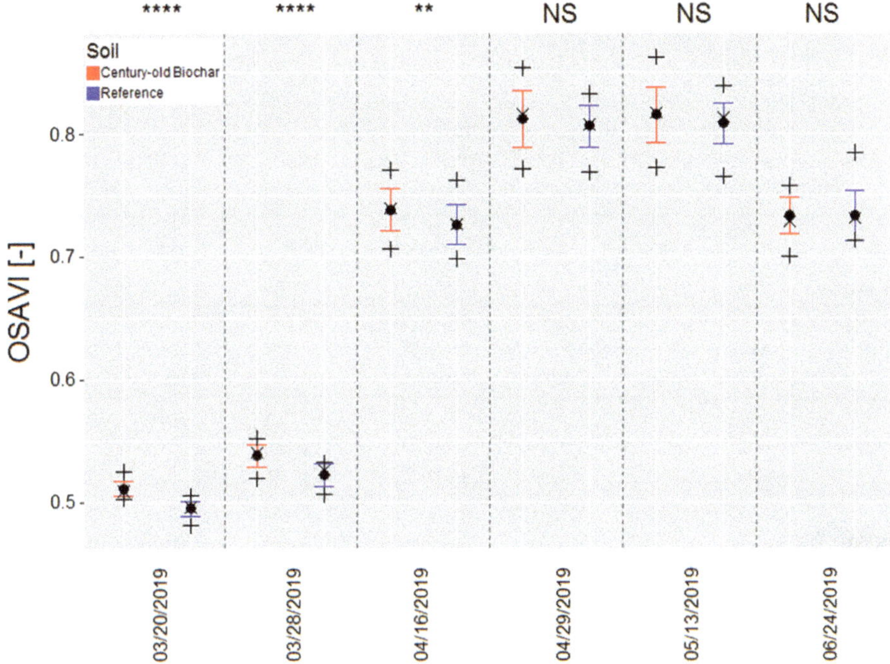

Figure 7. Comparison of the optimized soil adjusted vegetation index (OSAVI) of all of the 11 century-old biochar plots (red) versus the 11 reference plots for each acquisition date. The black circle and cross represent the mean and median OSAVI, respectively. The error-bar represents the corresponding standard deviation. The bottom and top black pluses (+) indicate the minimum and maximum OSAVI, respectively. Outside of the plot, asterisks *, **, ***, and **** reveal the statistical levels of significance and the acronym NS stands for non-significant.

The impact of century-old biochar on OSAVI considering all of the 11 century-old biochar plots versus the reference plots for each acquisition date is shown in Figure 7. The mean OSAVI (mean value of the 11 plots) was higher in the century-old biochar plots than the reference plots at each acquisition date except on 24 June when the mean OSAVI was 0.73 for both reference and century-old biochar plots (Figure 7). This finding is in accordance with the decreased plant greenness at the end of growing season for chicory in century-old biochar patches of the same experimental farm [26]. The observed OSAVI difference was non-significant between the reference and century-old biochar plots towards the end of the season, while it was highly significant from the beginning of the season until mid-April (Figure 7). This could be partially explained by the sparseness of the wheat canopies at the beginning of the season, increasing the interactions between plants and dark background soils in the century-old biochar plots.

Contrast between reference and century-old biochar plots for each vegetation index is presented in Table 3 for all the acquisition dates. Century-old biochar significantly affected the crop spectral information derived from NDVI, NDRE, OSAVI, CI-red, and s-CCCI on 20 and 28 March. However, WDVI, CVI, and EVI did not exhibit significant differences in crop spectral status between reference and century-old biochar plots during March. All of the examined vegetation indices showed a significant impact of century-old biochar on crop spectral status on 16 April except WDVI, EVI, and s-CCCI (Table 3). On 29 April only NDVI, NDRE, and CI-red, and on 13 May only NDVI and CI-red, exhibited a significant impact of century-old biochar on crop spectral status. All of the examined vegetation indices exhibited non-significant differences between reference and century-old biochar plots on 24 June (Table 3). This finding is in accordance with [26] who reported no significant difference of the multispectral vegetation indices between the century-old biochar and reference plots for chicory crop at the beginning of July.

Table 3. Mean vegetation index values over the study area comparing the century-old biochar and reference plots during the growing season of 2019. The statistical significance of the difference is expressed as the p-value of the paired t-test. Next to the p-values asterisks, *, **, ***, and **** reveal the statistical levels of significance and acronym NS stands for non-significant.

Index	Date	03/20	03/28	04/16	04/29	05/13	06/24
NDVI	Biochar	0.43 ± 0.02	0.72 ± 0.02	0.88 ± 0.02	0.93 ± 0.01	0.95 ± 0.01	0.86 ± 0.01
	Reference	0.64 ± 0.01	0.67 ± 0.01	0.85 ± 0.02	0.92 ± 0.01	0.95 ± 0.01	0.86 ± 0.01
	p-value	0.0000 ****	0.0000 ****	0.0000 ****	0.0175 *	0.0126 *	0.3863 NS
WDVI	Biochar	0.16 ± 0.00	0.18 ± 0.01	0.34 ± 0.02	0.44 ± 0.03	0.43 ± 0.04	0.35 ± 0.02
	Reference	0.16 ± 0.00	0.18 ± 0.01	0.35 ± 0.02	0.43 ± 0.02	0.42 ± 0.02	0.35 ± 0.01
	p-value	0.5290 NS	0.9388 NS	0.6326 NS	0.8048 NS	0.1820 NS	0.5469 NS
NDRE	Biochar	0.30 ± 0.02	0.30 ± 0.02	0.43 ± 0.03	0.55 ± 0.04	0.63 ± 0.03	0.55 ± 0.02
	Reference	0.27 ± 0.01	0.27 ± 0.01	0.42 ± 0.02	0.53 ± 0.02	0.62 ± 0.02	0.54 ± 0.01
	p-value	0.0002 ***	0.0000 ****	0.0050 **	0.0496 *	0.0563 NS	0.3518 NS
OSAVI	Biochar	0.51 ± 0.01	0.54 ± 0.01	0.74 ± 0.02	0.82 ± 0.02	0.82 ± 0.02	0.73 ± 0.02
	Reference	0.50 ± 0.01	0.53 ± 0.01	0.73 ± 0.01	0.81 ± 0.01	0.81 ± 0.01	0.73 ± 0.01
	p-value	0.0000 ****	0.0000 ****	0.0046 **	0.2313 NS	0.0948 NS	1.0000 NS
CVI	Biochar	2.27 ± 0.21	2.03 ± 0.14	2.47 ± 0.11	2.74 ± 0.29	4.47 ± 0.34	5.40 ± 0.28
	Reference	2.39 ± 0.11	2.20 ± 0.06	2.77 ± 0.11	2.84 ± 0.16	4.61 ± 0.14	5.33 ± 0.12
	p-value	0.0520 NS	0.0042 **	0.0000 ****	0.0755 NS	0.4628 NS	0.2260 NS
EVI	Biochar	0.38 ± 0.01	0.42 ± 0.01	0.65 ± 0.03	0.79 ± 0.04	0.79 ± 0.04	0.65 ± 0.03
	Reference	0.38 ± 0.01	0.42 ± 0.01	0.65 ± 0.03	0.78 ± 0.03	0.78 ± 0.02	0.66 ± 0.02
	p-value	0.66374 NS	0.89354 NS	0.75669 NS	0.68177 NS	0.25339 NS	0.59717 NS
CI-red	Biochar	4.55 ± 0.60	5.15 ± 0.48	14.47 ± 3.20	28.13 ± 5.66	39.79 ± 7.11	12.76 ± 1.14
	Reference	3.53 ± 0.20	4.05 ± 0.19	11.11 ± 1.66	23.03 ± 2.36	36.65 ± 4.08	12.14 ± 0.97
	p-value	0.00022 ***	0.0000 ****	0.00040 ***	0.01113 *	0.00726 **	0.39363 NS
s-CCCI	Biochar	0.43 ± 0.02	0.42 ± 0.01	0.49 ± 0.02	0.59 ± 0.03	0.67 ± 0.03	0.63 ± 0.01
	Reference	0.41 ± 0.01	0.40 ± 0.01	0.48 ± 0.01	0.58 ± 0.01	0.66 ± 0.01	0.63 ± 0.01
	p-value	0.01007 *	0.00324 **	0.17634 NS	0.08249 NS	0.08488 NS	0.32757 NS

It is worth mentioning that the WDVI and the EVI performed the worst in discriminating spectral difference between century-old biochar and reference plots throughout the season. The latter could be attributed to the use of the blue band in EVI's computation which is not of particular importance in the vegetation spectral signature [64].

Figure 8 shows the outcome of the OSAVI analysis for each experimental plot. Most of the experimental plots exhibited higher OSAVI values in century-old biochar plots compared to their adjacent reference plots for most of the season (Figure 8). Graphs of the other examined multispectral vegetation indices (Table 2) are not presented here. No clear trend was observed in terms of the OSAVI difference between reference and century-old biochar plots (Figure 8). The contrast between reference and century-old biochar plots was highly significant at several acquisitions (e.g., for plot 1 on 29 April), whereas some acquisitions exhibited a non-significant difference (e.g., plot 1 on 28 March).

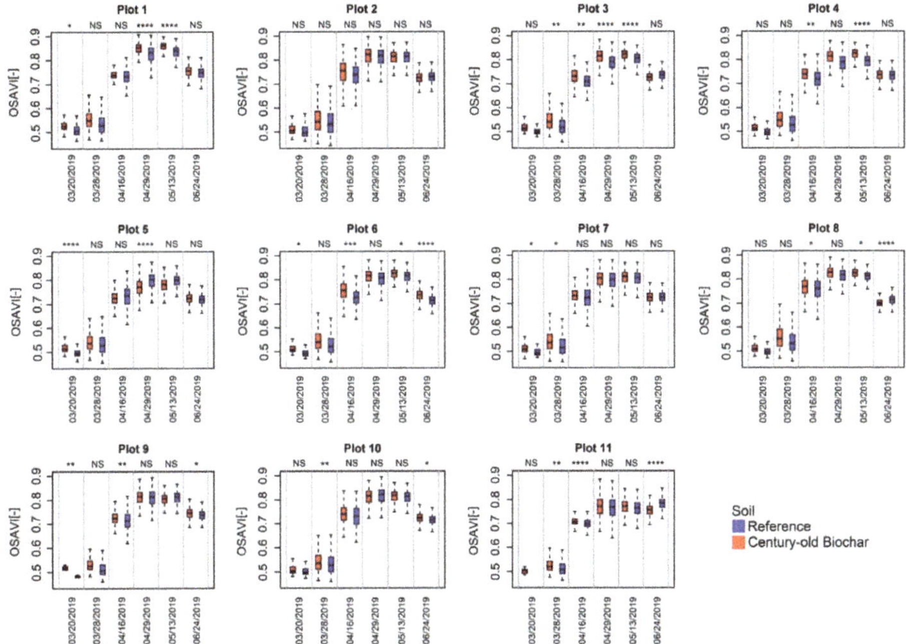

Figure 8. Box plot visualization of optimized soil adjusted vegetation index for the 11 experimental pairs of century-old biochar versus reference plots over the growing season of 2019. The horizontal black line displays the median value, surrounded by box edges representing the 25th and 75th percentiles. Outside of the plot, asterisks *, **, ***, and **** reveal the statistical levels of significance and the acronym NS stands for non-significant. No spectral information is available for reference plot 1 on 20 March because of a flight planning constraint.

3.1.4. Crop Health

Both canopy cover and OSAVI were characterized by a decreasing trend on 24 June (Figures 4 and 7, respectively). The impact of century-old biochar on crop health during the senescence phase is presented in Figure 9. In addition, Figure 9 presents the two different crop health maps, obtained from the two different sensors, including the associated general methodological workflows as represented in Figure 9a,b for the multispectral sensor and Figure 9c,d for the RGB sensor. An unsupervised k-means clustered crop health map of OSAVI including "good", "moderate", and "poor" classes, corresponding to 24 June, is presented in Figure 9b. In addition, an unsupervised k-means clustered crop health map derived from the first component (PC1) from the principal component analysis of the red and green bands of the RGB orthomosaic image corresponding to 24 June is shown in Figure 9d, clustering the study area into "good", "moderate", and "poor" classes. It is worth mentioning that 99.06% of the observed variance was explained by the first component (PC1) and only 0.94% of the variance retained by the second component (PC2) (Figure 9c). Moreover, there was a positive significant correlation between the red and green bands (Kendall's τ coefficient = 0.89 and p-value$_{(Kendall)}$ < 0.0001). The experimental field was completely covered by vegetation on 24 June enabling a proper PCA analysis over the vegetation pixels, without any soil background effects, to retrieve an accurate crop health map.

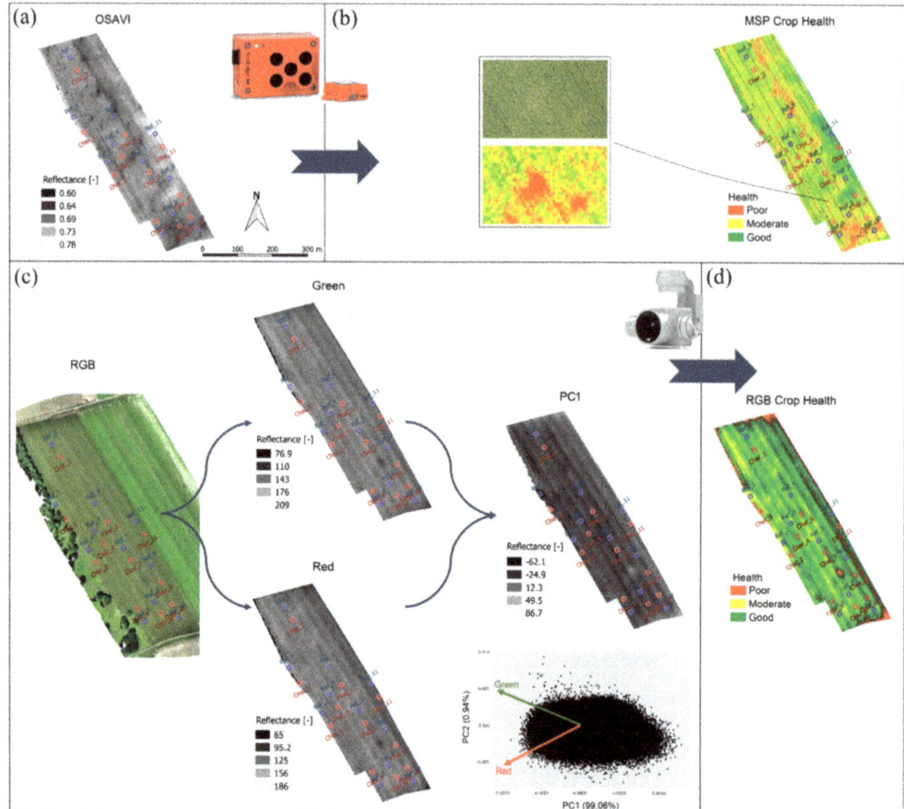

Figure 9. Data from 24 June 2019: (**a**) Map of optimized soil adjusted vegetation index (OSAVI). (**b**) Crop health map derived from the k-means clustering of OSAVI; the magnified window represents an example of an area characterized by poor crop health in the study field. (**c**) RGB orthomosaic image (left), the green and red spectral responses of the visible RGB orthomosaic image (middle), and the first component (PC1) raster of the principal component analysis (PCA) of the green and red bands (right). (**d**) Crop health map derived based on the k-means clustering of the PC1 raster of the PCA of green and red spectral channels.

Categorical classification of the experimental plots into the crop health clusters is presented in Table 4 for both of the multispectral and RGB crop health maps (Figure 9). Based on the multispectral crop health map, the average health was 42.5% for the century-old biochar plots compared to 39.4% in the reference plots. Similarly, the average health based on the RGB crop health map was higher for the century-old biochar plots (54.6%) than the reference plots (51.6%). This finding shows that there is only a small positive impact of century-old biochar on crop health status during the senescence phase. The present study paves the way for further research focusing on the effects of biochar on crop health during the senescence phase.

Table 4. Categorical classification of the crop health into good, moderate, and poor classes comparing the century-old biochar versus reference plots. The classification is based on the multispectral (MSP) crop health (left) and the first component (PC1) of the principal component analysis (PCA) of the green and red spectral channels of the RGB orthomosaic image (right).

Plot ID	MSP Crop Health		RGB Crop Health	
	Biochar	Reference	Biochar	Reference
1	Good	Good	Moderate	Moderate
2	Moderate	Moderate	Good	Good
3	Moderate	Good	Good	Good
4	Moderate	Moderate	Good	Moderate
5	Moderate	Moderate	Good	Moderate
6	Good	Poor	Good	Good
7	Moderate	Moderate	Good	Good
8	Poor	Moderate	Poor	Moderate
9	Good	Good	Good	Good
10	Moderate	Poor	Good	Good
11	Good	Good	Moderate	Moderate
Average health	42.5%	39.4%	54.6%	51.6%

The cross-tabulation analysis of the computed crop health maps is presented in Table 5 comparing the number of all of the pixels within the study area classified into good, moderate, and poor health classes. The result yielded a clustering agreement of 74.82% (Table 5). Out of a total of 564,760 pixels across the entire study area, 36.06% were classified as good, 31.54% as moderate, and 7.21% as poor by both methods (Table 5). Ground-based optical imagery, such as from the Fabry-Perot interferometer (FPI) camera system [39], is an essential step for future research focusing on the evaluation of crop health because of its improved spatial and spectral resolutions, enabling an accurate monitoring of the development of various plant diseases at the leaf level.

Table 5. Cross-tabulation table of the computed MSP and RGB crop health maps comparing the clustering accuracy of all of the pixels within the study area into good, moderate, and poor classes.

	Clusters	Pixel Class Types Determined from MSP Crop Health			
		Good	Moderate	Poor	Total
Pixel class types determined from RGB crop health	Good	**203,674**	39,326	15,756	258,756
	Moderate	38,757	**178,178**	11,568	228,503
	Poor	18,997	17,784	**40,720**	77,501
	Total	261,428	235,288	68,044	**564,760**
Clustering agreement					**74.82%**

3.2. Ground-Based Crop Evaluation and Harvested Crop Yield

Ground-based measurements of crop traits are shown in Figure 10. Estimates of the general crop growth stages taken on 28 May 2019, scaled according to the BBCH, reported a more advanced crop stage for the century-old biochar plots (median BBCH = 43) compared to the reference plots (median BBCH of 40.5) as shown in Figure 10a. The observed difference in crop growth stages (BBCH) was statistically significant (p-value = 0.01265) between the reference and century-old biochar plots. This could be explained by the dark color of the century-old biochar plots, resulting in higher soil and plant temperatures stimulating plant development.

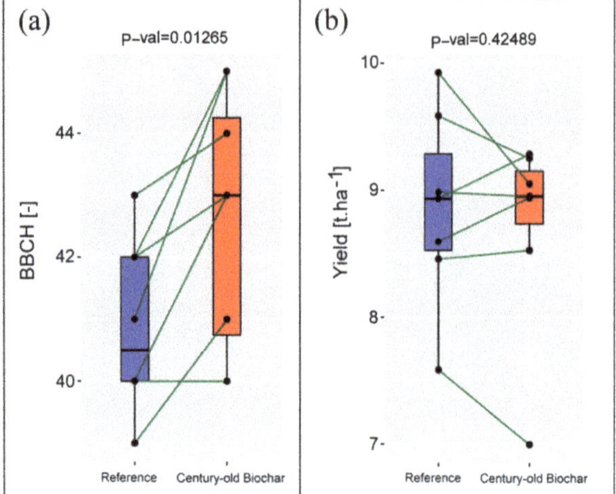

Figure 10. Impact of century-old biochar on ground-based crop traits including crop growth stages, expressed as the biologische bundesanstalt, bundessortenamt and chemical industry (BBCH) scale, on 28 May 2019 (**a**), and yield on 20 July 2019 (**b**). The horizontal black line displays the median value, surrounded by box edges representing the 25th and 75th percentiles. The black circles show the experimental plots of each pair and the dark green lines relate the corresponding reference and century-old biochar plots of each pair.

The mean harvested crop yield (taken on 20 July 2019) value was 8.69 ± 0.69 $t.ha^{-1}$ for the century-old biochar plots compared to 8.84 ± 0.66 $t.ha^{-1}$ in the reference plots; the observed contrast was not significant with a *p*-value of 0.42489 (Figure 10b). This finding contrasts with the positive promotion of the aboveground productivity for long-term biochar enriched soils [6,23]. This can be explained by the fact that our study area is situated in a quite rich soil type and hence the influence of century-old biochar on the harvested crop yield is limited. However, other studies considering poorer soil types have shown that the difference in yield due to the long-term biochar enrichment is more substantial [6].

3.3. UAV-Based Yield Map Generation

Relationships between harvested crop yields on 20 July (Section 3.2) and the UAV-based multispectral vegetation indices on the last acquisition date (i.e., 24 June which was the closest to the harvest date) are presented in Table 6. The s-CCCI, NDRE, NDVI, and CI-red exhibited a good relationship with the harvested crop yield as shown in Table 6. The obtained relationships were statistically significant with $p\text{-value}_{(\text{F-test})}$ of 0.002, 0.002, 0.008, and 0.008 for s-CCCI, NDRE, NDVI, and CI-red, respectively. The aforementioned indices showed a strong correlation with the harvested crop yield with Kendall's τ coefficient of 0.60, 0.61, 0.54, and 0.54 for the s-CCCI, NDRE, NDVI, and CI-red, respectively (Table 6). The obtained correlations were statistically significant with $p\text{-value}_{(\text{Kendall})}$ of 0.001, 0.0008, 0.004, and 0.004 for the s-CCCI, NDRE, NDVI, and CI-red, respectively (Table 6) which is in line with the previous finding of Das and Singh (2012).

Table 6. Results of correlation analysis of the multispectral vegetation indices, contained in Table 2, on 24 June and their relationships to the harvested crop yield on 20 July. The coefficient of determination, p-value of the statistical F-test, non-parametric Kendall's rank correlation tau, and its significance are expressed as R^2, $p\text{-value}_{(F\text{-test})}$, τ, $p\text{-value}_{(Kendall)}$, respectively.

Index	R^2	$p\text{-Value}_{(F\text{-test})}$	τ	$p\text{-Value}_{(Kendall)}$
NDVI	0.42	0.008	0.54	0.004
WDVI	0.04	0.424	0.33	0.092
NDRE	0.50	0.002	0.61	0.0008
OSAVI	0.13	0.172	0.44	0.020
CVI	0.007	0.763	0.16	0.435
EVI	0.08	0.289	0.39	0.460
CI-red	0.40	0.008	0.54	0.004
s-CCCI	0.52	0.002	0.60	0.001

The predicted UAV-based multispectral crop yield map based on the identified linear regression fit between the s-CCCI and crop harvested yield, which reported the strongest relationship (Table 6), is shown in Figure 11a. The result of the predicted UAV-based MSP crop yield is shown in Figure 11b. The western part of the experimental field clearly shows lower yield values. However, the eastern part of the field exhibited rather high yield values (Figure 11b). The impact of century-old biochar on the predicted UAV-based MSP crop yield is shown in Figure 11c. The mean predicted MSP crop yield (mean value of the 11 plots) was not statistically different (p-value of 0.437) between the century-old biochar plots (8.82 ± 0.39 $t.ha^{-1}$) and the reference plots (8.74 ± 0.42 $t.ha^{-1}$) as shown in Figure 11c. This finding balances the previous works reporting a higher aboveground productivity as a consequence of biochar presence [6,23]. However, our results showed a highly significant difference ($p < 0.0001$) of the predicted MSP crop yield between the good and moderate classes derived from the MSP sensor (Figure 11d). Similarly, the observed difference of the predicted MSP crop yield was highly significant ($p < 0.0001$) between the moderate and poor classes based on the MSP sensor (Figure 9d). The mean predicted MSP crop yields were 8.96 ± 0.85, 8.50 ± 0.79, and 7.84 ± 0.83 $t.ha^{-1}$ for the good, moderate, and poor classes derived from the MSP sensor (Figure 11d). In addition, the observed difference of the predicted MSP crop yield was highly significant ($p < 0.0001$) between the good and moderate classes, and between the moderate and poor classes, derived from the RGB sensor (Figure 11e). The mean predicted MSP crop yields were 8.76 ± 0.91, 8.40 ± 0.86, and 7.54 ± 0.89 $t.ha^{-1}$ for the good, moderate, and poor classes based on the RGB sensor (Figure 11e). This result indicates that other factors (such as fertilization inputs, agricultural practices, and inherent possible pedological variations) are of higher importance than the presence of century-old biochar in determining the crop health and yield variability at the within-field scale.

This paper provides a proof of concept of using high-resolution UAV images in assessing the impacts of century-old biochar on crop performance at the canopy scale. As such, the results of this study reveal important findings regarding the impact of century-old biochar on crop health and yield (in particular over the senescence period). However, since the approach described in this paper is only suitable for small-sized fields of a few hectares, further research should consider extending this analysis across a wider area, including a range of crops under contrasting weather conditions, using available fine resolution optical (e.g., Sentinel-2) and radar (e.g., Sentinel-1 Synthetic-aperture radar (SAR)) satellite images.

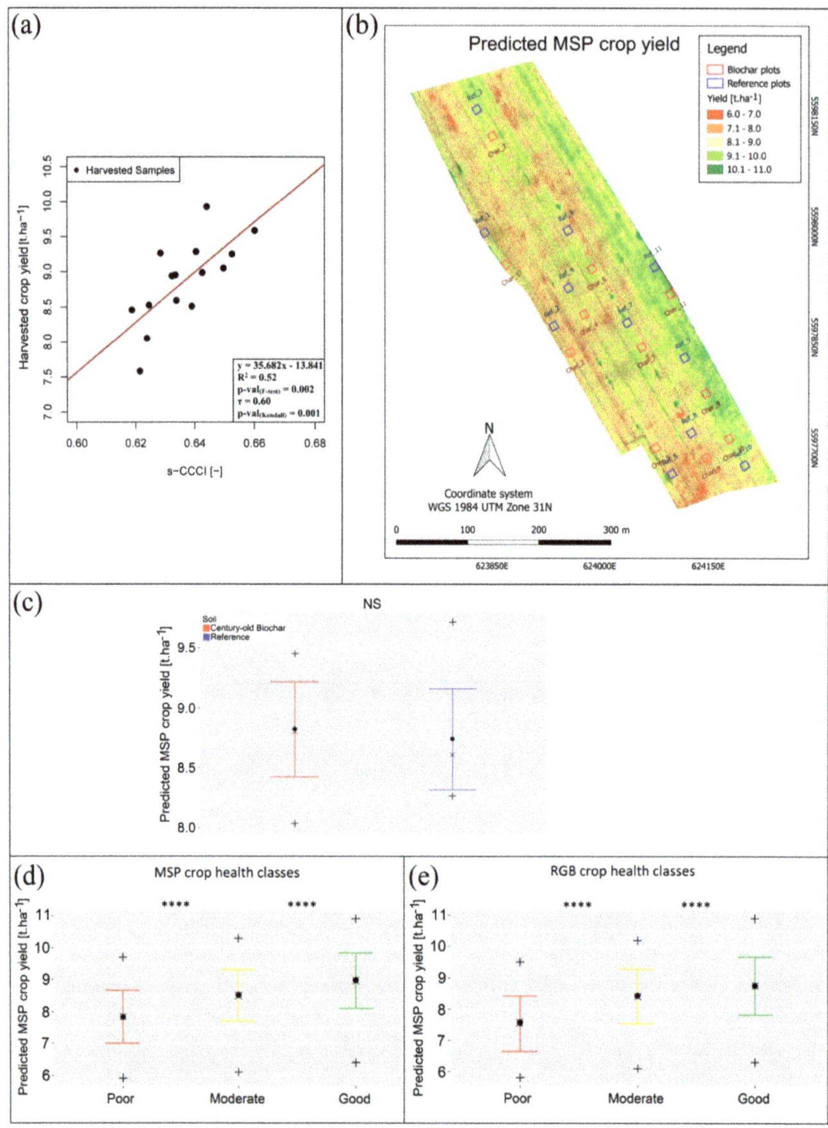

Figure 11. (**a**) Relationship between the multispectral (MSP) simplified canopy chlorophyll content index (s-CCCI) on 24 June and the harvested crop yield on 20 July. The black circles represent the harvested samples and the brown line indicates the linear regression fit. (**b**) Map of the predicted MSP crop yield computed from the relationship between the MSP s-CCCI and the harvested crop yield. (**c**) A comparison of the predicted MSP crop yield of all of the 11 century-old biochar plots (red) versus the 11 reference plots. The black circle and cross represent the mean and median MSP crop yield, respectively. The error-bar represents the corresponding standard deviation. The bottom and top black pluses (+) indicate the minimum and maximum MSP crop yield, respectively. A comparison of the predicted MSP

crop yield of the good (green), moderate (yellow), and poor (red) crop health classes derived from the MSP (**d**) and RGB (**e**) sensors. The black circle and cross represent the mean and median MSP crop yield, respectively. The error-bar represents the corresponding standard deviation. The bottom and top black pluses (+) indicate the minimum and maximum MSP crop yield, respectively. Outside of the plot, asterisks *, **, ***, and **** reveal the statistical levels of significance and the acronym NS stands for non-significant.

4. Conclusions

This paper explored the potential of using high-resolution UAV-based imagery to assess the impacts of century-old biochar on winter wheat crop performance. The combination of crop growth and health monitoring, through RGB and multispectral images, provided new insights into the alteration in crop dynamics at the canopy level associated with century-old biochar presence. Based on the RGB data, a significant positive impact (p-value = 0.00007) of century-old biochar on the evolution of winter wheat canopy cover was highlighted. Moreover, century-old biochar was found to slightly increase wheat height throughout the growing season. Multispectral OSAVI imagery underlined a significant positive effect on crop performance as a consequence of century-old biochar only at the beginning of the season (p-values < 0.01), although the impact was not remarkable at the end of the monitoring period.

Century-old biochar showed a significant positive impact on winter wheat crop growth stages (p-value of 0.01265). However, the effect of century-old biochar on harvested crop yield was not significant. Harvested crop yield exhibited a significant correlation with the multispectral vegetation indices, particularly s-CCCI and NDRE ($p\text{-value}_{(\text{Kendall})}$ of 0.001 and 0.0008, respectively). The predicted UAV-based multispectral crop yield was not significantly different between the century-old biochar plots compared to the reference plots. Hence, other factors, such as inherent pedological variations and agricultural practices, appear to be of higher importance than the presence of century-old biochar in determining the crop health and crop yield variability at the within-field scale.

Author Contributions: Conceptualization, R.H.D., A.D., J.M.; Investigation, R.H.D.; data curation, R.H.D., V.B., J.F. and E.P.G.; resources, R.H.D.; methodology, R.H.D. and J.M.; software, R.H.D.; formal analysis, R.H.D.; validation, R.H.D.; writing—original draft preparation, R.H.D.; writing—review and editing, R.H.D. and J.M.; supervision, J.M.; project administration, J.-T.C., J.M., F.N. and A.D.; funding acquisition, J.-T.C. and J.M.; conceptualization, R.H.D., J.M. and A.D. All authors have read and approved the published version of the manuscript.

Funding: This research was funded through the ARC grant 17/21-03 for Concerted Research Actions by the French Community of Belgium within the framework of the CHAR project at the University of Liège.

Acknowledgments: The authors thank the landowner for providing access to the CHAR project experimental farm. Thanks to Aurore Houtart within the framework of the CHAR project. Thanks also to Lammert Kooistra and Gilles Colinet for the helpful suggestions and supporting our research. The authors also acknowledge the editor and the anonymous reviewers for their insightful comments.

Conflicts of Interest: The authors declare no conflict of interest.

References

1. Malghani, S.; Gleixner, G.; Trumbore, S.E. Chars produced by slow pyrolysis and hydrothermal carbonization vary in carbon sequestration potential and greenhouse gases emissions. *Soil Biol. Biochem.* **2013**, *62*, 137–146. [CrossRef]
2. Trupiano, D.; Cocozza, C.; Baronti, S.; Amendola, C.; Vaccari, F.P.; Lustrato, G.; Di Lonardo, S.; Fantasma, F.; Tognetti, R.; Scippa, G.S. The effects of biochar and its combination with compost on lettuce (*Lactuca sativa* L.) growth, soil properties, and soil microbial activity and abundance. *Int. J. Agron.* **2017**, *2017*, 3158207. [CrossRef]
3. Yamato, M.; Okimori, Y.; Wibowo, I.F.; Anshori, S.; Ogawa, M. Effects of the application of charred bark of Acacia mangium on the yield of maize, cowpea and peanut, and soil chemical properties in South Sumatra, Indonesia. *Soil Sci. Plant Nutr.* **2006**, *52*, 489–495. [CrossRef]

4. Montanarella, L.; Lugato, E. The Application of Biochar in the EU: Challenges and Opportunities. *Agronomy* **2013**, *3*, 462–473. [CrossRef]
5. Stewart, C.E.; Zheng, J.; Botte, J.; Cotrufo, M.F. Co-generated fast pyrolysis biochar mitigates green-house gas emissions and increases carbon sequestration in temperate soils. *Gcb Bioenergy* **2013**, *5*, 153–164. [CrossRef]
6. Kerré, B.; Willaert, B.; Cornelis, Y.; Smolders, E. Long-term presence of charcoal increases maize yield in Belgium due to increased soil water availability. *Eur. J. Agron.* **2017**, *91*, 10–15. [CrossRef]
7. Lehmann, J.; Da Silva, J.P.; Steiner, C.; Nehls, T.; Zech, W.; Glaser, B. Nutrient availability and leaching in an archaeological Anthrosol and a Ferralsol of the Central Amazon basin: Fertilizer, manure and charcoal amendments. *Plant Soil* **2003**, *249*, 343–357. [CrossRef]
8. Biederman, L.A.; Stanley Harpole, W. Biochar and its effects on plant productivity and nutrient cycling: A meta-analysis. *Gcb Bioenergy* **2013**, *5*, 202–214. [CrossRef]
9. Glaser, B.; Lehmann, J.; Zech, W. Ameliorating physical and chemical properties of highly weathered soils in the tropics with charcoal—A review. *Biol. Fertil. Soils* **2002**, *35*, 219–230. [CrossRef]
10. Jeffery, S.; Meinders, M.B.J.; Stoof, C.R.; Bezemer, T.M.; van de Voorde, T.F.J.; Mommer, L.; van Groenigen, J.W. Biochar application does not improve the soil hydrological function of a sandy soil. *Geoderma* **2015**, *251–252*, 47–54. [CrossRef]
11. De la Rosa, J.M.; Paneque, M.; Miller, A.Z.; Knicker, H. Relating physical and chemical properties of four different biochars and their application rate to biomass production of *Lolium perenne* on a Calcic Cambisol during a pot experiment of 79 days. *Sci. Total Environ.* **2014**, *499*, 175–184. [CrossRef] [PubMed]
12. Gray, M.; Johnson, M.G.; Dragila, M.I.; Kleber, M. Water uptake in biochars: The roles of porosity and hydrophobicity. *Biomass Bioenergy* **2014**, *61*, 196–205. [CrossRef]
13. Hardie, M.; Clothier, B.; Bound, S.; Oliver, G.; Close, D. Does biochar influence soil physical properties and soil water availability? *Plant Soil* **2014**, *376*, 347–361. [CrossRef]
14. Liu, X.; Zhang, A.; Ji, C.; Joseph, S.; Bian, R.; Li, L.; Pan, G.; Paz-Ferreiro, J. Biochar's effect on crop productivity and the dependence on experimental conditions-a meta-analysis of literature data. *Plant Soil* **2013**, *373*, 583–594. [CrossRef]
15. Glaser, B.; Wiedner, K.; Seelig, S.; Schmidt, H.P.; Gerber, H. Biochar organic fertilizers from natural resources as substitute for mineral fertilizers. *Agron. Sustain. Dev.* **2015**, *35*, 667–678. [CrossRef]
16. Lehmann, J.; Rillig, M.C.; Thies, J.; Masiello, C.A.; Hockaday, W.C.; Crowley, D. Biochar effects on soil biota—A review. *Soil Biol. Biochem.* **2011**, *43*, 1812–1836. [CrossRef]
17. Major, J.; Rondon, M.; Molina, D.; Riha, S.J.; Lehmann, J. Maize yield and nutrition during 4 years after biochar application to a *Colombian savanna* oxisol. *Plant Soil* **2010**, *333*, 117–128. [CrossRef]
18. Liang, B.; Lehmann, J.; Solomon, D.; Kinyangi, J.; Grossman, J.; O'Neill, B.; Skjemstad, J.O.; Thies, J.; Luizão, F.J.; Petersen, J.; et al. Black Carbon Increases Cation Exchange Capacity in Soils. *Soil Sci. Soc. Am. J.* **2006**, *70*, 1719–1730. [CrossRef]
19. Crane-Droesch, A.; Abiven, S.; Jeffery, S.; Torn, M.S. Heterogeneous global crop yield response to biochar: A meta-regression analysis. *Environ. Res. Lett.* **2013**, *8*. [CrossRef]
20. Jeffery, S.; Abalos, D.; Prodana, M.; Bastos, A.C.; Van Groenigen, J.W.; Hungate, B.A.; Verheijen, F. Biochar boosts tropical but not temperate crop yields. *Environ. Res. Lett.* **2017**, *12*. [CrossRef]
21. Güereña, D.; Lehmann, J.; Hanley, K.; Enders, A.; Hyland, C.; Riha, S. Nitrogen dynamics following field application of biochar in a temperate North American maize-based production system. *Plant Soil* **2013**, *365*, 239–254. [CrossRef]
22. Van Zwieten, L.; Kimber, S.; Morris, S.; Chan, K.Y.; Downie, A.; Rust, J.; Joseph, S.; Cowie, A. Effects of biochar from slow pyrolysis of papermill waste on agronomic performance and soil fertility. *Plant Soil* **2010**, *327*, 235–246. [CrossRef]
23. Hernandez-Soriano, M.C.; Kerré, B.; Goos, P.; Hardy, B.; Dufey, J.; Smolders, E. Long-term effect of biochar on the stabilization of recent carbon: Soils with historical inputs of charcoal. *Gcb Bioenergy* **2016**, *8*, 371–381. [CrossRef]
24. Mastrolonardo, G.; Calderaro, C.; Cocozza, C.; Hardy, B.; Dufey, J.; Cornelis, J.T. Long-term effect of charcoal accumulation in hearth soils on tree growth and nutrient cycling. *Front. Environ. Sci.* **2019**, *7*, 1–15. [CrossRef]
25. Mikan, C.J.; Abrams, M.D. Mechanisms Inhibiting the Forest Development of Historic Charcoal Hearths in Southeastern Pennsylvania. *Can. J. For. Res.* **1996**, *26*, 1893–1898. [CrossRef]

26. Heidarian Dehkordi, R.; Denis, A.; Fouche, J.; Burgeon, V.; Cornelis, J.T.; Tychon, B.; Placencia, E.; Meersmans, J. Remotely-sensed assessment of the impact of century-old biochar on chicory crop growth using high-resolution UAV-based imagery. *Int. J. Appl. Earth Obs. Geoinf.* **2020**, *91*, 102147. [CrossRef]
27. Aragon, B.; Houborg, R.; Tu, K.; Fisher, J.B.; McCabe, M. Cubesats enable high spatiotemporal retrievals of crop-water use for precision agriculture. *Remote Sens.* **2018**, *10*, 1867. [CrossRef]
28. Segarra, J.; Buchaillot, M.L.; Araus, J.L.; Kefauver, S.C. Remote sensing for precision agriculture: Sentinel-2 improved features and applications. *Agronomy* **2020**, *10*, 641. [CrossRef]
29. Matese, A.; Toscano, P.; Di Gennaro, S.F.; Genesio, L.; Vaccari, F.P.; Primicerio, J.; Belli, C.; Zaldei, A.; Bianconi, R.; Gioli, B. Intercomparison of UAV, aircraft and satellite remote sensing platforms for precision viticulture. *Remote Sens.* **2015**, *7*, 2971–2990. [CrossRef]
30. Berni, J.A.J.; Zarco-Tejada, P.J.; Sepulcre-Cantó, G.; Fereres, E.; Villalobos, F. Mapping canopy conductance and CWSI in olive orchards using high resolution thermal remote sensing imagery. *Remote Sens. Environ.* **2009**, *113*, 2380–2388. [CrossRef]
31. Gago, J.; Douthe, C.; Coopman, R.E.; Gallego, P.P.; Ribas-Carbo, M.; Flexas, J.; Escalona, J.; Medrano, H. UAVs challenge to assess water stress for sustainable agriculture. *Agric. Water Manag.* **2015**, *153*, 9–19. [CrossRef]
32. Xia, T.; Kustas, W.P.; Anderson, M.C.; Alfieri, J.G.; Gao, F.; McKee, L.; Prueger, J.H.; Geli, H.M.E.; Neale, C.M.U.; Sanchez, L.; et al. Mapping evapotranspiration with high-resolution aircraft imagery over vineyards using one-and two-source modeling schemes. *Hydrol. Earth Syst. Sci.* **2016**, *20*, 1523–1545. [CrossRef]
33. Roosjen, P.P.J.; Suomalainen, J.M.; Bartholomeus, H.M.; Kooistra, L.; Clevers, J.G.P.W. Mapping reflectance anisotropy of a potato canopy using aerial images acquired with an unmanned aerial vehicle. *Remote Sens.* **2017**, *9*, 417. [CrossRef]
34. Zhang, C.; Kovacs, J.M. The application of small unmanned aerial systems for precision agriculture: A review. *Precis. Agric.* **2012**, *13*, 693–712. [CrossRef]
35. Yuan, L.; Huang, Y.; Loraamm, R.W.; Nie, C.; Wang, J.; Zhang, J. Spectral analysis of winter wheat leaves for detection and differentiation of diseases and insects. *Field Crop. Res.* **2014**, *156*, 199–207. [CrossRef]
36. Schils, R.; Olesen, J.E.; Kersebaum, K.C.; Rijk, B.; Oberforster, M.; Kalyada, V.; Khitrykau, M.; Gobin, A.; Kirchev, H.; Manolova, V.; et al. Cereal yield gaps across Europe. *Eur. J. Agron.* **2018**, *101*, 109–120. [CrossRef]
37. Su, J.; Liu, C.; Coombes, M.; Hu, X.; Wang, C.; Xu, X.; Li, Q.; Guo, L.; Chen, W.H. Wheat yellow rust monitoring by learning from multispectral UAV aerial imagery. *Comput. Electron. Agric.* **2018**, *155*, 157–166. [CrossRef]
38. Franke, J.; Menz, G.; Oerke, E.-C.; Rascher, U. Comparison of multi-and hyperspectral imaging data of leaf rust infected wheat plants. In *Remote Sensing for Agriculture, Ecosystems, and Hydrology VII*; SPIE: Bellingham, WA, USA, 2005; Volume 5976, p. 59761D.
39. Franceschini, M.H.D.; Bartholomeus, H.; van Apeldoorn, D.F.; Suomalainen, J.; Kooistra, L. Feasibility of unmanned aerial vehicle optical imagery for early detection and severity assessment of late blight in Potato. *Remote Sens.* **2019**, *11*, 224. [CrossRef]
40. Baxter, S. World Reference Base for Soil Resources. In *World Soil Resources Report 103. Rome: Food and Agriculture Organization of the United Nations*; Cambridge University Press: Cambridge, UK, 2007; p. 132. [CrossRef]
41. Witzenberger, A.; Lancashire, P. Phenological growth stages and BBCH-identification keys of cereals. In *Growth Stages Mono- and Dicotyledonous Plants BBCH Monograph*; Federal Biological Research Centre for Agriculture and Forestry: Berlin/Braunschweig, Germany, 2001; pp. 14–18. [CrossRef]
42. Clevers, J.G.P.W.; Van Leeuwen, H.J.; Sensing, R.; Verhoef, W. Estimanting apar by means of vegetation indeces: A sensitivity analysis. In *XXIX ISPRS Congress Technical Commission VII: Interpretation of Photographic and Remote Sensing Data*; ISPRS: Christian Heipke, Germany, 1989; pp. 691–698.
43. Bouman, B.A.M.; Van Kasteren, H.W.J.; Uenk, D. Standard relations to estimate ground cover and LAI of agricultural crops from reflectance measurements. *Eur. J. Agron.* **1992**, *1*, 249–262. [CrossRef]
44. Clevers, J.G.P.W.; Kooistra, L.; van den Brande, M.M.M. Using Sentinel-2 data for retrieving LAI and leaf and canopy chlorophyll content of a potato crop. *Remote Sens.* **2017**, *9*, 405. [CrossRef]
45. Siegmann, B.; Jarmer, T.; Lilienthal, H.; Richter, N.; Selige, T.; Höfled, B. Comparison of narrow band vegetation indices and empirical models from hyperspectral remote sensing data for the assessment of wheat nitrogen concentration. In Proceedings of the 8th EARSeL Workshop on Imaging Spectroscopy, Nantes, France, 8–10 April 2013; pp. 1–2.

46. Kooistra, L.; Clevers, J.G.P.W. Estimating potato leaf chlorophyll content using ratio vegetation indices. *Remote Sens. Lett.* **2016**, *7*, 611–620. [CrossRef]
47. Gao, X.; Huete, A.R.; Ni, W.; Miura, T. Optical-biophysical relationships of vegetation spectra without background contamination. *Remote Sens. Environ.* **2000**, *74*, 609–620. [CrossRef]
48. Gitelson, A.A.; Gritz, Y.; Merzlyak, M.N. Relationships between leaf chlorophyll content and spectral reflectance and algorithms for non-destructive chlorophyll assessment in higher plant leaves. *J. Plant Physiol.* **2003**, *160*, 271–282. [CrossRef] [PubMed]
49. Fitzgerald, G.J.; Rodriguez, D.; Christensen, L.K.; Belford, R.; Sadras, V.O.; Clarke, T.R. Spectral and thermal sensing for nitrogen and water status in rainfed and irrigated wheat environments. *Precis. Agric.* **2006**, *7*, 233–248. [CrossRef]
50. Venkateswaran, K.; Kasthuri, N.; Balakrishnan, K.; Prakash, K. Performance Analysis of K-Means Clustering For Remotely Sensed Images. *Int. J. Comput. Appl.* **2013**, *84*, 23–27. [CrossRef]
51. Pearson, K. On lines and planes of closest fit to systems of points in space. *Lond. Edinb. Dublin Philos. Mag. J. Sci.* **1901**, *2*, 559–572. [CrossRef]
52. Zhang, M.; Liu, X.; O'Neill, M. Spectral discrimination of Phytophthora infestants infection on tomatoes based on principal component and cluster analyses. *Int. J. Remote Sens.* **2002**, *23*, 1095–1107. [CrossRef]
53. Das, S.; Singh, T.P. Correlation analysis between biomass and spectral vegetation indices of forest ecosystem. *Int. J. Eng. Res. Technol.* **2012**, *1*, 1–13.
54. Rouse, J.W., Jr.; Hass, R.H.; Schell, J.A.; Deering, D.W.; Harlan, J.C. *Monitoring the Vernal Advancement and Retrogradation (Green Wave Effect) of Natural Vegetation*; Texas a&m University Remote Sensing Center: College Station, TX, USA, 1974.
55. Clevers, J.G.P.W. The application of a weighted infrared-red vegetation index for estimating leaf area index by correcting for soil moisture. *Remote Sens. Environ.* **1989**, *29*, 25–37. [CrossRef]
56. Rondeaux, G.; Steven, M.; Baret, F. Optimization of soil-adjusted vegetation indices. *Remote Sens. Environ.* **1996**, *55*, 95–107. [CrossRef]
57. Vincini, M.; Frazzi, E.; D'Alessio, P. A broad-band leaf chlorophyll vegetation index at the canopy scale. *Precis. Agric.* **2008**, *9*, 303–319. [CrossRef]
58. Huete, A.; Didan, K.; Miura, T.; Rodriguez, E.P.; Gao, X.; Ferreira, L.G. Overview of the radiometric and biophysical performance of the MODIS vegetation indices. *Remote Sens. Environ.* **2002**, *83*, 195–213. [CrossRef]
59. Kim, T.K. T test as a parametric statistic. *Korean J. Anesthesiol.* **2015**, *68*, 540. [CrossRef] [PubMed]
60. Kendall, A.M.G. A New Measure of Rank Correlation. *Biometrika* **1938**, *30*, 81–93. [CrossRef]
61. Ten Harkel, J.; Bartholomeus, H.; Kooistra, L. Biomass and Crop Height Estimation of Different Crops Using UAV-Based Lidar. *Remote Sens.* **2020**, *12*, 17. [CrossRef]
62. Carter, S.; Shackley, S.; Sohi, S.; Suy, T.; Haefele, S. The Impact of Biochar Application on Soil Properties and Plant Growth of Pot Grown Lettuce (*Lactuca sativa*) and Cabbage (*Brassica chinensis*). *Agronomy* **2013**, *3*, 404–418. [CrossRef]
63. Schulz, H.; Dunst, G.; Glaser, B. Positive effects of composted biochar on plant growth and soil fertility. *Agron. Sustain. Dev.* **2013**, *33*, 817–827. [CrossRef]
64. Parece, T.E.; Campbell, J.B. *Advances in Watershed Science and Assessment*; Springer: Berlin, Germany, 2015; Volume 33.

Article

A Plant-by-Plant Method to Identify and Treat Cotton Root Rot Based on UAV Remote Sensing

Tianyi Wang [1,*], J. Alex Thomasson [1,2], Thomas Isakeit [3], Chenghai Yang [4] and Robert L. Nichols [5]

1 Department of Biological and Agricultural Engineering, Texas A&M University, College Station, TX 77843, USA; athomasson@abe.msstate.edu
2 Department of Agricultural and Biological Engineering, Mississippi State University, Starkville, MS 39762, USA
3 Department of Plant Pathology and Microbiology, Texas A&M University, College Station, TX 77843, USA; t-isakeit@tamu.edu
4 USDA-Agricultural Research Service, Aerial Application Technology Research Unit, 3103 F&B Road, College Station, TX 77845, USA; chenghai.yang@usda.gov
5 Cotton Incorporated, Agricultural & Environmental Research, 6399 Weston Parkway, Cary, NC 27513, USA; bnichols@cottoninc.com
* Correspondence: timothywangty@tamu.edu; Tel.: +1-814-441-5300

Received: 25 June 2020; Accepted: 28 July 2020; Published: 30 July 2020

Abstract: Cotton root rot (CRR), caused by the fungus *Phymatotrichopsis omnivora*, is a destructive cotton disease that mainly affects the crop in Texas. Flutriafol fungicide applied at or soon after planting has been proven effective at protecting cotton plants from being infected by CRR. Previous research has indicated that CRR will reoccur in the same regions of a field as in past years. CRR-infected plants can be detected with aerial remote sensing (RS). As unmanned aerial vehicles (UAVs) have been introduced into agricultural RS, the spatial resolution of farm images has increased significantly, making plant-by-plant (PBP) CRR classification possible. An unsupervised classification algorithm, PBP, based on the Superpixel concept, was developed to delineate CRR-infested areas at roughly the single-plant level. Five-band multispectral data were collected with a UAV to test these methods. The results indicated that the single-plant level classification achieved overall accuracy as high as 95.94%. Compared to regional classifications, PBP classification performed better in overall accuracy, kappa coefficient, errors of commission, and errors of omission. The single-plant fungicide application was also effective in preventing CRR.

Keywords: precision agriculture; UAV; disease detection; cotton root rot; plant-level; single-plant; plant-by-plant; classification; image analysis; machine learning

1. Introduction

The United States (U.S.) produced 20.9 million 218-kg (480-lb) bales of cotton in the 2017–2018 season with a production value of $7.2 billion (USD), ranking 3rd after India and China, and it is the largest cotton-exporting country in the world [1]. The state of Texas produced 9.5 million bales, approximately 44% of U.S. cotton production, ranking 1st in the U.S. [1]. While Texas is by far the largest producing state, a major obstacle to cotton production in Texas is a disease called cotton root rot (CRR) or Texas root rot. The disease is caused by the soilborne fungus, *Phymatotrichopsis omnivora*, a destructive plant disease throughout the southwestern U.S. The first documented study of CRR was in the 19th century by Pammel [2]. The disease rots the root, disrupting the vascular system and preventing water and nutrients from being transported from the roots to the rest of the plant, eventually killing the plant. An infected cotton plant usually dies within 10 days. If the disease develops in an early stage of growth, the plants will die before bearing fruit. If it develops after flowering, the disease will reduce yield and lower the quality of the cotton lint by stopping transport of nutrients to the maturing bolls.

The fungus spreads within a field by direct root contact between plants and the growth of its mycelia through the soil [3]. Infested field areas are commonly circular in shape [4], providing an indication of the cause of plant death. One study reported that the overall area of plants infected with CRR in a particular field increased from 10% to 50% from August to September [4], during a later growth stage of the crop. Until the recent advent of soil-applied flutriafol fungicide, several control practices (e.g., crop rotation, soil fumigation, host resistance) were tested but found to be either not economical or not effective [4]. As CRR generally occurs at the same place in a field from year to year, its position can be mapped, allowing for targeted spot treatments of flutriafol fungicide. Both multispectral and hyperspectral imagery can distinguish infected areas accurately, but three-band multispectral is a preferred technique, being effective, less expensive, more widely available, and simpler to use than hyperspectral [5].

The large sizes of typical cotton fields make it impractical to map CRR position from ground-based platforms. Remote sensing, on the other hand, can provide such data quickly and relatively inexpensively, and it is thus an important technology for practical CRR sensing [5]. Visible, near-infrared (NIR), and thermal remote sensing data have been studied widely with aerial and satellite platforms to understand many crop phenomena [6]. Biotic and abiotic stresses, including insects, pathogens, weeds, drought, and nutrition deficiencies have been widely studied with remote sensing [7–12], in addition to applications such as yield prediction and general crop management [13–16]. In cotton, remote sensing has been used to evaluate the effectiveness of defoliation and regrowth control strategies [17] as well as in other applications.

Remote sensing has also been used to map CRR in cotton fields dating back to 1929 [18]. Nixon et al. later introduced color-infrared (CIR) technology to document the distribution of CRR infestation and to detect the effect of chemical treatment for CRR [19]. Multispectral video imagery of CRR was evaluated and reported as early as 1987 [20], and Yang et al. reported using manned-aircraft based remote sensing and high precision global positioning system (GPS) technology to map CRR in 2015 [21].

Satellite images typically have relatively low resolution but can be acquired periodically at a reasonable cost. However, there is a risk that clouds may cover the view when the images are taken. Aerial images commonly have higher spatial resolution that may be advantageous as well as some flexibility in timing, but the cost of acquisition is relatively high [22]. Previous research indicates that airborne and satellite multispectral imagery data can be used to successfully detect the CRR-infested area in both dryland and irrigated fields [23–26]. The resolution of such imagery limits CRR mapping to field zones, but unmanned aerial vehicles (UAVs) have recently emerged as remote-sensing tools that provide resolution high enough to potentially enable even single-plant level prescription map creation. Compared to manned aircraft, UAVs have a limited payload capacity, but they can fly lower and slower than manned aircraft. The above-ground level (AGL) with UAVs is commonly 20–120 m, providing for spatial resolution at the cm level. The temporal resolution of UAVs is commonly improved as well [27] because UAVs can be flown any time the weather permits. In addition, UAV flights generally cost less than traditional manned aircraft remote sensing.

Rotary-wing UAVs are much more common than fixed-wing UAVs for agricultural remote sensing. Rotary-wing UAVs are slower and more stable in flight and thus are able to generate higher-quality mosaicked images. However, the slower flying speed also generally leads to smaller coverage area. On the other hand, fixed-wing UAVs flying at the legally allowable limit of 120 m AGL are commonly able to cover a 24–40 ha (60–100 acre) area on one battery charge (about 20 min), depending on the weather conditions. Therefore, fixed-wing UAVs appear to be well-suited to the large-area farming that is commonly seen in cotton production.

UAV remote sensing has been increasingly used in agricultural research in recent years and has been considered for general production management, yield prediction, and disease detection [16,17,28–32]. RGB (red, green, and blue) and other multispectral sensors are commonly used, but the frequently used normalized difference vegetation index (NDVI) requires the NIR band. Without the NIR band, the normalized difference photosynthetic vigor ratio (NDPVR) and the visible atmospherically resistant index (VARI) can be calculated, and they have been used to estimate crop yield [28,29]. Both RGB and other multispectral images have been used for rice growth and yield estimation. However, the vegetation indices (VIs) derived from multispectral images including NIR correlate better with grain yield than VIs derived from RGB images [29]. Albetis et al. used the support vector machine (SVM) classifier on UAV images to differentiate diseased from non-diseased areas of vineyards. Due to the high spatial resolution, they could distinguish grapevine vegetation from bare soil, shadow, and inter-row vegetation. A high classification accuracy of 97–99% was achieved in four vineyards [30]. Furthermore, artificial neural networks (ANNs) have been used to estimate water potential in vineyards based on UAV data [31]. While a great deal of recent agricultural research has involved UAV-based remote sensing, there is scant research about UAV-based remote sensing for the delineation of CRR.

In general, classification procedures are a way to categorize data according to various characteristics. Classification in RS means categorizing or mapping an image into different classes depending on the features of the data such as tone, texture, pattern, etc. Unsupervised and supervised classification are common and differ according to whether human-guided training is involved in classifying the data.

The Superpixel algorithm segments images into many multi-pixel pieces (superpixels) based on shape, color, texture, etc. In essence, the Superpixel method converts images from pixel-level to district-level, and thus belongs to the image segmentation category in image processing. The Superpixel algorithm keeps the main features of the aggregated pixels, resulting in a sharp reduction in the number of data-containing units. As a result, it improves the image processing speed significantly. Sultani et al. used the Superpixel algorithm to detect objects in pavement images [33]. Different shaped objects such as patches, maintenance hole covers, and markers could be detected efficiently. After dividing the images into many small segments, features like histogram of oriented gradients (HOG), co-occurrence matrix (COOC), intensity histogram (IH), and mean intensity (MI) of each superpixel were calculated. HOG and COOC are texture and shape characteristics, while IH and MI are spectral intensity variations. Then, SVM was used to generate classifications based on each feature. The Superpixel algorithm has also been used to detect disease in agricultural crops. Zhang et al. developed a new method based on the Superpixel algorithm to detect cucumber diseases [34,35]. Leaf images were divided into superpixels, and then the expectation maximization (EM) method was applied to obtain lesion images. After feature extraction, SVM was used to detect the disease. The result indicated that the proposed method had the highest recognition rate and fastest processing speed compared to four other methods that have been used for cucumber disease recognition. Zhang et al. later proposed a new leaf recognition method based on the Superpixel algorithm, k-means, and pyramid of histograms of orientation gradients (PHOG) algorithms [34,35]. First, the RGB leaf image was divided into segments with the Superpixel algorithm. Then, k-means clustering was applied to segment the lesion section of the leaf. PHOG features were extracted and used to recognize the disease. Three apple and three cucumber leaf diseases were used to assess the method. The result indicated that the proposed method was effective and usually achieved the highest recognition rate compared to other methods that had been used for cucumber disease recognition.

Conventional CRR identification methods developed for 1-m resolution aerial images can only detect the CRR-infested area at the regional level, leading to the application of a large amount of fungicide to field areas that do not need it. UAV remote sensing makes high-resolution data collection possible, meaning that fungicide treatments could conceivably be applied at the level of individual plants. To take advantage of these high-resolution data, a novel high-precision CRR identification method is proposed to enable high-precision CRR detection and treatment. The objectives of this research were thus to (1) develop and evaluate a plant-by-plant (PBP) CRR detection and classification method; (2) compare the PBP classification method to common regional classification methods; (3) examine the effectiveness of PBP fungicide treatment to validate the necessity for the method.

2. Materials and Methods

2.1. Study Sites

This study involved five selected regions in four dryland cotton fields (Figure 1) in central Texas, USA, near the town of Thrall: Chase Section 1 (CH1), Chase Section 2 (CH2), West Poncho (WP), School House (SH), and a field for a plot test (PL) to examine the effectiveness of various fungicide treatments. All of these fields and regions have a history of CRR infestation.

Figure 1. The study was conducted on Stiles farm located in Williamson County, central Texas. The five areas of field experiments are marked on the map.

2.2. Data Collection

A fixed-wing UAV (Tuffwing Mapper, Tuffwing LLC, Boerne, TX, USA; Figure 2a) was used to acquire image data of all the fields and regions on a cloud-free day, August 20, 2017. This UAV is equipped by the manufacturer with a global navigation satellite system (GNSS) receiver and an inertial measurement unit (IMU). A multispectral camera (RedEdge, Micasense, Seattle, WA, USA; Figure 2b) mounted on the UAV collected images at 120-m above ground level (AGL). The images had a pixel resolution of 7.64 cm and contained five spectral bands: blue (≈475–500 nm), green (≈550–565 nm), red (≈665–675 nm), red edge (≈715–725 nm), and NIR (≈825–860 nm). The images were taken between 11:00 and 13:00 local time with an optimized fixed-exposure. Eight ground control points (GCPs) were used during each flight to improve the geographical accuracy of the mosaicked image of each field area. The GCPs were placed in each field at the four corners and four midpoints of each side. Ground-truth data were collected on August 25 2017, and involved using a GPS receiver to record boundary locations of some CRR zones (Figure 3). A total of about 20 plants from these zones were removed and evaluated for the presence of the fungus on the roots in order to validate the presence of CRR in the zone.

2.3. Data Preprocessing

A 0.95-ha area was covered in each image with the AGL and camera used. An 80% forward overlap and 70% side overlap flight plan was used for image acquisition. The raw images were collected in tiff format with data from the GNSS receiver and IMU stored as image metadata. Image mosaicking was conducted with Pix4D software (Pix4D S.A, Lausanne, Switzerland). A Geoexplorer 6000 (Trimble, Sunnyvale, CA, USA) GNSS receiver was used in the fields to collect the coordinates of the GCP centroids for geo-referencing of the images, also conducted in Pix4D. The centers of the GCPs in each raw image were manually linked to the corresponding ground-truth GNSS coordinates during the geo-referencing process.

Three spectrally flat reference tiles were used for radiometric calibration: dark gray (≈3% reflectance) medium gray (≈20% reflectance), and light gray (≈45% reflectance). The actual reflectance spectra of the calibration tiles were collected with a portable spectroradiometer (Figure 4) (PSR+ 3500, Spectral Evolution, Haverhill, MA, USA). A linear relationship between known reflectance and image-pixel digital numbers (DNs) was established for each band. All DN images were converted to reflectance images in ENVI software (Harris geospatial solution, Boulder, CO, USA) based on these relationships.

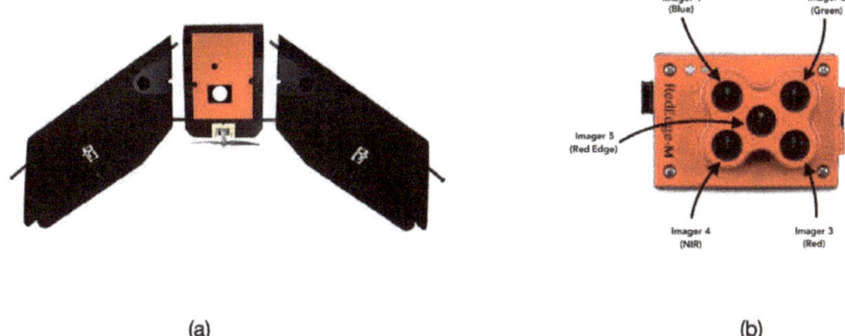

Figure 2. The images were captured from (**a**) a fixed-wing unmanned aerial vehicle (UAV) "TuffWing UAV Mapper" with (**b**) MicaSense RedEdge camera.

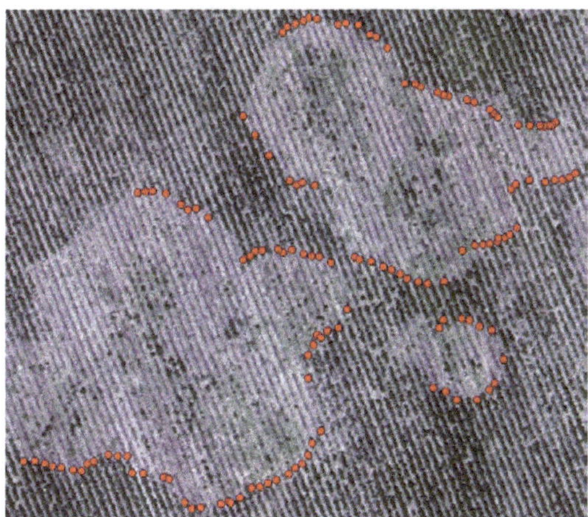

Figure 3. A portion of ground-truth data collected on August 25 2017. Totally, 627 ground-truth data points (in red) were recorded to delineate the boundary of some cotton root rot (CRR)-infested zones.

Figure 4. The PSR+ 3500 Spectroradiometer was used to collect reflectance data.

2.4. Regional Classification

Previous studies have used regional classification for CRR detection [4,5,21,24,36,37]. In a related prior study involving UAV remote sensing of CRR [38], each CRR-infested zone was identified as a region of plants rather than individual plants. The image data were classified into healthy and CRR-infested regions with unsupervised, semi-supervised, and supervised classification methods. The unsupervised and semi-supervised methods were based on k-means clustering and included one two-class method (unsupervised) and three multi-class (3, 5, and 10 classes) methods that combined more classes to form two classes based on user knowledge and judgment (semi-supervised). The supervised methods, which required selection of training data by a human operator, were two-class methods and included support vector machine (SVM), minimum distance (MD), maximum likelihood (ML), and Mahalanobis distance (MHD) classifiers. The CIR images of four of the five regions described above (the other region in the current study was used for a fungicide application test as described below) were used to evaluate the classification methods by comparing their overall accuracies, errors of omission and commission, and kappa coefficients. These images covered 5.68, 0.15, 0.42, and 0.34 ha of field regions CH1, CH2, SH, and WP, respectively (Figure 5). In the current study, the regional classification results from the related prior study were used for comparison with plant-by-plant classification.

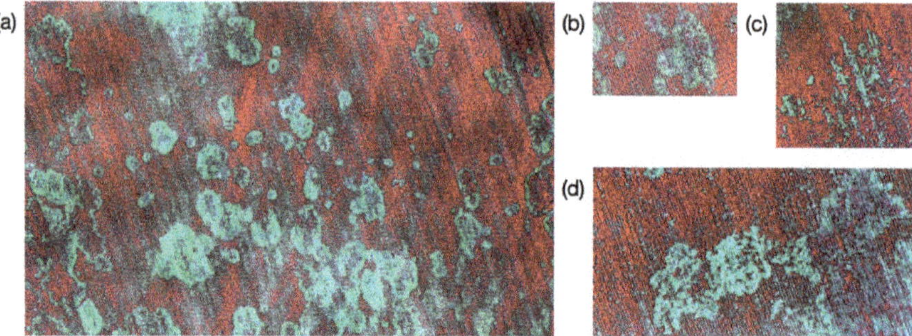

Figure 5. Multispectral color-infrared (CIR) images for (**a**) Region 'CH1' (Scale 1:3000), (**b**) Region 'CH2' (Scale 1:2000), (**c**) Region 'WP' (Scale 1:2400), and (**d**) Region 'SH' (Scale 1:3800). The regions are shown in different scales for a better visualization of details.

2.5. Plant-by-Plant Classification

The high spatial resolution of UAV RS images makes it possible to detect CRR infection at a plant-by-plant (PBP) level of precision. The 7.64-cm resolution in the current study provides for roughly 120 pixels per plant zone at full canopy cover, assuming 76-cm (30-in.) row spacing and average seeding distance of 12 cm (4.6 in.) at 45,000 seeds per 0.405 ha (1.0 ac.). This number of pixels per plant should be adequate for identifying specific features to enable discrimination of plants based on spectral and spatial information. A new PBP classification method was thus proposed and based on Superpixel and k-means algorithms. This combination of algorithms was selected because the Superpixel algorithm, used appropriately, should have the capability to identify single plant zones, and k-means has been demonstrated to distinguish CRR-infected from healthy plants.

The simple linear iterative clustering (SLIC) Superpixel algorithm [39,40] is based on visual color converted to the three-dimensional (3D) spherical CIE-Lab color space. CIE-Lab expresses colors in numeric terms and deals with the issue that colorimetric distance in measurements does not correspond with the color difference perceived by humans. In this study, CIR images were used based on the fact that healthy plants are more easily visually differentiated from unhealthy plants with CIR instead of visual color images. In CIR images, the image spectral bands are converted from green to blue, red to green, and NIR to red for visual display. Other researchers have used NIR to enhance the SLIC Superpixel algorithm results based on RGB [41], but in this study the SLIC Superpixel algorithm was used on the RGB-channel outputs of the CIR images, so CIR was directly converted to an artificial CIE-Lab by way of a CIR-based XYZ color space.

Figure 6 is the flowchart of the PBP classification algorithm. The SLIC Superpixel algorithm was first applied to a CIR image, and a number (k) of superpixel "seeds" were then generated and distributed uniformly across the image. The imported image was then divided into superpixels (small, rather homogenous, areas in the image) based on spectral and shape information around each "seed." The mean of the DN values within each superpixel was calculated and assigned to the superpixel so that it had a single DN value. The number (k) of superpixels was user-determined and provided to the algorithm based on the field planting rate; i.e., the number of superpixels was expected to be similar to the number of cotton plants in the image, multiplied by a scaling factor to account for bare soil areas in the image. K-means clustering was applied to the superpixel image to achieve a two-class regional classification, with "1" and "0" to represent CRR and healthy zones, respectively. At the same time, planting rows were detected by calculating the gradient of the raw image. A binary plant row image was generated, with "0" and "1" representing plant rows and the gaps between them, respectively. The centroids of each superpixel were identified and marked as potential cotton plant centroids, with "1" and "0" representing cotton plant and bare soil, respectively. The centroids of superpixels located in the CRR zone (as determined by k-means) and the planting row were regarded as locations of CRR-infected plants. The centroids of superpixels located in healthy zones and within the planting row were regarded as locations of healthy plants.

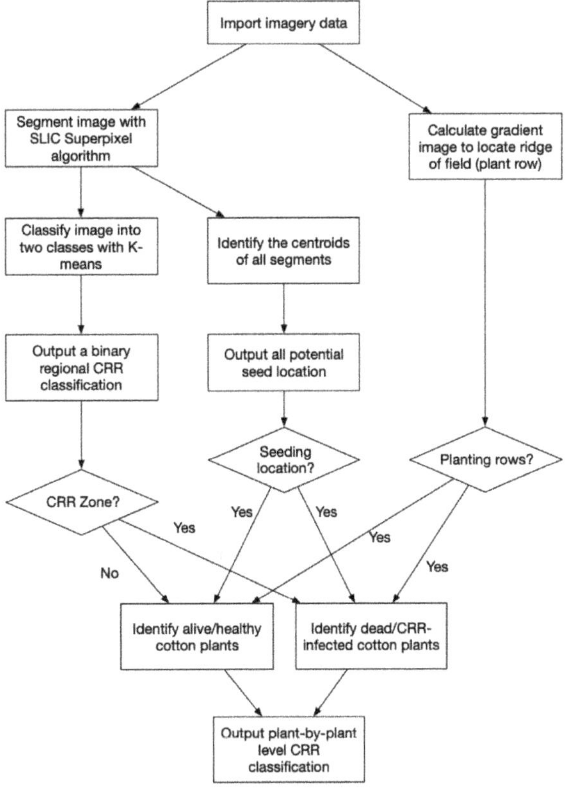

Figure 6. Flow chart of unsupervised plant-by-plant (PBP) classification algorithm. CRR = cotton root rot; SLIC = simple linear iterative clustering.

The classifier logic was applied to all image pixels and can be expressed with the following equation:

$$C = \begin{cases} 0 | \ S_n \cap P_n = 1, \ Z_n = 0 \\ 1 | \ S_n \cap P_n = 1, \ Z_n = 1 \\ 2 | \ \textit{other pixels} \end{cases} \quad (1)$$

where Z_n is the regional classification based on k-means, in which value "0" = healthy zone, and value "1" = CRR zone; S_n is the status of superpixel centroid location or not superpixel centroid location, in which value "0" = not superpixel centroid location, and value "1" = superpixel centroid location; P_n is the status of planting row or gap between rows, in which value "0" = gap, and value "1" = row; and $C = \{\psi|\psi : \mathbb{R} \rightarrow \{0, 1, 2\}\}$ is an overall class containing all pixel classes.

While C = 0, the superpixel centroid location is marked as an individual healthy cotton plant. While C = 1, the superpixel centroid location is marked as a CRR-infected plant. C = 2 represents no superpixel centroid at this location, no matter whether the pixel is classified as healthy or infested.

2.6. Accuracy Assessment

Accuracy assessment involves specific means of evaluating the performance of classifications [42,43]. A ground-truth map was drawn manually on the original high-resolution UAV images. In this process, the ground-truth data collected on August 25 2017, were used as a visual reference when applying the following protocol. Zones in the field with more than approximately 10 immediately adjacent infected plants were categorized as CRR-infested zones. In larger CRR-infested zones, more than 10 immediately adjacent healthy plants were categorized as healthy zones. The regional classification maps were resampled to the higher resolution of the ground-truth map and compared to it on a pixel-by-pixel basis. This method is common for accuracy assessment of raster-based classification maps. On the other hand, the PBP classification maps, which were vector point maps, were compared to the ground-truth map at only the locations of the superpixel centroids; i.e., the classification of a superpixel was compared to the pixel at its centroid location on the ground-truth map. This method was selected to enable comparison to the ground-truth map at the plant level instead of the pixel level. It should be noted that the number of comparisons between a regional classification and the ground-truth map was much higher than the number for PBP classification. However, the methods used are considered reasonable for the type of data being evaluated; e.g., the regional classification maps did not have adequate resolution for plant-level comparison. Confusion matrices were developed based on the individual comparisons within these zones (Table 1).

Table 1. Confusion matrix to evaluate classification methods.

		Class Types Determined from Reference Source (Ground-Truth)		
Class types determined from classified map		Healthy plant	Infested plant	Totals
	Healthy plant	A	B	A + B
	Infested plant	C	D	C + D
	Totals	A + C	B + D	A + B + C + D

The kappa coefficient, which indicates the agreement between the "predicted" and "true" values, was calculated from the derived confusion matrices and the following formula:

$$k = \frac{N \sum_{i=1}^{n} t_{i,i} - \sum_{i=1}^{n} (G_i P_i)}{N^2 - \sum_{i=1}^{n} (G_i P_i)} \quad (2)$$

where N is the total number of pixels; i is the class number; $t_{i,i}$ is the correctly classified number of pixels in Class i; G_i is the total number of pixels classified as Class i in ground-truth data; and P_i is the total number of pixels classified as Class i in the predicted data.

A kappa value of 1 indicates that the classification has perfect agreement with the true value, and a value of 0 indicates no agreement between the classification and ground truth (Figure 7). The errors of commission, representing a measure of false-positives, and errors of omission, representing a measure of false-negatives, were also calculated to evaluate the classifiers. Regional classification methods including k-means, SVM, MD, ML, and MHD were compared to the PBP classification method based on overall classification accuracy, the kappa coefficient, and errors of commission and omission.

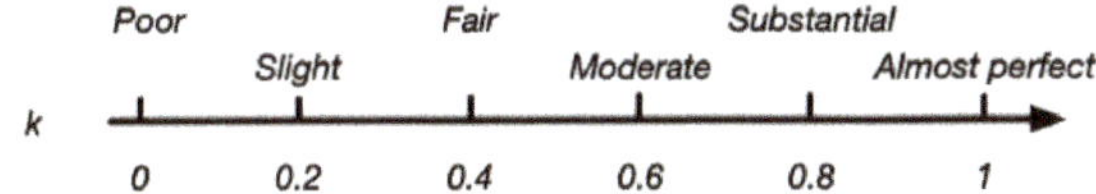

Figure 7. The interpretation of the kappa coefficient.

2.7. Test of PBP Fungicide Treatment in the Field

An in-furrow, at-planting spray application is the most common way to apply the Topguard Terra (FMC Agricultural Solutions, Philadelphia, PA, USA) fungicide (flutriafol) that is licensed for treatment of CRR. The continuous application of the fungicide over the top of seeds as they are planted treats not only soil close to the seed, but also a length of soil between seeds that may not need treatment. This process may result in applying more product than necessary, so it is important to determine whether applying the fungicide to individual seeds or plants is effective. To test whether PBP fungicide treatment is effective in protecting cotton plants from CRR infection, a stem-drench treatment—also proven in research trials to be an effective application method—was used in place of the at-planting application method. Specifically, the fungicide spray solution was applied to the stem of the cotton plant and a small amount of soil surrounding the stem. An 18.3 × 30.5 m (60 × 100 ft) test plot in field region PL was used to conduct this experiment.

Four treatments were applied: (1) a conventional at-planting treatment of in-furrow continuous spray over the top of planted seeds, applied with a tractor-pulled planter; (2) a stem-drench continuous spray, applied manually with a backpack sprayer; (3) a stem-drench pulsed spray on individual plants, applied manually with a backpack sprayer; (4) a no-fungicide control. The experiment had 24 rows (Figure 8) of 100 ft (30.48 m) in length, with an adjacent pair of rows receiving a treatment, and three replications per treatment, arranged in a randomized complete block design. For the two continuous-spray treatments, Topguard Terra was applied at 0.675 g active ingredient in 216 mL per 30.5 m (100 ft) of row length. This rate is equivalent to the labeled application rate of 237 mL (8 fluid oz.) formulation in 37.9 L (10 gal.) of water applied to 0.405 ha (1.0 ac.).

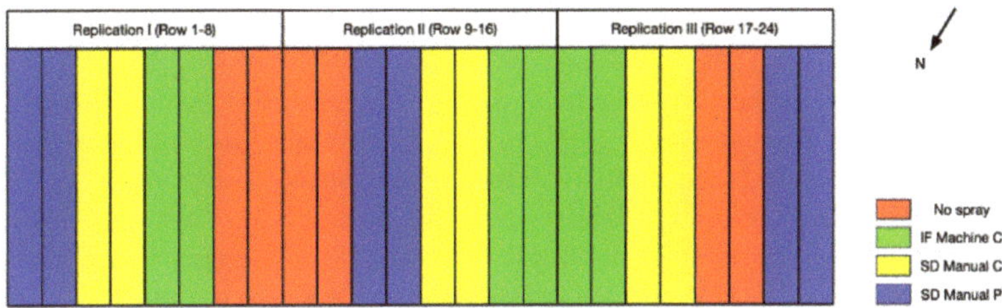

Figure 8. Plot design for testing the effectiveness of fungicide on plant-by-plant treatment. Note: IF Machine C = in-furrow machinery continuous spray, SD manual C = stem-drench manual continuous spray, and SD Manual P = stem-drench manual pulsed spray.

A CASE IH 1230 12-row Early Riser planter (Case Corp., Wisconsin, USA) with 76.2 cm (30-inch) row spacing was used for seeding of all treatments and the application of the in-furrow fungicide treatment (Figure 9a). The seeding rate of Phytogen 490 cotton seed was roughly 45,000 per 0.405 ha

(1.0 ac.). For the manual continuous-spray and pulsed stem-drench treatments, a pressurized CO_2 backpack sprayer was used (Figure 9b) with a pressure of 241 kPa (35.0 psi). Plots were planted and the in-furrow treatment was applied at planting on May 10, 2018. The soil temperature was 34 °C (94 °F).

(a)

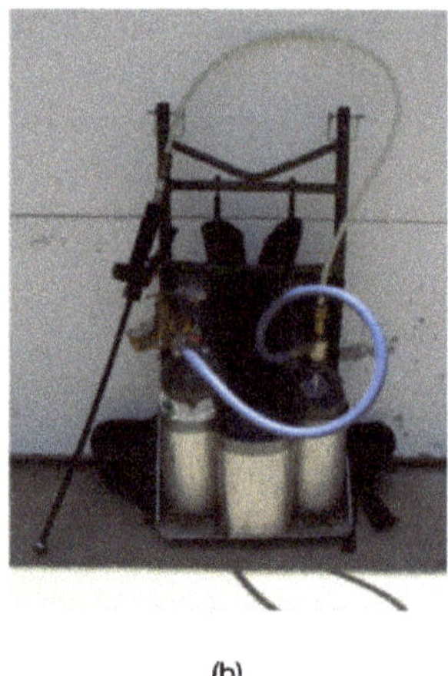

(b)

Figure 9. (**a**) A Case IH 1230 planter was used for in-furrow application; and (**b**) a CO_2-pressurized backpack sprayer was used for manual treatments.

The manual continuous and pulsed stem-drench spray treatments were applied when the cotton plants were at the four true leaf growth stage, on June 6, 2018. Spray was applied close to the stems at a height of 2–3 cm above the ground. With the pulse treatment, the volume of water containing the fungicide applied to the rows varied depending upon plant stand and spacing, with an average of 268 mL per 30.5 m (100 ft) of row; the application range was 160–450 mL. In other words, 1.67 mL of flutriafol (active ingredient) was used on average per 30.5-m-length of cotton plant row, and the actual flutriafol application range was 1.00–2.80 mL. For comparison, in-furrow planting consumed 1.68 mL flutriafol per 30.5-length of row. The experiment was repeated in 2019 with a reduction in fungicide application rate. An average of 122 mL of fungicide (0.67 mL flutriafol) was used per row for stem drench pulse spray treatment in 2019 (range: 60–130 mL).

3. Results

3.1. Plant-by-Plant Classification

One example (Region CH2) of the progression of the image processing results from each step in the PBP classification method is shown in Figure 10. The raw image (Figure 10a) gradient was calculated to identify the planting rows (Figure 10b). SLIC Superpixel segmentation was applied to the gradient map to determine possible locations of individual plants (Figure 10c). The original pixels of the raw image were aggregated into larger superpixels (Figure 10d) to identify individual plant locations. The k-means algorithm was applied to the superpixels to generate a two-class regional classification (Figure 10e). The final result of PBP classification is shown in Figure 10f, in which each individual healthy plant is marked with a yellow point, and each CRR-infected plant is marked with a blue point.

The accuracy assessment of PBP classification showed that it is a highly accurate method of differentiating between healthy and CRR-infected plants at the individual-plant level. In Region CH2, the PBP classification had the highest overall accuracy of 95.94%, as well as the highest kappa coefficient of 0.8617, which indicated very strong agreement between classification and ground-truth data (Table 2). Table 2 is the confusion matrix for PBP classification applied to Region CH2. Over 11,000 plants identified by the PBP algorithm in CH2 were evaluated, and about 82% of those were identified as healthy according to ground-truth data. About 13.1% of the healthy plants were misclassified as CRR-infected (overclassification), while about 9.5% of the actually CRR-infected plants were misclassified as healthy (underclassification).

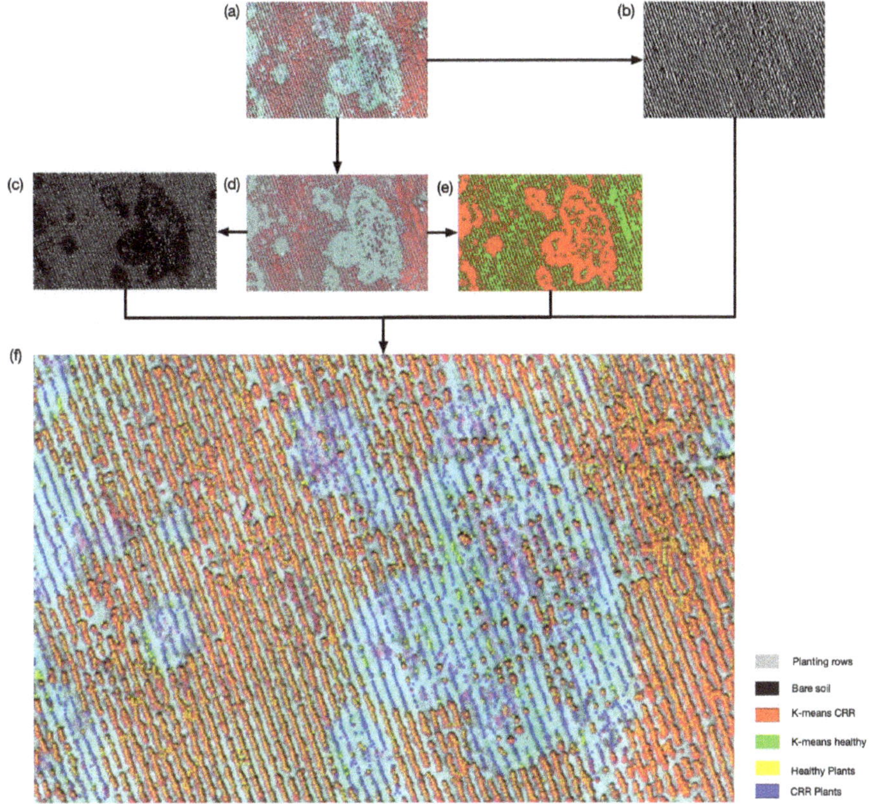

Figure 10. The imagery results getting from each step of unsupervised PBP classification algorithm. each image indicated: (**a**) CIR raw image, (**b**) the location of planting row, (**c**) the position of each individual plant, (**d**) the result of Superpixel segmentation, (**e**) the regional distribution of CRR-infested areas, and (**f**) the final result of PBP classification. In (**b**), planting rows are shown in light grey, and bare soil is shown in black. In (**e**), healthy regions are shown in green, and CRR-infested regions are shown in red. In (**f**), each yellow dot represents a healthy cotton plant, and each blue dot represents and a CRR-infected plant.

Table 2. Confusion matrix of PBP classification for Region CH2.

		Class Types Determined from Reference Source (Ground-Truth)			Commission	Omission
Class types determined from classified map		Healthy plant	Infected plant	Totals		
	Healthy plant	8963	186	9149	186/9149 2.03%	268/9231 2.9%
	Infected plant	268	1771	2039	268/2039 13.14%	186/1957 9.5%
	Totals	9231	1957	11,188		

The PBP classifications were also generated for CH1, WP, and SH. In CH1, the PBP algorithm achieved an overall accuracy of 93.5%, a kappa coefficient of 0.7848, an error of commission of 16.1%, and an error of omission of 18.6%. In SH, the PBP algorithm also had a high accuracy of 90.6% with a kappa coefficient of 0.7494. The errors of commission and omission were 12.2% and 8.5%, respectively. Compared to the other regions, WP had the lowest accuracy of 88.4%, but even this level would typically be acceptable for field application. The kappa coefficient, error of commission, and error of omission for WP were 0.6048, 20.9%, and 26.8%, respectively.

3.2. Comparison to Regional Classifications

Thirty-six confusion matrices were generated from the results of the nine overall classification methods as applied to the four field regions. These confusion matrices are summarized in Table 3. The two-class k-means classifier identified CRR-infected cotton plants in the image automatically, but the overall accuracy averaged only 77.5%. The kappa coefficient of 0.491 also indicated relatively weak agreement between the classification and ground-truth data. The error of commission of 46% indicated that almost half the plants classified as CRR-infected were overclassified. Manually combining three-class, five-class, and 10-class k-means classifications improved the overall accuracy to 83.5%, 84.4%, and 84.1%, respectively. The kappa coefficients also increased to 0.547, 0.552, and 0.576, respectively, indicating moderate agreement between classification and ground truth. However, it must be noted that combining classes required expertise from and implementation by the user, meaning that the ideal of automated processing was not realized.

Using supervised classifiers including SVM, MD, ML, and MHD increased the overall accuracy to 86.3%, 85.7%, 86.5%, and 87.7%, respectively. All of these classifiers performed significantly better ($\alpha = 0.05$) than two-class k-means in overall accuracy. The kappa coefficients for these classifiers were 0.659, 0.636, 0.667, and 0.786, respectively. While supervised classifiers performed better than the unsupervised and semi-supervised classifiers, they did not perform as well as the PBP classifier, and it must be noted that these also need human intervention, specifically for selection of training data based on subjective judgment.

The PBP classification method averaged 92.1% overall accuracy, by far the best among all classifiers considered. This accuracy level was significantly higher ($\alpha = 0.05$) than that of the unsupervised and combined unsupervised classifiers. The average kappa coefficient was 0.786, indicating strong agreement between the classifications and ground truth, and this value was significantly better ($\alpha = 0.05$) than that of all the other classifiers considered. The average errors of commission and omission, 15.56% and 15.85%, were also the lowest in the overall comparison group.

A comparison chart of the errors of commission and omission is shown in Figure 11. Theoretically, the ideal classifier, which has 100% accuracy and thus no errors of commission or omission, should be located at the origin of this coordinate system. The PBP classifier is the one closest to the origin by far, indicating that it clearly performed the best in terms of overall accuracy. It is worth noting here that in the aforementioned related prior study [38], two methods proposed to take advantage of the high resolution of UAV images, k-means plus support vector machine (KMSVM) and k-means segmentation (KMSEG), were evaluated with only regions CH1, WP, and SH. The two methods had approximately 22% and 16% error of commission and 18% and 11% of error of omission, respectively. The KMSEG classifier, which is a fully automated regional classifier, generated similar results to the PBP classifier on a somewhat different data set. Both KMSEG and KMSVM are regional classifiers that were designed to take advantage of the morphological information available in high-resolution UAV images. Thus, like the PBP classifier, they were meant as improvements over traditional regional classifiers. The added advantage of the PBP classifier is that it is designed to classify individual plants, a tremendous advantage when subsequently applying fungicide on a PBP basis.

Table 3. The accuracy comparison between unsupervised, combined-unsupervised, supervised classifications, and proposed automatic regional classifications.

		Overall Accuracy (%)						Kappa Coefficient					
		CH1	CH2	WP	SH	Mean	Std. Dev.	CH1	CH2	WP	SH	Mean	Std. Dev.
U	2-class KM	71.11	78.60	78.76	81.44	77.48 [a]	4.42	0.3826	0.5106	0.4527	0.6162	0.4905 [a]	0.0988
S-S	3 to 2-class KM	78.28	88.89	87.26	79.45	83.47 [ab]	5.38	0.3875	0.6868	0.5751	0.5392	0.5471 [ab]	0.1236
	5 to 2-class KM	76.50	88.67	88.81	83.49	84.37 [ab]	5.80	0.4232	0.6085	0.5293	0.6452	0.5516 [ab]	0.0983
	10 to 2-class KM	76.36	90.97	88.01	81.14	84.12 [ab]	6.61	0.4264	0.6986	0.5885	0.5911	0.5762 [ab]	0.1122
S	SVM	87.04	92.02	78.66	87.48	86.30 [bc]	5.57	0.6962	0.7587	0.4481	0.7345	0.6594 [ab]	0.1432
	MD	85.65	88.12	86.14	82.79	85.68 [bc]	2.20	0.6721	0.6753	0.5604	0.6346	0.6356 [ab]	0.0534
	ML	88.55	91.71	77.92	87.65	86.46 [bc]	5.95	0.7342	0.7498	0.4419	0.7422	0.6670 [ab]	0.1502
	MHD	87.89	89.60	87.13	86.27	87.72 [bc]	1.42	0.7238	0.7076	0.5764	0.7144	0.6806 [bc]	0.0698
P	PBP	93.52	95.94	88.43	90.64	92.13 [c]	3.28	0.7848	0.8617	0.6048	0.7494	0.7855 [c]	0.0746
		Errors of Commission (%)						**Errors of Omission (%)**					
		CH1	CH2	WP	SH	Mean	Std. Dev.	CH1	CH2	WP	SH	Mean	Std. Dev.
U	2-class KM	49.89	50.43	56.88	27.16	46.09 [a]	13.01	26.72	7.84	14.10	18.00	16.67 [a]	7.90
S-S	3 to 2-class KM	26.46	30.04	40.17	17.43	28.53 [ab]	9.40	60.87	17.22	28.28	41.43	36.95 [b]	18.77
	5 to 2-class KM	40.18	14.26	24.23	19.36	24.51 [ab]	11.21	42.32	44.58	51.63	25.30	40.96 [b]	11.17
	10 to 2-class KM	40.83	10.85	37.51	21.20	27.60 [ab]	14.08	40.57	34.95	29.83	30.73	34.02 [b]	4.90
S	SVM	25.38	18.50	57.07	16.18	29.28 [ab]	18.93	16.21	19.66	15.04	16.69	16.90 [a]	1.97
	MD	29.17	32.90	43.70	22.15	31.98 [ab]	9.00	14.13	14.48	24.71	23.19	19.13 [a]	5.60
	ML	23.85	19.38	57.88	18.53	29.91 [ab]	18.79	11.92	20.17	13.23	12.39	14.43 [a]	3.87
	MHD	26.13	28.74	40.79	20.73	29.10 [ab]	8.48	9.91	15.18	26.86	13.25	16.30 [a]	7.37
P	PBP	16.06	13.14	20.88	12.16	15.56 [b]	3.91	18.59	9.5	26.80	8.51	15.85 [a]	8.60

Note: U = unsupervised, S-S = semi-supervised, S = supervised, P = proposed method, KM = k-means, MD = minimum distance, ML = maximum likelihood, MHD = Mahalanobis distance, PBP = plant-by-plant. Letters a, b, and c in Column Mean indicate statistical different groups ($\alpha = 0.05$, Duncan test).

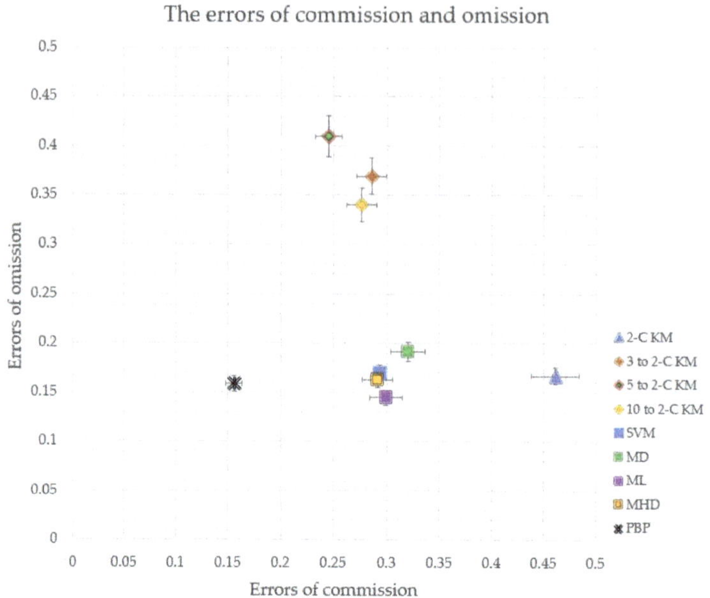

Figure 11. Comparison of the errors of commission and omission among classifiers.

3.3. Test of Method of Fungicide Application for CRR Control

In the study on fungicide application methods, the application of Topguard Terra generally reduced the incidence of CRR compared to the control (Table 4), as expected. The manually pulsed stem-drench treatment had the lowest plant mortality among the treatments, but the difference from the other two fungicide treatments was not significant. While in most years, the portion of the field (PL) used in this experiment eventually approaches 100% mortality from CRR, the dry weather in 2018, the first year this study was conducted, resulted in low severity of CRR. Assuming a 5% significance level, there was no statistically significant difference among all the treatments. However, the pulsed stem-drench (i.e., PBP) treatment would be considered significantly better than the no-spray control treatment if a 15% level of significance were assumed. While this is an uncommonly weak significance level, it is reasonable to believe that all three methods of applying fungicide, including the PBP method, offered some protection against CRR. Due to even drier conditions in 2019 than 2018, no CRR development was observed during the experiment, so efficacy could not be assessed in 2019.

Table 4. Effect of Topguard Terra application method on CRR (% mortality) in the 2018 experiment.

Method *	Replication 1	Replication 2	Replication 3	Mean **
No spray	4.79	0.32	7.69	4.27 [a]
In-furrow	3.92	0.57	0.98	1.83 [ab]
Stem Drench C	0.00	2.45	2.17	1.54 [ab]
Stem Drench P	0.62	1.42	1.05	1.03 [b]

* C = continuous application, P = pulse application. ** The different letters indicate statistically different groups ($\alpha = 0.15$, Duncan test).

4. Discussion

In this study, the errors of commission represent the percentage of plants over-classified into the CRR category. The errors of omission represent the percentage of CRR-infected plants misclassified as healthy plants. From an economics perspective, omission plays a more important role than commission for CRR detection, because over-spraying of fungicide caused by over-classification would likely cost less than the loss of CRR-infected plants that could have been protected. The zones with weeds growing on bare soil, very possibly next to a dead cotton plant, contributed to the errors of omission. While not necessarily critical, it should be noted that the errors of commission were commonly observed at zones where bare soil was evident where there was no CRR-infected plant. The mixed pixels of soil, plant leaves, and shadow of plants, which were commonly present at the boundaries between healthy and infested zones, also caused a large number of errors of commission and omission with the regional classifiers, because the mixed pixels do not represent the reflectance information from a single object.

The homogeneity of the field could also affect the classification accuracy. Regions CH1 and CH2 were from the same cotton field, and the images of them were from the same flight mission. The patterns of planting and disease as well as the reflectance information were similar, and the lighting conditions were the same. The main difference between the regions was that CH2 was smaller than CH1. The results of classification in CH2 were better than in CH1 in most cases, especially for unsupervised classification. One reason is that unsupervised classification clusters data into different classes based on the "otherness" of data. Once the sample size becomes larger, more diverse data besides healthy and infected cotton are introduced into the field of view, such as a concrete road, power line, pond, or other objects. All of this "noise" can reduce the accuracy of classifiers. A prerequisite for an accurate automated classification is to have images consisting of only rows of cotton plants. All the classifiers had relatively low accuracies on Region WP compared to the other three regions. It is possible this was because the planter experienced mis-seeding during planting, causing a narrow and long "dead zone" consisting mainly of bare soil. Manual manipulation can be used as postprocessing to correct the misclassification from mis-seeding, but it violates the intent of automation in classification. Morphological image processing tools such as erosion and dilation could be introduced in the future to improve the performance of differentiating mis-seeding from cotton root rot while maintaining the automation of classification [38].

Comparing regional and PBP classifications is challenging because regional classification is based on pixels while PBP classification is based on individual plants. To make the comparison even more convincing, the classifications should ideally be evaluated with the same protocols. Comparing pixel differences between all classifications (PBP and regional classifications) and the same ground-truth map is a fair way to evaluate and compare classifications, but it is not readily done when the classifiers produce different types of maps as results. The PBP classifier output a vector point map, whereas the regional classifiers output raster maps, so the comparisons to the ground-truth map had to be done with different methods appropriate to each form of data.

In the comparison of fungicide application methods, the results should be validated by further study, as the disease pressure was low in both years due to dry weather conditions. However, the experiment did substantially support the concept of pulse application to reduce fungicide use. In 2018, the manual application of the pulse spray method was inefficient and actually resulted in the application of a greater amount of fungicide than the continuous spray, but in 2019, an average of 43% less fungicide was used because of improvements in the application technique. If efficacy of the fungicide holds up with that method and application rate, the overall concept of PBP detection, mapping, and fungicide application will be validated.

Considering computational requirements, the PBP algorithm required more computing time than the regional classification methods, because segmenting and locating seed positions is computationally intensive. For the 0.15-ha CH2 image, about 30 s was required to generate the classification on a 2016 Macbook Pro computer with an Intel i7-6920HQ central processing unit (CPU) and Radeon Pro 460 graphic processing unit (GPU). The PBP classification algorithm is slower than other conventional regional classification methods, but it is still acceptably fast. While a larger field might require a few hours to complete the classification, these classifications do not need to be done in real time. Rather, they can be performed between growing seasons.

As discussed previously, UAVs provide much higher resolution (decimeter level to centimeter-level) remote sensing data than manned aircraft or satellites (meter level). This study intended to explore how to make use of the high-resolution data in CRR detection in the creation of PBP prescription maps. The application of fungicide at the PBP level is clearly possible from a technological standpoint. Wilkerson et al. tested a seed-specific in-furrow fungicide application system and found that the system could achieve as high as 95% accuracy for seed-specific treatment in cotton [44]. However, identifying, predicting, locating, and treating the disease at the PBP level are the obstacles to high-precision treatment of CRR, as well as other plant diseases. This study shows that CRR-infected plants in the current season can be individually identified with high accuracy, and PBP fungicide treatment appears to be effective in controlling CRR. The remaining challenges to be investigated are (1) whether the precise location of individual CRR-infected plants is predictive for the following year, and (2) whether previously developed precision-spray technology [44] enables fungicide to be practically applied at these locations on a seed-by-seed basis.

There is no evidence to suggest that CRR can be cured once a plant is infected; the fungicide must be applied prior to disease development. Thus, PBP application of fungicide requires previous years' data to predict CRR-infested areas in future years. The entire PBP classification process can be conducted automatically if an appropriate seeding rate is known beforehand. Management of other crop diseases that can be treated during the growing season could potentially benefit from this type of high-resolution classifier.

5. Conclusions

This study involved development and evaluation of a PBP classifier that is able to detect CRR-infected plants at the single-plant level automatically. The PBP classifier is mainly based on the Superpixel segmentation and k-means clustering algorithms. Eight conventional regional and PBP classification methods, based on UAV remote sensing image mosaics of cotton, were evaluated in four field sections with a history of CRR. Among the regional classification methods, the unsupervised two-class k-means classifier achieved an overall accuracy of 77.48%, lower than the semi-supervised (83.47–84.12%) and supervised (85.68–87.72%) classifiers but requiring less human involvement. Compared to these conventional regional classification methods, the PBP method achieved the highest overall accuracy of 92.1%, the highest kappa coefficient of 0.786, the lowest errors of commission of 15.6%, and the second-lowest errors of omission of 15.9%. Furthermore, the PBP method is able to classify the image mosaics automatically. The PBP-based fungicide treatment in the field appeared to be effective in controlling CRR infection. These results generally validate the idea of plant-level CRR treatment and suggest the likelihood of major advances in high-resolution precision agriculture practices in the future.

Author Contributions: Conceptualization, T.W. and J.A.T.; methodology, T.W., J.A.T., T.I. and C.Y.; software, T.W.; validation, T.W., J.A.T., T.I. and C.Y.; formal analysis, T.W.; investigation, T.W.; resources, J.A.T.; data curation, T.W.; writing—original draft preparation, T.W.; writing—review and editing, T.W., J.A.T., C.Y., T.I. and R.L.N.; visualization, T.W.; funding acquisition, R.L.N. All authors have read and agreed to the published version of the manuscript.

Funding: This research was funded by Cotton Incorporated, Cooperative Research Agreement number 16-233.

Acknowledgments: We thank Cody Bagnall, Lantian Shangguan, Xiongzhe Han, Xiwei Wang, and Roy Graves for helping in data collection. We thank Ryan M. Collett for helping in selection of survey fields.

Conflicts of Interest: The authors declare no conflict of interest.

References

1. USDA-NASS Agricultural Statistics. 2017. Available online: https://quickstats.nass.usda.gov (accessed on 30 July 2020).
2. Pammel, L.H. Root rot of cotton or "cotton blight". *Tex. Agric. Exp. Stn. Bull.* **1888**, *4*, 450–465.
3. Streets, R.B.; Bloss, H.E. Phymatotrichum root rot. *Phytopathol. Monogr.* **1973**, *8*, 1–38.

4. Yang, C.; Odvody, G.N.; Fernandez, C.J.; Landivar, J.A.; Minzenmayer, R.R.; Nichols, R.L.; Thomasson, J.A. Monitoring cotton root rot progression within a growing season using airborne multispectral imagery. *J. Cotton Sci.* **2014**, *93*, 85–93.
5. Yang, C.; Everitt, J.H.; Fernandez, C.J. Comparison of airborne multispectral and hyperspectral imagery for mapping cotton root rot. *Biosyst. Eng.* **2010**, *107*, 131–139. [CrossRef]
6. Mahlein, A.K.; Oerke, E.C.; Steiner, U.; Dehne, H.W. Recent advances in sensing plant diseases for precision crop protection. *Eur. J. Plant Pathol.* **2012**, *133*, 197–209. [CrossRef]
7. Gogoi, N.K.; Deka, B.; Bora, L.C. Remote sensing and its use in detection and monitoring plant diseases: A review. *Agric. Rev.* **2018**, *39*, 307–313. [CrossRef]
8. Qin, Z.; Zhang, M. Detection of rice sheath blight for in-season disease management using multispectral remote sensing. *Int. J. Appl. Earth Obs. Geoinf.* **2005**, *7*, 115–128. [CrossRef]
9. Chen, X.; Ma, J.; Qiao, H.; Cheng, D.; Xu, Y.; Zhao, Y. Detecting infestation of take-all disease in wheat using landsat thematic mapper imagery. *Int. J. Remote Sens.* **2007**, *28*, 5183–5189. [CrossRef]
10. Barton, C.V.M. Advances in remote sensing of plant stress. *Plant Soil* **2012**, *354*, 41–44. [CrossRef]
11. Hazaymeh, K.; Hassan, Q.K. Remote sensing of agricultural drought monitoring: A state of art review. *AIMS Environ. Sci.* **2016**, *3*, 604–630. [CrossRef]
12. Gerhards, M.; Schlerf, M.; Mallick, K.; Udelhoven, T. Challenges and future perspectives of multi-/Hyperspectral thermal infrared remote sensing for crop water-stress detection: A review. *Remote Sens.* **2019**, *11*, 1240. [CrossRef]
13. Cook, C.G.; Escobar, D.E.; Everitt, J.H.; Cavazos, I.; Robinson, A.F.; Davis, M.R. Utilizing airborne video imagery in kenaf management and production. *Ind. Crops Prod.* **1999**, *9*, 205–210. [CrossRef]
14. Calvão, T.; Pessoa, M.F. Remote sensing in food production—A review. *Emir. J. Food Agric.* **2015**, *27*, 138–151. [CrossRef]
15. Campos, I.; González-Gómez, L.; Villodre, J.; Calera, M.; Campoy, J.; Jiménez, N.; Plaza, C.; Sánchez-Prieto, S.; Calera, A. Mapping within-field variability in wheat yield and biomass using remote sensing vegetation indices. *Precis. Agric.* **2019**, *20*, 214–236. [CrossRef]
16. Matese, A.; Toscano, P.; Di Gennaro, S.F.; Genesio, L.; Vaccari, F.P.; Primicerio, J.; Belli, C.; Zaldei, A.; Bianconi, R.; Gioli, B. Intercomparison of UAV, aircraft and satellite remote sensing platforms for precision viticulture. *Remote Sens.* **2015**, *7*, 2971–2990. [CrossRef]
17. Yang, C.; Greenberg, S.M.; Everitt, J.H.; Fernandez, C.J. Assessing cotton defoliation, regrowth control and root rot infection using remote sensing technology. *Int. J. Agric. Biol. Eng.* **2011**, *4*, 1–11.
18. Taubenhaus, J.J.; Ezekiel, W.N.; Neblette, C.B. Airplane photography in the study of cotton root rot. *Phytopathology* **1929**, *19*, 1025–1029.
19. Nixon, P.R.; Lyda, S.D.; Heilman, M.D.; Bowen, R.L. Incidence and control of cotton root rot observed with color infrared photography. *MP Tex. Agric. Exp. Stn.* **1975**, *1241*, 4.
20. Nixon, P.R.; Escobar, D.E.; Bowen, R.L. A multispectral false-color video imaging system for remote sensing applications. In Proceedings of the 11th Biennial Workshop on Color Aerial Photography and Videography in the Plant Sciences and Related Fields, Weslaco, TX, USA, 1 September 1987; pp. 295–305.
21. Yang, C.; Odvody, G.N.; Fernandez, C.J.; Landivar, J.A.; Minzenmayer, R.R.; Nichols, R.L. Evaluating unsupervised and supervised image classification methods for mapping cotton root rot. *Precis. Agric.* **2015**, *16*, 201–215. [CrossRef]
22. Gogineni, S.; Thomasson, J.A.; Iqbal, J.; Wooten, J.R.; Kolla, B.M.; Sui, R. Remote sensing input to GIS-integrated cotton growth model: Preliminary results. In Proceedings of the Optical Science and Technology, the SPIE 49th Annual Meeting, Denver, CO, USA, 2 August 2004; Volume 5544, p. 186. [CrossRef]
23. Yang, C.; Fernandez, C.J.; Everitt, J.H. Mapping phymatotrichum root rot of cotton using airborne three-band digital imagery. *Trans. ASAE* **2005**, *48*, 1619–1626. [CrossRef]
24. Song, X.; Yang, C.; Wu, M.; Zhao, C.; Yang, G.; Hoffmann, W.C.; Huang, W. Evaluation of Sentinel-2A satellite imagery for mapping cotton root rot. *Remote Sens.* **2017**, *9*, 906. [CrossRef]
25. Huang, Y.; Brand, H.J.; Sui, R.; Thomson, S.J.; Furukawa, T.; Ebelhar, M.W. Cotton yield estimation using very high-resolution digital images acquired with a low-cost small unmanned aerial vehicle. *Trans. ASABE* **2016**, *59*, 1563–1574.

26. Yang, C.; Odvody, G.N.; Thomasson, J.A.; Isakeit, T.; Minzenmayer, R.R.; Drake, D.R.; Nichols, R.L. Site-specific management of cotton root rot using airborne and high-resolution satellite imagery and variable-rate technology. *Trans. ASABE* **2018**, *61*, 849–858. [CrossRef]
27. Bagheri, N. Development of a high-resolution aerial remote-sensing system for precision agriculture. *Int. J. Remote Sens.* **2017**, *38*, 2053–2065. [CrossRef]
28. Huang, Y.; Thomson, S.J.; Brand, H.J.; Reddy, K.N. Development and evaluation of low-altitude remote sensing systems for crop production management. *Int. J. Agric. Biol. Eng.* **2016**, *9*, 1–11.
29. Zhou, X.; Zheng, H.B.; Xu, X.Q.; He, J.Y.; Ge, X.K.; Yao, X.; Cheng, T.; Zhu, Y.; Cao, W.X.; Tian, Y.C. Predicting grain yield in rice using multi-temporal vegetation indices from UAV-based multispectral and digital imagery. *ISPRS J. Photogramm. Remote Sens.* **2017**, *130*, 246–255. [CrossRef]
30. Albetis, J.; Duthoit, S.; Guttler, F.; Jacquin, A.; Goulard, M.; Poilvé, H.; Féret, J.B.; Dedieu, G. Detection of Flavescence dorée grapevine disease using unmanned aerial vehicle (UAV) multispectral imagery. *Remote Sens.* **2017**, *9*, 308. [CrossRef]
31. Romero, M.; Luo, Y.; Su, B.; Fuentes, S. Vineyard water status estimation using multispectral imagery from an UAV platform and machine learning algorithms for irrigation scheduling management. *Comput. Electron. Agric.* **2018**, *147*, 109–117. [CrossRef]
32. Mattupalli, C.; Moffet, C.A.; Shah, K.N.; Young, C.A. Supervised classification of RGB Aerial imagery to evaluate the impact of a root rot disease. *Remote Sens.* **2018**, *10*, 917. [CrossRef]
33. Sultani, W.; Mokhtari, S.; Yun, H.B. Automatic pavement object detection using Superpixel segmentation combined with conditional random field. *IEEE Trans. Intell. Transp. Syst.* **2017**, *19*, 2076–2085. [CrossRef]
34. Zhang, S.; Zhu, Y.; You, Z.; Wu, X. Fusion of superpixel, expectation maximization and PHOG for recognizing cucumber diseases. *Comput. Electron. Agric.* **2017**, *140*, 338–347. [CrossRef]
35. Zhang, S.; Wang, H.; Huang, W.; You, Z. Plant diseased leaf segmentation and recognition by fusion of superpixel, K-means and PHOG. *Optik* **2018**, *157*, 866–872. [CrossRef]
36. Yang, C.; Odvody, G.N.; Thomasson, J.A.; Isakeit, T.; Nichols, R.L. Change detection of cotton root rot infection over 10-year intervals using airborne multispectral imagery. *Comput. Electron. Agric.* **2016**, *123*, 154–162. [CrossRef]
37. Song, H.; Yang, C.; Zhang, J.; He, D.; Thomasson, J.A. Combining fuzzy set theory and nonlinear stretching enhancement for unsupervised classification of cotton root rot. *J. Appl. Remote Sens.* **2015**, *9*, 96013. [CrossRef]
38. Wang, T.; Thomasson, J.A.; Yang, C.; Isakeit, T.; Nichols, R.L. Automatic Classification of Cotton Root Rot Disease Based on UAV Remote Sensing. *Remote Sens.* **2020**, *12*, 1310. [CrossRef]
39. Achanta, R.; Shaji, A.; Smith, K.; Lucchi, A.; Fua, P.; Susstrunk, S. *SLIC Superpixels*; EPFL Technical Report No. 149300; EPFL: Lausanne, Switzerland, 2010; Available online: https://infoscience.epfl.ch/record/149300?ln=en (accessed on 30 July 2020).
40. Achanta, R.; Shaji, A.; Smith, K.; Lucchi, A.; Fua, P.; Süsstrunk, S. SLIC Superpixels compared to state-of-the-art superpixel methods. *IEEE Trans. Pattern Anal. Mach. Intell.* **2012**, *34*, 2274–2282. [CrossRef]
41. Haccius, C.; Hariharan, H.P.; Herfet, T.; Hach, T.; Cine, R.; Gmbh, T. Infrared-aided superpixel segmentation. In Proceedings of the 9th International Workshop on Video Processing and Quality Metrics for Consumer Electronics (VPQM), Chandler, AZ, USA, 5–6 February 2015.
42. Congalton, R.G. A review of assessing the accuracy of classifications of remotely sensed data. *Remote Sens. Environ.* **1991**, *37*, 35–46. [CrossRef]
43. Huang, C.; Davis, L.S.; Townshend, J.R.G. An assessment of support vector machines for land cover classification. *Int. J. Remote Sens.* **2002**, *23*, 725–749. [CrossRef]
44. Wilkerson, J.B.; Hancock, J.H.; Moody, F.H.; Newman, M.A. Design of a seed-specific application system for in-furrow chemicals. *Trans. ASAE* **2004**, *47*, 537–645. [CrossRef]

Article

Comparison of Object Detection and Patch-Based Classification Deep Learning Models on Mid- to Late-Season Weed Detection in UAV Imagery

Arun Narenthiran Veeranampalayam Sivakumar [1], Jiating Li [1], Stephen Scott [2], Eric Psota [3], Amit J. Jhala [4], Joe D. Luck [1] and Yeyin Shi [1,*]

1 Department of Biological Systems Engineering, University of Nebraska-Lincoln, Lincoln, NE 68583, USA; arun-narenthiran@huskers.unl.edu (A.N.V.S.); jiatingli@huskers.unl.edu (J.L.); jluck2@unl.edu (J.D.L.)
2 Department of Computer Science and Engineering, University of Nebraska-Lincoln, Lincoln, NE 68588, USA; sscott2@unl.edu
3 Department of Electrical and Computer Engineering, University of Nebraska-Lincoln, Lincoln, NE 68588, USA; epsota@unl.edu
4 Department of Agronomy and Horticulture, University of Nebraska-Lincoln, Lincoln, NE 68583, USA; amit.jhala@unl.edu
* Correspondence: yshi18@unl.edu

Received: 30 April 2020; Accepted: 29 June 2020; Published: 3 July 2020

Abstract: Mid- to late-season weeds that escape from the routine early-season weed management threaten agricultural production by creating a large number of seeds for several future growing seasons. Rapid and accurate detection of weed patches in field is the first step of site-specific weed management. In this study, object detection-based convolutional neural network models were trained and evaluated over low-altitude unmanned aerial vehicle (UAV) imagery for mid- to late-season weed detection in soybean fields. The performance of two object detection models, Faster RCNN and the Single Shot Detector (SSD), were evaluated and compared in terms of weed detection performance using mean Intersection over Union (IoU) and inference speed. It was found that the Faster RCNN model with 200 box proposals had similar good weed detection performance to the SSD model in terms of precision, recall, f1 score, and IoU, as well as a similar inference time. The precision, recall, f1 score and IoU were 0.65, 0.68, 0.66 and 0.85 for Faster RCNN with 200 proposals, and 0.66, 0.68, 0.67 and 0.84 for SSD, respectively. However, the optimal confidence threshold of the SSD model was found to be much lower than that of the Faster RCNN model, which indicated that SSD might have lower generalization performance than Faster RCNN for mid- to late-season weed detection in soybean fields using UAV imagery. The performance of the object detection model was also compared with patch-based CNN model. The Faster RCNN model yielded a better weed detection performance than the patch-based CNN with and without overlap. The inference time of Faster RCNN was similar to patch-based CNN without overlap, but significantly less than patch-based CNN with overlap. Hence, Faster RCNN was found to be the best model in terms of weed detection performance and inference time among the different models compared in this study. This work is important in understanding the potential and identifying the algorithms for an on-farm, near real-time weed detection and management.

Keywords: CNN; Faster RCNN; SSD; Inception v2; patch-based CNN; MobileNet v2; detection performance; inference time

---

1. Introduction

Weeds are unwanted plants that grow in the field and compete with the crops for water, light, nutrients, and space. If uncontrolled, weeds can have several negative consequences, such as crop yield loss, production of a large number of seeds thereby creating a weed seed bank in the field, and contamination of grain during harvesting [1,2]. Traditionally, weed management programs involve the control of weeds through chemical or mechanical means such as the uniform application of herbicides throughout the field. However, the spatial density of weeds is not uniform across the field, thereby leading to overuse of chemicals which results in environmental concerns and evolution of herbicide-resistant weeds. To overcome this issue, a concept of site-specific weed management (SSWM), which refers to detecting weed patches and spot spraying or removal by mechanical means, was proposed in the early 1990s [3–5]. Weed control early in the season is critical, since otherwise the weeds would compete with the crops for resources during the critical growth stage of the crops resulting in possible yield loss [6,7]. Therefore, in addition to the application of pre-emergence herbicides, the early application of post emergence herbicides is preferred for effective weed control and also to reduce the damage to crops. The effectiveness of weed control from post-emergence herbicides depends on the timing of application [8,9]. Detection of early season weeds in an accurate and timely manner helps in the creation of prescription maps for the site-specific application of post-emergence herbicides [10–12]. Prescription maps for post-emergence application can also be created from late-season weeds detected during the previous seasons [13–17]. Compared to early season weeds, late-season weeds do not directly affect the yield of the crop, since it is not competing for resources during the critical growth period of the crop. However, if unattended, late-season weeds can produce large numbers of seeds creating problems in the subsequent growing seasons. Therefore, the detection and control of late-season weeds can be complementary to early season weed control.

Earlier studies on weed detection often used Color Co-occurrence Matrix-based texture analysis for digital images [18,19]. Following this, there were several studies on combining optical sensing, image processing algorithms, and variable rate application implements for real-time site-specific herbicide application on weeds. However, the speed of these systems was limited by computational power constraints for real-time detection, which in turn limited their ability to cover large areas of fields [20]. Unmanned aerial vehicles (UAVs) with their ability to cover large areas in a short amount of time and payload capacity to carry optical sensors provide an alternative. UAVs have been studied for various applications in precision farming such as weed, disease, pest, biotic and abiotic stress detection using high-resolution aerial imagery [21–24]. Several studies have investigated the potential of using remote sensing to discriminate between crops and weeds for weed mapping at different phenological stages and found that results differ based on the phenology [2,10,25–33]. The similar spectral signature of the crops and the weeds, occurrence of weeds as small patches and interference of soil pixels in detection are the major challenges for remote sensing in early season weed detection [2,12]. A common approach is to use vegetation indices to segment the vegetation pixels from the soil pixels, followed by crop row detection for weed classification using techniques such as object-based image analysis (OBIA) and Hough Transform [29,32,34]. However, crop row detection-based approaches cannot detect intra-row weeds. Hence, machine learning based classifiers using features computed from OBIA were used to detect intra-row weeds as well [10]. However, the performance of OBIA is sensitive to the segmentation accuracy and so optimal parameters for the segmentation step in OBIA have to be found for different crops and field conditions [35].

With advancements in parallel computing and the availability of large datasets, convolutional neural networks (CNN) were found to perform very well in computer vision tasks such as classification, prediction, and object detection [36]. In addition to performance, another principal advantage of CNN is that the network learns the features by itself during the training process, and hence manual feature engineering is not necessary. CNNs have been studied for various image-based applications in agriculture such as weed detection, disease detection, fruit counting, crop yield estimation, obstacle detection for autonomous farm machines, and soil moisture content estimation [37–41]. CNNs have been used for weed detection using data obtained in three different ways—using UAVs, using the autonomous ground robot, and high-resolution images obtained manually in the field. A simple CNN binary classifier was trained to classify manually collected small high-resolution images of maize and weeds [42,43]. The performance of the classifier with transfer learning on various pre-trained networks such as LeNet and AlexNet was compared, but this study was limited in variability in the obtained dataset and on the evaluation of the classification approach with large images. Dyrmann et al. [23] used a pre-trained VGG-16 network and replaced the fully connected layer with a deconvolution layer to output a pixel-wise classification map of maize, weeds, and soil. The training images were simulated by overlapping a small number of available images of soil, maize, and weeds with various sizes and orientations. The use of an encoder-decoder architecture for real-time output of pixel-wise classification maps for site-specific spraying was studied. It was found that by adding hand-crafted features such as vegetation indices, different color spaces, and edges as input channels to CNN, the model's ability to generalize to different locations and at the different growth stages of the crop improved [44–46]. Furthermore, to improve the generalization performance of the CNN-based weed detection system, Lottes et al. [25] studied the use of fully-convolutional DenseNet with spatiotemporal fusion and spatiotemporal decoder with sequential images to learn the local geometry of crops in fixed straight lines along the path of a ground robot. In the case of overlapping crop and weed objects, Lottes et al. [15] proposed a key point based feature extraction approach that was used to detect weed objects that overlap with the crop. In addition to weed detection, for effective removal of weeds using mechanical or laser-based methods, it is necessary to detect the stem location of weeds prior to actuation. A fully-convolutional DenseNet was trained to output the stem location as well as a pixel-wise segmentation map of crops and weeds [47,48].

In the case of weed detection using UAV imagery, similar to OBIA approaches mentioned above, dos Santos Ferreira et al. [3] used a Superpixel segmentation algorithm to segment objects and trained a CNN to classify these clusters. They then compared the performance with other machine learning classifiers which use handcrafted features. Sa et al. [27] studied the use of an encoder-decoder architecture, Segnet, for the pixel-wise classification of multispectral imagery and followed up with a performance evaluation of this detection system using different UAV platforms and multispectral cameras [49–51]. Bah et al. [29] used the Hough transform along with a patch-based CNN to detect weeds from UAV imagery and found that overlapping weed and crop objects led to some errors in this approach. It is to be noted that, in this approach, the patches are sliced from the large image in a non-overlapping manner. Huang et al. [30] studied the performance of various deep learning architectures for pixel-wise classification of rice and weeds and found that the fully-convolutional network architecture outperformed other architectures. Yu et al. [52] studied the use of CNN for multispecies weed detection in rye grass.

From the literature reviewed, it can be seen that automated weed detection has been primarily focused on early season weeds, since that is found to be the critical period for weed management and to prevent crop yield loss. However, it should be noted that mid- to late-season weeds that escape from the routine early-season management also threaten production by creating a large number of seeds which creates problems for several future growing seasons. With herbicide resistance, escaped weeds can proliferate and become difficult to manage. Studies on early season weeds can use vegetation segmentation as a preprocessing step to reduce the memory requirements; however, this does not apply to mid- to late-season weed imaging with no soil pixels due to canopy closure. Furthermore, because of the significant overlap between crops and weeds, it is challenging to find the optimal scale and other parameters of segmentation in OBIA to achieve the maximum performance. With deep learning-based object detection methods proving successful for tasks such as fruit counting—another situation with a cluttered background—it is hypothesized that such methods would be able to detect mid- to late-season weeds from UAV imagery. Hence, the objective of this study was to evaluate deep learning-based object detection models on detecting mid- to late-season weeds and compare their performance with patch-based CNN method for near-real time weed detection. Near-real time refers to on-farm processing of the aerial imagery on the edge device as it is collected. We refer to this as near-real time rather than real-time because there is no real time control output generated from the collected imagery and so we refer to near-real time as the completion of processing shortly after completion of data collection. The specific objectives of the study are:

1. Evaluate the performance of two object detection models with different detection performance and inference speed—Faster RCNN and the Single Shot Detector (SSD) models—in detecting mid- to late-season weeds from UAV imagery using precision, recall, f1 score, and mean IoU as the evaluation metrics for their detection performance and inference time as the metric for their speed;
2. Compare the performance of object detection CNN models with the patch-based CNN model in terms of weed detection performance using mean IoU and inference time.

2. Materials and Methods

2.1. Study Site

The study sites were located in the South Central Agricultural Laboratory of the University of Nebraska, Lincoln at Clay Center, NE, USA (40.5751, -98.1309). The two study sites were located adjacent to each other. They were different soybean weed management research plots. Figure 1 shows the stitched maps of the study sites.

2.2. UAV Data Collection

A DJI Matrice 600 pro unmanned aerial vehicle (UAV) platform (Figure 2) was used with a Zenmuse X5R camera to capture aerial imagery. In order to collect data with varying growth stages of the crop as well as variations in illumination conditions, the images from study site 1 (shown at the top in Figure 1) were collected on 2 July 2018 whereas the images from study site 2 (shown at the bottom in Figure 1) were collected on 12 July 2018. The flight altitude in both the cases was 20m above ground level. The Zenmuse X5R camera used is a 16 megapixel camera with 4/3″ sensor and 72 degree diagonal field of view. The dimension of the captured images is 4608 × 3456 pixels in three bands—Red, Green, and Blue. To develop an economical solution, this study focuses on only using RGB imagery. At a 20-m altitude, for the given sensor specifications, the spatial resolution of the output image is 0.5 cm/pixel. DJI Ground Station pro software was used for flight control. Common weed species at the experimental site were waterhemp (Amaranthus tuberculatus), Palmer amaranthus (Amaranthus palmeri), common lambsquarters (Chenopodiam album), velvetleaf (Abutilon theophrasti), and foxtail species such as yellow and green foxtails. The weeds were naturally infesting the crop and were forming patches. The two data collections were performed after 45 to 50 days after soybean planting and 15 to 20 days after post-emergence herbicides were applied in most treatments, except in plots where only pre-emergence herbicides were applied and in non-treated control plots. Soybean was at V6 (six trifoliate stage) to R2 (full flowering) growth stage.

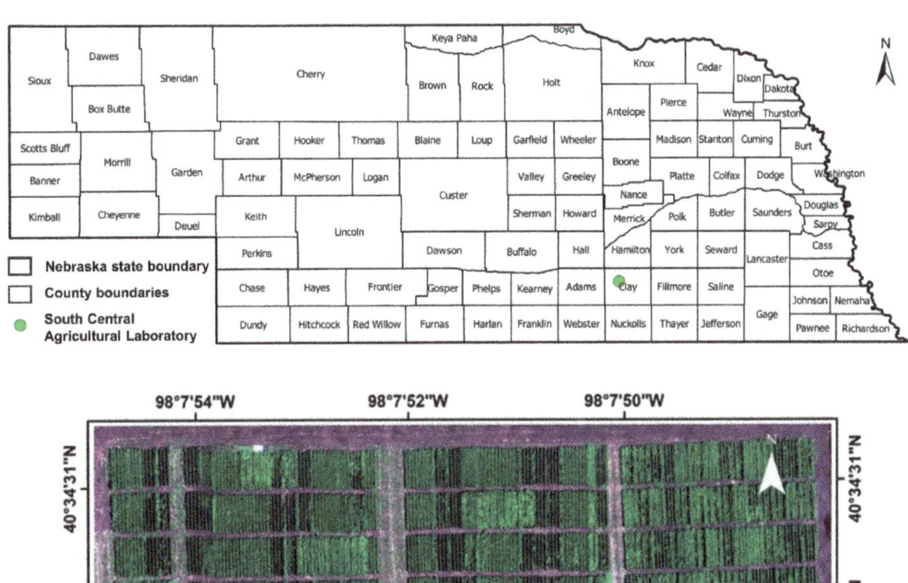

Figure 1. Study area at South Central Ag Laboratory in Clay Center, NE.

Figure 2. DJI Matrice 600 pro UAV platform with Zenmuse X5R camera.

2.3. Data Annotation and Processing

The objective of the study is to develop a weed detection system with on-farm data processing capability. Since the mosaicking of overlapping aerial images is the time-consuming process in the workflow and is not required in this case, overlapping images were removed, and only the non-overlapping raw images were retained. The original dimension of the raw image is too large to fit in the memory for processing so each raw image of size 4608 × 3456 pixels was sliced into 12 sub-images of size 1152 × 1152 pixels. The weed areas in each sub-image were annotated as rectangular bounding boxes using the python labeling tool LabelImg [53]. Only one annotator was involved in the labeling process. The annotator was trained to draw rectangular bounding boxes around weed patches. In case of weed patches of complex shapes, multiple rectangular bounding boxes were drawn to cover such patches. A total of 450 sub-images were annotated manually and were then randomly split into 90% training images and 10% test images

2.4. Patch Based CNN

Convolutional neural networks (CNNs) are feedforward artificial neural networks with the fully connected layers in the input hidden layers replaced with convolutional filters. This reduces the number of filters in each layer and enables CNNs to learn spatial patterns in images and other two-dimensional data. The advantage of a CNN is its ability to learn the features by itself, thereby preventing the need for time-consuming hand engineering of features needed in case of other Computer Vision algorithms. CNN architectures have been proposed, and its use in applications, such as document recognition by using backpropagation for training, has been studied much earlier [54]. However, their applications were limited because of the need for very large datasets to train a large number of parameters in deep networks, and also the computational needs for training. In the last decade, with advancements in parallel processing capabilities using graphical processing units and increases in the availability of large datasets, Krizhevsky et al. [36] showed the potential of CNNs in complex multiclass image classification tasks. However, in most cases, it was found that there were not enough data available to train a deep CNN from scratch. Transfer learning helped overcome this limitation. Transfer learning is the technique of using the weights of pre-trained networks trained on very large datasets such as Alexnet or GoogleNet and retraining them with small datasets for other applications [55]. This has been found to lead to exceptional classification performance and one hypothetical explanation is that the features learned in the initial convolutional layers are global features common across various image classification tasks. Several studies have looked at the application of neural networks for weed detection, such as [28,56].

In this study, a pre-trained network called Mobilenet v2 has been used for transfer learning [57]. Mobilenet v2 was developed primarily for use in mobile devices with limited memory capabilities. Hence, in order to reduce the number of parameters, each convolutional block of Mobilenet v2 consists of an expansion layer with a convolutional kernel of window size 1. This layer increases the number of channels in the input. This is followed by a depthwise convolutional layer which is then followed by a projection layer that consists of a convolutional kernel of window size 1. The depthwise convolution layer applies a single convolutional filter per input channel. The 1 × 1 convolutional layer that follows is called point wise layer. It reduces the number of channels in the output, thereby reducing the number of parameters in the next convolutional block. Hence in each block, feature maps are projected to a high dimensional space followed by learning higher dimensional features in the depthwise convolutional layer which are then encoded using a pointwise convolutional projection layer. The Mobilenet v2 network was trained on the ImageNet dataset containing 1.4 million images belonging to 1000 classes [57]. This network was then fine-tuned using the training patches belonging to both the classes in this study. Initially, for the first 10 epochs, only the classifier layer of the network were trained by freezing the weights of all other layers. This was performed to use the global features learned on the ImageNet dataset and fine-tune the classifier for this specific application. After this, fine-tuning was performed in which all the top layers were unfrozen and to allow the network to adapt to this specific application. The fine tuning was performed for 10 epochs and, hence, the model was only trained for 20 epochs in total [58].

2.5. Object Detection Models

An object in Computer Vision refers to a connected, single element present in the image. Object detection is defined as the problem of finding the class of an object, and also localizing it in the image [59]. Hence, for every object in the image, the model is expected to regress the coordinates of the bounding box of the object in addition to the class probabilities for classification. Two different models have been investigated—Faster RCNN and SSD, both with Inception v2 as a feature extractor. Faster RCNN and SSD were chosen since Faster RCNN was found to have better performance, whereas SSD was found to have better speed [60]. Several different models trained on Imagenet dataset such as Inception v2 [61], Mobilenet v2 [57], Resnet 101 [62], VGG 16 [63] can be used as feature extractors for transfer learning. Of these, Inception v2 and Mobilenet v2 have been found to be the fastest in terms of inference speed [60]. The objective was to develop a weed detection system with on-farm real-time data processing capabilities. Since with similar inference speed, Inception v2 has better performance than Mobilenet v2 for object detection tasks, Inception v2 was chosen as the feature extractor [60].

2.5.1. Faster RCNN

Faster RCNN is a region proposal method-based object detection algorithm. Region-based CNN (R-CNN) was the first region proposal method-based model [64]. However, it was computationally expensive since CNN based feature extraction has to be performed for each proposed region. Fast RCNN was proposed to reduce the computational time by sharing convolutional features across the region proposals [65]. To improve the speed, Faster RCNN was proposed with fully convolutional Region Proposal Networks (RPN) that are trained to propose better object regions [66]. The Faster RCNN model consists of four sections: the feature extractor, the region proposal network, Region of Interest (RoI) pooling, and classification (as shown in Figure 3).

For feature extraction, the convolutional layers from Inception v2 were used. The advantage of the Inception v2 network is its use of wider networks with filters of different kernel sizes in each layer which makes it translation and scale invariant. Hence, the Inception v2 architecture outputs a reduced-dimensional feature map for the region proposal layer. The region proposal network is defined by anchors or fixed boundary boxes at each location. At each location, anchors of different scale and aspect ratio are defined, thereby enabling the region proposal network to make scale invariant proposals. The region proposal layer uses a convolutional filter on the feature map to output a confidence score for two classes; object and background. This is called the objectness score. Furthermore, the convolutional filter outputs regression offsets for anchor boxes. Hence, assuming there are k anchors at a location, the convolutional filter in the region proposal network outputs 6k values, namely 4k coordinates and 2k scores. Two losses are calculated from this output—classification loss and bounding box regression loss. The bounding box coordinates of anchors classified as objects are then combined with the feature map from feature extractor. In the RoI pooling layer, bounding box regions of different sizes and aspect ratios are resized to fixed size outputs using max pooling. Pooling layer refers to a down sampling layer and in case of max pooling, the down sampling is done by maximum of pixels [36]. The max-pooled feature map of a fixed size corresponding to each output is then classified, and its bounding box offsets with respect to ground truth boxes are regressed. Hence, as in the region proposal layer, two losses are computed at this output, namely the classification loss and bounding box regression loss.

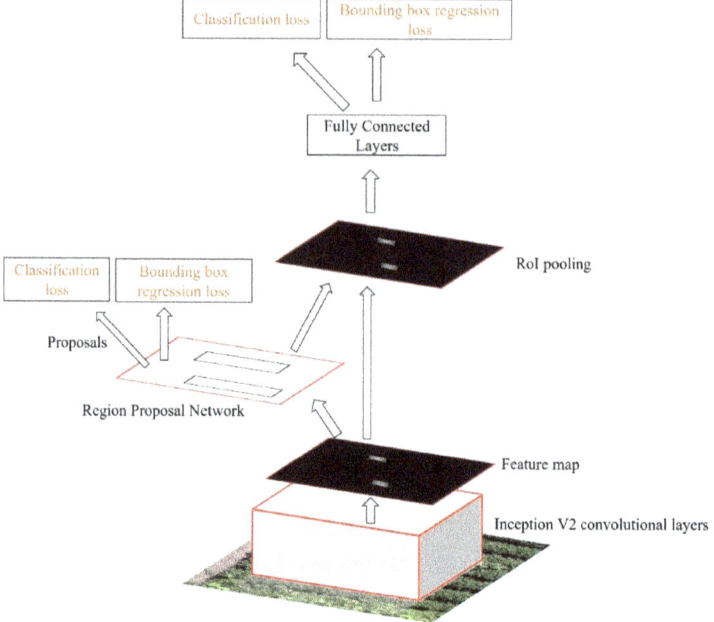

Figure 3. Faster RCNN architecture.

2.5.2. Hyperparameters of the Architecture

In the framework that was used, the input images to the Faster RCNN network were resized to images of fixed size 1024 × 1024 pixels. At each location in the region proposal layer, 4 different scales namely 0.25, 0.5, 1.0, 2.0 and 3 different aspect ratios namely 0.5, 1.0 and 2.0 were used. Hence, in total, there were 12 anchors at each location. The model was trained for 25,000 epochs with a batch size of 1 using stochastic gradient descent with momentum optimizer. The training dataset was split into training and validation datasets and the performance of the model on validation data was continuously monitored during training to check if the model starts to overfit. Random horizontal flip and random crop operations were performed to augment the training data. The data collected had the crop rows always parallel to the horizontal axis of the image, therefore random horizontal flip and crop operations augment the training data.

2.5.3. Single Shot Detector

The Single Shot Detector (SSD) (Figure 4) model was proposed to improve the inference time of objection detection models with region proposal network such as Faster RCNN. The main difference in SSD compared to Faster RCNN is the generation of detection outputs without a separate region proposal layer. Similar to Faster RCNN, SSD uses a feature extractor which is the Inception v2 architecture in this case. At each location of the feature map output, the model outputs a set of bounding boxes of different scales and aspect ratios. This is very similar to Faster RCNN but the difference being the convolutional filter on the feature map directly outputs the confidence scores corresponding to the output classes along with regression box offsets. Hence, the class and bounding box offsets are output in a single shot as the name suggests. For the model to be scale and translation invariant, rather than outputting bounding boxes from only the feature map, extra feature layers are added to the feature map output and detection boxes are output at different scales from each output. Hence, in total, the SSD model has 6 layers that output detection boxes at different scales [67].

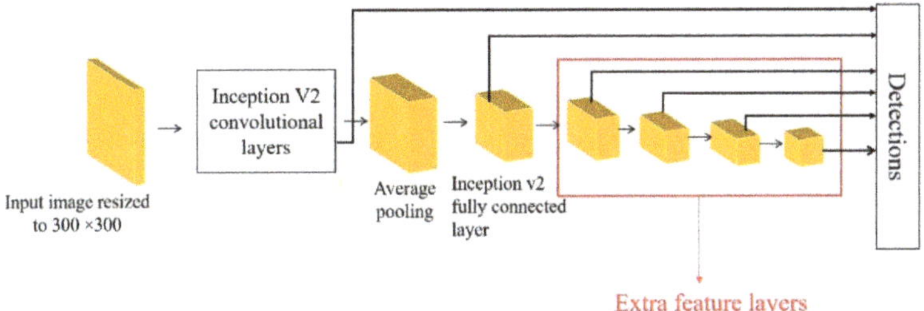

Figure 4. Single Shot Detector (SSD) architecture.

2.5.4. Hyperparameters of the Architecture

In the case of SSD, in the framework that has been used, the input images are always reshaped to a fixed dimension of 300 × 300 pixels. After the feature extraction, in 6 different layers that output detection boxes, 6 different scales in the range 0.2–0.95 were used. Five different aspect ratios namely 1.0, 2.0, 0.5, 3.0 and 0.333 were generated at each location. The model was trained for 25,000 epochs as in the case of Faster RCNN. A batch size of 24 was used in training and the RMS prop optimizer was used. Data augmentation was applied with random horizontal flipping and random cropping of images. Validation images were, again, evaluated periodically during the training to check if the model is overfitting.

2.6. Hardware and Software Used

The models were trained and evaluation of the models was performed on a computer with Intel i9 processor with 18 cores and 64 GB of RAM and NVIDIA GeForce RTX 2080 Ti graphics card. Tensorflow object detection API [61] in Python was used to train and evaluate Faster RCNN and SSD. Tensorflow tutorial on transfer learning [58] was used to train the MobileNet v2 architecture for patch-based CNN.

2.7. Evaluation Metrics

Precision, recall, f1 score, and Intersection over Union (IoU) are the evaluation metrics used in this study.

$$\text{Precision} = \frac{\text{TP}}{\text{TP} + \text{FP}} \tag{1}$$

$$\text{Recall} = \frac{\text{TP}}{\text{TP} + \text{FN}} \tag{2}$$

$$\text{F1 score} = \frac{2 \times \text{Precision} \times \text{Recall}}{\text{Precision} + \text{Recall}} \tag{3}$$

Here TP refers to True Positive, FP refers to False Positive, and FN refers to False negative. Moreover, mean Average Precision (mAP) is another metric that is commonly used in object detection problems [59,68]. It is the mean of the average precision at all recall values at different IoUs for prediction and ground truth thresholds from 0.5 to 0.95. It should be noted that these metrics were primarily formulated for object detection. Even though, in this study, we use object detection models, the objective is not to find weed objects rather all the area covered by weeds for management purposes. In case of a deep learning-based object detection model, multiple objects with their bounding box are predicted. Of these, only the boxes which have IoU with the ground truth greater than threshold and class score (probability of that object being in each class) greater than confidence threshold are considered positive prediction boxes. Among these, only the box with highest class score is considered as the true positive and other positive boxes are considered as false positives. In our case, for a weed patch that is marked as a ground truth box, the model might have multiple positive weed boxes corresponding to that one ground truth box. However, only one of those would be considered as true positive and other boxes are false positives. As can be seen in the following Figure 5, the output of this image has two prediction boxes covering the weed area in the left but in the ground truth it was marked as one bounding box. Hence, if precision is used as the evaluation metric, the box on the bottom will be regarded as False Positive even though that box adds to more weed area being detected. Therefore, the Intersection over Union (IoU) of binary output image representing weed and background pixels with the ground truth binary image is used as the primary evaluation metric. The binary output images corresponding to prediction outputs and ground truth are obtained by considering pixels representing weed objects as 1 and other areas as 0. The intersection and union of the two binary images obtained are then used to find the IoU ratio. Hence, IoU here represents the ratio between the intersection of all positive prediction boxes (true positive and false positives in object detection terms) and all ground truth boxes in an image.

$$\text{IoU} = \frac{\text{Area of overlap}}{\text{Area of union}} \tag{4}$$

Figure 5. Example output image showing a weed patch annotated with single box in ground truth image detected as two boxes in output. This will lead to lesser precision as only the bigger box is considered true positive and therefore IoU is a better evaluation metric for this problem.

To evaluate the patch-based CNN on the sub-image, an overlap slicing approach is used. The sub-image of size 1152 × 1152 pixels is sliced into patches of size 128 × 128 pixels with a stride of 32 on the horizontal and vertical. Therefore, the sliced patches have 75% horizontal and vertical overlap. Hence, each small area of size 32 × 32 is part of 8 patches and the class with maximum votes from the 4 patches is assigned as the class of the small area. To evaluate this result with ground truth and to compare with the results of Faster RCNN and SSD, IoU is used as the evaluation metric.

3. Results and Discussion

3.1. Training of Faster RCNN and SSD

Figure 6 shows the training graph for Faster RCNN and SSD. The decrease in training loss and the increase in mAP of the validation data with training epochs can be seen. By the end of the training, very little difference in the mAP of Faster RCNN and the SSD validation data was obtained. Faster RCNN converged faster than SSD. The training process of Faster RCNN might appear to oscillate more than SSD, which could be due to the different batch sizes and optimizers being used by the two models. However, it should be noted that the scale of the two loss plots was different. The different batch size and optimizer could also be the reason for the Faster RCNN model converging to high validation mAP earlier than SSD, since a batch size of 1 for Faster RCNN leads to 24 times more gradient updates than SSD with a batch size of 24.

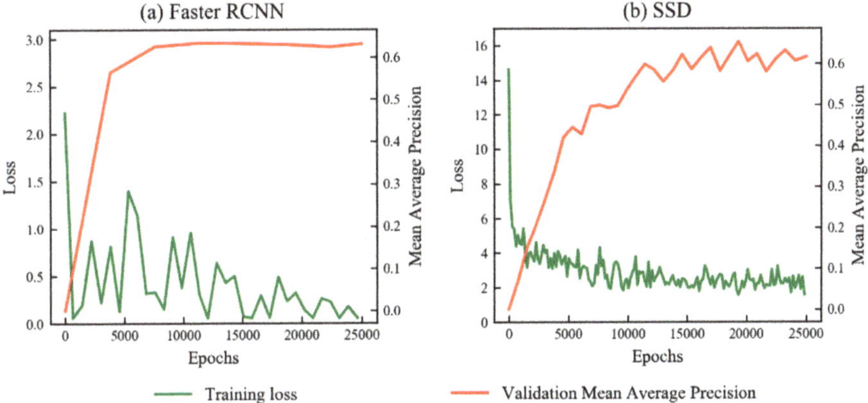

Figure 6. Change in training loss and Validation Mean Average Precision with number of epochs of (**a**) Faster RCNN and (**b**) SSD.

3.2. Optimal IoU and Confidence Thresholds for Faster RCNN and SSD

In order to find the optimal threshold for IoU of the prediction boxes and ground truth boxes that would result in best performance of the model, precision recall curves were drawn using various confidence thresholds from 0 to 1 at various IoU thresholds ranging from 0.5 to 0.95 (Figure 7).

It can be seen that the area under the precision-recall curve is almost the same in case of Faster RCNN and SSD which explains the fact that the validation mAP during the final epochs as seen from the training graph was very similar (0.63 in Faster RCNN and 0.62 in SSD). Furthermore, both Faster RCNN and SSD achieved the maximum area under the precision-recall curve at an IoU threshold of 0.5 for the prediction box and ground truth box. Hence, for each ground truth box, among all prediction boxes with a confidence score greater than the threshold for confidence score, the prediction box with the highest value of IoU with the ground truth box and also whose IoU with ground truth box is greater than the threshold for IoU was considered a true positive. All prediction boxes that were not a true positive with any ground truth box are regarded as false positives. The number of false negatives is equal to the number of ground truth boxes that do not have a corresponding true positive. With the optimal IoU threshold found for Faster RCNN and SSD, the following graph (Figure 8) was plotted to find the optimal confidence threshold for Faster RCNN and SSD that results in the best performance.

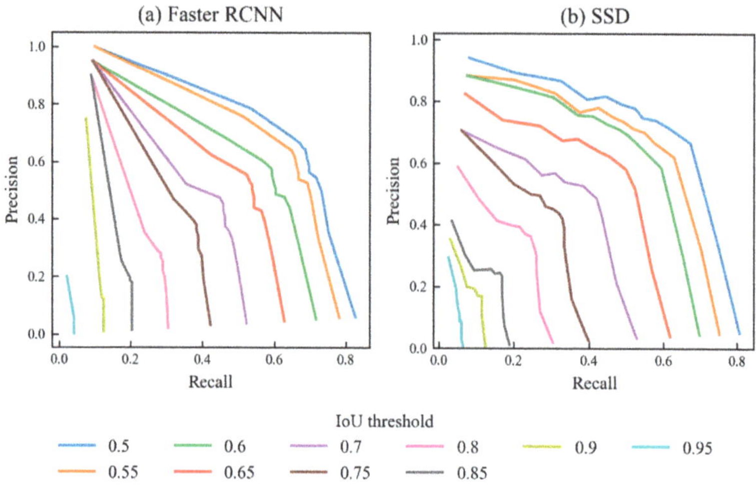

Figure 7. Precision-recall curve at different thresholds for IoU of the predicted box and ground truth box (**a**) Faster RCNN and (**b**) SSD.

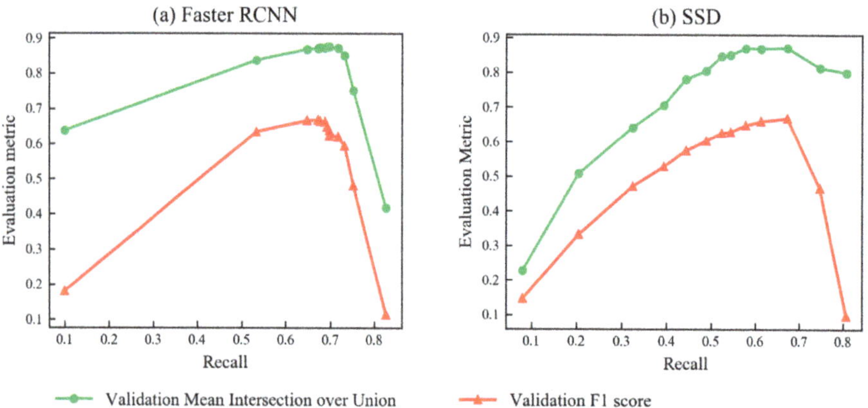

Figure 8. Change in IoU of output binary image and ground truth binary image as well as f1 score with change in recall.

Figure 8 shows the change in f1 score and the mean IoU of the output binary image of the model with the ground truth binary image with change in recall. From the figure, the recall value which results in the best IoU and F1 score was found using the peak. The recall at which the best mean IoU and f1 score were observed was around 0.7 and its corresponding confidence threshold for class scores was 0.6 in the case of Faster RCNN, and 0.1 in the case of SSD. It is to be noted that mean IoU here refers to the Intersection over Union of the whole binary model output image with the ground truth binary image whereas the IoU mentioned earlier was the Intersection over Union of individual prediction bounding boxes with individual ground truth bounding boxes.

3.3. Comparison of Performance of Faster RCNN and SSD

Table 1 shows the precision, recall, f1 score, and mean IoU of the model output binary image and the ground truth binary along with the inference time for a 1152 × 1152 image. The precision, recall, f1 score, and mean IoU of both the models were similar but the SSD model was slightly faster in execution than Faster RCNN. It should be noted that the above performance was in the case that the Faster RCNN network outputs 300 proposals from the region proposal network. However, Huang et al. [36] found that by reducing the number of proposals output by Faster RCNN, the inference time of Faster RCNN can be improved with a slight cost in precision, recall, and f1 score. Therefore, experiments were conducted to study the change in inference time, precision, recall, f1 score and mean IoU, by varying the number of proposal boxes from the Faster RCNN network from 50 to 300 and the results are plotted in Figure 9.

Table 1. Performance of test data in Faster RCNN and Single Shot Detector (SSD).

Model	Precision	Recall	F1 Score	Mean IoU	Inference Time of 1152 × 1152 Image in Seconds
Faster RCNN	0.65	0.68	0.66	0.85	0.23
SSD	0.66	0.68	0.67	0.84	0.21

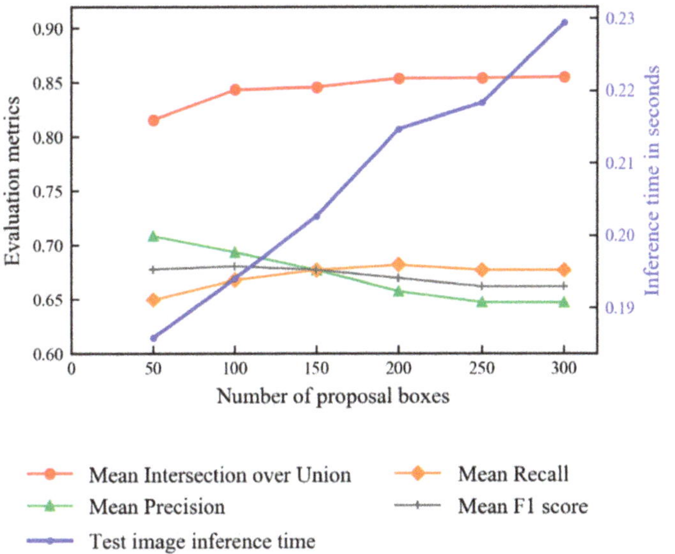

Figure 9. Change in evaluation metrics and inference time of Faster RCNN model with increase in number of proposals.

The inference time of Faster RCNN had a linear time complexity with the number of proposal boxes output from the region proposal network. It can be seen that, from 200 to 300 proposals, there was no change in performance of the model but the inference time decreased. Hence, 200 proposals was selected as the optimal number of proposals for this dataset. At 200 proposals, the inference time of Faster RCNN was 0.21 seconds, which was the same as SSD. In the case of constraints in computational power, using 100 proposal boxes would result in significant compute savings with minimal loss in mean IoU. Hence, no difference in performance was found between Faster RCNN with 200 proposals and SSD in terms of the evaluation metrics used in this study. However, it is to be noted that, even with the same performance metric, Faster RCNN output weed objects with high confidence compared to SSD, since the confidence threshold being used for Faster RCNN was 0.6, whereas it was a very low 0.1 for SSD. Though this threshold might result in the best performance with the current validation test, it might affect the generalization performance of the model in the case of a test dataset from a different location or from a field with different management practices. In such cases, the low threshold might lead to reduced precision.

On visual observation of the outputs of all the 44 test images, it was found that in 41 images, both the networks detected all the weed areas. Hence, in these images, the difference in IoU between the model output and the ground truth is only because of the slight displacements of the boundaries of the bounding boxes from each other. As mentioned in Section 2.7, the low values of precision, recall, and f1 score obtained are primarily because of the way these metrics are calculated, since only one bounding box is considered as a true positive for one ground truth box, whereas the model in case of some weed areas with slight discontinuities outputs multiple prediction boxes to detect those areas. Therefore, the mean IoU of the binary output image with the binary image of the ground truth is the appropriate metric. In three of the test images (shown in Figure 10), there was a difference in the output of Faster RCNN and SSD. In the output image 1, Faster RCNN failed to detect a small strip of weed between the crop rows, but this was detected by SSD. However, by looking at the confidence score of the weed object from SSD, it can be understood that SSD was only able to detect this weed object because of the very low confidence threshold set for it. Whereas in output image 2, SSD misclassified a row of soybean crops with herbicide drift injury as weeds. Moreover, in case of output image 3, SSD could not detect the weeds on the left vertical border of the image. With both the failure areas being present in the border of the images, this might show the susceptibility of the SSD model in the image border. This could be due to the architecture of SSD that does detection of objects and classification into its class in a single shot, unlike Faster RCNN. Another possible reason could be that, by default, the API used to train both the models was resizing the input images of Faster RCNN to 600 × 600 whereas in case of SSD it was resized to 300 × 300. Therefore, this further loss of detail in the input image compared to the Faster RCNN input image might have led to the misclassifications in the border. Hence, further study with the same input image resolution is needed for a fair comparison.

Other than the above-mentioned three images, Faster RCNN, as well as SSD, performed exceptionally well in detecting weed objects of various scales as seen in Figure 11. As mentioned earlier, it can be seen that though SSD detected all the weed objects that were detected by Faster RCNN, the confidence of many of those predictions were very low and ended up as true positive because of the low confidence threshold. Since, by reducing the number of proposals to 200, Faster RCNN can be as fast SSD in terms of inference time, it can be concluded that Faster RCNN has better speed performance tradeoff.

3.4. Comparison of Performance of Faster RCNN and Patch-Based CNN

The Mobilenet v2 network trained on the training patches showed very high performance in classifying test patches with an f1 score of 0.98. However, in order to evaluate its performance in detecting the weed objects in the sub-image and compare its performance with the Faster RCNN object detection model, the overlapping approach explained earlier was used. Table 2 shows the mean IoU of the output binary image from Faster RCNN and patch-based CNN with the ground truth binary image. Furthermore, the table shows the time taken to evaluate one sub-image by both the models.

Figure 10. Output images with discrepancies between Faster RCNN and SSD and their corresponding ground truth.

Table 2. Performance of Faster RCNN and patch-based CNN in test sub-images.

Model	Mean IoU	Inference Time in Seconds for Each Sub-Image (1152 × 1152)
Faster RCNN with 200 proposals	0.85	0.21
Patch based CNN sliced with overlap	0.61	1.03
Patch based CNN sliced without overlap	0.6	0.22

Figure 11. Example output images with good model performance.

Faster RCNN had better performance than patch-based CNN with overlap, both in terms of mean IoU and inference time. However, patch-based CNN without overlap has an inference time which is almost the same as Faster RCNN. The low values of IoU of patch-based CNN without overlap were because of the coarse nature of this algorithm. Since each sub-image was split into 81 patches in this approach, weeds that were smaller in size would not be detected in this approach. Furthermore, because of the way the patches were sliced, there could be a lot of patches with weeds and background in equal proportion, whereas the Mobilenet v2 model had only been trained with patches that contained only weed or only background, and hence the model was prone to error in this approach. To reduce this error, the slicing with overlap approach was tested. Since, for each small block within a patch, the class was determined by majority vote in eight patches, the problem of mixed patches was solved to some extent. Still, the similar IoU of slicing with overlap and without overlap is because the ground truth binary image represents weed objects as rectangular boxes whereas output binary images from the patch-based overlap approach consist of weed objects, which are polygonal in nature because of the majority vote as can be seen in Figure 12. Therefore, patch-based CNN with overlap has better performance than the IoU value with ground truth image suggests. However, the drawback of this approach is the very high inference time compared to Faster RCNN and patch-based RCNN without overlap. Further studies can be done with different levels of horizontal and vertical overlap and its influence on the inference time of this approach. However, with the inference time of Faster RCNN being the same as the patch-based CNN without overlap, any amount of overlap would lead to more patches to be evaluated than the non-overlap approach and hence greater inference time. Therefore, among the approaches investigated in this study, Faster RCNN had the best overall performance. It would be interesting to study a modified Fast RCNN architecture with the region proposal part replaced with an image analysis method that selects polygons. This could achieve faster computational speed as well as better performance for a patch-based CNN method.

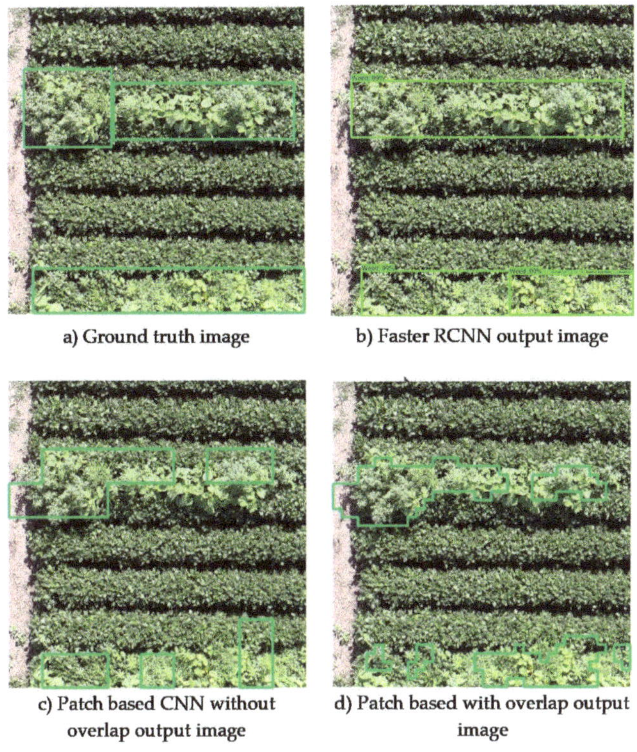

Figure 12. Output images of patch-based CNN and Faster RCNN.

In order to implement this system for on-farm detection, further evaluation of the performance of these approaches at higher altitudes is needed. At the altitude of 20m at which these data were collected, it is practically impossible to cover the large soybean fields with the current limitations on the battery capacity of UAV systems. Therefore, the evaluation of the performance of these models at low-resolution images from high altitude is needed for practical adoption of these systems. Like SSD, it can be seen that there is a higher misclassification rate of patches in the border of the images. In this case, it is suggested to collect images with some overlap, such as 15%, so that weed objects present in the border of one image end up in the interior of the next image. Furthermore, it is to be noted that the dataset used to train the models in the study was only collected on two different days. Therefore, the differences in phenological stage of the crop and the weed and lighting conditions are limited within the dataset. Further experiments with wide variations in lighting conditions, flight altitudes, different phenological stages are needed to analyze and compare the generalizability of performance of these models in varying conditions in the field. In addition, since the manual labeling of bounding boxes used in this study was labeled by one annotator, it is possible that there is error due to bias of the observer. Therefore, further studies using multiple annotators for labeling data with more variations as mentioned above is needed to remove bias and study the generalizability of the model. With the manual annotation of images being a time-consuming process, use of multiresolution segmentation approaches from OBIA could help in automating this. In that case, OBIA could help generate polygon labels from which rectangular bounding box labels can be generated for object detection tasks.

4. Conclusions

In this study, Faster RCNN and SSD object detection models were trained and evaluated over UAV imagery for mid- to late-season weed detection in soybean fields. The performance of two object detection models, Faster RCNN and the Single Shot Detector (SSD) models, as well as the performance of object detection CNN models with the patch-based CNN model, were evaluated and compared in terms of weed detection performance using mean IoU and inference speed.

It was found that the Faster RCNN model with 200 box proposals had a similar weed detection performance to the SSD model in terms of precision, recall, f1 score, and IoU as well as similar inference time. The precision, recall, f1 score and IoU were 0.65, 0.68, 0.66 and 0.85 for Faster RCNN with 200 proposals, and 0.66, 0.68, 0.67 and 0.84 for SSD respectively. However, the optimal confidence threshold of SSD was found to be 0.1, indicating the lower confidence of this model in the case of weed objects detected, whereas the optimal confidence threshold was found to be 0.6 in the case of Faster RCNN, meaning higher confidence in the weed objects detected. In addition, SSD was susceptible to misclassification in the border of some test images. These findings indicate that SSD might have lower generalization performance than Faster RCNN for mid- to late-season weed detection in soybean fields using UAV imagery. Hence, Faster RCNN was determined to be the better performing model among the two in this study. Between Faster RCNN and patch-based CNN, Faster RCNN had better weed detection performance than patch-based CNN with overlap as well as without overlap. The inference time of Faster RCNN was similar to patch-based CNN without overlap, but significantly less than patch-based CNN with overlap. Hence, Faster RCNN was found to be the best model in terms of weed detection performance and inference time among the different models compared in this study.

Future work can evaluate the performance variation of models in different weed species. In addition, the performance of Faster RCNN at different altitudes by resampling high-resolution images to low-resolution images can be studied. Furthermore, the inference time experiments at different altitudes should be performed on low computational power devices such as regular laptops and mini-PCs used for the flight control of UAV systems. Inference time experiments should also be performed on low cost hardware accelerators available for edge computing such as the Intel Neural Compute Stick or Google Coral. This would help understand the potential of using such devices for on-farm, near real-time data processing and actuation. In addition, the effect of model compression techniques and approximation algorithms developed for neural networks can be studied to understand the limit of edge computing for in-field near real-time weed detection. Moreover, further work can be performed on using the RTK GPS data of individual images and their corresponding IMU data to orthorectify the image and find the geolocation of the weed patches detected by the object detection models. In addition, the performance of object detection models for weed detection can be compared between raw individual images as used in this study and stitched mosaic maps. With the manual annotation of images being a laborious part of the process, using techniques such as self-supervised learning [69] and active learning [70] to reduce the amount of manual labeling for this task can be studied. Furthermore, few-shot learning algorithms can be studied to investigate the transfer learning of this algorithm to other crops and weed species by training with a few labeled instances from those crops and weed species.

Author Contributions: Conceptualization, Y.S. and E.P.; methodology, A.N.V.S., Y.S. and S.S.; data acquisition, A.N.V.S. and J.L.; software, analysis and evaluation, A.N.V.S.; writing—original draft preparation, A.N.V.S.; writing—review and editing, Y.S., E.P., S.S., A.J.J., J.D.L. and J.L.; project administration, Y.S.; funding acquisition, Y.S., E.P. and A.J.J. All authors have read and agreed to the published version of the manuscript.

Funding: This research was funded by the Nebraska Research Initiative (NRI) Collaboration Initiative Seed Grant 2132250011, the Nebraska Corn Board, and the Nebraska Agricultural Experiment Station through the Hatch Act capacity funding program (Accession Number 1011130) from the USDA National Institute of Food and Agriculture.

Acknowledgments: Thanks to Jonathan Forbes for their assistance in data collection.

Conflicts of Interest: The authors declare no conflict of interest. The funders had no role in the design of the study; in the collection, analyses, or interpretation of data; in the writing of the manuscript, or in the decision to publish the results.

References

1. Dos Santos Ferreira, A.; Matte Freitas, D.; Gonçalves da Silva, G.; Pistori, H.; Theophilo Folhes, M. Weed detection in soybean crops using ConvNets. *Comput. Electron. Agric.* **2017**, *143*, 314–324. [CrossRef]
2. Thorp, K.R.; Tian, L.F. A Review on Remote Sensing of Weeds in Agriculture. *Precis. Agric.* **2004**, *5*, 477–508. [CrossRef]
3. Weis, M.; Gutjahr, C.; Rueda Ayala, V.; Gerhards, R.; Ritter, C.; Schölderle, F. Precision farming for weed management: Techniques. *Gesunde Pflanz.* **2008**, *60*, 171–181. [CrossRef]
4. Christensen, S.; SØgaard, H.T.; Kudsk, P.; NØrremark, M.; Lund, I.; Nadimi, E.S.; JØrgensen, R. Site-specific weed control technologies. *Weed Res.* **2009**. [CrossRef]
5. Zhang, N.; Wang, M.; Wang, N. Precision agriculture—A worldwide overview. *Comput. Electron. Agric.* **2002**, *36*, 113–132. [CrossRef]
6. O'Donovan, J.T.; De St. Remy, E.A.; O'Sullivan, P.A.; Dew, D.A.; Sharma, A.K. Influence of the Relative Time of Emergence of Wild Oat (Avena fatua) on Yield Loss of Barley (Hordeum vulgare) and Wheat (Triticum aestivum). *Weed Sci.* **1985**. [CrossRef]
7. Swanton, C.J.; Mahoney, K.J.; Chandler, K.; Gulden, R.H. Integrated Weed Management: Knowledge-Based Weed Management Systems. *Weed Sci.* **2008**. [CrossRef]
8. JUDGE, C.A.; NEAL, J.C.; DERR, J.F. Response of Japanese Stiltgrass (Microstegium vimineum) to Application Timing, Rate, and Frequency of Postemergence Herbicides 1. *Weed Technol.* **2005**. [CrossRef]
9. Chauhan, B.S.; Singh, R.G.; Mahajan, G. Ecology and management of weeds under conservation agriculture: A review. *Crop. Prot.* **2012**, *38*, 57–65. [CrossRef]
10. De Castro, A.I.; Torres-Sánchez, J.; Peña, J.M.; Jiménez-Brenes, F.M.; Csillik, O.; López-Granados, F. An automatic random forest-OBIA algorithm for early weed mapping between and within crop rows using UAV imagery. *Remote Sens.* **2018**, *10*, 285. [CrossRef]
11. Fernández-Quintanilla, C.; Peña, J.M.; Andújar, D.; Dorado, J.; Ribeiro, A.; López-Granados, F. Is the current state of the art of weed monitoring suitable for site-specific weed management in arable crops? *Weed Res.* **2018**, *58*, 259–272. [CrossRef]
12. López-Granados, F. Weed detection for site-specific weed management: Mapping and real-time approaches. *Weed Res.* **2011**. [CrossRef]
13. Barroso, J.; Fernàndez-Quintanilla, C.; Ruiz, D.; Hernaiz, P.; Rew, L.J. Spatial stability of Avena sterilis ssp. ludoviciana populations under annual applications of low rates of imazamethabenz. *Weed Res.* **2004**. [CrossRef]
14. Koger, C.H.; Shaw, D.R.; Watson, C.E.; Reddy, K.N. Detecting Late-Season Weed Infestations in Soybean (Glycine max) 1. *Weed Technol.* **2003**. [CrossRef]
15. De Castro, A.I.; Jurado-Expósito, M.; Peña-Barragán, J.M.; López-Granados, F. Airborne multi-spectral imagery for mapping cruciferous weeds in cereal and legume crops. *Precis. Agric.* **2012**. [CrossRef]
16. De Castro, A.I.; López-Granados, F.; Jurado-Expósito, M. Broad-scale cruciferous weed patch classification in winter wheat using QuickBird imagery for in-season site-specific control. *Precis. Agric.* **2013**. [CrossRef]
17. Castillejo-González, I.L.; López-Granados, F.; García-Ferrer, A.; Peña-Barragán, J.M.; Jurado-Expósito, M.; de la Orden, M.S.; González-Audicana, M. Object- and pixel-based analysis for mapping crops and their agro-environmental associated measures using QuickBird imagery. *Comput. Electron. Agric.* **2009**. [CrossRef]
18. Meyer, G.E.; Mehta, T.; Kocher, M.F.; Mortensen, D.A.; Samal, A. Textural imaging and discriminant analysis for distinguishing weeds for spot spraying. *Trans. Am. Soc. Agric. Eng.* **1998**. [CrossRef]
19. Burks, T.F.; Shearer, S.A.; Payne, F.A. Classification of weed species using color texture features and discriminant analysis. *Trans. Am. Soc. Agric. Eng.* **2000**, *43*, 411. [CrossRef]
20. Wang, A.; Zhang, W.; Wei, X. A review on weed detection using ground-based machine vision and image processing techniques. *Comput. Electron. Agric.* **2019**, *158*, 226–240. [CrossRef]

21. Sankaran, S.; Khot, L.R.; Espinoza, C.Z.; Jarolmasjed, S.; Sathuvalli, V.R.; Vandemark, G.J.; Miklas, P.N.; Carter, A.H.; Pumphrey, M.O.; Knowles, N.R.; et al. Low-altitude, high-resolution aerial imaging systems for row and field crop phenotyping: A review. *Eur. J. Agron.* **2015**, *70*, 112–123. [CrossRef]
22. Rasmussen, J.; Nielsen, J.; Streibig, J.C.; Jensen, J.E.; Pedersen, K.S.; Olsen, S.I. Pre-harvest weed mapping of Cirsium arvense in wheat and barley with off-the-shelf UAVs. *Precis. Agric.* **2019**. [CrossRef]
23. Casa, R.; Pascucci, S.; Pignatti, S.; Palombo, A.; Nanni, U.; Harfouche, A.; Laura, L.; Di Rocco, M.; Fantozzi, P. UAV-based hyperspectral imaging for weed discrimination in maize. In Proceedings of the Precision Agriculture 2019—Papers Presented at the 12th European Conference on Precision Agriculture, ECPA 2019, Montpellier, France, 8–11 July 2019; Wageningen Academic Publishers: Wageningen, The Netherlands, 2019; pp. 365–371.
24. Sánchez-Sastre, L.F.; Casterad, M.A.; Guillén, M.; Ruiz-Potosme, N.M.; Veiga, N.M.S.A.; da Navas-Gracia, L.M.; Martín-Ramos, P. UAV Detection of Sinapis arvensis Infestation in Alfalfa Plots Using Simple Vegetation Indices from Conventional Digital Cameras. *AgriEngineering* **2020**, *2*, 12. [CrossRef]
25. Peña-Barragán, J.M.; López-Granados, F.; Jurado-Expósito, M.; García-Torres, L. Spectral discrimination of Ridolfia segetum and sunflower as affected by phenological stage. *Weed Res.* **2006**. [CrossRef]
26. Gray, C.J.; Shaw, D.R.; Gerard, P.D.; Bruce, L.M. Utility of Multispectral Imagery for Soybean and Weed Species Differentiation. *Weed Technol.* **2008**. [CrossRef]
27. Martin, M.P.; Barreto, L.; RiañO, D.; Fernandez-Quintanilla, C.; Vaughan, P. Assessing the potential of hyperspectral remote sensing for the discrimination of grassweeds in winter cereal crops. *Int. J. Remote Sens.* **2011**. [CrossRef]
28. De Castro, A.I.; Jurado-Expósito, M.; Gómez-Casero, M.T.; López-Granados, F. Applying neural networks to hyperspectral and multispectral field data for discrimination of cruciferous weeds in winter crops. *Sci. World J.* **2012**. [CrossRef]
29. Peña, J.M.; Torres-Sánchez, J.; de Castro, A.I.; Kelly, M.; López-Granados, F. Weed Mapping in Early-Season Maize Fields Using Object-Based Analysis of Unmanned Aerial Vehicle (UAV) Images. *PLoS ONE* **2013**, *8*, e77151. [CrossRef]
30. Torres-Sánchez, J.; López-Granados, F.; De Castro, A.I.; Peña-Barragán, J.M. Configuration and Specifications of an Unmanned Aerial Vehicle (UAV) for Early Site Specific Weed Management. *PLoS ONE* **2013**, *8*, e58210. [CrossRef]
31. Torres-Sánchez, J.; Peña, J.M.; de Castro, A.I.; López-Granados, F. Multi-temporal mapping of the vegetation fraction in early-season wheat fields using images from UAV. *Comput. Electron. Agric.* **2014**. [CrossRef]
32. Pérez-Ortiz, M.; Peña, J.M.; Gutiérrez, P.A.; Torres-Sánchez, J.; Hervás-Martínez, C.; López-Granados, F. A semi-supervised system for weed mapping in sunflower crops using unmanned aerial vehicles and a crop row detection method. *Appl. Soft Comput.* **2015**, *37*, 533–544. [CrossRef]
33. Castaldi, F.; Pelosi, F.; Pascucci, S.; Casa, R. Assessing the potential of images from unmanned aerial vehicles (UAV) to support herbicide patch spraying in maize. *Precis. Agric.* **2017**. [CrossRef]
34. López-Granados, F.; Torres-Sánchez, J.; Serrano-Pérez, A.; de Castro, A.I.; Mesas-Carrascosa, F.-J.; Peña, J.-M. Early season weed mapping in sunflower using UAV technology: Variability of herbicide treatment maps against weed thresholds. *Precis. Agric.* **2016**, *17*, 183–199. [CrossRef]
35. Liu, D.; Xia, F. Assessing object-based classification: Advantages and limitations Assessing object-based classification: Advantages and limitations. *Remote Sens. Lett.* **2010**. [CrossRef]
36. Krizhevsky, A.; Sutskever, I.; Hinton, G.E. ImageNet classification with deep convolutional neural networks. In Proceedings of the Advances in Neural Information Processing Systems, Lake Tahoe, NV, USA, 3–6 December 2012; pp. 1097–1105.
37. Steen, K.; Christiansen, P.; Karstoft, H.; Jørgensen, R.; Steen, K.A.; Christiansen, P.; Karstoft, H.; Jørgensen, R.N. Using Deep Learning to Challenge Safety Standard for Highly Autonomous Machines in Agriculture. *J. Imaging* **2016**, *2*, 6. [CrossRef]
38. Song, X.; Zhang, G.; Liu, F.; Li, D.; Zhao, Y.; Yang, J. Modeling spatio-temporal distribution of soil moisture by deep learning-based cellular automata model. *J. Arid Land* **2016**, *8*, 734–748. [CrossRef]
39. Mohanty, S.P.; Hughes, D.P.; Salathé, M. Using Deep Learning for Image-Based Plant Disease Detection. *Front. Plant. Sci.* **2016**, *7*, 1419. [CrossRef]

40. Kuwata, K.; Shibasaki, R. Estimating crop yields with deep learning and remotely sensed data. In Proceedings of the 2015 IEEE International Geoscience and Remote Sensing Symposium (IGARSS), Milan, Italy, 26–31 July 2015; pp. 858–861.
41. Rahnemoonfar, M.; Sheppard, C.; Rahnemoonfar, M.; Sheppard, C. Deep Count: Fruit Counting Based on Deep Simulated Learning. *Sensors* **2017**, *17*, 905. [CrossRef]
42. Andrea, C.-C.; Mauricio Daniel, B.B.; Jose Misael, J.B. Precise weed and maize classification through convolutional neuronal networks. In Proceedings of the 2017 IEEE Second Ecuador Technical Chapters Meeting (ETCM), Salinas, Ecuador, 16–20 October 2017; pp. 1–6.
43. Dyrmann, M.; Mortensen, A.; Midtiby, H.; Jørgensen, R. Pixel-wise classification of weeds and crops in images by using a fully convolutional neural network. In Proceedings of the International Conference on Agricultural Engineering, Aarhus, Denmark, 26–29 June 2016; pp. 26–29.
44. Milioto, A.; Lottes, P.; Stachniss, C. Real-Time Semantic Segmentation of Crop and Weed for Precision Agriculture Robots Leveraging Background Knowledge in CNNs. In Proceedings of the 2018 IEEE International Conference on Robotics and Automation (ICRA), Brisbane, Australia, 21–25 May 2018; pp. 2229–2235.
45. Lottes, P.; Behley, J.; Milioto, A.; Stachniss, C. Fully Convolutional Networks With Sequential Information for Robust Crop and Weed Detection in Precision Farming. *IEEE Robot. Autom. Lett.* **2018**, *3*, 2870–2877. [CrossRef]
46. Lottes, P.; Khanna, R.; Pfeifer, J.; Siegwart, R.; Stachniss, C. UAV-based crop and weed classification for smart farming. In Proceedings of the 2017 IEEE International Conference on Robotics and Automation (ICRA), Singapore, 29 May–3 June 2017; pp. 3024–3031.
47. Lottes, P.; Behley, J.; Chebrolu, N.; Milioto, A.; Stachniss, C. Robust joint stem detection and crop-weed classification using image sequences for plant-specific treatment in precision farming. *J. Field Robot.* **2020**, *37*, 20–34. [CrossRef]
48. Sa, I.; Chen, Z.; Popovic, M.; Khanna, R.; Liebisch, F.; Nieto, J.; Siegwart, R. WeedNet: Dense Semantic Weed Classification Using Multispectral Images and MAV for Smart Farming. *IEEE Robot. Autom. Lett.* **2018**, *3*, 588–595. [CrossRef]
49. Sa, I.; Popović, M.; Khanna, R.; Chen, Z.; Lottes, P.; Liebisch, F.; Nieto, J.; Stachniss, C.; Walter, A.; Siegwart, R.; et al. WeedMap: A Large-Scale Semantic Weed Mapping Framework Using Aerial Multispectral Imaging and Deep Neural Network for Precision Farming. *Remote Sens.* **2018**, *10*, 1423. [CrossRef]
50. Bah, M.D.; Dericquebourg, E.; Hafiane, A.; Canals, R. Deep Learning Based Classification System for Identifying Weeds Using High-Resolution UAV Imagery. In *Intelligent Computing. SAI 2018*; Advances in Intelligent Systems and Computing: Cham, Switzerland, 2019; Volume 857, pp. 176–187.
51. Huang, H.; Deng, J.; Lan, Y.; Yang, A.; Deng, X.; Zhang, L. A fully convolutional network for weed mapping of unmanned aerial vehicle (UAV) imagery. *PLoS ONE* **2018**, *13*, e0196302. [CrossRef]
52. Yu, J.; Schumann, A.W.; Cao, Z.; Sharpe, S.M.; Boyd, N.S. Weed Detection in Perennial Ryegrass With Deep Learning Convolutional Neural Network. *Front. Plant Sci.* **2019**, *10*. [CrossRef] [PubMed]
53. Tzutalin LabelImg. LabelImg 2015. Available online: https://github.com/tzutalin/labelImg (accessed on 30 June 2020).
54. LeCun, Y.; Bottou, L.; Bengio, Y.; Haffner, P. Gradient-based learning applied to document recognition. *Proc. IEEE* **1998**, *86*, 2278–2324. [CrossRef]
55. Torrey, L.; Shavlik, J. Transfer learning. In *Handbook of Research on Machine Learning Applications and Trends: Algorithms, Methods, and Techniques*; IGI Global: Hershey, PA, USA, 2010; pp. 242–264.
56. Karimi, Y.; Prasher, S.O.; McNairn, H.; Bonnell, R.B.; Dutilleul, P.; Goel, P.K. Classification accuracy of discriminant analysis, artificial neural networks, and decision trees for weed and nitrogen stress detection in corn. *Trans. Am. Soc. Agric. Eng.* **2005**. [CrossRef]
57. Sandler, M.; Howard, A.; Zhu, M.; Zhmoginov, A.; Chen, L.C. MobileNetV2: Inverted Residuals and Linear Bottlenecks. In Proceedings of the IEEE Computer Society Conference on Computer Vision and Pattern Recognition, Salt Lake City, UT, USA, 18–23 June 2018; pp. 4510–4520.
58. Chollet, F. Transfer Learning Using Pretrained ConvNets. Available online: https://www.tensorflow.org/alpha/tutorials/images/transfer_learning (accessed on 30 June 2020).

59. Lin, T.Y.; Maire, M.; Belongie, S.; Hays, J.; Perona, P.; Ramanan, D.; Dollár, P.; Zitnick, C.L. Microsoft COCO: Common objects in context. In *Computer Vision—ECCV 2014*; Lecture Notes in Computer Science: Cham, Switzerland, 2014; pp. 740–755. [CrossRef]
60. Huang, J.; Rathod, V.; Sun, C.; Zhu, M.; Korattikara, A.; Fathi, A.; Murphy, K. Speed/accuracy trade-offs for modern convolutional object detectors. In Proceedings of the IEEE Conference on Computer Vision and Pattern Recognition, Honolulu, HI, USA, 21–26 July 2017; pp. 7310–7311.
61. Szegedy, C.; Vanhoucke, V.; Ioffe, S.; Shlens, J.; Wojna, Z. Rethinking the Inception Architecture for Computer Vision. In Proceedings of the IEEE Computer Society Conference on Computer Vision and Pattern Recognition, Las Vegas, NV, USA, 27–30 June 2016; Volume 2016, pp. 2818–2826.
62. He, K.; Zhang, X.; Ren, S.; Sun, J. Deep residual learning for image recognition. In Proceedings of the IEEE Computer Society Conference on Computer Vision and Pattern Recognition, Las Vegas, NV, USA, 27–30 June 2016; Volume 2016, pp. 770–778.
63. Liu, S.; Deng, W. Very deep convolutional neural network based image classification using small training sample size. In Proceedings of the 3rd IAPR Asian Conference on Pattern Recognition, ACPR 2015, Kuala Lumpur, Malaysia, 3–6 November 2015; pp. 730–734.
64. Girshick, R.; Donahue, J.; Darrell, T.; Malik, J. Region-Based Convolutional Networks for Accurate Object Detection and Segmentation. *IEEE Trans. Pattern Anal. Mach. Intell.* **2016**, *38*, 142–158. [CrossRef]
65. Girshick, R. Fast R-CNN. In Proceedings of the IEEE International Conference on Computer Vision, Santiago, Chile, 7–13 December 2015.
66. Ren, S.; He, K.; Girshick, R.; Sun, J. Faster R-CNN: Towards Real-Time Object Detection with Region Proposal Networks. *IEEE Trans. Pattern Anal. Mach. Intell.* **2017**, *39*, 1137–1149. [CrossRef]
67. Liu, W.; Anguelov, D.; Erhan, D.; Szegedy, C.; Reed, S.; Fu, C.-Y.; Berg, A.C. SSD: Single Shot MultiBox Detector. In *Computer Vision—ECCV 2016*; Lecture Notes in Computer Science: Cham, Switzerland, 2015. [CrossRef]
68. Everingham, M.; Van Gool, L.; Williams, C.K.I.; Winn, J.; Zisserman, A. The pascal visual object classes (VOC) challenge. *Int. J. Comput. Vis.* **2010**, *88*, 303–338. [CrossRef]
69. Doersch, C.; Gupta, A.; Efros, A.A. Unsupervised visual representation learning by context prediction. In Proceedings of the IEEE International Conference on Computer Vision, Santiago, Chile, 7–13 December 2015.
70. Brust, C.A.; Käding, C.; Denzler, J. Active learning for deep object detection. In Proceedings of the VISIGRAPP 2019—14th International Joint Conference on Computer Vision, Imaging and Computer Graphics Theory and Applications, Prague, Czech, 25–27 February 2019; SciTePress: Setúbal, Portugal, 2019; Volume 5, pp. 181–190.

Article

Assessing the Operation Parameters of a Low-altitude UAV for the Collection of NDVI Values Over a Paddy Rice Field

Rui Jiang [1,2,3,4], Pei Wang [1,3,4], Yan Xu [1], Zhiyan Zhou [1,3,4,*], Xiwen Luo [1,3,4], Yubin Lan [3,5], Genping Zhao [6], Arturo Sanchez-Azofeifa [2] and Kati Laakso [2]

1 College of Engineering, South China Agricultural University/Guangdong Engineering Research Center for Agricultural Aviation Application (ERCAAA), Guangzhou 510642, China; rjiang3@ualberta.ca (R.J.); wangpei@scau.edu.cn (P.W.); autoxuyan@stu.scau.edu.cn (Y.X.); xwluo@scau.edu.cn (X.L.)
2 Centre for Earth Observation Sciences (CEOS), Department of Earth and Atmospheric Sciences, University of Alberta, Edmonton, AB T6G 2E3, Canada; gasanche@ualberta.ca (A.S.-A.); laakso@ualberta.ca (K.L.)
3 National Center for International Collaboration Research on Precision Agricultural Aviation Pesticides Spraying Technology (NPAAC), Guangzhou 510642, China; ylan@scau.edu.cn
4 Key Laboratory of Key Technology on Agricultural Machine and Equipment (South China Agricultural University), Ministry of Education, Guangzhou 510642, China
5 College of Electronic Engineering, South China Agricultural University, Guangzhou 510642, China
6 School of Computer Science and Technology, Guangdong University of Technology, Guangzhou 510006, China; genping.zhao@gdut.edu.cn
* Correspondence: zyzhou@scau.edu.cn; Tel.: +86-135-6002-6139

Received: 5 May 2020; Accepted: 4 June 2020; Published: 8 June 2020

Abstract: Unmanned aerial vehicle (UAV) remote sensing platforms allow for normalized difference vegetation index (NDVI) values to be mapped with a relatively high resolution, therefore enabling an unforeseeable ability to evaluate the influence of the operation parameters on the quality of the thus acquired data. In order to better understand the effects of these parameters, we made a comprehensive evaluation on the effects of the solar zenith angle (SZA), the time of day (TOD), the flight altitude (FA) and the growth level of paddy rice at a pixel-scale on UAV-acquired NDVI values. Our results show that: (1) there was an inverse relationship between the FA (≤100 m) and the mean NDVI values, (2) TOD and SZA had a greater impact on UAV–NDVIs than the FA and the growth level; (3) Better growth levels of rice—measured using the NDVI—could reduce the effects of the FA, TOD and SZA. We expect that our results could be used to better plan flight campaigns that aim to collect NDVI values over paddy rice fields.

Keywords: NDVI; solar zenith angle; flight altitude; time of day; operating parameters; UAV

---

1. Introduction

Paddy rice is a staple crop for more than half the world's population, especially in Asia [1]. China is the largest paddy rice-producing country, but still faces challenges from a rapid increase in global food demand and food security [2]. Hence, it is of utmost importance to increase food productivity in a sustainable manner. Successful and timely estimations of paddy-rice growth are useful means of cultivation management, agricultural decision-making and tillage improvement [3]. Cultivating paddy rice scientifically can contribute to disease prevention, optimal nutrient use, yield increments and the reduction of environmental pollution arising due to excessive nitrogen use [4–6]. One essential aspect of effective field management practices is to monitor crop growth with higher efficiency.

For decades, optical satellite systems have been an effective means of monitoring crop growth and estimating yields [7,8]. To this end, Landsat, MODIS, AVHRR, SPOT and other satellites have been commonly used [9]. Data from these satellites can be effectively used to monitor crop growth over a vast region, but most freely available satellite data are still untimely, and the spatial resolution is insufficient for precision agriculture [10]. For example, Landsat 8 has a temporal resolution of 16 days and a spatial resolution of 30 m [11]. As timeliness and high resolution are crucial for agricultural production and management, such data have limited applicability for agricultural production monitoring and management. Additionally, satellite-derived data are often affected by atmospheric effects induced by the absorption and scattering of aerosols and molecules. Conducting the required atmospheric and spatial corrections before satellite data can be effectively analyzed adds to the time required for data analysis.

Moreover, there are many commercial field sensors that can be used to infer information on the structural and biochemical properties of vegetation in a nondestructive manner. Examples on these sensors include, but are not limited to, the GreenSeeker handheld crop sensor (Trimble, Inc., Sunnyvale, CA, USA), Crop Circle™ (Holland Scientific, Inc., Lincoln, NE, USA), N-Sensor ALS (Precision Decisions, Ltd., Shipton, UK), Crop Spec (Topcon Positioning Systems, Inc., Livermore, CA, USA) and FieldSpec 2500© (Malvern Panalytical, Malvern, UK) [12]. However, these sensors also have disadvantages, such as short sensory distance and thus, a small spatial coverage [13]. One must usually distribute a large number of sensors in the field for data collection, but even then, handheld sensor are inefficient to cover large areas [14].

Recently, UAV-based remote sensing has emerged as a promising solution for monitoring crop growth in a flexible and widely applicable manner and with a high throughput. UAVs typically have an excellent maneuverability, low cost and high safety, which gives them an advantage over some other platforms [15]. Compared to field sensors, optical imaging sensors mounted on UAVs can be used to efficiently acquire data due to their broader spatial coverage [16–18]. Furthermore, unlike satellite data, UAVs offer means for real-time image analysis based on high spatial resolution data. Due to the multitude of advantages of UAVs, many researchers have applied such data to study agricultural production and practices. For instance, UAV-based optical sensing has been successfully used to infer information on soybeans [19], wheat breeding, yield estimation [20], paddy-rice diseases [21] and yield evaluation [22,23].

Due to the many benefits of UAVs, they have been identified [24] as an established tool for precision agriculture [25]. In many studies, normalized difference vegetation index (NDVI), the most commonly used index in vegetation studies [26], has been successfully applied to UAV-based data [23]. Flight parameters such as the flight altitude (FA) need to be considered because they determine the quality of the remote sensing data. For instance, the ground sampling distance (GSD = size of pixel × flight altitude/focal length) [27] and the spatial coverage of the survey [28,29] determine to which extent patterns and objects can be detected. On the other hand, collecting high-quality remote-sensing data can become challenging if the bidirectional reflectance distribution function (BRDF), resulting from a large solar zenith angle (SZA), affects the results [30]. The time of day (TOD), also related to the SZA, is also an important factor that can influence optical remote sensing data and hence needs to be taken into account. In addition, in the context of rice crops, the different growth levels can influence the spectral discrimination between rice and other image components (soil and water). However, there are few studies to have comprehensively evaluated these factors. Because of these reasons, it is essential to make a comprehensive assessment of the above discussed parameters (TOD, SZA, FA and growth level) on UAV data acquired for precision agriculture.

Here, we present comparison trials of UAV-borne NDVI values, called UAV–NDVIs from herein, acquired using different flight parameters and crop growth levels. We conducted our study in a paddy rice field that was treated with different fertilizers to induce variability in crop growth levels. The specific goals of this study were to assess (1) the effects of the TOD, SZA and FA on the UAV–NDVIs; and (2) the susceptibility of the UAV–NDVIs to the variability in the growth levels of crops.

2. Materials and Methods

2.1. Study Site

A 1500 m^2 (30 m × 50 m) paddy rice field (Figure 1a) (variety: '*Meixiang zhan*') in the Ningxi Teaching and Research Bases (NTRB) at the South China Agricultural University, Guangzhou, China (23°14′22″N, 113°38′31″E), was selected for the survey. Plant spacing and row spacing in the field were both 0.2 m. The rice field was divided into five comparison zones: F1, F2, F3, F4 and F5. Each zone consisted of three fields, F1: A1, B1 and C1; F2: A2, B2 and C2; F3: A3, B3 and C3; F4: A4, B4 and C4; F5: A5, B5 and C5 (Figure 1b). The size of each field was 10 m × 10 m. Additionally, five zones were fertilized with different nitrogen application rates (Figure 2). It should be noted that the fields of each comparison zone were treated with the same fertilization strategy. This strategy involved applying different amounts of base fertilizer, tillering fertilizer and panicle fertilizer to create differences in crop growth.

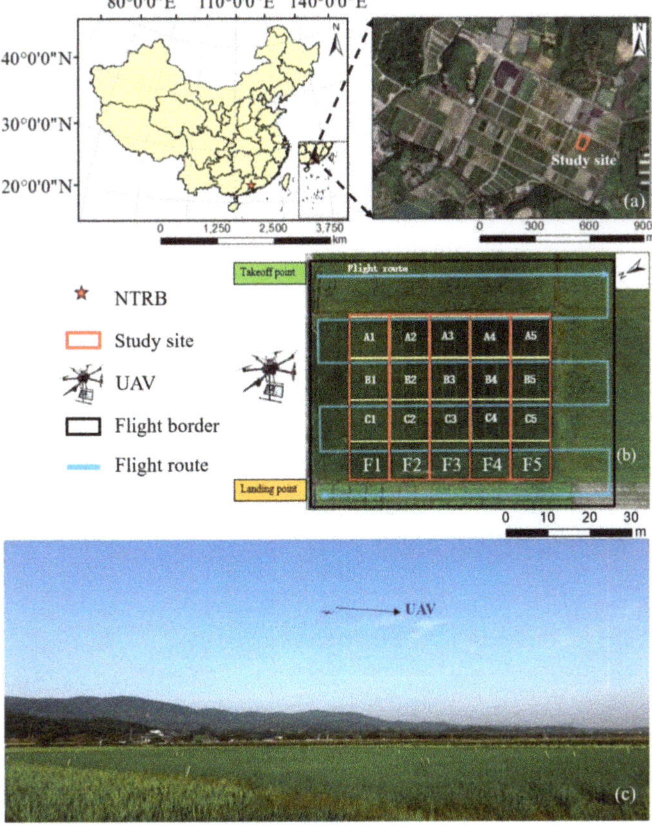

Figure 1. Study site. (**a**) The study site was located at a paddy-rice field, marked by the pink rectangle in Ningxi Teaching and Research Bases (NTRB) at the South China Agricultural University, Guangzhou, China; (**b**) five field zones were present: F1, F2, F3, F4 and F5, marked by red rectangles; blue flight route is a schematic diagram that represents the 4 different routes at different flight altitudes: 40 m, 60 m, 80 m and 100 m; (**c**) unmanned aerial vehicle (UAV)-based multispectral camera acquiring raw multispectral images of the study site at a low flight altitude (40–100 m).

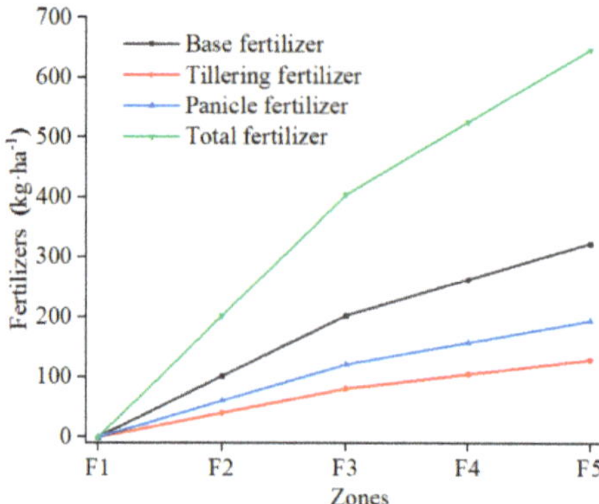

Figure 2. Fertilization program of the test field.

2.2. UAV Data Collection

A Micasense RedEdge-M camera (MicaSense, Inc., Seattle, WA, USA) was attached to the Matrice-600 Pro UAV (DJI, Inc., Shenzhen, China), a six-rotor UAV, for the aerial imagery collection. This camera has five independent imagers, each with a custom narrowband filter that enables the imager to receive a precise spectrum. The RedEdge-M camera also includes a downwelling light sensor (DLS) and a GPS module. The DLS is a light sensor with five bands, which need to connect directly to the host for radiation calibration. The specifications of the RedEdge-M camera are given in Table 1.

Table 1. The specifications of the RedEdge-M camera. Acronyms: DLS—downwelling light sensor; FOV—field of view; FWHM—full width at half maximum; NIR—near infrared.

Specification	Parameters
Weight	170 g (including DLS)
Dimensions	9.4 cm × 6.3 cm × 4.6 cm
Power	4.2–15.8 V DC, 4 W nominal, 8 W peak
Spectral bands	Blue (475 nm, FWHM: 20 nm); green (560 nm, FWHM: 20 nm); red (668 nm, FWHM: 10 nm); red edge (717 nm, FWHM: 10 nm); NIR (840 nm, FWHM: 40 nm)
Maximum capture speed	1 capture per second (all bands)
Storage format	16 bits TIFF
Field of view (FOV)	46°

The flight plans were designed using the Atlas Flight application (Micasense, Inc., Fremont, CA, USA). The flight altitudes were set to 40 m, 60 m, 80 m and 100 m and the flight speed was set to 5 m/s. In addition. the side overlap, and forward overlap of the camera were set to 80% to ensure effective orthorectification and mosaicking of the resulting images. Following the instructions of the Micasense manual (https://support.micasense.com), a calibrated Teflon panel was imaged by the RedEdge-M camera for radiometric calibration before and after each flight. Using these settings, the image acquisition was conducted above the study area from 6 am to 6 pm on 11 June 2018. A total of 48 flights were thus conducted and 31,980 images were collected by performing each flight mission at an interval of 15 min. Detailed flight parameters are given in Table 2.

2.3. Ground Data Collection

The ground NDVI data were obtained using the GreenSeeker handheld crop sensor (Trimble, Inc., Sunnyvale, CA, USA), which has two independent LEDs that can emit 774 nm near-infrared (NIR) light and 656 nm red light. Due to its own light source, the sensor is relatively unaffected by solar conditions. In this study, the GreenSeeker sensor was used to acquire NDVI ground data (called GS–NDVIs from here forward), to determine the different growth levels of the five zones (F1–F5) described above. The GreenSeeker sensor calculates the NDVI values using the following formula:

Table 2. Flight parameters of the study.

Start Time of Each Hour (min)	Flight Altitude (m)	Image Overlap (%)	Speed ($m{\cdot}s^{-1}$)	Ground Sampling Distance (cm)
<15	40	80	5	2.88
15–30	60			4.32
30–45	80			5.81
>45	100			7.22

$$GS - NDVI(656\,nm, 774\,nm) = \frac{R_{774} - R_{656}}{R_{774} + R_{656}} \tag{1}$$

where R_{774} is the reflectance of near-infrared light (774 nm) and R_{656} is the reflectance of red light (656 nm).

In practice, the GS–NDVI values were acquired by positioning the sensor at a 0.8-m height above the canopy. These data were acquired simultaneously with the UAV data to minimize potential discrepancies caused by asynchronous data acquisition. A total of 60 randomly distributed GS–NDVI values were collected from each zone, resulting in a total of 300 measurements across the study area.

2.4. Data Processing and Analysis

The 60 GS–NDVI measurements acquired from each of the five zones were first averaged to create a single representative value for each zone. A map was then compiled of these five GS–NDVI values using the ArcMap software (ESRI, Redlands, CA, USA).

The UAV images were processed in three steps as shown in Figure 3. First, the original images were stitched and geometrically and radiometrically corrected using the Pix4D mapper software (Pix 4D, Inc., Lausanne, Switzerland). Using these images, the UAV–NDVI values were computed using the red and NIR bands of the RedEdge-M camera:

$$UAV - NDVI(668\,nm, 860\,nm) = \frac{R_{840} - R_{668}}{R_{840} + R_{668}} \tag{2}$$

where R_{840} is the reflectance of near-infrared light (840 nm) and R_{668} is the reflectance of red light (668 nm).

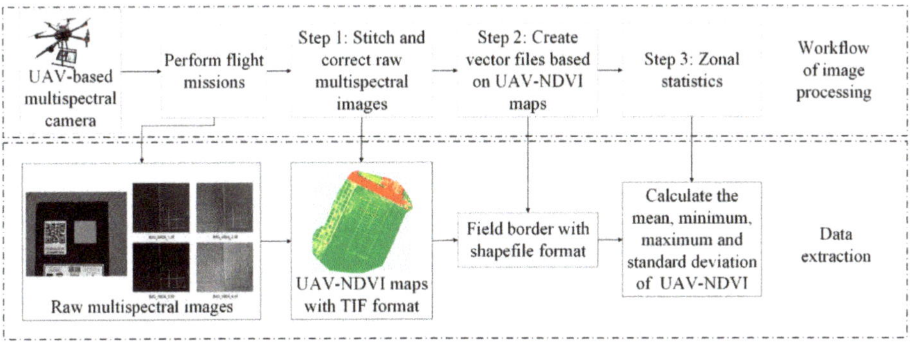

Figure 3. Image processing workflow.

Second, five shapefiles were created to represent each zone using ArcMap. Finally, the Zonal Statistics tool of ArcMap was used to calculate the mean, minimum, maximum and standard deviation (STDEV) of the pixel-scale UAV–NDVI data. These statistics were calculated separately for the five zones, represented by the five shapefiles created in step two. Of the collected values, the mean UAV–NDVI reflects the mean productivity and biomass, and the standard deviation indicates the spatial variability in productivity. The minimum and maximum UAV–NDVIs can provide information on pixel mixing. The count of pixels (COP) of a certain NDVI can give indicators of the NDVI distribution and is determined by the ground sampling distance (GSD), which in turn is influenced by the FA. Therefore, the COPs of each NDVI value were computed to analyze the effect of the FA on the UAV–NDVIs.

In addition to the aforementioned, a total of 48 SZAs were calculated using the NOAA Solar Calculator tool (https://www.esrl.noaa.gov) that was given the location of the study site and the TOD of each flight mission. From herein, identifiers M6–M17 are used to represent each TOD. For example, M6 stands for TOD from 6 a.m. to 7 a.m., and M13 represents the TOD from 1 p.m. to 2 p.m., and so on. It should be noted that this experiment was only conducted for the UAV data because the GreenSeeker instrument has its internal light source and hence, the effects of the TOD are negligent in the ground data. The significance of the effects of the FA, TOD, SZA and growth level on the UAV–NDVIs was determined using the Standardized Regression Coefficients [31] analysis which reflects the relative importance of different variables. This analysis was conducted using the SPSS® software package (IBM, NY, USA) to evaluate the order of influence.

3. Results

3.1. The Rice Growth Levels Determined by the GS–NDVIs

In Figure 4, the mean GS–NDVI values, obtained through ground measurements, are shown. As can be seen in figure, there is a trend from the highest NDVI values in zone F5 toward the lowest NDVI values in zone F1. This trend is consistent with fertilizer use, shown in Figure 2, that increases from zone F1 toward zone F5. Therefore, it is obvious that the differences in the NDVI values are due to the fertilizer strategy and the zones thus represent different growth levels of paddy rice.

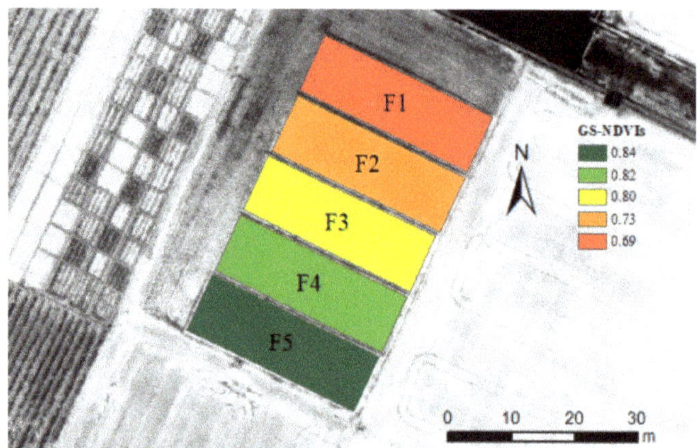

Figure 4. Map of the five zones (F1–F5) and their average normalized difference vegetation index (NDVI) values (GS–NDVIs) based on ground data.

3.2. Effects of the Flight Altitude (FA) on the NDVI Values

In this experiment, we analyzed the influence of the FA on the UAV–NDVIs. The resulted, shown in Figure 5, suggest five trends: (i) the mean and maximum UAV–NDVIs increase with decreasing FA

(Figure 5a,c); (ii) the minimum UAV–NDVIs increase with increasing FA (Figure 5b); (iii) the standard deviation values decrease with increasing FA (Figure 5d); (iv) the mean, minimum and maximum UAV–NDVIs increase from zone F1 to zone F5 and (v) the standard deviation values decrease from zone F1 to zone F5. These trends were based on the obtained UAV–NDVI values throughout the day.

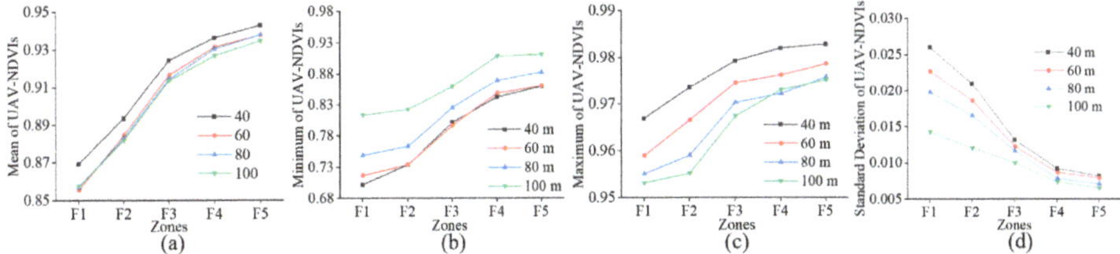

Figure 5. Mean (**a**), minimum (**b**), maximum (**c**) and standard deviation values (**d**) of the UAV–NDVIs of the five zones (F1–F5) at different flight altitudes.

In order to further analyze the COP differences of the UAV–NDVIs in the field-scale, the COP distribution curves of the UAV–NDVIs were analyzed under different FAs for fields B1–B5 (Figure 6). It could be seen that, on the whole, the higher the FA, the lower the COP, the smaller the NDVI value of the peak COP and the smaller the NDVI value overall. For instance, in B1 (Figure 6a), the peak COP NDVI values were 0.785, 0.775, 0.755 and 0.755 for 40 m, 60 m, 80 m, 100 m, respectively, and the COPs of these peak NDVI values were 7820, 3808, 2162 and 1596, respectively. The mean UAV–NDVIs of the FAs from 40 m to 100 m were 0.766, 0.756, 0.742 and 0.733. The same tendency could be observed in the other fields.

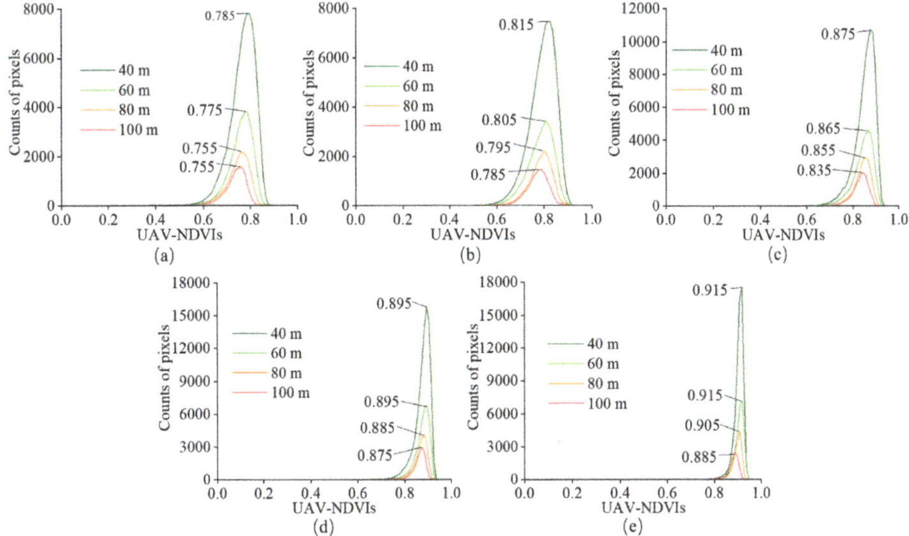

Figure 6. Count of pixels (COP) distribution curves of the UAV–NDVIs of the B1–B5 (**a**–**e**) fields under different flight altitudes at the time of day (TOD) of 11 a.m.–12 p.m. (M11).

3.3. *Effects of the Time of Day (TOD) on the NDVI Values*

Figure 7 showed the trends of the UAV–NDVIs of the five zones (F1–F5) as a function of the TOD and FA. As can be seen in figure, the general trend of the mean UAV–NDVIs was nearly uniform across all zones and FAs. This suggests that the mean, minimum and maximum UAV–NDVIs were highest during the morning (TODs: M6–M9) and late afternoon (TODs: M13–M17) and lowest around the midday (TODs: M10 to M12). In contrast, the standard deviation values were higher around midday than at other times.

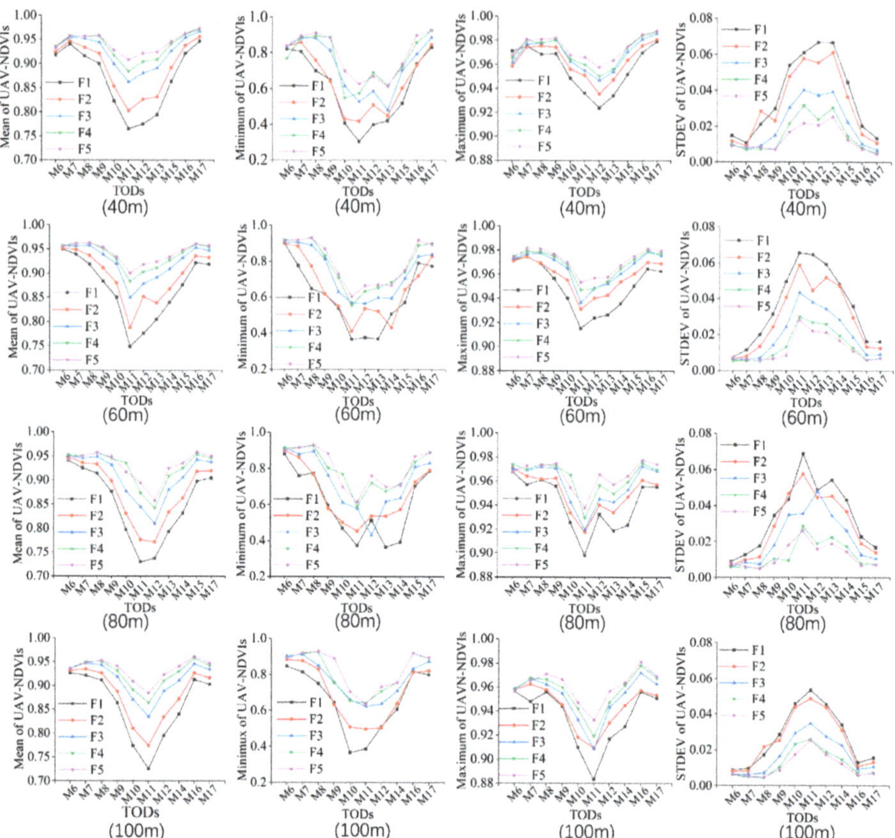

Figure 7. Mean, minimum, maximum and standard deviation of the UAV–NDVIs shown for different zones (F1–F5), times of day (TODs; M6–M17) and flight altitudes (FAs; 40, 60, 80 and 100 m).

3.4. Dependence of the UAV–NDVIs on the Solar Zenith Angle (SZA)

As can be seen in Figure 8, two general trends can be observed in the SZA values of the study: (i) the mean, minimum and maximum UAV–NDVI values generally increased as a function of increasing SZA and (ii) the standard deviation values decreased as a function of increasing SZA. The trends were similar across all zones (F1–F5) and FAs. As the lowest SZAs occur around the solar noon this means that the UAV–NDVI values were lower around noon than in the morning or in the afternoon. The reverse was true for the standard deviation values. These results are consistent with those discussed in Section 3.3, which is logical, since the SZAs depend on the TOD.

3.5. Effects of the Growth Levels of Rice on the UAV–NDVIs

The differences between the mean UAV–NDVI values of the five zones followed those acquired using the ground data (F1: lowest, F5: highest, see Figure 4). In contrast, the standard deviation values had a downward trend from zone F1 to zone F5. With some slight variation, these differences could be detected irrespective of the FA, TOD and SZA (see Figures 5, 7 and 8). Thus, better growth levels induced higher NDVI values and lower standard deviation values. Furthermore, as could be seen in

Figure 6, the curves of each field (B1–B5) consistently show a right shift (toward higher UAV–NDVI values) and a compressed value distribution with increasing growth level (B1: lowest, B5: highest). Hence, the better the growth level, the denser the UAV–NDVI distribution at the pixel scale.

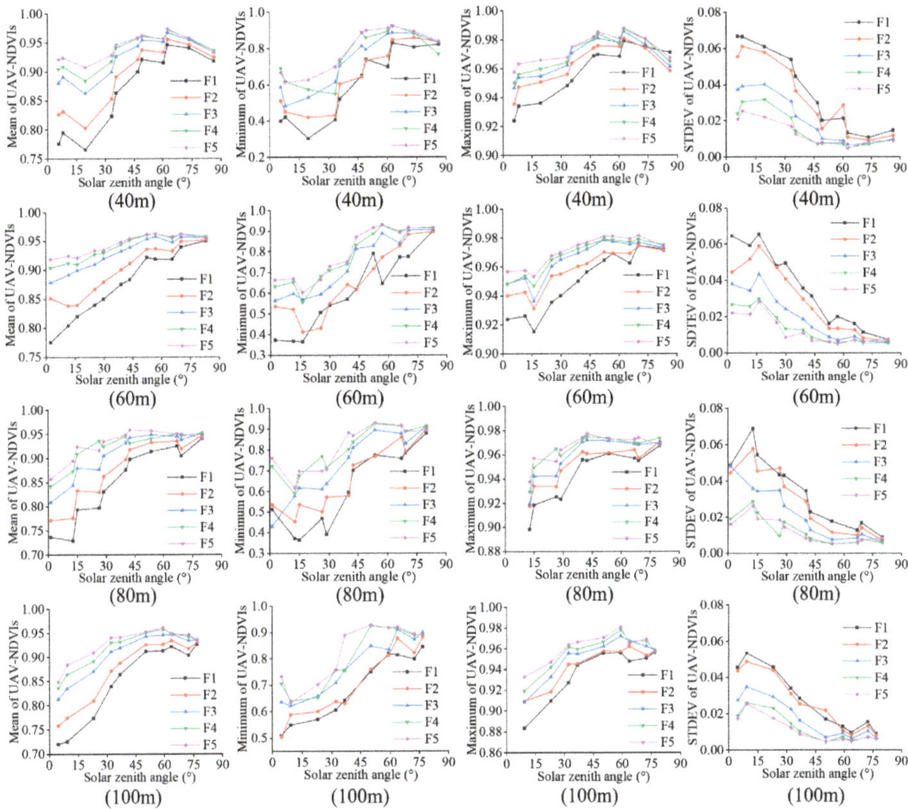

Figure 8. Mean, minimum, maximum and standard deviation of the UAV–NDVIs shown for different zones (F1–F5), solar zenith angles and flight altitudes (40, 60, 80 and 100 m).

To further study the trends associated with the growth levels, the averaged range of the mean UAV–NDVIs are shown individually for the FAs, TODs and SZAs in Table 3. As can be seen, the maximum averaged range of the FAs was 0.031 and that of TODs and SZAs was 0.200 in zone F1. In zone 5, the range was from 0.019 to 0.085. Based on the resulted, shown on Table 3, indicating that the better growth level could reduce the effect of FAs, TODs and SZAs.

Table 3. Average range of means UAV–NDVIs of F1, F2, F3, F4 and F5 fields.

Impactors	F1	F2	F3	F4	F5
FAs	0.031	0.031	0.025	0.022	0.019
TODs(h)/SZAs (°)	0.200	0.168	0.122	0.100	0.085

3.6. Relative Importance of Different Impactors: FAs, SZAs/TODs, Growth Levels

In order to further analyze the contribution of the above factors, the standardized regression coefficients (SRC) of the linear model are shown in Table 4. According to the positive relationship between the contribution and absolute values of the SRC, the descending contribution rank regarding the mean and STDEV UAV–NDVIs can be listed as: SZAs/TODs, growth levels and FAs. It can be found that the FAs affect the values less than the other parameters and the SZAs/TODs had a significant effect on the mean UAV–NDVIs and STDEV UAV–NDVIs.

Table 4. Standardized regression coefficients (SRC) of FAs, SZAs/TODs and growth levels.

Impactors	Mean UAV–NDVIs	STDEV UAV–NDVIs
FAs	−0.148 **	−0.089 **
TODs(h)/SZAs (°)	0.698 **	−0.728 **
Growth levels	0.509 **	−0.510 **

Note: ** represents the $p \leq 0.01$.

4. Discussion

Remote sensing has been widely used to assess crop growth in different environments. UAV-based imaging technology has the potential to provide high spatial resolution (up to centimeter-scale) maps for this end, providing instant feedback needed for crop management and decision making. In this context, increasing the understanding of how the flight parameters affect the NDVI values acquired using UAV-systems can help improve the image quality of the thus acquired data. In this study, we used a lightweight UAV equipped with a multispectral camera to collect NDVI values over a paddy rice under different flight parameters as well as different crop growth levels. The results suggest that the flight parameters and growth levels have a significant effect on the UAV-based data, thus highlighting the importance of careful flight planning. Nevertheless, it is essential to mention that the parameters and environmental factors (FA, TOD, SZA and growth level) discussed in this study do not fully account for the quality of the UAV-acquired data: Factors such as the flight speed can have a significant impact on the quality of the remote sensing data.

4.1. Sensitivity of UAV–NDVIs to the Flight Altitude (FA)

Our resulted, discussed in Section 3.2, suggest that the mean, minimum, maximum and standard deviation values of the UAV–NDVIs were highly related to the FA. This was particularly true for the COP of the peak UAV–NDVI values, shown in Figure 6. Additionally, as can be seen in the Appendix A Table A1 and Figure A1, under the same growth level, the higher the FA, the smaller the individual UAV–NDVIs and the narrower the range of the UAV–NDVIs. Regarding the effects of the FA on the NDVI values, previous studies have reported different results. Stow et al. [32] evaluated the effects of the FA on the NDVI values through field experiments, finding that the relationship between them was inconclusive. In studies by Rasmussen et al. [33] and Yu et al. [34], no significant associations were found between the FA and the NDVI values. Easterday et al. [35] concluded that of the NDVI values acquired under different FAs (30 m, 60 m, 100 m and 120 m), the lowest FAs were better for detecting water deficit in plants. Similar to our study, a report by Mesas-Carrascosa et al. [36] concluded that even though the mean and maximum UAV–NDVIs were negatively related to the FA, the reverse was true for the minimum and STDEV UAV–NDVIs.

Notwithstanding the relationship between the NDVIs and the FA, the selection of the FAs has important implications from the point of view of the spatial resolution of the remote sensing imagery. In the field of imaging technology, a higher GSD means that each pixel represents a larger area. However, with the resulting lower spatial resolution of the image, the spectral information included in each pixel becomes more mixed and hence, detecting objects and phenomena can become more difficult. At high FAs associated with low spatial resolutions, some pixels could be mixed, and the average NDVIs may become diluted [37].

Looking at the relationship between the FA and the number of pixels of this study, Figure A1 shows four subsets of the UAV–NDVI maps of the B2 field at different FAs at TOD M11. In this analysis, four subsets were selected to represent the FAs of 40, 60, 80 and 100 m. These FAs resulted in GSDs values of 2.88, 4.32, 5.81 and 7.22 cm×pixel^{-1} that had NDVI values between 0.5 and 0.86. As can be seen in Figure A1, increasing the GSD resulted in fewer pixels in the same sampled area and narrower COP distributions. Under normal circumstances, the higher the FA of the UAV platform, the larger the area represented by a single pixel, and thus, the lesser the ability of the UAV system to detect small features. In the case of a paddy rice field, despite their commonly high canopy densities, there are still many canopy gaps. These gaps have other elements such as soil and water, which, depending on the FA, can lead to different degrees of spectral mixing. This, in turn, can lead to overall smaller mean NDVI and standard deviation values as suggested by Figure 5a,b, respectively. Therefore, as far as the NDVI monitoring of rice is concerned, the average NDVI was inversely proportional to the FAs. This further supports the conclusion that the FA was an important factor affecting the image quality for the crops. In general, it can be concluded that FAs do impact the mean UAV–NDVIs, but it is not conclusive that the mean UAV–NDVIs will keep decreasing with increasing FA. A possible extreme flight altitude needs further exploration.

Another noticeable factor regarding the effect of FA on vegetation indexes was atmospheric effects. Yu et al. suggested normalized excess green index (ExG) and the normalized green–red difference index (NGRDI) were relatively susceptible to FAs than the NDVI values (FAs: 10 m, 30 m, 50 m and 100 m) considering of the atmospheric effects [34]. The atmospheric attenuation was small in the low-altitude atmospheres and the effect of path radiance in their model was negligible for red and infrared wavelengths. Thus, we did not make an analysis of the atmospheric effects upon NDVIs when the FAs were under 100 m. In fact, the atmospheric effects were more significant at higher FAs (manned aircrafts and satellites) [38].

4.2. Influence of the time of Day (TOD) and Solar Zenith Angle (SZA) on the UAV–NDVIs

Although the SZA values vary as a function of the time of day, date and location [39], in this study, SZAs and TODs were discussed independently, because there is the spatiotemporal difference. To the knowledge of the authors, our study is one of the first ones to report and discuss the effects of the TODs and SZAs on UAV-based NDVI values continuously from sunrise to sunset. Based on our results, the mean, minimum, maximum and standard deviation of the UAV–NDVIs are all highly related to the TODs and SZAs (see Sections 3.3 and 3.4). In previous studies, Ishihara et al. [28] suggests that the response of vegetation indices to the SZA was evident under a clear sky. Rahman [29] reported NDVIs from ground-based observations that decreased with decreasing SZA at a pasture site, but did not provide explanations to this finding. Their conclusions were directly based on the average assessment of ground-NDVIs, and hence, not remote sensing data. Although the results in this study show a similar phenomenon, we have further analyzed the UAV–NDVI distribution on a pixel-scale (including mean, minimum, maximum, standard deviation and COP of the UAV–NDVIs), an experimental setup that can reveal the relationships between the UAV–NDVIs and the flight parameters (SZAs/TODs) more clearly. Ishihara et al. suggested that surveys involving NDVI measurements be performed at a SZA of 60° to be effective for the accurate assessment of the canopy structure and function. However, our results, shown in Figure 8, suggest no obvious differences between the mean UAV–NDVIs and standard deviation values of the UAV–NDVIs with different growth levels at a SZA of 60°. Therefore, we do not follow the recommendation by Ishihara et al. to acquire UAV–NDVI values at a SZA of 60°. In practice, the TODs and SZAs were the direct reasons for changing the angle between the view direction and incident light. An important phenomenon in this context is the widely discussed hot spot effect, also called the hot spot directional signature in the backscattering direction [40]. In recent studies, this phenomenon was also called the bidirectional reflectance distribution function (BRDF) effects. Generally, the radiance of the reflecting medium will decrease with an increasing angle between the view direction and incident light because of the decreased probability of seeing illuminated particles [41]. Therefore, because of the vertical downward view of the RedEdge camera, the strongest radiance and relatively complete reflectivity from the rice canopy was collected when the SZAs were relatively small. By definition, crop reflectance is the ratio of the amount of light leaving the canopy to the amount of incoming light. We suggest the reflectivity was relatively accurate at a smaller SZAs, in turn, resulting in reliable NDVI values.

4.3. Effect of the Growth Level on the UAV–NDVIs

It can also be seen in the data that the application rate of fertilizers can have a significant impact on the NDVI values (Figure 4). In the early stages of rice growth or in the case of weak growth that results from inadequate fertilization, the canopy is distributed sparsely and there are more gaps with water between seedlings [42]. These potential gaps would influence the NDVI values. For instance, looking at the image acquired at M11 (11 a.m.–12 p.m.) at 40 m (Figure A1), the image has a relatively high GSD of 2.88 cm at a FA of 40 m. Due to the resulting high spatial resolution, the image contains many details, including those related to non-foliar areas, and thus, small NDVI values that result from pixels with soil or water. Similarly, areas with lesser amounts of fertilizers and thus, less vigorous growth levels will have more gaps with smaller NDVI values and higher standard deviation values than result from spatial heterogeneity (vegetation, water and soil). The higher the flight altitude, the lower the spatial resolution and the lesser the amount of details in the image. Therefore, in the UAV image acquired at 100 m, the standard deviation is lower and the range of NDVI values is narrower than in the image that was acquired at 40 m (Figures 7 and A1). On the other hand, the main limitation in our growth level analysis was the lack of observations to cover entire growth stages. Such an analysis would have provided a more comprehensive quantitative framework to interpret the interrelationships between the growth levels and the flight parameters. In fact, in the early stages of rice growth, due to the sparse rice canopy, aerial remote sensing will collect a large area of ground water surface (mirror reflectance), which is a growth period that is not suitable for UAV remote sensing monitoring. Hence, the available growth stages for reliable UAV remote sensing are limited.

5. Conclusions

In this study, a paddy rice field was used as a test area to assess the effects of FA, TOD, SZA and growth level on the UAV-acquired NDVI values. Based on the SRC values of the linear regression resulted, all of the former parameters had a significant effect on the UAV–NDVIs. More specifically, our results suggest that (1) the SZA/TOD had the largest impact on the mean, minimum, maximum and STDEV of UAV–NDVI values, followed by the growth level and the FA; (2) the mean, maximum and STDEV of UAV–NDVIs were inversely proportional to the FA (≤100 m) and the minimum of UAV–NDVIs decreased with increasing FA; (3) the mean, minimum and maximum of UAV–NDVIs were proportional to the SZA, but it was contrary with the STDEV of UAV–NDVIs; (4) the effect of TOD on UAV–NDVIs could refer to the SZA;(5) Furthermore, according to our results, the UAV–NDVIs close to the smaller SZAs show the highest signal-to-noise ratios which we infer to reflect the most realistic growth status values. We expect that our resulted and the recommendations could provide a reference to the operating parameters of UAVs in the context of precision agriculture.

Author Contributions: R.J. was responsible for the framework design of the entire system in this research and arranged tests, conducted several verification tests and wrote the article. P.W. and Y.X. assisted in collected all the test data and checked the study carefully. Z.Z. proposed the main plans, ideas and guidance for the work and reviewed the study, as well as acquired the funding. X.L., Y.L., G.Z., A.S.-A. and K.L. provided guidance and advice. All authors have read and agreed to the published version of the manuscript.

Funding: This work was supported by the National Natural Science Foundation of China (31871520), the National Key R&D Program of China (2018YFD0200301), Science and Technology Plan of Guangzhou of China (201807010111), Science and Technology Plan of Guangdong Province of China (2017B090903007) and Innovative Research Team of Agricultural and Rural Big Data in Guangdong Province of China (2019KJ138).

Acknowledgments: We would like to thank the flying permission of the administration of Ningxi Teaching and Research Bases, at the South China Agricultural University, Guangzhou, China. Furthermore, the authors would like to thank the reviewer for their helpful comments.

Conflicts of Interest: The authors declare no conflict of interest.

Appendix A

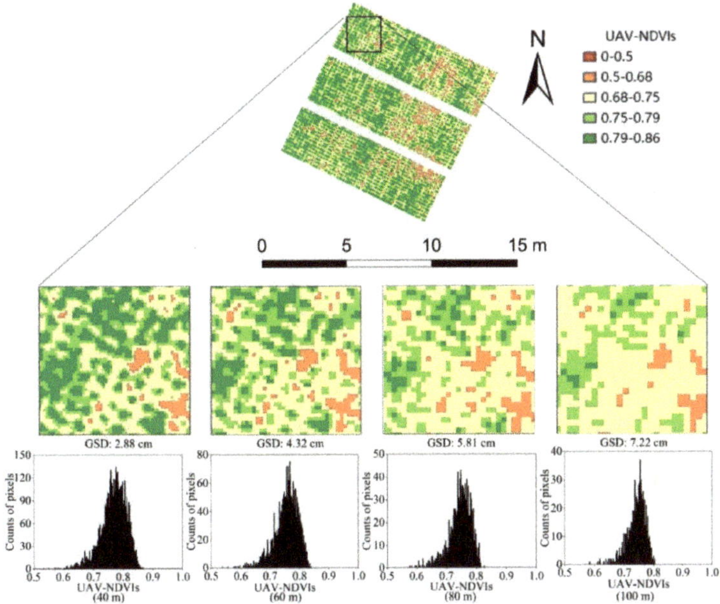

Figure A1. Subset of the B2 NDVI map at 40, 60, 80, 100 m with TOD of M11 (11 a.m.–12 p.m.).

Table A1. Minimum, maximum and range of UAV–NDVIs of the B1–B5 fields.

Fields	Flight Altitude/m	Minimum of UAV–NDVIs	Maximum of UAV–NDVIs	Range
B1	40	0.38	0.88	0.50
	60	0.48	0.88	0.40
	80	0.49	0.86	0.37
	100	0.52	0.84	0.32
B2	40	0.43	0.92	0.49
	60	0.44	0.91	0.47
	80	0.46	0.91	0.45
	100	0.50	0.90	0.40
B3	40	0.53	0.94	0.41
	60	0.57	0.93	0.36
	80	0.61	0.92	0.31
	100	0.65	0.90	0.25
B4	40	0.59	0.94	0.35
	60	0.62	0.93	0.31
	80	0.68	0.92	0.24
	100	0.71	0.91	0.20
B5	40	0.74	0.94	0.20
	60	0.77	0.94	0.17
	80	0.77	0.93	0.16
	100	0.78	0.92	0.14

References

1. Aruva, S.; Sayantani, D.; Vijayalakshmi, S.; Moses, J.A.; Anandharamakrishnan, C. Ageing of rice: A review. *J. Cereal Sci.* **2018**, *81*, 161–170.
2. Zhang, X.; Liu, Y.; Liu, Y.; Cui, Q.; Yang, L.; Hu, X.; Guo, J.; Zhang, J.; Yang, S.; Abou-Amer, I. Impacts of climate change on self-sufficiency of rice in China: A CGE-model-based evidence with alternative regional feedback mechanisms. *J. Clean. Prod.* **2019**, *230*, 150–161. [CrossRef]
3. Sharma, S.; Rout, K.K.; Khanda, C.M.; Tripathi, R.; Shahid, M.; Nayak, A.; Satpathy, S.; Banik, N.C.; Iftikar, W.; Parida, N.; et al. Field-specific nutrient management using Rice Crop Manager' decision support tool in Odisha, India. *Field Crop. Res.* **2019**, *241*, 107578. [CrossRef] [PubMed]
4. Deng, L.; Mao, Z.; Li, X.; Hu, Z.; Duan, F.; Yan, Y. UAV-based multispectral remote sensing for precision agriculture: A comparison between different cameras. *ISPRS J. Photogramm.* **2018**, *146*, 124–136. [CrossRef]
5. Agueera Vega, F.; Carvajal Ramirez, F.; Perez Saiz, M.; Orgaz Rosua, F. Multi-temporal imaging using an unmanned aerial vehicle for monitoring a sunflower crop. *Biosyst. Eng.* **2015**, *132*, 19–27. [CrossRef]
6. Khanal, S.; Fulton, J.; Shearer, S. An overview of current and potential applications of thermal remote sensing in precision agriculture. *Comput. Electron. Agric.* **2017**, *139*, 22–32. [CrossRef]
7. Brinkhoff, J.; Dunn, B.W.; Robson, A.J.; Dunn, T.S.; Dehaan, R.L. Modeling mid-season rice nitrogen uptake using multispectral satellite data. *Remote Sens.* **2019**, *11*, 1837. [CrossRef]
8. Nutini, F.; Confalonieri, R.; Crema, A.; Movedi, E.; Paleari, L.; Stavrakoudis, D.; Boschetti, M. An operational workflow to assess rice nutritional status based on satellite imagery and smartphone apps. *Comput. Electron. Agric.* **2018**, *154*, 80–92. [CrossRef]
9. Fan, X.; Liu, Y. A global study of NDVI difference among moderate-resolution satellite sensors. *ISPRS J. Photogramm.* **2016**, *121*, 177–191. [CrossRef]
10. Matese, A.; Toscano, P.; Di Gennaro, S.F.; Genesio, L.; Vaccari, F.P.; Primicerio, J.; Belli, C.; Zaldei, A.; Bianconi, R.; Gioli, B. Intercomparison of UAV, Aircraft and satellite remote sensing platforms for precision viticulture. *Remote Sens.* **2015**, *7*, 2971–2990. [CrossRef]
11. Wu, M.; Huang, W.; Niu, Z.; Wang, C.; Li, W.; Yu, B. Validation of synthetic daily Landsat NDVI time series data generated by the improved spatial and temporal data fusion approach. *Inform. Fusion* **2018**, *40*, 34–44. [CrossRef]
12. Ali, A.M.; Abou-Amer, I.; Ibrahim, S.M. Using GreenSeeker active optical sensor for optimizing maize nitrogen fertilization in calcareous soils of Egypt. *Arch. Agron. Soil Sci.* **2018**, *64*, 1083–1093. [CrossRef]
13. Barker, D.W.; Sawyer, J.E. Using active canopy sensors to quantify corn nitrogen stress and nitrogen application rate. *Agron. J.* **2010**, *102*, 964–971. [CrossRef]
14. Zheng, H.; Cheng, T.; Yao, X.; Deng, X.; Tian, Y.; Cao, W.; Zhu, Y. Detection of rice phenology through time series analysis of ground-based spectral index data. *Field Crop. Res.* **2016**, *198*, 131–139. [CrossRef]
15. Ostos-Garrido, F.J.; de Castro, A.I.; Torres-Sanchez, J.; Piston, F.; Pena, J.M. High-throughput phenotyping of bioethanol potential in cereals using UAV-based multi-spectral imagery. *Front. Plant Sci.* **2019**, *10*, 948. [CrossRef] [PubMed]
16. Sankaran, S.; Khot, L.R.; Espinoza, C.Z.; Jarolmasjed, S.; Sathuvalli, V.R.; Vandemark, G.J.; Miklas, P.N.; Carter, A.H.; Pumphrey, M.O.; Knowles, N.R.; et al. Low-altitude, high-resolution aerial imaging systems for row and field crop phenotyping: A review. *Eur. J. Agron.* **2015**, *70*, 112–123. [CrossRef]
17. Candiago, S.; Remondino, F.; De Giglio, M.; Dubbini, M.; Gattelli, M. Evaluating multispectral images and vegetation indices for precision farming applications from UAV images. *Remoe Sens.* **2015**, *7*, 4026–4047. [CrossRef]
18. Xiang, H.; Tian, L. Development of a low-cost agricultural remote sensing system based on an autonomous unmanned aerial vehicle (UAV). *Biosyst. Eng.* **2011**, *108*, 174–190. [CrossRef]
19. Yu, N.; Li, L.; Schmitz, N.; Tian, L.F.; Greenberg, J.A.; Diers, B.W. Development of methods to improve soybean yield estimation and predict plant maturity with an unmanned aerial vehicle based platform. *Remote Sens. Environ.* **2016**, *187*, 91–101. [CrossRef]
20. Duan, T.; Chapman, S.C.; Guo, Y.; Zheng, B. Dynamic monitoring of NDVI in wheat agronomy and breeding trials using an unmanned aerial vehicle. *Field Crop. Res.* **2017**, *210*, 71–80. [CrossRef]

21. Liu, T.; Li, R.; Zhong, X.; Jiang, M.; Jin, X.; Zhou, P.; Liu, S.; Sun, C.; Guo, W. Estimates of rice lodging using indices derived from UAV visible and thermal infrared images. *Agric. For. Meteorol.* **2018**, *252*, 144–154. [CrossRef]
22. Reza, M.N.; Na, I.S.; Baek, S.W.; Lee, K. Rice yield estimation based on K-means clustering with graph-cut segmentation using low-altitude UAV images. *Biosyst. Eng.* **2019**, *177*, 109–121. [CrossRef]
23. Zhou, X.; Zheng, H.B.; Xu, X.Q.; He, J.Y.; Ge, X.K.; Yao, X.; Cheng, T.; Zhu, Y.; Cao, W.X.; Tian, Y.C. Predicting grain yield in rice using multi-temporal vegetation indices from UAV-based multispectral and digital imagery. *ISPRS J. Photogramm.* **2017**, *130*, 246–255. [CrossRef]
24. Guan, S.; Fukami, K.; Matsunaka, H.; Okami, M.; Tanaka, R.; Nakano, H.; Sakai, T.; Nakano, K.; Ohdan, H.; Takahashi, K. Assessing Correlation of High-Resolution NDVI with fertilizer application level and yield of rice and wheat crops using small UAVs. *Remote Sens.* **2019**, *11*, 112. [CrossRef]
25. Maes, W.H.; Steppe, K. Perspectives for remote sensing with unmanned aerial vehicles in precision agriculture. *Trends Plant Sci.* **2019**, *24*, 152–164. [CrossRef]
26. Al-Bakri, J.T.; Suleiman, A.S. NDVI response to rainfall in different ecological zones in Jordan. *Int. J. Remote Sens.* **2004**, *25*, 3897–3912. [CrossRef]
27. Seifert, E.; Seifert, S.; Vogt, H.; Drew, D.; van Aardt, J.; Kunneke, A.; Seifert, T. Influence of drone altitude, image overlap, and optical sensor resolution on multi-view reconstruction of forest images. *Remote Sens.* **2019**, *11*, 1252. [CrossRef]
28. Ishihara, M.; Inoue, Y.; Ono, K.; Shimizu, M.; Matsuura, S. The impact of sunlight conditions on the consistency of vegetation indices in croplands—Effective usage of vegetation indices from continuous ground-based spectral measurements. *Remote Sens.* **2015**, *7*, 14079–14098. [CrossRef]
29. Rahman, M.M.; Lamb, D.W.; Samborski, S.M. Reducing the influence of solar illumination angle when using active optical sensor derived NDVIAOS to infer fAPAR for spring wheat (Triticum aestivum L.). *Comput. Electron. Agric.* **2019**, *156*, 1–9. [CrossRef]
30. Wang, H.; Zhang, W.; Dong, A. Measurement and modeling of Bidirectional Reflectance Distribution Function (BRDF) on material surface. *Measurement* **2013**, *46*, 3654–3661. [CrossRef]
31. Bring, J. How to Standardize Regression Coefficients. *Am. Stat.* **1994**, *48*, 209–213.
32. Stow, D.; Nichol, J.C.; Wade, T.; Assmann, J.J.; Simpson, G.; Helfter, C. Illumination geometry and flying height influence surface reflectance and NDVI derived from multispectral UAS imagery. *Drones* **2019**, *3*, 55. [CrossRef]
33. Rasmussen, J. Are vegetation indices derived from consumer-grade cameras mounted on UAVs sufficiently reliable for assessing experimental plots? *Eur. J. Agron.* **2016**, *74*, 75–92. [CrossRef]
34. Yu, X.; Liu, Q.; Liu, X.; Liu, X.; Wang, Y. A physical-based atmospheric correction algorithm of unmanned aerial vehicles images and its utility analysis. *Int. J. Remote Sens.* **2017**, *38*, 3101–3112. [CrossRef]
35. Easterday, K.; Kislik, C.; Dawson, T.E.; Hogan, S.; Kelly, M. Remotely sensed water limitation in vegetation: Insights from an experiment with Unmanned Aerial Vehicles (UAVs). *Remote Sens.* **2019**, *11*, 1853. [CrossRef]
36. Mesas-Carrascosa, F.; Torres-Sanchez, J.; Clavero-Rumbao, I.; Garcia-Ferrer, A.; Pena, J.; Borra-Serrano, I.; Lopez-Granados, F. Assessing optimal flight parameters for generating accurate multispectral orthomosaicks by UAV to support site-specific crop management. *Remote Sens.* **2015**, *7*, 12793–12814. [CrossRef]
37. Tu, Y.; Phinn, S.; Johansen, K.; Robson, A.; Wu, D. Optimising drone flight planning for measuring horticultural tree crop structure. *ISPRS J. Photogramm.* **2020**, *160*, 83–96. [CrossRef]
38. Khaliq, A.; Comba, L.; Biglia, A.; Ricauda Aimonino, D.; Chiaberge, M.; Gay, P.; Abou-Amer, I.P. Comparison of Satellite and UAV-based multispectral imagery for vineyard variability assessment. *Remote Sens. Environ.* **2019**, *11*, 436. [CrossRef]
39. Wei, S.; Fang, H. Estimation of canopy clumping index from MISR and MODIS sensors using the normalized difference hotspot and darkspot (NDHD) method: The influence of BRDF models and solar zenith angle. *Remote Sens. Environ.* **2016**, *187*, 476–491. [CrossRef]
40. Bréon, F.; Maignan, F.; Leroy, M.; Grant, I. Analysis of hot spot directional signatures measured from space. *J. Geophys. Res. Atmos.* **2002**, *107*, 1–15. [CrossRef]

41. Kuusk, A. The hot spot effect in plant canopy reflectance. In *Photon-Vegetation Interactions: Applications in Optical Remote Sensing and Plant Ecology*; Myneni, R.B., Ross, J., Eds.; Springer: Berlin/Heidelberg, Germany, 1991; pp. 139–159.
42. Stanton, C.; Starek, M.; Elliott, N.; Brewer, M.; Maeda, M.; Chu, T. Unmanned aircraft system-derived crop height and normalized difference vegetation index metrics for sorghum yield and aphid stress assessment. *J. Appl. Remote Sens.* **2017**, *11*, 26035. [CrossRef]

Article

Closing the Phenotyping Gap: High Resolution UAV Time Series for Soybean Growth Analysis Provides Objective Data from Field Trials

Irene Borra-Serrano [1,2], Tom De Swaef [1], Paul Quataert [1], Jonas Aper [1], Aamir Saleem [1], Wouter Saeys [3], Ben Somers [2], Isabel Roldán-Ruiz [1,4] and Peter Lootens [1,*]

1 Flanders Research Institute for Agriculture, Fisheries and Food (ILVO), Plant Sciences Unit, Caritasstraat 39, 9090 Melle, Belgium; Irene.Borra-Serrano@ilvo.vlaanderen.be (I.B.-S.); tom.deswaef@ilvo.vlaanderen.be (T.D.S.); Paul.Quataert@ilvo.vlaanderen.be (P.Q.); Jonas.Aper@ilvo.vlaanderen.be (J.A.); aamir.saleem@ilvo.vlaanderen.be (A.S.); isabel.roldan-ruiz@ilvo.vlaanderen.be (I.R.-R.)

2 KU Leuven, Division of Forest, Nature and Landscape, Celestijnenlaan 200E, 3001 Leuven, Belgium; ben.somers@kuleuven.be

3 KU Leuven, Department of Biosystems, MeBios, Kasteelpark Arenberg 30, 3001 Leuven, Belgium; wouter.saeys@kuleuven.be

4 Ghent University, Department of Plant Biotechnology and Bioinformatics, Technologiepark 927, 9052 Ghent, Belgium

* Correspondence: peter.lootens@ilvo.vlaanderen.be

Received: 28 February 2020; Accepted: 15 May 2020; Published: 20 May 2020

Abstract: Close remote sensing approaches can be used for high throughput on-field phenotyping in the context of plant breeding and biological research. Data on canopy cover (CC) and canopy height (CH) and their temporal changes throughout the growing season can yield information about crop growth and performance. In the present study, sigmoid models were fitted to multi-temporal CC and CH data obtained using RGB imagery captured with a drone for a broad set of soybean genotypes. The Gompertz and Beta functions were used to fit CC and CH data, respectively. Overall, 90.4% fits for CC and 99.4% fits for CH reached an adjusted $R^2 > 0.70$, demonstrating good performance of the models chosen. Using these growth curves, parameters including maximum absolute growth rate, early vigor, maximum height, and senescence were calculated for a collection of soybean genotypes. This information was also used to estimate seed yield and maturity (R8 stage) (adjusted $R^2 = 0.51$ and 0.82). Combinations of parameter values were tested to identify genotypes with interesting traits. An integrative approach of fitting a curve to a multi-temporal dataset resulted in biologically interpretable parameters that were informative for relevant traits.

Keywords: *Glycine max*; RGB; canopy cover; canopy height; close remote sensing; growth model; curve fitting

1. Introduction

It has been demonstrated that close remote sensing approaches in combination with appropriate experimental designs and data integration can increase accuracy, precision, and throughput of on-field phenotyping experiments while also reducing cost and labor requirements [1]. These high throughput phenotyping approaches can also be implemented in crop breeding to determine architectural/morphological and physiological traits to aid early detection of desirable genotypes [2]. In these areas, the use of unmanned aerial vehicles (UAVs) is increasing steadily, as UAVs are suitable for use in complex field environments (e.g., muddy soil), are non-invasive, flexible in use (e.g., not restricted in use to particular field plots), and are available at a low cost [3]. UAV flights can be scheduled in function of the needs of the breeding program or experiment, using combinations of sensors that are suitable to investigate the traits of interest. In one of the most straightforward applications, canopy cover (CC) and canopy height (CH) data can be derived from images obtained using an RGB camera mounted on an UAV [4]. Canopy cover, defined as the ratio of the projected plant area to the total area considered, goes through changes from crop establishment to senescence. CC yields information regarding seedling emergence, early vigor, or senescence [5] and is therefore an important criterion in a breeding context [6]. Furthermore, CC data can be used to estimate canopy photosynthesis and transpiration [7] or to evaluate crop responses to abiotic stresses (e.g., drought stress; [8]). Canopy height (CH) is also dynamic and linked to growth, crop development, and yield [9,10].

One of the main advantages of UAV-based imaging approaches is the high precision and throughput for data acquisition and processing in comparison to the manual ruler measurements or scoring systems that have been traditionally used to analyze crop growth and development. By performing UAV flights throughout the entire growing season, accurate crop data can be obtained at a high spatio-temporal resolution, delivering dynamic information on crop growth and performance [11]. For example, in Yu et al. (2016) [12], data of five time points (using RGB and modified RGB-NIR imaging) were combined to classify soybean populations in different maturity groups and to estimate final yield. In Holman et al. (2016) [13], a method for deriving crop height and growth rate from multi-temporal 3D digital surface models of winter wheat field trials was developed. In Pugh et al. (2018) [14], UAV-based approaches to estimate height in sorghum and maize were determined and growth curves were fitted for different groups of breeding material. Similarly, in Chang et al. (2017) [15] sigmoidal curves were fitted to UAV-derived multi-temporal height data of sorghum. Multi-temporal height data (four flights) have also been used in combination with clustering analysis methods to detect different growth patterns in maize varieties [16].

These examples illustrate the value of multi-temporal data to describe crop growth and to predict specific parameters such as maturity, growth rate (in the above examples this was restricted to estimations considering two consecutive measurements) or yield. Up to now, however, little attention was paid to the biological interpretation of the growth curves built using UAV-based information. Nevertheless, the possibility for such interpretation represents one of the main advantages of this integrative approach. Multi-temporal data at high resolution can be of great value for sophisticated plant and crop modeling approaches. This is because crop models typically contain a large number of parameters that need to be estimated in an identifiable manner [17] based on preferentially large quantities of objective experimental (field) data. Furthermore, mathematical models can be fitted to multi-temporal data to link growth patterns to relevant environmental and developmental clues for a precise description of crop performance from sowing to harvest. This might also enable derivation of 'hidden' variables as phenome descriptors that are useful in a breeding context (e.g., determinacy).

In this study, we demonstrate this concept using a soybean (Glycine max (L.) Merr.) experiment with 454 field plots comprising a diverse set of genotypes. More specifically, we fit mathematical models to multi-temporal data of CC and CH obtained for each plot and explore the possibilities of a global description of crop growth and development throughout the season in the context of breeding applications. First, CC and CH from each flight are estimated. Second, curves are fitted considering multi-temporal data per plot to extract biologically relevant information. Third, the correlation/complementarity of the model-based parameters obtained are evaluated. Fourth, parameters are examined for a possible relation to seed yield and maturity. Finally, an example to illustrate how the data could be used in a breeding context is developed to select interesting plots or to depict variability in the collection of genotypes evaluated.

2. Materials and Methods

2.1. Field Trial

An experimental field located in Melle (ILVO, Belgium, latitude 50.59 N, longitude 3.47 E) was used. The trial was established in 2018 in the context of the EU project EUCLEG (http://www.eucleg.eu/) and comprised 359 soybean genotypes randomized in a row-column-block augmented design in which three reference genotypes were replicated nine times, nine reference genotypes were replicated six times, and 13 reference genotypes were replicated three times. The remaining 334 genotypes were not replicated, rendering a total of 454 plots of 0.50 x 1.20 m (0.60 m^2) or 0.80 x 0.75 m (0.60 m^2), depending on the group to which the genotype belongs (Figure 1A). Within the 454 plots, there were 12 border plots (located between groups sections in Figure 1A) that were not considered in our analysis. This design was used to achieve the highest possible genotypic variation for an association mapping study in which genomic and phenotypic data will be combined. The genotypes used were classified beforehand in four groups that were sown on four different dates in an attempt to synchronize flowering of early maturing and late maturing types: April 20 (GP1), May 3 (GP2), May 8 (GP3), and May 11 (GP4) (Figure 1C). Each plot consisted of three rows; only the middle row was considered for the measurements. The seeds were inoculated with rhizobium bacteria (BiodozTM, De Sangosse, France) and the field was fertilized with 36 kg N ha^{-1}, 20 kg P2O5 ha^{-1}, 140 kg K2O ha^{-1}, and 20 kg MgO ha^{-1}. For weed control, 910 g ha^{-1} pendimethaline and 72 g ha^{-1} clomazone was applied directly after sowing (on the four sowing dates). A manual correction on June 13 was applied to ensure that no weeds developed that could hamper the growth of the crop or influence UAV-based measurements.

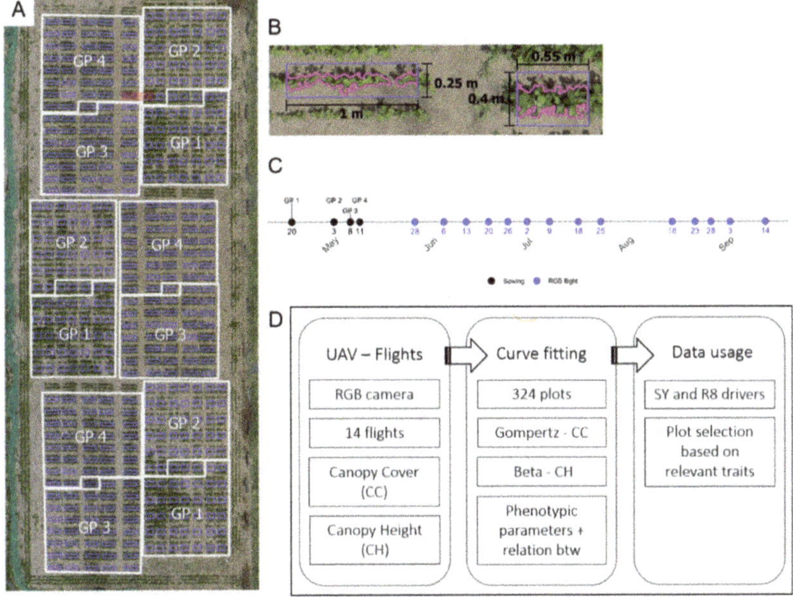

Figure 1. Schematic overview of (**a**) the trial with group locations (white boxes, GP1-4), (**b**) zoom of two plots, marked by a red box in A, with the mask applied to the images to select pixels corresponding to vegetation, (**c**) timeline of the sowing dates (per GP) and UAV flights and (**d**) workflow followed in the manuscript.

2.2. Data Acquisition

2.2.1. UAV Flights

Between May 28 and September 14, UAV flights were performed once per week when weather conditions allowed (Figure 1C), using a dodecacopter model Onyxstar HYDRA-12 (AltiGator, Belgium) equipped with a RGB (visible spectral range) camera (α6000, Sony Corporation, Japan), with 6000 × 4000 pixels and a 35 mm lens (f/1.8 Sony Corporation, Japan). Per flight, a set of nadir images was collected from a height of 30 m above ground level (raw image resolution of 3.35 mm/pix), ensuring an overlap of 70% in both directions (forward-longitudinal-lap and side-lap) at a flight speed of 2.5 m/s. The RGB camera settings (shutter speed, aperture, and ISO) were adjusted before each trial, set in manual mode, and triggered automatically according to the designed flight route. Depending on the weather conditions (light intensity and wind), shutter speed was ≤1/1000 s and ISO was ≤800.

In total, 14 flights were executed around 13:00h (local time) and the required images were captured in less than 20 minutes. A graph thermal time (°C d) versus time for the growing season is included in the Supplementary Materials (Figure S1). Correction for white balance and exposure was applied to the images. Processing of the 14 sets of images and building of geo-referenced orthophotos and digital elevation models (DEMs) were done using Agisoft Photoscan v1.2.6 Professional Edition (Agisoft LLC, Russia) based on 10 Ground Control Points (GCPs) geo-referenced with an RTK GPS (Stonex S10 GNSS, Stonex SRL, Italy). The resulting RGB ortho-mosaicked images had a planimetric spatial resolution of 0.5 cm; the DEMs had an altimetric resolution of 1.25 cm. Spatial resolutions were defined when both products (orthophotos and DEMs) were exported to ensure that products from different flights had the same resolution. Digital elevation models were used to build a Canopy Height Model (CHM) for each date. A triangulated irregular network (TIN) interpolation of 25 GCPs distributed all over the trial was used as Digital Terrain Model (DTM). The CHM was obtained by subtracting the DTM from the DEM.

2.2.2. On-Ground Measurements

Traits commonly recorded by breeders to track growth and reproductive development were monitored during the growing season by experienced UAV operators. The percentage of emergence was calculated by dividing the number of emerged plants by the total seed density. Plots with a percentage of emergence lower than 50% were discarded. The reproductive stage (from R1 to R6) was recorded at five fixed dates for each plot, as described in Fehr and Caviness (1977) [18]. The date of maturity (R8) was recorded separately by frequent visits to the field trial at the end of the growing season. R1 corresponds to beginning bloom (one open flower at any node on the main stem) and R8 is full maturity (95% of the pods have reached their mature pod color). As the R1 stage of each of the plots did not always coincide with one of the five time points considered, we estimated it by linear regression of the reproductive stage (five time points) and thermal time at time of scoring. Thermal time was calculated as the daily accumulation of the difference between the mean daily temperature and a base temperature (6 °C) over a period of time (°C d). Plots for which R^2 was lower than 0.7 were discarded, as the determination of R1 stage was considered inaccurate. In total, 324 plots remained for further analysis after the discards related to emergence rate and R1 stage.

At maturity, five plants of the middle row of each plot were harvested by hand and were threshed together to determine average seed yield and number of seeds per plant.

2.3. Data Handling and Statistical Analysis

The different analysis steps are presented schematically in Figure 1D. First, canopy cover (CC) and canopy height (CH) data were extracted per plot. Second, sigmoid growth curves were fitted to CC and CH data using the 14 time points available and variables were estimated (see Section 2.3.2). Third, a principal component analysis (PCA) was carried out to investigate the complementarity of parameters describing the genotypes. Last, predictive models were built for seed yield (SY) and R8 using multiple linear regression (MLR) models.

QGIS v2.18.26-Las Palmas (QGIS Geographic Information System; Open Source Geospatial Foundation Project) was used to extract information from the images. Georeferenced polygons with the dimension of the plot middle row were defined in QGIS to ensure that data were extracted from the same location for each and every flight and that only the information of the middle row was captured (Figure 1). The curve fitting and the statistical analyses were carried out in R v3.5.1 using RStudio v1.1.456 (RStudio: Integrated Development Environment for R, RStudio Inc., USA) coding customized functions for fitting the curves and using packages such as FactoMineR and optimx for the PCA and optimization of the functions, respectively. These customized functions can be shared, upon request to the corresponding author.

2.3.1. Canopy Cover and Canopy Height

The canopy cover was calculated per time point and plot using the Excess Green vegetation index (ExG, [19]). Pixels corresponding to vegetation were differentiated from pixels corresponding to soil (Figure 1B) using a simple thresholding, and the proportion of pixels classified as vegetation was then calculated.

CHMs were built for each date (described in Section 2.2.1) and used to estimate the 90th percentile (P90) of the canopy height for each plot. Only pixels corresponding to vegetation (the pixel classification described in the previous paragraph was used to build masks) were considered for this calculation.

2.3.2. Model Fitting and Parameter Estimation

Canopy cover and canopy height data extracted for the 14 time points considered (timeline Figure 1C) were used to fit curves describing their temporal evolution in 324 plots. After reaching the maximum value, subsequent time points were only considered for fitting if the difference in value was not greater than 10% from the maximum canopy cover, and 20% from the maximum plant height. When plants get older and start to senesce, they gradually lose leaves and the CC and/or CH is reduced. These criteria prevented any excess influence of these changes on the CC and/or CH fits, as our interest was to fit the sigmoid part of canopy development. The last two time points were excluded by default, because senescence was already observed in a majority of plots. This implies that a maximum of 12 time points were included in each fit. For the fits, we chose curves that have parameters that are biologically interpretable [20] and that reach a "plateau" that defines growth stop.

For the canopy cover (CC) data the Gompertz function [21] was used. In this function, an initial value (at the beginning of the growing season) is assumed. In the present experiment, the first drone flight was carried out when the plants had already emerged (canopy cover >0%). From the Gompertz fits, two parameters were extracted (Figure 2A, formula 1): maximum cover (CC_Cmax) and thermal time when growth rate was maximal (CC_tm). The Gompertz function used for CC has the following equation:

$$CC = CC\_C_{max}e^{e^{-k(t-CC\_t_m)}}, \tag{1}$$

where CC_Cmax [%] is the maximum CC value attained, t is the thermal time [°C d], k [°C-1 d-1] is a constant that determines the curvature of the growth pattern, and CC_tm [°C d] is the inflection point at which the growth rate reaches its maximum value.

For the canopy height (CH) data, the Beta function defined in Yin et al. (2003) [22] was chosen because it has more flexibility than Gompertz at the beginning of the growing season. In addition, for the Beta function the limit is not known in advance, while the CC data values are restricted by definition to the range 0–100. The Beta function was used to estimate the following parameters (Figure 2B, formula 2): maximum height (CH_Hmax), thermal time when maximum canopy height was achieved (CH_te), and thermal time when the growth rate was maximal (CH_tm). The Beta function used for CH has the following equation:

$$CH = CH\_H_{max}\left(1 + \frac{CH\_t_e - t}{CH\_t_e - CH\_t_m}\right)\left(\frac{t}{CH\_t_e}\right)^{\frac{CH\_t_e}{CH\_t_e - CH\_t_m}} \tag{2}$$

where CH_Hmax [m] is the maximum value of CH reached at CH_te [°C d], t is the thermal time [°C d], and CH_tm [°C d] is the inflection point at which the growth rate reaches its maximum value.

The accuracy of the fitting was evaluated by adjusted R^2 (CC_RsqA and CH_RsqA). Plots with an adjusted R^2 lower than 0.70 were discarded, resulting in 291 remaining plots (when both datasets were merged) for further analysis.

Table 1. Overview of the phenotypic parameters.

Data Source		Abbreviation	Description
UAV Orthophoto	Canopy Cover (CC)	CC_tm	Thermal time when maximum growth rate is achieved
		CC_Cmax	Maximum cover
		CC_AGRmax	Maximum absolute growth rate
		CC_6W	Cover percentage after six weeks emergence
		CC_75PCT	Thermal time when 75% of cover is achieved
		CC_SNC	Senescence rate calculated with CC data (CC_Cmax – average cover two last time points)/CC_Cmax
Digital Elevation Model	Canopy Height (CH)	CH_tm	Thermal time when maximum growth rate is achieved
		CH_Hmax	Maximum height
		CH_te	Thermal time when CC_Hmax occurred
		CH_AGRmax	Maximum absolute growth rate
		CH_DET	Determinacy calculated as CH_te – R1
		CH_SNC	Senescence rate derived from CH data (CC_Hmax—average height two last time points)/CC_Hmax
Breeders	Field observations	EP	Percentage of emergence
		SY	Seed yield
		R1	Thermal time when R1 stage starts (beginning bloom)
		R8	Thermal time when R8 stage is reached (full maturity)

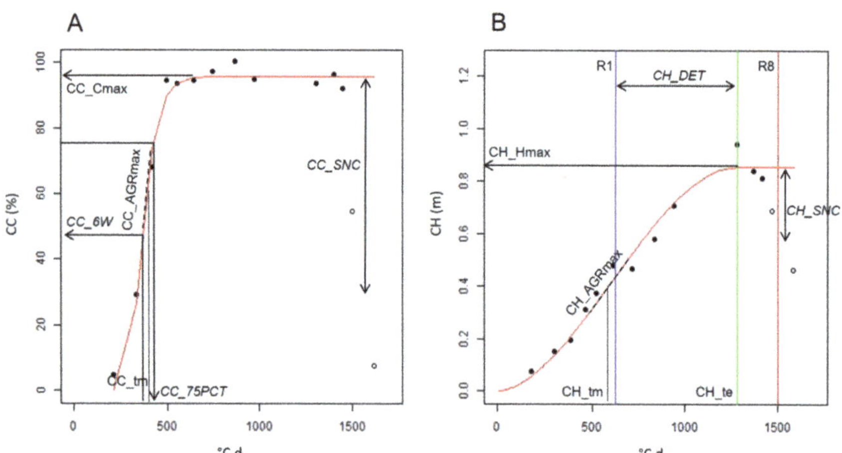

Figure 2. Schematic representation of the phenotypic parameters derived from the curve fitting for (**a**) canopy cover using the Gompertz function and (**b**) canopy height using the Beta function. For abbreviations and explanations of the parameters, see Table 1.

2.3.3. Derived Phenotypic Parameters

Using the fitted growth curves, additional phenotypic parameters were derived (Figure 2). These phenotypes are more difficult to measure in the field. From the CC growth function, we calculated the maximum absolute growth rate (CC_AGRmax) as the first derivative of the function (which is the slope), the canopy cover six weeks after emergence (CC_6W; related to the capacity of the genotype to suppress weeds) and the thermal time needed to reach 75% of full canopy cover (CC_75PCT; related to canopy closure).

From the CH growth function, maximum absolute growth rate (CH_AGRmax) was calculated and the degree of determinacy (CH_DET) was estimated by calculating the thermal time difference between the CH_te and the R1 stage (based on visual scores). According to this calculation, a high CH_DET value indicates a longer period of growth and a lower level of determinacy.

Finally, the senescence rate was calculated as the difference between the maximum value of the fitted curve and the average value of the last two time points. Calculations were performed for CC and CH (CC_SNC and CH_SNC, respectively).

To determine whether a significant difference is present between the groups, a multiple pairwise comparison using Wilcoxon rank sum was carried out for all phenotypic variables.

2.3.4. Relations Among Phenotypic Parameters

A PCA analysis was performed to investigate the relation between the different phenotypic parameters and to find out which ones were correlated or complementary. Twelve phenotypic parameters obtained from the UAV imagery were included (Table 1) and three on-ground determined variables (as quantitative supplementary variables). The four groups (GP1 to GP4) were considered together. To prevent a specific GP from determining the results, an equal number of plots per group were chosen at random (avoiding plots that corresponded to replicates of the same genotype) to ensure a balanced dataset. This resulted in a set of 148 plots. Pairwise correlations were also calculated between the parameters for the same set of plots.

In addition, pairwise correlations per GP were estimated separately. When one genotype was represented by more than one plot (see Section 2.1), only one was retained to prevent the replicates from influencing the correlation output.

2.3.5. Seed Yield and R8 Stage Prediction from UAV Data

From a breeding perspective it is important to know which variables act as driver for seed yield (SY) or influence the thermal time needed to reach maturity (R8). Therefore, two backward multiple linear regression (MLR) models were built for SY and R8 with different sets of variables. The first MLR model was built using the values of the 12 phenotypic parameters that were determined from the curve fitting, whereas the second MLR model was built on the raw data of the 14 time points (flights). In both cases, the group classification was included as a dummy variable for the basic model to be used as starting point. To avoid collinearity between the phenotypic variables in building the first MLR model, variables were selected based on expert knowledge. Consequently, the variables CC_6W and CH_DET were discarded as they are highly correlated with CC_75PCT and CH_te. For the second MLR model, no expert variable selection was done since the raw data set is difficult to interpret in terms of redundancy. The MLR models were built using subset selection regression including a 10 k-fold cross validation. The variables for both datasets were normalized. In each subsequent step, the number of variables included was increased by one, while considering all possible combinations of variables and the best model was chosen based on the lowest RMSE.

3. Results

3.1. Model Fitting for Canopy Cover and Canopy Height

In an exploratory data analysis of CC and CH data, after discarding plots with emergence <50% or $R^2 < 0.7$ in the estimation of R1 stage, different growth patterns per group were observed (data not shown), which was confirmed in the subsequent analysis.

Fitting of CC data using the Gompertz function resulted in 90.4% fits with an adjusted $R^2 > 0.70$. For the CH data to which the Beta model was fitted, 99.4% fits had an adjusted $R^2 > 0.70$. Combining the 0.7 limits for both fits, rendered 291 plots that fulfilled both criteria (89.8%), with a median CC_RsqA of 0.95 and CH_RsqA of 0.98 (Figure 3).

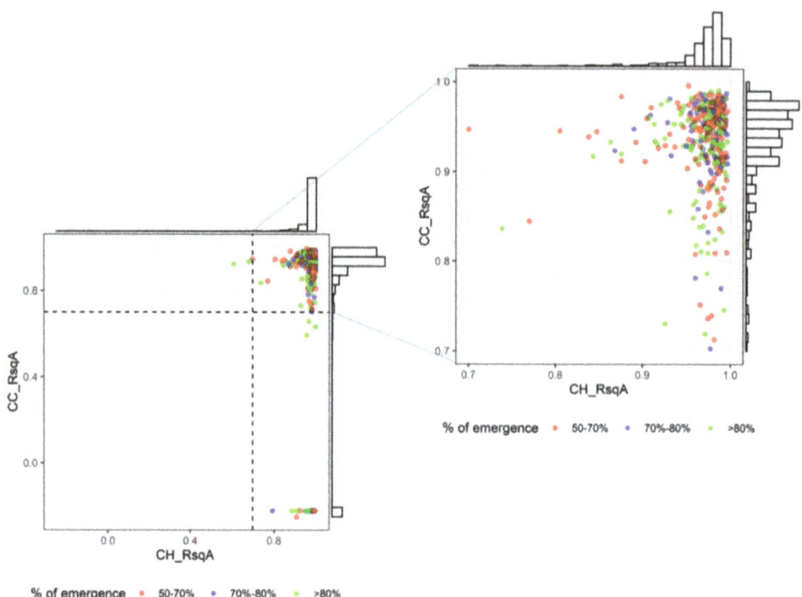

Figure 3. Adjusted R^2 of the curve fitting for canopy cover (CC_RsqA) and canopy height (CH_RsqA) data of all plots. A color code is used to show the percentage of emergence. Zoom of the plot distribution with an adjusted R^2 higher than 0.70 (n = 291 plots remaining for further analysis) is included.

The parameter values for CC are presented in Table 2. For CC_tm (°C d) the median values ranged from 299 to 446; for CC_Cmax (%) the values ranged from 94.2 to 95.6. On average, CC_tm was lower for the later-sown groups GP3 and GP4 than for the groups GP1 and GP2 (Figure 4A).

Table 2. Mean, median, standard deviation (Stdev) and coefficient of variation (cv) of the phenotypic parameters derived from fitting the canopy cover data. Data are presented per group (GP1-4). For the abbreviations, see Table 1.

		UAV Parameters					
		Estimated Parameters		Derived Parameters			
		CC_tm (°C d)	CC_Cmax (%)	CC_AGRmax (% °C^{-1} d^{-1})	CC_6W (%)	CC_75PCT (°C d)	CC_SNC (-)
GP1	Mean	390	94.1	0.518	79.5	509	0.315
	Median	446	94.2	0.660	65.8	525	0.074
	Stdev	69	2.6	0.300	21.8	64	0.230
	cv	15.9	2.8	45.3	36.7	11.9	129.5
GP2	Mean	390	94.1	0.518	79.5	509	0.315
	Median	378	94.2	0.475	81.0	499	0.238
	Stdev	51	2.3	0.217	11.5	65	0.300
	cv	13.1	2.4	41.9	14.5	12.8	95.1
GP3	Mean	329	95.3	0.807	93.3	401	0.493
	Median	324	95.6	0.761	94.6	372	0.483
	Stdev	59	1.8	0.293	5.2	57	0.262
	cv	17.9	1.9	36.6	5.5	14.3	53.1
GP4	Mean	306	94.6	0.689	90.6	399	0.715
	Median	299	94.8	0.577	93.4	379	0.815
	Stdev	56	3.1	0.348	9.1	81	0.235
	cv	18.2	3.3	50.4	10.1	20.3	32.9

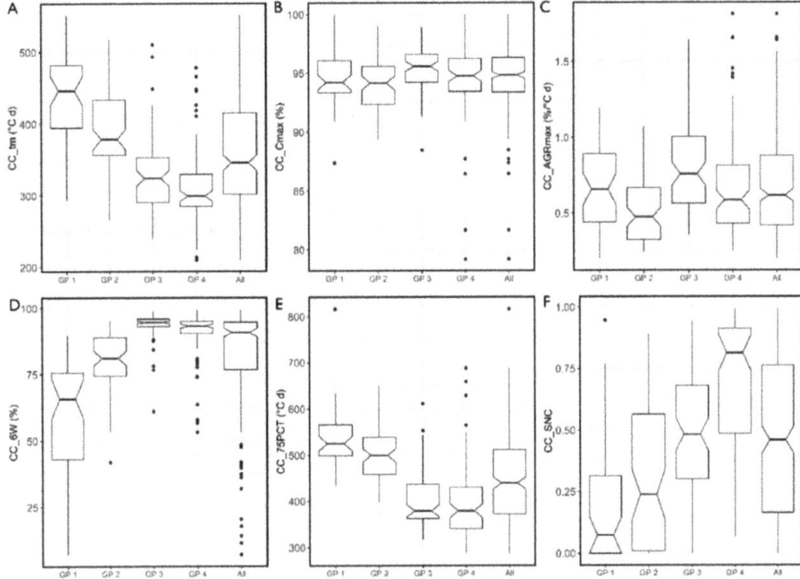

Figure 4. Boxplots of the phenotypic parameters (**a–f**) derived from the curve fitting of canopy cover data. The data is presented per group (GP 1–4) and all groups together (All). For abbreviations see Table 1.

For CH, the parameter values are presented in Table 3. For CH_tm (°C d) this resulted in a range from 547 to 735, for CH_Hmax (m) from 0.76 to 0.96 and for CH_te (°C d) from 901 to 1312.

Table 3. Mean, median, standard deviation (Stdev) and coefficient of variation (cv) of the phenotypic parameters derived from fitting the canopy height data. Data are presented per group (GP1-4). For the abbreviations, see Table 1.

		UAV Parameters					
		Estimated Parameters			Derived Parameters		
		CH_tm (°C d)	CH_Hmax (m)	CH_te (°C d)	CH_AGRmax (m °C^{-1} d^{-1})	CH_DET (°C d)	CH_SNC (-)
GP1	Mean	753	0.96	1333	0.00118	636	0.274
	Median	735	0.93	1312	0.00118	595	0.245
	Stdev	113	0.21	258	0.00025	293	0.139
	cv	15.1	21.5	19.3	20.9	46.0	50.9
GP2	Mean	695	0.95	1280	0.00121	675	0.336
	Median	699	0.96	1272	0.00119	644	0.261
	Stdev	112	0.18	248	0.00025	262	0.181
	cv	16.1	19.2	19.4	21.0	38.8	53.9
GP3	Mean	627	0.91	1029	0.00151	402	0.427
	Median	609	0.91	989	0.00149	389	0.379
	Stdev	72	0.14	150	0.00026	156	0.173
	cv	11.6	15.0	14.6	16.9	38.7	40.4
GP4	Mean	535	0.77	897	0.00149	340	0.637
	Median	547	0.76	901	0.00155	339	0.639
	Stdev	92	0.18	170	0.00035	181	0.186
	cv	17.2	23.3	19.0	23.5	53.2	29.1

To complete the results presented in Figure 4; Figure 5, we checked whether the differences between groups were significant. For CC_Cmax, the only pair with a significant difference was GP2-GP3. All pairwise comparisons were significant for CC_tm.

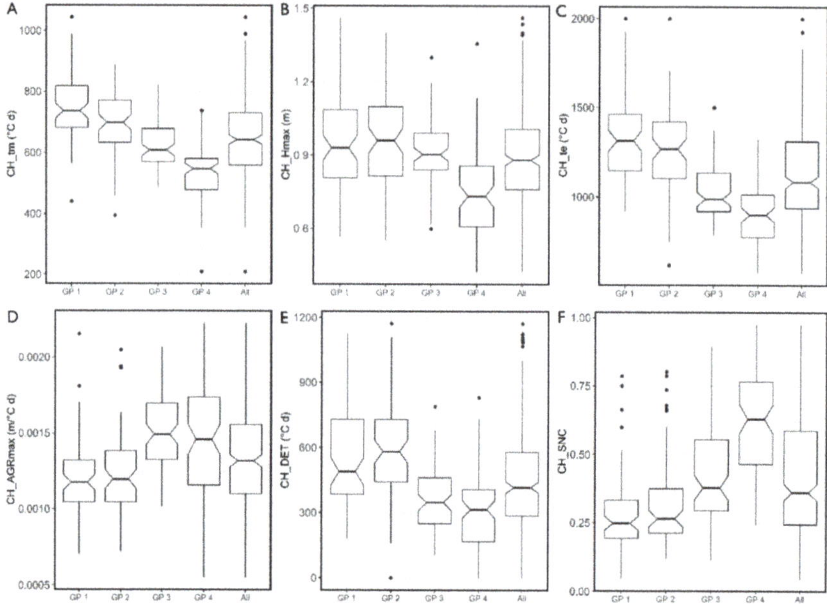

Figure 5. Notched boxplots of the phenotypic parameters (**a**–**f**) derived from the curve fitting of canopy height data. The data is presented per group (GP 1–4) and all groups together (All). For abbreviations see Table 1.

In the case of CH_Hmax only GP4 showed significant differences with the other three groups. All six pairwise combinations were significant for CH_tm. Almost the same results were found for CH_te— except GP1-GP2, which was not significantly different.

3.2. Extra Phenotypic Parameters Derived from the Fitted Curves

As models were fitted to the time series, we could derive phenotypic parameters that correspond to traits that are hard to observe or measure in the field such as growth rate, weed suppression, determinacy or senescence.

The median of the CC_AGRmax (% $°C^{-1}$ d^{-1}) ranged from 0.475 to 0.761 when different GPs were considered. Although the four groups differed in CC_AGRmax (Figure 4C) almost all their plots reached high CC_Cmax > 90–95% (Figure 4B) as can be expected because standard deviations are <3.1%. Based on the canopy cover growth curve, a value related to weed suppression (Table 2 and Figure 4D) was derived by estimating the percentage of canopy cover reached six weeks after emergence (CC_6W (%)). The median values varied between 65.8 and 94.6 for the four groups considered. Strikingly, a large spread of values in GP1 and GP2 was observed, indicating a slow canopy development for some of the genotypes contained in these groups. Thermal time values when 75% of canopy coverage was reached (CC_75PCT; °C d) ranged from 372 to 525, meaning that GP1 and GP2 needed more time (were slower) to close canopy. A group effect was also present here, with considerable variation found within each group (Figure 4D,E).

The median values for CH_AGRmax (m $°C^{-1}$ d^{-1}) ranged from 0.00118 to 0.00155. The two groups with the highest CH_AGRmax were GP3 and GP4; they reached their maximum height earlier (lower CH_te), but the other two groups showed a higher CH_Hmax (Figure 5B–D). The determinacy values are summarized in Table 3 and Figure 5E. This variable indicates how long the plants kept growing after reaching R1. The median CH_DET values ranged from 339 to 644 (°C d). Figure 5E shows that GP3 and GP4 contained more determinate genotypes than GP1 and GP2, but again with a large variation.

Finally, based both on the canopy cover and the height fits, senescence was estimated (Table 2; Table 3 and Figures 4F and 5F). These senescence parameters describe the ratio of decrease for CC and CH between the maximum value derived from the model fitting and the average value of the two last flights. In other words, the higher the values, the more senescence was observed. For CC_SNC median values ranged from 0.074 to 0.815, and in the case of CH_SNC, the median values ranged from 0.245 to 0.639. GP4 showed the highest CC_SNC and CH_SNC values followed by GP3 (Figures 4F and 5F). Taken together with results for determinacy, we can conclude that GP3 and GP4 contained more determinate genotypes that senesce earlier in the environment tested. The histograms of all parameters are available in the Supplementary Materials (Figure S2 and Figure S3).

The analysis to identify significant differences between groups resulted in the following. In the case of CC_AGRmax, all six pairwise comparisons were significant, except GP1-GP4. For CC_6W and CC_SNC, all pairwise comparisons were significant. For CC_75PCT, all pairwise comparisons significant except for the comparison GP3-GP4. In the case of CH_AGRmax the two pairs that were not significantly different were GP1-GP2 and GP3-GP4. For CH_DET and CH_SNC all comparisons were significant except for GP1-GP2.

3.3. Relations between Phenotypic Parameters

To visualize the relations and complementarity among all phenotypic data, a PCA was carried out (Figure 6) and pair-wise correlations were calculated (Figure 7). The PCA (for all GP together) explained 81.7% of the variance present using the first four dimensions. Positively correlated variables were CC_75PCT with CC_tm, CH_DET with CH_tm and CH_SNC with CC_SNC. Examples of negatively correlated variables were CC_75PCT and CC_6W (Figures 6A and 7E). Variables that show an orthogonal relation in the PCA graphs provide complementary data, for example CH_Hmax and CC_6W, indicating that the maximum height achieved is not correlated with the canopy development in early stages. Based on the contributions of the variables to the two first principal components (Figure 6A), the most important variables to explain the variability in the data set were CH_tm, CH_te and CC_75PCT for the first principal component and CH_Hmax and CC_AGRmax for the second principal component. The total variance in the PCA is further explained by the CH derived parameters than the CC ones (57.9% versus 42.1%, data not shown). A large overlap between GPs was observed when the plots were plotted in the two first dimensions of the PCA (Figure 6B) and similar patterns were observed when correlations were calculated per GP (Figure 7), indicating no clear-cut differentiation among GPs.

SY, R, and R8 values obtained using ground-based methods were plotted on top of the PCA carried out with the phenotypic variables derived from UAV-based data (Figure 6, red arrows). SY and R8 are positively correlated, both with each other as well as largely with CH_te, CH_tm, and CH_DET. SY and R8 are negatively correlated with CC_SNC and CH_SNC. This is consistent with the results shown in Figure 7. The low correlation of CC_max with all variables was expected, as all plots reached a high maximum canopy cover, resulting in a small variance. Lower correlations between SY, R1, and R8 with all the other variables were observed for GP4 compared to the other groups (Figure 7D).

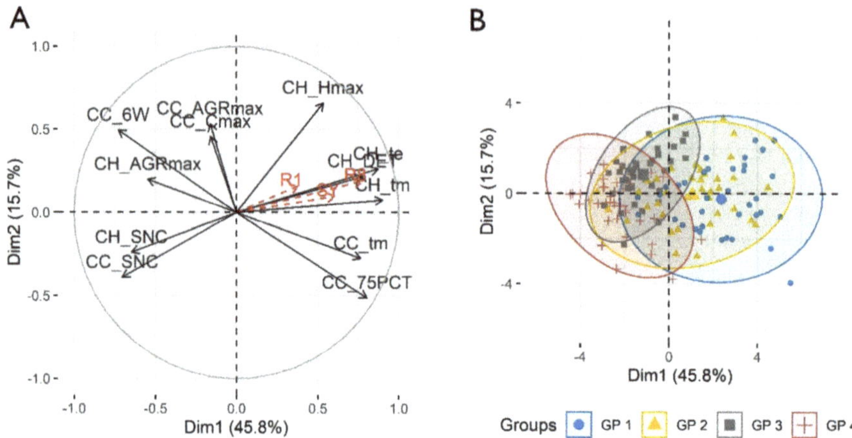

Figure 6. PCA of (**a**) UAV derived phenotypic parameters and field observations (in red) and (**b**) individual plots (n = 148) with color code related to the group/sowing date. For the abbreviations, see Table 1.

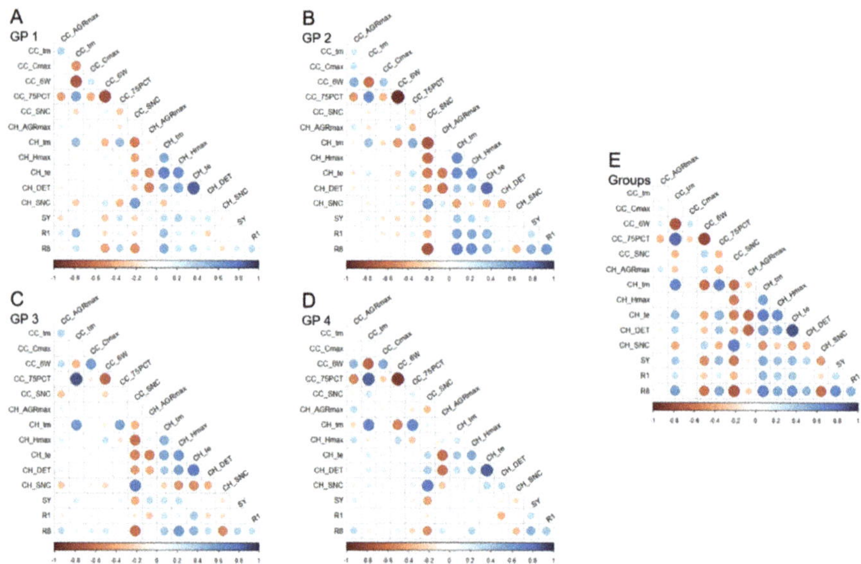

Figure 7. Correlograms between the UAV derived phenotypic parameters and field observations per group (**a–d**, GP1-4) and grouped (**e**) (n = 148). For the abbreviations, see Table 1.

3.4. Linking UAV-Based and Breeder Information

In Table 4, the results of the four MLR models built for SY and R8 with different sets of variables are presented. Higher adjusted R^2 value and lower RMSE were obtained for R8 compared to SY. The similar adjusted R^2 between the MLR models based on different datasets indicated that SY and R8 could be estimated with similar performance either using the phenotypic parameters (from curve fitting) or the 14 time points (flights), with a trend towards a better prediction based on the raw data, but resulting in models that are less interpretable.

Table 4. Statistics and coefficients of the normalized variables for the MLRs for seed yield (SY) and R8 stage based on the phenotypic parameters and the raw flight data.

	RMSE	Adj R^2	Int.	GP2	GP3	GP4	Variables Included in Model		
SY	4.650	0.486	17.68	−2.94	−7.55	−6.13	CC_SNC −1.70	CH_te 1.74	CH_AGRmax 1.77
	4.579	0.501	18.31	−3.96	−9.10	−6.09	CC_246 2.05 CC_157 −0.92	CC_228 0.93 CC_199 −0.59	CH_199 1.05
R8	63.463	0.804	1777.51	−112.73	−169.99	−208.54	CC_SNC −44.05 CC_Cmax 10.29	CH_Hmax 14.69	CH_tm 17.60
	56.944	0.842	1790.79	−122.79	−200.07	−221.49	CC_246 43.234 CC_228 11.07 CH_164 −28.45	CH_235 30.626 CC_177 −24.76 CH_177 16.67	CC_206 14.227 CC_171 30.40

When UAV phenotypic variables were used in the MLR building, the best result for SY was achieved when using three variables: CC_SNC (senescence ratio), CH_te (thermal time when maximum height is reached) and CH_AGRmax (maximum growth rate) with an RMSE of 4.65 g (9% improvement compared to the basic model). In Figure 8A, the RMSE evolution in response to adding a variable is visualized, whereas Figure 8B shows the intercept, the coefficients for each GP and the variables included in each model. The red dashed line indicates the best model with three extra variables compared to the basic model. Figure 8 and Table 4 demonstrate that the models for SY show stable coefficients for the groups, independently of the number of variables added. This indicates a separate effect explained by the dummy variables. The coefficients for CC_SNC, CH_te and CH_AGRmax in the final model have the same magnitude and have consequently comparable importance. The negative coefficient for CC_SNC indicates that an earlier senescence reduces SY. A positive coefficient for CH_te and CH_AGRmax suggests that a later date to achieve the maximum canopy height and a higher AGRmax are linked with a higher SY In the case of R8, a similar figure is presented and a best model with four phenotypic variables resulted in a RMSE of 63.46 (23% improvement compared to the basic model). The variables included in the model were CC_SNC, CH_Hmax, CH_tm and CC_Cmax. Figures for the other three MLR models are included in the Supplementary Materials (Figures S4–S6).

In the case of SY, using the 14 time points, five variables were selected as the best model obtaining a RMSE of 4.58 g (11% improvement compared to the basic model). Four of these variables corresponded to CC data. For R8, eight variables were selected including CC and CH data and RMSE was 56.94 (improvement of 31% compared to the basic model).

In a next step, we explored the information obtained in a breeding context to detect any plots displaying interesting traits. For example, by plotting genotypes/plots in terms of relevant complementary trait values (e.g., CH_Hmax, CC_6W and CC_SNC; maximum height reached, fast soil coverage and the rate of senescence, respectively) (Figure 9A). As apparent in Figure 9, taller genotypes have a tendency to senesce later. Based on the values obtained, breeders could select a plot with a high CC_6W (good weed suppression), not too tall (low sensitivity to lodging), not too small (strong correlation with seed yield) and a fast senescence. As indicated in Figure 9B, plot number 30 fulfils these criteria (plots 110, 131, 177, and 392 are also not too tall and display a good weed suppression, but they senesce later).

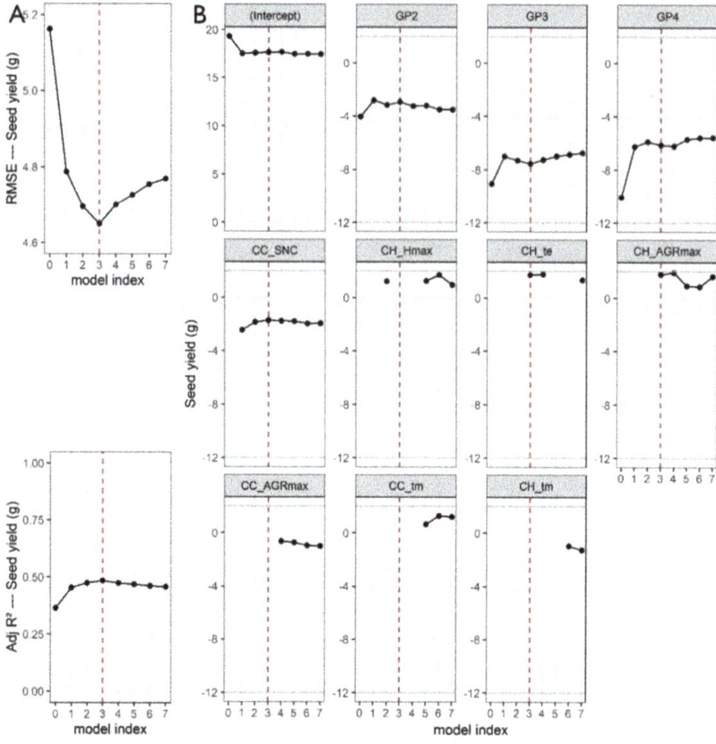

Figure 8. MLR model building overview for SY with RMSE and Adj. R^2 evolution (**a**). The red dashed line indicates the selected (best) model with the lowest RMSE. The subset selection regression with normalized UAV phenotypic variables included a 10 k-fold cross validation. In each subsequent step one variable was added, and all possible combinations were evaluated. The model index = 0 included only the groups (GP1-4, GP1 is the reference group) as dummy variables. For the abbreviations, see Table 1. Graphs are provided per variable with the value of their respective coefficients (**b**).

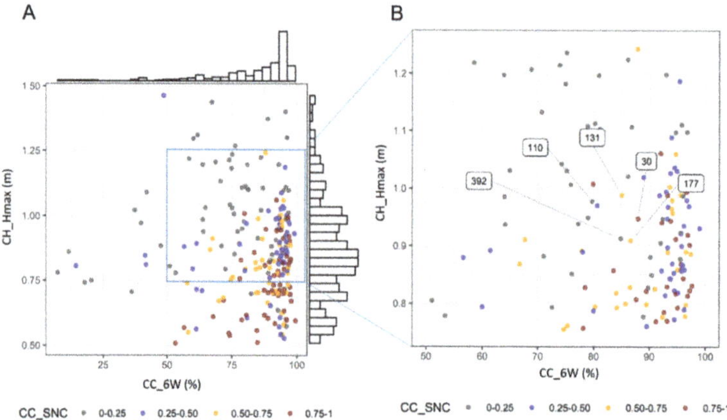

Figure 9. Representation of how data could be used to select interesting plots in a breeding program based on phenotypic data using three parameters: CC_6W, CH_Hmax, and CC_SNC (**a**). Zoom of the plot distribution based on a applicable values range for selection (**b**). For the abbreviations, see Table 1.

4. Discussion

4.1. General Assessment of the Methodology Developed

Weekly flights (n = 14) of a UAV equipped with an RGB sensor allowed to phenotype a large field trial during the entire growing season in terms of the dynamics of canopy cover based on the calculation of the vegetation index Excess Green (ExG), and canopy height based on the Structure from Motion (SfM) photogrammetry with high temporal and spatial resolution. These data were used to fit sigmoid curves (Gompertz and Beta function) in an automated fashion using R scripts. Model parameters to objectively describe the different soybean plots/genotypes were extracted. The methodology presented here is suitable for the efficient phenotyping of a large number of genotypes under field conditions and can provide breeders with trait information that may be difficult to obtain using other methods, e.g., maximum absolute growth rate, weed suppressive ability, onset and evolution of senescence or determinacy. Large differences were observed among the four groups considered, and high variability was present within the groups (~genotypes). The full canopy cover was reached starting from day of the year (DOY) 165 (mid-June). A decrease in cover and height was observed starting from DOY 200 (mid/end July) and was related to the onset of senescence.

By fitting a sigmoid growth function such as the Gompertz or the Beta function to multi-temporal data, it is possible to determine reliable parameters with a straightforward biological meaning enabling characterization of between-plot differences in growth and developmental processes [22]. An extra advantage of curve fitting to multi-temporal data is that possible small errors at one time point are smoothed out using the entire time series. However, model choice is critical because use of the wrong model can introduce bias [23]. The success rate reached for model fitting was 91.0% for canopy cover and 99.6% for canopy height, hence the models chosen were appropriate for fitting these data. The majority of the cases that did not reach a good fit, originated from CC data mainly from genotypes of GP1, the group first sown. Thus, a bad fit was most likely a combination of having only two data points in the increasing phase of the Gompertz model fit for CC data, the first data point having already a canopy cover of >15%, and the reduced flexibility of the Gompertz function to properly fit the first time points. These first time points are important for the model fit, as these present bigger differences because of rapid growth. This suggests that for UAV based phenotyping using time series, flights should start early enough, at plant emergence, and should last long enough to enable full senescence of all treatments. This sometimes conflicts with traditional experimental methods. In this case, bird netting was used to cover the field trial at the time of emergence to prevent bird damage. This of course interferes with UAV imaging, requiring removal and replacement of large areas of netting at the starting phase of the experiment. Near the end of the experiment, full senescence is not always required in a breeding trial before harvests are performed. Therefore, before starting the experiment, these issues should be discussed carefully between the breeders and the phenotyping researchers. A gap in the dataset can happen too—in this study, no flights were possible between July 25 and August 16. This was solved by the curve fitting, as the values during the two weeks gap were estimated. On July 25 the canopy was fully closed, and on August 16 senescence has been initiated only for some plots. Potentially, this could affect the parameters CC_SNC and CH_SNC, but not all other parameters, as full canopy and maximum height were reached before that gap. Scripting the models in R proved appropriate to automatize the fitting step for large sets of plots. Nevertheless, care should be taken to define appropriate initialization values for the Gompertz model as an optimization was not always achieved with the optimx function in R. In our case different values needed to be tested. Regarding canopy height, some datasets contained close to zero or even negative values for some plots in the first and the second flight. This can be explained by the intrinsic altimetric error and the difficulty of detecting small objects in the photogrammetry process of a flight [10]. This "noise" cannot be corrected or removed, but is in our opinion acceptable (x-y-z average error is 0.07 m).

4.2. Describing General Growth Patterns in the Different Groups

Using the derived and estimated phenotypic parameters (Figure 4; Figure 5), it was possible to detect and quantify differences among the groups. A relevant trait is the canopy cover at six weeks after emergence (CC_6W) as this is related to weed suppression capacity and row closure. In this study, most of the plots of GP3 and GP4 had CC_6W values >90% and closed the row faster than the other two groups. Senescence was also different among groups, with fast senescence for GP4. For CH_AGRmax comparable values were observed for GP3 and GP4 and for GP1 and GP2, but at different thermal times (CH_tm). By combining the results obtained for CH_Hmax and CH_te, we can conclude that plots of GP4 did not grow tall (relatively lower CH_Hmax) but grew faster (lower CH_te in combination with higher CH_AGRmax) than plots in other groups. This confirms our expectations, as GP4 comprised genotypes expected to develop quickly and mature early, according to the information available beforehand. We can thus conclude that the pipeline developed to characterize the seasonal growth and development of soybean is functional and suitable for implementation in real-time breeding applications.

Determinacy describes the extent to which a plant continues to produce extra internodes and consequently extra possibilities to produce pods and seeds after entering the generative phase (R1). Indeterminacy or semi-determinacy can be a favorable trait in soybean for cultivation in Northwest Europe as it represents an 'insurance policy' if a period of unfavorable weather conditions occurs, such as low temperature during flowering or drought in the summer months. Based on experience, determinate genotypes perform less well under Northwestern European conditions [24,25]. In contrast, a high level of determinacy contributes to a synchronous seed maturation, avoiding variable moisture content in the harvested seed that could cause quality problems. In soybean, determinate cultivars are used in areas with long growing seasons while indeterminate cultivars are more suited to short growing seasons [26]. Detailed information on this trait is therefore very useful for soybean breeders. Here, we present a method to estimate determinacy in a straightforward way that could be implemented routinely in soybean breeding. Overall, genotypes included in groups 1 and 2 had a higher level of indeterminacy and kept growing during a longer thermal time.

4.3. Estimating SY and Maturity Using UAV-Derived Information

In Northwest Europe, early maturity (quick progress to R8) in combination with high seed yield and a high protein content are important breeding targets. We evaluated if seed yield (SY) and the R8 stage could be predicted by an MLR with the individual time series data points or with the estimated and derived phenotypic parameters as variables. Similar adjusted R^2 values were obtained using the data of the different flights and using the derived phenotypic parameters. This demonstrates that the phenotypic parameters captured all variance present. Seed yield is influenced by canopy cover because a longer lasting canopy cover most likely results in more photosynthesis for optimal SY at the end of the growing season. This is supported by the presence of CC_SNC in the first MLR model and because four out of five time points included were CC data for the second model. Canopy height also plays a role in terms of growth rate and the moment when maximum height is reached. An early senescence decreases SY and the later the maximum height is reached, the higher the chances for a higher SY.

Maturity (R8) was determined by more CC time points (five dates, DOY range 164–246) than CH time points (three dates were included). The phenotypic variable CC_SNC has the most influence on R8 followed by CH_tm. Similar results related to maturity prediction classification (93% accuracy) were achieved in Yu et al. 2016 [12] using RGB and NIR drone data. In the present study we obtained similar results using higher resolution RGB imagery but by using simpler statistical methods (MLR). Soybean agronomic traits such as yield, maturity, and height were predicted in Yuan et al. (2019) [27] using six flights over four locations by color and texture features. Resulting R^2 values were 0.68, 0.76, and 0.27, respectively, using five regression modeling techniques, i.e., Partial Least Squares Regression, Random Forests, Cubist, Artificial Neural Networks, and Support Vector Regression. In our case, seed yield and R8 estimated by MLR achieved an adjusted R^2 of 0.51 and 0.85, respectively, which is slightly lower for SY but higher for maturity.

4.4. Perspectives for Practical Applications

The UAV-estimated and derived phenotypic data can be used directly in a breeding context, as they are related to relevant traits, e.g., weed suppression capacity, fast canopy establishment, early vigor, growth rates, maximum plant height, risk for lodging, determinacy, etc., for each genotype. For example, maximum plant height contributes negatively to lodging resistance, while it is positively correlated with the number of internodes, which in turn is positively correlated with the number of pods. Therefore, breeders should find a compromise considering different traits. For example, the combination of CC_6W, CH_Hmax, and CC_SNC makes it possible to identify genotypes that combine appropriate values for the different traits.

Besides the use in a breeding context, multi-temporal data can be highly valuable for plant and crop modeling. Time series of different traits have been found to be essential for proper and identifiable parameter estimation [17]. Large amounts of data can be gathered using UAV with a high temporal resolution.

4.5. Practical Lessons Learned and Future Perspectives

The high spatial resolution achieved from the UAV flights (0.5 cm pixel size for the orthophotos and 1.25 cm for the DEM) revealed the variability present within the plot and yielded a higher number of measurements in comparison to manual data-gathering. High resolution imagery does require a longer processing time, however. To reduce the processing time, a lower spatial resolution could be defined if the aim of the study does not require highly detailed information. Future research could investigate which resolution is sufficient without influencing the predictability of traits.

In this study, we considered 14 flights spread over the entire growing season at regular intervals. This time interval was short in comparison to similar studies, which included an average of only four to 10 dates (e.g., [13,16]). During data processing and interpretation, we noticed that a different flight scheme may have been more appropriate. For example, in some cases, as senescence occurred in a short period of time, it could have been appropriate to add to extra flights at the end of the growing season. Similarly, when rapid changes (e.g., fast growth) are expected, extra time points are welcome to ensure a good fit. In this regard, the date of the first flight should also be carefully chosen to capture the entire growing season. In conclusion, planning the flights according to the expected development of the crop may yield more useful data than regular intervals between flights.

5. Conclusions

In this study we collected UAV data of a soybean field trial at an unprecedented temporal resolution. The time series data allowed us to fit growth curves with high accuracy (>90%) and to derive relevant traits. The trait data that became available in this study by combining multi-temporal UAV with curve fitting will allow for a better biological interpretation of the differential response of a large set of soybean genotypes. In addition, we developed a methodology to predict seed yield and to estimate the moment at which each accession reached the R8 stage.

Furthermore, this developed strategy may be applied in the future to the analysis of multi-temporal data obtained using other sensors, e.g., multispectral, hyperspectral and thermal, to evaluate responses to abiotic and biotic stresses. This would enable the development of more resilient plant varieties at a faster pace.

Supplementary Materials: The following are available online at http://www.mdpi.com/2072-4292/12/10/1644/s1. **Figure S1**: Evolution of thermal time (°C d) during the growing season of the trial. First flight and the two weeks gap are indicated. **Figure S2**: Histograms with median values (dashed lines) of the phenotypic parameters (a-f) derived from the curve fitting of canopy cover data. The data is presented per group (GP 1-4) and all groups together (All). For the abbreviations see Table 1. **Figure S3**: Histograms with median values (dashed lines) of the phenotypic parameters (a-f) derived from the curve fitting of canopy height. The data is presented per group (GP 1-4) and all groups together (All). For the abbreviations see Table 1. **Figure S4**: MLR model building overview for SY stage with RMSE and Adj.R^2 evolution. The red dashed line indicates the selected (best) model with the lowest RMSE. The subset selection regression with normalized UAV time points included a 10 k-fold cross validation. In each subsequent step one variable was added and all possible combinations were evaluated. The model index = 0 included only the groups (GP1-4, GP1 is the reference group) as dummy variables. Graphs are provided per variable with the value of their respective coefficients. **Figure S5**: MLR model building overview for R8 stage with RMSE and Adj.R^2 evolution. The red dashed line indicates the selected (best) model with the lowest RMSE. The subset selection regression with normalized UAV phenotypic variables included a 10 k-fold cross validation. In each subsequent step one variable was added and all possible combinations were evaluated. The model index = 0 included only the groups (GP1-4, GP1 is the reference group) as dummy variables. For the abbreviations see Table 1. Graphs are provided per variable with the value of their respective coefficients. **Figure S6**: MLR model building overview for R8 stage with RMSE and Adj.R^2 evolution. The red dashed line indicates the selected (best) model with the lowest RMSE. The subset selection regression with normalized UAV time points included a 10 k-fold cross validation. In each subsequent step one variable was added and all possible combinations were evaluated. The model index = 0 included only the groups (GP1-4, GP1 is the reference group) as dummy variables. Graphs are provided per variable with the value of their respective coefficients.

Author Contributions: Conceptualization, I.B.-S., P.L., T.D.S., J.A., I.R.-R., W.S. and B.S.; methodology, I.B.-S., P.L., T.D.S., P.Q., J.A., I.R.-R., W.S. and B.S.; formal analysis, I.B.-S., T.D.S., P.Q., P.L.; investigation, I.B.-S., A.S. and J.A.; writing—original draft preparation, I.B-S.; writing—review and editing, I.B.-S., P.L., T.D.S., I.R.-R., J.A., P.Q., A.S., W.S. and B.S; visualization, I.B.-S.; supervision, P.L., I.R.-R., B.S. and W.S. All authors have read and agreed to the published version of the manuscript.

Funding: This research received no external funding.

Acknowledgments: The authors would like to thank Thomas Vanderstocken and Filip De Brouwer for performing the UAV flights and Thomas Vanderstocken for the help with pre-processing the UAV data. We would also like to thank the technical teams that helped to set up the soybean field trial. Special thanks for Miriam Levenson for English language corrections. The field trial was part of the trials within the EUCLEG project (EU-H2020, http://www.eucleg.eu/).

Conflicts of Interest: The authors declare no conflict of interest.

References

1. Mir, R.R.; Reynolds, M.; Pinto, F.; Khan, M.A.; Bhat, M.A. High-throughput phenotyping for crop improvement in the genomics era. *Plant Sci.* **2019**, *282*, 60–72. [CrossRef] [PubMed]
2. Shakoor, N.; Lee, S.; Mockler, T.C. High throughput phenotyping to accelerate crop breeding and monitoring of diseases in the field. *Curr. Opin. Plant Biol.* **2017**, *38*, 184–192. [CrossRef] [PubMed]
3. Yang, G.; Liu, J.; Zhao, C.; Li, Z.; Huang, Y.; Yu, H.; Xu, B.; Yang, X.; Zhu, D.; Zhang, X.; et al. Unmanned Aerial Vehicle Remote Sensing for Field-Based Crop Phenotyping: Current Status and Perspectives. *Front. Plant Sci.* **2017**, *8*, 1111. [CrossRef] [PubMed]
4. Araus, J.L.; Kefauver, S.C. Breeding to adapt agriculture to climate change: Affordable phenotyping solutions. *Curr. Opin. Plant Biol.* **2018**, *45*, 237–247. [CrossRef] [PubMed]
5. Walter, A.; Liebisch, F.; Hund, A. Plant phenotyping: From bean weighing to image analysis. *Plant Methods* **2015**, *11*, 14. [CrossRef]
6. Yu, K.; Kirchgessner, N.; Grieder, C.; Walter, A.; Hund, A. An image analysis pipeline for automated classification of imaging light conditions and for quantification of wheat canopy cover time series in field phenotyping. *Plant Methods* **2017**, *13*, 15. [CrossRef]
7. Mullan, D.J.; Reynolds, M.P. Quantifying genetic effects of ground cover on soil water evaporation using digital imaging. *Funct. Plant Biol.* **2010**, *37*, 703. [CrossRef]
8. Sayed, M.A.; Schumann, H.; Pillen, K.; Naz, A.A.; Léon, J. AB-QTL analysis reveals new alleles associated to proline accumulation and leaf wilting under drought stress conditions in barley (Hordeum vulgare L.). *BMC Genet.* **2012**, *13*, 61. [CrossRef]
9. Viljanen, N.; Honkavaara, E.; Näsi, R.; Hakala, T.; Niemeläinen, O.; Kaivosoja, J. A Novel Machine Learning Method for Estimating Biomass of Grass Swards Using a Photogrammetric Canopy Height Model, Images and Vegetation Indices Captured by a Drone. *Agriculture* **2018**, *8*, 70. [CrossRef]
10. Borra-Serrano, I.; De Swaef, T.; Muylle, H.; Nuyttens, D.; Vangeyte, J.; Mertens, K.; Saeys, W.; Somers, B.; Roldán-Ruiz, I.; Lootens, P. Canopy height measurements and non-destructive biomass estimation of *Lolium perenne* swards using UAV imagery. *Grass Forage Sci.* **2019**, *74*, 356–369. [CrossRef]

11. Zhang, C.; Kovacs, J.M. The application of small unmanned aerial systems for precision agriculture: A review. *Precis. Agric.* **2012**, *13*, 693–712. [CrossRef]
12. Yu, N.; Li, L.; Schmitz, N.; Tian, L.F.; Greenberg, J.A.; Diers, B.W. Development of methods to improve soybean yield estimation and predict plant maturity with an unmanned aerial vehicle based platform. *Remote Sens. Environ.* **2016**, *187*, 91–101. [CrossRef]
13. Holman, F.H.; Riche, A.B.; Michalski, A.; Castle, M.; Wooster, M.J.; Hawkesford, M.J. High throughput field phenotyping of wheat plant height and growth rate in field plot trials using UAV based remote sensing. *Remote Sens.* **2016**, *8*, 1031. [CrossRef]
14. Pugh, N.A.; Horne, D.W.; Murray, S.C.; Carvalho, G.; Malambo, L.; Jung, J.; Chang, A.; Maeda, M.; Popescu, S.; Chu, T.; et al. Temporal Estimates of Crop Growth in Sorghum and Maize Breeding Enabled by Unmanned Aerial Systems. *Tppj* **2018**, *1*. [CrossRef]
15. Chang, A.; Jung, J.; Maeda, M.M.; Landivar, J. Crop height monitoring with digital imagery from Unmanned Aerial System (UAS). *Comput. Electron. Agric.* **2017**, *141*, 232–237. [CrossRef]
16. Han, L.; Yang, G.; Yang, H.; Xu, B.; Li, Z.; Yang, X. Clustering Field-Based Maize Phenotyping of Plant-Height Growth and Canopy Spectral Dynamics Using a UAV Remote-Sensing Approach. *Front. Plant Sci.* **2018**, *9*, 1–18. [CrossRef]
17. De Swaef, T.; Bellocchi, G.; Aper, J.; Lootens, P.; Roldán-Ruiz, I. Use of identifiability analysis in designing phenotyping experiments for modelling forage production and quality. *J. Exp. Bot.* **2019**, *70*, 2587–2604. [CrossRef]
18. Fehr, W.; Caviness, C. *Stages of Soybean Development*; Iowa State University: Aldrich Ames, IA, USA, 1977; p. 87.
19. Woebbecke, D.M.; Meyer, G.E.; Von Bargen, K.; Mortensen, D.A. Color indices for weed identification under various soil, residue, and lighting conditions. *Trans. ASAE* **1995**, *38*, 259–269. [CrossRef]
20. Zeide, B. Analysis of Growth Equations. *For. Sci.* **1993**, *39*, 594–616. [CrossRef]
21. Gompertz, B. On the nature of the function expressive of the law of human mortality, and on a new mode of determining the value of life contingencies. In a letter to Francis Baily, Esq. F. R. S. &c. *Philos. Trans. R. Soc. Lond.* **1825**, *182*, 513–583.
22. Yin, X.; Goudriaan, J.; Lantinga, E.A.; Vos, J.; Spiertz, H.J. A flexible sigmoid function of determinate growth. *Ann. Bot.* **2003**, *91*, 361–371. [CrossRef] [PubMed]
23. Thorney, J.H.M. A New Formulation of the Logistic Growth Equation and Its Application to Leaf Area Growth. *Ann. Bot.* **1990**, *66*, 309–311. [CrossRef]
24. Schori, A.; Charles, R.; Peter, D. Soja: Sélection, agronomie et production en Suisse. *Rev. Suisse d'Agronomie* **2003**, *35*, 69–75.
25. Aper, J.; De Clercq, H.; Baert, J. Agronomic characteristics of early-maturing soybean and implications for breeding in Belgium. *Plant Genet. Resour. Characterisation Util.* **2016**, *14*, 142–148. [CrossRef]
26. Ping, J.; Liu, Y.; Sun, L.; Zhao, M.; Li, Y.; She, M.; Sui, Y.; Lin, F.; Liu, X.; Tang, Z.; et al. Dt2 is a gain-of-function MADS-domain factor gene that specifies semideterminacy in soybean. *Plant Cell* **2014**, *26*, 2831–2842. [CrossRef]
27. Yuan, W.; Wijewardane, N.K.; Jenkins, S.; Bai, G.; Ge, Y.; Graef, G.L. Early Prediction of Soybean Traits through Color and Texture Features of Canopy RGB Imagery. *Sci. Rep.* **2019**, *9*, 1–18. [CrossRef]

Article

Segmenting Purple Rapeseed Leaves in the Field from UAV RGB Imagery Using Deep Learning as an Auxiliary Means for Nitrogen Stress Detection

Jian Zhang [1,2,†], Tianjin Xie [1,2,†], Chenghai Yang [3], Huaibo Song [4], Zhao Jiang [1,2], Guangsheng Zhou [5], Dongyan Zhang [6], Hui Feng [7] and Jing Xie [8,*]

1 Macro Agriculture Research Institute, College of Resource and Environment, Huazhong Agricultural University, 1 Shizishan Street, Wuhan 430070, China; Jz@mail.hzau.edu.cn (J.Z.); purple@webmail.hzau.edu.cn (T.X.); jzh@webmail.hzau.edu.cn (Z.J.)
2 Key Laboratory of Arable Land Conservation (Middle and Lower Reaches of Yangtze River), Ministry of Agriculture, Wuhan 430070, China
3 Aerial Application Technology Research Unit, USDA-Agricultural Research Service, College Station, TX 77845, USA; chenghai.yang@ars.usda.gov
4 College of Mechanical and Electronic Engineering, Northwest A&F University, 22 Xinong Road, Yangling 712100, China; songhuaibo@nwafu.edu.cn
5 College of Plant Science and Technology, Huazhong Agricultural University, Wuhan 430070, China; zhougs@mail.hzau.edu.cn
6 Anhui Engineering Laboratory of Agro-Ecological Big Data, Anhui University, 111 Jiulong Road, Hefei 230601, China; zhangdy@ahu.edu.cn
7 College of Informatics, Huazhong Agricultural University, Wuhan 430070, China; fenghui@mail.hzau.edu.cn
8 College of Science, Huazhong Agricultural University, Wuhan 430070, China
* Correspondence: xiejing625@mail.hzau.edu.cn
† These authors contributed equally to this work.

Received: 6 March 2020; Accepted: 27 April 2020; Published: 29 April 2020

Abstract: Crop leaf purpling is a common phenotypic change when plants are subject to some biotic and abiotic stresses during their growth. The extraction of purple leaves can monitor crop stresses as an apparent trait and meanwhile contributes to crop phenotype analysis, monitoring, and yield estimation. Due to the complexity of the field environment as well as differences in size, shape, texture, and color gradation among the leaves, purple leaf segmentation is difficult. In this study, we used a U-Net model for segmenting purple rapeseed leaves during the seedling stage based on unmanned aerial vehicle (UAV) RGB imagery at the pixel level. With the limited spatial resolution of rapeseed images acquired by UAV and small object size, the input patch size was carefully selected. Experiments showed that the U-Net model with the patch size of 256 × 256 pixels obtained better and more stable results with a F-measure of 90.29% and an Intersection of Union (IoU) of 82.41%. To further explore the influence of image spatial resolution, we evaluated the performance of the U-Net model with different image resolutions and patch sizes. The U-Net model performed better compared with four other commonly used image segmentation approaches comprising support vector machine, random forest, HSeg, and SegNet. Moreover, regression analysis was performed between the purple rapeseed leaf ratios and the measured N content. The negative exponential model had a coefficient of determination (R^2) of 0.858, thereby explaining much of the rapeseed leaf purpling in this study. This purple leaf phenotype could be an auxiliary means for monitoring crop growth status so that crops could be managed in a timely and effective manner when nitrogen stress occurs. Results demonstrate that the U-Net model is a robust method for purple rapeseed leaf segmentation and that the accurate segmentation of purple leaves provides a new method for crop nitrogen stress monitoring.

Keywords: purple rapeseed leaves; unmanned aerial vehicle; U-Net; plant segmentation; nitrogen stress

1. Introduction

Biotic and abiotic stresses such as plant diseases, low temperature, and deficiencies of mineral elements can greatly affect the crop production and yield [1,2]. Crops can usually exhibit phenotypic and physiological changes to respond to adverse conditions. Leaf purpling is a common phenotypic change caused by environmental stresses in many crops such as rapeseed, wheat, corn, and rice [3]. Leaf purpling is mainly due to lack of nitrogen (N), phosphorus (P), potassium (K), or other nutrients causing the stagnation and accumulation of anthocyanin, which is usually red or purple in the leaf tissues [4]. Anthocyanins as water-soluble pigments found in all plant tissues can be used as a protective layer to improve the resistance to cold, drought, and disease, thereby minimizing the permanent damage to the leaves [5], but stresses can cause a temporary inhibition of growth [6]. Previous studies of purple leaves mostly focused on their ecological and physiological significance [1,7] and genomics research [3,4]. However, compared with green leaves, purple leaves have attracted little attention in phenotype monitoring, crop management, and yield estimation in agriculture, due to shorter appearance periods and smaller areas. As a phenotypic trait, purple leaves can intuitively respond to stresses, and crops could be treated by timely site-specific application of remedies. The ability to accurately and efficiently segment purple leaves in crops may greatly facilitate crop phenotype analysis, monitoring, and management.

Purple leaf segmentation is difficult due to the complexity of the field environment and differences in leaf size, shape, texture, and color gradation. Plant segmentation using images based on the color space is a traditional target segmentation method [8]. For example, Tang et al. used the HSeg method based on the H-component of the hue-saturation-intensity (HSI) color space to separate tassels from corn plants [9]. However, the segmentation accuracy was affected greatly by the illumination. Bai et al. proposed a morphology modeling method to establish a crop color model based on lightness in the Lab color space for crop image segmentation [10]. The experimental results showed that the method was robust to the lighting conditions, but its dependence on color information alone could lead to incomplete extraction.

In the past few decades, machine learning methods such as support vector machine (SVM) [11,12], random forest (RF) [13], and artificial neural network (ANN) [14] have been used widely in plant segmentation. Deep learning is a relatively new area of machine learning where multiple processing layers are employed to learn representations of complex data, and it has dramatically achieved state-of-the-art results in agriculture [15–17]. The fully convolutional neural network (FCNN) for semantic segmentation work has demonstrated its great capacity in plant segmentation due to rich feature representations and end-to-end structure [18,19]. Nevertheless, the FCNN-based models are limited by low-resolution prediction because of the sequential max-pooling and down-sampling structure. Subsequently, some sophisticated network architecture emerged, such as SegNet [20], U-Net [21], and DeepLab [22], which can carefully address the issue through the encoder and decoder operation and have shown huge potential in plant segmentation [23–25]. For example, Sa et al. used a complete pipeline for semantic sugar beet and weed segmentation using multispectral images obtained by UAV. The model with nine channels significantly outperformed a baseline SegNet architecture with RGB input [19]. Milioto et al. proposed an end-to-end encoder-decoder network, which can accurately perform the pixel-wise prediction task for real-time semantic segmentation of crop and weed [26].

Low-altitude remote sensing using unmanned aerial vehicles (UAVs) has developed rapidly for applications in precision agriculture in recent years [27–29], due to their ability to capture high-resolution imagery with low-cost portable sensors and their flexibility for quick image acquisition. However, UAVs have been mostly used for object detection [30,31] and image classification [32,33] in deep learning. Pixel-wise plant segmentation based on UAV images is a big challenge due to limited spatial resolution, small object size, and complex background features, compared with images obtained from the ground platform.

In this study, we used a U-Net convolutional neural network architecture as a binary semantic segmentation task of purple rapeseed leaves during the seedling stage based on UAV imagery. The U-Net model has shown significant outperformance over other deep architectures in some segmentation tasks, which is more suitable for target segmentation with uncertain size, small resolution, and complicated background [21,34,35]. The specific objectives of this study were to (1) develop a novel method for efficiently assessing and quantifying the crop stress based on UAV imagery and compare the performance with commonly used deep learning algorithms (SegNet), traditional machine learning architecture (RF and SVM), and color space method (HSeg), (2) explore the regulation of image spatial resolution on the selection of sample size, and (3) verify the ability of using purple leaves to monitor crop nitrogen status.

2. Materials and Methods

2.1. Study Area

The study area was located at the experimental base of Huazhong Agricultural University (30°28′10′′N, 114°21′21′′E) in Wuhan, Hubei, China, which lies approximately in the center of the Yangtze River basin. The study site has a humid sub-tropical monsoon climate and the main physicochemical properties of the soil are presented in Table 1. In the Yangtze River region, winter oilseed rape (*Brassica napus* L.) is widely planted, making up 89% of total rapeseed yields in China [36]. In this experiment, the conventional winter rapeseed hybrid variety "Huayouza No. 9" was used and the leaves are green in the natural state. The rapeseeds were first sown in the prepared seedbeds in September using high fertility soils, then the seedlings were transplanted into the tilled field at a density of 100,000 plants/ha in October. The experimental site with a total area of 0.29 ha was divided into three areas shown in Figure 1b. Area 1 contained 36 plots with a plot size of 11 × 2.7 m and was used for a substitution experiment where crop residues provided nutrients in a rice–rapeseed rotation system. Rice–rapeseed and corn–rapeseed rotation experiments were conducted in Area 2 and Area 3, respectively, where each area was divided into 30 plots with a plot of 4 × 5 m. In the rapeseed season, four different N fertilizer treatments (0, 75, 150, and 225 kg/ha) were randomly assigned with three replicates in Area 2 and Area 3. The samples used for the U-Net model comprised data from all three areas. In order to ensure the consistency of the experiment, the fitting model regarding the correlation between N content and area of purple rapeseed leaves only used the 60 sample plots from Area 2 and Area 3. The study area and plot arrangement are shown in Figure 1d. The table in Figure 1c shows the N fertilizer rates for each plot number.

Table 1. Main physicochemical properties of the soil in the study area.

Texture	pH	Organic Matter (g/kg)	Available N (mg/kg)	Available P (mg/kg)	Available K (mg/kg)
Silt clay loam	6.71	24.16	133.12	17.16	145.89

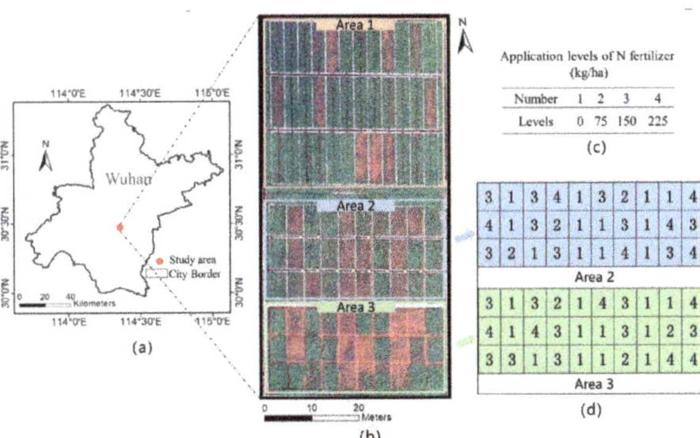

Figure 1. Study area and arrangement of the experimental sites: (**a**) geographic map showing the study area, (**b**) test fields for rapeseed, (**c**) nitrogen fertilizer application levels, and (**d**) application levels of nitrogen fertilizer in each plot in Area 2 and Area 3 according to the table.

2.2. Field Data Acquisition

In this study, nitrogen content was measured in the field by a GreenSeeker handheld crop sensor (Trimble Navigation Limited, Sunnyvale, CA, USA). Many studies have verified that the values obtained by GreenSeeker have a strong correlation with nitrogen content [37–39]. The GreenSeeker sensor used two LEDs as a light source to detect the reflection in the visible and NIR spectral regions. This study placed the instrument 1.2 m above the canopy and the measurement area was a circular area with a radius of 0.25 m. A total of 66 samples were collected in Area 2 and Area 3.

2.3. UAV Image Acquisition

The UAV flight task was carried out under clear and sunny weather conditions between 12:00 and 14:00 local time on December 25, 2016, when rapeseed was at the seedling stage. We employed a Matrice 600 UAV (DJI, Shenzhen, China) with a maximum payload of 6 kg, 16 min of continuous flight, and a maximum horizontal flight speed of 18 m/s in a windless environment. The platform carried a full frame digital single lens reflex camera Nikon D800 (Nikon, Inc., Tokyo, Japan) with a sensor size of 35.9 × 24 mm. The camera was equipped with a Nikon AF 50 mm f/1.4D fixed focal length lens which was vertical to the ground. The aperture, shutter speed, and ISO of the camera were set at f/5.6, 1/1000 s, and 100, respectively. A Global Position System (GPS) module and a wireless trigger were also integrated with the camera. The flight time was 10 min and flight height was set to 20 m with a forward overlap of 80% and a side overlap of 70%. During the flight, 369 images were captured every 1.0 s automatically with 36 million pixels (7360 × 4912 pixels). All the images were stored in the 24-bit TIFF format of true color.

2.4. Image Pre-Processing

Pre-processing of the 369 RGB images mainly involved image corrections and mosaicking. Image corrections included a geometric correction to reduce the effect of lens distortion and a vignetting correction to reduce the progressive radial decrease in radiance strength towards the image periphery [40–42]. All 369 images were used to create the orthomosaic to cover the whole study and meanwhile to reduce geometric distortion with large image overlapping [42]. The above steps were carried out using PIE-UAV software (Beijing Piesat Information Technology Co., Ltd., Beijing, China) software. In order to improve the accuracy of the mosaicking results, eight ground control points (GCPs) collected by a global navigation satellite system real-time kinematic (GNSS RTK) receiver (UniStrong Science and Technology Co., Ltd, Beijing, China) were imported into the software. The spatial resolution of the mosaicked orthoimage was 1.86 mm/pixel and the image was stored in TIFF format.

2.5. Dataset Preparation

The mosaicked orthoimage was clipped using a window with a fixed size in order to prepare the dataset for the model. The window size greatly influenced the final detection results for the purple rapeseed leaves [28] and an appropriate size provided a good trade-off in terms of the overall information regarding the purple rapeseed leaves. For exploring the optimal patch size for purple rapeseed leaf segmentation with the U-Net model, the datasets were finally resized to four different sizes of 64 × 64, 128 × 128, 256 × 256, and 512 × 512 pixels, and the information about sample capacities and clip stride is reported in Table 2. The clipped dataset was randomly divided into a training set, a validation set, and a test set according to the ratio of 3:1:1. To ensure the fairness of the assessment results, the test set was only used for the final model evaluation. Two categories, with and without purple rapeseed leaves, had the same area in images for different patch sizes and the ratio of purple leaf pixels to the total pixels was 14.25%. In this study, since there is no accepted standard to identify purple leaves, three experts created the mask of purple leaves by visual interpretation through prior experience using ArcMap 10.3 (Esri Inc., Redlands, CA, USA), and then the ground truth was obtained by taking the intersection of three mask results. Additionally, the corresponding labels were cut to the same size as image patches. The labels contained two classes, corresponding to the categories with and without purple rapeseed leaves.

Table 2. Statistics for sample capacities and clip strides with all four patch sizes.

Patch Size	Training Set	Validation Set	Test Set	Stride
64 × 64	62,759	20,920	20,920	64
128 × 128	25,313	8438	8438	128
256 × 256	8825	2941	2941	128
512 × 512	2841	947	947	128

In this study, the flight height of the UAV was set at 20 m, which was the lowest altitude possible in practice and the resolution of the rapeseed images was 1.86 mm/pixel. However, in agricultural production, it is more practical to obtain coarser resolution images at higher altitudes, considering the cost and efficiency. Therefore, to explore the influence of image resolution on sample size selection with the specific network architecture (U-Net), the original orthoimage was re-sampled to resolutions of 3.72, 5.58, and 7.44 mm/pixel, to mimic the image acquisition at different flight heights of a UAV based platform. The resampled images were obtained through the cubic interpolation algorithm in ArcMap. Degraded images can effectively simulate the images acquired at different heights [43,44]. The three degraded images and the original orthoimage were respectively cropped into four data sets for training and test of the model.

2.6. Network Architecture

The U-Net model has shown good performance of semantic segmentation, especially in biomedical image segmentation [34,45,46]. In this study, we extracted purple leaves from UAV imagery by a U-Net model. Figure 2 shows the overall structure of the U-Net used in this study, which was a typical encoder-decoder architecture. The part of the encoder consists of repeated convolutional layer with a kernel size of 3 × 3 and each follows by a 2 × 2 max pooling with stride 2, which gradually reduces the patch sizes. The part of the decoder contains the up-sampling operations by a 2 × 2 convolution, which halves the number of feature channels. As the main difference of the U-Net model compared with other encoder-decoder architectures, the skip connections could concatenate feature maps in up-sampling with feature maps of the same level in the encoder, which could contribute the decoder to better recovering and optimizing the details of the object. Moreover, it could output images of the same sizes as the input due to the usage of the same padding for filters. In this study, Adam optimizer [47] and an initial learning rate of 10^{-4} were used. The training batch size was set to 16 and the epoch was set to 100. For the model, an adjustment was carried out in the layer weight initialization procedure using He normal initializer proposed by [48]. For binary classification in this study, the U-Net was trained with a pixel-wise binary cross-entropy loss [26], which can handle the imbalanced number of pixels for each class, and a Sigmoid function as the activation layer.

The implementation of the model employed the Keras library with TensorFlow as its backend. All the experiments were carried on a Linux platform with a 4.00-GHz Intel(R) Core i7-8086k CPU, 64 GB of memory, and an NVIDIA Quadro GV100 graphics processing unit (32 GB).

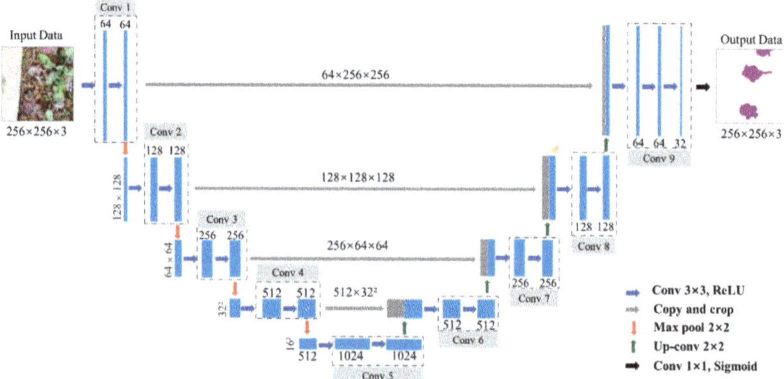

Figure 2. The architecture of the U-Net model used in this study with an input sample size of 256 × 256 pixels as an example. Each box represents the feature maps with the dimension of width × height on the left. For Conv 1, it performs two convolution operations with a 3 × 3 convolutional kernel, the number of feature maps is 64 with the size of 256 × 256 pixels, and then the max-pooling with a kernel size of 2 × 2 is performed.

2.7. Evaluation Metrics

In order to quantitatively evaluate the performance of the proposed method for segmenting purple rapeseed leaves, we calculated the precision, recall, F-measure, and Intersection of Union (IoU) [23] based on comparisons with the ground truth. The precision and recall were defined using the true positives (*TP*), false positives (*FP*), and false negatives (*FN*). The F-measure score was calculated using precision and recall as a powerful evaluation metric for the harmonic mean of precision and recall. IoU calculated the ratio of the intersection and union of the area in prediction and the ground truth. These measures were calculated as follows:

$$\text{Precision} = \frac{TP}{TP + FP} \quad (1)$$

$$\text{Recall} = \frac{TP}{TP + FN} \quad (2)$$

$$\text{F} - \text{measure} = \frac{2 \times \text{Precision} \times \text{Recall}}{(\text{Precision} + \text{Recall})} \quad (3)$$

$$\text{IoU} = \frac{TP}{TP + FP + FN} \quad (4)$$

where the precision is the percentage of matched pixels in the extraction results and the recall measures the proportion of matched pixels in the ground truth. The F-measure and IoU indicate the final exponent for evaluating the accuracy of purple rapeseed leaf segmentation with the U-Net model.

2.8. Segmentation of Purple Rapeseed Leaves with Four Other Methods

To further evaluate our proposed purple rapeseed leaf segmentation method, we implemented four other commonly used image segmentation methods comprising HSeg, RF, SVM, and SegNet. In this work, all methods used the same training set and test set. In particular, the accuracy of the models using only two categories of labels for RF and SVM was low. Therefore, the original class "without purple leaves" was subdivided into five new categories: green leaves, light soil, dark soil, cement, and withered leaves as the training set for RF and SVM. The HSeg method is a traditional target segmentation method of unsupervised classification. In this study, the original image was converted to the H-component of the HSI color space, which can make a big difference between the purple leaf and the background, and then the Otsu approach [49] was used to find an optimal threshold for purple rapeseed leaf segmentation. RF and SVM are both machine learning methods that have been widely used in agriculture for yield prediction and crop classification [15]. RF is a nonparametric and nonlinear classification method with a low computational burden [50,51] and the SVM classifier divides the classes with a decision surface by using a kernel function that maximizes the margin between the classes [52]. In this study, we found the optimal parameters in terms of accuracy obtained by five-fold cross-validation through continuous adjustment. Finally, we used RF containing 80 trees which were all grown fully and unpruned, and the number of features at each node was set to the square root of the total number of features. For the SVM classifier, this study employed the Radial Basis Function as the kernel, and the penalty factor and gamma parameter value were 2 and 0.33, respectively. The structure of the SegNet model is also divided into the encoder and the decoder [20]. This model records the feature locations by pooling indices computed in the max-pooling of the encoder to optimize the production of dense feature maps in the decoder. The process of parameter tuning and samples with four sizes for the SegNet were the same as those for the U-Net. After adjusting the parameters, the optimal input size of SegNet in this study was 256 × 256 pixels.

3. Results

3.1. Segmentation Results Obtained with the U-Net Model

Figure 3 shows the loss curves for training and validating datasets in four different patch sizes, which drop sharply at the beginning and then all reach the convergence states. Figure 4 summarizes the test results of the U-Net model for the patch sizes. Comparisons of the F-measure and IoU showed that as the patch sizes increased, the values tended to increase initially and then decrease. The peak value was obtained with a sample size of 256 × 256 pixels, which provided the best F-measure of 90.29% and IoU of 82.41%. In theory, larger patches should provide more information and yield more accurate predictions. However, the finer resolution also requires a longer training time, and the added information may be redundant and then influence the model training process [34]. The clipped patches with an excessively small window size would lack sufficient features for the model and then increase the risk of over-fitting. Moreover, the patches with a small size could not complete all down-sampling processes of the U-Net model. Thus, excessively small or large sizes made it difficult to distinguish between the purple leaf and the other categories with the U-Net model. Therefore, a patch size of 256 × 256 pixels was selected as an appropriate window size for the U-Net model in this study.

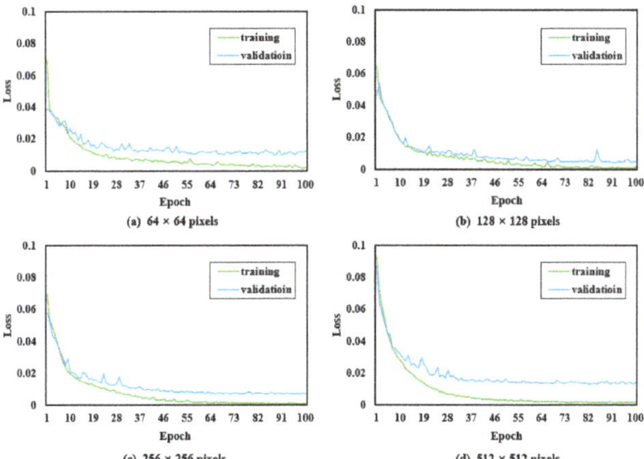

Figure 3. Loss curves of the proposed model for training and validating the datasets in four input patch sizes.

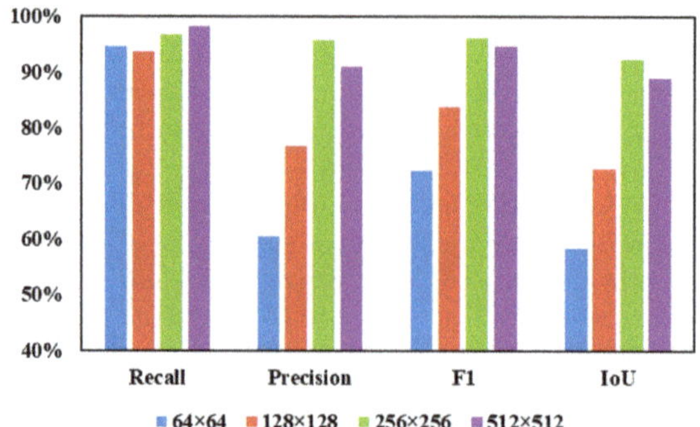

Figure 4. Four evaluation metrics results obtained for four patch sizes.

3.2. *Accuracy Evaluation for the U-Net Model and Four Other Image Segmentation Methods*

To compare the differences between the U-Net method and the other four methods (RF, HSeg, SVM, and SegNet), the four methods employed the same training set and test set as the U-Net model. The results are shown in Figure 5. The U-Net model outperformed each of the other four methods in terms of F-measure and IoU, which were about 6% and 9%, respectively, higher than those for SegNet, the second-best method. RF and SVM had a relatively low recall, and the precision with HSeg was clearly low. However, U-Net had a good balance between the two metrics.

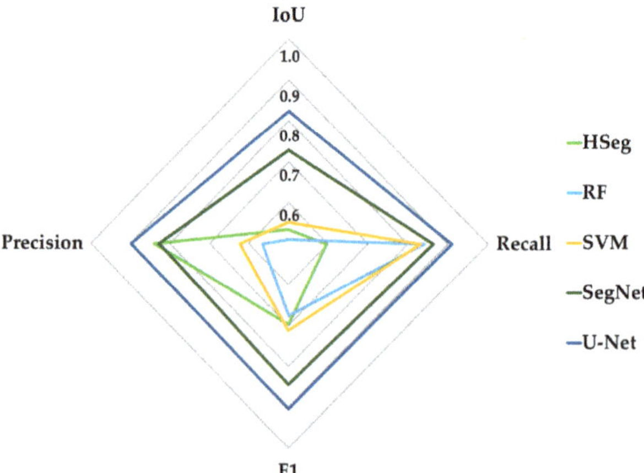

Figure 5. Summary of the precision, recall, F-measure (F1), and Intersection of Union (IoU) results obtained with all five segmentation methods.

4. Discussion

4.1. *Comparison of the Proposed Method and Four Commonly Used Methods*

To facilitate more intuitive comparisons of the proposed method and the four commonly used image segmentation methods, we selected four representative test images to analyze the segmentation results obtained with the different approaches. In Figure 6, the first row represents the original RGB rapeseed images, the second row represents the corresponding ground truth, and the remaining five rows show the segmentation results obtained by the five methods. The field environment was very complex with some uncontrollable factors such as illumination and many non-target objects.

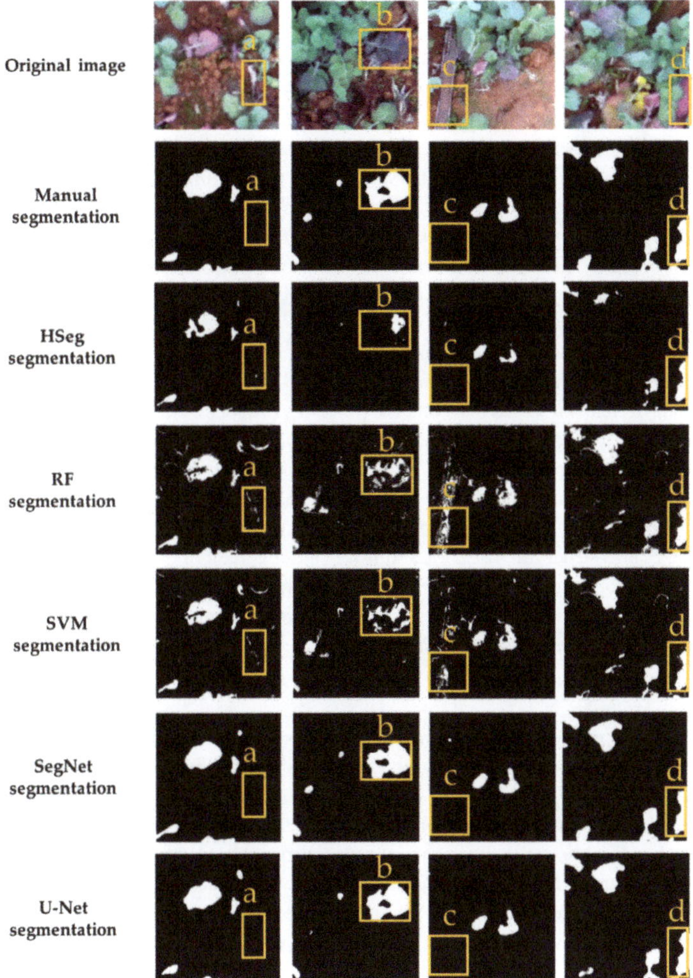

Figure 6. Purple rapeseed leaf extraction results obtained using five methods. Four representative rapeseed field images were selected to illustrate the segmentation effects. The first row shows the original rapeseed images in the field. The second row shows the manual segmentation results obtained using ArcMap software. The third to seventh rows show the purple rapeseed leaf segmentation results obtained using HSeg, random forest (RF), support vector machine (SVM), SegNet, and U-Net, respectively. Objects of interest (a withered leaf in region a, a darker purple leaf in region b, a ground object in region c, and a purple leaf overlapping with a green leaf in region d) were marked with orange boxes.

Region a contained a withered leaf that resembled a purple leaf. HSeg, SegNet, and U-Net could distinguish between them. However, there were a few incorrectly identified pixels with RF and SVM. For region b, the color gradation of the purple leaves was very different from that of general purple leaves, which led to many misclassified pixels except for SegNet and U-Net. Additionally, the appearance of some ground objects in region c interfered with the segmentation processes by SVM and RF. In region d, when the purple leaves were close to the green leaves, RF and SVM tended to misclassify the mixed pixels at the boundary of the purple leaf.

In the complex background conditions, the U-Net model had a higher capacity for purple rapeseed leaf segmentation and it obviously performed better than the other four methods. HSeg is only based on color information [8]. The color of the purple leaves could vary with changes in illumination and plant growth stage, which severely affected the segmentation results obtained using HSeg. In addition to these problems, the excessive dependence on color information could also lead to incomplete extraction and low recall values [8]. Both traditional machine learning methods (i.e., RF and SVM) create a serious salt-and-pepper phenomenon in the segmentation images. RF is time-consuming, but it is insensitive to the parameters and outliers [50,51]. RF had a difficulty distinguishing similar targets such as various leaves, shadows, and dark soil. Moreover, due to the influence of mixed pixels containing soil and leaves, RF often incorrectly identified the edges of the green leaves as purple leaves. SVM required a long time for model training and it had a problem with incomplete extraction. In general, the main problem with supervised methods is that their accuracy relies mainly on the quality of the training data set and the dimensionality of the input sample, which makes sample labeling a time-consuming process [52]. In this study, the field environment contained many non-target objects and it was difficult to obtain samples from all classes. Therefore, the number of classes in the image was much larger than the sample number, which led to low segmentation accuracy with SVM and RF. By contrast, the U-Net method exhibited higher generalizability and greater robustness during purple leaf segmentation because of its capacity to build a hierarchy of local and sparse features, whereas the traditional image segmentation methods are based on the global transformation of features [53]. Moreover, U-Net could maintain the edges and structural integrity, and it was independent of the color and shape information, thereby obtaining higher accuracy. Similarity, SegNet also had these traits, but the segmented purple leaves had non-sharp boundaries. For SegNet, the coarse outputs were produced after five consecutive down-samplings, which would have had an effect on the process of feature recovery in the up-sampling.

4.2. Influence of Image Resolution on Sample Size Selection

The images used in this study were acquired by the UAV platform. In practice, if the distance between the UAV and rapeseed plants is too close, the strong downdraft produced by the UAV makes the leaves sway and thus affects the image registration process. Thus, the restriction on the flight height made it difficult to obtain ultrahigh-resolution images. The flight height of the UAV was 20 m, which was almost the limit. In order to mimic other higher flight heights, the original orthoimage was re-sampled to resolutions of 3.72, 5.58, and 7.44 mm/pixel. Figure 7 presents the IoU results obtained for each image resolution and patch size. As the image resolution becomes coarser, the optimal accuracy tends to move towards a small patch size. On the one hand, for the same patch size, the number of background pixels was much more than the target (purple rapeseed leaf), which tended to make the trained network divide the pixels into backgrounds. To prevent this from happening, patch size could be reduced appropriately. Nevertheless, the patches with too small a size could easily lead to over-fitting of the model. On the other hand, if the target sizes were too small, the down-sampling process of the U-Net model could not complete, causing the loss of target feature extraction. Especially, the coarser the image spatial resolution, the smaller the area of the target in the image. Figure 8 shows the sizes of all purple rapeseed leaves in the original image used in this experiment, and most purple leaf sizes were between 11 × 11 pixels and 20 × 20 pixels. When the original image was re-sampled to the resolution of 7.44 mm/pixel, the mean and median of all purple leaf sizes were smaller than 8 × 8 pixels, which was not sufficient to obtain more features in Conv 4 (Figure 2). Usually, to obtain new and more complex features, a deep convolutional neural network is added with convolution layers to increase the receptive field. However, the deeper network requires a relatively larger image patch and target size. For pixel-level plant segmentation based on UAV imagery, it is difficult to meet such sample requirements. Excessive pursuit of network depth could lead to counterproductive results. For future plant segmentation work based on UAV images, this discussion can help to make an appropriate decision on sample size selection.

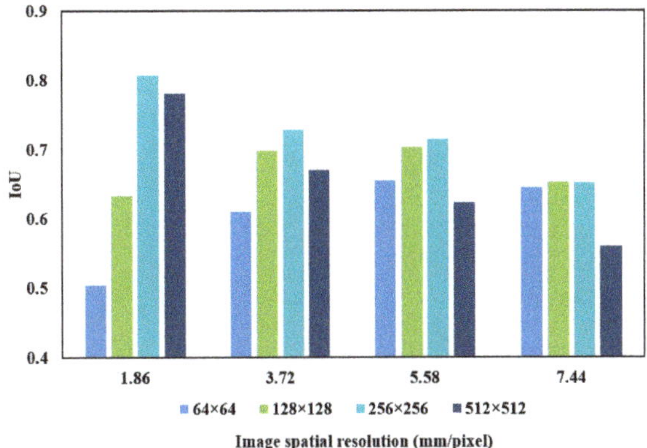

Figure 7. IoU results obtained for four image spatial resolutions and four patch sizes.

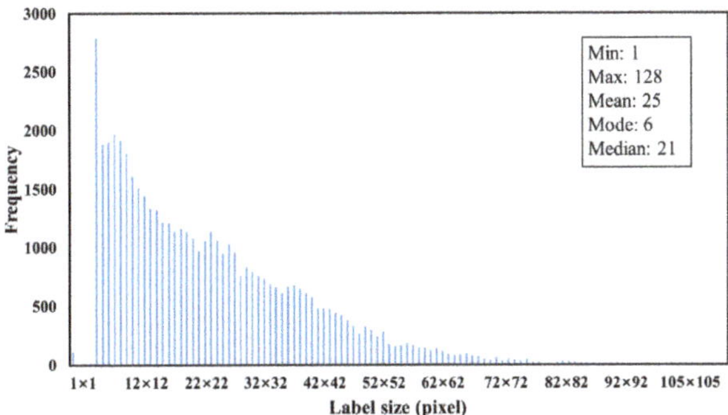

Figure 8. Frequency diagram and descriptive statistics of purple leaf size in unmanned aerial vehicle (UAV) imagery with resolution of 1.86 mm/pixel.

4.3. Relationship Between Nitrogen Content and Area Ratios of Purple Rapeseed Leaf

Nitrogen nutrition has a great effect on the growth and yield of rapeseed [37,54,55]. In this study, four N fertilizer rates were randomly applied to plots as shown in Figure 1 (Area 2 and Area 3). In order to accurately obtain the real-time N content, GreenSeeker (GS) values were measured as the ground truths of N content in 66 field samples. Table 3 presents the descriptive statistics for analyzing the differences in the GS values and purple leaf ratios under the four N application levels. As the ground truths of N content in this experiment, GS values increased steadily with the increase of N application level and maintained the similar range and coefficient of variation (CV). Purple leaf ratios have an increasing increment with the decrease of N application level, which led to a greater range and CV. At the same N application level, the nutrient uptake of rapeseed also varied. To further analyze the differences in GS values and purple leaf ratios among the four N treatments, the analysis of variance (ANOVA) was carried out, the *p* values of GS and purple leaf ratios were less than 0.05, indicting the decrease of N application induced different level of the N stress in rapeseed on the whole. Multiple comparison results using the least significant difference (LSD) method results (Table 3) showed that the GS means were significantly different among the four N levels, while the purple leaf ratio means were significantly different among all four N levels except for the two high N levels. These results indicate that the extraction method of purple leaves in this experiment was sensitive to detect N stress from 0 to 150 kg/ha. However, this method had a low sensitivity to nitrogen application at relatively high levels (over 150 kg/ha).

Table 3. Descriptive statistics and the differences of GreenSeeker (GS) values and purple leaf ratios among four N application levels.

	N Rate (kg/ha)	Number of Samples	Max	Min	Range	Mean	CV (%)	ANOVA Results
GS values	0	23	0.59	0.405	0.185	0.474[d]	9.4	$p < 0.05$
	75	13	0.665	0.52	0.145	0.568[c]	6.1	
	150	15	0.705	0.54	0.165	0.624[b]	8.4	
	225	14	0.72	0.605	0.115	0.678[a]	5.8	
Purple Leaf Ratios	0	23	0.536	0.167	0.369	0.308[a]	33.2	$p < 0.05$
	75	13	0.165	0.048	0.105	0.056[b]	18.3	
	150	15	0.05	0	0.05	0.016[c]	11.2	
	225	14	0.002	0	0.002	0.001[c]	1.0	

Max = maximum, Min = minimum, and CV = coefficient of variation; mean values followed by the same letter within a column of GS values or purple leaf ratios are not significantly different from one another at 0.05 probability level.

Results in Table 3 clearly indicate that leaf purpling was associated with N deficiency. However, there are few studies on the relationship between crop leaf area and nitrogen stress. Therefore, we performed a regression analysis to explore the specific relationship between the area of rapeseed purple leaves and N content. For rapeseed planting, the input of N is fitted to the law of diminishing returns, for which the increments of the rapeseed output will decrease while the increments of the N application input remain constant [56]. Additionally, some studies have indicated that an exponential function can measure the relationship between nitrogen content and green leaf canopy cover well [57,58]. As shown in Table 3, the area of purple leaves and the N fertilizer level had opposite trends, and thus it was reasonable to use a negative exponential model to fit the relationship between the area ratios of the purple rapeseed leaf and the N application rate. Figure 9a illustrates a linear relationship between nitrogen application level and GS values in the range of 0.3–0.8, which are effective for N content detection according to [59]. A negative exponential regression model between N content and purple leaf ratios based on the U-Net model was established in Figure 9b with the R^2 value 0.858. When more N was applied, the purple leaf ratio was reduced. Additionally, when the N application rate reached a certain level, no purple rapeseed leaves were observed, as shown in Figure 9c. However, monitoring crop growth by calculating the leaf area index or canopy coverage using only the green leaves is not always accurate. In particular, when the area of green leaves reached the maximum, as shown in Figure 9c–e corresponding to three different N fertilizer treatments (0, 75, and 150 kg/ha), it was difficult to assess the crop state by observing the green leaves. However, purple leaf extraction as a visible trait can be an effective auxiliary means for nitrogen stress detection.

As for the relationship between N rate and yield, low N application could lead to low photosynthetic assimilation and thus reduce grain yield, and excessive N application could cause environmental pollution and even yield decrease [60,61]. Additionally, this experiment illustrated that the exponential function could explain the relationship between N content and purple leaf ratios. For the commonly used nitrogen detection instruments in the field such as GS, it is time-consuming and laborious compared to the UAV-based method in this study. In addition, these sensors can only collect data at discrete points. Moreover, compared with the scientific-grade sensors including near-infrared and red-edge bands, which are sensitive for crops, the consumer-grade camera used in this experiment can only obtain RGB images, however, it is low-cost, easy to operate, and can obtain images with high spatial resolution [62]. In view of the above differences, the purple leaf extraction method based on UAV provided an efficient alternative method for nitrogen stress monitoring. In future experiments, we will increase the number and range of nitrogen levels for exploring the relationship between N stress and the purple leaves ratios comprehensively to find the best threshold for balancing the yield and environmental impact.

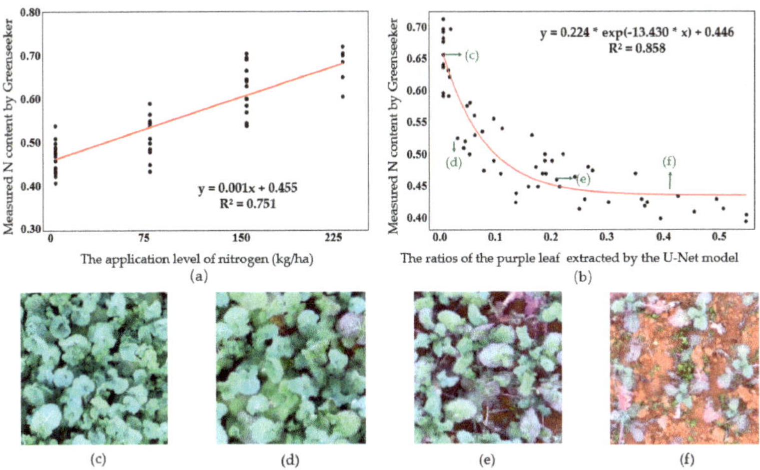

Figure 9. The results evaluation of N stress detected by the ratios of purple rapeseed leaves (purple leaf area to total leaf area). (**a**) Scatter plot and fitted curve between four N application level and N content measured by GS, (**b**) scatter plot and fitted curve between the ratios of purple rapeseed leaves extracted by the U-Net model and GS values, and (**c–f**) the images corresponded to the points in the (**b**).

5. Conclusions

In this study, we developed a novel method, which can efficiently monitor the crop nitrogen stress as an auxiliary means based on UAV RGB imagery. We designed and implemented a U-Net method for purple rapeseed leaf segmentation at the pixel level, and identified the optimum parameters by adjusting the sample size. The experimental results demonstrated the effectiveness of our proposed method for extracting purple rapeseed leaves, with a F-measure of 91.56% and IoU of 83.19%. Compared with four other commonly used image segmentation approaches (i.e., HSeg, RF, SVM, and SegNet), the U-Net model performed better.

The resolution of the rapeseed images obtained using the low-altitude remote sensing platform was much lower than that acquired with a high-resolution camera from close range. Therefore, in this study, we discussed the importance of patch size selection and evaluated the influence of image resolution on sample size selection, and the result showed that as the spatial resolution became coarser, a relatively smaller patch size had a higher accuracy.

We found that there was a negative exponential relationship (R^2 = 0.858) between the area of purple rapeseed leaf ratios and the corresponding N content in this experiment. Additionally, using the low-cost consumer-grade camera based on UAV platform is more suitable for agricultural practical application. Therefore, purple leaf extraction was a feasible and effective auxiliary means for monitoring nitrogen stress. Nitrogen is a significant parameter determining the photosynthetic functioning and productivity in crops. In further research, additional nitrogen levels will be added, and purple leaf area will be assessed as a visual trait to find the optimal nitrogen threshold for balancing crop yield and environmental impact. Moreover, other crops and stress types (e.g., water stress) will be studied based on purple leaves.

Author Contributions: J.Z. and T.X. designed the method, conducted the experiment, analyzed the data, discussed the results, and wrote the majority of the manuscript. C.Y. guided the study design, advised on data analysis, and revised the manuscript. H.S., H.F., and J.X. contributed to the method design and discussed the results. Z.J., D.Z., and G.Z. were involved in the process of the experiment, ground data collection, or manuscript revision. All authors reviewed and approved the final manuscript.

Funding: This work was financially supported by the National Key Research and Development Program of China (2018YFD1000900) and the Fundamental Research Funds for the Central Universities (Grant No. 2662018PY101 and 2662018JC012).

Acknowledgments: Special thanks go to the field staff of Huazhong Agricultural University for their daily management of the field experiments. We are grateful to the editor and reviewers for their valuable comments and recommendations.

Conflicts of Interest: The authors declare no conflict of interest.

Data Availability: The rapeseed datasets of this experience are available at https://figshare.com/s/e7471d81a1e35d5ab0d1.

Disclaimer: Mention of trade names or commercial products in this publication is solely for the purpose of providing specific information and does not imply recommendation or endorsement by the authors and their employer.

References

1. Gitelson, A.A.; Merzlyak, M.N.; Chivkunova, O.B. Optical properties and nondestructive estimation of anthocyanin content in plant leaves. *Photochem. Photobiol.* **2007**, *74*, 38–45. [CrossRef]
2. Van den Ende, W.; El-Esawe, S.K. Sucrose signaling pathways leading to fructan and anthocyanin accumulation: A dual function in abiotic and biotic stress responses? *Environ. Exp. Bot.* **2014**, *108*, 4–13. [CrossRef]
3. Sakamoto, W.; Ohmori, T.; Kageyama, K.; Miyazaki, C.; Saito, A.; Murata, M.; Noda, K.; Maekawa, M. The purple leaf (Pl) locus of rice: The plw allele has a complex organization and includes two genes encoding basic helix-loop-helix proteins involved in anthocyanin biosynthesis. *Plant Cell Physiol.* **2001**, *42*, 982–991. [CrossRef]
4. Chin, H.S.; Wu, Y.P.; Hour, A.L.; Hong, C.Y.; Lin, Y.R. Genetic and evolutionary analysis of purple leaf sheath in rice. *Rice* **2016**, *9*, 8. [CrossRef] [PubMed]
5. Ithal, N.; Reddy, A.R. Rice flavonoid pathway genes, OsDfr and OsAns, are induced by dehydration, high salt and ABA, and contain stress responsive promoter elements that interact with the transcription activator, OsC1-MYB. *Plant Sci.* **2004**, *166*, 1505–1513. [CrossRef]
6. Oren-Shamir, M.; Levi-Nissim, A. Temperature effects on the leaf pigmentation of cotinus coggygria 'Royal Purple'. *J. Hortic. Sci.* **1997**, *72*, 425–432. [CrossRef]
7. Hughes, N.M.; Lev-Yadun, S. Red/purple leaf margin coloration: Potential ecological and physiological functions. *Environ. Exp. Bot.* **2015**, *119*, 27–39. [CrossRef]
8. Xiong, X.; Duan, L.; Liu, L.; Tu, H.; Yang, P.; Wu, D.; Chen, G.; Xiong, L.; Yang, W.; Liu, Q. Panicle-SEG: A robust image segmentation method for rice panicles in the field based on deep learning and superpixel optimization. *Plant Methods* **2017**, *13*, 104. [CrossRef]
9. Tang, W.; Zhang, Y.; Zhang, D.; Yang, W.; Li, M. Corn tassel detection based on image processing. In Proceedings of the 2012 International Workshop on Image Processing and Optical Engineering, Harbin, China, 15 November 2012.
10. Bai, X.D.; Cao, Z.G.; Wang, Y.; Yu, Z.H.; Zhang, X.F.; Li, C.N. Crop segmentation from images by morphology modeling in the CIE L*a*b* color space. *Comput. Electron. Agric.* **2013**, *99*, 21–34. [CrossRef]
11. Tian, Y.; Li, T.; Li, C.; Piao, Z.; Sun, G.; Wang, B. Method for recognition of grape disease based on support vector machine. *Trans. Chin. Soc. Agric. Eng.* **2007**, *23*, 175–180.
12. Guerrero, J.M.; Pajares, G.; Montalvo, M.; Romeo, J.; Guijarro, M. Support Vector Machines for crop/weeds identification in maize fields. *Expert Syst. Appl.* **2012**, *39*, 11149–11155. [CrossRef]
13. Tatsumi, K.; Yamashiki, Y.; Torres, M.A.C.; Taipe, C.L.R. Crop classification of upland fields using Random forest of time-series Landsat 7 ETM+ data. *Comput Electron. Agric.* **2015**, *115*, 171–179. [CrossRef]
14. Jeon, H.Y.; Tian, L.F.; Zhu, H. Robust crop and weed segmentation under uncontrolled outdoor illumination. *Sensors* **2011**, *11*, 6270–6283. [CrossRef]
15. Liakos, K.; Busato, P.; Moshou, D.; Pearson, S.; Bochtis, D. Machine learning in agriculture: A review. *Sensors* **2018**, *18*, 2674. [CrossRef]
16. Romera-Paredes, B.; Torr, P.H.S. Recurrent instance segmentation. In Proceedings of the 14th European Conference, Amsterdam, The Netherlands, 11–14 October 2016; pp. 312–329.

17. Pound, M.P.; Atkinson, J.A.; Townsend, A.J.; Wilson, M.H.; Griffiths, M.; Jackson, A.S.; Bulat, A.; Tzimiropoulos, G.; Wells, D.M.; Murchie, E.H.; et al. Deep Machine Learning provides state-of-the-art performance in image-based plant phenotyping. *Gigascience* **2017**, *6*, 1–10. [CrossRef]
18. Long, J.; Shelhamer, E.; Darrell, T. Fully convolutional networks for semantic segmentation. *IEEE Trans. Pattern Anal. Mach. Intell.* **2015**, *79*, 1337–1342.
19. Sa, I.; Chen, Z.; Popović, M.; Khanna, R.; Liebisch, F.; Nieto, J.; Siegwart, R. weednet: Dense semantic weed classification using multispectral images and mav for smart farming. *IEEE Robot. Autom. Lett.* **2017**, *3*, 588–595. [CrossRef]
20. Badrinarayanan, V.; Kendall, A.; Cipolla, R. Segnet: A deep convolutional encoder-decoder architecture for image segmentation. *IEEE Trans. Pattern Anal. Mach. Intell.* **2017**, *39*, 2481–2495. [CrossRef]
21. Ronneberger, O.; Fischer, P.; Brox, T. U-net: Convolutional networks for biomedical image segmentation. In Proceedings of the 18th Medical Image Computing and Computer-Assisted Intervention, Munich, Germany, 5–9 October 2015.
22. Chen, L.C.; Papandreou, G.; Kokkinos, I.; Murphy, K.; Yuille, A.L. Deeplab: Semantic image segmentation with deep convolutional nets, atrous convolution, and fully connected crfs. *IEEE Trans. Pattern Anal. Mach. Intell.* **2018**, *40*, 834–848. [CrossRef]
23. Sa, I.; Popović, M.; Khanna, R.; Chen, Z.; Lottes, P.; Liebisch, F.; Nieto, J.; Stachniss, C.; Walter, A.; Siegwart, R. WeedMap: A large-scale semantic weed mapping framework using aerial multispectral imaging and deep neural network for precision farming. *Remote Sens.* **2018**, *10*, 1423. [CrossRef]
24. Barth, R.; IJsselmuiden, J.; Hemming, J.; Van Henten, E.J. Synthetic bootstrapping of convolutional neural networks for semantic plant part segmentation. *Comput. Electron. Agric.* **2019**, *161*, 291–304. [CrossRef]
25. De Brabandere, B.; Neven, D.; Van Gool, L. Semantic instance segmentation with a discriminative loss function. In Proceedings of the IEEE Conference on Computer Vision and Pattern Recognition, Honolulu, HI, USA, 21–26 July 2017; pp. 478–480.
26. Milioto, A.; Lottes, P.; Stachniss, C. Real-time semantic segmentation of crop and weed for precision agriculture robots leveraging background knowledge in CNNs. In Proceedings of the IEEE International Conference on Robotics and Automation, Brisbane, Australia, 21–25 May 2018.
27. Gómez-Candón, D.; De Castro, A.I.; López-Granados, F. Assessing the accuracy of mosaics from unmanned aerial vehicle (UAV) imagery for precision agriculture purposes in wheat. *Precis. Agric.* **2014**, *15*, 44–56. [CrossRef]
28. Rango, A.; Laliberte, A.; Steele, C.; Herrick, J.E.; Bestelmeyer, B.; Schmugge, T.; Roanhorse, A.; Jenkins, V. Research article: Using unmanned aerial vehicles for rangelands: Current applications and future potentials. *Environ. Pract.* **2006**, *8*, 159–168. [CrossRef]
29. Gong, Y.; Duan, B.; Fang, S.; Zhu, R.; Wu, X.; Ma, Y.; Peng, Y. Remote estimation of rapeseed yield with unmanned aerial vehicle (UAV) imaging and spectral mixture analysis. *Plant Methods* **2018**, *14*, 70. [CrossRef] [PubMed]
30. Konoplich, G.V.; Putin, E.O.; Filchenkov, A.A. Application of deep learning to the problem of vehicle detection in UAV images. In Proceedings of the 2016 IEEE International Conference on Soft Computing and Measurements, Adygeya, Russia, 25–27 May 2016; pp. 4–6.
31. Benjamin, K.; Diego, M.; Devis, T. Detecting mammals in UAV images: Best practices to address a substantially imbalanced dataset with deep learning. *Remote Sens. Environ.* **2018**, *216*, 139–153.
32. Zhang, X.; Chen, G.; Wang, W.; Wang, Q.; Dai, F. Object-based land-cover supervised classification for very-high-resolution UAV images using stacked denoising autoencoders. *IEEE J. Sel. Top. Appl. Earth Obs. Remote Sens.* **2017**, *10*, 3373–3385. [CrossRef]
33. Yang, M.D.; Huang, K.S.; Kuo, Y.H.; Tsai, H.P.; Lin, L.M. Spatial and spectral hybrid image classification for rice lodging assessment through UAV imagery. *Remote Sens.* **2017**, *9*, 583. [CrossRef]
34. Milletari, F.; Navab, N.; Ahmadi, S.A. V-Net: Fully convolutional neural networks for volumetric medical image segmentation. In Proceedings of the 14th 3D Vision, Stanford, CA, USA, 25–28 October 2016; pp. 565–571.
35. Yuan, M.; Liu, Z.; Wang, F. Using the wide-range attention U-Net for road segmentation. *Remote Sens. Lett.* **2019**, *10*, 506–515. [CrossRef]

36. Yao, Y.; Miao, Y.; Huang, S.; Gao, L.; Ma, X.; Zhao, G.; Jiang, R.; Chen, X.; Zhang, F.; Yu, K.; et al. Active canopy sensor-based precision N management strategy for rice. *Agron. Sustain. Dev.* **2012**, *32*, 925–933. [CrossRef]
37. Ozer, H. Sowing date and nitrogen rate effects on growth, yield and yield components of two summer rapeseed cultivars. *Eur. J. Agron.* **2003**, *19*, 453–463. [CrossRef]
38. Singh, I.; Srivastava, A.K.; Chandna, P.; Gupta, R.K. Crop sensors for efficient nitrogen management in sugarcane: Potential and constraints. *Sugar Tech.* **2006**, *8*, 299–302. [CrossRef]
39. Teal, R.K.; Tubana, B.; Girma, K.; Freeman, K.W.; Arnall, D.B.; Walsh, O.; Raun, W.R. In-season prediction of corn grain yield potential using normalized difference vegetation index. *Agron. J.* **2006**, *98*, 1488–1494. [CrossRef]
40. Ma, N.; Yuan, J.; Li, M.; Li, J.; Zhang, L.; Liu, L.; Naeem, M.S.; Zhang, C. Ideotype population exploration: Growth, photosynthesis, and yield components at different planting densities in winter oilseed rape (*Brassica napus* L.). *PLoS ONE* **2014**, *9*, e114232. [CrossRef]
41. Wang, M.; Li, Q.; Hu, Q.; Yuan, H. A parallel interior orientation method based on lookup table for UAV images. In Proceedings of the IEEE International Conference on Spatial Data Mining and Geographical Knowledge Services, Fuzhou, China, 29 June–1 July 2011; pp. 397–400.
42. Kelcey, J.; Lucieer, A. Sensor correction of a 6-band multispectral imaging sensor for UAV remote sensing. *Remote Sens.* **2012**, *4*, 1462–1493. [CrossRef]
43. Hu, P.; Guo, W.; Chapman, S.C.; Guo, Y.; Zheng, B. Pixel size of aerial imagery constrains the applications of unmanned aerial vehicle in crop breeding. *ISPRS J. Photogramm. Remote Sens.* **2019**, *154*, 1–9. [CrossRef]
44. Zhou, K.; Cheng, T.; Zhu, Y.; Cao, W.; Ustin, S.L.; Zheng, H.; Yao, X.; Tian, Y. Assessing the impact of spatial resolution on the estimation of leaf nitrogen concentration over the full season of paddy rice using near-surface imaging spectroscopy data. *Front. Plant Sci.* **2018**, *9*, 964. [CrossRef]
45. Feng, W.; Sui, H.; Huang, W.; Xu, C.; An, K. Water body extraction from very high-resolution remote sensing imagery using deep U-Net and a superpixel-based conditional random field model. *IEEE Geosci. Remote Sens. Lett.* **2018**, *16*, 618–622. [CrossRef]
46. Langner, T.; Hedström, A.; Mörwald, K.; Weghuber, D.; Forslund, A.; Bergsten, P.; Ahlström, H.; Kullberg, J. Fully convolutional networks for automated segmentation of abdominal adipose tissue depots in multicenter water–fat MRI. *Magn. Reson. Med.* **2019**, *81*, 2736–2745. [CrossRef]
47. Kingma, D.P.; Ba, J. Adam: A method for stochastic optimization. In Proceedings of the 2014 International Conference on Learning Representations, Banff, AB, Canada, 14–16 April 2014.
48. He, K.; Zhang, X.; Ren, S.; Sun, J. Delving deep into rectifiers: Surpassing human-level performance on imageNet classification. In Proceedings of the IEEE International Conference on Computer Vision (ICCV), Santiago, Chile, 13–16 December 2015; pp. 1026–1034.
49. Otsu, N. A threshold selection method from gray-level histograms. *IEEE Trans. Syst. Man Cybern.* **1979**, *9*, 62–66. [CrossRef]
50. Puissant, A.; Rougier, S.; Stumpf, A. Object-oriented mapping of urban trees using Random Forest classifiers. *Int. J. Appl. Earth Obs. Geoinf.* **2014**, *26*, 235–245. [CrossRef]
51. Kang, L.; Ye, P.; Li, Y.; Doermann, D. Convolutional neural networks for no-reference image quality assessment. In Proceedings of the 2014 IEEE Computer Vision and Pattern Recognition, Columbus, OH, USA, 23–28 June 2014.
52. Camps-Valls, G.; Gómez-Chova, L.; Calpe-Maravilla, J.; Soria-Olivas, E.; Martín-Guerrero, J.D.; Moreno, J. Support vector machines for crop classification using hyperspectral data. In Proceedings of the 1st Pattern Recognition and Image Analysis, Puerto de Andratx, Mallorca, Spain, 4–6 June 2003; pp. 134–141.
53. Kussul, N.; Lavreniuk, M.; Skakun, S.; Shelestov, A. Deep learning classification of land cover and crop types using remote sensing data. *IEEE Geosci. Remote Sens. Lett.* **2017**, *14*, 778–782. [CrossRef]
54. Inoue, Y.; Sakaiya, E.; Zhu, Y.; Takahashi, W. Diagnostic mapping of canopy nitrogen content in rice based on hyperspectral measurements. *Remote Sens. Environ.* **2012**, *126*, 210–221. [CrossRef]
55. Zhang, F.; Cui, Z.; Fan, M.; Zhang, W.; Chen, X.; Jiang, R. Integrated soil–crop system management: Reducing environmental risk while increasing crop productivity and improving nutrient use efficiency in China. *J. Environ. Qual.* **2011**, *40*, 1051–1057. [CrossRef] [PubMed]
56. Wright, G.C.; Smith, C.J.; Woodroofe, M.R. The effect of irrigation and nitrogen fertilizer on rapeseed (Brassica napes) production in South-Eastern Australia: I. Growth and seed yield. *Irrig. Sci.* **1988**, *9*, 1–13. [CrossRef]

57. Jia, B.; He, H.; Ma, F.; Diao, M.; Jiang, G.; Zheng, Z.; Cui, J.; Fan, H. Use of a digital camera to monitor the growth and nitrogen status of cotton. *Sci. World J.* **2014**, *1*, 1–2. [CrossRef]
58. Lee, K.J.; Lee, B.W. Estimation of rice growth and nitrogen nutrition status using color digital camera image analysis. *Eur. J. Agron.* **2013**, *48*, 57–65. [CrossRef]
59. Walsh, O.S.; Klatt, A.R.; Solie, J.B.; Godsey, C.B.; Raun, W.R. Use of soil moisture data for refined green seeker sensor based nitrogen recommendations in winter wheat. *Precis. Agric.* **2013**, *14*, 343–356. [CrossRef]
60. Jay, S.; Rabatel, G.; Hadoux, X.; Moura, D.; Gorretta, N. In-field crop row phenotyping from 3D modeling performed using Structure from Motion. *Comput. Electron. Agric.* **2015**, *110*, 70–77. [CrossRef]
61. Kaushal, S.S.; Groffman, P.M.; Band, L.E.; Elliott, E.M.; Shields, C.A.; Kendall, C. Tracking nonpoint source nitrogen pollution in human-impacted watersheds. *Environ. Sci. Technol.* **2011**, *45*, 8225–8232. [CrossRef]
62. Zhang, J.; Yang, C.; Song, H.; Hoffmann, W.C.; Zhang, D.; Zhang, G. Evaluation of an airborne remote sensing platform consisting of two consumer-grade cameras for crop identification. *Remote Sens.* **2016**, *8*, 257. [CrossRef]

Letter

Recognition of Banana Fusarium Wilt Based on UAV Remote Sensing

Huichun Ye [1,2,3], Wenjiang Huang [1,2,3,*], Shanyu Huang [4], Bei Cui [1,2,3], Yingying Dong [1,2], Anting Guo [1,5], Yu Ren [1,5] and Yu Jin [6]

1. Key Laboratory of Digital Earth Science, Institute of Remote Sensing and Digital Earth, Chinese Academy of Sciences, Beijing 100094, China; yehc@radi.ac.cn (H.Y.); cuibei@radi.ac.cn (B.C.); dongyy@radi.ac.cn (Y.D.); guoat@aircas.ac.cn (A.G.); renyu@aircas.ac.cn (Y.R.)
2. Aerospace Information Research Institute, Chinese Academy of Sciences, Beijing 100094, China
3. Key Laboratory of Earth Observation of Hainan Province, Sanya 572029, China
4. Chinese Academy of Agricultural Engineering Planning & Design, Beijing 100125, China; s08020406@cau.edu.cn
5. University of Chinese Academy of Sciences, Beijing 100049, China
6. School of Electronics and Information Engineering, Anhui University, Hefei 230601, China; ahuJingy@163.com

* Correspondence: huangwj@radi.ac.cn

Received: 22 January 2020; Accepted: 12 March 2020; Published: 13 March 2020

Abstract: Fusarium wilt (Panama disease) of banana currently threatens banana production areas worldwide. Timely monitoring of Fusarium wilt disease is important for the disease treatment and adjustment of banana planting methods. The objective of this study was to establish a method for identifying the banana regions infested or not infested with Fusarium wilt disease using unmanned aerial vehicle (UAV)-based multispectral imagery. Two experiments were conducted in this study. In experiment 1, 120 sample plots were surveyed, of which 75% were used as modeling dataset for model fitting and the remaining were used as validation dataset 1 (VD1) for validation. In experiment 2, 35 sample plots were surveyed, which were used as validation dataset 2 (VD2) for model validation. An UAV equipped with a five band multispectral camera was used to capture the multispectral imagery. Eight vegetation indices (VIs) related to pigment absorption and plant growth changes were chosen for determining the biophysical and biochemical characteristics of the plants. The binary logistic regression (BLR) method was used to assess the spatial relationships between the VIs and the plants infested or not infested with Fusarium wilt. The results showed that the banana Fusarium wilt disease can be easily identified using the VIs including the green chlorophyll index (CI_{green}), red-edge chlorophyll index (CI_{RE}), normalized difference vegetation index (NDVI), and normalized difference red-edge index (NDRE). The fitting overall accuracies of the models were greater than 80%. Among the investigated VIs, the CI_{RE} exhibited the best performance both for the VD1 (OA = 91.7%, Kappa = 0.83) and VD2 (OA = 80.0%, Kappa = 0.59). For the same type of VI, the VIs including a red-edge band obtained a better performance than that excluding a red-edge band. A simulation of imagery with different spatial resolutions (i.e., 0.5-m, 1-m, 2-m, 5-m, and 10-m resolutions) showed that good identification accuracy of Fusarium wilt was obtained when the resolution was higher than 2 m. As the resolution decreased, the identification accuracy of Fusarium wilt showed a decreasing trend. The findings indicate that UAV-based remote sensing with a red-edge band is suitable for identifying banana Fusarium wilt disease. The results of this study provide guidance for detecting the disease and crop planting adjustment.

Keywords: Fusarium wilt; crop disease; banana; multispectral remote sensing; UAV

1. Introduction

Banana (*Musa spp.*) is a widely grown cash crop in the tropics and subtropics. Fusarium wilt (Panama disease) of banana, which is caused by the fungus *Fusarium oxysporum* f. sp. *cubense* race 4 (*Foc* 4), is a serious soilborne fungal disease [1]. This disease currently threatens the banana planting areas worldwide, including areas in Southeast Asia, Jordan, Australia, Lebanon, Pakistan, Mozambique, and Oman [2]. Fusarium wilt disease may have affected approximately 100,000 ha of banana plantations, and it is likely to spread further either through infected plant materials, contaminated soil, or farm machinery or due to flowing water and inappropriate sanitation measures [2]. Externally, the first signs of this disease are wilted banana plants with yellowing of the older leaves around the margins. As the disease advances, the plant leaves finally droop, forming a 'skirt' around the pseudo-stem before falling off. The new leaves may present pale margins and irregular and wrinkled blades [3]. Currently, there are no efficient chemical treatment for Fusarium wilt control. Once a diseased plant has been found, 'timely removal' is the best way to avoid the formation of a disease center [4]. Therefore, timely monitoring of banana Fusarium wilt disease is important for the disease treatment and crop planting adjustment.

Real-time monitoring and identification of crop disease are the basis of timely prevention and control [5]. Traditionally, ground surveys have been the only effective approach to monitor and discriminate crop disease, but these investigations are time-consuming and often very expensive. Remote sensing has become a feasible technology for disease detection and assessment over the last several decades. Diseases that have been detected using remote sensing include rust infection [6–8], Fusarium head blight [9,10], and powdery mildew [9–12] in wheat, bacterial leaf blight in rice [13,14], grey leaf spot in maize [15], and late blight disease and bacterial spot in tomato [16,17]. When plants are infected with diseases, the leaf water, pigment content and internal structure undergo changes, and these biochemical and biophysical changes are also reflected in the spectral characteristics of plants [18]. Many studies have successfully applied sensitive spectral bands or vegetation indices (VIs) to the identification and monitoring of crop diseases in the leaf and canopy scales. Bravo et al. [19] calculated the normalized difference vegetation index (NDVI) using wavelengths of 740–760 nm and 620–640 nm to extract powdery mildew wheat patches. Devadas et al. [20] showed that yellow rust wheat and healthy wheat could be distinguished by the anthocyanin reflection index (ARI). Huang et al. [7] suggested that the red-edge position can be used as an indicator for disease detection. However, spectral bands and VIs exhibit different sensitivity to different diseases and it is necessary to determine which spectral bands and VIs are suitable for the identification of specific diseases.

Satellite-based imagery is an affordable source of data for large-scale agricultural monitoring. There are a few previous studies that have shown successful detection of crop disease using high-resolution satellite multispectral images. For example, Oumar and Mutanga [21] demonstrated the applicability of Worldview-2 image for disease monitoring in a study on the prediction of bronze bug damage in plantation forests. Zhang et al. [22] established a multitemporal, modified soil-adjusted vegetation index for HJ-CCD images, and detected and mapped the outbreak of armyworm. Shi et al. [5] successfully used PlanetScope imagery to identify rice blast, rice dwarf, and glume blight. However, canopy structural characteristics and the biological effects induced by diseases often vary at fine spatial scales. Thus, in practice, the use of satellite-based imagery to monitoring diseases at field or subfield scales must address the constraint that different objects with similar spectral properties are affected by a mixed pixel effect from low-to-moderate resolution satellites (e.g., Landsat OLI-8, Sentinel-2) [5]. Furthermore, the use of high-resolution imageries acquired from satellite platforms is deficient for the long revisit period due to high cost and unfavorable weather conditions. In recent years, the development of unmanned aerial vehicles (UAVs) has provided new imagery acquisition platforms that can collect very high-resolution imagery and data in a short period of time in a cost-effective manner [23]. Therefore, UAVs provide a new technical means from which the in-season growth information of crops can be extracted in a timely and nondestructive manner [24]. Significant progress has been made in crop classification, growth monitoring, and pest and disease identification using

UAV-based multispectral and hyperspectral imagery [23,25,26]. A few studies also applied UAV-based imagery to map spatial patterns of photosynthetic activity in banana plantations [27]. However, studies using UAV-based remote sensing technology to monitor Fusarium wilt of banana are scarce [28].

Moreover, due to the scale effects, the scaling topic has become one of the hotspots in remote sensing research [29]. Although higher spatial resolution images show more landscape details and more accurate estimates [30], due to expensive costs and processing difficulties, it is unnecessary and unrealistic to seek very high-resolution data for the agriculture application. Therefore, it is better to select an appropriate spatial resolution image for agricultural monitoring after considering various factors. In addition, choosing an appropriate method for data analysis is very important, as it directly affects the reliability and accuracy of the results. Many approaches or models have been used to determine bands and features that are sensitive to crop disease detection and discrimination [5,18,31]. Binary logistic regression (BLR) is one of the most commonly used multivariate analysis approaches to describe the relationship between a dependent variable and multiple independent variables, where the dependent variable is a binary variable that indicates whether an event exists [32]. Logistic regression has advantages over linear regression and log-linear linear regression because logistic regression does not need to assume normality [33].

The objectives of this study were to (i) develop an identification method for Fusarium wilt of banana using UAV-based multispectral imagery, (ii) determine the optimal VI for establishing an optimal identification model, and (iii) assess the effect of different image resolution on the identification accuracy of banana Fusarium wilt disease to provide a reference for large-scale applications of satellite-based data.

2. Materials and Methods

2.1. Study Area

The experiments were conducted at two experimental locations: The Guangxi site and Hainan site (Figure 1).

The Guangxi site is located in Long'an County, Guangxi Province, China (23°7′53.2″–23°8′4.0″ N, 107°43′44.9–07°44′7.2″ E) (Figure 1). The region has a subtropical monsoon climate characterized by year-round sufficient sunshine and rainfall. The mean annual temperature is 20.8–22.4 °C. The average rainfall is 1200 mm a year. The soil type is a Ferralsol according to the IUSS Working Group WRB soil classification system [34]. The field crops were the banana variety "Williams B6." The plant height was about 2.4–3 m, the leaf number was 34–36, and the growth period was 10–12 months. The farm was developed in September 2015 and was harvested for the first time in November 2016. By August 2018 (the time of field investigation in this study), the third generation of bananas was present in the fields. The planting distance was 2.0 m by 2.6 m with a planting density of 1950 plants per hectare. In this study area, more than 40% of banana plants were infected with Fusarium wilt disease of different severity.

The Hainan site is located in Chengmai County, Hainan Province, China (19°49′4.4″–19°49′15.8″ N, 109°54′40.0″–109°54′53.0″ E). The region has a tropical monsoon climate characterized by year-round sufficient sunshine and rainfall. The mean annual temperature is 23.1–24.5 °C. The average rainfall is 1750 mm a year. The soil type is a Humic Acrisol according to the IUSS Working Group WRB soil classification system [34]. This experimental site was divided into two fields (left field and right field) along the boundary of the middle road. The left field was planted the banana variety "Baxijiao." The plant height was about 2.6–3.2 m and the growth period was 9–12 months. This field was developed in June 2017 and was harvested for the first time in July 2018. By December 2018, the second generation of bananas was present in the field. The planting distance was 2.0 m by 2.3 m with a planting density of 2100 plants per hectare. In this field, about 10% of banana plants were infected with Fusarium wilt disease of different severity. The right field was developed in August 2018 and the banana variety was "Nantianhuang." The plant height was about 2.5–3.0 m and the growth period was 10–13 months.

The planting distance was the same as the left field. At the time of field investigation in December 2018, there were no plants infected with Fusarium wilt found in this field.

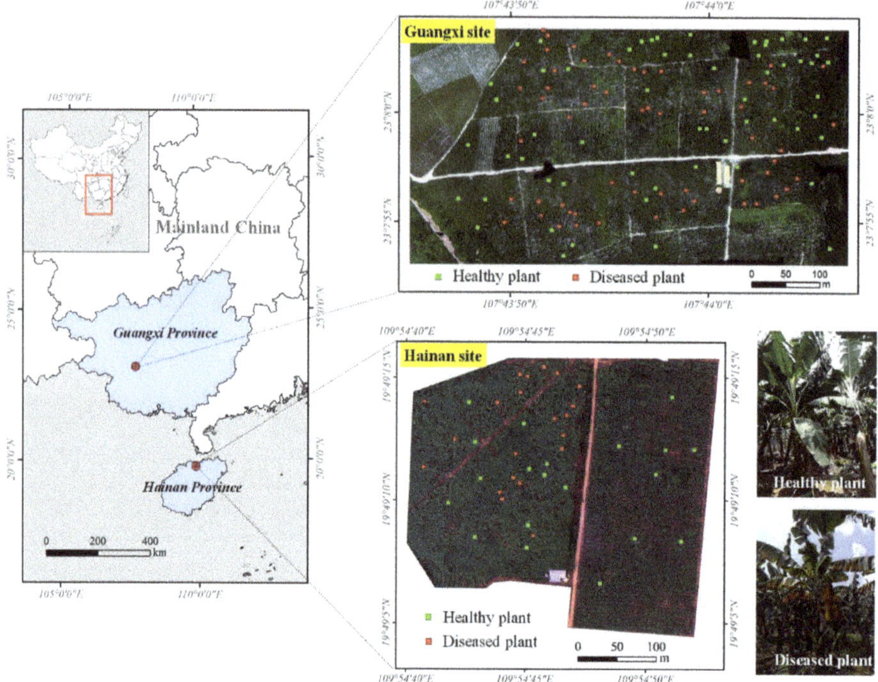

Figure 1. Location of the experimental sites with the distribution of survey sites in the banana plantations.

In this study, the Guangxi site was used for Fusarium wilt identifying model calibration and validation and the Hainan site was used for model validation.

2.2. Field Investigation

The Guangxi experiment was conducted on 7 August 2018. A total of 120 sample plots were surveyed to assess the occurrence of banana Fusarium wilt disease as ground truth data (Figure 1). The size of each sample point covered one banana plant. These samples were classified into two categories: Healthy samples (total of 57) and diseased samples (total of 63), representing the occurrence or nonoccurrence of Fusarium wilt as reflected by the external characteristics. The classification standard adopted in this paper was based on the percent of the yellowing leaf area to the total leaf area of the plant. If the percent of the yellowing leaf area to the total leaf area of the plant was less than 1%, the plant was considered to be healthy. Otherwise, it was considered to be diseased. Finally, 75% of the samples were randomly chosen and used as modeling dataset for model fitting, and the remaining 25% were used as validation dataset 1 (VD1) for validation.

The Hainan site experiment was conducted on 11 December 2018. The investigation scheme was consistent with that of Guangxi site experiment. A total of 35 sample plots were investigated, of which 16 were healthy and 19 were diseased. All the sample plots were used as validation dataset 2 (VD2) for model validation.

2.3. *UAV Multi-Spectral Imagery Acquisition*

The surveys were done using a DJI Phantom 4 quadcopter (DJI Innovations, Shenzhen, China). This UAV was equipped with a five-band multispectral camera MicaSense RedEdge M^{TM} (MicaSense, Inc., Seattle, WA, USA), which has five narrow bands: Blue (465–485 nm), green (550–570 nm), red (653–673 nm), red edge (712–722 nm), and near-infrared (800–880 nm). The flight at Guangxi site was conducted between 12:30 p.m.–13:30 p.m. on 7 August 2018 and covered an area of 21 ha. The flight at Hainan site was conducted between 11:00 a.m.–12:00 p.m. on 11 December 2018 and covered an area of 11 ha. Both the flight plans were developed to ensure greater than 80% cross-track and along-track overlap rates. The multispectral imagery was acquired from a flying height of 120 m above the ground with a ground sample distance of 0.08 m. Pre- and post-flight images of MicaSense calibrated reflectance panels with known reflectance were also captured using the RedEdge sensor to aid in radiometric conversion.

2.4. *Data Analysis*

In this study, different VIs were used to identify the infestation status of Fusarium wilt in banana. BLR was used to assess the spatial relationship between the VIs and the plants infested or not infested with Fusarium wilt. In order to assess the classification accuracy of images with different spatial resolutions, we chose to resample the original UAV imagery of 0.08 m to generate images with 0.5-m, 1-m, 2-m, 5-m, and 10-m resolutions using the nearest neighbor resampling algorithm. These resolutions were selected because they were similar to those of several mainstream and easily accessible satellite imagery products (i.e., WorldView series with a resolution of 0.5 m, GF-2 with a resolution of 1 m, GF-1 and GF-6 with a resolution of 2 m, RapidEye with a resolution of 5 m, and Sentinel-2 with a resolution of 10 m) for agricultural applications.

2.4.1. Vegetation Indices

Considering the potential pathological characteristics of the Fusarium wilt disease infestations, eight VIs related to pigment absorption and plant growth were selected to characterize the biochemical and biophysical variations caused by individual infestations. The VIs included the NDVI, normalized difference red edge index (NDRE), green chlorophyll index (CI_{green}), red-edge chlorophyll index (CI_{RE}), structural independent pigment index (SIPI), red-edge structural independent pigment index ($SIPI_{RE}$), carotenoid index (CARI), and anthocyanin reflectance index (ARI). The definitions of these VIs are listed in Table 1.

Table 1. List of eight vegetation indices (Vis) used in this study.

Vegetation Index	Description	Formulation	Sensitive Parameter	Reference
NDVI	Normalized difference vegetation index	$(R_{NIR}-R_{red})/(R_{NIR}+R_{red})$	Leaf area index, green biomass	[35]
NDRE	Normalized difference red edge index	$(R_{NIR}-R_{RE})/(R_{NIR}+R_{RE})$	Leaf area index, green biomass	[36]
CI_{green}	Green chlorophyll index	$R_{NIR}/R_{green}-1$	Chlorophyll content	[37]
CI_{RE}	Red-edge chlorophyll index	$R_{NIR}/R_{RE}-1$	Chlorophyll content	[38]
SIPI	Structural independent pigment index	$(R_{NIR}-R_{blue})/(R_{NIR}-R_{red})$	Pigment content	[39]
$SIPI_{RE}$	Red-edge structural independent pigment index	$(R_{RE}-R_{blue})/(R_{RE}-R_{red})$	Pigment content	[40]
CARI	Carotenoid index	$R_{RE}/R_{green}-1$	Carotenoid content	[41]
ARI	Anthocyanin reflectance index	$1/R_{green}-1/R_{RE}$	Anthocyanin content	[42]

2.4.2. Binary Logistic Regression

BLR is one of the most frequently used multivariate analysis methods, where the dependent variable is a binary variable representing the presence or absence of an event. The dependent variable in the BLR method is a function of the probability and is expressed as [32]:

$$p = 1/(1 + e^{-y}) \tag{1}$$

where p is the probability of disease occurrence, which ranges from 0 to 1, y is the linear combination, and e is the numerical constant. The y can be expressed by formula as:

$$y = \beta_0 + \beta_1 x_1 + \beta_2 x_2 + \ldots + \beta_n x_n \tag{2}$$

where β_0 is the intercept and β_i and x_i (i = 0, 1, 2,..., n) are the slope coefficients independent variables, respectively. In this study, the BLR method was used to establish the spatial relationships between the plants infested or not infested with Fusarium wilt and the VIs extracted from different resolution images. The modelling dataset were used to fit the logistic regression models through SPSS 20.0 software (SPSS Inc., Chicago, Illinois, USA).

2.4.3. Accuracy Assessment

After the model fitting, two validation datasets (VD1 and VD2) were used to quantitatively evaluate the disease identification accuracy, respectively, with indicators such as the overall accuracy (OA) and the Kappa coefficient [43,44]. The OA is the sum of the correctly classified plots divided by the total number of plots. The Kappa value ranges between −1 and 1 with a larger value indicating better model performance. Model performance can be judged as excellent if kappa ≥ 0.75, good if 0.75 > kappa ≥ 0.4, or poor if kappa < 0.4 [45].

3. Results

3.1. Statistical Characteristics of Samples

We analyzed the differences in the VI values between the healthy and diseased samples obtained from the Guangxi site and Hainan site, and conducted independent t-test analyses for each sample. Table 2 shows the statistical characteristics of the VI values of the healthy and diseased samples. The results showed that there were significant differences in the values of CARI, CI_{green}, CI_{RE}, NDVI, NDRE, and ARI between the healthy and diseased samples ($p < 0.01$), but no significant differences in the SIPI and $SIPI_{RE}$ values ($p > 0.05$). Therefore, CI_{green}, CI_{RE}, NDVI, NDRE, CARI, and ARI were selected for the subsequent analysis.

3.2. Model Fitting with Different Vegetation Indices

In this study, the modeling dataset was used to fit the logistic regression models describing the relationship between the VIs and the plants infested or not infested with Fusarium wilt. Both the validation dataset 1 (VD1) from the Guangxi site and validation dataset 2 (VD2) from the Hainan site were used to verify the classification accuracy of the fitted models. The results showed that the use of the CI_{green}, CI_{RE}, NDVI, and NDRE resulted in relatively good fitting models with an OA greater than 80% (Table 3). Of all VIs, the CI_{RE} obtained the highest validation OA and highest validation Kappa coefficient both for VD1 (OA = 91.7%, Kappa = 0.83) and VD2 (OA = 80.0%, Kappa = 0.59), indicating that CI_{RE} had the best performance for Fusarium wilt identification. For the same type of VI, higher validation OA and Kappa coefficient were obtained for VIs that included the red-edge band (e.g., CI_{RE} vs. CI_{green}, and NDRE vs. NDVI). However, the validation OA and Kappa coefficients based on the CARI and ARI were relatively low.

Table 2. Statistical characteristics of VIs values of the healthy and diseased samples.

Experiment	VI	Sample Category	No. of Samples	Mean of VI Value	Std. Deviation	*p* Value (*t*-test)
Guangxi site	NDVI	Healthy Diseased	57 63	0.54 0.34	0.11 0.14	0.00
	NDRE	Healthy Diseased	57 63	0.20 0.02	0.08 0.09	0.00
	CI_{green}	Healthy Diseased	57 63	1.08 0.43	0.32 0.33	0.00
	CI_{RE}	Healthy Diseased	57 63	0.56 0.09	0.22 0.22	0.00
	SIPI	Healthy Diseased	57 63	0.88 1.68	0.36 5.26	0.24
	$SIPI_{RE}$	Healthy Diseased	57 63	0.58 2.07	0.71 9.77	0.25
	CARI	Healthy Diseased	57 63	0.34 0.30	0.04 0.06	0.00
	ARI	Healthy Diseased	57 63	0.85 0.62	0.15 0.16	0.00
Hainan site	NDVI	Healthy Diseased	16 19	0.44 0.36	0.05 0.06	0.00
	NDRE	Healthy Diseased	16 19	0.35 0.12	0.10 0.09	0.00
	CI_{green}	Healthy Diseased	16 19	0.92 0.49	0.26 0.26	0.00
	CI_{RE}	Healthy Diseased	16 19	0.35 0.12	0.10 0.09	0.00
	SIPI	Healthy Diseased	16 19	1.07 1.18	0.07 0.12	0.06
	$SIPI_{RE}$	Healthy Diseased	16 19	1.11 1.23	0.11 0.16	0.04
	CARI	Healthy Diseased	16 19	0.43 0.33	0.16 0.19	0.01
	ARI	Healthy Diseased	16 19	0.87 0.61	0.30 0.35	0.03

Table 3. The logistic regression models for different vegetation indices.

VI	Logistic Regression Equation	OA* of the Fitting (%)	Validation Dataset 1		Validation Dataset 2	
			OA (%)	Kappa	OA (%)	Kappa
NDVI	$y = -11.851 \times NDVI + 5.373$	86.3	83.3	0.66	62.9	0.22
NDRE	$y = -15.775 \times NDRE + 1.802$	90.5	87.5	0.75	65.7	0.39
CI_{green}	$y = -4.144 \times CI_{green} + 3.118$	89.5	87.5	0.74	74.3	0.47
CI_{RE}	$y = -6.110 \times CI_{RE} + 1.935$	91.6	91.7	0.83	80.0	0.59
CARI	$y = -9.966 \times CARI + 3.172$	62.1	66.7	0.35	60.0	0.21
ARI	$y = -7.247 \times ARI + 5.326$	75.8	83.3	0.66	68.6	0.37

* Overall accuracy.

3.3. Model Fitting for Different Resolution Imagery

The effect of resolution on the identification accuracy of banana Fusarium wilt disease was assessed to provide reference information for large-scale applications of satellite-based data. The UAV imagery was resampled to represent five resolutions (0.5 m, 1 m, 2 m, 5 m, and 10 m) to monitor the occurrence of Fusarium wilt. In order to consider satellite imagery with red-edge bands, both the optimal VI with a red-edge band (CI_{RE}) and the optimal VI without a red-edge band (CI_{green}) were calculated for the images with different spatial resolutions. Table 4 lists the results of logistic regression fitting between locations of infested or noninfested plants and the optimal VIs (CI_{RE} and CI_{green}) at different resolutions. The results showed that the logistic regression models for the CI_{RE} for the 0.5-m, 1-m, and 2-m resolution imagery had an acceptable fitting accuracy with the fitting OA greater than 80% (OA = 90.5%, 83.2% and 81.1% for 0.5-m, 1-m, and 2-m resolution, respectively). Verification results also showed that the CI_{RE} for the 0.5-m, 1-m, and 2-m resolution obtained the acceptable validation OA (over 70%) and Kappa coefficient (over 0.40). For the VD1, the validation OA for the 0.5-m, 1-m, and 2-m resolution were 91.7%, 79.2%, and 75.0%, respectively, and the Kappa coefficients were 0.83, 0.60, and 0.53, respectively. For the VD2, the validation OA for the 0.5-m, 1-m, and 2-m resolution were 85.7%, 74.3%, and 71.4%, respectively, and the Kappa coefficients were 0.71, 0.48, and 0.41, respectively. However, the OA and Kappa coefficient for the 5-m and 10-m resolutions were relatively low. As the resolution decreased, the OA and Kappa coefficient showed a decreasing trend. Moreover, at the same spatial resolution, the CI_{green} resulted in lower accuracy of the identification models of Fusarium wilt than the CI_{RE}. For the CI_{green}, the result was only acceptable at 0.5-m resolution.

Table 4. The logistic regression models for the CI_{RE} and CI_{green} VIs for images with different resolutions.

Resolution	Logistic Regression Equation	OA* of the Fitting (%)	Validation Dataset 1		Validation Dataset 2	
			OA (%)	Kappa	OA (%)	Kappa
CI_{RE}						
0.5 m	$y = -5.826\times CI_{RE}+1.987$	90.5	91.7	0.83	85.7	0.71
1 m	$y = -4.896\times CI_{RE}+1.645$	83.2	79.2	0.60	74.3	0.48
2 m	$y = -4.178\times CI_{RE}+1.475$	81.1	75.0	0.53	71.4	0.41
5 m	$y = -2.854\times CI_{RE}+1.027$	76.8	70.8	0.42	65.7	0.30
10 m	$y = -1.817\times CI_{RE}+0.761$	69.5	62.5	0.25	62.9	0.24
CI_{green}						
0.5 m	$y = -3.946\times CI_{green}+3.166$	87.4	87.5	0.75	74.3	0.48
1 m	$y = -3.266\times CI_{green}+2.633$	83.2	75.0	0.51	65.7	0.32
2 m	$y = -2.936\times CI_{green}+2.421$	78.9	75.0	0.51	62.9	0.26
5 m	$y = -1.862\times CI_{green}+1.552$	70.5	66.7	0.35	48.6	0.01
10 m	$y = -1.158\times CI_{green}+1.044$	61.1	58.3	0.18	45.7	−0.01

* Overall accuracy.

3.4. Mapping Disease Distribution using Imagery with Different Resolutions

In order to further understand the visual effect of resolution, the distributions of banana Fusarium wilt infested or non-infested regions at Guangxi site were mapped at different resolutions (including 0.5-m, 1-m, 2-m, 5-m, and 10-m resolutions). CI_{RE} and CI_{green} were used as input variables to create disease distribution maps based on their identification models of banana Fusarium (Figures 2 and 3). The maps with 0.08-m, 0.5-m, 1-m and 2-m resolutions appeared quite similar with regard to the occurrence of Fusarium wilt disease (Figures 2a–d and 3a–d), whereas the maps with 5-m and 10-m resolutions showed little detail (Figure 2e,f and Figure 3e,f). We also calculated the area and percentage of the Fusarium wilt infected regions based on different resolution maps (see Table 5). For the CI_{RE}-based maps, the areas of Fusarium wilt disease regions were in the range of 5.69–6.59 ha, accounting for 38.2%–44.3% of the total planting area of bananas at different resolutions. Within the 2-m resolution, the percentage of the Fusarium wilt infected regions were in the range of 40.8%–43.6%. For the CI_{green}-based maps, the areas of Fusarium wilt disease regions were in the range of 5.09–6.63 ha, accounting for 34.2%–44.6% of the total planting area of bananas. At 0.08-m and 0.5-m resolutions, the percentage of the Fusarium wilt infected regions were 40.1% and 44.6%, respectively.

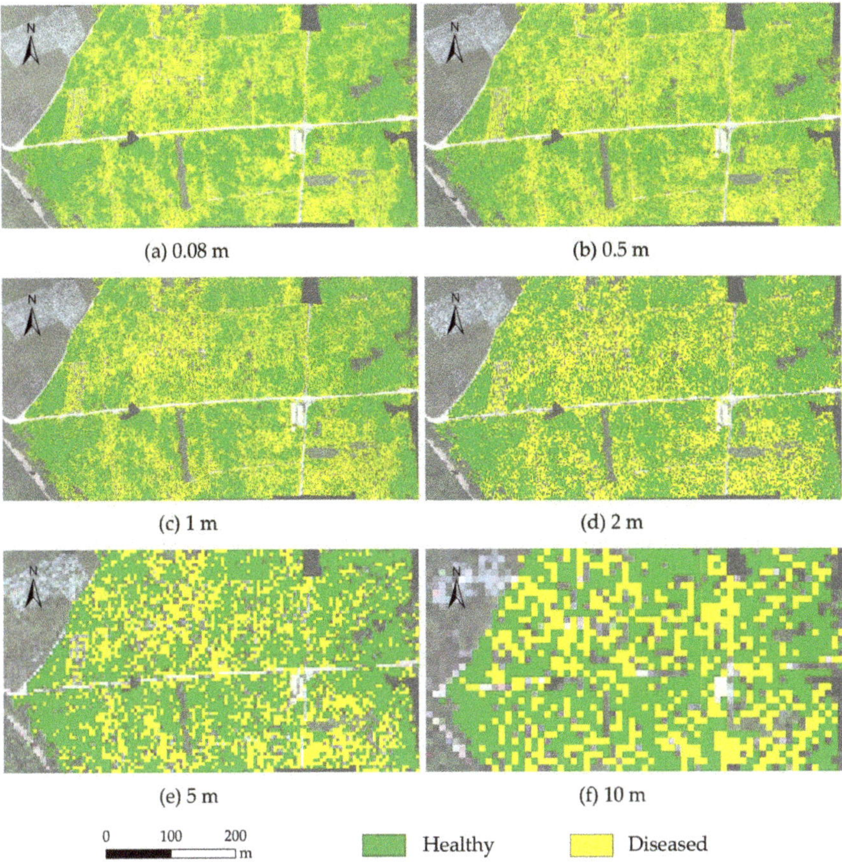

Figure 2. Maps of the distribution of banana Fusarium wilt at Guangxi site based on the CI_{RE} for different resolutions. (**a**) 0.08-m resolution; (**b**) 0.5-m resolution; (**c**) 1-m resolution; (**d**) 2-m resolution; (**e**) 5-m resolution; (**f**) 10-m resolution.

Table 5. Area and percentage of the Fusarium wilt infected regions based on different resolution maps.

Resolution	Healthy Area (ha)	Diseased Area (ha)	Percentage of Diseased Area (%)
CI_{RE}			
0.08 m	8.78	6.04	40.8
0.5 m	8.28	6.59	44.3
1 m	8.60	6.28	42.2
2 m	8.38	6.47	43.6
5 m	9.11	5.70	38.5
10 m	9.19	5.69	38.2
CI_{green}			
0.08 m	8.87	5.95	40.1
0.5 m	8.24	6.63	44.6
1 m	8.44	6.44	43.3
2 m	8.22	6.63	44.6
5 m	9.12	5.69	38.4
10 m	9.79	5.09	34.2

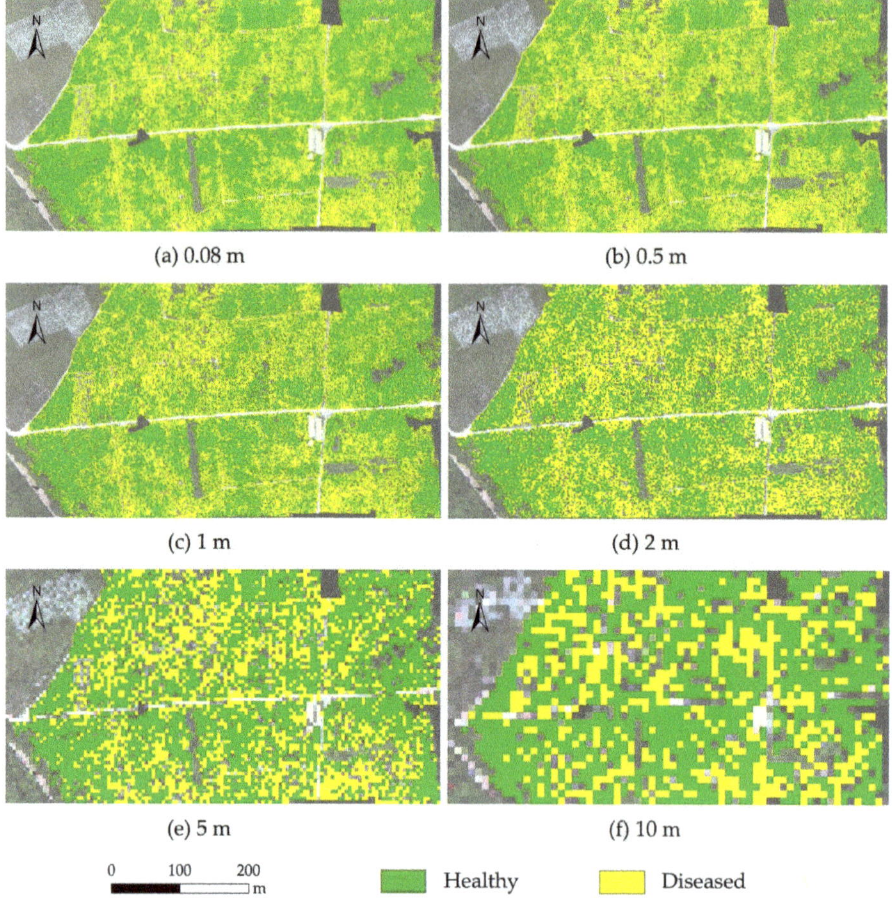

Figure 3. Maps of the distribution of banana Fusarium wilt at Guangxi site based on the CI_{green} for different resolutions. (**a**) 0.08-m resolution; (**b**) 0.5-m resolution; (**c**) 1-m resolution; (**d**) 2-m resolution; (**e**) 5-m resolution; (**f**) 10-m resolution.

4. Discussion

The results of this study indicate that the CI_{RE} was the optimal red-edge VI and the CI_{green} was the optimal non-red-edge VI for developing identification models for banana Fusarium wilt. This is attributed to the fact that as the infection of Fusarium wilt progresses, the chlorophyll content decreases significantly [46], and the CI_{RE} and CI_{green} values are sensitive to small variations in the chlorophyll content [37,38]. Furthermore, for the same type of VI, higher OA and Kappa coefficients were obtained for VIs that included the red-edge band than for those without a red-edge band (i.e., CI_{RE} vs. CI_{green} and NDRE vs. NDVI). Many studies have demonstrated that the red-edge region is highly sensitive to changes in chlorophyll, and bands in this region are well-suited for estimating the chlorophyll content [47,48], which decreased significantly as the infection of Fusarium wilt progressed. Huang et al. [7] also proved that the red-edge band can be used for disease detection. However, the UAV-based multispectral images used in this study only had five spectral bands, which may not fully reflect the differences in spectral characteristic between healthy and diseased plants. It is necessary to conduct further studies on the sensitivity of certain bands to Fusarium wilt using hyperspectral data.

The results also demonstrated the potential of BLR combined with VIs for the accurate identification of banana Fusarium wilt. This approach provides an ideal framework for using spectral features to determine pathological mechanisms. In this study, the dependent variable was the infection or non-infection of banana Fusarium wilt. BLR is a suitable approach when the predicted variable has a binary nature [32]. Moreover, when the predictor variables are continuous, categorical, or a combination of the two, its performance is better than discriminant analysis [49]. Because BLR is very efficient and highly interpretable and does not require large computational resources, it is a widely used technique to describe the relationship between a dependent variable and multiple independent variables [32]. However, logistic regression is not one of the most powerful algorithms, and some more complex algorithms may easily perform better. Moreover, nonlinear problems cannot be solved with logistic regression due to the linear decision surface. With the development of artificial intelligence, pattern recognition and machine learning methods will become more prevalent for monitoring and forecasting of crop diseases using remote sensing [50].

In this study, VD1 from the Guangxi site and VD2 from the Hainan site were used to verify the Fusarium wilt detection models. The verification results at two locations showed that CI_{RE} and CI_{green} had good performances for Fusarium wilt identification with the OA all greater than 70% and Kappa values all greater than 0.4, indicating a good transferability of the Fusarium wilt detection methodology in other areas. However, in Tables 3 and 4, it can be seen that the Kappa values of VD2 were lower than those of VD1. For example, in Table 3, the Kappa value of CI_{RE} was 0.83 in VD1 and 0.59 in VD2, and the Kappa value of CI_{green} was 0.74 in VD1 and 0.47 in VD2. This shows that the application of the Fusarium wilt detection methodology in other areas would cause some loss of precision. This situation may be caused by the following factors. First of all, the different varieties of the two experimental sites may be one of the most important reasons affecting the verification results. The variety for VD1 was "Williams B6" and for VD2 was "Baxijiao." There were differences in biophysical and biochemical characteristics between those two varieties. Differences in these variety characteristics can lead to differences in spectral information. Second, due to the differences in planting systems between these two experimental sites, their growth stages differ greatly. When acquiring images in this study, the two experiments were at different growth stages. Besides, it is also better to consider factors, such as planting density, soil types, and crop growth environmental conditions, that could affect the applicability of the Fusarium wilt detection methodology. Therefore, when applying the method in different regions, it is suggested to optimize the parameters of BLR if there is a large difference between the application and the modeling area of banana planting and growth.

This study demonstrates that UAV-based multispectral imagery is well-suited for the identification of banana Fusarium wilt disease. We also simulated the resolutions of satellite-based imagery (i.e., WorldView series with a resolution of 0.5 m, GF-2 with a resolution of 1 m, GF-1 and GF-6 with a resolution of 2 m, RapidEye with a resolution of 5 m, and Sentinel-2 with a resolution of 10 m) to assess the effects of imagery with different spatial resolution on the identification of disease. The results showed that imagery with a spatial resolution higher than 2 m had good identification accuracy of Fusarium wilt, which might be related to the plant spacing of bananas. As the resolution decreased, the mixed pixel effect worsened, and the monitoring accuracy decreased. However, the resolution was not the only difference among the UAV and satellites. The satellites captured information in wavelengths that was not the same as the ones used in the UAV sensors. Hence, the simulated results of the different resolutions need to be further verified while applied with the actual satellite data. In this study, single date multispectral imagery was used, which represents limitations with regard to determining the spectral response mechanism of the changes in the biophysical and chemical parameters caused by Fusarium wilt. In the future, multitemporal and hyperspectral imagery should be investigated. Moreover, the differences in the spectral response characteristic between Fusarium wilt and other yellowing phenomena caused by other stresses (i.e., nutrition deficiency and drought stress) should also be examined.

5. Conclusions

This study used VIs derived from UAV-based multispectral imagery and BLR to develop an identification method for detecting banana Fusarium wilt. The results showed that Fusarium wilt of banana can be identified with this method. The fitting OA of the CI_{green}, CI_{RE}, NDVI, and NDRE were all higher than 80%. Among the investigated VIs, the CI_{RE} exhibited the best performance both for the verification dataset 1 (OA = 91.7%, Kappa = 0.83) and verification dataset 2 (OA = 80.0%, Kappa = 0.59). For the same type of VI, the VIs including a red-edge band obtained a better performance than those excluding a red-edge band. The simulation of imagery with different spatial resolutions (i.e., 0.5-m, 1-m, 2-m, 5-m, and 10-m resolutions) showed that good identification accuracy of Fusarium wilt was obtained when the resolution was higher than 2 m. As the resolution decreased, the identification accuracy of Fusarium wilt showed a decreasing trend. The results of this study indicate that UAV-based remote sensing imagery with a red-edge band is well-suited for the identification of banana Fusarium wilt disease, providing guidance for disease treatment and crop planting adjustment.

Author Contributions: H.Y. performed the data analysis and wrote the manuscript. W.H. guided the study and discussed the methods and results. S.H., B.C. and Y.D. provided suggestions for the study, reviewed and edited the manuscript. H.Y., A.G., Y.R. and Y.J. conducted the field experiments. All authors have read and agreed to the published version of the manuscript.

Funding: This research was funded by the Hainan Provincial Key R&D Program of China (ZDYF2018073), Hainan Provincial Major Science and Technology Program of China (ZDKJ2019006), National Natural Science Foundation of China (41571354), Agricultural Science and Technology Innovation Program of Sanya, China (2019NK17), National Special Support Program for High-level Personnel Recruitment (Ten-thousand Talents Program) (Wenjiang Huang).

Acknowledgments: We gratefully acknowledge the National Meteorological Information Center of China, Guangxi Jiejiarun Technology Co., Ltd. and Guangxi Jinsui Agriculture Group Co., Ltd. for the experiments.

Conflicts of Interest: The authors declare no conflict of interest.

References

1. Shen, Z.; Xue, C.; Penton, C.R.; Thomashow, L.S.; Zhang, N.; Wang, B.; Ruan, Y.; Li, R.; Shen, Q. Suppression of banana Panama disease induced by soil microbiome reconstruction through an integrated agricultural strategy. *Soil. Biol. Biochem.* **2019**, *128*, 164–174. [CrossRef]
2. Ordonez, N.; Seidl, M.F.; Waalwijk, C.; Drenth, A.; Kilian, A.; Thomma, B.P.H.J.; Ploetz, R.C.; Kema, G.H.J. Worse comes to worst: Bananas and Panama disease-when plant and pathogen clones meet. *PLoS Pathog.* **2015**, *11*, e1005197. [CrossRef] [PubMed]
3. Van den Berg, N.; Berger, D.K.; Hein, I.; Birch, P.R.; Wingfield, M.J.; Viljoen, A. Tolerance in banana to Fusarium wilt is associated with early up-regulation of cell wall-strengthening genes in the roots. *Mol. Plant Pathol.* **2007**, *8*, 333–341. [CrossRef] [PubMed]
4. Lin, B.; Shen, H. Fusarium oxysporum f. sp. cubense. In *Biological Invasions and Its Management in China: Volume 2*; Wan, F., Jiang, M., Zhan, A., Eds.; Springer: Singapore, 2017.
5. Shi, Y.; Huang, W.; Ye, H.; Ruan, C.; Xing, N.; Geng, Y.; Dong, Y.; Peng, D. Partial least square discriminant analysis based on normalized two-stage vegetation indices for mapping damage from rice diseases using PlanetScope datasets. *Sensors* **2018**, *18*, 1901. [CrossRef]
6. Huang, W.; Lamb, D.W.; Niu, Z.; Zhang, Y.; Liu, L.; Wang, J. Identification of yellow rust in wheat using in-situ spectral reflectance measurements and airborne hyperspectral imaging. *Precis. Agric.* **2007**, *8*, 187–197. [CrossRef]
7. Huang, W.; Guan, Q.; Luo, J.; Zhang, J.; Zhao, J.; Liang, D.; Huang, L.; Zhang, D. New optimized spectral indices for identifying and monitoring winter wheat diseases. *IEEE J. Sel. Top. Appl. Earth Observ. Remote Sens.* **2014**, *7*, 2516–2524. [CrossRef]
8. Shi, Y.; Huang, W.; Gonzalez-Moreno, P.; Luke, B.; Dong, Y.; Zheng, Q.; Ma, H.; Liu, L. Wavelet-based rust spectral feature set (WRSFs): A novel spectral feature set based on continuous wavelet transformation for tracking progressive host-pathogen interaction of yellow rust on wheat. *Remote Sens.* **2018**, *10*, 525. [CrossRef]

9. Jin, X.; Jie, L.; Wang, S.; Qi, H.J.; Li, S.W. Classifying wheat hyperspectral pixels of healthy heads and Fusarium head blight disease using a deep neural network in the wild field. *Remote Sens.* **2018**, *10*, 395. [CrossRef]
10. Mahlein, A.K.; Alisaac, E.; Al Masri, A.; Behmann, J.; Dehne, H.W.; Oerke, E.C. Comparison and Combination of Thermal, Fluorescence, and Hyperspectral Imaging for Monitoring Fusarium Head Blight of Wheat on Spikelet Scale. *Sensors* **2019**, *19*, 2281. [CrossRef]
11. Yuan, L.; Pu, R.; Zhang, J.; Wang, J.; Yang, H. Using high spatial resolution satellite imagery for mapping powdery mildew at a regional scale. *Precis. Agric.* **2016**, *17*, 332–348. [CrossRef]
12. Zhao, J.; Xu, C.; Xu, J.; Huang, L.; Zhang, D.; Liang, D. Forecasting the wheat powdery mildew (Blumeria graminis f. Sp tritici) using a remote sensing-based decision-tree classification at a provincial scale. *Australas Plant Path.* **2018**, *47*, 53–61. [CrossRef]
13. Huang, J.; Liao, H.; Zhu, Y.; Sun, J.; Sun, Q.; Liu, X. Hyperspectral detection of rice damaged by rice leaf folder (Cnaphalocrocis medinalis). *Comput. Electron. Agric.* **2012**, *82*, 100–107. [CrossRef]
14. Yang, C.M. Assessment of the severity of bacterial leaf blight in rice using canopy hyperspectral reflectance. *Precis. Agric.* **2010**, *11*, 61–81. [CrossRef]
15. Dhau, I.; Adam, E.; Mutanga, O.; Ayisi, K.; Abdel-Rahman, E.M.; Odindi, J.; Masocha, M. Testing the capability of spectral resolution of the new multispectral sensors on detecting the severity of grey leaf spot disease in maize crop. *Geocarto Int.* **2018**, *33*, 1223–1236. [CrossRef]
16. Zhang, M.; Qin, Z.; Liu, X.; Ustin, S.L. Detection of stress in tomatoes induced by late blight disease in California, USA, using hyperspectral remote sensin. *Int. J. Appl. Earth. Obs. Geoinf.* **2003**, *4*, 295–310. [CrossRef]
17. Jones, C.D.; Jones, J.B.; Lee, W.S. Diagnosis of bacterial spot of tomato using spectral signatures. *Comput. Electron. Agric.* **2010**, *74*, 329–335. [CrossRef]
18. Zheng, Q.; Huang, W.; Cui, X.; Shi, Y.; Liu, L. New spectral index for detecting wheat yellow rust using Sentinel-2 multispectral imagery. *Sensors* **2018**, *18*, 868. [CrossRef]
19. Bravo, C.; Moshou, D.; West, J.; McCartney, A.; Ramon, H. Early disease detection in wheat fields using spectral reflectance. *Biosyst. Eng.* **2003**, *84*, 137–145. [CrossRef]
20. Devadas, R.; Lamb, D.W.; Simpfendorfer, S.; Backhouse, D. Evaluating ten spectral vegetation indices for identifying rust infection in individual wheat leaves. *Precis. Agric.* **2009**, *10*, 459–470. [CrossRef]
21. Oumar, Z.; Mutanga, O. Using WorldView-2 bands and indices to predict bronze bug (Thaumastocoris peregrinus) damage in plantation forests. *Int. J. Remote. Sens.* **2013**, *34*, 2236–2249. [CrossRef]
22. Zhang, J.; Huang, Y.; Yuan, L.; Yang, G.; Chen, L.; Zhao, C. Using satellite multi-spectral imagery for damage mapping of armyworm (Spodoptera frugiperda) in maize at a regional scale. *Pest Manag. Sci.* **2016**, *72*, 335–348. [CrossRef] [PubMed]
23. Dash, J.P.; Watt, M.S.; Pearse, G.D.; Heaphy, M.; Dungey, H.S. Assessing very high resolution UAV imagery for monitoring forest health during a simulated disease outbreak. *ISPRS J. Photogramm. Remote Sens.* **2017**, *131*, 1–14. [CrossRef]
24. Deng, L.; Mao, Z.; Li, X.; Hu, Z.; Yan, Y. UAV-based multispectral remote sensing for precision agriculture: A comparison between different cameras. *ISPRS J. Photogramm. Remote Sens.* **2018**, *146*, 124–136. [CrossRef]
25. Liu, B.; Shi, Y.; Duan, Y.; Wu, W. UAV-based crops classification with joint features from orthoimage and DSM data. *Int. Arch. Photogram. Remote Sens. Spat. Inform. Sci.* **2018**, *XLII-3*, 1023–1028. [CrossRef]
26. Liu, K.; Zhou, Q.B.; Wu, W.B.; Xia, T.; Tang, H.J. Estimating the crop leaf area index using hyperspectral remote sensing. *J. Integr. Agr.* **2016**, *15*, 475–491. [CrossRef]
27. Machovina, B.L.; Feeley, K.J.; Machovina, B.J. UAV remote sensing of spatial variation in banana production. *Crop Pasture Sci.* **2016**, *67*, 1281–1287. [CrossRef]
28. Ye, H.; Cui, B.; Huang, S.; Dong, Y.; Huang, W.; Ren, Y.; Guo, A.; Jin, Y. Identification of banana Fusarium wilt disease based on UAV multi-spectral imagery. In Proceedings of the International Conference on Intelligent Agriculture, Beijing, China, 18–21 October 2019.
29. Li, N.; Xie, G.; Zhou, D.; Zhang, C.; Jiao, C. Remote sensing classification of marsh wetland with different resolution images. *J. Resour. Ecol.* **2016**, *7*, 107–114.
30. Fisher, J.R.; Acosta, E.A.; Dennedy-Frank, P.J.; Kroeger, T.; Boucher, T.M. Impact of satellite imagery spatial resolution on land use classification accuracy and modeled water quality. *Remote Sens. Ecol. Conserv.* **2018**, *4*, 137–149. [CrossRef]

31. Mahlein, A.K.; Oerke, E.C.; Steiner, U.; Dehne, H.W. Recent advances in sensing plant diseases for precision crop protection. *Eur. J. Plant Pathol.* **2012**, *133*, 197–209. [CrossRef]
32. Lee, S.; Pradhan, B. Landslide hazard mapping at Selangor, Malaysia using frequency ratio and logistic regression models. *Landslides* **2007**, *4*, 33–41. [CrossRef]
33. Ozdemir, A. Using a binary logistic regression method and GIS for evaluating and mapping the groundwater spring potential in the Sultan Mountains (Aksehir, Turkey). *J. Hydrol.* **2011**, *405*, 123–136. [CrossRef]
34. IUSS Working Group WRB. *World Reference Base for Soil Resources 2006*; FAO: Rome, Italy, 2006.
35. Rouse, J.W.; Haas, R.H.; Schell, J.A.; Deering, D.W. Monitoring vegetation systems in the great plains with ERTS. In Proceedings of the Third ERTS-1 Symposium NASA SP-351, Greenbelt, MD, USA, 10–14 December 1973.
36. Gitelson, A.; Merzlyak, M.N. Spectral reflectance changes associated with autumn senescence of aesculus–hippocastanum L and acer-platanoides L leaves—spectral features and relation to chlorophyll estimation. *J. Plant Physiol.* **1994**, *143*, 286–292. [CrossRef]
37. Gitelson, A.A.; Gritz, Y.; Merzlyak, M.N. Relationships between leaf chlorophyll content and spectral reflectance and algorithms for non-destructive chlorophyll assessment in higher plant leaves. *J. Plant Physiol.* **2003**, *160*, 271–282. [CrossRef] [PubMed]
38. Gitelson, A.A.; Vina, A.; Ciganda, V.; Rundquist, D.C.; Arkebauer, T.J. Remote estimation of canopy chlorophyll content in crops. *Geophys. Res. Lett.* **2005**, *32*, L08403. [CrossRef]
39. Peñuelas, J.; Inoue, Y. Reflectance indices indicative of changes in water and pigment contents of peanut and wheat leaves. *Photosynthetica* **1999**, *36*, 355–360. [CrossRef]
40. Ramoelo, A.; Skidmore, A.K.; Cho, M.A.; Schlerf, M.; Mathieu, R.; Heitkonig, I.M.A. Regional estimation of savanna grass nitrogen using the red-edge band of the spaceborne RapidEye sensor. *Int. J. Appl. Earth. Obs. Geoinf.* **2012**, *19*, 151–162. [CrossRef]
41. Zhou, X.; Huang, W.; Zhang, J.; Kong, W.; Casa, R.; Huang, Y. A novel combined spectral index for estimating the ratio of carotenoid to chlorophyll content to monitor crop physiological and phenological status. *Int. J. Appl. Earth. Obs. Geoinf.* **2019**, *76*, 128–142. [CrossRef]
42. Gitelson, A.A.; Merzlyak, M.N.; Chivkunova, O.B. Optical properties and nondestructive estimation of anthocyanin content in plant leaves. *Photochem. Photobiol.* **2001**, *74*, 38–45. [CrossRef]
43. Congalton, R.G. A review of assessing the accuracy of classifications of remotely sensed data. *Remote Sens. Environ.* **1991**, *37*, 35–46. [CrossRef]
44. Foody, G.M. Classification accuracy comparison: Hypothesis tests and the use of confidence intervals in evaluations of difference, equivalence and non-inferiority. *Remote Sens. Environ.* **2009**, *113*, 1658–1663. [CrossRef]
45. Tuanmu, M.N.; Viña, A.; Bearer, S.; Xu, W.; Ouyang, Z.; Zhang, H.; Liu, J. Mapping understory vegetation using phenological characteristics derived from remotely sensed data. *Remote Sens. Environ.* **2010**, *114*, 1833–1844. [CrossRef]
46. Dong, X.; Wang, M.; Ling, N.; Shen, Q.; Guo, S. Potential role of photosynthesis-related factors in banana metabolism and defense against Fusarium oxysporum f. sp cubense. *Environ. Exp. Bot.* **2016**, *129*, 4–12. [CrossRef]
47. Clevers, J.G.P.W.; de Jong, S.M.; Epema, G.F.; van der Meer, F.; Bakker, W.H.; Skidmore, A.K.; Addink, E.A. MERIS and the red-edge position. *Int. J. Appl. Earth. Obs. Geoinf.* **2001**, *3*, 313–320. [CrossRef]
48. Dash, J.; Curran, P.J. The MERIS terrestrial chlorophyll index. *Int. J. Remote Sens.* **2004**, *25*, 5403–5413. [CrossRef]
49. Mathew, J.; Jha, V.K.; Rawat, G.S. Application of binary logistic regression analysis and its validation for landslide susceptibility mapping in part of Garhwal Himalaya, India. *Int. J. Remote Sens.* **2007**, *28*, 2257–2275. [CrossRef]
50. Lu, J.; Sun, L.; Huang, W. Research progress in monitoring and forecasting of crop pests and diseases by remote sensing. *Remote Sens. Technol. Appl.* **2019**, *34*, 21–32.

Article

Influence of Model Grid Size on the Estimation of Surface Fluxes Using the Two Source Energy Balance Model and sUAS Imagery in Vineyards

Ayman Nassar [1,*], Alfonso Torres-Rua [1], William Kustas [2], Hector Nieto [3], Mac McKee [1], Lawrence Hipps [4], David Stevens [1], Joseph Alfieri [2], John Prueger [5], Maria Mar Alsina [6], Lynn McKee [2], Calvin Coopmans [7], Luis Sanchez [6] and Nick Dokoozlian [6]

1 Department of Civil and Environmental Engineering, Utah State University, Logan, UT 84322, USA; Alfonso.Torres@usu.edu (A.T.-R.); mac.mckee@usu.edu (M.M.); david.stevens@usu.edu (D.S.)
2 U. S. Department of Agriculture, Agricultural Research Service, Hydrology and Remote Sensing Laboratory, Beltsville, MD 20705, USA; bill.kustas@usda.gov (W.K.); Joe.Alfieri@ars.usda.gov (J.A.); lynn.mckee@ars.usda.gov (L.M.)
3 Complutum Tecnologías de la Información Geográfica (COMPLUTIG), 28801 Madrid, Spain; hector.nieto@complutig.com
4 Plants, Soils and Climate Department, Logan, UT 84322, USA; Lawrence.Hipps@usu.edu
5 U. S. Department of Agriculture, Agricultural Research Service, National Laboratory for Agriculture and the Environment, Ames, IA 50011, USA; john.prueger@usda.gov
6 E & J Gallo Winery Viticulture Research, Modesto, CA 95354, USA; MariadelMar.Alsina@ejgallo.com (M.M.A.); Luis.Sanchez@ejgallo.com (L.S.); Nick.Dokoozlian@ejgallo.com (N.D.)
7 Department of Electrical Engineering, Utah State University, Logan, UT 84322, USA; cal.coopmans@usu.edu
* Correspondence: aymnassar@aggiemail.usu.edu

Received: 25 October 2019; Accepted: 10 January 2020; Published: 21 January 2020

Abstract: Evapotranspiration (*ET*) is a key variable for hydrology and irrigation water management, with significant importance in drought-stricken regions of the western US. This is particularly true for California, which grows much of the high-value perennial crops in the US. The advent of small Unmanned Aerial System (*sUAS*) with sensor technology similar to satellite platforms allows for the estimation of high-resolution *ET* at plant spacing scale for individual fields. However, while multiple efforts have been made to estimate *ET* from *sUAS* products, the sensitivity of *ET* models to different model grid size/resolution in complex canopies, such as vineyards, is still unknown. The variability of row spacing, canopy structure, and distance between fields makes this information necessary because additional complexity processing individual fields. Therefore, processing the entire image at a fixed resolution that is potentially larger than the plant-row separation is more efficient. From a computational perspective, there would be an advantage to running models at much coarser resolutions than the very fine native pixel size from *sUAS* imagery for operational applications. In this study, the Two-Source Energy Balance with a dual temperature (*TSEB2T*) model, which uses remotely sensed soil/substrate and canopy temperature from *sUAS* imagery, was used to estimate *ET* and identify the impact of spatial domain scale under different vine phenological conditions. The analysis relies upon high-resolution imagery collected during multiple years and times by the Utah State University *AggieAir*^{TM} *sUAS* program over a commercial vineyard located near Lodi, California. This project is part of the USDA-Agricultural Research Service Grape Remote Sensing Atmospheric Profile and Evapotranspiration eXperiment (*GRAPEX*). Original spectral and thermal imagery data from *sUAS* were at 10 cm and 60 cm per pixel, respectively, and multiple spatial domain scales (3.6, 7.2, 14.4, and 30 m) were evaluated and compared against eddy covariance (*EC*) measurements. Results indicated that the *TSEB2T* model is only slightly affected in the estimation of the net radiation (R_n) and the soil heat flux (*G*) at different spatial resolutions, while the sensible and latent heat fluxes (*H* and *LE*, respectively) are significantly affected by coarse grid sizes. The results indicated overestimation of

H and underestimation of *LE* values, particularly at Landsat scale (30 m). This refers to the non-linear relationship between the land surface temperature (*LST*) and the normalized difference vegetation index (*NDVI*) at coarse model resolution. Another predominant reason for *LE* reduction in *TSEB2T* was the decrease in the aerodynamic resistance (R_a), which is a function of the friction velocity (u_*) that varies with mean canopy height and roughness length. While a small increase in grid size can be implemented, this increase should be limited to less than twice the smallest row spacing present in the *sUAS* imagery. The results also indicated that the mean *LE* at field scale is reduced by 10% to 20% at coarser resolutions, while the with-in field variability in *LE* values decreased significantly at the larger grid sizes and ranged between approximately 15% and 45%. This implies that, while the field-scale values of *LE* are fairly reliable at larger grid sizes, the with-in field variability limits its use for precision agriculture applications.

Keywords: evapotranspiration (*ET*); *GRAPEX*; *sUAS*; remote sensing; Two Source Energy Balance model (*TSEB*); contextual spatial domain/resolution; data aggregation; eddy covariance (*EC*)

1. Introduction

Evapotranspiration (*ET*) is a key factor in the hydrologic cycle and in irrigation demand. Conventional methods for estimating *ET*, such as lysimeters and flux towers, are limited to sampling small areas on the order of 10^1 to 10^3 m^2. For that, a more efficient method is needed as *ET* varies spatially under different micro-meteorological and vegetative conditions. Accordingly, spatially distributed data are important for mapping *ET* variations over large areas, particularly in agricultural regions containing many of crop types and growth stages. In recent decades, remote sensing products from various platforms and at various spatial resolutions have been applied in modeling different environmental processes (e.g., surface energy fluxes, water and carbon balance, net primary productivity) [1]. Improved sensor systems and methods in remote sensing, and particularly the advent of small unmanned aerial systems (*sUAS*), have made these technologies a valuable source of spatial information for *ET* estimation at the canopy level. *sUAS* can offer spatial coverage with sub-meter-resolution imagery for mapping canopy and soil temperature, which are the key surface states for estimating *ET* [2]. While satellites are characterized by either coarse resolution and high temporal frequency or by high spatial resolution and low repeatability [3], *sUAS* technology, in addition to offering high-resolution data [4–6], can be described as "flexible on timing" [7]. This means that remotely sensed information can be obtained when needed or on demand using *sUAS*. For these reasons, various methods are under development to employ *sUAS* data for *ET* estimation [2].

Remote sensing is a valuable source for accessing land surface spatial information [8]. Nonetheless, spatial scaling is recognized as a challenging issue, particularly in surface-atmosphere exchange [8,9], environmental modeling, and agricultural management [10] applications and research. Previous studies by Brunsell and Gillies [11] and Giorgi [12] indicated that spatial scaling becomes more complex in cases of heterogeneous land surfaces, and homogeneity is less likely to be met in reality [13]. Various models have been developed to describe aerodynamic or energy balance fluxes, but these models assume homogeneity in terms of agricultural type, surface roughness, surface temperature, and meteorological condition [13,14]. Heat fluxes, including latent heat flux (*LE*) and sensible heat flux (*H*), are highly influenced by land surface heterogeneity [15]. Therefore, the variability in land cover within a pixel or model grid size can result in significant error in the mean pixel or grid heat flux estimation [16]. Vegetated areas with partial canopy cover will have underlying soil/substrate affecting the remotely sensed data, and hence, require models that explicitly consider the different effects of these two sources on energy exchange and sensor integration [2]. Typically, remotely sensed data at different resolutions are employed as an approximation to describe the spatial variability of the interaction between surface and atmosphere [11]. Current and future developments in remote sensing, with information spanning from sub-meters to kilometers, are making upscaling (data aggregation) a crucial issue in scientific and methodological advances. This is particularly true for understanding the physics behind climate, weather, and the surface energy balance [13,17].

In general, spatial aggregation can be performed under two different procedures: forcing inputs to a coarser resolution or aggregating the derived fluxes from initial high-resolution data (contextual spatial domain). Long et al. [18] pointed out that forcing spatial data aggregation from Landsat bands to MODIS (Moderate Resolution Imaging Spectroradiometer) resolution results in different statistical and spatial properties in *ET* estimates than at the original Landsat resolution. Study cases of *LE* resulted in inaccuracies [19,20] due to a reduction in surface variability at MODIS scale [11]. Moreover, the structure of vegetation and aerodynamic roughness influence the aggregation of turbulent fluxes and produce bias when MODIS data is used [15]. On the other hand, Bian and Butler [21] showed that low-resolution data could retain the statistical characteristics of the original data using specific aggregation techniques such as average and median. In addition, the spatial aggregation of *ET* inputs removes the effects of heterogeneity on the land surface. Still, scaling up energy fluxes from Landsat to MODIS scale is necessary in large-scale environmental models [22]. However, Landsat resolution is needed for validating modeled outputs using flux towers [23].

Several methods exist for spatial aggregation of *ET* data, but they are in the exploratory stage [24]. Ershadi et al. [14] demonstrated that *ET* results reduced by 15% when aggregating Landsat TM (Thematic Mapper) imagery by 50% using the Surface Energy Balance System (*SEBS*) model. The *ET* reduction was caused by the decrease in roughness parameterization [14]. This outcome was also supported by Brunsell and Gillies [11], who indicated that the land surface heterogeneity is highly influenced by the input forcing aggregation of Landsat TM data affecting the surface heat fluxes. In contrast, French et al. [25] found no significant difference in daily *ET* estimates when they used *METRIC* (Mapping EvapoTranspiration at high Resolution with Internalized Calibration) model and upscaled data acquired by aircraft to Landsat resolution. However, another study by Kustas and Norman [16] that used a detailed soil-vegetation atmosphere simulation model along with the thermal-based two-source energy balance model found that varying the degree of heterogeneity within a pixel, either in terms of surface roughness, moisture status, or a combination thereof, can have a significant impact on the pixel aggregated flux.

A key question related to data aggregation was raised by Su et al. [26]: "How does the level of aggregation affect surface energy fluxes as fluxes are aggregated from the resolution at which they are observed to the coarse grid cell size of the atmospheric model?". The study conducted by Guzinskia and Nieto [27] aimed to estimate *ET* using a Two Source Energy Balance (*TSEB*) model. They reported that sharpening Sentinel 3 thermal imagery at 1-km pixel resolution to higher resolution (20 m) visible/near-infrared is indicative of the main issue of the lack of fine resolution thermal-IR (InfraRed) data for input to remote sensing-based *ET* models, particularly when applied to agricultural areas. In addition, Niu et al. [28] indicated that the *TSEB* model *ET* output using *sUAS* imagery gives more reliable estimates compared to coarse-resolution data because the model can separate between canopy and soil components. Moreover, most previous studies exploring the effects of sensor resolution on modeled *ET* have used semi-empirical models (e.g., Surface Energy Balance Algorithm for Land (*SEBAL*) model) [14], while physically-based *ET* models are required to quantify changes in the water and energy exchange due to changes in fractional vegetation cover, roughness, canopy structure, phenology, etc. that are occurring at plant scale [29]. In addition, it is common knowledge that vineyards and orchard fields do not have the same row spacing. The spacing varies from 6 ft to 12 ft for vineyards [30] and from 8 ft to 18 ft for orchards [31].

In the same context as the investigations discussed above on spatial resolution and surface heterogeneity, this study investigates the impact of grid-size resolution on *LE* outputs from *TSEB* model using the component soil/substrate and canopy temperature version (*TSEB2T*) model applied to a complex agricultural canopy, namely a vineyard in California's Central Valley. The study directly quantifies the effect of sensor resolution on key *TSEB* model inputs (i.e., land surface temperature

(*LST*), leaf area index (*LAI*), canopy height (h_c), canopy width-to-height ratio (w_c/h_c), and fractional cover (f_c)) for estimating surface energy balance/*ET*. High-resolution optical and thermal data were acquired by an *sUAS* platform for vine and cover crop phenological stages at several different times during the day. In this research effort, the topics investigated include determining (a) whether the separation between canopy and soil/substrate temperature (T_c and T_s, respectively) using *TSEB2T* is valid for coarse spatial domains (e.g., towards Landsat scale); (b) the effect of spatial resolution of *TSEB2T* inputs on the magnitude and spatial variation of *LE*; (c) if the different spatial domain scales/pixel resolutions under study (3.6, 7.2, 14.4 and 30 m) have an impact on the magnitude of the *LE* and quantify the discrepancies as a function of resolution.

1.1. TSEB2T Model

TSEB2T is a physically based approach developed by Norman et al. [32] that explicitly accommodates the difference between aerodynamic and radiometric surface temperature that affect the radiative and convective exchange of energy between soil and canopy systems and the lower atmosphere. The main concept underpinning the *TSEB2T* approach is modeling of the partitioning of radiative and turbulent energy fluxes between canopy and soil systems. In this case, *H* is partitioned between soil and canopy, which is dependent mainly on T_c and T_s differences with the overlying atmosphere and their respective aerodynamic coupling.

As shown in the Figure 1, the *TSEB2T* model separates the surface energy balance between soil and vegetation as follows:

$$R_n = LE + H + G, \tag{1}$$

$$R_{nc} = H_c + LE_c, \tag{2}$$

$$R_{ns} = H_s + LE_s + G, \tag{3}$$

where R_n is the net radiation, *H* is the sensible heat flux, *LE* is the latent heat flux, and *G* is the soil heat flux. All units of fluxes are in W/m^2. Subscripts of *c* and *s* represent the canopy and soil components, respectively. Because T_s and T_c can be derived from the *LST* with a high enough resolution of optical data, energy fluxes (R_n, *H*) can be calculated directly from the component temperatures (T_c and T_s) and estimated aerodynamic resistances of canopy and soil components, while *G* is parametrized as a portion of soil net radiation (R_{ns}). LE_c and LE_s are solved as residuals when (T_c and T_s) observations are available.

$$G = c_G R_{ns} \tag{4}$$

where c_G is an empirical coefficient changing over the daytime [2].

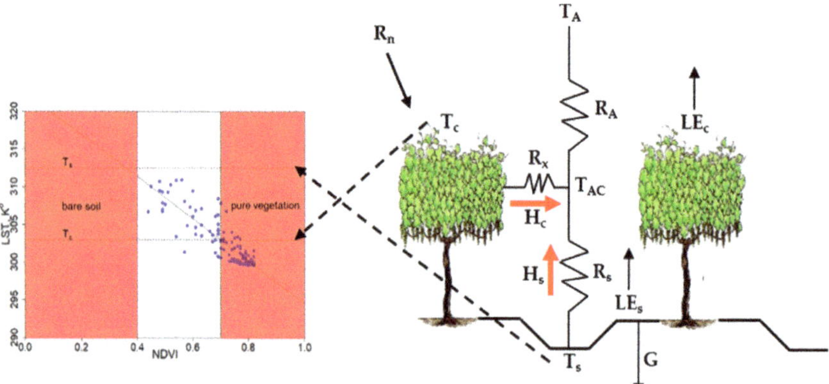

Figure 1. Schematic representation of *TSEB2T* model.

To estimate the sensible heat flux for vegetation and canopy, Norman et al. [32] proposed a series of soil-vegetation resistance network as illustrated in Figure 1:

$$H = H_c + H_s = \rho_{air} C_p \frac{T_{AC} - T_A}{R_A} = \rho_{air} C_p \left[\frac{T_C - T_{AC}}{R_x} + \frac{T_s - T_{AC}}{R_s} \right] \tag{5}$$

$$R_A = \frac{ln\left(\frac{z_T - d_0}{z_{0H}}\right) - \Psi_h\left(\frac{z_T - d_0}{L}\right) + \Psi_h\left(\frac{z_{0H}}{L}\right)}{\kappa' u_*} \tag{6}$$

where ρ_{air} is the air density (kg/m^3); C_p is the heat capacity of the air at constant pressure (J/(kg·K)); T_c and T_s are canopy and soil temperature (K), respectively; T_{AC} is the temperature of canopy-air space (K); and T_A is the temperature of air (K). R_A is the aerodynamic resistance to heat transport from the soil/canopy system (s/m), R_x is the boundary layer resistance of the canopy leaves (s/m), R_s is the aerodynamic resistance to heat transport in the boundary layer close to the soil surface (s/m), z_T is the measurement height for T_A, z_{0H} is the roughness length for heat transport, d_0 is the zero-plane displacement height (m), L is the Monin-Obukhov length (m), $\kappa' = 0.4$ is the von Karman's constant, u_* is the friction velocity (m/s), and Ψ_h is the adiabatic correction factor for the momentum.

Key factors, including T_s and T_c, *LAI*, f$_c$, w_c/h_c, and h_c, are required as inputs for the *TSEB* model to parameterize the radiative and convective flux exchanges between soil/substrate and canopy. Other parameters related to micro-meteorological data are also needed to run the model. In the study conducted by Chirouze et al. [33] comparing different remote sensing *ET* models, results indicated that *TSEB* is a better model for *ET* estimation compared to others, being less sensitive to roughness parameters. This lack of sensitivity to roughness parameters was also recently verified for vineyards by Alfieri et al. [34]. The *TSEB* model has been extensively tested for years over agroecosystems [35–37], natural ecosystems [38,39], and wetlands [40,41].

TSEB2T was originally developed and evaluated by Kustas and Norman [42] using multiple thermal-IR radiometer viewing angles and was further refined and tested by Nieto et al. [2] applied to high resolution imagery from *sUAS* or other airborne sources. They found that *TSEB2T* gave better agreement with tower fluxes compared to other versions of *TSEB*, including *TSEB-PT* (Priestly-Taylor), *TSEB-DTD* (Dual-time-difference), and *TSEB-2T-DMS* (Data-mining sharpening of temperature). *TSEB-PT* is one version of the *TSEB* model that assumes a composite radiometric temperature (T_{rad}) containing temperature contribution from the canopy and soil/substrate, which is typically provided by the radiometer. The decomposition of radiometric temperature (T_{rad}) between plant canopy and soil/substrate is based on f_c. *TSEB-DTD* is a further development of the *TSEB-PT* model described by Norman et al. [43]. The *TSEB-DTD* model is similar to the *TSEB-PT* model in that it divides the composite T_{rad} into T_c and T_s. However, *TSEB-DTD* uses two observations of T_{rad}: the first observation obtained 1.5 h after the sunrise ($T_{rad,0}$) and the second one during the daytime ($T_{rad,1}$). This version is less sensitive to errors in absolute radiometric surface temperature or the use of non-local air temperature observations. *TSEB-2T-DMS* partitions T_s and T_c using a data-mining fusion algorithm [44] to sharpen the original *LST* to be similar to the optical data, which would allow a better discrimination between T_s and T_c.

The Nieto et al. [2] *TSEB2T* approach is a contextual *TSEB* that estimates T_s and T_c from composite *LST* imagery using the relationship between vegetation index (*VI*) and *LST* for extracting T_s and T_c within a spatial domain. T_s and T_c are calculated by averaging the temperature of pixels that are considered pure soil/substrate and pure canopy in a contextual spatial domain, namely, a two-dimensional plot of *LST* versus *VI*, such as Normalized Difference Vegetation Index (*NDVI*) (see Figure 1). That is, each pixel of the spatial domain is assigned based on T_c and T_s corresponding to the average temperature of the 0.6-m grids that are considered pure vegetation and bare soil, respectively. Both soil/substrate or canopy features are determined using *NDVI* threshold values (or any other vegetation index). The selection criterion for detecting the *NDVI* threshold of pure soil for bare soil interrows or, for most of the growing season, a soil senescent and cover crop stubble mixture (substrate) ($NDVI_s$) can be further supported by other sources such as *NDVI* value from a *NDVI-LAI* curve when *LAI* in the interrows is nearly zero. The pure vine canopy *NDVI* threshold ($NDVI_v$) can be calculated as the mean value of pixels identified as pure vegetation in a binary (soil-vegetation) classification of a multispectral image. In cases of very dense vegetation where pure soil pixels do not exist or sparse vegetation lacking pure vegetation pixels inside the spatial domain, a linear fit between *LST* and *NDVI* can be developed where T_s and T_c can be estimated by previously defining the *NDVI* thresholds of canopy and bare soil (Figure 1).

1.2. TSEB2T Main Inputs

1.2.1. Leaf Area Index (*LAI*)

LAI is one of the key inputs in *TSEB* influencing the computation of *ET* as leaves distribution is the driving factor in energy and mass exchange in this model. *LAI* is also difficult to acquire using ground-based leaf-scale measurements, due to the time-intensive effort required [45], complications using indirect methods in complex canopies, and lack of any spatial extent for mapping, even at the field scale [46]. Therefore, considerable efforts have been devoted to developing remote sensing approaches to estimate *LAI* [47].

Estimating spatial distribution of *LAI* is challenging in vineyards, with their rows of vines and interrows with little to no vegetation. A previous study conducted by Johnson [48] evaluated the *LAI-NDVI* relationship in vineyards using IKONOS satellite imagery with 1-m pixel resolution and comparing *NDVI* to ground-based *LAI* measurements. They concluded that *LAI* can be computed from *NDVI* using simple linear regression for the vineyard they studied planted with red grape in six blocks of different planting density, trellis, age, and cultivar. In addition, Johnson et al. [48] and Dobrowski et al. [49] showed that remotely sensed indices of soil and vegetation can be used to estimate *LAI*. However, a study by Fang [50] indicated that limitations exit when using vegetation indices (*VIs*) to describe the spatially distributed *LAI* due to sensitivity of the *LAI-VIs* relationship to vegetation type and substrate/soil type, and hence, will not be stable or applicable over large areas. Indeed, operational satellite retrievals of *LAI*, particularly for vineyards [51], have a level of uncertainty that could affect modeling fluxes using *TSEB*. Furthermore, canopy phenological properties (i.e., chlorophyll content and average leaf angle), along with other factors such as atmospheric scattering, soil reflectance, and the effects of mixed pixel due to a composite of soil and vegetation that changes with time and from one place to another, affect the accuracy of *LAI* estimation [47]. To improve the *LAI-VIs* relationships, numerous studies have been conducted to estimate *LAI* using statistical approaches. Artificial Neural Network (*ANN*) was very promising and is simple to use [50]; however, this method does not allow for standardization of the *LAI* estimation [52]. As described by Gonsamo and Pellikka [53], there is currently no standard or arbitrary characteristic parameters, specific vegetation types, or data sources can be employed for *LAI* estimation. Thus, researchers must develop custom models by considering the sensitivity of parameters to *LAI* within an expected range [53].

1.2.2. Canopy Height (h_c)

The h_c value is representative (mean) over the area of interest, but it can also be incorporated from spatial sources. An estimate of h_c can be produced using high-resolution images from *sUAS* and other airborne sources processed with structure-from-motion (*SfM*) methods in Agisoft or Pix4D, among others, along with digital elevation models (*DEM*) and point clouds (*LiDAR*). The value of h_c is required for the *TSEB2T* model to estimate surface aerodynamic roughness and radiation transmission in row crops and to calculate the foliage density, which are all required for the canopy wind attenuation model (Figure 2).

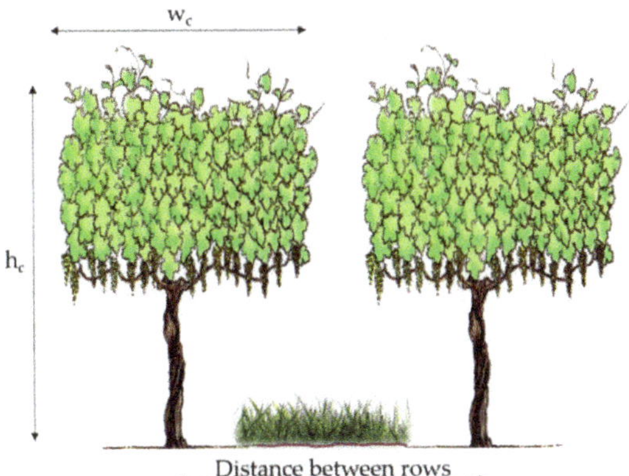

Figure 2. Schematic diagram for canopy w_c/h_c ratio.

1.2.3. Fractional Cover (f_c) and Canopy Width (w_c):

Fractional cover (f_c) is the proportional area of vine for each spatial domain under analysis, where values vary from 0 through 1. f_c is used to estimate w_c and the clumping index, which is a factor to adjust the remotely sensed *LAI* value, which is assumed to be uniformly distributed (homogeneous) over the landscape instead of being clumped [54]. These are used to estimate the actual canopy gap fraction, which is greater than the homogenous case. It is required as input for the radiation transmission and wind extinction algorithms through the canopy layer. The magnitude of w_c is a length scale representing the area occupied by vine leaves along the vine row, which varies spatially and temporally based on phenology and management (i.e., vine manipulation via the trellis system and pruning) (Figure 2).

1.2.4. w_c/h_c Ratio

In *TSEB* and *TSEB2T* models, the w_c/h_c ratio is required as input to the radiation transmission and wind extinction algorithms through the canopy layer developed for vineyards [2,55]. The w_c/h_c ratio value is obtained by simply calculating canopy width over canopy height (Figure 2).

2. Materials and Methods

The methodology to assess the impact of changes in the contextual spatial domain for the *TSEB2T* model is graphically presented in Figure 3. The analysis was performed for wine grape growing seasons (May–August) using different spatial domain scales.

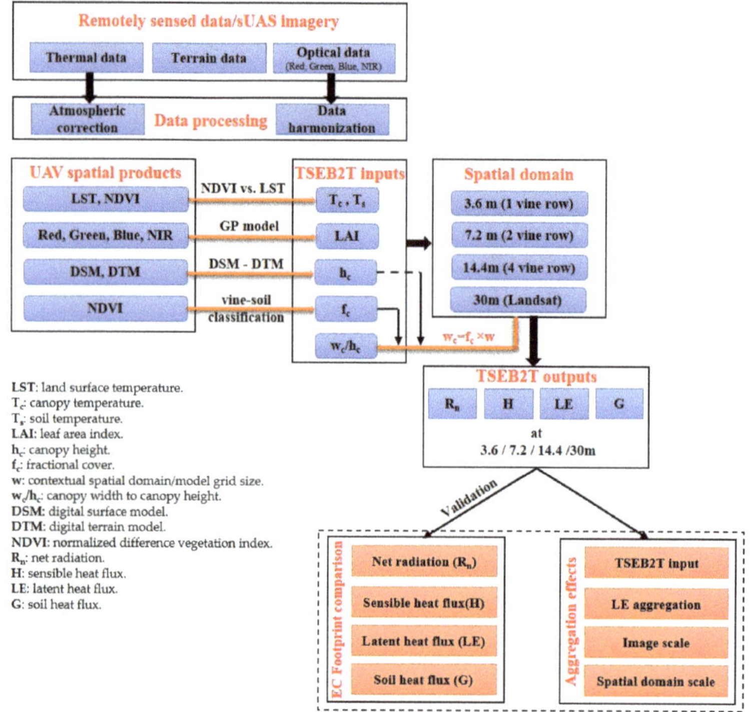

Figure 3. Study methodology for assessing the impact of the *TSEB2T* contextual spatial domain.

2.1. Study Area and Data Sources

The study site is located near Lodi, California (38.29°N, 121.12°W) with an area of approximately 150 ha. The two vineyard blocks (north and south) are part of the Sierra Loma vineyard ranch (Figure 4). The north block was planted in 2009, while the south block was implemented in 2011, leading to different levels of vine maturity, and hence, biomass and grape production. Both vineyards are managed cooperatively by Pacific Agri-Lands Management. The plantation structure in both fields is the same, with vine rows having east–west orientation with a row width of 3.35 m (11 feet). A cover crop grows in the interrows, occupying ~2 m, with bare soil strips along the vine rows spanning ~0.7 m. The purpose of the cover crop is to deplete plant available water in the interrows from the fall and winter precipitation in order to control vine growth in the spring by irrigation. Typically, the vine height varies between 2 m and 2.5 m above ground level (*agl*) and vine biomass is concentrated mainly in the upper half of the vine canopy height. The actual vine canopy width varies spatially and temporally due to vine management practices. This study site is a part of the Grape Remote Sensing Atmospheric Profile and Evapotranspiration eXperiment (*GRAPEX*) project run by the USDA Agricultural Research Service in collaboration with E&J Gallo Winery, Utah State University, University of California in Davis, and others [56].

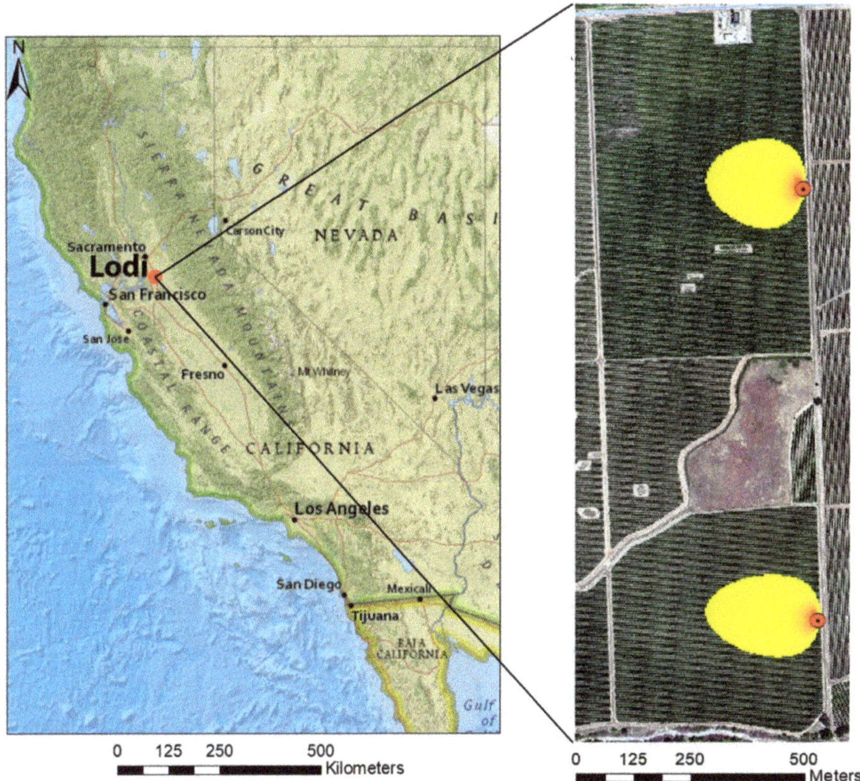

Figure 4. Layout of study area in Lodi, California, locations of *EC* towers and example of 90% of *EC* footprint at afternoon for 02 June 2015.

Flights campaigns were conducted by the *AggieAir sUAS* program at Utah State University (https://uwrl.usu.edu/aggieair/). Optical and thermal high-resolution imagery of the study site were collected from different flights in 2014, 2015, and 2016. Vegetative and soil conditions changed between the field campaigns. The 2016 flight imagery represents the early part of the growing season, around the time phenologically of fruit set, while other flights in 2014 and 2015 represent full vine canopy development and grape vine phenology in the pre- and post-veraison stages. Table 1 lists information concerning the different flights. The pixel resolution of the *sUAS* imagery collected is 10 cm and 60 cm for the optical and thermal bands, respectively. The spectral range of the optical data is similar to Landsat and includes visible bands (red, green, and blue) as well as near-infrared. However, the thermal band is different than Landsat, having a bandwidth spanning from 7 to 14 μm [57]. Thermal data, acquired using a lightweight micro-bolometer camera, were radiometrically calibrated [58].

Table 1. Dates and times of AggieAir *GRAPEX* flights used in this study.

Flight Date	Landsat Time PST	Afternoon PST	Midafternoon PST
09 August 2014	10:41 am		-
02 June 2015	10:43 am	14:07 pm	-
11 July 2015	10:35 am	14:14 pm	
02 May 2016	-	12:05 pm	15:04 pm
03 May 2016	-	12:48 pm	-

To evaluate the *ET* performance at different spatial domain scales, two eddy covariance (*EC*) flux systems were deployed for the measurements of turbulent fluxes, including *LE* and *H*, and the available energy terms of R_n and *G*. Both towers are located at the eastern edge of the fields, due to predominant winds from the west. Ground measurements, including soil temperature and soil moisture were also collected. A complete listing of all measurements on the towers is given by Kustas et al. [56]. Details of the post processing of the *EC* data as well as the available energy measurements are provided by Alfieri et al. and Agam et al. [59,60].

EC micrometeorological data also included wind speed, air temperature, vapor pressure, air pressure, and shortwave radiation. Hourly average values of these atmospheric forcing variables, as well as the components of the surface energy balance, were computed. Table 2 illustrates the in-situ micrometeorological parameters and the name of the instruments used for the measurements.

Table 2. Description of in-situ micrometeorological measurements in this study.

ID	Micrometeorological Parameters	Instrument Name [1]	Elevation
1	Water vapor concentration	Infrared gas analyzer (EC150, Campbell Scientific, Logan, Utah)	5 m agl
2	Wind velocity	Sonic anemometer (CSAT3, Campbell Scientific)	5 m agl
3	Net radiation	4-way radiometer (CNR-1, Kipp and Zonen, Delft, The Netherlands)	6 m agl
4	Air temperature	Gill shielded temperature (Vaisala, Helsinki, Finland)	5 m agl
5	Water vapor pressure	Humidity probe (HMP45C, Vaisala, Helsinki, Finland)	5 m agl
6	Soil heat flux	Five plates (HFT-3, Radiation Energy Balance Systems, Bellevue, Washington)	−8 cm
7	Soil temperature	Thermocouples	−2 cm
8	Soil moisture	Soil moisture probe (HydraProbe, Stevens Water Monitoring Systems, Portland, Oregon)	−5 cm

[1] The use of trade, firm, or corporation names in this article is for the information and convenience of the reader. Such use does not constitute official endorsement or approval by the US Department of Agriculture or the Agricultural Research Service of any product or service to the exclusion of others that may be suitable.

Given the high fluctuation of atmospheric conditions during the daytime, the flux footprint or contributing source area of each *EC* tower was estimated for the hourly period encompassing *sUAS* flight campaigns using the two-dimensional (*2D*) flux footprint model developed recently by Kljun et al. [61]. Because a 100% *EC* footprint fetch could extend over the study area, a 90% footprint area (90% cutoff) was used for analysis. Then, the weighted footprint area was divided by 0.9.

2.2. Data Processing

In this study, images were acquired remotely by *sUAS*, and the data were terrain corrected using georeferencing based on ground control points (*GCPs*). Furthermore, both thermal and optical data were atmospherically corrected.

2.2.1. Thermal Data

Torres-Rua [57] indicated that the thermal data obtained from the *sUAS* thermal sensors in this study are adversely affected by changes in transmissivity and atmospheric radiance. For this reason, ground measurements of temperature were collected in the same timeframe as the *sUAS* flight and compared with the imagery to calibrate the thermal image data. More details about the calibration of temperature imagery related to this study can be found in Torres-Rua [57].

2.2.2. Optical Data

Radiometric agreement between remotely sensed data from different platforms constitutes one of the major challenges in image processing. Therefore, in this research, the images acquired by *sUAS* were upscaled and harmonized with Landsat using the point spread function (*PSF*). More details related to *sUAS* data harmonization can be found in Hassan-Esfahani et al. [62].

2.3. Energy Balance Closure Adjustment Methods for EC

While the *EC* technique provides measurements of turbulent fluxes H and LE, a lack of energy balance closure with the available energy terms R_n and G [63] is well documented. This results in $R_n - G > LE + H$ [64,65], and the computed closure ratio (*CR*) evaluates the energy balance discrepancy, $CR = (H + LE)/(R_n - G)$. This ratio varies during the daytime, but for the *sUAS* flights [55] it was found to be above 0.8, except for the May 2 afternoon flight where it fell to around 0.7.

To avoid any bias when comparing the energy balance models with *EC* field measurements, the energy closure issue needs to be handled and resolved. Twine et al. [66] suggested a method for energy balance closure that assumes the Bowen ratio (*H/LE*) before and after adjustment are the same, while considering both R_n and G as reliable measurements. A modified H and LE can be calculated as:

$$LE^* = \frac{(R_n - G)}{(B+1)} \tag{7}$$

$$H^* = \frac{(R_n - G)}{\left(\frac{1}{B} + 1\right)} \tag{8}$$

where LE^* and H^* denotes the closure adjusted latent and sensible heat flux, respectively.

2.4. Contextual Spatial Domain

The representative *TSEB2T* modeling grid size for the vineyard blocks was taken at 3.6 m, which corresponds to encompassing 6 × 6 grid or 36 *sUAS* thermal pixels having a resolution of 0.6 m. At this grid size, the inputs to *TSEB2T* incorporate the thermal-IR and optical bands of a vine row and adjusted interrows having a length scale of 3.35 m. Larger spatial domain scales were considered in this study, including 7.2 m, 14.4 m, and 30 m, to investigate the influence of domain size on the *TSEB2T* estimates. These selected values correspond to multiple vine rows spacing of 7.2 m (two rows), 14.4 m (four rows), and 30 m (Landsat scale—nine rows).

2.5. TSEB2T Inputs

The *TSEB2T* model developed by Nieto et al. [2] and implemented in Python language and is available at https://github.com/hectornieto/pyTSEB.

2.5.1. Leaf Area Index (*LAI*)

To assess the spatial heterogeneity of *LAI*, an approach was developed in this study to calculate *LAI* using a genetic programming (*GP*) model using the Eureqa software. The *GP* model associated *sUAS* imagery and *LAI* ground measurements collected with an indirect method using (*LAI-2200*, *LI-COR*, Lincoln, Nebraska) plant canopy analyzer measurements at several locations within the northern and southern vineyards with additional validation using destructive *LAI* sampling at several locations [46]. Before performing the *GP* model calculations, imagery features were classified into two categories, vine and interrow, and then statistical calculations were separately carried out for the optical properties of each category. The main optical reflectance used in this analysis comprise the original bands (red (*R*), green (*G*), blue (*B*), and near-infrared (*NIR*)), along with two conventional *VIs* (*NDVI* and *NIR*/*R*). Statistical computations were performed using the fine-resolution data inside the spatial domain scales (3.6 m, 7.2 m, 14.4 m, and 30 m), which included the maximum, minimum, area, mean, standard deviation, and sum. The *GP* model integrates all of these corresponding statistics to construct a relationship to *LAI* observations.

2.5.2. Canopy Height (h_c)

Spatial data from the digital terrain model (*DTM*) [67] and digital surface model (*DSM*) were aggregated into multiple spatial scales by employing a simple averaging method; then, h_c was calculated using the expression: $h_c = DSM\text{-}DTM$. For example, in the case of a 7.2-m grid, the average values of *DSM* and *DTM*, $DSM_{(7.2)}$, and $DTM_{(7.2)}$, respectively, were computed inside the grid window, then the height of the canopy was computed as: $h_{c_{(7.2)}} = DSM_{(7.2)} - DTM_{(7.2)}$.

2.5.3. Fractional Cover (f_c) and Canopy Width (w_c)

The north and south vineyard blocks were classified into two categories, vine and interrow, based on *NDVI*. Then the vine area inside each spatial domain was calculated and divided by the total area of the grid to calculate the f_c. w_c inside each spatial domain was computed using f_c and the width of the grid (w) under analysis, i.e., $w_c = f_c \times w$. To calculate the representative width of the vine canopy, the total width was rescaled and standardized at multiple spatial domain scales, depending on the number of rows inside each grid. For example, in the case of a 3.6-m grid, one vine row was counted inside, while in a 7.2-m grid, the number of rows was doubled.

2.5.4. w_c/h_c Ratio

w_c/h_c was calculated by simply dividing canopy width by canopy height at each contextual spatial domain.

2.6. Goodness-of-Fit Statistics

Evaluating the performance of the *TSEB2T* model with the *sUAS* imagery for the four different modeling grid resolutions involved comparing the estimated fluxes with measurements from the *EC* towers. Computed statistical metrics included the root mean square error (*RMSE*), the normalized root mean square error (*NRMSE*), mean absolute error (*MAE*), mean absolute percentage error (*MAPE*), and Nash–Sutcliffe efficiency coefficient (*NSE*). A value of $NSE = 1$ indicates perfect agreement between modeled and observed flux, while *NSE* approaching 0 means that the agreement is very poor, and $NSE < 0$ indicates unacceptable performance [68]. These statistical measurements are calculated as follows using *LE* as the flux:

$$RMSE = \sqrt{\frac{1}{N}\sum_{i=1}^{N}(LE_{m,i} - LE_{o,i})^2} \tag{9}$$

$$NRMSE = \frac{RMSE}{\sigma_0} \tag{10}$$

$$MAE = \frac{\sum_{i=1}^{n}\left|LE_{m,i} - LE_{o,i}\right|}{n} \tag{11}$$

$$MAPE = \frac{\sum_{i=1}^{N}\left|\frac{LE_{m,i}-LE_{o,i}}{LE_{o,i}}\right| * 100}{n} \tag{12}$$

$$NSE = 1 - \frac{\sum_{i=1}^{n}\left(LE_{m,i} - LE_{o,i}\right)^2}{\sum_{i=1}^{n}\left(LE_{m,i} - \overline{LE}_{o,i}\right)^2} \tag{13}$$

where LE_m. denotes the modeled latent heat flux obtained from the *TSEB2T* aggregated up for the estimated flux footprint/source area, LE_o denotes the observed values from the *EC* tower, and n represents the number of observations, σ_o denotes the standard deviation of observed values.

LE was used for evaluating the impact of spatial resolution or grid size on modeled fluxes. At field scale, the evaluation is done using the spatial mean and coefficient of variation (*CV*) statistics. For *LE* statistical characteristics, frequency and cumulative distribution curves were used. Finally, to evaluate the effect of aggregating *LE* at 3.6 m, 7.2 m, 14.4 m, and Landsat scale, relative difference (relative error) was used. Relative difference (relative error) is defined as the root mean square error (*RMSE*) between the aggregated resolution and its reference grid size resolution of 3.6 × 3.6 m divided by the spatial mean (μ) value computed from the reference grid size (3.6 m × 3.6 m), i.e., $E_r = RMSE/\mu$ [14]. Each grid value of aggregated data was compared to the $n \times n$ set of reference scale or resolution (3.6 m) grid using E_r.

3. Results and Discussion

3.1. TSEB2T Contextual Spatial Domains Validation

3.1.1. EC Footprint Estimation

The results of footprint analysis using the *2D* flux model developed by Kljun et al. [61] and described in Section 2.1 are shown in Figure 5 for the different *sUAS* flights.

The orientation and size of each flux footprint/source area depends on the micro-meteorological conditions at the site measured by the *EC* towers, which include the turbulence fluxes, friction velocity (u_*), and wind speed, which affect atmospheric stability, and canopy and *EC* measurement height, which affect the effective sampling height and wind direction that affects the orientation of the footprint. The total statistical weight of the footprint is taken to equal unity, although the actual area computed by the footprint model represents 90% of the contribution since the additional 10% essentially makes no measurable contribution. To compare the fluxes computed by the *TSEB2T* model at the different spatial resolutions with the *EC* measurements, the source area estimated by the footprint model was multiplied by the corresponding modeled fluxes (R_n, H, LE, G) using *ArcGIS10.6*. Then, a comparison between the weighted fluxes at the different spatial resolutions or grid sizes from the *TSEB2T* version of *TSEB* and *EC* measurements was performed to assess model performance.

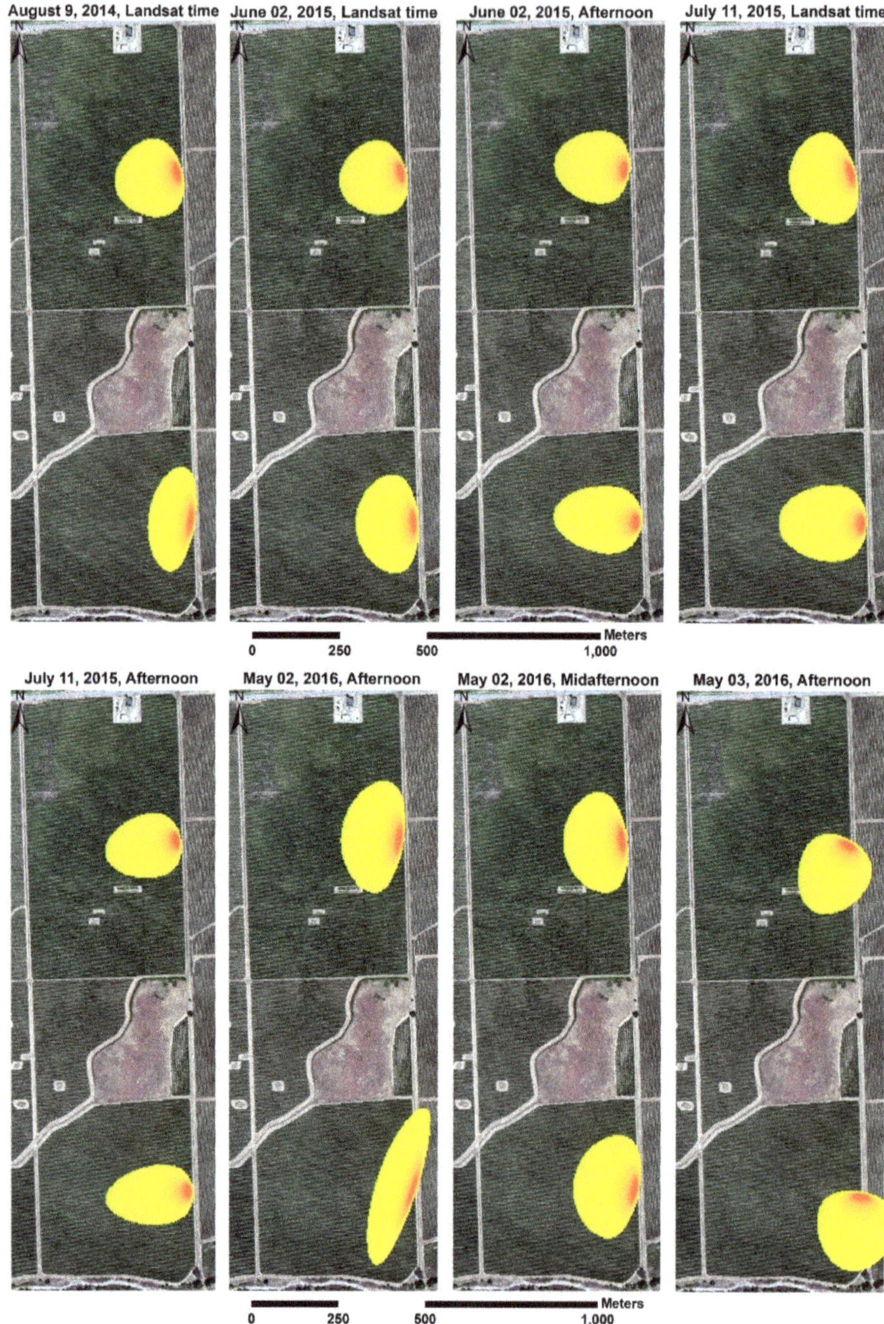

Figure 5. Layout of 90% *EC* footprints for two towers at different times considered by this study.

3.1.2. Statistical Performance

Table 3 lists the goodness-of-fit statistics between the energy fluxes using *TSEB2T* at different spatial resolutions and *EC* tower observations, while Figure 6 shows the relationship between the modeled and measured fluxes. The results indicate a significant deterioration in model performance at the 30-m grid size. A major factor that may be responsible for this poor performance in the *TSEB2T* model at 30-m resolution is that the size and dimension of the *EC* source area estimated by the footprint model cannot incorporate a representative range in the spatial variability in the fluxes at 30-m resolution. This conclusion agrees with a previous study conducted by Song et al. [69] that showed a major problem in comparing modeled and measured fluxes when there is a mismatch in pixel resolution or model grid size in the remotely sensed *ET* output and in the source area contributing to the *EC* tower measurements in a heterogeneous landscape.

Results in Table 3 also indicate that R_n and G across multiple aggregated grids demonstrated a close agreement between the *TSEB2T* output and observed measurements, as indicated by lower *MAE* and *MAPE* with quite constant correlation (R^2). The *MAE* and *MAPE* in the R_n estimate at grid sizes of 3.6 m, 7.2 m, and 14.4 m accounted for less than 25 W/m^2 and 5%, respectively. However, at Landsat scale the *MAE* increased slightly to 29 W/m^2. A similar result was obtained for *H*, where *MAE* at the finer resolutions yielded values less than 45 W/m^2, while the coarser grid size of 30 m yielded a larger *MAE* of nearly 80 W/m^2. As shown in Table 3, the correlation of *H* is higher than *G* and *LE*, except for 30-m resolution/model grid. This implies that the performance of the 30-m resolution is different compared to the 3.6-m, 7.2-m, and 14.4-m resolutions. The results for *LE* indicated good agreement with the flux measurements at 3.6-m, 7.2-m, and 14.4-m modeling grid sizes, while at the 30-m resolution, the *MAE* value was around 85 W/m^2. As demonstrated in Figure 6d, all values of *LE* are underestimated (below 1:1 line) with an *NSE* coefficient of 0.2. Furthermore, the highest *NRMSE* values were observed for *LE*, compared with other surface fluxes, particularly at 30-m resolution. The lowest *NRMSE* was obtained for R_n across different spatial domains/model grids.

Table 3. Goodness-of-fit statistics between the eddy covariance and the *TSEB2T* fluxes at different spatial scales (3.6 m, 7.2 m, 14 m, and 30 m).

Spatial Domain	Fluxes	RMSE (W/m^2)	NRMSE	MAE (W/m^2)	MAPE (%)	NSE	R^2
3.6 m	R_n	28	0.3	25	5	0.9	0.94
	LE	69	1.2	58	20	0.5	0.49
	H	54	0.8	41	26	0.7	0.67
	G	34	0.9	30	51	0.6	0.56
7.2 m	R_n	27	0.3	24	4	0.9	0.94
	LE	66	1.2	56	19	0.5	0.53
	H	51	0.7	36	24	0.7	0.67
	G	33	0.8	30	50	0.6	0.58
14.4 m	R_n	25	0.3	20	4	0.9	0.95
	LE	79	1.4	56	18	0.1	0.21
	H	48	0.7	35	26	0.6	0.69
	G	32	0.8	29	49	0.6	0.59
30 m	R_n	34	0.4	29	5	0.9	0.96
	LE	101	1.8	86	30	0.2	0.53
	H	93	1.3	78	67	−0.1	0.23
	G	31	0.8	28	48	0.6	0.60

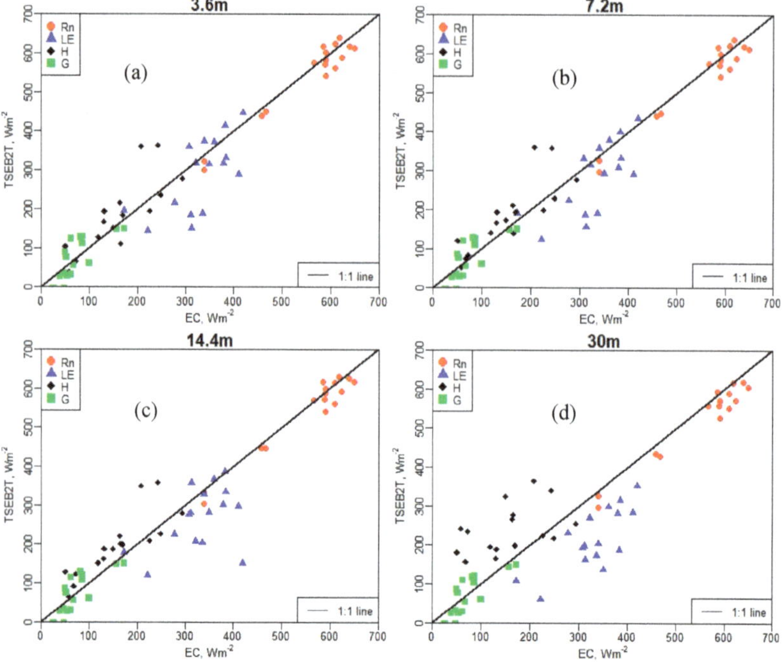

Figure 6. Scatterplot of observed versus estimated surface fluxes using different model grid sizes/resolution with the *TSEB2T* model; (**a**) 3.6 m, (**b**) 7.2 m, (**c**) 14.4 m, and (**d**) 30 m.

With the *TSEB2T* model and other remote sensing-based models using thermal-IR as the boundary condition, *LE* is solved as the residual component of the surface energy balance, $LE = R_n - H - G$. Therefore, an error in the calculation of energy fluxes (R_n, *H*, and *G*) adversely affects the estimation of *LE*. Based on Figure 6, the *LE* estimation (or bias) is mainly influenced by the estimation of *H*. This conclusion was also reached by Kustas et al. [70], who showed the discrepancies between modeled and measured *LE* is due in large part, up to approximately 90%, to errors in modeled *H*.

3.2. Contextual Spatial Domain Aggregations Effects

3.2.1. The Effect of Model Grid Size on *TSEB2T* Inputs

(a) Canopy and Soil Temperatures (T_c, T_s)

T_c and T_s were estimated based on a linear *LST-NDVI* relationship as described by Nieto et al. [2]. However, this relationship does not fulfill the homoscedasticity criterion when the spatial domain/resolution reaches a certain size (i.e., 30-m) as shown in Figure 7. For example, in the case of a 30-m grid size, a higher variability is observed in the *LST-NDVI* data compared with finer resolutions (3.6 m, 7.2 m, and 14.4 m). At micro-scale (e.g., 3.6 m), there are small number of pixels inside the spatial domain compared with others (7.2 m, 14.4 m, and 30 m), and exhibit an apparent linear relationship between *LST* and *NDVI*. However, at coarse resolution (e.g., 30 m), there are many more pixels, more rows of vineyard are included, and large vegetated and bare soil pixels exist inside the spatial domain. The result is a partially filled triangular shape. This indicates the relationship between *LST* and *NDVI* starts to resemble the "triangle method" [71] to estimate *ET* as the sampling domain increases.

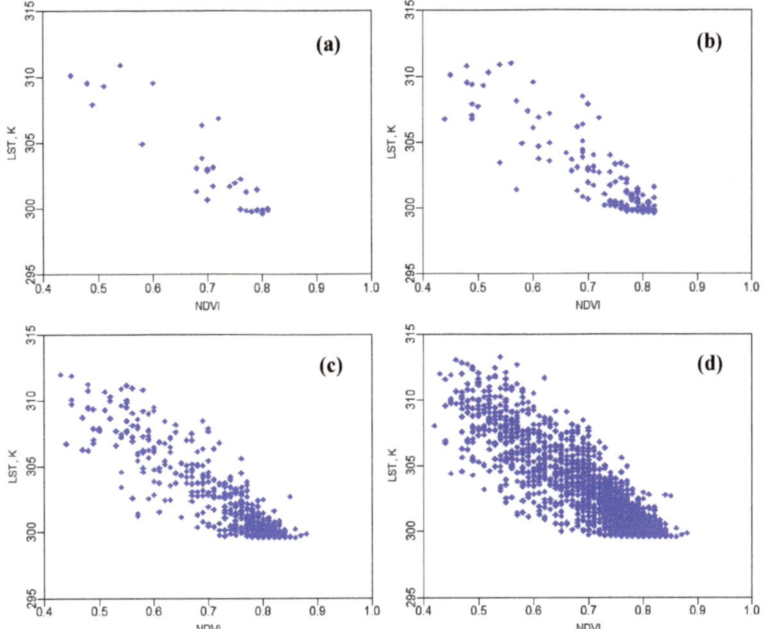

Figure 7. The *LST-NDVI* relationship used for finding T_c and T_s as proposed by the *TSEB2T* model at different spatial domains (09 August 2014). (**a**) 3.6 m, (**b**) 7.2 m, (**c**) 14.4 m, (**d**) 30 m.

Figure 8 illustrates the T_c and T_s maps at different resolutions, which provide an indication of the loss in spatial variability due to spatial aggregation. The ranges of T_c and T_s were between 290 K (16.85 °C) and 320 K (46.85 °C) for the *sUAS* flight in 2014.

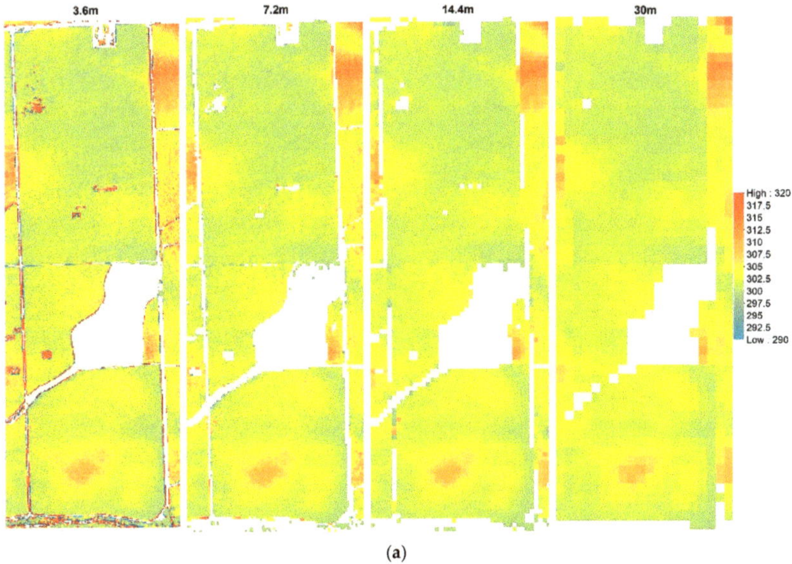

(**a**)

Figure 8. *Cont.*

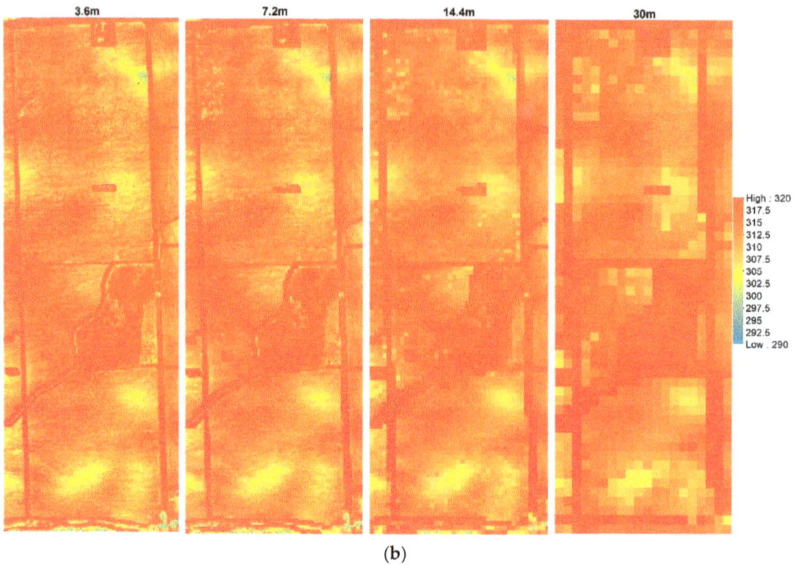

(**b**)

Figure 8. Example of (**a**) canopy temperature (T_c) and (**b**) soil temperature (T_s) in Kelvin (K) at different spatial domains for 09 August 2014.

(b) Leaf Area Index (*LAI*)

With the *GP* model results, it was found that the main estimators for computing *LAI* are the mean of *NIR/R* ratio of the vine, area of the vine, sum of *NDVI* of the vine, standard deviation of *NIR* of the interrow, and standard deviation of *NIR/R* ratio of the vine. The *GP* model (Equation (14)) was applied to the remote-sensing imagery to map spatial *LAI* distribution across the study area.

$$LAI = 0.21NDVI_{v\_area} - 0.004NDVI_{v\_sum} + 0.34\left(\frac{NIR}{R}\right)_{v\_mean} - \frac{0.94}{exp\left(0.23(NDVI_{v\_area})^2\right)} - 2.8NIR_{i\_STD}\left(\frac{NIR}{R}\right)_{vSTD} - 0.7 \quad (14)$$

LAI values from the *GP* model compared with the actual *LAI* field measurements showed good agreement with an R^2 of 0.73.

To evaluate the difference between multiple model grid sizes of *LAI* for each flight, *LAI* maps at different resolutions were estimated (see Figure 9) and statistics including the spatial mean, standard deviation, and coefficient of variation (*CV*) were calculated as shown in Table 4. Figure 9 provides an indication of the loss in spatial variability in *LAI* images due to spatial aggregation. *LAI* at each contextual spatial domain/resolution was calculated using the *LAI* model (Equation (14)). Each parameter in that equation was calculated based on the pixel values inside the model grid. The ranges of *LAI* were between 0 and 2.5 for the *sUAS* flight in 2014. As illustrated in Table 4, the spatial mean value (μ) is the same across different scales, with a slight decrease in *CV*. The exception is the flight on 2 May 2016, which represents the early growing stage of the vine canopy with active/live interrow cover crop, showing a higher *CV*. Hardin and Jensen [72] also found greater uncertainty in estimating *LAI* under low *LAI* conditions using *VIs*. The frequency histogram in Figure 10 indicates the distribution of values is skewed such that the lower values are more pronounced for the flight of 2 May 2016, with a non-significant change between curves from the different grid sizes, except the 30-m resolution spatial domain, which shows a higher variation compared with other scales. This behavior aligns with the decreasing *CV* values due to loss in internal or pixel variability of the *LAI* values. A similar trend of lower *CV* toward large scale (30 m) has been observed for other *TSEB2T* inputs including h_c, f_c, and w_c/h_c.

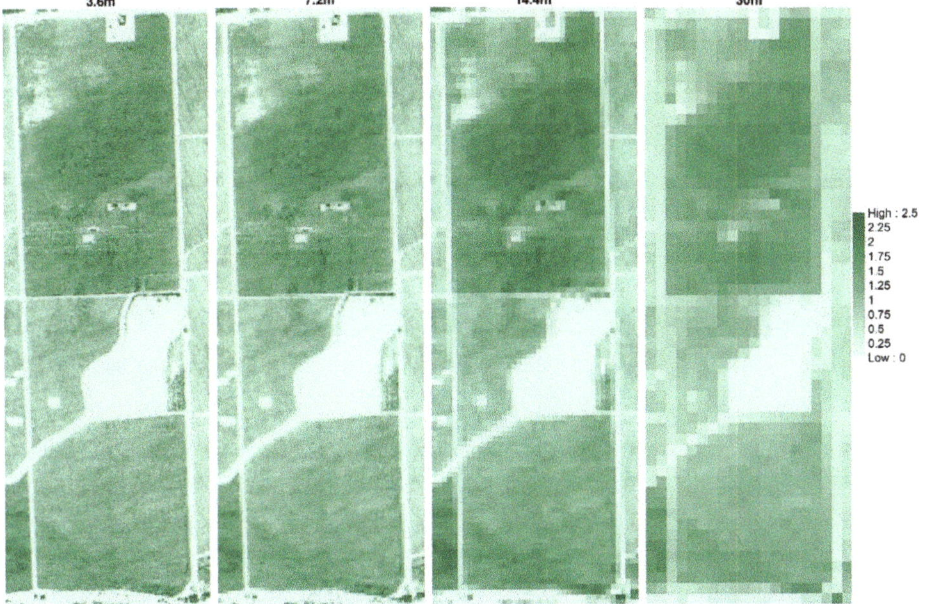

Figure 9. Example of modeled *LAI* (unitless) across different spatial domains for 09 August 2014.

Table 4. Spatial domain effect on *LAI* estimation.

Flight	Spatial Domain	μ	σ	*CV*
09 August 2014	3.6 m	0.91	0.56	0.61
	7.2 m	0.91	0.54	0.59
	14.4 m	0.91	0.52	0.57
	30.0 m	0.91	0.48	0.53
02 June 2015	3.6 m	0.57	0.38	0.66
	7.2 m	0.57	0.33	0.58
	14.4 m	0.57	0.30	0.52
	30.0 m	0.57	0.27	0.47
11 July 2015	3.6 m	0.52	0.39	0.75
	7.2 m	0.52	0.36	0.69
	14.4 m	0.52	0.34	0.65
	30.0 m	0.52	0.31	0.60
02 May 2016	3.6 m	0.06	0.11	1.90
	7.2 m	0.06	0.10	1.75
	14.4 m	0.06	0.10	1.66
	30.0 m	0.06	0.09	1.59

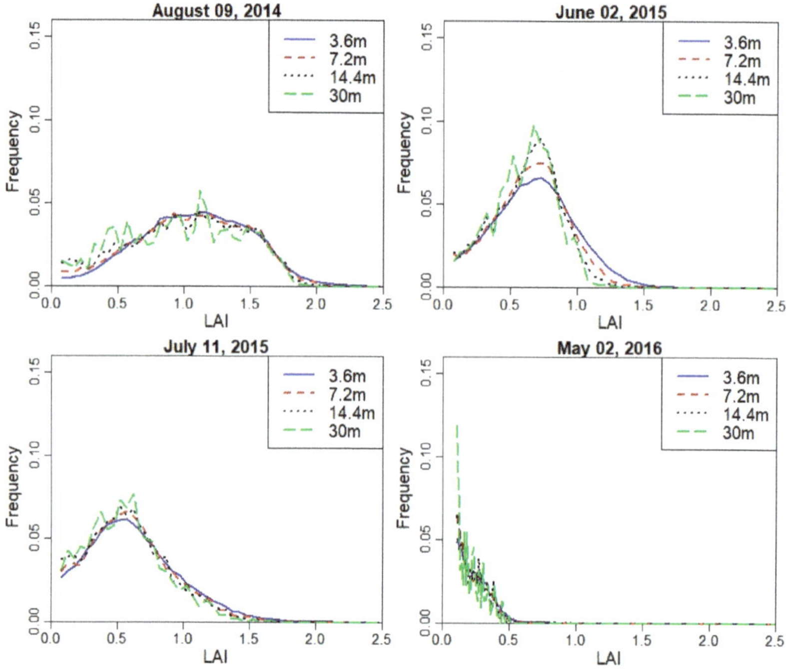

Figure 10. Frequency curve of *LAI* at different times from 3.6 m and 7.2 m, 14.4 m and 30 m.

3.2.2. Contextual Spatial Domain Effect on Field-Scale *LE* Estimation

An example of the maps of *LE* across different model grid sizes is shown in Figure 11. The maps of the energy balance components for 2014 flight at different resolutions are shown in Appendix A. The statistics (mean (μ) and coefficient of variation (*CV*)) for the *LE* maps at the different modeling resolutions are illustrated as bar graphs in Figures 12 and 13, respectively. For *LE*, the highest mean value is on 02 May 2016, at midafternoon. Although the grapevine canopy is fully developed by June, *LE* in May at both overpass times is higher than the acquisition in June, July, and August. However, on 3 May, the model yields the lowest *LE* values due to overcast conditions that day significantly reducing incoming solar radiation, and hence, the energy fluxes. The phenocam data (https://hrsl.ba.ars.usda.gov/awhite/CAM/) indicate the high rate of *LE* on 2 May is the result of a rapidly developing vine canopy, together with a transpiring cover crop.

At a contextual spatial domain level, the magnitude of *LE* is degraded as shown in Figure 12 due to the data aggregation from the 3.6-m grid to Landsat scale (30 m). For example, the mean *LE* value from *TSEB2T* on 02 May 2016 at midafternoon was 315 W/m^2 for the 3.6-m grid decreasing to 304 W/m^2 for the 7.2-m grid, then decreases further to 293 W/m^2 and 278 W/m^2, respectively, for 14.4-m and 30-m grids. As shown in Figure 13, *CV* value slightly increases as the model grid scale/resolution size increases despite a decrease in variation of *LAI* and *LST* distribution as seen in Section 3.2.1. While *LE* degrades, the *CV* values do not show significant differences. This can be due to internal *TSEB2T* compensation of the energy balance components at the different evaluated scales.

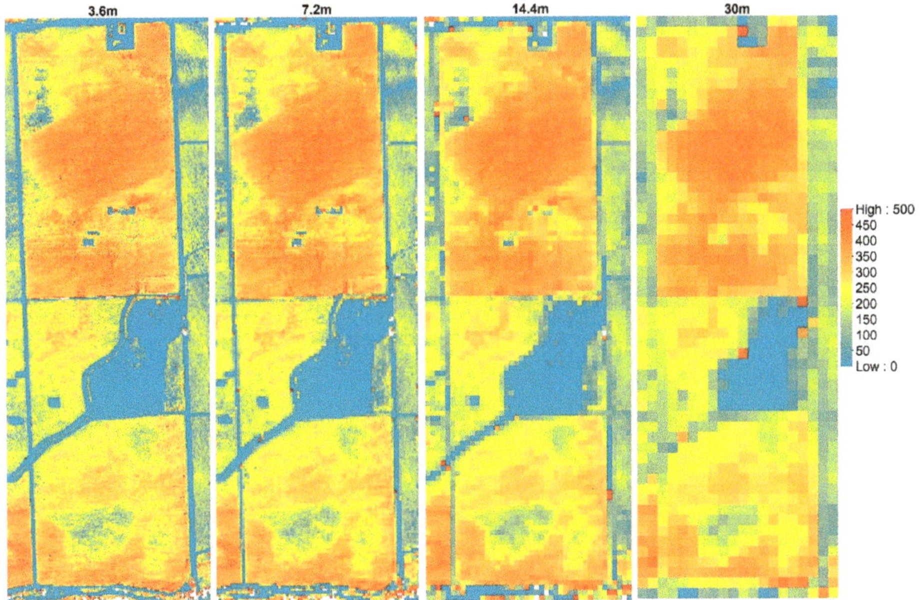

Figure 11. *LE* (W/m^2) aggregation at 3.6 m, 7.2 m, 14.4 m and 30 m for 09 August 2014.

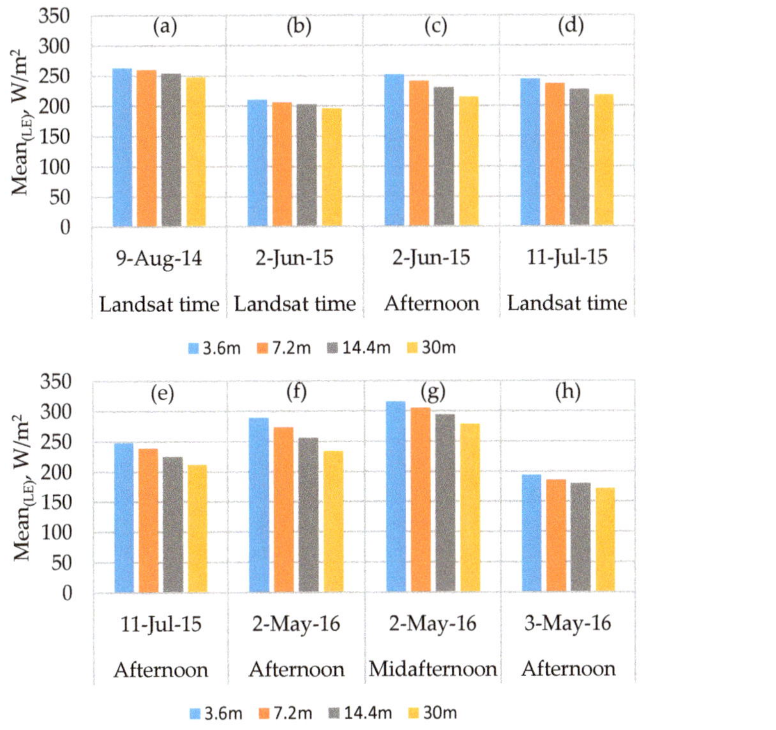

Figure 12. Spatial domain effect on the mean of *LE* spatial data at different times.

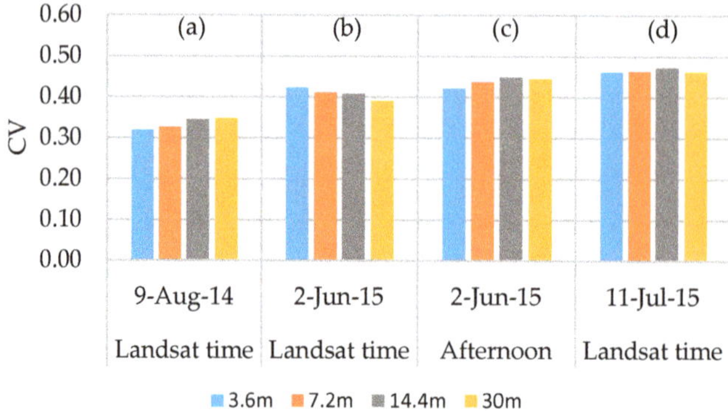

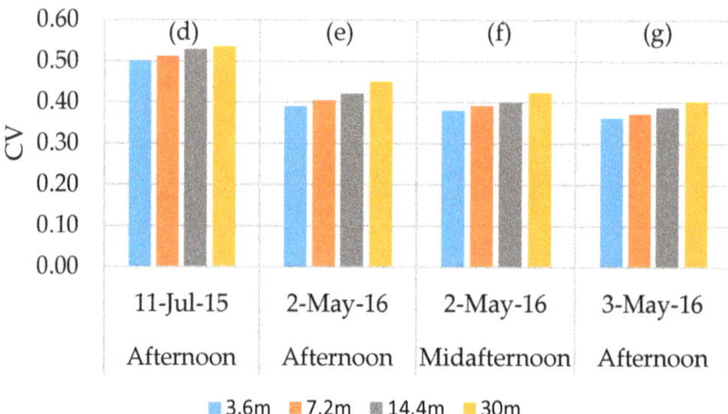

Figure 13. Spatial domain effect on the coefficient of variation (*CV*) of *LE* spatial data at different times.

3.2.3. Contextual Spatial Domain Effect on *LE* Statistical Characteristics

To provide quantitative evaluation of the impact of spatial aggregation of inputs on *LE* estimation for the resulting pixel values, frequency and cumulative distribution plots for the *LE* maps are illustrated in Figure 14. This figure shows that *LE* varies at different grid sizes. The cumulative frequency distribution curves indicate that, especially at the 30-m grid size, *LE* distribution tends to have the highest cumulative values at lower *LE* range (below 300 W/m^2). A magnitude shift towards lower *LE* persists across different times, with one exception. In the case of a 30-m grid on 02 June 2015, the frequency moved up then decreased below the frequency curves of other grid sizes (3.6 m, 7.2 m, and 14.4 m). In general, the results in Figure 14 show a reduction in *LE* distribution as the scale becomes coarser. Hong et al. [22] indicated that an increase in the peak of the *LE* histogram curve spans as much 10% to 20% as a response to spatial data aggregation using *SEBAL*. In the *TSEB* model, the soil and vegetation components of the scene are treated separately, while the *SEBAL* model uses a single source approach using the composite soil/canopy temperature and is contextual defining wet and dry *ET* limits based on the hot and cold extremes in the *LST* field within the image [73]. Moreover, Ershadi et al. [14] pointed out three possible reasons behind the different results obtained from *ET* models: (a) the approach (e.g., contextual hot/cold surface temperature limits versus using absolute surface-atmosphere temperature differences) of each model to estimate *ET*, (b) the study area and eco-hydrological conditions of the surface, which may favor certain *ET* model parameterizations over others, or (c) the different models of aerodynamic resistance formulations and sensitivity to the roughness parameters.

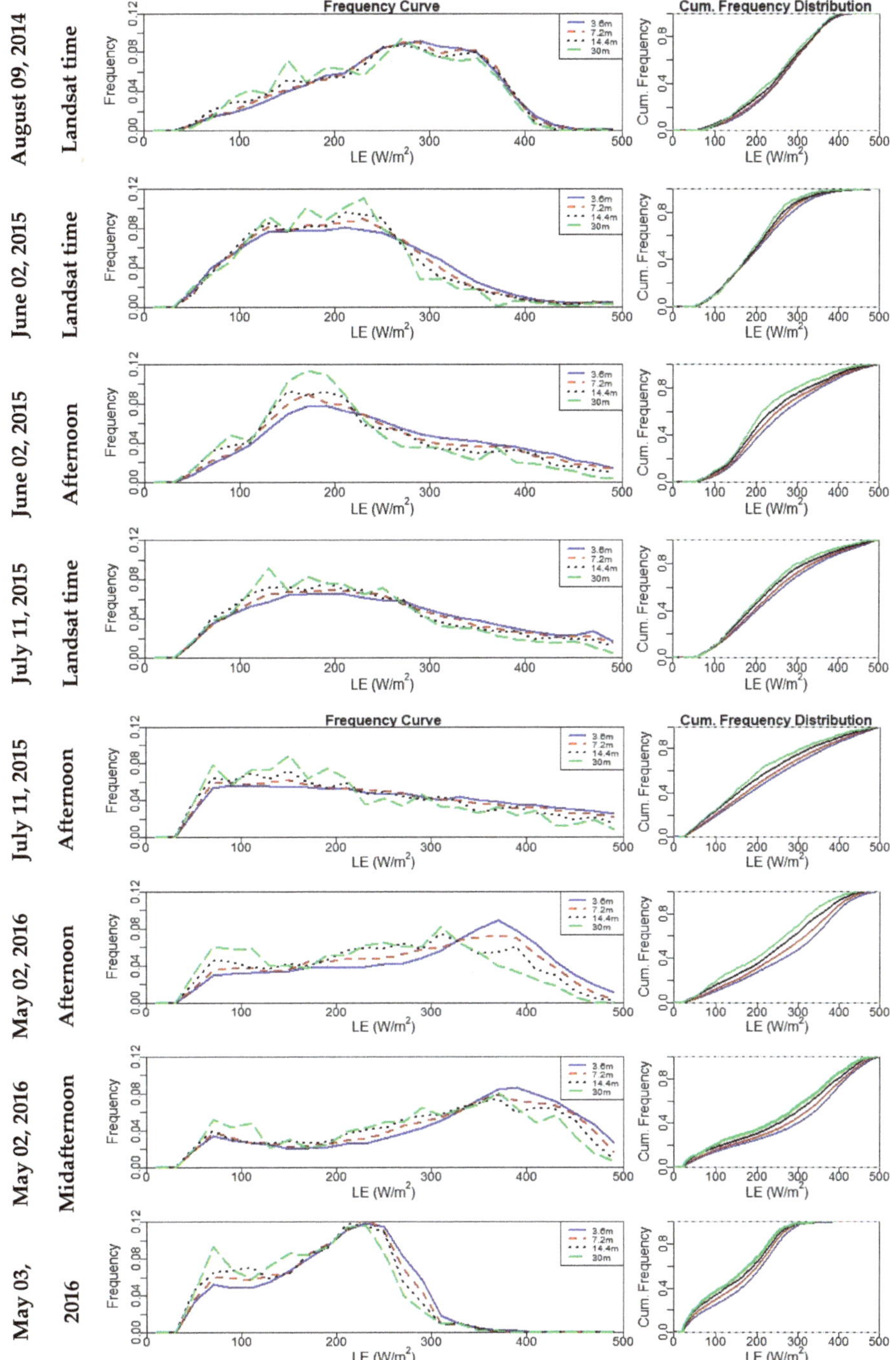

Figure 14. Frequency curve (**left**) and cumulative frequency distribution (**right**) plots of instantaneous *LE* for all *sUAS* flights at 3.6 m, 7.2 m, 14.4 m, and 30 m.

Increasing the spatial domain/resolution affects the estimation of *TSEB2T* parameters as the fine details of the surface disappear. To test these claims, R_a (s/m) and *LST-NDVI* relationship were evaluated at different spatial domain/resolution; the latter is shown in Section 3.2.1 (a). As shown in Figure 15, there is a decreasing trend in the relative spatial mean (μ_r) of R_a for all flights, ranging approximately from 20% to 60%. The high variability in R_a is related mainly to the variables that affect the friction velocity (u_*), which the mean canopy height and roughness length (z_{oH}), which are derived from the imagery at different resolution/spatial domain. This finding is in agreement with Ershadi et al. [14] and Moran et al. [15], who indicated that the reduction of R_a value at coarse spatial domain/resolution is a key factor behind the underestimation of *LE*.

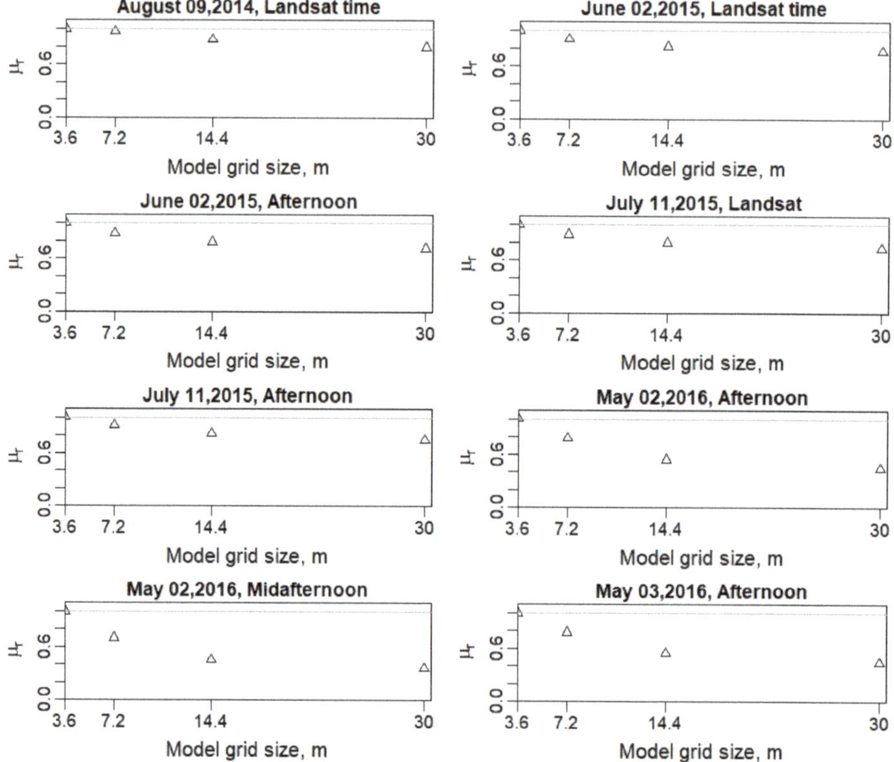

Figure 15. Variation of the relative spatial mean (μ_r) of R_a for different flights.

3.2.4. Effects of Model Grid Size on *LE*

To evaluate quantitatively the impact of model grid size via the resolution of key input data, the relative difference (relative error) (E_r) was computed using as the reference the *LE* at 3.6-m model grid size/resolution. For example, the *LE* value at the 7.2-m grid is compared to the *LE* at the 3.6-m grid size by resampling the 7.2-m grid to a 4 × 4 set of 3.6-m *LE* output which will have a uniform *LE*-value at the finer resolution, and taking the difference. As illustrated in Figure 16, E_r is calculated with the mean and percentiles (25th and 75th) for the coarser grid sizes used in the *TSEB2T* model for the different *sUAS* acquisitions. The plots demonstrate an increasing trend in E_r as the model grid size/resolution increases/decreases. The largest E_r value was computed for the imagery on 11 July 2015 at afternoon at nearly 45% for the Landsat resolution. In contrast to 11 July 2015, the lowest range of relative error was observed on 09 August 2014, where the E_r ranged approximately between 15% for the 7.2-m grid and 25% for the 30-m grid. On an average, E_r value ranged from approximately 25% using the 7.2-m model grid size to 40% with the 30-m model resolution.

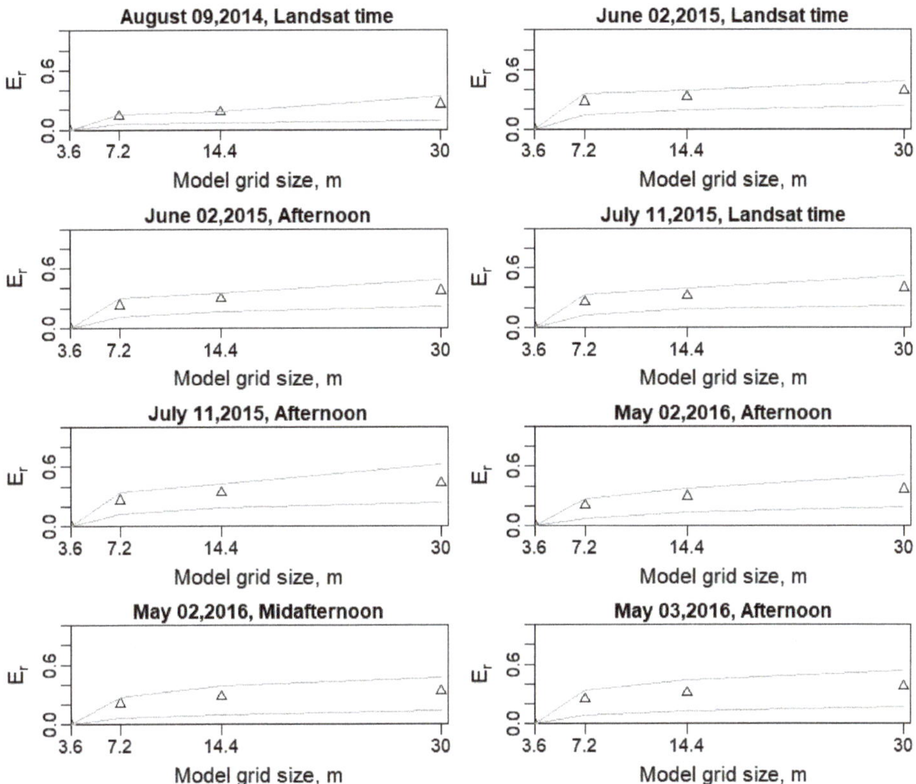

Figure 16. Relative error (E_r) at different spatial resolutions for *LE* with the triangle symbols indicating mean and light lines indicating the 25th and 75th percentiles for the coarse grid sizes.

These results are supported by an Ershadi et al. [14] study that found the E_r of *LE* varied between 20% and 40% when aggregating the Landsat data incrementally from 120 m to 960 m and using the *SEBS* model to calculate surface heat fluxes. Furthermore, Moran et al. [15] indicated that a larger error could appear (larger than 50%) in *H* estimation over a heterogeneous area due to a mix of stable and unstable conditions and the variation in aerodynamic roughness, especially for highly unstable conditions. As previously mentioned in Section 3.1, the underestimated *LE* could be influenced by overestimation in *H*, which implies that a large error is expected in the residual flux (*LE*) estimate at coarse spatial domains [70]. Furthermore, the effect of model grid size on *LE* is also visible at the 25th and 75th percentiles, which immediately increases at the 7.2-m grid size and continues increasing towards the 30-m resolution, providing a clear indication of increasing discrepancy with the reference grid (3.6 m) *LE* estimates.

4. Conclusions

The objective of this study was to assess high-resolution *LE* estimation in vineyards at different model grid sizes or resolutions, specifically 3.6 m, 7.2 m, 14.4 m, and 30 m (Landsat scale), using a physically-based *ET* model known as *TSEB2T*. The reference grid size of 3.6 m represents the finest pixel resolution that includes both vine canopy and interrow conditions, which is the resolution where the

TSEB model algorithms of soil/substrate and canopy temperature partitioning radiation and convective energy exchange are applicable [2]. Multiple statistical measures were used to assess the effect of decreasing the spatial resolution or increasing the model grid size 2, 4, and nearly 10 times the original 3.6 m resolution. These included validation of *TSEB2T* fluxes at the different model grid sizes with the *EC* measurements, comparing *LE* spatial statistics (mean and coefficient of variation, frequency distributions) and *LE* differences over the imaged domain at the different resolutions using *LE* at 3.6 m grid size as the reference. The results showed that separation of T_c and T_s, required in *TSEB2T*, affects the *LST-NDVI* linear trend as a function of resolution of the pixels. The validation results with the flux tower measurements indicate that R_n and *G* discrepancies do not change across different model grid sizes, while for *H* and *LE* there is an increase in model-measurement differences, particularly at the 30-m resolution. This is largely caused by an overestimation in *H*, causing an underestimation in *LE* (bias), particularly at the coarsest resolution (30-m grid size). This refers mainly to the non-linear relationship of *LST-NDVI* and the variability of R_a due to the variables that affect the u_* which are the mean canopy height and roughness length, which are derived from remote sensing imagery at different spatial domain/resolution.

The effects of model grid size were evaluated at field and at grid scale using the spatial mean and coefficient of variation and relative difference, respectively. At field scale, the results show small decreases in the spatial mean over the image, ranging approximately from 10% to 20%, as the data aggregated for model grid size increased from 3.6 m to 30 m. However, the relative differences with resolution indicate a significant decrease in *LE*, ranging approximately from 25% to 45%, when aggregating the data from 3.6 m to Landsat scale (30 m). This means that, while field values of *LE* may be adequate to use, the field variability reduction limits its use for precision agriculture applications, such as identifying areas within the field under actual stress conditions or being over irrigated. These results suggest that *TSEB2T* is only applicable using imagery with high enough resolution that can readily distinguish plant canopy and soil/substrate temperatures and the modeling grid size is at a resolution where it is appropriate to apply *TSEB2T* algorithms for modeling the radiative and convective energy exchange from both the vegetation and soil substrate systems. Aggregating inputs to *TSEB2T* to multiple grid sizes of the interrow/row spacings for vineyards is not advisable, since it is likely the accuracy of surface fluxes, particularly *LE*, will deteriorate. While this study was limited to evaluating different modeling grid sizes, a future comparison with Landsat and ECOstress *ET* products is also planned, which would provide a more comprehensive scaling assessment of *ET* estimates for *sUAS*-Satellite *ET* integration. Furthermore, the effect of remote sensing resolution on the output of other *TSEB* versions such as *TSEB-PT* may be less affected and will be evaluated in a future study.

Author Contributions: A.N. designed, analyzed, and wrote the paper. A.T.-R. supervised the research, contributed with ideas during the interpretation of results, and reviewed the paper. W.K. supervised the research and reviewed the paper. M.M. contributed with ideas during the interpretation of results and reviewed the paper. D.S. contributed in the statistical analysis. L.H. contributed with ideas and reviewed the paper. H.N., J.A., J.P., L.M., M.M.A., C.C., L.S., and N.D. helped in collecting micrometeorological and ground measurements. All authors have read and agreed to the published version of the manuscript.

Funding: Funding provided by E.&J. Gallo Winery. Utah Water Research Laboratory contributed towards the acquisition and processing of the ground truth and *UAV* imagery data collected during *GRAPEX IOPs*. We would like to acknowledge the financial support for this research from NASA Applied Sciences-Water Resources Program [NNX17AF51G] and the USDA Non Assistance Cooperative Agreement 58-8042-5-092 funding. USDA is an equal opportunity provider and employer.

Acknowledgments: We would like to thank Aggieair Service Center team (Ian Gowing, Mark Winkelaar, and Shannon Syrstad) for their extraordinary support in this research, whose cooperation greatly improved the data collection and data processing, and the staff of Viticulture, Chemistry and Enology Division of E.&J. Gallo Winery for the assistance in the collection and processing of field data during *GRAPEX IOPs*. This project would not have been possible without the cooperation of Ernie Dosio of Pacific Agri Lands Management, along with the Sierra Loma vineyard staff, for logistical support of *GRAPEX* field and research activities. The authors would like to thank Carri Richards for editing this paper.

Conflicts of Interest: The authors declare no conflict of interest.

Appendix A

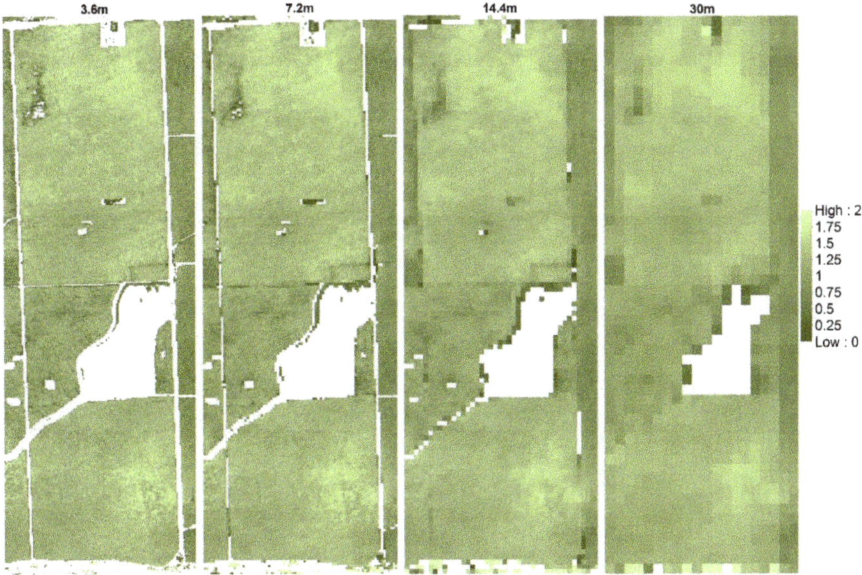

Figure A1. Example of modeled h_c (m) across different spatial domains for 09 August 2014.

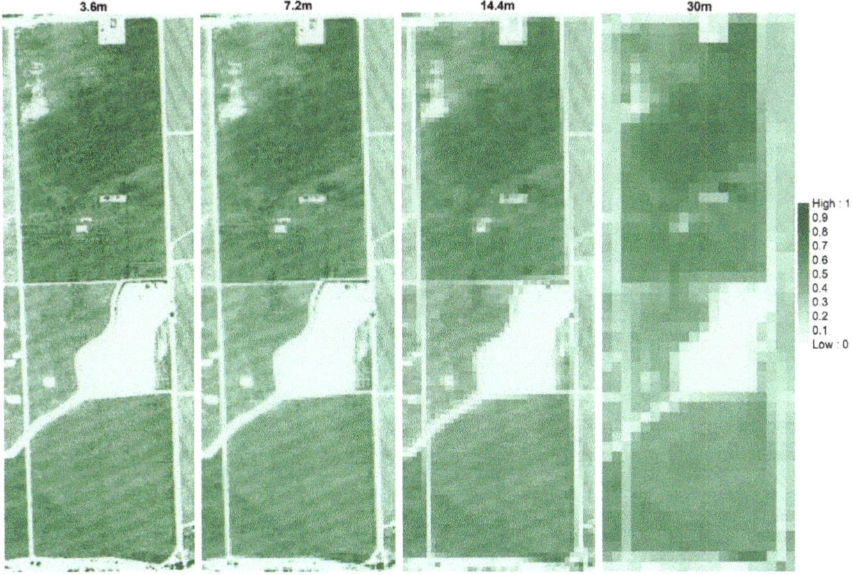

Figure A2. Example of modeled f_c across different spatial domains for 09 August 2014.

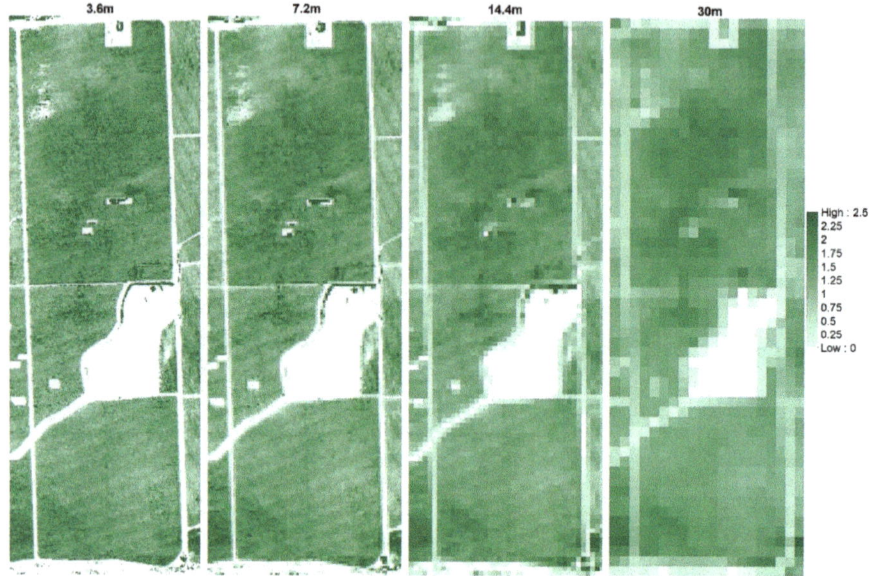

Figure A3. Example of Modeled w_c/h_c different spatial domains for 09 August 2014.

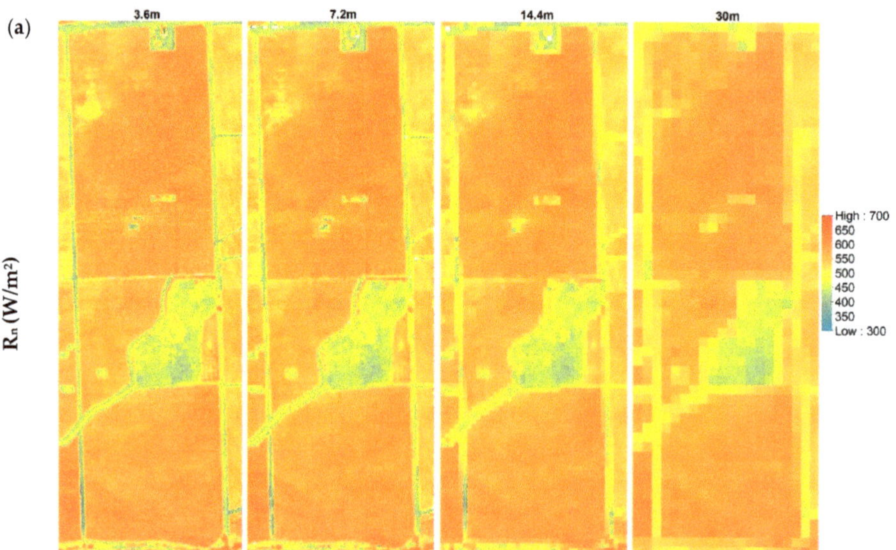

Figure A4. *Cont.*

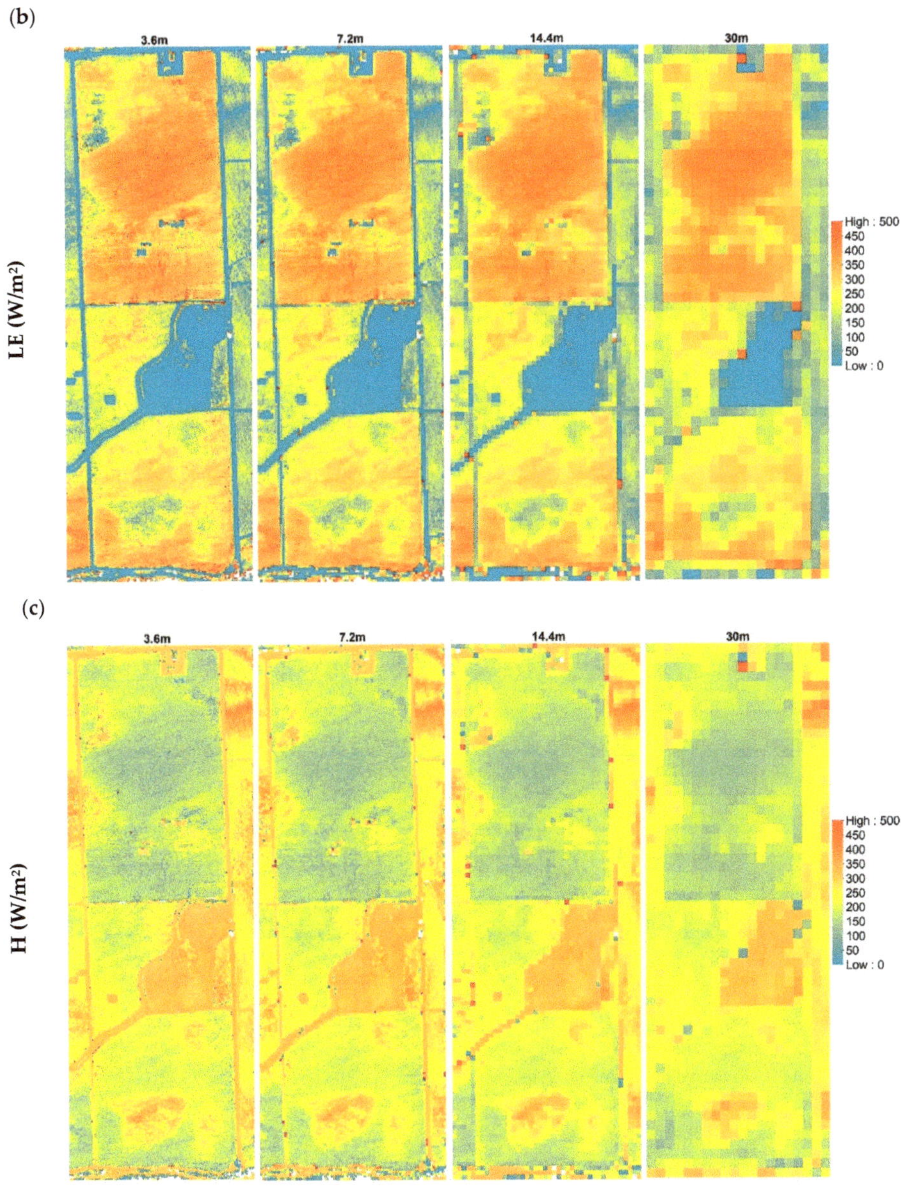

Figure A4. *Cont.*

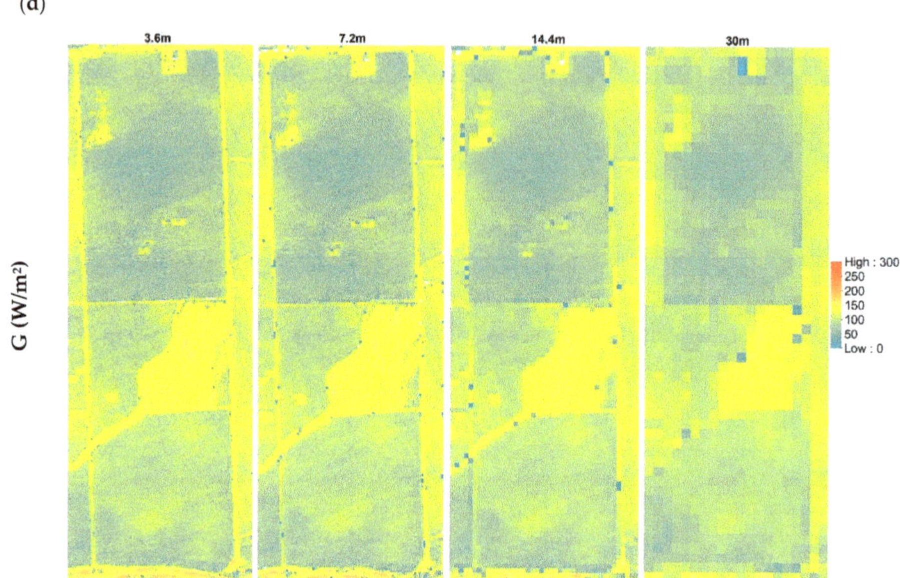

Figure A4. Aggregation of surface energy fluxes across different spatial domain (3.6 m, 7.2 m, 14.4 m, and 30 m) on 09 August 2014: (**a**) R_n, (**b**) *LE*, (**c**) *H*, and (**d**) *G*.

References

1. Melesse, A.M.; Weng, Q.; Thenkabail, P.S.; Senay, G.B. Remote Sensing Sensors and Applications in Environmental Resources Mapping and Modelling. *Sensors* **2007**, *7*, 3209–3241. [CrossRef] [PubMed]
2. Nieto, H.; Kustas, W.P.; Torres-Rúa, A.; Alfieri, J.G.; Gao, F.; Anderson, M.C.; White, W.A.; Song, L.; Alsina, M.D.M.; Prueger, J.H.; et al. Evaluation of TSEB turbulent fluxes using different methods for the retrieval of soil and canopy component temperatures from UAV thermal and multispectral imagery. *Irrig. Sci.* **2018**, *37*, 389–406. [CrossRef]
3. Cammalleri, C.; Anderson, M.; Gao, F.; Hain, C.R.; Kustas, W.P. A data fusion approach for mapping daily evapotranspiration at field scale. *Water Resour. Res.* **2013**, *49*, 4672–4686. [CrossRef]
4. McKee, M.; Torres-Rua, A.F.; Aboutalebi, M.; Nassar, A.; Coopmans, C.; Kustas, W.P.; Gao, F.; Dokoozlian, N.; Sanchez, L.; Alsina, M.M. Challenges that beyond-visual-line-of-sight technology will create for UAS-based remote sensing in agriculture (Conference Presentation). In *Autonomous Air and Ground Sensing Systems for Agricultural Optimization and Phenotyping IV*; SPIE-The International Society for Optical Engineering: Bellingham, WA, USA, 2019; Volume 11008, p. 110080J.
5. Torres-Rue, A.F.; Aboutalebi, M.; Wright, T.; Nassar, A.; Guillevic, P.; Hipps, L.; Gao, F.; Jim, K.; Alsina, M.; Coopmans, C.; et al. Estimation of Surface Thermal Emissivity In A Vineyard For UAV Microbolometer Thermal Cameras Using NASA Hytes Hyperspectral Thermal, And Landsat And Aggieair Optical Data. In *Autonomous Air and Ground Sensing Systems for Agricultural Optimization and Phenotyping IV*; SPIE: Baltimore, MD, USA, 2019; pp. 15–17.
6. Nassar, A.; Nieto, H.; Aboutalebi, M.; TorresRue, A.F.; McKee, M.; Kustas, W.P.; Prueger, J.H.; McKee, L.; Alfieri, J.G.; Hipps, L.; et al. *Pixel Resolution Sensitivity Analysis for the Estimation of Evapotranspiration Using the Two Source Energy Balance Model and sUAS Imagery under Agricultural Complex Canopy Environments*; American Geophysical Union (AGU): Washington, DC, USA, 2018.
7. Nassar, A.; Torres-Rue, A.F.; McKee, M.; Kustas, W.P.; Coopmans, C.; Nieto, H.; Hipps, L. *Assessment of UAV Flight Times for Estimation of Daily High Resolution Evapotranspiration in Complex Agricultural Canopy Environments*; Universities Council in Water Resources (UCOWR): Snowbird, UT, USA, 2019.

8. Wu, H.; Li, Z.-L. Scale Issues in Remote Sensing: A Review on Analysis, Processing and Modeling. *Sensors* **2009**, *9*, 1768–1793. [CrossRef]
9. El Maayar, M.; Chen, J.M. Spatial scaling of evapotranspiration as affected by heterogeneities in vegetation, topography, and soil texture. *Remote. Sens. Environ.* **2006**, *102*, 33–51. [CrossRef]
10. Miu, M.; Zhang, X.; Dewan, M.A.A.; Wang, J. Aggregation and Visualization of Spatial Data with Application to Classification of Land Use and Land Cover. *Geoinformatics Geostat. Overv.* **2017**, *5*. [CrossRef]
11. Brunsell, N.; Gillies, R. Scale issues in land–atmosphere interactions: Implications for remote sensing of the surface energy balance. *Agric. For. Meteorol.* **2003**, *117*, 203–221. [CrossRef]
12. Giorgi, F. An Approach for the Representation of Surface Heterogeneity in Land Surface Models. Part I.; Theoretical Framework. *Mon. Weather Rev.* **1997**, *125*, 1885–1899. [CrossRef]
13. Sharma, V.; Kilic, A.; Irmak, S. Impact of scale/resolution on evapotranspiration from Landsat and MODIS images. *Water Resour. Res.* **2016**, *52*, 1800–1819. [CrossRef]
14. Ershadi, A.; McCabe, M.; Evans, J.; Walker, J. Effects of spatial aggregation on the multi-scale estimation of evapotranspiration. *Remote. Sens. Environ.* **2013**, *131*, 51–62. [CrossRef]
15. Moran, M.S.; Humes, K.S.; Pinter, P.J. The scaling characteristics of remotely-sensed variables for sparsely-vegetated heterogeneous landscapes. *J. Hydrol.* **1997**, *190*, 337–362. [CrossRef]
16. Kustas, W.; Norman, J. Evaluating the Effects of Subpixel Heterogeneity on Pixel Average Fluxes. *Remote Sens. Environ.* **2000**, *74*, 327–342. [CrossRef]
17. Verburg, P.H.; Crossman, N.; Ellis, E.C.; Heinimann, A.; Hostert, P.; Mertz, O.; Nagendra, H.; Sikor, T.; Erb, K.-H.; Golubiewski, N.; et al. Land system science and sustainable development of the earth system: A global land project perspective. *Anthropocene* **2015**, *12*, 29–41. [CrossRef]
18. Long, D.; Singh, V.P.; Li, Z. How sensitive is SEBAL to changes in input variables, domain size and satellite sensor? *J. Geophys. Res. Space Phys.* **2011**, *116*, 116. [CrossRef]
19. Goodchild, M.F.; Gopal, S. *The Accuracy of Spatial Databases*; CRC Press: Boca Raton, FL, USA, 1989.
20. Fritz, S.; See, L.; Rembold, F. Comparison of global and regional land cover maps with statistical information for the agricultural domain in Africa. *Int. J. Remote Sens.* **2010**, *31*, 2237–2256. [CrossRef]
21. Bian, L.; Butler, R. Comparing effects of aggregation methods on statistical and spatial properties of simulated spatial data. *Photogramm. Eng. Remote Sens.* **1999**, *65*, 73–84.
22. Hong, S.-H.; Hendrickx, J.M.; Borchers, B. Up-scaling of SEBAL derived evapotranspiration maps from Landsat (30m) to MODIS (250m) scale. *J. Hydrol.* **2009**, *370*, 122–138. [CrossRef]
23. Singh, R.K.; Senay, G.B.; Velpuri, N.M.; Bohms, S.; Verdin, J.P. On the Downscaling of Actual Evapotranspiration Maps Based on Combination of MODIS and Landsat-Based Actual Evapotranspiration Estimates. *Remote Sens.* **2014**, *6*, 10483–10509. [CrossRef]
24. Li, X.; Liu, S.; Li, H.; Ma, Y.; Wang, J.; Zhang, Y.; Xu, Z.; Xu, T.; Song, L.; Yang, X.; et al. Intercomparison of Six Upscaling Evapotranspiration Methods: From Site to the Satellite Pixel. *J. Geophys. Res. Atmos.* **2018**, *123*, 6777–6803. [CrossRef]
25. French, A.N.; Hunsaker, D.J.; Thorp, K.R. Remote sensing of evapotranspiration over cotton using the TSEB and METRIC energy balance models. *Remote Sens. Environ.* **2015**, *158*, 281–294. [CrossRef]
26. Su, Z.; Pelgrum, H.; Menenti, M. Aggregation effects of surface heterogeneity in land surface processes. *Hydrol. Earth Syst. Sci.* **1999**, *3*, 549–563. [CrossRef]
27. Guzinski, R.; Nieto, H. Evaluating the feasibility of using Sentinel-2 and Sentinel-3 satellites for high-resolution evapotranspiration estimations. *Remote Sens. Environ.* **2019**, *221*, 157–172. [CrossRef]
28. Niu, H.; Zhao, T.; Wang, D.; Chen, Y. Evapotranspiration Estimation with UAVs in Agriculture: A Review. *Preprints* **2019**. [CrossRef]
29. Mauser, W.; Schädlich, S. Modelling the spatial distribution of evapotranspiration on different scales using remote sensing data. *J. Hydrol.* **1998**, *212*, 250–267. [CrossRef]
30. The California Garden Web, University of California. Available online: http://cagardenweb.ucanr.edu/Growing_Grapes_in_the_California_Garden/?uid=1&ds=436 (accessed on 16 September 2019).
31. Growing Fruit Trees in Maine, the University of Maine. Available online: https://extension.umaine.edu/fruit/growing-fruit-trees-in-maine/spacing/ (accessed on 1 October 2019).
32. Norman, J.; Kustas, W.; Humes, K. Source approach for estimating soil and vegetation energy fluxes in observations of directional radiometric surface temperature. *Agric. For. Meteorol.* **1995**, *77*, 263–293. [CrossRef]

33. Chirouze, J.; Boulet, G.; Jarlan, L.; Fieuzal, R.; Rodríguez, J.C.; Ezzahar, J.; Er-Raki, S.; Bigeard, G.; Merlin, O.; Garatuza-Payan, J.; et al. Intercomparison of four remote-sensing-based energy balance methods to retrieve surface evapotranspiration and water stress of irrigated fields in semi-arid climate. *Hydrol. Earth Syst. Sci.* **2014**, *18*, 1165–1188. [CrossRef]
34. Alfieri, J.G.; Kustas, W.P.; Nieto, H.; Prueger, J.H.; Hipps, L.E.; McKee, L.G.; Gao, F.; Los, S. Influence of wind direction on the surface roughness of vineyards. *Irrig. Sci.* **2019**, *37*, 359–373. [CrossRef]
35. Kustas, W.P.; Alfieri, J.G.; Nieto, H.; Wilson, T.G.; Gao, F.; Anderson, M.C. Utility of the two-source energy balance (TSEB) model in vine and interrow flux partitioning over the growing season. *Irrig. Sci.* **2019**, *37*, 375–388. [CrossRef]
36. Bigeard, G.; Coudert, B.; Chirouze, J.; Er-Raki, S.; Boulet, G.; Ceschia, E.; Jarlan, L. Evapotranspiration monitoring based on thermal infrared data over agricultural landscapes: Comparison of a simple energy budget model and a SVAT model. *Hydrol. Earth Syst. Sci. Discuss.* **2018**, *2018*, 1–44.
37. Kustas, W.; Anderson, M. Advances in thermal infrared remote sensing for land surface modeling. *Agric. For. Meteorol.* **2009**, *149*, 2071–2081. [CrossRef]
38. Yang, Y.; Qiu, J.; Zhang, R.; Huang, S.; Chen, S.; Wang, H.; Luo, J.; Fan, Y. Intercomparison of Three Two-Source Energy Balance Models for Partitioning Evaporation and Transpiration in Semiarid Climates. *Remote Sens.* **2018**, *10*, 1149. [CrossRef]
39. Andreu, A.; Kustas, W.P.; Polo, M.J.; Carrara, A.; González-Dugo, M.P. Modeling Surface Energy Fluxes over a Dehesa (Oak Savanna) Ecosystem Using a Thermal Based Two-Source Energy Balance Model (TSEB) I. *Remote Sens.* **2018**, *10*, 567. [CrossRef]
40. Yao, W.; Han, M.; Xu, S. Estimating the regional evapotranspiration in Zhalong wetland with the Two-Source Energy Balance (TSEB) model and Landsat7/ETM+ images. *Ecol. Inform.* **2010**, *5*, 348–358. [CrossRef]
41. Anderson, M.; Gao, F.; Knipper, K.; Hain, C.; Dulaney, W.; Baldocchi, D.; Eichelmann, E.; Hemes, K.; Yang, Y.; Medellin-Azuara, J.; et al. Field-Scale Assessment of Land and Water Use Change over the California Delta Using Remote Sensing. *Remote Sens.* **2018**, *10*, 889. [CrossRef]
42. Kustas, W.P.; Norman, J.M. A two-source approach for estimating turbulent fluxes using multiple angle thermal infrared observations. *Water Resour. Res.* **1997**, *33*, 1495–1508. [CrossRef]
43. Norman, J.M.; Kustas, W.P.; Prueger, J.H.; Diak, G.R. Surface flux estimation using radiometric temperature: A dual-temperature-difference method to minimize measurement errors. *Water Resour. Res.* **2000**, *36*, 2263–2274. [CrossRef]
44. Gao, F.; Kustas, W.P.; Anderson, M.C. A Data Mining Approach for Sharpening Thermal Satellite Imagery over Land. *Remote Sens.* **2012**, *4*, 3287–3319. [CrossRef]
45. Zheng, G.; Moskal, L.M. Retrieving Leaf Area Index (LAI) Using Remote Sensing: Theories, Methods and Sensors. *Sensors* **2009**, *9*, 2719–2745. [CrossRef]
46. White, W.A.; Alsina, M.M.; Nieto, H.; McKee, L.G.; Gao, F.; Kustas, W.P. Determining a robust indirect measurement of leaf area index in California vineyards for validating remote sensing-based retrievals. *Irrig. Sci.* **2019**, *79*, 269–280. [CrossRef]
47. Liu, K.; Zhou, Q.-B.; Wu, W.-B.; Xia, T.; Tang, H.-J. Estimating the crop leaf area index using hyperspectral remote sensing. *J. Integr. Agric.* **2016**, *15*, 475–491. [CrossRef]
48. Johnson, L.F. Temporal stability of an NDVI-LAI relationship in a Napa Valley vineyard. *Aust. J. Grape Wine Res.* **2003**, *9*, 96–101. [CrossRef]
49. Dobrowski, S.; Ustin, S.; Wolpert, J. Remote estimation of vine canopy density in vertically shoot-positioned vineyards: Determining optimal vegetation indices. *Aust. J. Grape Wine Res.* **2002**, *8*, 117–125. [CrossRef]
50. Fang, H. Retrieving leaf area index using a genetic algorithm with a canopy radiative transfer model. *Remote Sens. Environ.* **2003**, *85*, 257–270. [CrossRef]
51. Sun, L.; Gao, F.; Anderson, M.C.; Kustas, W.P.; Alsina, M.M.; Sanchez, L.; Sams, B.; McKee, L.; Dulaney, W.; White, W.A.; et al. Daily Mapping of 30 m LAI and NDVI for Grape Yield Prediction in California Vineyards. *Remote Sens.* **2017**, *9*, 317. [CrossRef]
52. Kimes, D.; Knyazikhin, Y.; Privette, J.; Abuelgasim, A.; Gao, F. Inversion methods for physically-based models. *Remote Sens. Rev.* **2000**, *18*, 381–439. [CrossRef]
53. Gonsamo, A.; Pellikka, P. The sensitivity based estimation of leaf area index from spectral vegetation indices. *ISPRS J. Photogramm. Remote Sens.* **2012**, *70*, 15–25. [CrossRef]

54. Kustas, W.P.; Norman, J.M. Evaluation of soil and vegetation heat flux predictions using a simple two-source model with radiometric temperatures for partial canopy cover. *Agric. For. Meteorol.* **1999**, *94*, 13–29. [CrossRef]
55. Nieto, H.; Kustas, W.P.; Alfieri, J.G.; Gao, F.; Hipps, L.E.; Los, S.; Prueger, J.H.; McKee, L.G.; Anderson, M.C. Impact of different within-canopy wind attenuation formulations on modelling sensible heat flux using TSEB. *Irrig. Sci.* **2019**, *37*, 315–331. [CrossRef]
56. Kustas, W.P.; Anderson, M.C.; Alfieri, J.G.; Knipper, K.; Torres-Rua, A.; Parry, C.K.; Nieto, H.; Agam, N.; White, W.A.; Gao, F.; et al. The Grape Remote Sensing Atmospheric Profile and Evapotranspiration Experiment. *Bull. Am. Meteorol. Soc.* **2018**, *99*, 1791–1812. [CrossRef]
57. Torres-Rua, A. Vicarious Calibration of sUAS Microbolometer Temperature Imagery for Estimation of Radiometric Land Surface Temperature. *Sensors* **2017**, *17*, 1499. [CrossRef]
58. McKee, M.; Nassar, A.; Torres-Rua, A.F.; Aboutalebi, M.; Kustas, W. Implications of sensor inconsistencies and remote sensing error in the use of small unmanned aerial systems for generation of information products for agricultural management. In *Autonomous Air and Ground Sensing Systems for Agricultural Optimization and Phenotyping III*; International Society for Optics and Photonics: Bellingham, WA, USA, 2018; Volume 10664, p. 1066402.
59. Alfieri, J.G.; Kustas, W.P.; Prueger, J.H.; McKee, L.G.; Hipps, L.E.; Gao, F. A multi-year intercomparison of micrometeorological observations at adjacent vineyards in California's Central Valley during GRAPEX. *Irrig. Sci.* **2019**, *37*, 345–357. [CrossRef]
60. Agam, N.; Kustas, W.P.; Alfieri, J.G.; Gao, F.; McKee, L.M.; Prueger, J.H.; Hipps, L.E. Micro-scale spatial variability in soil heat flux (SHF) in a wine-grape vineyard. *Irrig. Sci.* **2019**, *37*, 253–268. [CrossRef]
61. Kljun, N.; Calanca, P.; Rotach, M.W.; Schmid, H.P. A simple two-dimensional parameterisation for Flux Footprint Prediction (FFP). *Geosci. Model Dev.* **2015**, *8*, 3695–3713. [CrossRef]
62. Hassan-Esfahani, L.; Ebtehaj, A.M.; Torres-Rua, A.; McKee, M. Spatial Scale Gap Filling Using an Unmanned Aerial System: A Statistical Downscaling Method for Applications in Precision Agriculture. *Sensors* **2017**, *17*, 2106. [CrossRef] [PubMed]
63. Moorhead, J.E.; Marek, G.W.; Colaizzi, P.D.; Gowda, P.H.; Evett, S.R.; Brauer, D.K.; Marek, T.H.; Porter, D.O. Evaluation of Sensible Heat Flux and Evapotranspiration Estimates Using a Surface Layer Scintillometer and a Large Weighing Lysimeter. *Sensors* **2017**, *17*, 2350. [CrossRef]
64. Imukova, K.; Ingwersen, J.; Hevart, M.; Streck, T. Energy balance closure on a winter wheat stand: Comparing the eddy covariance technique with the soil water balance method. *Biogeosci. Discuss.* **2015**, *12*, 6783–6820. [CrossRef]
65. Mauder, M.; Genzel, S.; Fu, J.; Kiese, R.; Soltani, M.; Steinbrecher, R.; Zeeman, M.; Banerjee, T.; De Roo, F.; Kunstmann, H. Evaluation of energy balance closure adjustment methods by independent evapotranspiration estimates from lysimeters and hydrological simulations. *Hydrol. Process.* **2017**, *32*, 39–50. [CrossRef]
66. Twine, T.; Kustas, W.; Norman, J.; Cook, D.; Houser, P.; Meyers, T.; Prueger, J.; Starks, P.; Wesely, M. Correcting eddy-covariance flux underestimates over a grassland. *Agric. For. Meteorol.* **2000**, *103*, 279–300. [CrossRef]
67. Al-Juaidi, A.E.M.; Nassar, A.M.; Al-Juaidi, O.E.M. Evaluation of flood susceptibility mapping using logistic regression and GIS conditioning factors. *Arab. J. Geosci.* **2018**, *11*, 765. [CrossRef]
68. Sun, H.; Yang, Y.; Wu, R.; Gui, D.; Xue, J.; Liu, Y.; Yan, D. Improving Estimation of Cropland Evapotranspiration by the Bayesian Model Averaging Method with Surface Energy Balance Models. *Atmosphere* **2019**, *10*, 188. [CrossRef]
69. Song, L.; Liu, S.; Kustas, W.P.; Zhou, J.; Xu, Z.; Xia, T.; Li, M. Application of remote sensing-based two-source energy balance model for mapping field surface fluxes with composite and component surface temperatures. *Agric. For. Meteorol.* **2016**, *230*, 8–19. [CrossRef]
70. Kustas, W.P.; Nieto, H.; Morillas, L.; Anderson, M.C.; Alfieri, J.G.; Hipps, L.E.; Villagarcía, L.; Domingo, F.; Garcia, M. Revisiting the paper "Using radiometric surface temperature for surface energy flux estimation in Mediterranean drylands from a two-source perspective". *Remote Sens. Environ.* **2016**, *184*, 645–653. [CrossRef]
71. Carlson, T. An Overview of the "Triangle Method" for Estimating Surface Evapotranspiration and Soil Moisture from Satellite Imagery. *Sensors* **2007**, *7*, 1612–1629. [CrossRef]

72. Hardin, P.J.; Jensen, R.R. Neural Network Estimation of Urban Leaf Area Index. *GIScience Remote Sens.* **2005**, *42*, 251–274. [CrossRef]
73. Timmermans, W.J.; Kustas, W.P.; Anderson, M.C.; French, A.N. An intercomparison of the Surface Energy Balance Algorithm for Land (SEBAL) and the Two-Source Energy Balance (TSEB) modeling schemes. *Remote Sens. Environ.* **2007**, *108*, 369–384. [CrossRef]

Article

A Comparative Study of RGB and Multispectral Sensor-Based Cotton Canopy Cover Modelling Using Multi-Temporal UAS Data

Akash Ashapure [1], Jinha Jung [1,*], Anjin Chang [2], Sungchan Oh [1], Murilo Maeda [3] and Juan Landivar [4]

1 Lyles School of Civil Engineering, Purdue University, West Lafayette, IN 47907, USA; aashapur@purdue.edu (A.A.); oh231@purdue.edu (S.O.)
2 School of Engineering & Computing Science, Texas A&M University– Corpus Christi, Corpus Christi, TX 78412, USA; anjin.chang@tamucc.edu
3 Texas A&M AgriLife Extension, Lubbock, TX 79403, USA; mmaeda@ag.tamu.edu
4 Texas A&M AgriLife Research, Corpus Christi, TX 78406, USA; jalandivar@ag.tamu.edu
* Correspondence: jinha@purdue.edu; Tel.: +1-765-496-1267

Received: 22 October 2019; Accepted: 21 November 2019; Published: 23 November 2019

Abstract: This study presents a comparative study of multispectral and RGB (red, green, and blue) sensor-based cotton canopy cover modelling using multi-temporal unmanned aircraft systems (UAS) imagery. Additionally, a canopy cover model using an RGB sensor is proposed that combines an RGB-based vegetation index with morphological closing. The field experiment was established in 2017 and 2018, where the whole study area was divided into approximately 1 x 1 m size grids. Grid-wise percentage canopy cover was computed using both RGB and multispectral sensors over multiple flights during the growing season of the cotton crop. Initially, the normalized difference vegetation index (NDVI)-based canopy cover was estimated, and this was used as a reference for the comparison with RGB-based canopy cover estimations. To test the maximum achievable performance of RGB-based canopy cover estimation, a pixel-wise classification method was implemented. Later, four RGB-based canopy cover estimation methods were implemented using RGB images, namely Canopeo, the excessive greenness index, the modified red green vegetation index and the red green blue vegetation index. The performance of RGB-based canopy cover estimation was evaluated using NDVI-based canopy cover estimation. The multispectral sensor-based canopy cover model was considered to be a more stable and accurately estimating canopy cover model, whereas the RGB-based canopy cover model was very unstable and failed to identify canopy when cotton leaves changed color after canopy maturation. The application of a morphological closing operation after the thresholding significantly improved the RGB-based canopy cover modeling. The red green blue vegetation index turned out to be the most efficient vegetation index to extract canopy cover with very low average root mean square error (2.94% for the 2017 dataset and 2.82% for the 2018 dataset), with respect to multispectral sensor-based canopy cover estimation. The proposed canopy cover model provides an affordable alternate of the multispectral sensors which are more sensitive and expensive.

Keywords: precision agriculture; canopy cover; UAS; image analysis; multispectral; crop mapping

---

1. Introduction

Numerous studies are being conducted on cotton crop growth monitoring for precision agriculture. Cotton is an important crop in the state of Texas, which produces more than 50% of the total cotton produced by the entire country, comprising a spatial coverage of around six million acres [1]. Recent advances in genetic engineering and genomics have significantly accelerated the breeding process of cotton [2]. There is a growing need for phenotyping to match this high pace breeding process. Consequently, plant breeders and agriculture scientists have recognized the need for a high-throughput phenotyping (HTP) system that can efficiently measure phenotypic traits such as crop height, volume, canopy cover, and vegetation indices (VIs) with reasonable accuracy [3]. An accurate phenotyping process is very critical for the reliable quantification of phenotypical traits to select the genotypes of interest. HTP is an extensively discussed phenomenon; however, until recently, its implementation has been rather fragmentary [4]. The change in this situation has been mainly attributed to the recent developments in unmanned aircraft systems (UAS). Lightweight platforms combined with consumer grade imaging sensors have provided an affordable system to perform the necessary remote sensing activities for precision agriculture, especially with low altitude flights that provide high temporal and spatial resolution data [5–8].

In this paper, canopy cover (CC), which is commonly expressed as the percentage of total ground areal coverage by the vertical projection of plant canopy, is studied. Plant canopy cover is strongly related to crop growth, development, water use, and photosynthesis, which makes it an important trait to be observed throughout the growing season [9]. In addition, CC is an important ancillary variable in the estimation of the leaf area index (LAI) [10]. Various remote sensing techniques have been employed in the literature to compute CC, and these include satellite imagery with varying degree of resolutions [11–15], airborne imagery [16] and light detection and ranging (LiDAR) data [17,18]. Satellite imagery has the advantage of providing large spatial coverage. However, coarser spatial resolution limits its application in computing CC over small breeding fields where genotype screening is the objective. Moreover, the temporal resolution of satellite imagery is also not enough for phenotypic applications. Furthermore, satellite imagery is highly affected by cloud cover and other atmospheric conditions [19]. On the other hand, aerial imagery usually has a higher spatial resolution, but it has fewer spectral bands as compared to satellite imagery [20]. CC estimation using LiDAR data can be slightly biased in visual interpretation; however, in general, it is particularly useful in the estimation of vertical canopy cover and angular canopy closure which is otherwise difficult to compute [21]. Terrestrial and airborne LiDAR data have been successfully used to compute CC in the literature [22,23]. However, data collection frequency has remained a significant issue, as LiDAR sensors and airborne imaging sensors are relatively expensive compared to UAS. Recently, UAS have emerged as an alternate to the satellite, airborne imaging sensors or LiDAR sensors to estimate CC, and this approach is more affordable and could provide higher temporal and spatial resolution [24–28]. UAS-based CC measurements have been efficiently used to estimate LAI [29,30] and have been used as one of the comparison parameters to quantify the difference between various crop management practices throughout the growing season [31]. Moreover, a recent study conducted over maize field indicated that UAS-based CC is significantly correlated with the grain yield [32].

CC computation using multispectral (MS) sensors has gained more popularity over RGB(red, green, and blue)-based CC, with the primary reason being that the MS sensor is more stable over time and remains relatively unaffected by changes in environmental conditions (e.g., sunlight angle and cloud cover) throughout the crop growing season due to its irradiance sensor [3,7,33,34]. However, MS sensors are more sensitive and expensive compared to RGB sensors. RGB-based CC estimation methods can be divided into two categories, namely the thresholding method and the pixel classification method. Thresholding methods require the specification of the color thresholds or the ratios to identify canopy pixels. Pixel classification methods use a supervised or unsupervised pixel-wise classification method to identify canopy pixels. Though pixel classification methods are highly accurate, they are time consuming and computationally extensive. Supervised classification methods require training samples to be collected, which is expensive and prone to human error. However, pixel classification methods are particularly useful to calibrate thresholding methods [35]. There is an ample amount of work in the literature that has used RGB sensors to compute CC. Early work in this direction includes the quantification of turfgrass cover using digital image analysis by Richarson et al., (2001) [36]. Lee and Lee, (2011), estimated canopy cover over the rice field using an RGB sensor [37]. Patrignani and Ochsner,

(2015), developed the Canopeo algorithm to extract fractional green canopy cover [38]. Despite having a significant amount of previous literature exploring RGB-based CC estimation, there is a scarcity of work that compares different CC estimations throughout the crop growing season. Torres-Sánchez et al., (2014), [39] developed a multitemporal CC framework for a wheat field using UAS-based RGB images. However, it was limited to early season CC estimation only. Moreover, the highest accuracy that they achieved in mapping CC was less than 92%. Fang et al., (2016), [40] presented a case study of CC estimation using UAS-based MS sensor data over an oilseed rape. However, their study was aimed to provide CC estimation and flower fraction for the crop species that have conspicuous non-green flowers or fruits. Moreover, they primarily used MS sensor-based CC estimation methodology in their study, with only one RGB-based CC estimation approach that only worked efficiently during the vegetative period. Marcial-Pablo et al., (2019), [41] compared CC estimation using RGB and MS sensor-based vegetation indices over a maize field. Their results suggested that RGB-based CC estimation can be useful in the early-season growth stage of the crop, while later in the season CC estimation, using MS sensor-based indices was more accurate. Moreover, the accuracy of the CC estimation was also dependent on automatic thresholding using the Otsu method. Lima-Cueto et al., (2019), [42] used 11 VIs to quantify vegetation cover in olive groves, and they suggested that MS sensor-based CC had better accuracy as compared to RGB-based CC. A consistent observation in the aforementioned case studies was that RGB-based CC estimation was not efficient in the late season. Therefore, the objective of this study was not only to compare various RGB-based CC estimation methods with MS sensor-based CC estimation but also to improve RGB-based CC estimation to provide a more affordable option to breeders and agriculture scientists, particularly in late season.

2. Materials and Methods

2.1. Study Area and Sensors

A field experiment was established at the Texas A&M AgriLife Research and Extension Center in Corpus Christi, TX (Latitude 27°46′59″ N and longitude 97°34′13″ W). The trial consisted of 5 cotton genotypes from the Texas A&M AgriLife Cotton Breeding Program (presented in Figure 1). Genotypes were planted 22 March 2017 in skip and solid row patterns (i.e., one- or two-row plots, respectively), and each was replicated four times. For canopy cover estimation, another field experiment was established at the same location in 2018. The trial consisted of 10 cotton genotypes from the Texas A&M AgriLife Cotton Breeding Program. Genotypes were planted in the first week of April in skip and solid row patterns. To maintain the integrity of the experiment, only part of the field which is highlighted by yellow boxes in Figure 1, was considered for both 2017 and 2018, as the selected area was the only area which had an alternate pattern of skip and non-skip rows and a common variety. The selected area was divided into 1 x 1 m size grids. The number of grids in 2017 and 2018 experiments was 300 and 600, respectively.

RGB and MS sensors were used for this study, as presented in Figure 2. A DJI Phantom 4 Pro (SZ DJI Technology Co., Ltd., Shenzhen, China) was used for RGB data collection. It comprised a 3-axis gimbal-stabilized RGB sensor with a resolution of 20 mega pixels. MS data were collected using the DJI Matrice 100 platform (SZ DJI Technology Co., Ltd., Shenzhen, China) with a SlantRange 3p sensor (Slantrange Inc, San Diego, CA) equipped with integrated solar spectrometer for frame-to-frame, radiometrically accurate reflectance measurements. The sensor has a spatial resolution of 4.8 cm/pixel at 120 m above ground level. It collects data using four spectral bands, namely green, red, red-edge, and near infrared bands, for which the peak wavelengths are presented in Table 1.

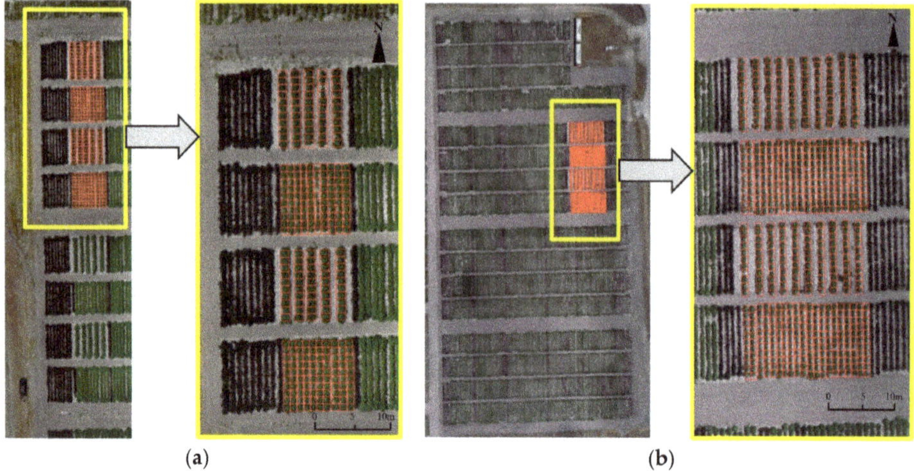

Figure 1. Experimental field setup consisted of cotton in skip and solid row patterns in: (**a**) 2017 and (**b**) 2018. The experimental field setup is presented with an RGB (red, green, and blue) orthomosaic of the study area on June 7, 2017, and June 6, 2018.

Figure 2. RGB and multispectral sensors used for data collection. (**a**) The DJI Phantom 4 Pro for RGB and (**b**) the DJI Matrice 100 platform with the SlantRange 3p sensor for multispectral data collection.

Table 1. Peak wavelength and FWHM (full width at half maximum) for SlantRange 3p sensor bands.

SlantRange 3p Sensor Band	Peak Wavelength (nm)	FWHM (nm)
Green	560	40
Red	655	35
Red-edge	710	20
Near infrared	830	110

2.2. Data Collection and Preprocessing

Data collection and preprocessing was followed from the method of Ashapure et al. (2019) [31]. UAS data (both MS and RGB) were collected over the experimental field on a weekly basis. Table 2 presents the flight specifications for both RGB and MS data collection. A total of eleven and ten flights were conducted in 2017 and 2018, respectively, using RGB and MS sensors. The overlap for MS sensor data collection was 70%, and it was 80%–85% for RGB sensor data collection. In 2017, the altitude was about 20 m for the RGB sensor and 25 m for the MS sensor. In 2018, the flight altitude for RGB and MS sensors was 35 and 47 m, respectively. Provided the experimental field was in a coastal area, wind speed and rain were the potential factors to be considered before every flight. Most flights were conducted between 10:00AM and 2:00PM, except under unfavorable weather conditions, such as a wind speed greater than 15 mph or raining. Moreover, the temperature variation throughout the growing season varied between 79 and 96 °F.

Table 2. UAS data collection timeline and sensor-wise flight specification.

Date	Flight Altitude (m)		Overlap (%)		Spatial Resolution (cm)	
	RGB	Multispectral	RGB	Multispectral	RGB	Multispectral
24 April 2017	20	30	85	75	0.51	0.93
5 May 2017	20	25	85	70	0.50	0.85
12 May 2017	20	25	85	70	0.51	0.81
20 May 2017	20	25	85	70	0.52	0.82
30 May 2017	20	25	85	70	0.51	0.85
7 June 2017	20	25	85	70	0.51	0.83
19 June 2017	20	25	85	70	0.52	0.81
5 July 2017	20	25	85	70	0.51	0.81
10 July 2017	20	25	85	70	0.50	0.83
18 July 2017	20	25	85	70	0.51	0.82
23 July 2017	20	25	85	70	0.51	0.82
23 April 2018	35	47	80	70	0.73	1.61
7 May 2018	35	47	80	70	0.69	1.65
14 May 2018	35	47	80	70	0.71	1.61
23 May 2018	35	47	80	70	0.71	1.64
1 June 2018	37	47	80	70	0.73	1.62
6 June 2018	35	47	80	70	0.72	1.61
13 June 2018	35	47	80	70	0.71	1.63
3 July 2018	35	47	80	70	0.71	1.61
9 July 2018	35	47	80	70	0.72	1.63
19 July 2018	35	47	80	70	0.70	1.62

Generally, UAS are equipped with a consumer grade global positioning system (GPS) that do not have satisfactory location accuracy for aerial mapping applications. To overcome this problem, semi-permanent ground control points (GCPs) with a high reflectance were installed over the study area. The GCPs were surveyed using a dual frequency, post processed kinematic (PPK) GPS system, 20 Hz V-Map Air model (Micro Aerial Project L.L.C., Gainesville, FL). Images obtained from the UAS platform with significant overlaps along with the 3D coordinates of the GCPs were imported to Agisoft Photoscan Pro (Agisoft LLC, St. Petersburg, Russia), which uses structure from motion (SfM) photogrammetry algorithms to derive high dens 3D point clouds, fine spatial resolution 2D orthomosaics, and digital surface models (DSM). The SfM refers to the process of finding the three-dimensional structure of an object by analyzing local motion signals over time [43].

2.3. Canopy Cover Computation

The percentage CC was computed as the ratio of canopy area to the total area of the grid computed using Equation (1), where GSD is ground sampling distance. An RGB orthomosaic image was converted into a binary image, where zero represents non-canopy pixels and one represents canopy pixels. As the whole field was divided into a square meter grid by using Equation (1), a grid-wise percentage CC was computed (As shown in Figure 3).

$$CC = \left(\frac{\Sigma\left(GCD^2\right)\ if\ Canopy\ Pixel}{\Sigma\left(GCD^2\right)} \right) \times 100, \tag{1}$$

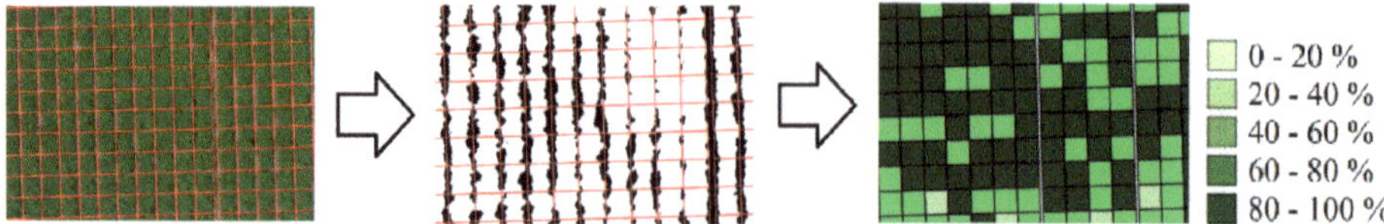

Figure 3. Canopy cover estimation from the orthomosaic images (Red square: individual crop grid, and each grid is 1 × 1 m): an RGB orthomosaic image collected using the unmanned aerial systems (UAS) platform, followed by the binary classification results of the orthomosaic image, where white represents the canopy class and black represents the non-canopy class; the last image represents the grid-wise estimated canopy cover (CC).

As mentioned earlier, MS sensor-based CC estimation is considered, in the literature, as the most reliable estimation technique that uses a normalized difference vegetation index (NDVI) to separate the canopy from the non-canopy areas (computed using Equation (2) [44]). SlantView, the software developed for the SlantRange 3p MS sensor, was used for the radiometric calibration in order to accurately compare the crop conditions across datasets collected in varying lighting conditions throughout the day and growing season. A detailed visual inspection was performed to find a threshold NDVI value to separate the canopy area from the non-canopy area in the image throughout the growing season regardless of the growth stage.

$$\text{NDVI} = \left(\frac{NIR - Red}{NIR + Red}\right), \tag{2}$$

To investigate the accurate CC estimation using the RGB-based sensor, which is equivalent to CC estimation using NDVI, a pixel-wise classification method was implemented; this is presented in Figure 4. As found in the literature, pixel classification methods are considered highly accurate for separating the canopy and non-canopy classes, and they are mainly used to calibrate RGB-based methods [35,38,45]. A pixel classification method based on K-means clustering was used to compare the RGB-based methods that use vegetation indices to separate canopy areas from non-canopy areas. Initially, K-means clustering with five classes was applied to the RGB orthomosaics, considering five potential classes representing soil, shadow, cotton bolls, green canopy and brown canopy, as presented in Figure 4. After assigning the class labels to the clustered map, it was validated using ground-truth samples collected over RGB orthomosaics using visual inspection, and the overall classification accuracy was found to be at least 97%. Later, soil, shadow and cotton bolls were merged and assigned as non-canopy, and green and brown canopy were merged and identified as canopy classes. However, as pixel-wise classification-based CC estimation is computationally extensive, its implementation was solely done with an intention to investigate the maximum achievable performance using the RGB-based sensors and to compare it with MS sensor-based CC estimation.

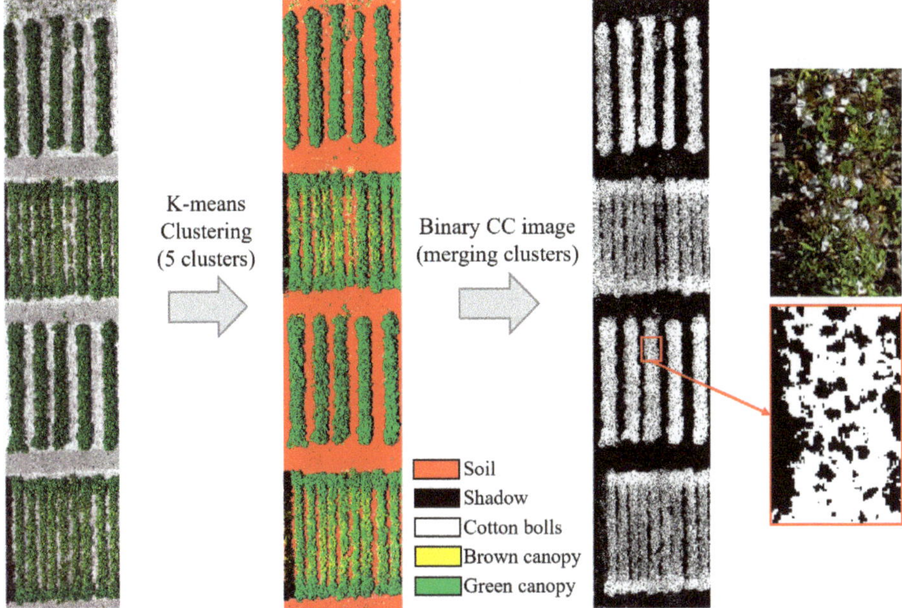

Figure 4. K-means clustering-based pixel classification method workflow where the orthomosaic is classified into five classes, and, later, classes are merged into two clusters, namely canopy and non-canopy. The RGB orthomosaic presented was captured on 19th June 2017.

In this study, four different RGB-based methods were used, namely Canopeo, the excessive greenness index (ExG), the modified green red vegetation index (MGRVI) and the red green blue vegetation index (RGBVI), to generate the binary images to separate canopy areas from non-canopy areas (Table 3).

Table 3. RGB image-based vegetation indices and their formula.

Vegetation Index	Formula	Reference
Canopeo	$canopy = (i_1 < \theta_2) \times (i_2 < \theta_1) \times (i_3 > \theta_3)$ $i_1 = \frac{red}{green}, i_2 = \frac{blue}{green}, \; i_3 = 2 \times green - blue - red$ $\theta_1 = 0.95, \; \theta_2 = 0.95, \; \theta_3 = 20$	[38]
ExG	$2G_n - R_n - B_n$ $R_n = \frac{R}{R+G+B}, G_n = \frac{G}{R+G+B}, B_n = \frac{B}{R+G+B}$	[46]
MGRVI	$\frac{G^2-R^2}{G^2+R^2}$	[47]
RGBVI	$\frac{G^2-R\times B}{G^2+R\times B}$	[47]

The overall procedure to compute the CC binary map using RGB vegetation indices is presented in Figure 5. The Canopeo algorithm resulted in a binary map that separated canopy areas from non-canopy areas; however, applying vegetation indices over the RGB mosaics resulted in a grayscale image. Similar to the NDVI, an empirical evaluation was performed over all the grey scale vegetation index maps to decide a threshold value for the ExG, the MGRVI and the RGBVI that could separate canopy areas from non-canopy areas. A detailed visual inspection was performed to determine a threshold value to separate canopy areas from non-canopy areas in the image throughout the growing season, regardless of the growth stage. Considering the homogeneity of the crop, only a subset of the image was used to determine the threshold, and the threshold chosen for each VI is presented in Table 4. The demonstration of visual inspection is presented in Figure 6, where a subset of an early stage and a mature stage RGB image of the same area is considered. Originally for all the RGB images in the growing season, a range of threshold values with a step size of 0.01 was applied to a subset area in the images to generate binary map using the grayscale VI map of the subset area. However, for the demonstration, only one VI (ExG) image was considered, and this was generated using one early stage (image taken on 7 June 2017) and one mature stage (image taken on 10 July 2017) RGB image with a range of threshold values and a step size of 0.02. It can be observed from Figure 6 that variation in threshold values did not affect the binarization of the early stage image, as most of the canopy was green. However with the higher threshold (0.22), the binary image had some canopy pixels not classified as canopy due to their darker color. The effect of variation in threshold was more significant in the mature stage image. A lower threshold value of 0.18 resulted in a lot of non-canopy pixels classified as canopy, especially the shadow pixels. A slightly higher threshold value of 0.22 resulted in more conservative classification that omitted a substantial number of canopy pixels. Visual inspection suggested that the threshold value of 0.2 resulted in a most appropriate classification for the ExG. Similarly, other VIs were also examined under visual inspection to select a single threshold value for all the images in the growing season.

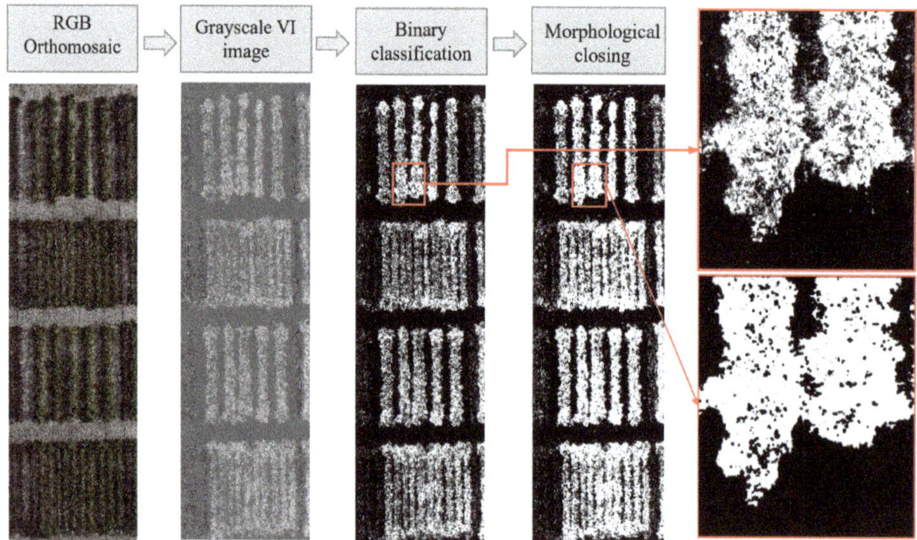

Figure 5. Procedure to generate a binary map that indicates canopy and non-canopy areas. Applying the vegetation index over the RGB orthomosaic resulted in a grayscale image. By applying thresholding, a binary image was generated. Lastly, morphological closing was applied over binary image to improve binary classification. The presented RGB orthomosaic was captured on 10 July 2017, and the excessive greenness index (ExG) was the VI used for the demonstration of the methodology.

Table 4. Threshold chosen for vegetation indices (Vis) to separate canopy and non-canopy areas.

VI	Threshold	Range
NDVI	0.6	0 to 1
ExG	0.2	−2 to 2
MGRVI	0.15	−1 to 1
RGBVI	0.15	−1 to 1

RGB-based vegetation indices accurately identified healthy green canopy; however, later in the season as the canopy started to change the color, their ability to identify canopy deteriorated. To further improve the binary map (indicating canopy and non-canopy areas), a morphological closing operation was performed. The morphological closing operation is a combination of dilation and erosion, and it helps to remove small holes while keeping the separation boundary intact [48]. For this experiment, a 3 × 3 kernel window over one iteration was used to perform closing operation.

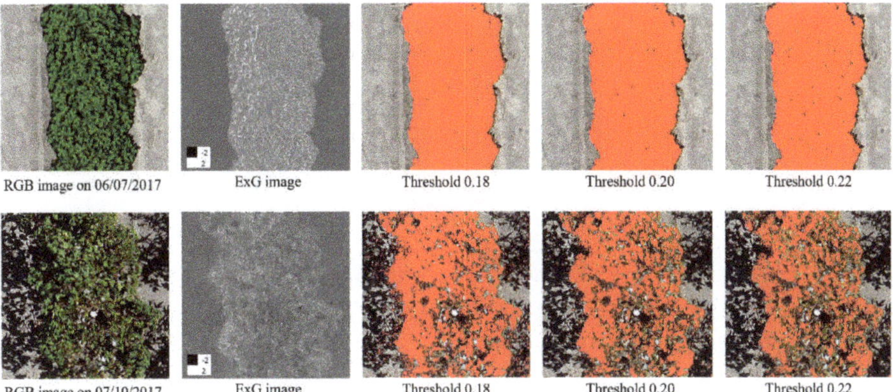

Figure 6. Procedure to select appropriate threshold value to generate a binary map that indicates canopy and non-canopy areas. The first images are a subset of RGB images captured on 7 June 2017 and 10 July 2017. Second images are the result of applying the vegetation index (ExG) over RGB images. The next three images are the result of applying varying threshold values that were superimposed on the original RGB images (the canopy classified pixels are represented by the red color, and non-canopy pixels were set to transparent).

3. Results

The CC grid maps at each flight for both 2017 and 2018 are presented in Figures 7 and 8, respectively. The visual inspection of the grid maps revealed that the percentage canopy cover increased in the growing season and reached its plateau right after the middle of the season (19 June for the 2017 experiment and 13 June for the 2018 experiment). Later in the season, it was observed that the percentage CC started to slightly decrease with the canopy senescence. For the 2017 experiment, a rapid decay in CC was observed between 18 July and 23 July due to a common practice in the cotton fields known as defoliation, which prepares the crop for harvesting. A similar effect was observed in the 2018 experiment between 9 July and 19 July.

Following the methodology presented in Figure 4 using RGB images, K-means clustering-based classification maps were generated considering five clusters which were later labeled to represent soil, cotton boll, shadow, green canopy, and brown canopy. Later, binary maps were generated by merging soil, cotton boll and shadow classes to indicate non-canopy, while brown canopy and green canopy classes were merged to indicate canopy pixels. The comparison of the average CC per grid using the NDVI and K-means (also referred as RGB reference) is presented in Figures 9 and 10 for the 2017 and 2018 experiments, respectively. From both the 2017 and 2018 experiments, it was observed that the NDVI-based and RGB reference-based average CC per grid followed the same trend throughout the growing season, and there was a one-to-one correspondence between the two when plotted as a straight line at an intercept of one with very high R^2 values (0.98 for 2017 and 0.97 for 2018). The results indicated that it is possible to achieve the same level of performance using the RGB-based sensor as that of the MS sensors using NDVI. However, the purpose of the RGB reference CC estimation was only to provide a comparison reference for the MS sensor-based CC estimation.

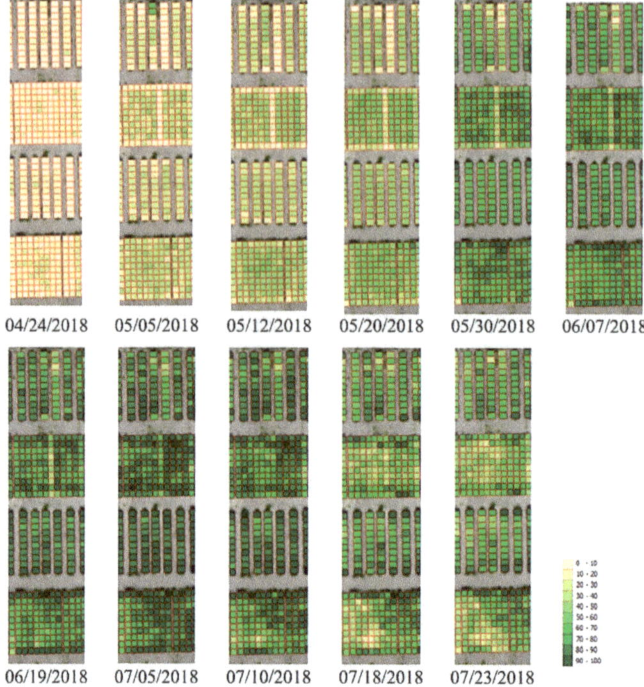

Figure 7. CC grid maps generated at each flight in growing season using normalized difference vegetation index (NDVI) maps for the 2017 dataset.

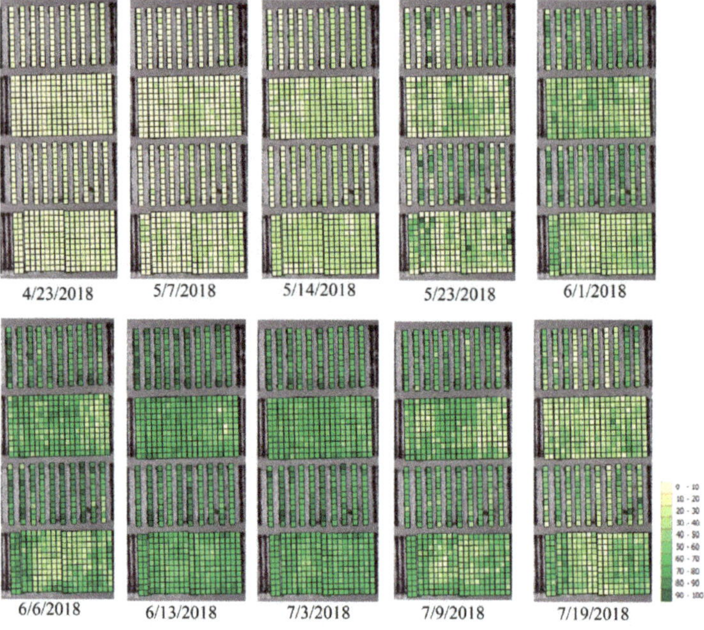

Figure 8. CC grid maps generated at each flight in growing season using NDVI maps for the 2018 dataset.

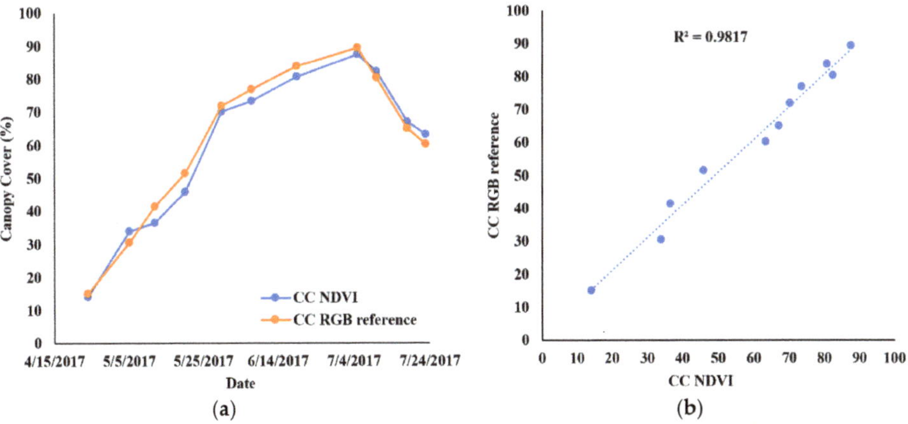

Figure 9. For the 2017 experiment: (**a**) the average NDVI and RGB reference-based percentage CC for each flight in the growing season; (**b**) a comparison of the NDVI and RGB reference-based percentage CC techniques with R^2.

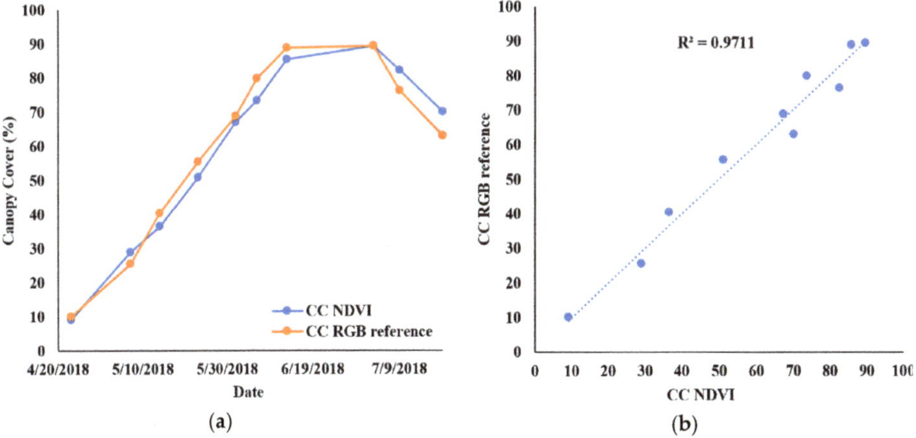

Figure 10. For the 2018 experiment: (**a**) the average NDVI and RGB reference-based percentage CC for each flight in the growing season; (**b**) a comparison of the NDVI and RGB reference-based percentage CC techniques with R^2.

Using RGB images, four thresholding-based CC estimation methods were implemented, namely Canopeo, the ExG, the MGRVI and the RGBVI. Along with the NDVI-based average CC estimation per grid, the average CC estimation per grid in the growing season for each RGB-based CC estimation method before and after applying morphological closing (MC) is presented in Figures 11 and 12 for the 2017 and 2018 experiments, respectively. It can be observed from Figure 11 that, early in the growing season, the average CC estimated by all the methods was in agreement with very less variation and followed the same increasing trend as the NDVI-based estimation. However, the Canopeo and ExG methods reached their peak early in the season (7th June), as compared to other methods. The main reason for this was that they were accurately identifying healthy green canopy, but later, as the canopy started to change color from green to yellow and eventually to brown, they were not as efficient as other methods to identify the matured canopy. Moreover, after they reached their peak, they rapidly started to decrease later in the season, commensurate with the rate of change of color in the canopy in the later season. As can be observed from Table 5, the average root mean square error (RMSE) of the percentage CC turned out to be highest amongst all (17.87 for Canopeo and 16.97 for the ExG). The MGRVI showed a slightly better performance over Canopeo and the ExG, and it was able to identify mature canopy. However, in comparison to the NDVI, it also reached its peak relatively early. The RGBVI turned out to be the most efficient method to estimate CC, as, especially later in the season, it outperformed the other RGB-based methods. However, the RGBVI still could not match the NDVI-based CC estimation. It was noticed that morphological closing significantly improved the CC estimation, and the average RMSE with the NDVI-based CC estimation was substantially reduced (Table 5).

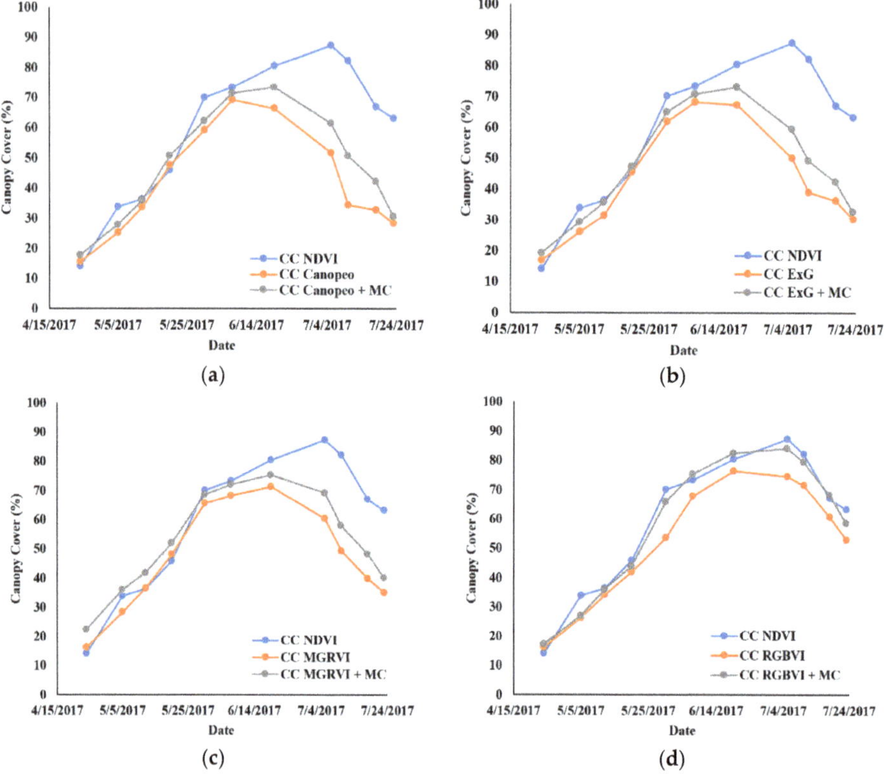

Figure 11. For the 2017 experiment, the average CC estimation per grid using the NDVI-based CC estimation throughout the growing season, along with the average CC estimation using (**a**) Canopeo, (**b**) the ExG, (**c**) the modified green red vegetation index (MGRVI) and (**d**) the red green blue vegetation index (RGBVI), before and after applying the morphological closing (MC) operation.

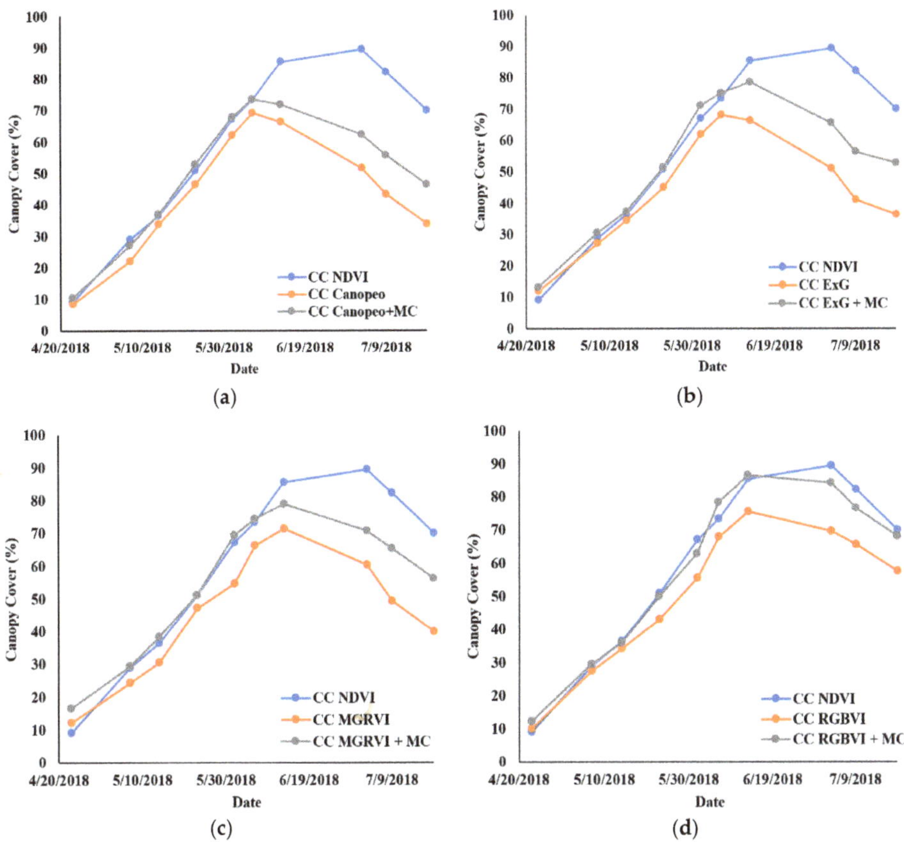

Figure 12. For the 2018 experiment, the average CC estimation per grid using the NDVI-based CC estimation throughout the growing season, along with the average CC estimation using (**a**) Canopeo, (**b**) the ExG, (**c**) the MGRVI and (**d**) the RGBVI, before and after applying the morphological closing (MC) operation.

Table 5. Average RMSE of the thresholding-based CC estimation methods with respect to the NDVI-based CC estimation (%) throughout the growing season.

RGB-Based Method	Average RMSE with Respect to NDVI-Based CC Estimation (%) 2017 Experiment		2018 Experiment	
	Before MC	After MC	Before MC	After MC
Canopeo	17.87	13.34	15.56	9.73
ExG	16.97	13.00	15.51	8.67
MGRVI	13.11	10.35	14.34	6.95
RGBVI	7.44	2.94	8.85	2.82

A similar trend was observed in the performance of thresholding-based CC estimation methods in the 2018 experiment (Figure 12). Canopeo and the ExG method had the higher RMSE, as compared to other methods (Table 5). However, CC estimation values of 2018 experiment were slightly higher compared to CC estimation values of 2017 experiment. The RGBVI turned out to be the most efficient CC estimation method amongst all. The application of the morphological closing operation significantly reduced the RMSE. It was observed that morphological closing operation resulted in slightly better performance in the 2018 experiment as compared to the 2017 experiment. It was noticed that in the 2018 experiments, the spatial resolution of the RGB images was slightly lower than the 2017 experiments. However, the difference in the spatial resolution was not much (difference of ~0.2cm), and a slight decrease of the GSD might contribute to the better classification and later filtering performance because a very high resolution leads to more details being observed, which could be undesirable when the information class under consideration is more general. Furthermore, a little difference in the CC growth pattern in two years was observed, which was a function of weather conditions (such as daily temperature and precipitation) and the type of genotype planted.

4. Discussion

Recent years have witnessed an upsurge in UAS and sensor technology, an upsurge which has made it possible to collect high temporal and spatial resolution data over crops throughout the growing season. The main objective of this study was to provide a comparison framework between MS sensor-based CC estimation and RGB-based CC estimation, as very scarce attention has been paid to explore different VIs generated using UAS-based sensors to compute canopy cover in the literature. As mentioned in the literature, MS sensor-based CC estimation is more accurate and stable because it accounts for the live canopy based on the chlorophyll content present in the canopy, and this content is highly reflected in the near-infrared (NIR) band; hence, the canopy varieties which are not green in color can be correctly accounted for. Moreover, changes in the color as the season progresses can also be identified accurately. However, the accuracy of the NDVI is a function of the type and quality of the multispectral sensor used to collect the image. In this study, temporal NDVI maps were comparable, despite changes in lighting conditions over different flights throughout the season, as they were generated from multispectral data collected using the SlantRange 3p sensor that performed radiometric calibration. The RGB-based CC estimation performed inadequately if the plant color deviated from green, which was confirmed by the experiments performed in both 2017 and 2018.

Except for the Canopeo method, all the other RGB VIs considered in this study required a threshold to be applied in order to separate canopy areas from non-canopy areas. In previous studies, the Otsu method has mostly been used for automatic thresholding [41]. The Otsu method resulted in an accurate thresholding early in the season, as the image had a mostly bimodal histogram and the variances of the spectral clusters were small compared to the mean difference. However, later in the season, thresholding by the Otsu method was questionable, because, as the season progressed, the variance in the spectral signature of the canopy increased, and, closer to senescence, the image no longer possessed a bimodal histogram due to the emergence of new spectral classes such as open cotton bolls. Consequently, in this study, the VIs were examined under visual inspection to select a single threshold value for all the images in the growing season. It was observed that the selected single threshold value successfully classified canopy pixels. Though they were less affected due to senescence as compared to the RGB-based VIs, the overall NDVI values still decreased in the late season; however, the selected single threshold value successfully classified canopy pixels with reasonable accuracy based on visual interpretation. Since this study was limited to the cotton crop, the threshold value might differ for other crops, and there might be a different trend observed in response to the senescence in the growing season. In future, an efficient thresholding method that can classify canopy regardless of growth stage would help automate the process.

As previous studies have affirmed that RGB-based CC estimation efficiently works early in the season [40,41], it was also observed in this study that early in the season, both the MS- and RGB-based CC estimations were in agreement and followed a similar increasing trend. Moreover, in previous studies, MS sensor-based CC estimation has been found to be more accurate in the later season as compared to RGB-based CC estimation [41]. In this study, it was observed that as the season progressed, MS sensor-based CC estimation kept on increasing, but RGB-based CC estimation peaked early and started to drop relatively rapidly for both the 2017 and 2018 experiments. Nevertheless, the RGBVI outperformed other RBG-based CC estimation methods, though still not close enough to match the estimation by MS sensor-based CC estimation. That led to the question "is it possible to achieve the same level of accuracy by using RGB-based CC estimation as that of the NDVI-based CC estimation?" With the aforementioned objective, a K-means clustering-based CC estimation method was implemented, and this was tested by using ground truth samples for the canopy and non-canopy classes. It was observed that K-means clustering-based approach matched the accuracy level of the MS sensor-based CC estimation. However, there was a requirement to investigate any scope for an improvement to the RGB-based CC estimation approach, as the K-means clustering-based approach, or any classification-based approach is accurate but computationally extensive, especially in its parameter tuning and demand for ground truth sample collection, which is labor-intensive and time consuming. The objective of this study was to improve the RGB-based CC estimation approach. It was noticed that, later in the season, RGB-based indices were not able to capture the canopy pixels that were not green, which resulted in CC maps with a lot of small holes. The morphological closing operation proved to be a solution to this problem and helped to fill very small holes and keep the boundary of the canopy intact. In both the 2017 and 2018 experiments, it was noticed that applying the morphological closing operation significantly improved the performance of the RGB-based CC estimation. The RGBVI with morphological closing applied was found to have a CC estimation very close to the NDVI-based CC estimation. With the proposed approach, a more affordable alternate to the MS sensor can be provided to estimate CC.

5. Conclusions

A comparative study was performed to evaluate CC estimation using an RGB sensor. With a multi-year CC analysis, MS sensor-based CC estimation was used as a reference, as it is considered a stable and accurate form of estimation. The correlation of RGB reference-based CC estimation with MS sensor-based CC estimation ensured the feasibility of using an RGB sensor to match the MS sensor-based CC estimation. An analysis of RGB-based methods suggested that the RGBVI was more tolerant to changes in the color of the canopy when the canopy started to senescence. Moreover, when applied with the morphological closing operation, the RGBVI-based CC estimation was found to be as accurate as MS sensor-based CC estimation, with an average RMSE of less than three percent. CC is a good predictor variable for plant growth parameters. As multispectral sensors are more sensitive and expensive, the proposed RGB-based CC estimation could provide an affordable alternate to agriculture scientists and breeders. In the future, this methodology will be investigated on other crops, as there might be a different trend observed in response to senescence in the growing season.

Author Contributions: conceptualization, A.A. and J.J.; data curation, A.A., A.C. and S.O.; formal analysis, A.A.; investigation, M.M.; methodology, A.A. and J.J.; project administration, J.J., M.M. and J.L.; resources, J.L.; software, A.A.; supervision, J.J.; validation, A.A.; visualization, A.A.; writing—original draft, A.A.; writing—review & editing, J.J., A.C., S.O. and J.L.

Funding: This research received no external funding.

Acknowledgments: This research was supported by Texas A&M AgriLife Research, Corpus Christi.

Conflicts of Interest: The authors declare no conflict of interest.

References

1. Adhikari, P.; Gowda, P.; Marek, G.; Brauer, D.; Kisekka, I.; Northup, B.; Rocateli, A. Calibration and validation of csm-cropgro-cotton model using lysimeter data in the texas high plains. *J. Contemp. Water Res. Educ.* **2017**, *162*, 61–78. [CrossRef]
2. Phillips, R.L. Mobilizing science to break yield barriers. *Crop Sci.* **2010**, *50*, 99–108. [CrossRef]
3. Xu, R.; Li, C.; Paterson, A.H. Multispectral imaging and unmanned aerial systems for cotton plant phenotyping. *PLoS ONE* **2019**, *14*, e0205083. [CrossRef] [PubMed]
4. Pierpaoli, E.; Carli, G.; Pignatti, E.; Canavari, M. Drivers of precision agriculture technologies adoption: A literature review. *Procedia Technol.* **2013**, *8*, 61–69. [CrossRef]

5. Tokekar, P.; Vander Hook, J.; Mulla, D.; Isler, V. Sensor planning for a symbiotic uav and ugv system for precision agriculture. *IEEE Trans. Robot.* **2016**, *32*, 1498–1511. [CrossRef]
6. Singh, K.K.; Frazier, A.E. A meta-analysis and review of unmanned aircraft system (uas) imagery for terrestrial applications. *Int. J. Remote Sens.* **2018**, *39*, 5078–5098. [CrossRef]
7. Roth, L.; Streit, B. Predicting cover crop biomass by lightweight uas-based rgb and nir photography: An applied photogrammetric approach. *Precis. Agric.* **2018**, *19*, 93–114. [CrossRef]
8. Gevaert, C.M.; Suomalainen, J.; Tang, J.; Kooistra, L. Generation of spectral–temporal response surfaces by combining multispectral satellite and hyperspectral uav imagery for precision agriculture applications. *IEEE J. Sel. Top. Appl. Earth Obs. Remote Sens.* **2015**, *8*, 3140–3146. [CrossRef]
9. Trout, T.J.; Johnson, L.F.; Gartung, J. Remote sensing of canopy cover in horticultural crops. *HortScience* **2008**, *43*, 333–337. [CrossRef]
10. Nielsen, D.C.; Miceli-Garcia, J.J.; Lyon, D.J. Canopy cover and leaf area index relationships for wheat, triticale, and corn. *Agron. J.* **2012**, *104*, 1569–1573. [CrossRef]
11. Chopping, M. Canapi: Canopy analysis with panchromatic imagery. *Remote Sens. Lett.* **2011**, *2*, 21–29. [CrossRef]
12. Halperin, J.; LeMay, V.; Chidumayo, E.; Verchot, L.; Marshall, P. Model-based estimation of above-ground biomass in the miombo ecoregion of zambia. *For. Ecosyst.* **2016**, *3*, 14. [CrossRef]
13. Hansen, M.C.; DeFries, R.S.; Townshend, J.R.; Carroll, M.; DiMiceli, C.; Sohlberg, R.A. Global percent tree cover at a spatial resolution of 500 meters: First results of the modis vegetation continuous fields algorithm. *Earth Interact.* **2003**, *7*, 1–15. [CrossRef]
14. Korhonen, L.; Hovi, A.; Rönnholm, P.; Rautiainen, M. The accuracy of large-area forest canopy cover estimation using landsat in boreal region. *Int. J. Appl. Earth Obs. Geoinf.* **2016**, *53*, 118–127.
15. Chemura, A.; Mutanga, O.; Odindi, J. Empirical modeling of leaf chlorophyll content in coffee (coffea arabica) plantations with sentinel-2 msi data: Effects of spectral settings, spatial resolution, and crop canopy cover. *IEEE J. Sel. Top. Appl. Earth Obs. Remote Sens.* **2017**, *10*, 5541–5550. [CrossRef]
16. Melin, M.; Korhonen, L.; Kukkonen, M.; Packalen, P. Assessing the performance of aerial image point cloud and spectral metrics in predicting boreal forest canopy cover. *ISPRS J. Photogramm. Remote Sens.* **2017**, *129*, 77–85. [CrossRef]
17. Korhonen, L.; Ali-Sisto, D.; Tokola, T. Tropical forest canopy cover estimation using satellite imagery and airborne lidar reference data. *Silva Fenn.* **2015**, *49*, 1–18. [CrossRef]
18. Li, W.; Niu, Z.; Huang, N.; Wang, C.; Gao, S.; Wu, C. Airborne lidar technique for estimating biomass components of maize: A case study in zhangye city, northwest china. *Ecol. Indic.* **2015**, *57*, 486–496. [CrossRef]
19. Tattaris, M.; Reynolds, M.P.; Chapman, S.C. A direct comparison of remote sensing approaches for high-throughput phenotyping in plant breeding. *Front. Plant Sci.* **2016**, *7*, 1131. [CrossRef]
20. Chen, A.; Orlov-Levin, V.; Elharar, O.; Meron, M. Comparing satellite and high-resolution visible and thermal aerial imaging of field crops for precision irrigation management and plant biomass forecast. In *Precision Agriculture'19*; Wageningen Academic Publishers: Wageningen, The Netherlands, 2019; pp. 37–44.
21. Korhonen, L.; Korpela, I.; Heiskanen, J.; Maltamo, M. Airborne discrete-return lidar data in the estimation of vertical canopy cover, angular canopy closure and leaf area index. *Remote Sens. Environ.* **2011**, *115*, 1065–1080. [CrossRef]
22. Anderson, K.E.; Glenn, N.F.; Spaete, L.P.; Shinneman, D.J.; Pilliod, D.S.; Arkle, R.S.; McIlroy, S.K.; Derryberry, D.R. Estimating vegetation biomass and cover across large plots in shrub and grass dominated drylands using terrestrial lidar and machine learning. *Ecol. Indic.* **2018**, *84*, 793–802. [CrossRef]
23. Ma, Q.; Su, Y.; Guo, Q. Comparison of canopy cover estimations from airborne lidar, aerial imagery, and satellite imagery. *IEEE J. Sel. Top. Appl. Earth Obs. Remote Sens.* **2017**, *10*, 4225–4236. [CrossRef]
24. Holman, F.; Riche, A.; Michalski, A.; Castle, M.; Wooster, M.; Hawkesford, M. High throughput field phenotyping of wheat plant height and growth rate in field plot trials using uav based remote sensing. *Remote Sens.* **2016**, *8*, 1031. [CrossRef]
25. Chianucci, F.; Disperati, L.; Guzzi, D.; Bianchini, D.; Nardino, V.; Lastri, C.; Rindinella, A.; Corona, P. Estimation of canopy attributes in beech forests using true colour digital images from a small fixed-wing uav. *Int. J. Appl. Earth Obs. Geoinf.* **2016**, *47*, 60–68. [CrossRef]

26. Fernandez-Gallego, J.A.; Kefauver, S.C.; Kerfal, S.; Araus, J.L. Remote Sensing for Agriculture, Ecosystems, and Hydrology XX. In *Comparative Canopy Cover Estimation Using RGB Images from UAV and Ground*; International Society for Optics and Photonics: Berlin, Germany, 2018; p. 107830J.
27. Ashapure, A.; Oh, S.; Marconi, T.G.; Chang, A.; Jung, J.; Landivar, J.; Enciso, J. Autonomous Air and Ground Sensing Systems for Agricultural Optimization and Phenotyping IV. In *Unmanned Aerial System Based Tomato Yield Estimation Using Machine Learning*; International Society for Optics and Photonics: Baltimore, MD, USA, 2019.
28. Chu, T.; Chen, R.; Landivar, J.A.; Maeda, M.M.; Yang, C.; Starek, M.J. Cotton growth modeling and assessment using unmanned aircraft system visual-band imagery. *J. Appl. Remote Sens.* **2016**, *10*, 036018. [CrossRef]
29. Ballesteros, R.; Ortega, J.; Hernández, D.; Moreno, M. Applications of georeferenced high-resolution images obtained with unmanned aerial vehicles. Part II: Application to maize and onion crops of a semi-arid region in spain. *Precis. Agric.* **2014**, *15*, 593–614. [CrossRef]
30. Ballesteros, R.; Ortega, J.F.; Hernandez, D.; Moreno, M.A. Onion biomass monitoring using uav-based rgb imaging. *Precis. Agric.* **2018**, *19*, 840–857. [CrossRef]
31. Ashapure, A.; Jung, J.; Yeom, J.; Chang, A.; Maeda, M.; Maeda, A.; Landivar, J. A novel framework to detect conventional tillage and no-tillage cropping system effect on cotton growth and development using multi-temporal uas data. *ISPRS J. Photogramm. Remote Sens.* **2019**, *152*, 49–64. [CrossRef]
32. Makanza, R.; Zaman-Allah, M.; Cairns, J.; Magorokosho, C.; Tarekegne, A.; Olsen, M.; Prasanna, B. High-throughput phenotyping of canopy cover and senescence in maize field trials using aerial digital canopy imaging. *Remote Sens.* **2018**, *10*, 330. [CrossRef]
33. Clevers, J.; Kooistra, L.; Van Den Brande, M. Using sentinel-2 data for retrieving lai and leaf and canopy chlorophyll content of a potato crop. *Remote Sens.* **2017**, *9*, 405. [CrossRef]
34. Pauly, K. Applying conventional vegetation vigor indices to uas-derived orthomosaics: Issues and considerations. In Proceedings of the International Conference of Precision Agriculture (ICPA), Sacramento, CA, USA, 20–23 July 2014.
35. Booth, D.T.; Cox, S.E.; Berryman, R.D. Point sampling digital imagery with 'samplepoint'. *Environ. Monit. Assess.* **2006**, *123*, 97–108. [CrossRef] [PubMed]
36. Richardson, M.; Karcher, D.; Purcell, L. Quantifying turfgrass cover using digital image analysis. *Crop Sci.* **2001**, *41*, 1884–1888. [CrossRef]
37. Lee, K.J.; Lee, B.W. Estimating canopy cover from color digital camera image of rice field. *J. Crop Sci. Biotechnol.* **2011**, *14*, 151–155. [CrossRef]
38. Patrignani, A.; Ochsner, T.E. Canopeo: A powerful new tool for measuring fractional green canopy cover. *Agron. J.* **2015**, *107*, 2312–2320. [CrossRef]
39. Torres-Sánchez, J.; Peña, J.M.; de Castro, A.I.; López-Granados, F. Multi-temporal mapping of the vegetation fraction in early-season wheat fields using images from uav. *Comput. Electron. Agric.* **2014**, *103*, 104–113. [CrossRef]
40. Fang, S.; Tang, W.; Peng, Y.; Gong, Y.; Dai, C.; Chai, R.; Liu, K. Remote estimation of vegetation fraction and flower fraction in oilseed rape with unmanned aerial vehicle data. *Remote Sens.* **2016**, *8*, 416. [CrossRef]
41. Marcial-Pablo, M.D.J.; Gonzalez-Sanchez, A.; Jimenez-Jimenez, S.I.; Ontiveros-Capurata, R.E.; Ojeda-Bustamante, W. Estimation of vegetation fraction using rgb and multispectral images from uav. *Int. J. Remote Sens.* **2019**, *40*, 420–438. [CrossRef]
42. Lima-Cueto, F.J.; Blanco-Sepúlveda, R.; Gómez-Moreno, M.L.; Galacho-Jiménez, F.B. Using vegetation indices and a uav imaging platform to quantify the density of vegetation ground cover in olive groves (*Olea europaea* L.) in southern spain. *Remote Sens.* **2019**, *11*, 2564. [CrossRef]
43. Westoby, M.J.; Brasington, J.; Glasser, N.F.; Hambrey, M.J.; Reynolds, J.M. 'Structure-from-motion'photogrammetry: A low-cost, effective tool for geoscience applications. *Geomorphology* **2012**, *179*, 300–314. [CrossRef]
44. Rouse, J.W., Jr.; Haas, R.; Schell, J.; Deering, D. *Monitoring Vegetation Systems in the Great Plains with ERTS*; Texas A&M Univ.: College Station, TX, USA, 1974.
45. Hulvey, K.B.; Thomas, K.; Thacker, E. A comparison of two herbaceous cover sampling methods to assess ecosystem services in high-shrub rangelands: Photography-based grid point intercept (gpi) versus quadrat sampling. *Rangelands* **2018**, *40*, 152–159. [CrossRef]

46. Woebbecke, D.M.; Meyer, G.E.; Von Bargen, K.; Mortensen, D. Color indices for weed identification under various soil, residue, and lighting conditions. *Trans. ASAE* **1995**, *38*, 259–269. [CrossRef]
47. Bendig, J.; Yu, K.; Aasen, H.; Bolten, A.; Bennertz, S.; Broscheit, J.; Gnyp, M.L.; Bareth, G. Combining uav-based plant height from crop surface models, visible, and near infrared vegetation indices for biomass monitoring in barley. *Int. J. Appl. Earth Obs. Geoinf.* **2015**, *39*, 79–87. [CrossRef]
48. Dougherty, E.R. *An Introduction to Morphological Image Processing*; SPIE: Bellingham, WA, USA, 1992.

Letter

Watson on the Farm: Using Cloud-Based Artificial Intelligence to Identify Early Indicators of Water Stress

Daniel Freeman [1], Shaurya Gupta [2], D. Hudson Smith [3], Joe Mari Maja [4,*], James Robbins [5], James S. Owen [6], Jose M. Peña [7] and Ana I. de Castro [8]

1 Department of Biochemistry and Genetics, Clemson University, Poole Agricultural Center, Clemson, SC 29634, USA; freemaf@g.clemson.edu
2 Department of Electrical and Computer Engineering, Clemson University, Riggs Hall, Clemson, SC 29634, USA; shauryg@g.clemson.edu
3 Watt Family Innovation Center, Clemson University, 405 S Palmetto Blvd., Clemson, SC 29634, USA; dane2@clemson.edu
4 Department of Agricultural Science, Clemson University, 240 McAdams Hall, Clemson, SC 29634, USA
5 Division of Agriculture, University of Arkansas, Little Rock, AR 72204, USA; jrobbins@uaex.edu
6 School of Plant and Environmental Sciences, Virginia Tech, Hampton Roads Agricultural Research and Extension Center, Virginia Beach, VA 23455, USA; jsowen@vt.edu
7 Institute of Agricultural Sciences, CSIC, 28006 Madrid, Spain; jmpena@ica.csic.es
8 Institute for Sustainable Agriculture, CSIC, 14004 Cordoba, Spain; anadecastro@ias.csic.es
* Correspondence: jmaja@clemson.edu

Received: 27 September 2019; Accepted: 8 November 2019; Published: 13 November 2019

Abstract: As demand for freshwater increases while supply remains stagnant, the critical need for sustainable water use in agriculture has led the EPA Strategic Plan to call for new technologies that can optimize water allocation in real-time. This work assesses the use of cloud-based artificial intelligence to detect early indicators of water stress across six container-grown ornamental shrub species. Near-infrared images were previously collected with modified Canon and MAPIR Survey II cameras deployed via a small unmanned aircraft system (sUAS) at an altitude of 30 meters. Cropped images of plants in no, low-, and high-water stress conditions were split into four-fold cross-validation sets and used to train models through IBM Watson's Visual Recognition service. Despite constraints such as small sample size (36 plants, 150 images) and low image resolution (150 pixels by 150 pixels per plant), Watson generated models were able to detect indicators of stress after 48 hours of water deprivation with a significant to marginally significant degree of separation in four out of five species tested ($p < 0.10$). Two models were also able to detect indicators of water stress after only 24 hours, with models trained on images of as few as eight water-stressed Buddleia plants achieving an average area under the curve (AUC) of 0.9884 across four folds. Ease of pre-processing, minimal amount of training data required, and outsourced computation make cloud-based artificial intelligence services such as IBM Watson Visual Recognition an attractive tool for agriculture analytics. Cloud-based artificial intelligence can be combined with technologies such as sUAS and spectral imaging to help crop producers identify deficient irrigation strategies and intervene before crop value is diminished. When brought to scale, frameworks such as these can drive responsive irrigation systems that monitor crop status in real-time and maximize sustainable water use.

Keywords: sUAS; water stress; ornamental; container-grown; artificial intelligence; machine learning; deep learning; neural network; visual recognition; precision agriculture

1. Introduction

Freshwater is a finite resource that is required for the daily production of container crops to be used for food, ecosystem services, urban development, and other purposes. The United Nations Education, Scientific, and Cultural Organization (UNESCO) has indicated that the combined expansion of manufacturing, agriculture, and urban populations has created excessive strain on the existing fresh water supply and has called for more sustainable water management [1]. One opportunity to reduce water consumption lies in the development of intelligent irrigation systems that can optimize water use in real-time [2]. Crop producers routinely provide an excess of water to container-grown plants to mitigate plant stress and subsequent economic loss, resulting in inefficient use of agrichemicals, energy, and freshwater. Site-specific irrigation systems minimize these losses by using sensors to allocate water to plants as needed, improving crop production while minimizing operating costs [3]. Sensor-based irrigation is not a new concept [1,4–6]. Kim et al. [5] developed software for an in-field wireless sensor network (WSN) to implement site-specific irrigation management in greenhouse containers. Coates et al. [7] developed site-specific applications using soil water status data to control irrigation valves.

In 2017, the U.S. nursery industry had sales of $5.9 billion and ornamental production accounted for 2.2 percent of all U.S. farms [8]. Plants grown in containers are the primary (73%) production method [9] and the majority (81%) of nursery production acreage is irrigated [10]. The largest production cost for nurseries is labor, which amounts to 39% of total costs [11], and labor shortages are linked to reduced production [12]. Adoption of appropriate technologies may offset increasing labor costs and labor shortages. Small unmanned aircraft systems (sUAS) have been suggested as an important tool in nursery production to help automate certain processes such as water resource management [13].

sUASs allow farmers to quickly survey large plots of land using aerial imagery. sUAS imagery has been used to detect diseases and weeds [14,15], predict cotton yield [16], measure the degree of stink bug aggregation [17], and identify water stress in ornamental plants [18]. Several thermal and spectral indices have been correlated to biophysical plant parameters based on sUAS imagery [19,20]. Analyses of sUAS imagery have been shown to be sensitive to time of day, cloud cover, light intensity, image pixel size, soil water buffering capacity, and atmospheric conditions at the canopy level [21,22]. Still, multispectral data collected with sUAS were shown to be more accurate than data collected using manned aircraft [23]. A variety of methodologies, including thermal and spectral imagery, have been used to assess water stress in conventional sustainable agriculture using sUAS [3]. Stagakis et al. [24] indicated that the high spatial and spectral resolution provided by sUAS-based imagery could be used to detect deficient irrigation strategies. Zovkoa et al. [25] reported difficulty measuring three levels of water stress of grape grown in soil; however, they were able to discern irrigated vs. non-irrigated plots via hyperspectral image analysis (409–988 nm and 950–2509 nm) when employing a support vector machine (SVM). de Castro et al. [18] successfully identified water-stressed and non-stressed containerized ornamental plants using two multispectral cameras aboard an sUAS, although the spectral separation was higher when information from the sensors was combined. Data being produced by de Castro and Zovkoa could be utilized as a roadmap for real-time, sustainable water management of specialty or container-grown crops using sUAS. Fulcher et al. [26] indicated that the adoption of sUAS to monitor crop water status will be useful in addressing the challenge of sustainable water use in container nurseries. Unlike conventional crops produced in soil systems, containerized soilless-based systems have low water buffering capacity, resulting in rapid physiological changes that may not be observed at the ground level visually, but can be monitored by reflected wavelengths captured by sUAS. To reduce size and cost, sUAS can collect and wirelessly transmit high-resolution image data to cloud providers that can perform analyses on offsite servers. Thus, the convergence of technologies—such as sUAS, Internet of Things (IoT), spectral imagery, and cloud-based computing—can be used to build intelligent irrigation systems that monitor crop status and optimize water allocation in real time.

In this study, images were analyzed with IBM Watson Visual Recognition, a cloud-hosted artificial intelligence service that allows users to train custom image classifiers using deep convolutional neural networks (CNNs). Unlike linear algorithms, CNNs model complex non-linear relationships between the independent variables (pixels comprising the image) and the dependent variable (plant health) by transforming data through layers of increasingly abstract representation (Figure 1). The first layer is an array of pixel values from the original image; nodes in subsequent layers represent local features such as color, texture, and shape; deeper layers encode semantic information such as leaf or branch morphology. Individual nodes become optimized to represent different features of the image through an iterative learning process that rewards nodes that amplify aspects of the image that are useful for classification and suppresses those that do not [27]. The convolutional relationship from one layer to the next allows CNNs to model complex relationships between input variables, making it particularly useful for analyzing image data that cannot be understood by examining pixels in isolation. Given a set of images of stressed and non-stressed plants, for example, individual nodes in the network may become optimized to represent spectral indices that are sensitive to water stress. Those nodes can affect the outcome directly, or they can feed forward into higher-order features such as the specific location and pattern of discoloration within the plant. Spectral indices may combine with other plant features such as the unique structure of sagging branches or the distinct texture created by the shadows from drooping leaves. All of these features culminate in a single output node that returns a value from zero to one representing the confidence that a given image belongs to the desired class (i.e., water stress).

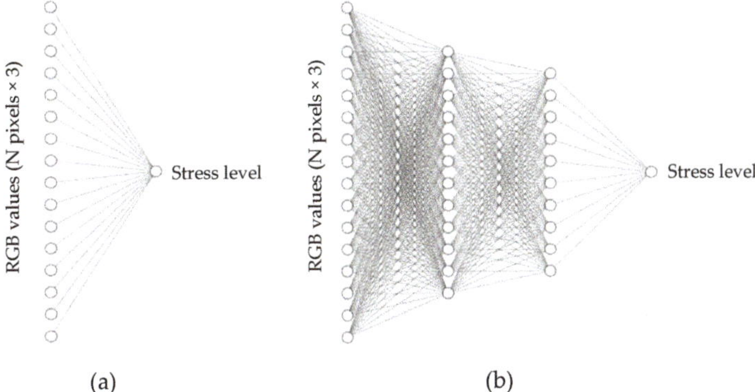

Figure 1. (**a**) Linear model in which each variable directly affects the outcome versus (**b**) a convolutional neural network (CNN) in which data is transformed through multiple layers.

While CNNs' layers allow networks to model complex nonlinear relationships that simpler algorithms might miss, they are also prone to overfitting. This occurs when CNNs learn patterns that are specific to the training set and do not generalize to the overall population. In one case study, for example, a model trained to predict a patient's age based on MRI images was found to have learned the shape of the head rather than the content of the scan itself [28]. The challenge of overfitting is compounded by CNNs' inherent 'black box' quality. Since information is passed through so many transformations, it is difficult to identify which input variables have the largest influence on the final outcome. While CNNs often must be trained with large datasets to overcome their tendency to overfit, transfer learning techniques allow fully trained networks to be repurposed for new classification tasks with much smaller datasets. A growing set of tools are also making it possible to introspect models to determine feature importance directly. Saliency heat maps, for example, can highlight regions of the image that are used for classification [29,30]. Overfitting can be tested with a cross-validation scheme in which models are trained with one set of images and then used to classify a new, previously unseen set of images. Performance metrics are based on how well the model's classification of unseen data matches a ground truth standard. A final limitation of CNNs is the significant amount of time and resources required to train them. To circumvent this, the computation may be outsourced to cloud computing providers that train models on large servers and offer a suite of tools for hyperparameter tuning, transfer learning, and cross-validation [31].

Despite their inherent limitations, CNNs have become popular for image recognition tasks ranging from Facebook photo-tagging to self-driving cars [32,33]. In agriculture, CNNs have been used to predict wheat yield based on soil parameters, diagnose diseases with simple images of leaves, and detect nitrogen stress using hyperspectral imagery [34,35]. CNNs' ability to learn complex nonlinear features makes them particularly useful for analyzing image data in which individual pixels form larger features such as shape or texture. Extensive research has demonstrated that CNNs perform image classification tasks with higher accuracy than traditional machine vision algorithms [36].

In our study, a small set of aerial images were used to train custom image classification models to detect water stress in ornamental shrubs. The objective was to evaluate the ability of IBM Watson's Visual Recognition service to detect early indicators of plant stress. These experiments provide a strong rationale for the deployment of cloud-based artificial intelligence frameworks that use larger datasets to monitor crop status and maximize sustainable water use.

2. Materials and Methods

This research was conducted at the Hampton Roads Agricultural Research and Extension Center (Hampton Roads AREC-Virginia Tech), located in Virginia Beach, VA, USA (36.8919N, 76.1787W). Six plots with container-grown ornamental plants across two experimental areas were studied. Containers were established outdoors on gravel. The species and number of plants in each experimental plot are shown in Table 1.

Table 1. Species and number of plants in each experimental plot.

Scientific Name	Common Name	Height (cm)	Width ±SD (cm)	Plants
Buddleia x 'Podaras #5' (BUD)	Buddleia Flutterby Grande® Peach Cobbler	82	109 ± 12	34
Cornus obliqua 'Powell Gardens' (CO)	Red Rover® silky dogwood	43	55 ± 6	42
Hydrangea paniculata 'Limelight' (HP)	Limelight panicle hydrangea	46	50 ± 5	42
Hydrangea quercifolia 'Queen of Hearts' (HQ)	Queen of Hearts oakleaf hydrangea	31	49 ± 7	21
Physocarpus opulifolius 'Seward' (PO)	Summer Wine® ninebark	61	100 ± 23	36
Spiraea japonica 'Neon Flash' (SJ)	Neon Flash spirea	83	68 ± 21	32

A subset of plants from each species was removed from the open-air nursery and transferred to a greenhouse where the plants experienced water stress due to the absence of overhead irrigation. High water stress (HWS) plants were transferred to the greenhouse on 8 Aug 2017 and low water stress (LWS) plants were transferred to the greenhouse on 9 Aug 2017. The plants were then returned to the open-air nursery on 10 Aug 2017 after non-stressed plants received overhead irrigation daily, including 10 Aug 2017. This process produced three levels of water stress for this experiment; high, low, and non-stressed (Table 2). At the time of flight, the soilless substrate of HWS plants contained ~19% less water (mL) than non-stress plants and soilless substrate of LWS plants contained ~13% less water (mL) than non-stress plants. There were no easily detectable visual symptoms of water stress in any of the treatment plants. After the data collection, all water-stressed plants were returned to normal irrigation on 10 August 2017 where they fully recovered and continued to grow. This strategy was part of a broader research program with the aim of studying the adaptation of ornamental species to stress conditions.

Table 2. Technical specifications of the sensors onboard the sUAS and the number of images taken at each field.

	Modified Canon	MAPIR Survey 2
Sensor resolution (pixels)	3246 × 2448	4608 × 3456
Focal length (mm)	28	23
Radiometric resolution (bit)	24	24
Image format	jpg	jpg
Image no. for Area 1	131	262
Image no. for Area 2	154	324
Wavebands	Red, Green, NIR	Red, Green, NIR
GSD (at 120 m)	4.05 cm	4.05 cm

2.1. Image Acquisition

Experimental plots were photographed with a quadcopter drone (DJI Inspire 2, DJI Science and Technology Co. Ltd., Shenzhen, China) (Figure 2a) mounted with two cameras: (1) a modified (Llewellyn Data Processing LLC, Carlstadt, NJ, USA) Canon (Canon, Tokyo, Japan) PowerShot ELPH 130 IS; and (2) a MAPIR Survey2 (MAPIR, Peau Productions, Inc., CA, USA) in Figure 2b. During each flight, the quadcopter took images using each camera at a height of 30 meters and a forward and side lap of 90% and 60%, respectively. The technical specifications of the two sensors are shown in Table 2. Figure 3 shows the data collected from both sensors.

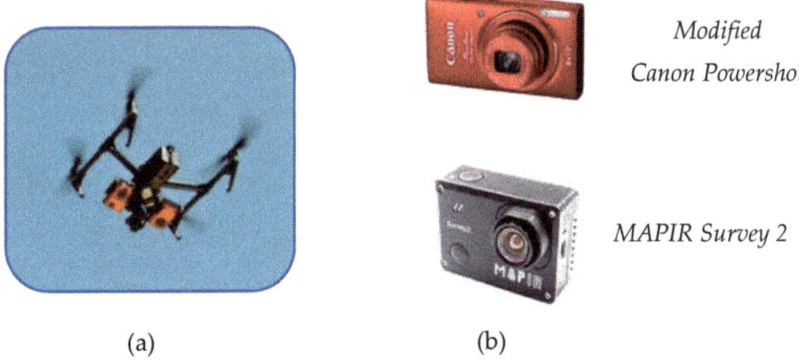

Figure 2. (**a**) DJI Inspire 2 with (**b**) sensors installed underneath. Images on (b) were from www.canon.com (Canon Powershot) and www.mapir.camera (MAPIR).

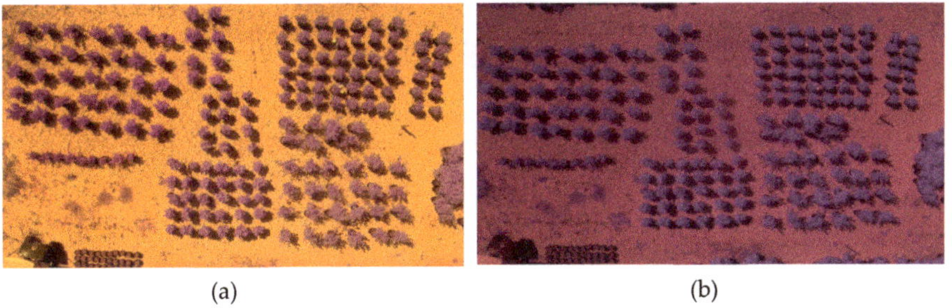

Figure 3. Examples of false-color near-infrared images taken by (**a**) modified Canon and (**b**) MAPIR cameras.

2.2. Image Pre-Processing

Images were cropped using a custom, browser-based interface in LabelBox (Figure 4). Data were annotated by dragging a bounding box across each plant and labeling it as 'high water stress', 'low water stress', or 'no stress' according to the key provided in Figure 5. The GraphQL application programming interface (API) was used to pull the pixel coordinates of each bounding box onto a local computer so that individual plants could be cropped from the original aerial images. The resolution of the cropped images was approximately 150 by 150 pixels. The number of cropped images for each condition is shown in Table 3.

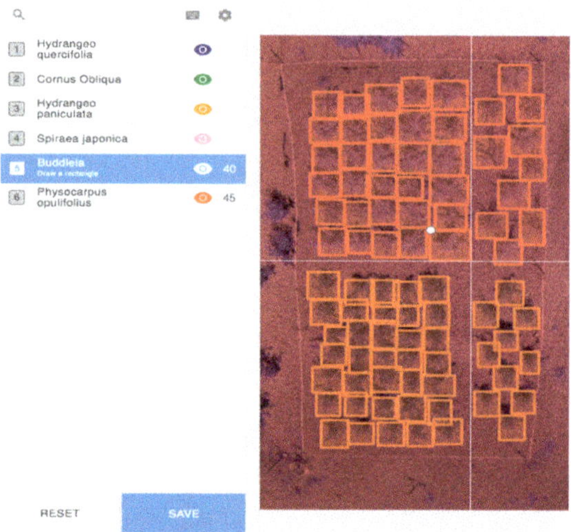

Figure 4. Custom LabelBox interface used to crop aerial images.

Figure 5. Example of the key used to identify plants according to species and condition. Stressed plants are labeled red (high water stress), yellow (low water stress), or blue (low phosphorus fertilizer, not used). Unmarked plants are non-stressed.

Since multiple photographs were taken of the same plots from different angles, cropped images of the same plants were grouped together so that they could be segregated into the training set or validation set as complete units. This procedure protected against overly optimistic performance estimates that would occur if photographs of the same plant appeared in both the training and validation datasets. For each species and treatment, the centers of each bounding box were calculated and normalized to a range of zero to one. Spatstat (http://spatstat.org), an open-source R package for analyzing point patterns, was then used to match plants from different aerial images based on the similarity of their pixel coordinates. For example, if there were eight plants in the HWS treatment of a certain species, all images of plants one through six would be used to train the model and all images of plants seven and eight would be used for validation. This allowed us to make full use of the data during the training phase without artificially inflating performance metrics by validating models with images of the same plants they were trained with. The successful grouping was confirmed by visual inspection (Figure 6).

Table 3. Number of images for each camera, species, and treatment.

Species	Water Treatment	Plants	Modified Canon	MAPIR
Buddleia (BUD)	HWS	8	24	24
	LWS	8	24	24
	NS	18	72	72
Cornus (CO)	HWS	8	25	25
	LWS	8	25	25
	NS	28	85	85
Hydrangea paniculata (HP)	HWS	8	40	40
	LWS	8	40	40
	NS	30	150	150
Hydrangea quercifolia (HQ)	HWS	6	30	30
	LWS	6	30	30
	NS	25	125	125
Physocarpus (PO)	HWS	8	24	24
	LWS	6	18	18
	NS	31	93	93
Spiraea (SJ)	HWS	10	0	10
	LWS	10	0	10
	NS	36	0	36

HWS = high water stress; LWS = low water stress; and NS = no stress treatment.

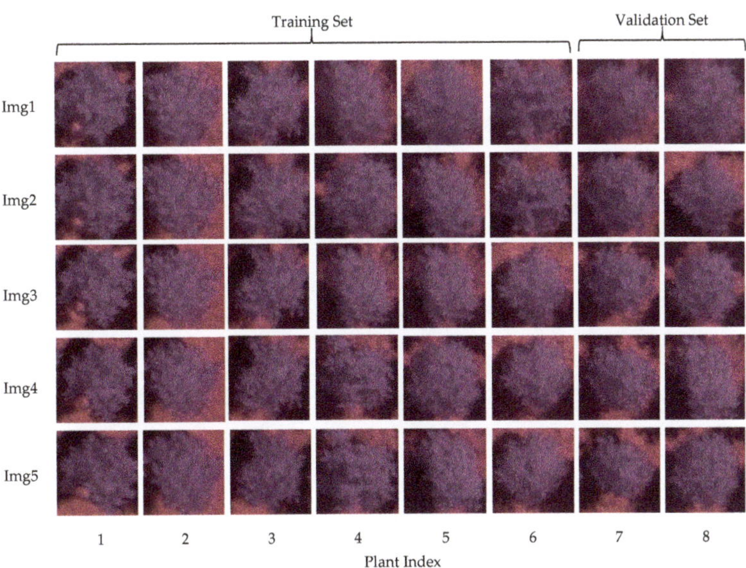

Figure 6. Images of the same plants cropped from multiple aerial images (Img1–Img5) were matched based on their pixel coordinates so that unique plants could be grouped into either the training set or the validation set. Brackets show one of four train/test splits.

2.3. Model Training and Testing

Cropped images were used to train models with the Watson Visual Recognition API, a cloud-hosted artificial intelligence service provided by IBM that uses CNNs to build custom image classifiers. Here, models were trained to predict water stress status using red, green, and near-infrared pixel values of the cropped images. A Python script was used to access the service and transfer images from a local computer to a cloud server for model training and testing. For each species and camera, three-quarters of NS and HWS images were used to train a model that was then used to classify the remaining quarter. The API returned a prediction between zero and one for each validation image with zero indicating no stress and one indicating water stress (Table 4). This process was repeated four times so that a prediction could be made for each image in the dataset and compared to the ground truth.

Table 4. Predictions returned by the Watson Visual Recognition API (Score) are compared to the ground truth (Stress).

Row	Image	Score	Stress
1	IMG_7696_bud_lws_729.JPG	0.906	TRUE
2	IMG_7695_bud_lws_815.JPG	0.875	TRUE
3	IMG_7694_bud_lws_644.JPG	0.916	TRUE
4	IMG_7696_bud_lws_730.JPG	0.798	TRUE
5	IMG_7694_bud_lws_645.JPG	0.777	TRUE
6	IMG_7695_bud_lws_814.JPG	0.856	TRUE
7	IMG_7694_bud_ns_658.JPG	0.031	FALSE
8	IMG_7696_bud_ns_747.JPG	0.001	FALSE
...	...	...	...
23	IMG_7696_bud_ns_746.JPG	0.062	FALSE
24	IMG_7696_bud_ns_732.JPG	0.852	FALSE

2.4. Statistical Analysis

A receiver operating characteristic area under the curve (AUC) score was used to quantify the degree of separation between treatments for each species and camera. A one-sample *t*-test was used to compare the AUC scores returned by the four-fold validation sets to a hypothesized mean of 0.5, corresponding to random classification.

3. Results

Of the 11 combinations of species and camera used in this study, four produced models that were able to discriminate images of NS and HWS plants with a statistically significant degree of separation ($p < 0.05$): Canon and MAPIR images of *Buddleia*, Canon images of *Physocarpus opulifolius*, and MAPIR images of *Hydrangea paniculata* (Table 5). Of these four, models trained with MAPIR or Canon images of NS and HWS *Buddleia* were also able to discriminate NS and LWS plants with high separation (Table 6). Four datasets produced models with a marginally significant degree of separation ($0.05 < p < 0.10$): Canon and MAPIR images of *Hydrangea quercifolia*, Canon images of *Hydrangea paniculata*, and MAPIR images of *Physocarpus opulifolius* (Table 5). Images of *Spiraea japonica* were not tested because the HWS class in the training set did not meet the minimum of 10 images required by the Visual Recognition API. Overall, models trained with four of five species tested achieved marginal significance or better ($p < 0.10$) in one or both cameras (Figures 7 and 8).

Results were compared to a previous study by de Castro et al. [18] that described the same dataset by masking the background and comparing mean pixel values in stressed and non-stressed plants. The three wavelengths detected by each camera were delineated and differences between treatments were evaluated by performing an analysis of variance (ANOVA) significance by a Tukey honestly significant difference (HSD) range test. Experiments that demonstrated a significant difference in mean pixel value between water stress treatments in one or more wavelengths ($p < 0.05$) are highlighted green in Table 5. Marginal significance is not shown because de Castro et al. [18] did not report specific p-values.

Table 5. Performance of models trained to classify HWS and NS images. Models achieving a statistically significant degree of separation (p-value < 0.05) are highlighted green and models achieving a marginal degree of separation (0.05 < p-value < 0.10) are highlighted yellow.

Species	Camera	Mean AUC	St. Dev	P-value	de Castro et al.
Buddleia (BUD)	Canon	0.9931	0.0046	1.10E−05	
	MAPIR	0.9907	0.0076	2.97E−05	
Cornus (CO)	Canon	0.463	0.0976	0.2635	
	MAPIR	0.5094	0.1661	0.4602	
Hydrangea (HP)	Canon	0.6448	0.1381	0.0854	
	MAPIR	0.7381	0.1036	0.0221	
Hydrangea (HQ)	Canon	0.6639	0.1284	0.0626	
	MAPIR	0.7113	0.1946	0.081	
Physocarpus (PO)	Canon	0.8177	0.1759	0.0344	
	MAPIR	0.5677	0.0503	0.0573	
Spiraea (SJ)	MAPIR	NA	NA	NA	

Table 6. Models that achieved a statistically significant degree of separation on HWS images were also used to classify LWS images.

Species	Camera	Mean AUC	St. Dev.	P-value	de Castro et al.
Buddleia (BUD)	Canon	0.9884	0.0139	1.01E−04	
Buddleia (BUD)	MAPIR	0.9653	0.0306	5.40E−04	
Hydrangea (HP)	MAPIR	0.6144	0.1126	0.0897	
Physocarpus (PO)	Canon	0.7167	0.1725	0.0643	

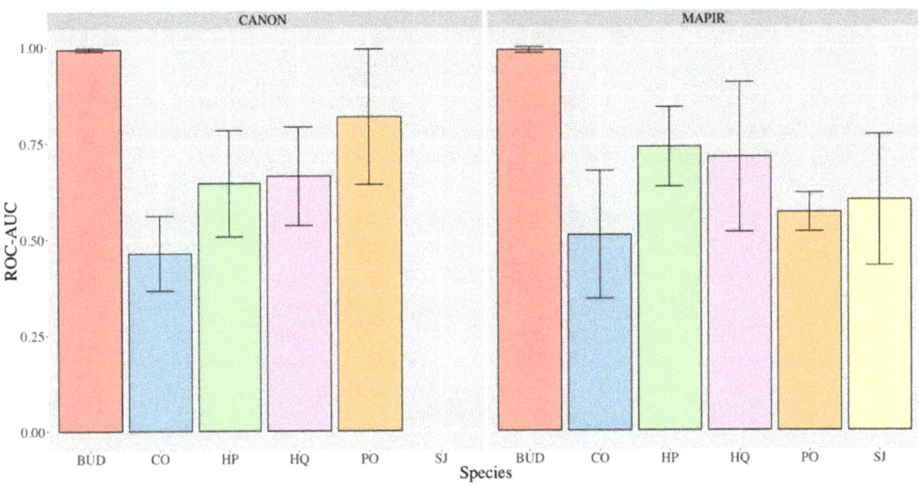

Figure 7. Performance of models trained to classify HWS and NS images.

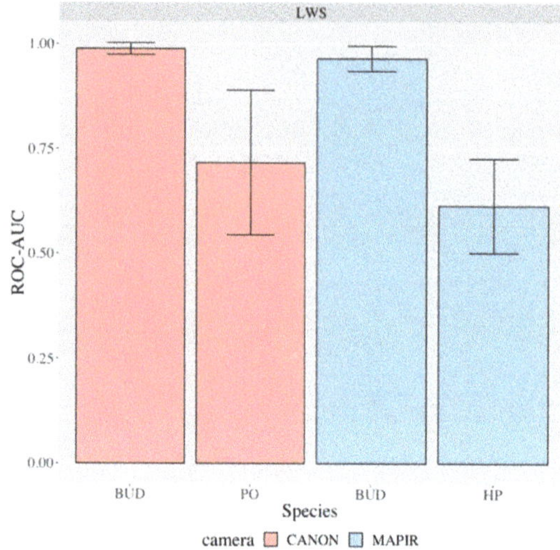

Figure 8. Models that achieved a statistically significant degree of separation on HWS images were also used to classify LWS images.

4. Discussions

Unlike traditional machine vision models that require users to manually select features, CNNs have layers of neurons that allow them to automatically learn relevant features from data. CNNs improve with each training example by iteratively rewarding neurons that amplify aspects of the image that are important for discrimination and suppressing those that do not. For example, in traditional techniques, the background must be manually segmented prior to analysis. By contrast, CNNs can automatically 'learn' to ignore the background because it is not relevant to the classification task. Similarly, rather than manually delineating spectral indices thought to be correlated with plant health, networks can infer relevant transformation of the input color channels from data. Low level features inferred by the network feed into higher-order features such as the specific location or pattern of discoloration within the plant. Information from spectral indices may combine with other features such as the unique structure of sagging branches or the distinct texture created by the shadows from wilted leaves. Thus, CNNs can learn multiple features of the training images and are not limited by a priori hypotheses.

Models tested in this study demonstrated significant variation in their ability to identify water stress in different species. Models trained on *Buddleia* achieved near-perfect separation while those trained on *Cornus* approximated random classification. Such variation is consistent with previous literature showing differences in morphological and physiological responses to water stress across genera, species, and even cultivar. In Michigan, Warsaw et al. [37] tracked daily water use and water use efficiency of 24 temperate ornamental taxa from 2006 and 2008. Daily water use varied from 12 to 24 mm per container and daily water use efficiency (increase in growth index per total liters applied) varied from 0.16 to 0.31. Of the similar taxa used, *Buddleia davidii* 'Guinevere' (24 mm per container) had the greatest water use followed by *Spirea japonica* 'Flaming Mound' (18 mm per container), *Hydrangea paniculata* 'Unique' (14 mm per container), and *Cornus sericea* 'Farrow' (12 mm per container) with estimated crop coefficients (KC) of 6.8, 5.0, 3.6, and 3.4, respectively. Low-water tolerant taxa such as *Cornus* may simply not have been demonstrating symptoms of water stress when they were photographed. Models that achieved moderate performance were likely provided with too few examples to distinguish patterns relevant to the classification task from those specific to the training data, causing them to generalize poorly to new data during the testing phase. Such overfitting bias can be overcome by training models with a larger and more diverse set of training images. Varying the location, weather, and growing period in which images are taken, for example, can force models to learn features that generalize to all conditions. Future studies can also use images of plants with multiple degrees of water stress to train regression models that return a value along a numeric scale rather than a stressed or not-stressed binary.

While CNNs' complicated nature prevents us from knowing what features are driving the model, insight can be gained from the conditions in which classifiers succeed or fail. For example, classifiers trained by pooling images of all species had significantly lower performance than classifiers trained with images of just one species despite having a considerably larger training set. This suggests that symptoms of water stress differ from one species to the next. Subsequent studies can identify what features are driving the model by iteratively removing them from the image. For example, one experiment could train models with individual R, G, or near-infrared channels to determine if certain spectral indices are more sensitive to water stress than others. Another experiment could crop a rectangle circumscribed to the plant in order to see if plant shape or other peripheral features aid the classifier. Features that significantly reduce performance when removed may represent biologically relevant phenotypes that are worthy of further study.

5. Conclusions

Our findings confirm that the IBM Watson Visual Recognition service can be used to identify early indicators of water stress in ornamental shrubs despite constraints such as small sample size, low image resolution, and lack of clear visual differences. Watson-generated models were able to detect indicators of stress after 48 hours of water deprivation with a significant to marginally significant degree of separation in four out of five species tested ($p < 0.10$). Models trained on images of *Buddleia* achieved near-perfect separation after only 24 hours with a max AUC of 0.9884. Furthermore, unlike traditional algorithms that require users to manually select plant parameters believed to correlate with health status, CNNs were able to automatically infer relevant features from the training data and combine multiple types of visual information. Despite this, not all models were successful. Failure of models trained on images of *Cornus* was consistent with previous literature, suggesting higher water stress tolerance in *Cornus* compared to the other species tested. Because all plants were grown in the same experimental area, authors cannot be certain that these models will generalize well to new situations.

Future studies can focus on improving model accuracy and generalizability by increasing the number of training examples and varying the conditions in which images are taken. Fully trained networks can also be introspected to give biological backing to the most predictive features. Other studies can expand the application of this workflow by testing data collected with different sensors and on different species. These experiments provide a valuable case study for the use of CNNs to monitor plant health. Brought to scale, artificial intelligence frameworks such as these can drive responsive irrigation systems that monitor plant status in real time and maximize sustainable water use.

Author Contributions: Conceptualization, A.I.d.C., J.M.P., J.M.M., J.R., and J.S.O.J.; Methodology, D.F., S.G., and D.H.S.; Software, D.F. and S.G.; Investigation, A.I.d.C., J.M.P., J.M.M., J.R., and J.S.O.J.; Resources, J.S.O.J. and J.M.M.; Validation, A.I.d.C., and J.M.P.; Formal Analysis, D.H.S., D.F., and S.G.; Writing—original draft preparation, D.F.; Writing—review and editing, D.F., S.G., D.H.S., J.M.M., J.R., and J.S.O.J.; Supervision, J.M.M.; Project Administration, J.M.M.; Funding Acquisition, J.M.M. and J.R.

Funding: This work was partially supported by a grant from the J. Frank Schmidt Family Charitable Foundation and is based on work supported by NIFA/USDA under project numbers SC-1700540, SC-1700543 and 2014-51181-22372 (USDA-SCRI Clean WateR3). Research of Drs. Peña and de Castro was financed by the "Salvador de Madariaga" for Visiting Researchers in Foreign Centers Program (Spanish MICINN funds) and the Intramural-CSIC Project (ref. 201940E074), respectively.

Acknowledgments: The authors would like to thank Julie Brindley and Ana Birnbaum for their support and assistance in this project. Special thanks to IBM for supporting our research with access to their artificial intelligence services.

Conflicts of Interest: The authors declare no conflict of interest.

References

1. Maja, J.M.; Robbins, J. Controlling irrigation in a container nursery using IoT. *AIMS Agric. Food* **2018**, *3*, 205–215. [CrossRef]
2. Cohen, Y.; Alchanatis, V.; Meron, M.; Saranga, S.; Tsipris, J. Estimation of leaf water potential by thermal imagery and spatial analysis. *J. Exp. Bot.* **2005**, *56*, 1843–1852. [CrossRef] [PubMed]
3. Gago, J.; Douthe, C.; Coopman, R.E.; Gallego, P.P.; Ribas-Carbo, M.; Flexas, J.; Escalona, J.; Medrano, H. UAVs challenge to assess water stress for sustainable agriculture. *Agric. Water Manag.* **2015**, *153*, 9–19. [CrossRef]
4. Jacobson, B.K.; Jones, P.H.; Jones, J.W.; Paramore, J.A. Real-time greenhouse monitoring and control with an expert system. *Comput. Electron. Agric* **1989**, *3*, 273–285. [CrossRef]
5. Kim, Y.; Evans, A.R.G. Software design for wireless sensor-based site-specific irrigation. *Comput. Electron. Agric.* **2009**, *66*, 159–165. [CrossRef]
6. Stone, K.C.; Smajstrla, A.G.; Zazueta, F.S. Microcomputer-based data acquisition system for continuous soilwater potential measurements. *Soil Crop Sci. Soc. Fla. Proc.* **1985**, *44*, 49–53.
7. Coates, R.W.; Delwiche, J.M.; Brown, P.H. Design of a system for individual microsprinkler control. *Trans. ASABE* **2006**, *49*, 1963–1970. [CrossRef]
8. USDA; National Agricultural Service. *2017 Census of Agriculture, United States Summary and State Data*; Geographic Area Series, Part 51. Publ AC-17-A-51; USDA: Washington, DC, USA; National Agricultural Service: Washington, DC, USA, 2019; Volume1, p. 820.
9. Vilsack, T.; Reilly, A.J.T. *2012 Census of Agriculture. Census of Horticultural Specialties (2014)*; Special Studies. Part 3. AC-12-SS-3; 2015; Volume 3, p. 422. Available online: https://www.nass.usda.gov/Publications/AgCensus/2012/Online_Resources/Farm_and_Ranch_Irrigation_Survey/fris13.pdf (accessed on 8 November 2019).
10. Vilsack, T.; Reilly, A.J.T. *2012 Census of Agriculture. Farm and Ranch Irrigation Survey (2013)*; Special Studies. Part 1. AC-12-SS-1; USDA NASS; Volume 3, pp. 84–195. Available online: https://www.nass.usda.gov/Publications/AgCensus/2012/Online_Resources/Farm_and_Ranch_Irrigation_Survey/fris13.pdf (accessed on 8 November 2019).
11. USDA. USDA Agricultural Resource Management Survey (ARMS) 2016. Available online: https://www.ers.usda.gov/topics/farm-economy/farm-labor/ (accessed on 16 September 2019).
12. Posadas, B.C.; Knight, P.R.; Coker, C.H.; Coker, R.Y.; Langlois, S.A. Hiring Preferences of Nurseries and Greenhouses in Southern United States. *Hort Technol.* **2014**, *24*, 101–117.
13. Robbins, J.A. Small unmanned aircraft systems (sUAS): An emerging technology for horticulture. *Hortic. Rev.* **2018**, *45*, 33–71.
14. De Castro, A.I.; Ehsani, R.; Ploetz, R.; Crane, J.H.; Abdulridha, J. Optimum spectral and geometric parameters for early detection of laurel wilt disease in avocado. *Remote Sens. Environ.* **2015**, *171*, 33–44. [CrossRef]
15. De Castro, A.I.; Torres-Sánchez, J.; Peña, J.M.; Jiménez-Brenes, F.M.; Csillik, O.; López-Granados, F. An Automatic Random Forest-OBIA Algorithm for Early Weed Mapping between and within Crop Rows Using UAV Imagery. *Remote Sens.* **2018**, *10*, 285. [CrossRef]
16. Maja, J.M.; Campbell, T.; Camargo-Neto, J.; Astillo, P. Predicting cotton yield of small field plots in a cotton breeding program using UAV imagery data. In Proceedings of the Autonomous Air and Ground Sensing Systems for Agricultural Optimization and Phenotyping, Baltimore, MD, USA, 17 May 2016; p. 98660C. [CrossRef]
17. Reay-Jones, F.; Greene, J.; Maja, J.M.; Bauer, P. Using remote sensing to improve the management of stink bugs in cotton in South Carolina. In Proceedings of the XXV International Congress of Entomology, Orlando, FL, USA, 26 September 2016.
18. De Castro, A.; Maja, J.M.; Owen, J.; Robbins, J.; Peña, J.M. 2018. Experimental approach to detect water stress in ornamental plants using sUAS-imagery. In Proceedings of the Autonomous Air and Ground Sensing Systems for Agricultural Optimization and Phenotyphing, SPIE Commercial + Scientific Sensing and Imaging, Orlando, FL, USA, 21 May 2018. [CrossRef]
19. Salami, E.; Barrado, C.; Pastor, E. UAV flight experiments applied to remote sensing of vegetated areas. *Remote Sens.* **2014**, *6*, 11051–11081. [CrossRef]

20. Shahbazi, M.; Jérôme, T.; Ménard, P. Recent applications of unmanned aerial imagery in natural resource management. *GISci. Remote Sens. J.* **2014**, *51*, 1548–1603. [CrossRef]
21. Jackson, R.D. Canopy temperature and crop water stress. In *Advances in Irrigation*; Hillel, D., Ed.; Academic Press: New York, NY, USA, 1982; Volume 1, pp. 43–80.
22. Bellvert, J.; Zarco-Tejada, P.J.; Girona, J.; Fereres, E. Mapping crop water stress index in a 'Pinot-noir' vineyard: Comparing ground measurements with thermal remote sensing imagery from an unmanned aerial vehicle. *Precis. Agric* **2014**, *15*, 361. [CrossRef]
23. Garcia Ruiz, F.; Sankaran, S.; Maja, J.M.; Lee, W.S.; Rasmussen, J.; Ehsani, R. Comparison of two aerial imaging platforms for identification of Huanglongbing-infected citrus trees. *Comput. Electron. Agric.* **2013**, *91*, 106–115. [CrossRef]
24. Stagakis, S.; Gonzalez-Dugo, V.; Cid, P.; Guillen-Climent, M.L.; Zarco-Tejada, P.J. Monitoring water stress and fruit quality in an orange orchard under regulated deficit irrigation using narrow-band structural and physiological remote sensing indices. *ISPRS J. Photogramm. Remote Sens.* **2012**, *71*, 47–61. [CrossRef]
25. Zovkoa, M.; Žibrat, U.; Knapič, M.; Kovačić, M.B.; Romić, D. Hyperspectral remote sensing of grapevine drought stress. *Precis. Agric* **2019**, *20*, 335–347. [CrossRef]
26. Fulcher, A.; LeBude, A.V.; Owen, J.S., Jr.; White, S.A.; Beeson, R.C. The Next Ten Years: Strategic Vision of Water Resources for Nursery Producers. *HortTechnology* **2016**, *26*, 121–132. [CrossRef]
27. LeCun, Y.; Bengio, Y.; Hinton, G. Deep learning. *Nature* **2015**, *521*, 436–444. [CrossRef]
28. Herent, P.; Jegoue, S.; Wainrib, G.; Clozel, T. Brain age prediction of healthy subjects on anatomic MRI with deep learning: going beyond with an "explainable AI" mindset. *bioRxiv* **2018**, 413302. [CrossRef]
29. Zhou, B.; Khosla, A.; Lapedriza, A.; Aude, O.; Torralba, A. Learning deep features for discriminative localization. *arXiv* **2015**, arXiv:1512.04150.
30. Zhang, Q.; Zhu, S. Visual interpretability for deep learning: a survey. *Front. Inf. Technol. Electron. Eng.* **2018**, *19*, 27–39. [CrossRef]
31. Awad, M.; Khanna, R. *Efficient Learning Machines*; Springer Link Apress: Berkeley, CA, USA, 2015; pp. 167–184. ISBN 978-1-4302-5989-3.
32. Ahmad, S.; Abdou, M.; Perot, E.; Yogamani, S. Deep reinforcement learning framework for autonomous driving. *arXiv* **2017**, arXiv:1704.02532.
33. Taigman, Y.; Yang, M.; Ranzato, M.; Wolf, L. DeepFace: closing the gap to human-level performance in face verification. In Proceedings of the IEEE Conference on Computer Vision and Pattern Recognition (CVPR 2014), Columbus, OH, USA, 24–27 June 2014; pp. 1701–1708.
34. Knyazikhin, Y.; Schull, M.A.; Stenberg, P.; Mõttus, M.; Rautiainen, M.; Yang, Y.; Marshak, A.; Carmona, P.L.; Kaufmann, R.K.; Lewis, P. Hyperspectral remote sensing of foliar nitrogen content. *Proc. Natl. Acad. Sci. USA* **2013**, *110*, E185–E192. [CrossRef]
35. Gómez-Casero, M.T.; López-Granados, F.; Peña-Barragán, J.M.; Jurado-Expósito, M.; García Torres, L.; Fernández-Escobar, R. Assessing Nitrogen and Potassium Deficiencies in Olive Orchards through Discriminant Analysis of Hyperspectral Data. *J. Am. Soc. Hort. Sci.* **2007**, *132*, 611–618. [CrossRef]
36. Alom, M.Z.; Taha, T.M.; Yakopcic, C.; Westberg, S.; Sidike, P.; Nasrin, M.S.; Van Essen, B.C.; Awwal, A.A.S.; Asari, V.K. The history began from alexNet: A comprehensive survey on deep learning approaches. *arXiv* **2018**, arXiv:1803.01164.
37. Warsaw, A.L.; Fernandez, R.T.; Cregg, B.M.; Andersen, J.A. Water conservation, growth, water use efficiency of container grown woody ornamentals irrigated based on daily water use. *HortScince* **2009**, *44*, 1308–1318. [CrossRef]

Article

Using Vegetation Indices and a UAV Imaging Platform to Quantify the Density of Vegetation Ground Cover in Olive Groves (*Olea Europaea* L.) in Southern Spain

Francisco J. Lima-Cueto *, Rafael Blanco-Sepúlveda, María L. Gómez-Moreno and Federico B. Galacho-Jiménez

Geographic Analysis Research Group, Department of Geography, University of Malaga, Campus of Teatinos, s/n. 29071 Malaga, Spain; rblanco@uma.es (R.B.-S.); geolugom@uma.es (M.L.G.-M.); fbgalacho@uma.es (F.B.G.-J.)
* Correspondence: lima@uma.es

Received: 25 September 2019; Accepted: 29 October 2019; Published: 1 November 2019

Abstract: In olive groves, vegetation ground cover (VGC) plays an important ecological role. The EU Common Agricultural Policy, through cross-compliance, acknowledges the importance of this factor, but, to determine the real impact of VGC, it must first be quantified. Accordingly, in the present study, eleven vegetation indices (VIs) were applied to quantify the density of VGC in olive groves (*Olea europaea* L.), according to high spatial resolution (10–12 cm) multispectral images obtained by an unmanned aerial vehicle (UAV). The fieldwork was conducted in early spring, in a Mediterranean mountain olive grove in southern Spain presenting various VGC densities. A five-step method was applied: (1) generate image mosaics using UAV technology; (2) apply the VIs; (3) quantify VGC density by means of sampling plots (ground-truth); (4) calculate the mean reflectance of the spectral bands and of the VIs in each sampling plot; and (5) quantify VGC density according to the VIs. The most sensitive index was IRVI, which accounted for 82% ($p < 0.001$) of the variability of VGC density. The capability of the VIs to differentiate VGC densities increased in line with the cover interval range. RVI most accurately distinguished VGC densities > 80% in a cover interval range of 10% ($p < 0.001$), while IRVI was most accurate for VGC densities < 30% in a cover interval range of 15% ($p < 0.01$). IRVI, NRVI, NDVI, GNDVI and SAVI differentiated the complete series of VGC densities when the cover interval range was 30% ($p < 0.001$ and $p < 0.05$).

Keywords: UAV; vegetation ground cover; multispectral; vegetation indices; agro-environmental measures; olive groves; southern Spain

1. Introduction

The Mediterranean basin contains 93.44% of the 10.24 million hectares of global olive cultivation (*Olea europaea* L.). Spain is the world's leading producer, with an annual crop production of 6.56 million tons, obtained from a growing area of over 2.57 million hectares (24.40% of the worldwide surface area in this respect), of which 1.55 million hectares are in Andalusia [1,2]. In the 1980s, the surface area of land dedicated to olive cultivation in Andalusia was expanded and production methods intensified. These changes, together with the continued existence of inappropriate practices linked to traditional land management, such as deep and continuous tillage in areas with high slopes or the elimination of vegetation ground cover (VGC) that protects the soil from torrential rains, aggravated the unsustainability environmental in the Andalusian olive groves [3,4].

VGC is the vegetal cover that grows spontaneously on the ground surface. This provides significant benefits to agricultural soils in the olive groves of southern Spain. Maintaining this cover is the best management option in rain-fed conditions, facilitating greater biomass and more bacterial diversity than in soils lacking vegetation cover. The microbial diversity of the soil deserves special attention because it an accurate biological indicator of soil quality, since it is very sensitive to changes that occur as a result of a bad crop management (weeds chemical control and soil tillage) [5]. Furthermore, the reduction of the soil biological diversity leads to a reduction of the stability of the microbial community, which causes a malfunction of the soil biota and its associated functions relating to the decomposition of the soil organic matter, with the nutrient cycle and, ultimately, with the plant production [6]. It is also important to highlight the advantages of maintaining VGC, compared to conventional tillage practices, in terms of improving soil fertility and water retention [7], and reducing water erosion in olive groves [8].

The importance of vegetation cover is also acknowledged in the cross-compliance system of the EU Common Agricultural Policy (CAP) (Regulation No. 1306/2013), specifically in the Good Agricultural and Environmental Conditions (GAEC) related to soil and the carbon stock, i.e., "minimum soil cover" (GAEC 4) and the "maintenance of soil organic matter level through appropriate practices including ban on burning arable stubble, except for plant health reasons" (GAEC 6). The application of these rules, which are mandatory for recipients of CAP assistance, requires the VGC density on the farms to be quantified. However, the effective control and monitoring of compliance is hampered by methodological shortcomings.

At present, VGC is usually analysed by means of vegetation indices (VIs), especially those using the R and NIR bands [9]. As is known, in remote sensing, the relationship between the two bands is basic because they are a test of vigour of vegetation check. VIs can be a more precise plant cover detection element than other traditional techniques, such as supervised or unsupervised classification, as they have the advantage of maintaining detailed spectral information and the image with all its sharpness. The basic principles are widely known: the vigorous vegetation, in the visible spectrum, between 400 and 700 nm (Red) appears in dark colour due to the high absorption of the pigments of its leaves (mainly chlorophyll), with a small increase of the reflectance around 550 nm (Green), in contrast to its high reflectance between 700–1300 nm (Near-Infrared).

VIs are derived from arithmetical operations performed on spectral information from the radiation reflected by the vegetation at different wavelengths [10]. These indices are used to evaluate the biophysical characteristics of the plant canopy, such as the leaf area index, biomass and physiological activities [11,12]; to classify types of vegetation cover [13]; and to determine the proportion of each type of cover with respect to the total area studied [14–17]. To our knowledge, VIs have not previously been used to quantify VGC density.

In olive cultivation (*Olea europea* L.), VIs are usually employed to differentiate types of soil cover (cover crop, bare soil and tree areas) [18] and to determine the agronomic and environmental characteristics of tree orchards [19]. They have also been used to analyse the three-dimensional geometric characteristics of individual trees and tree-rows [20] and their physiological characteristics, according to the canopy reflectance for cultivar recognition in an olive grove [21].

Traditionally, satellites have provided the platforms used for vegetation analysis [22]. However, the quantification of land cover in woody crops, such as olive groves, is subject to an important limitation, because the trees partially cover the soil, which generates high error rates in remote sensing [23]. This problem is exacerbated when the VGC presents heterogeneous densities, due to climatic seasonality and soil management (tillage vs. no tillage). Thus, in situations of scant VGC, the resulting radiance depends on both the nature of the underlying soil and the current level of vegetation cover [24,25].

The use of unmanned aerial vehicles (UAVs) for low-altitude remote sensing provides an interesting alternative to traditional detection systems [26], presenting numerous advantages [27]. For the purposes of this study, the following are particularly important: (a) UAVs provide considerable temporal flexibility for image capture, which enables periodic monitoring of the vegetation cover [28], and also enables the operator to obtain images unaffected by clouds; and (b) ultra-high resolution images can be obtained, with a low proportion of mixed pixels (in which both soil and vegetation are reflected) [15].

The aim of this study was to analyse the capacity of VIs, applied to images obtained by UAV technology, to quantify the density of VGC in olive groves.

2. Materials and Methods

2.1. Study Site

The study was carried out in 57 privately-owned olive groves located in the centre-west of the province of Malaga (southern Spain), in the municipalities of Alozaina (334494,72N, 4064214,89W ETRS 1989 UTM Zone 30N) and Casarabonela (336842, 64N, 4068297.68W ETRS 1989 UTM Zone 30N) (Figure 1a). The relief is undulating (flysch) with slopes ranging 7–30% (Figure 1b).

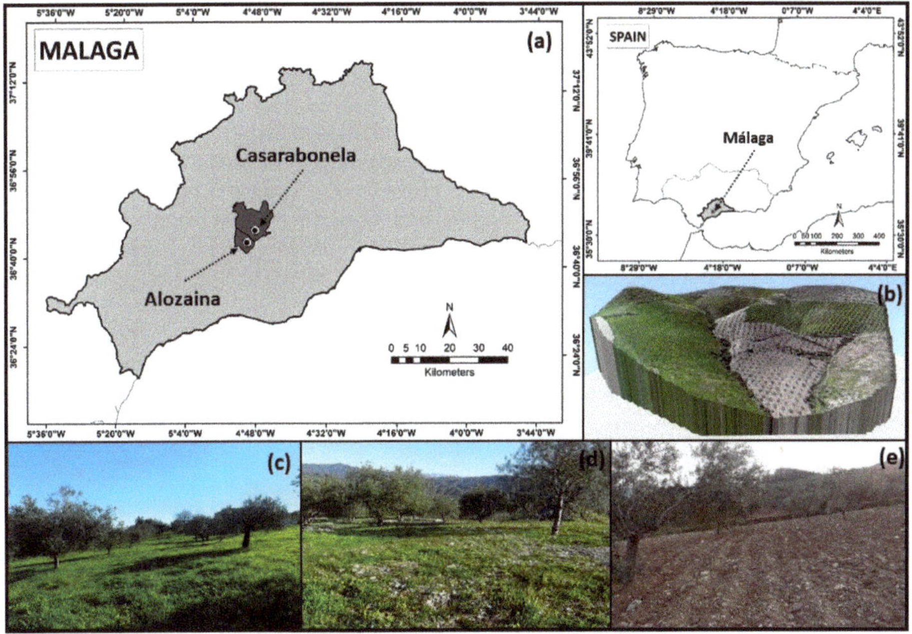

Figure 1. Study area: (**a**) location in the province of Malaga; (**b**) undulating relief and geometric planting framework; (**c**) areas with high VGC density; (**d**) areas with medium VGC density; and (**e**) no VGC (bare soil).

The predominant soils are Calcic Cambisols, Vertic Cambisols, Calcaric Regosols and Haplic Vertisols [29], with an average depth of 66.4 ± 30.9 cm and an average organic carbon content of 20.3 ± 13.5 g kg^{-1}.

The climate is temperate Mediterranean, with an average annual temperature of 18.4° C and mean annual precipitation of 636 mm. There is a period of prolonged water deficit, from April to September, which, together with the type of soil and its management, influence the development and seasonality of the VGC. These circumstances mean that the density of land cover is highly variable during the year (Figure 1c–e), with a vertical development that varies depending on the species, but in no case exceeds 1 m in height, as it is eliminated beforehand by the farmer. The VGC is mainly composed of grasses, together with weeds and ruderal nitrophytes of the Ruderali-Secalietea class. The Thero-Brometalia order is represented by species such as *Aegilops triuncialis, Bromus spp.* and *Inula viscosa*, while examples of the Chenopodietalia order (found in more nitrified environments) include species such as *Hordeum murinum* and *Malva sylvestris*, especially in compact surface soils, and *Chenopodium album* where there is more humidity and a high concentration of nitrogen [30].

Olive trees are cultivated in non-irrigated land, in geometric planting frameworks with 8–10 m between each tree, which provides an average density of 160 trees per hectare (Figure 1b). VGC is managed by means of deep (15–20 cm) and continuous mouldboard ploughing (2–3 times per year), beginning in January and continuing until June. As a result, the soil is bare for most of the year (Figure 1e), except for periods between ploughing operations, especially in the spring, which is when the VGC quickly returns (Figure 1c,d).

2.2. Quantifying the VGC Density

VGC density is the amount of vegetal cover present on the ground surface. As a preliminary step to quantifying the VGC, a field survey was carried out in each farm and the owners were interviewed to characterise the form of VGC management employed and determine the most appropriate time to perform the UAV flights and obtain the aerial images. The ideal time for this was considered to be mid-April (early spring), when the VGC presented greater variability of surface density, having been present in some cases for several months, giving rise to dense coverage (Figure 1c), while, in other plots, recently cleared, the soil was bare or had only minimal VGC (Figure 1e).

The flights were performed on 16 April at an altitude of 85 m in Alozaina and 90 m in Casarabonela. The difference in altitude was to ensure the safety of the device, because the relief is more abrupt in the Casarabonela farms. As a result of this difference, there was a slight variation in the spatial resolutions obtained (mean ground sampling distance), which were 10.11 cm/pixel and 11.18 cm/pixel in the Alozaina and Casarabonela images, respectively. This difference in altitude had no impact on the study results, as has been reported previously [16].

The method used to quantify VGC consisted of the following steps (Figure 2): (1) conduct the UAV flights and generate image mosaics; (2) apply the VIs; (3) quantify the VGC density by means of sampling plots (SP) (ground-truth); (4) calculate the mean reflectance value of the spectral bands and the VIs in the SPs; and (5) assess the VIs in order to quantify the density of the VGC.

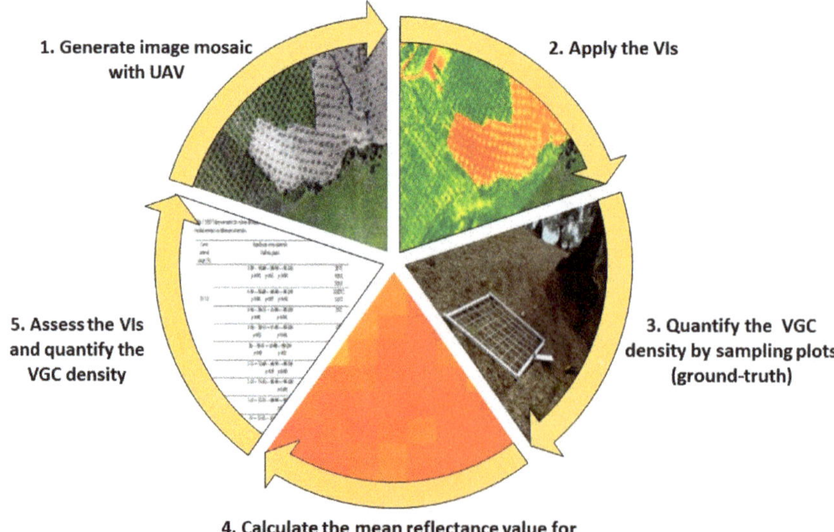

Figure 2. Stages of the method used to quantify the density of vegetation ground cover (VGC).

2.2.1. UAV Flights and Image Mosaicking

The images required for this study were obtained using a Parrot Bluegrass quadcopter (Parrot S. A, Paris, France) (Figure 3a), with vertical take off and landing. This device carries a Parrot Sequoia multispectral sensor (Parrot S. A, Paris, France) (Figure 3b), composed of four single-band global shutter cameras with a resolution of 1.2 Mpx (1280 × 960 pixels), capable of capturing four spectral bands in visible and invisible infrared light: Green (G), Red (R), Red Edge (RE), and Near Infrared (NIR). In addition, it has a brightness sensor that records the light conditions, the GPS location and inertial data.

Figure 3. UAV quadcopter (**a**) and multispectral sensor (**b**).

In each flight mission, a sequence of overlapping images (30% side-lap and 60% forward-lap) was taken of each farm in the study area. Pix4Dmapper Pro software, version 4.2.25 (Pix4D S. A, Prilly, Switzerland), was used for the mosaicking and for the radiometric calibration of the images. First, the overall orthomosaic reflectance was obtained for each band. These values were then spectrally corrected by applying an empirical linear relationship [31] from the light sensor data and from the reference photographs taken of the calibrated reflectance panel (Diana Parrot Sequoia 19 cm × 13.5 cm). The geometric calibration was obtained using the GPS parameters and the inertial measurement units from the sensor.

2.2.2. Application of the Vegetation Indices (VIs)

We selected the VIs based on two criteria. Firstly, if they used some of the four bands (G, R, RE and NIR) offered by the multispectral camera used in the flights in their calculations, they were selected. Secondly, the scientific literature has shown that the behaviour of the IVs is variable depending on the existing density of ground cover. Thus, some indices have provided better results when they have been applied in an area with high vegetation ground cover density, while others have responded better in areas with low cover. Taking into account the high heterogeneity of the VGC present in the study area, varied indices were selected to assess their responses to this situation.

The VIs used can be classified as conventional ratio or differential indices (IRVI, RVI, DVI, GVI, GRVI and VREI), indices corrected and derived from the traditional indicators (normalised difference vegetation indices) (NDVI, NRDE, NRVI and GNDVI) and soil reflectance adjusted indices (SAVI) (Table 1).

Table 1. Vegetation indices applied.

Index	Formula [a]	Reference
IRVI (Inverse Ratio Vegetation Index)	$\frac{R}{NIR}$	[32]
RVI (Ratio Vegetation Index)	$\frac{NIR}{R}$	[33]
DVI (Difference Vegetation Index)	$NIR - R$	[32]
NDVI (Normalised Difference Vegetation Index)	$\frac{NIR-R}{NIR+R}$	[34]
GNDVI (Green Normalised Difference Vegetation Index)	$\frac{NIR-G}{NIR+G}$	[35]
NDRE (Normalised Difference Red Edge)	$\frac{NIR-RE}{NIR+RE}$	[36]
GRVI (Green-Red Vegetation Index)	$\frac{G-R}{G+R}$	[37,38]
GVI (Green Vegetation Index)	$\frac{NIR}{G}$	[39]
NRVI (Normalised Ratio Vegetation Index)	$\frac{RVI-1}{RVI+1}$	[11]
SAVI (Soil-Adjusted Vegetation Index) [b]	$\frac{NIR-R}{NIR+R+L}(1+L)$	[40]
VREI (Vogelmann Rededge Index)	$\frac{NIR}{RE}$	[41]

[a] Wavelength band values G (550 nm centre, 40 nm bandwidth), R (660 nm centre, 40 nm bandwidth), RE (735 nm centre, 10 nm bandwidth), NIR (790 nm centre, 40 nm bandwidth). [b] L= Soil adjustment factor 0.5 for intermediate coverage values [42].

2.2.3. Quantifying VGC Density by Means of Sampling Plots (Ground-Truth)

VIs provide an abstract number that reproduces, in each pixel, the relationship between the bands used. These results must then be related to the real level of ground cover. To do so, a ground-truth field quantification survey of VGC density was performed in 115 sampling plots (SP), distributed randomly (1 SP per 0.5 ha of surface). No samples were taken in areas close to the tree canopies in order to avoid the influence of shade. With a surface area of 1 m^2, each SP was divided into quadrants measuring 12.5 cm × 12.5 cm, which provided 49 checkpoints per SP (Figure 4) in which the VGC was quantified by the binary classes "vegetation present" and "vegetation absent".

Figure 4. Quadrant-based method for quantifying VGC.

The SPs were located in the field using a Trimble Geo XH 6000 real-time decimetric precision GPS collector (10 cm DGNSS real-time accuracy) (Trimble GeoSpatial, Munich, Germany).

2.2.4. Calculating the Mean Reflectance Values of the Spectral Bands and the Vis in the Sampling Plots

The mean reflectance value of the spectral bands and the VIs in each SP was then calculated (Figure 5). Due to the above-discussed differences in spatial resolution, in the Alozaina farms, each SP contained 98 pixels, while in those in Casarabonela there were 80 pixels per SP.

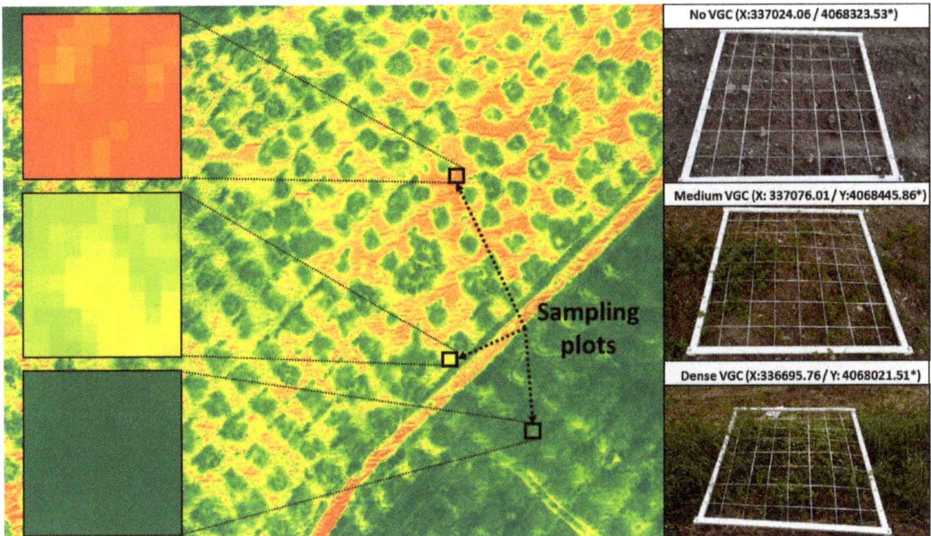

Figure 5. Random distribution of the sampling plots within the NDVI image (centre), detail of the pixels constituting the average NDVI reflectance value for each SP (left), ground-truth (right) (* Coordinate System: ETRS 1989 UTM Zone 30N).

2.2.5. Assessing the VIs to Quantify the VGC Density

The relationship among the spectral bands, the VIs and the ground-truth data was determined by linear regression analysis (stepwise method). We then determined how well the VIs differentiated VGC density at different ranges, using analysis of the variance (ANOVA) and Tukey's honestly-significant-difference (HSD) test. All statistical analyses were performed using IBM SPSS Statistics 25.0.

3. Results

Regression analysis between the reflectance of the spectral bands used alone and the VGC density shows that the R and G wavelengths have a greater explanatory capacity ($R^2 = 0.58$ and $R^2 = 0.50$, $p < 0.001$, respectively) than the NIR and RE bands ($R^2 = 0.33$ and $R^2 = 0.17$, $p < 0.001$, respectively) (Table 2). Multiple regression analysis significantly improves the results when the R and NIR bands are combined (adjusted $R^2 = 0.74$, $p < 0.001$). The equation with the R, NIR and G bands has the same explanatory capacity (Table 2).

Application of the VIs substantially improves the results. IRVI, NDVI and NRVI provide the most accurate estimates of VGC density ($R^2 > 0.81$, $p < 0.001$) (Figure 6).

Table 2. Estimating VGC density from spectral bands (n = 115).

Band	Regression Model	R	R^2	Adjusted R^2	P-Value
R	y = 108.05 − 413R	0.765	0.584	0.581	$p < 0.001$
G	y = 135.63 − 659.93G	0.710	0.504	0.500	$p < 0.001$
NIR	y = − 14.46 + 217.92NIR	0.571	0.326	0.320	$p < 0.001$
RE	y = − 6.08 + 237.197RE	0.409	0.167	0.160	$p < 0.001$
R, NIR	y = 49.53 − 358.58R + 155.76NIR	0.861	0.741	0.737	$p < 0.001$
R, NIR, G	y = 60.41−161.04R + 173.28NIR−346.05G	0.867	0.752	0.745	$p < 0.001$

G: Green; R: Red; RE: Red Edge; NIR: Near Infrared.

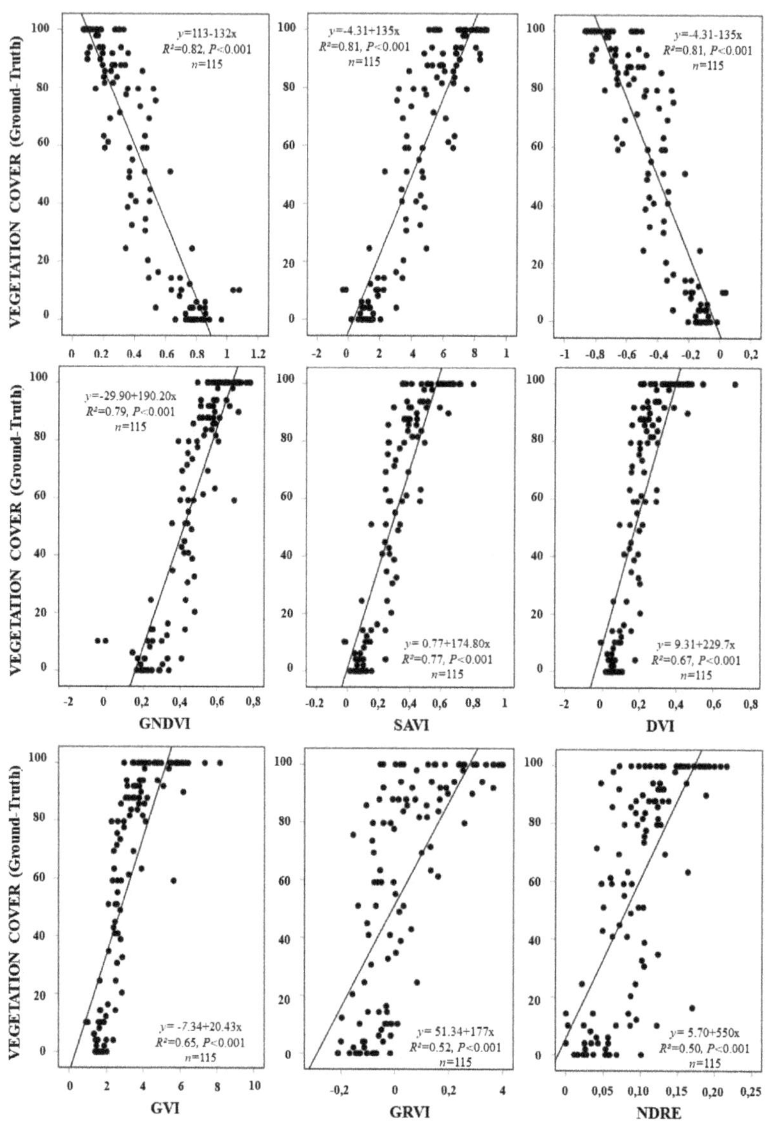

Figure 6. *Cont.*

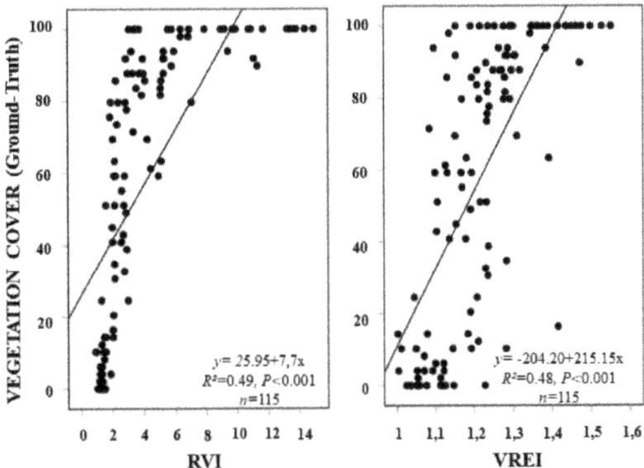

Figure 6. Scatter diagrams and regression curves of the relation between vegetation cover (Ground-Truth) and VIs.

Analysis of the capability of VIs to differentiate intervals of VGC densities highlighted the existence of important differences. Cover interval range is the reference value taken to establish the VGC density. The increase in the range in which the VGC density intervals are expressed was directly proportional to the separability of the VIs (Table 3). For a 10% cover interval range, Tukey's HSD test shows that RVI ($p < 0.001$); DVI, GVI and SAVI ($p < 0.01$); and NRVI and NDVI ($p < 0.05$) are the only VIs with the ability to differentiate VGC densities > 80%. Lower VGC densities are not differentiated by any of the VIs used.

Table 3. Tukey's HSD test of reflectance values for VIs and ground-truth VGC at different intervals. The table expresses in a range of increasing coverage interval (from 10% to 30%) the separability between coverage intervals for each VI applied. The coverage interval ranges within the parentheses correspond to VGC values greater than 90% because the entire series could not be divided into regular intervals. Bold letters indicate that there is statistically significant separability in coverage intervals, accompanied by the level of significance (p).

Cover Interval Range (%)	Significant Cover Intervals (Analysis Pairs)	VI
10	0–10 ↔ 10–20 ↔ 20–30 ↔ 30–40 ↔ 40–50 ↔ 50–60 ↔ 60–70 ↔ 70–80 ↔ **80–90 ↔ 90–100** **$p < 0.001$**	RVI
	0–10 ↔ 10–20 ↔ 20–30 ↔ 30–40 ↔ 40–50 ↔ 50–60 ↔ 60–70 ↔ 70–80 ↔ **80–90 ↔ 90–100** **$p < 0.01$**	DVI GVI SAVI
	0–10 ↔ 10–20 ↔ 20–30 ↔ 30–40 ↔ 40–50 ↔ 50–60 ↔ 60–70 ↔ 70–80 ↔ **80–90 ↔ 90–100** **$p < 0.05$**	NRVI NDVI
	0–10 ↔ 10–20 ↔ 20–30 ↔ 30–40 ↔ 40–50 ↔ 50–60 ↔ 60–70 ↔ 70–80 ↔ 80–90 ↔ 90–100	Other VIs
15 (10)	**0–15 ↔ 15–30** ↔ 30–45 ↔ 45–60 ↔ 60–75 ↔ **75–90 ↔ 90–100** **$p < 0.01$** **$p < 0.01$**	IRVI
	0–15 ↔ 15–30 ↔ 30–45 ↔ 45–60 ↔ 60–75 ↔ **75–90 ↔ 90–100** **$p < 0.05$** **$p < 0.001$**	NRVI NDVI GNDVI
	0–15 ↔ 15–30 ↔ 30–45 ↔ 45–60 ↔ 60–75 ↔ **75–90 ↔ 90–100** **$p < 0.001$**	RVI DVI GVI SAVI
	0–15 ↔ 15–30 ↔ 30–45 ↔ 45–60 ↔ 60–75 ↔ **75–90 ↔ 90–100** **$p < 0.05$**	VREI
	0–15 ↔ 15–30 ↔ 30–45 ↔ 45–60 ↔ 60–75 ↔ 75–90 ↔ 90–100	Other VIs

Table 3. *Cont.*

Cover Interval Range (%)	Significant Cover Intervals (Analysis Pairs)	VI
20	0–20 ↔ 20–40 ↔ 40–60 ↔ 60–80 ↔ 80–100 $p < 0.001$ $p < 0.001$	IRVI NRVI NDVI GNDVI
	0–20 ↔ 20–40 ↔ 40–60 ↔ 60–80 ↔ 80–100 $p < 0.01$ $p < 0.001$	SAVI
	0–20 ↔ 20–40 ↔ 40–60 ↔ 60–80 ↔ 80–100 $p < 0.05$ $p < 0.05$	NDRE
	0–20 ↔ 20–40 ↔ 40–60 ↔ 60–80 ↔ 80–100 $p < 0.001$	RVI DVI GVI
	0–20 ↔ 20–40 ↔ 40–60 ↔ 60–80 ↔ 80–100 $p < 0.05$	VREI
	0–20 ↔ 20–40 ↔ 40–60 ↔ 60–80 ↔ 80–100	Other VIs
25	0–25 ↔ 25–50 ↔ 50–75 ↔ 75–100 $p < 0.001$ $p < 0.001$	IRVI NRVI NDVI GNDVI SAVI
	0–25 ↔ 25–50 ↔ 50–75 ↔ 75–100 $p < 0.05$ $p < 0.001$	DVI NDRE
	0–25 ↔ 25–50 ↔ 50–75 ↔ 75–100 $p < 0.001$	RVI GVI GRVI VREI
	0–25 ↔ 25–50 ↔ 50–75 ↔ 75–100	Other VIs
30 (10)	0–30 ↔ 30–60 ↔ 60–90 ↔ 90–100 $p < 0.001$ $p < 0.01$ $p < 0.001$	IRVI NRVI NDVI
	0–30 ↔ 30–60 ↔ 60–90 ↔ 90–100 $p < 0.001$ $p < 0.05$ $p < 0.001$	GNDVI SAVI
	0–30 ↔ 30–60 ↔ 60–90 ↔ 90–100 $p < 0.001$ $p < 0.001$	DVI
	0–30 ↔ 30–60 ↔ 60–90 ↔ 90–100 $p < 0.01$ $p < 0.001$	GVI
	0–30 ↔ 30–60 ↔ 60–90 ↔ 90–100 $p < 0.05$ $p < 0.01$	NDRE
	0–30 ↔ 30–60 ↔ 60–90 ↔ 90–100 $p < 0.05$ $p < 0.001$	GRVI
	0–30 ↔ 30–60 ↔ 60–90 ↔ 90–100 $p < 0.001$	RVI
	0–30 ↔ 30–60 ↔ 60–90 ↔ 90–100 $p < 0.01$	VREI
	0–30 ↔ 30–60 ↔ 60–90 ↔ 90–100	Other VIs

Increasing the cover interval range to 15% raises the number of indices capable of differentiating VGC densities greater than 75% and, at the same time, improves the accuracy of those that were already significant in the previous interval (10%) (Table 3). Thus, in addition to the high level of significance already found for RVI ($p < 0.001$), NRVI, NDVI, SAVI, DVI and GVI are now significant. Moreover, new significant indices appear: GNDVI ($p < 0.001$), IRVI ($p < 0.01$) and VREI ($p < 0.05$). In this range, too, we observe the first indices capable of differentiating VGC densities below 30%, namely IRVI ($p < 0.01$) and NRVI, NDVI and GNDVI ($p < 0.05$). On the other hand, VGC densities between 30% and 75% remain undifferentiated.

The results obtained for VGC intervals of 20–25% are similar to those for the lower value (15%). New indices (NDRE and GRVI) are capable of differentiating VGC densities greater than 60% and 75%, respectively (Table 3). With VGC densities below 40% and 50%, respectively, the VIs found to be significant in the previous cover interval range (IRVI, NRVI, NDVI and GNDVI) become even more significant ($p < 0.001$), and further indices become significant, i.e., SAVI ($p < 0.01$) and NDRE ($p < 0.05$). VGC densities of 40–60% and 25–75%, respectively, continue to be undifferentiated. Only when the cover interval range reaches 30% do IRVI, NRVI and NDVI ($p < 0.01$) and GNDVI and SAVI ($p < 0.05$) discriminate the complete series of VGC density intervals (Table 3).

Figures 7 and 8 show the quantification of the VGC from the application of the regression equation obtained by IRVI. In both figures, the heterogeneity of VGC densities of the study area is clearly observed, which denotes the existence of different temporality in the management of the soil by farmers.

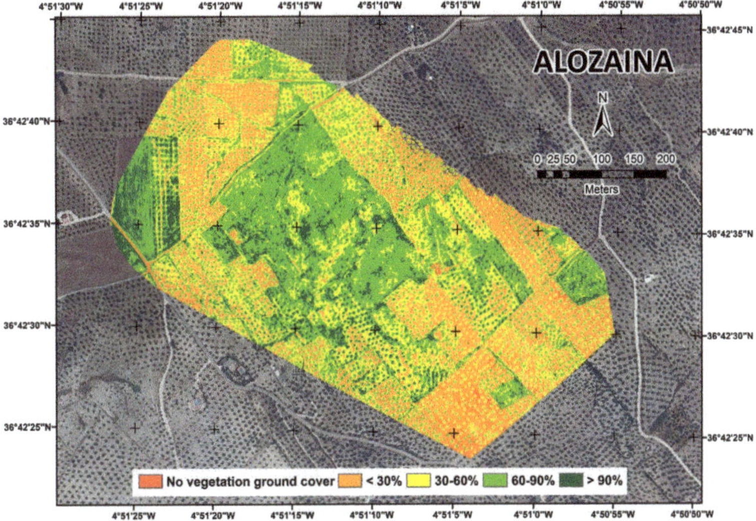

Figure 7. Quantification map of the VGC in Alozaina, made at a cover interval range of 30 % using IRVI index.

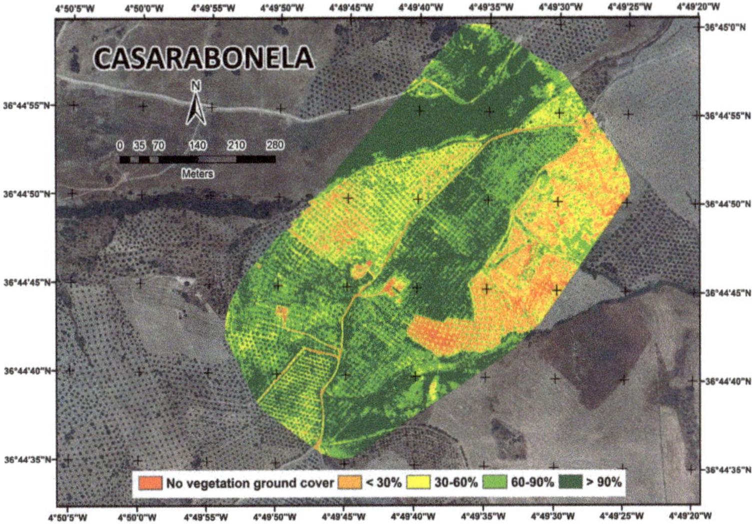

Figure 8. Quantification map of the VGC in Casarabonela, made at a cover interval range of 30% using IRVI index.

Lima et al. [8] showed for the study area that zones in which the VGC is non-existent or very scarce (with percentages below 30%) correspond to areas ploughed later (early April), when normally they are ploughed in March. For this reason, on the date of the flight (mid-April), the lands were recently ploughed and, therefore, the soil had little vegetation cover. However, the areas that present VGC intervals greater than 60% are those that have not yet were ploughed, which has allowed the ground cover development, taking into account that the last ploughing was approximately in the months of January and February. All these temporality differences in soil management, an aspect that is normal in situations of real use outside the controlled conditions of experimental farmland, increase the demands on the VGC detection and quantification system.

4. Discussion

According to the study results obtained, VIs can be used to quantify VGC density, increasing the vegetation reflectance information obtained by the spectral bands alone. The IRVI, NDVI and NRVI provide the most accurate estimates of VGC density ($R^2 > 0.81$, $p < 0.001$) in the study area. Although these VIs obtain very similar values, the best result is obtained with IRVI ($R^2 = 0.82$, $p < 0.001$). The ability of this index to measure green herbaceous biomass at the end of the rainy season was reported by Verbesselt et al. [43]. In the present analysis, we show that the results obtained by the inverse indices such as IRVI are substantially better than those from simple indices such as RVI ($R^2 = 0.49$, $p < 0.001$). The remaining simple VIs (DVI, GVI and VREI) present the lowest values, and only account for 48–67% of VGC variability.

The corrected indices derived from traditional indicators are also interesting. Normalisation substantially improves the results of the VIs. One example of this improvement is that of NDVI ($R^2 = 0.81$, $p < 0.001$). This same index, without normalisation (i.e., DVI), is less able to distinguish VGC ($R^2 = 0.67$, $p < 0.001$). The difference is even more apparent with NRVI ($R^2 = 0.81$, $p < 0.001$), which prior to normalisation (RVI) had a 32% poorer explanatory capability ($R^2 = 0.49$, $p < 0.001$). These results are consistent with those obtained by Carlson and Ripley [44] and by Hassan et al. [45], who obtained good results with normalised indices, such as NDVI, to estimate the fraction of vegetation cover. Index normalisation usually improves the results obtained because it provides a greater separation of the green vegetation from its background soil brightness [34] and reduces the effects produced by topographic, atmospheric and lighting factors [10].

The soil reflectance adjusted index (SAVI) is not among the best-performing indices ($R^2 = 0.77$, $p < 0.001$), because the study area considered has a predominance of areas with high cover density (in 44% of the SP the VGC was greater than 80%) and this VI was designed to analyse areas with little vegetation cover [40].

In our study, the VGC is best quantified by the VIs based on the R and NIR bands, due to the spectral behaviour of the vegetation; the chlorophyll absorbs a greater proportion of the electromagnetic waves in the R region, and high reflectance values are observed in the NIR region due to the microcellular structures of the leaf material [46].

However, this is not the case with the VIs derived from the G or RE bands. While the G band, used individually, obtains an acceptable result ($R^2 = 0.50$, $p < 0.001$), its incorporation into the VIs worsens their performance. This deterioration is apparent with GNDVI ($R^2 = 0.79$, $p < 0.001$) and even more so with GRVI ($R^2 = 0.52$, $p < 0.001$), with respect to NDVI ($R^2 = 0.81$, $p < 0.001$). Both of these VIs perform worse when the G band replaces the R band (GNDVI) or the NIR band (GRVI). These results corroborate those of Khajeddin [14] and Barati et al. [47], who reported that the use of the G band reduces the sensitivity of the VIs.

The RE band, used individually, does not obtain good results ($R^2 = 0.17$, $p < 0.001$). This is reflected in the VIs that incorporate this band, such as VREI ($R^2 = 0.48$, $p < 0.001$) and NDRE ($R^2 = 0.50$, $p < 0.001$). From these results, we conclude that VIs based on the RE band are relatively insensitive to the quantification of VGC, although Dong et al. [48] stated that RE-based VIs are more sensitive to chlorophyll and can be used to derive an empirical model for estimating the leaf area index in different crops.

As expected, the behaviour of the VIs in response to different VGC densities is not homogeneous, but improves in line with the increase in cover interval ranges. Tukey's HSD test shows that the most suitable indices to quantify areas with VGC densities greater than 80%, at a cover interval range of 10%, are RVI ($p < 0.001$); DVI, GVI and SAVI ($p < 0.01$); and NRVI and NDVI ($p < 0.05$). NDRE and VREI are expected to obtain very similar results, since these indices are normally more robust and perform better in areas of greater canopy density [49], and hence no saturation deficiencies [50–52]. These VIs begin to be significant ($p < 0.05$) at a cover range of 15% (VREI) and 20% (NDRE).

For the discrimination of areas with a VGC density of less than 30%, IRVI ($p < 0.01$) is the most significant at a cover interval range of 15%. Its estimation capacity decreases as the biomass increases, since the greater is the biomass, the lower is the reflectance of the R band and the greater is the reflectance of the NIR band [43].

IRVI, NRVI and NDVI ($p < 0.001$ and $p < 0.01$) and GNDVI and SAVI ($p < 0.001$ and $p < 0.05$) only achieve a significant differentiation at all coverage intervals within the same range when the cover interval range reaches 30%. From these results, we conclude that VIs are especially useful for detecting and quantifying homogeneous surfaces, such as areas that are either completely covered or have very little vegetation cover. However, when the VGC is slight or moderate, the reflectance measured does not depend exclusively on the vegetation cover, but also on other factors, such as the soil. The VI is then a less precise instrument. As mentioned above, the analysis of land cover in woody crops, such as olive groves, tends to generate high error rates in remote sensing. In the present study, the differentiation capacity of the VIs was severely tested by their use in a region of very heterogeneous soil cover; some of the olive groves had not been cleared for several months and so the VGC was quite dense, while in others the soil was bare, and subjected to diverse soil management regimes (tillage vs. no tillage). In this context, it can be considered normal that some of the VIs were only able to significantly differentiate all vegetation cover intervals with the same range above a cover interval range of 30%. Better results are to be expected for crops or areas in which the ground cover is more homogeneous.

5. Conclusions

In this paper, we show that UAV technology, together with image processing based on VIs, makes it possible to remotely quantify the density of VGC produced spontaneously in olive groves. Of the 11 VIs considered, IRVI was the most sensitive to quantify VGC density at intervals of 10–25%. Only when the cover interval range rose to 30% did IRVI, NRVI, NDVI, GNDVI and SAVI differentiate the complete series of densities.

The study described in this paper provides: (a) a better understanding of the behaviour of VIs in response to different soil cover densities; (b) a demonstration of the ability to remotely quantify the VGC in olive groves with heterogeneous soil cover; (c) a contribution to providing control and monitoring tools enabling recipients of CAP benefits to comply with cross-compliance requirements in terms of minimum soil cover; and (d) to know the temporality of the operations carried out in the soil, which is of great importance to adapt this to the rainfall conditions of the area, and avoid the existence of bare soils in the periods of more intense rainfall.

However, due to the dynamic, heterogeneous nature of the VGC in olive groves, further research is needed in this area, applying the method to images obtained in other seasons (i.e., in summer, autumn and winter), and in regions where there is less ground cover heterogeneity.

Author Contributions: F.J.L.-C. designed and processed the research and drafted the manuscript; F.J.L.-C., R.B.-S., and M.L.G.-M. worked on contextualising the document; F.J.L.-C. and R.B.-S. determined the approach taken to the study methodology, formal analysis and validation; and F.J.L.-C., R.B.-S., M.L.G.-M. and F.B.G.-J. played an active role in writing, reviewing and editing the manuscript. All authors contributed substantially to the paper.

Funding: This study was conducted within the framework of a predoctoral contract (A.2) under the I Research and Transfer Plan of the University of Málaga, and was also funded by the University of Málaga through the mode B3 of assistance for research projects.

Acknowledgments: Invaluable cooperation was provided by the farmers of the lower Guadalhorce Valley-Sierra de las Nieves and by the company TYC-GIS Soluciones Integrales (http://tycgis.com/), to whom we express our gratitude.

Conflicts of Interest: The authors declare they have no conflict of interest.

References

1. FAO. Food and Agriculture Organization of the United Nations, FAOSTAT. 2016. Available online: http://www.fao.org/faostat/es/#data/QC (accessed on 6 September 2018).
2. INE. National Institute of Statistics of Spain. Agrarian Census. 2009. Available online: https://www.ine.es/dyngs/INEbase/es/operacion.htm?c=Estadistica_C&cid=1254736176851&menu=resultados&secc=1254736194950&idp=1254735727106 (accessed on 6 September 2018).
3. Infante, J. The ecology and history of the Mediterranean olive grove: The Spanish great expansion, 1750–2000. *Rural Hist.* **2012**, *23*, 161–184. [CrossRef]
4. Lima-Cueto, F.J.; Gómez-Moreno, M.L.; Blanco-Sepúlveda, R. Mountain olive groves and soil preservation in the transition from an organic to an industrial economy: The case of Sierra de las Nieves (Málaga, Spain), 1940–1975. *J. Depopulation Rural Dev. Stud.* **2017**, *23*, 97–128.
5. Moreno, B.; García-Rodriguez, S.; Cañizares, R.; Castro, J.; Benítez, E. Rainfed olive farming in south-eastern Spain: Long-term effect of soil management on biological indicators of soil quality. *Agric. Ecosyst. Environ.* **2009**, *131*, 333–339. [CrossRef]
6. Brussaard, L.; Kuyper, T.W.; Didden, W.A.M.; de Goede, R.G.M.; Bloem, J. Biological soil quality from biomass to biodiversity–Importance and resilience to management stress and disturbance. In *Managing Soil Quality: Challenges in Modern Agriculture*; Schjønning, P., Elmholt, S., Christensen, B.T., Eds.; CAB International: Wallingford, UK, 2004; pp. 139–171.
7. Sastre, B.; Marques, M.J.; García-Díaz, A.; Bienesa, R. Three years of management with cover crops protecting sloping olive groves soils, carbon and water effects on gypsiferous soil. *Catena* **2018**, *171*, 115–124. [CrossRef]
8. Lima-Cueto, F.J.; Blanco-Sepúlveda, R.; Gómez-Moreno, M.L. Soil erosion and environmental regulations in the European Agrarian Policy for olive groves (*Olea europea* L.) of southern Spain. *Agrociencia* **2018**, *52*, 293–308.
9. Fang, H.; Liang, S. Leaf Area Index Models. Reference Module in Earth Systems and Environmental Sciences. *Encycl. Ecol.* **2014**, 2139–2148. Available online: https://www.sciencedirect.com/science/article/pii/B9780080454054001907 (accessed on 1 September 2018).
10. Broge, N.H.; Leblanc, E. Comparing prediction power and stability of broadband and hyperspectral vegetation indices for estimation of green leaf area index and canopy chlorophyll density. *Remote Sens. Environ.* **2000**, *76*, 156–172. [CrossRef]
11. Baret, F.; Guyot, G. Potentials and limits of vegetation indices for LAI and APAR assessment. *Remote Sens. Environ.* **1991**, *35*, 161–173. [CrossRef]
12. Richardson, A.J.; Wiegand, C.L.; Wajura, D.F.; Dusek, D.; Steiner, J.L. Multisite analyses of spectral-biophysical data for sorghum. *Remote Sens. Environ.* **1992**, *41*, 71–82. [CrossRef]
13. Song, W.; Mu, X.; Ruan, G.; Gao, Z.; Li, L.; Yan, G. Estimating fractional vegetation cover and the vegetation index of bare soil and highly dense vegetation with a physically based method. *Int. J. Appl. Earth Obs. Geoinf.* **2017**, *58*, 168–176. [CrossRef]
14. Xiao, J.; Moody, A. A comparison of methods for estimating fractional green vegetation cover within a desert-to-upland transition zone in central New Mexico, USA. *Remote Sens. Environ.* **2005**, *98*, 237–250. [CrossRef]
15. Barati, S.; Rayegani, B.; Saati, M.; Sharifi, A.; Nasri, M. Comparison the accuracies of different spectral indices for estimation of vegetation cover fraction in sparse vegetated areas. *Egypt. J. Remote Sens. Space Sci.* **2011**, *14*, 49–56. [CrossRef]
16. Torres, J.; Peña, J.M.; De Castro, A.I.; Lopez, F. Multi-temporal mapping of the vegetation fraction in early-season wheat fields using images from UAV. *Comput. Electron. Agric.* **2014**, *103*, 104–113. [CrossRef]

17. Yun, H.S.; Park, S.H.; Kim, H.J.; Lee, W.D.; Lee, K.D.; Hong, S.Y.; Jung, G.H. Use of unmanned aerial vehicle for multi-temporal monitoring of soybean vegetation fraction. *J. Biosyst. Eng.* **2016**, *41*, 126–137. [CrossRef]
18. Peña-Barragán, J.M.; Jurado-Expósito, M.; López-Granados, F.; Atenciano, S.; Sánchez-De la Orden, M.; García-Ferrer, A.; García-Torres, L. Assessing land-use in olive groves from aerial photographs. *Agric. Ecosyst. Environ.* **2004**, *103*, 117–122. [CrossRef]
19. García, L.; Peña-Barragán, J.M.; López-Granados, F.; Jurado-Expósito, M.; Fernández-Escobar, R. Automatic assessment of agro-environmental indicator from remotely sensed images of tree orchards and its evaluation using olive plantions. *Comput. Electron. Agric.* **2008**, *61*, 179–191. [CrossRef]
20. Torres-Sánchez, J.; López-Granados, F.; Serrano, N.; Arquero, O.; Peña, J.M. High-throughput 3-d monitoring of agricultural-tree plantations with unmanned aerial vehicle (UAV) technology. *PLoS ONE* **2015**, *10*, e0130479. [CrossRef] [PubMed]
21. Avola, G.; Di Gennaro, S.F.; Cantini, C.; Riggi Ezio Muratore, F.; Tornambe, C.; Matese, A. Remotely sensed vegetation indices to discriminate field-grown olive cultivars. *Remote Sens.* **2019**, *11*, 1242. [CrossRef]
22. De Castro, A.I.; Six, J.; Plant, R.E.; Peña, J.M. Mapping Crop Calendar Events and Phenology-Related Metrics at the Parcel Level by Object-Based Image Analysis (OBIA) of MODIS-NDVI Time-Series: A Case Study in Central California. *Remote Sens.* **2018**, *10*, 1745. [CrossRef]
23. Pinilla, C.; Ariza, F.J.; Sánchez, M.; Tovar, J. Improvement of the reliability in the identification of the olive grove using a geometric reflectance model. *J. Remote Sens.* **2001**, *16*, 11–15. (In Spanish)
24. Huete, A.R.; Jackson, R.D.; Post, D.F. Spectral response of a plant canopy with different soil backgrounds. *Remote Sens. Environ.* **1985**, *17*, 37–53. [CrossRef]
25. Iaquinta, J.; Fouilloux, A. Influence of the heterogeneity and topography of vegetated land surfaces for remote sensing applications. *Int. J. Remote Sens.* **1988**, *19*, 1711–1723. [CrossRef]
26. Saberioon, M.M.; Amin, M.S.M.; Anuar, A.R.; Gholizadeh, A.; Wayayok, A.; Khairunniza-Bejo, S. Assessment of rice leaf chlorophyll content using visible bands at different growth stages at both the leaf and canopy scale. *Int. J. Appl. Earth Obs. Geoinf.* **2014**, *32*, 35–45. [CrossRef]
27. De Castro, A.I.; Jiménez-Brenes, F.M.; Torres-Sánchez, J.; Peña, J.M.; Borra-Serrano, I.; López-Granados, F. 3-D Characterization of Vineyards Using a Novel UAV Imagery-Based OBIA Procedure for Precision Viticulture Applications. *Remote Sens.* **2018**, *10*, 584. [CrossRef]
28. Catur, A.R. The potential of UAV-based remote sensing for supporting precision agriculture in Indonesia. *Procedia Environ. Sci.* **2015**, *24*, 245–253.
29. FAO. *World Reference Base for Soil Resources 2014. International Soil Classification System for Naming Soils and Creating Legends for Soil Maps. Update 2015*; Food and Agriculture Organization of the United Nations: Rome, Italy, 2014; ISSN 0532-0488.
30. Foraster, L.; Gómez, E.J.; Iglesias, I.; Macías, F.J.; Ruiz, M. *Study for the Delimitation of Experimental Zones. Olive Grove of the Sierra de Las Nieves Region. Characterization and Recommendations for Ecological Management*; Mancomunidad de Municipios Sierra de las Nieves: Málaga, Spain, 2011; ISBN 978-84-930235-5-4. (In Spanish)
31. Hunt, E.R.; Hively, W.D.; Fujikawa, S.J.; Linden, D.S.; Daughtry, C.S.T.; McCarty, G.W. Acquisition of NIR-green–blue digital photographs from unmanned aircraft for crop monitoring. *Remote Sens.* **2010**, *2*, 290–305. [CrossRef]
32. Richardson, A.J.; Wiegand, C.L. Distinguishing vegetation from soil background information. *Photogramm. Eng. Remote Sens.* **1977**, *43*, 1541–1552.
33. Birth, G.S.; Birth, G.R. Measuring the color of growing turf with a reflectance spectrophotometer. *Am. Soc. Agron.* **1968**, *60*, 640–643. [CrossRef]
34. Rouse, J.W.; Haas, R.H.; Schell, J.A.; Deering, D.W. Monitoring vegetation systems in the Great Plains with ERTS. In *Third Earth Resources Technology Satellite–1 Syposium*; Freden, S.C., Mercanti, E.P., Becker, M., Eds.; Technical Presentations, NASA SP-351; NASA: Washington, DC, USA, 1973; Volume 1, pp. 309–317.
35. Gitelson, A.A.; Kaufman, Y.J.; Merzlyak, M.N. Use of a green channel in remote sensing of global vegetation from EOS-MODIS. *Remote Sens. Environ.* **1996**, *58*, 289–298. [CrossRef]
36. Gitelson, A.A.; Merzlyak, M.N. Spectral reflectance changes associated with autumn senescence of *Aesculus hippocatanum* I and *Hacer plantanoides* I. leaves. Spectral features and relation to chlorophyll estimation. *J. Plant Physiol.* **1994**, *143*, 286–292. [CrossRef]
37. Tucker, C.J. Red and photographic infrared linear combinations for monitoring vegetation. *Remote Sens. Environ.* **1979**, *8*, 127–150. [CrossRef]

38. Falkowski, M.J.; Gessler, P.E.; Morgan, P.; Hudak, A.T.; Smith, A.M.S. Characterizing and mapping forest fire fuels using ASTER imagery and gradient modeling. *For. Ecol. Manag.* **2005**, *217*, 129–146. [CrossRef]
39. Cruden, B.A.; Prabhu, D.; Martinez, R. Absolute radiation measurement in Venus and Mars entry conditions. *J. Spacecr. Rocket.* **2012**, *49*, 1069–1079. [CrossRef]
40. Huete, A.R. A soil vegetation adjusted index (Savi). *Remote Sens. Environ.* **1988**, *25*, 295–309. [CrossRef]
41. Vogelmann, B.N.; Rock, D.M. Moss Red edge spectral measurements from sugar maple leaves. *Int. J. Remote Sens.* **1993**, *14*, 1563–1575. [CrossRef]
42. Richardson, A.; Everitt, J. Using spectral vegetation indices to estimate rangeland productivity. *Geocarto. Int.* **1992**, *1*, 63–77. [CrossRef]
43. Verbesselt, J.; Somers, B.; Van Aardt, J.; Jonckheere, I.; Coppin, P. Monitoring herbaceous biomass and water content with spot vegetation time-series to improve fire risk assessment in savanna ecosystems. *Remote Sens. Environ.* **2006**, *101*, 399–414. [CrossRef]
44. Carlson, D.A.; Ripley, D.A. On the relation between NDVI, fractional vegetation cover, and leaf area index. *Remote Sens. Environ.* **1997**, *62*, 241–252. [CrossRef]
45. Hassan, M.A.; Yang, M.; Rasheed, A.; Yang, G.; Reynolds, M.; Xia, X.; Xiao, Y.; He, Z. A rapid monitoring of NDVI across the wheat growth cycle for grain yield prediction using a multi-spectral UAV platform. *Plant Sci.* **2019**, *282*, 95–103. [CrossRef]
46. Tucker, C.J. A spectral method for determining the percentage of green herbage material in clipped samples. *Remote Sens. Environ.* **1980**, *9*, 175–181. [CrossRef]
47. Khajeddin, S.J. *A Survey of the Plant Communities of the Jazmorian, Iran, Using Landsat MSS Data*; Reading University: Reading, UK, 1995.
48. Dong, T.; Lui, J.; Shang, J.; Qian, B.; Ma, B.; Kovacs, J.M.; Walters, D.; Jiao, X.; Geng, X.; Shi, Y. Assessment of red-edge vegetation indices for crop leaf area index estimation. *Remote Sens. Environ.* **2019**, *222*, 133–143. [CrossRef]
49. Mutanga, O.; Skidmore, A.K. Narrow band vegetation indices overcome the saturation problem in biomass estimation. *Int. J. Remote Sens.* **2004**, *25*, 3999–4014. [CrossRef]
50. Huete, A.; Didan, K.; Miura, T.; Rodriguez, E.P.; Gao, X.; Ferreira, L.G. Overview of the radiometric and biophysical performance of the MODIS vegetation indices. *Remote Sens. Environ.* **2002**, *83*, 195–213. [CrossRef]
51. Eitel, J.U.H.; Vierling, L.A.; Litvak, M.E.; Long, D.S.; Schulthess, U.; Ager, A.A.; Krofcheck, D.J.; Stoscheck, L. Broadband, red-edge information from satellites improves early stress detection in a New Mexico conifer woodland. *Remote Sens. Environ.* **2011**, *115*, 3640–3646. [CrossRef]
52. Marx, A. *The Impact of the RapidEye Red Edge Band in Mapping Defoliation Symptoms*; ESA Living Planet Symposium: Edinburgh, UK, 2013; Available online: http://seom.esa.int/LPS13/5405ddfb/ (accessed on 1 July 2017).

Article

Estimating and Examining the Sensitivity of Different Vegetation Indices to Fractions of Vegetation Cover at Different Scaling Grids for Early Stage *Acacia* Plantation Forests Using a Fixed-Wing UAS

Kotaro Iizuka [1,*], Tsuyoshi Kato [2], Sisva Silsigia [2], Alifia Yuni Soufiningrum [2] and Osamu Kozan [3,4]

1. Center for Spatial Information Science, The University of Tokyo, Kashiwa 277-8568, Japan
2. PT Mayangkara Tanaman Industri, Pontianak 787391/PT Wana Subur Lestari, Jakarta 10270, Indonesia
3. Research Institute for Humanity and Nature, Kyoto 603-8047, Japan
4. Center for Southeast Asian Studies, Kyoto University, Kyoto 606-8501, Japan

* Correspondence: kiizuka@csis.u-tokyo.ac.jp

Received: 12 July 2019; Accepted: 1 August 2019; Published: 3 August 2019

Abstract: Understanding the information on land conditions and especially green vegetation cover is important for monitoring ecosystem dynamics. The fraction of vegetation cover (FVC) is a key variable that can be used to observe vegetation cover trends. Conventionally, satellite data are utilized to compute these variables, although computations in regions such as the tropics can limit the amount of available observation information due to frequent cloud coverage. Unmanned aerial systems (UASs) have become increasingly prominent in recent research and can remotely sense using the same methods as satellites but at a lower altitude. UASs are not limited by clouds and have a much higher resolution. This study utilizes a UAS to determine the emerging trends for FVC estimates at an industrial plantation site in Indonesia, which utilizes fast-growing *Acacia* trees that can rapidly change the land conditions. First, the UAS was utilized to collect high-resolution RGB imagery and multispectral images for the study area. The data were used to develop general land use/land cover (LULC) information for the site. Multispectral data were converted to various vegetation indices, and within the determined resolution grid (5, 10, 30 and 60 m), the fraction of each LULC type was analyzed for its correlation between the different vegetation indices (Vis). Finally, a simple empirical model was developed to estimate the FVC from the UAS data. The results show the correlation between the FVC (acacias) and different Vis ranging from R^2 = 0.66–0.74, 0.76–0.8, 0.84–0.89 and 0.93–0.94 for 5, 10, 30 and 60 m grid resolutions, respectively. This study indicates that UAS-based FVC estimations can be used for observing fast-growing acacia trees at a fine scale resolution, which may assist current restoration programs in Indonesia.

Keywords: UAS; UAV; vegetation cover; multispectral; land cover; forest; *Acacia*; Indonesia; tropics

1. Introduction

Quantitative assessments of the green vegetative covers of terrestrial environments are essential for understanding ecosystem dynamics. The functions of green environments (e.g., vegetation, forests) provide important benefits to ecosystems, such as controlling air quality through photosynthesis, generating an energy supply from woody biomass, preventing soil erosion, improving water quality and balancing the heat fluxes of the earth [1–5]. The worldwide terrestrial environment is currently showing rapid changes in particular regions from anthropogenic activities, causing land degradations that engulf the natural environment [6]. Some areas show transitions away from green areas, which results in substantial impacts on local to global ecosystems, sociocultural and economic impacts [7,8].

To mitigate these impacts, known cooperative agencies and organizations are planning actions for the recovery of green areas [9,10]. Consistent monitoring of the rapid environmental changes in green vegetative coverages is important for conservation and maintaining the sustainability of the natural environment.

Indonesia's landmass includes approximately 24 million hectares (Mha) of peatlands, which represents approximately 83% of the peatlands found in Southeast Asia. Peatlands in Indonesia are distributed mainly among the four large islands of Sumatera (9.2 Mha), Kalimantan (4.8 Mha), Sulawesi (0.06 Mha) and Papua (6.6 Mha) [11]. One of the common uses of peatlands in Indonesia is for Industrial Forest Plantations (IFP). The area of IFP concessions in Indonesia, which are located on peatlands, is 2 Mha [12], or 54.79% of the total IFP area in 2006, which reached 3.65 Mha [13], and *Acacia crassicarpa* is a fast-growing species that has been developed as a staple plant for most IFPs on peatlands [14].

Indonesia peatland forests provide important local and global benefits. However, their drainage and conversion into agricultural lands without well-planned management has caused considerable and irreversible environmental, social and economic damage. The catastrophic 2015 fires in Indonesia [15] drew national and international attention. That event reinforced the commitments of the Indonesian government to both reduce peatland deforestation and fires and to rehabilitate and restore degraded peatlands via reforestation. Strategic and operational approaches for monitoring the peat ecosystem together with the conditions of the green vegetation are crucial.

Various researchers and institutions have performed related studies of quantitative analyzations of both local and global vegetation coverage. The products of MODIS Vegetation Continuous Fields or the fCover (fraction of vegetation cover, hereafter denoted as FVC) [16] were used, while other researchers utilized the MODIS reflectance data provided by Land Processes Distributed Active Archive Center (LP DAAC) for developing improved FVC data [17]. The remote sensing techniques for FVC development utilize the multispectral information observed from space and validates its product with ground truth information (e.g., field surveys). The estimation methods can vary depending on the model type used, including simple empirical models [18], linear spectral models [19], decision tree method [20], machine learning techniques [21], and so on. Although the input information is rather simple, remote sensing can use various spectral information or the computed vegetation indices (Vis) for its estimation, although correctly delineating the FVC for various regions of the world is still a challenge.

Many studies have indicated the capability of space-borne remotely sensed data for mapping and/or monitoring of regional to global vegetation cover. However, depending on the products or specific locations used, there can be constraints and challenges in processing or accurately estimating a fractional cover. One of the conventional issues observed is the cloud cover, which blocks otherwise available information for analyzing a terrestrial environment [22]. The missing information can be aided with a gap-filling technique [17] for including continuous land information. However, this technique can result in large differences in the spectral information of the area, and then the possibility of an incorrect FVC estimation rises. In tropical regions including Indonesia, obtaining sufficient land information from areas with lesser cloud cover can be challenging; even when cloud removal and gap filling are performed, correct land information for a certain period of time can be lacking. If a large area of peatlands in Indonesia is observed in IFPs with fast-growing trees, then even a small temporal gap of land information may show erroneous assumptions of the vegetation coverage (Figure 1). To accurately and effectively detect green areas, considering different platforms or sensors is an important step for effectively monitoring the tropical areas. This is especially true in Indonesia where rapid land transitions are occurring.

Figure 1. Example of temporal differences for fast-growing species. The left image shows the fast-growing *Acacia* trees in its early stages in August, 2018, while the right image shows its rapid growth in October, 2018. Even with small temporal differences, the situation of the land area would change dramatically.

In recent years, many studies using unmanned aerial systems (UASs) were carried out. The UAS platform provides alternatives to space-borne platforms since optical data can be observed in a clear-high spatial/temporal resolution for the region of interest [23]. This technique has been used in research on ecology [24], precision agriculture/forestry [25,26] and even analyses for estimating vegetation cover [27–29]. UAS was successful [27] in clearly estimating vegetation fractions and flower fractions in crop fields with the changing VIs, and work by Chen et al. [28] showed that utilizing UAS-captured imagery may clearly detect grassy vegetation covers due to its high-resolution data. Riihimäkia et al. [29] showed that the UAV-derived information can be aided by satellite-observed information in FVC estimations. As Indonesia is exposed to high and frequent cloud coverage nationwide, obtaining clear satellite information is often difficult. Even if this information is collected, radiometric corrections for both atmospheric and topographic data are mandatory, which is a difficult task [30]. Riihimäkia et al. [29] recently showed an approach for estimating the FVC at arctic vegetation using UAS and satellite data through multiple spatial scaling's and different indices. They have expressed that there is a strong correlation between the UAS-based FVC for validation data that can be used to bridge with the satellite data and noted that the sensitivity of VIs was better when using Red-edge or Short Wave Infrared (SWIR) information. The prior study of Riihimäkia et al. [29] shows a relationship analysis between the VIs and FVC that is based on only a single class that classified the area into vegetation/non-vegetation. Depending on regions where heterogeneous land use/land cover (LULC) types are seen, there may be more classes requiring further analysis and how those classes affect the VI response. Minimal research has been conducted in rapid changing environments such as Indonesia for estimating the fractional cover of green vegetation by utilizing UASs, especially in rapidly growing industrial forest plantations (IFP). Higher spatial/temporal resolution imagery may have a high potential to analyze where the changes in green vegetative cover are occurring.

The objective of this study is to develop a method for retrieving the FVC by utilizing UAS and multispectral sensors for the fast-growing *Acacia* plantation forests in Indonesia. Several VIs are computed using the raw band information to compare the sensitivity of the VIs to FVC, moreover, the result is also compared at different spatial resolutions and with other LULC types. The developed product is compared with the existing method for computing FVC by using satellite imagery, and it examines how the UAS observed product can compensate for conventional space-borne products. This work mainly focus on if UAV-based FVC can be obtained in the forested area, while it is out of the scope at the moment for generalizing the result which could be utilized for global estimations. This study may present advances in UAS research in developing FVC estimations and the possibility of utilizing the platform for collecting ground truth information to bridge airborne sensing with space-borne sensing.

2. Study Area

The study site is located in West Kalimantan and is the same area used in Iizuka et al. [31] (Figure 2). The large plantation area is managed by an industrial plantation company. The area was planted in January 2017 with *Acacia crassicarpa* as the main commercial species, which is one of the fast-growing species that can grow from saplings to up to a few meters in one year. Usually, the plantation site has a cycle of planting to logging in four-year intervals, which is a dramatic change rate. One section of the compartment site is considered for the test. Brief details of the area are explained in Iizuka et al. [31].

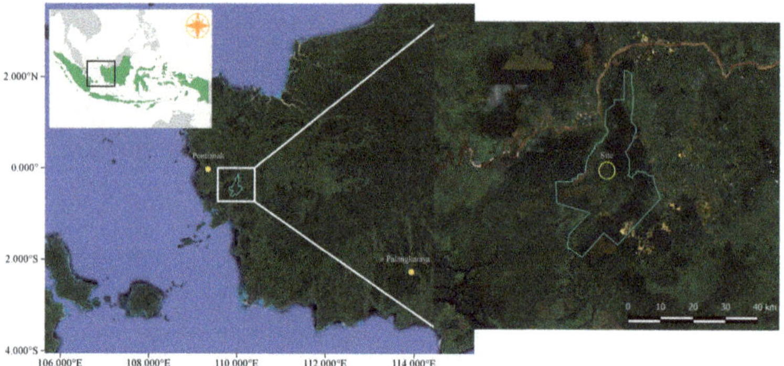

Figure 2. Location of the study site at West Kalimantan.

3. Materials and Methods

The overall workflow is shown in Figure 3. The data from the UAS were collected for RGB imagery and multispectral data. These data were used to produce a LULC map of the study area and to compute the fractions of the LULC classes. The multispectral data were used to further compute multiple VIs. Different scaled polygons were generated and within each grid, the fractional coverage and VI values were analyzed for their relationships. The data were further divided into training and validation sets, and an estimation of FVC was implemented for different spatial resolutions.

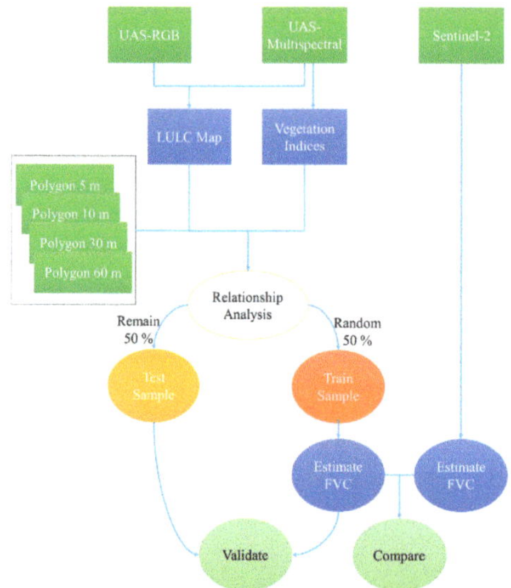

Figure 3. Overall flowchart of the methodology.

3.1. Coordinate Observation for Ground Control Points

3.1.1. GCP Collection Using Low Cost GNSS

For improving the 3D modeling and georeferencing, ground control points (GCPs) were placed and coordinates were collected before the flight campaign. The Reach (Emlid, Hong Kong, China) global navigation satellite system (GNSS) was utilized for collecting the XYZ geographical coordinates (Figure 4). A total of two stations were used. One was set on the tip of a long envy pipe and was fixed at 4 m high off the ground so the surrounding obstacles would not block the GNSS signals; this station was used as a base station. The other receiver was placed on a tripod and utilized as a rover station during the collection of coordinates at each GCP target. A total of six GCPs were observed and the GNSS signals at each GCP were recorded for 5 min. The signals were further processed with a post-processing kinematic (PPK) method to enhance the XYZ coordinate precision. Depending on the satellite signals and processing, the coordinate data can enhance the precision using the PPK method up to a centimeter-level error; when using non-processed GNSS data, the precision can result in a 10 m error (when only using the GPS [32]). The Reach GNSS was set to observe GNSS signals from GPS, QZSS, Galileo and Beidou at a logging frequency of 5 Hz.

Figure 4. Small and light-weight global navigation satellite system (GNSS) equipment for post-processing kinematic (PPK).

3.1.2. PPK Processing for Precise GCP Coordinates

For the postprocessing of the GNSS signals, the open source RTKLIB software [33] was used. RTKLIB is a program package for standard and precise positioning with GNSS and can perform a PPK analysis by using Receiver Independent Exchange Format (RINEX) files. First, using the RTKLIB, the log files for both the base and rover stations were opened and examined for the GNSS signal quality. The signals were checked for each satellite, and the only satellites that seemed to receive continuous L1 frequency signals during the recording were included for further processing, and the other satellites were omitted (Figure 5). Each observation of the GCP targets was postprocessed with the "static" option in the positioning mode, "combined" in the filter type, "Fixed and Held" for the integer ambiguity resolution and 15 degrees for the elevation mask, and 35 dB for SNR (signal-to-noise ratio) filtering were used. After the coordinate of each GCP was computed, the coordinate logs were averaged from the fixed solution (Q1) and the ratio factor of ambiguity validation ≧500. However, if there were no logs for Q1 then a float solution (Q2) was used.

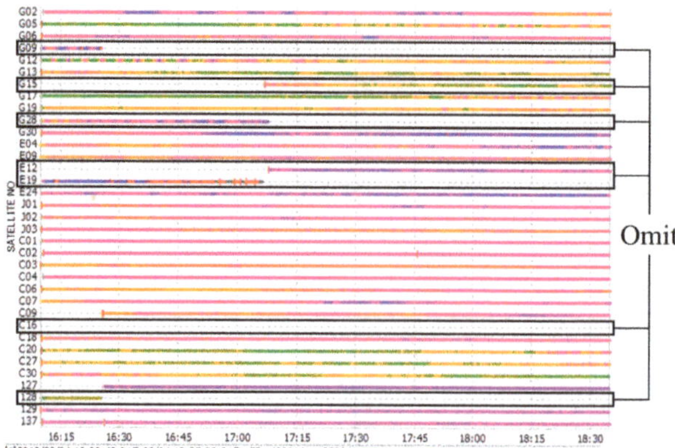

Figure 5. Example of the GNSS signal status from each satellite and the selection of the omitted satellites.

3.2. UAS Flights Data Collection

3.2.1. Fixed-Wing RGB Acquisition and Processing

On 17 October, 2018, a flight campaign was conducted at one of the compartment areas in the plantation site. A Firefly6 Pro (BirdsEyeView Aerobotics, New Hampshire, US) fixed-wing VTOL (vertical takeoff and landing) UAS was utilized to collect the aerial photography of the area. For the photo shoot, the SONY α6000 camera was embedded in the gimbal attached beneath the fixed-wing UAS and collected throughout the flight path. The flight path of the UAS was set to a south-north direction with the camera time-lapse set to two seconds, corresponding to a forward (side) overlap of 75% (70%), and flying at 140 m at altitude with an approximate 3 cm ground sampling distance (GSD). The flight course did not consider the wind direction. The collected aerial images were further processed with the software Photoscan Version 1.4.5 (Agisoft, St. Petersburg, Russia) for 3D modeling and mosaicking of the whole study area. The Photoscan parameter was considered with a "High" alignment with default tie and key points, and a dense point cloud of "High" with "Mild" depth filtering. A digital surface model (DSM) was generated using the dense-point cloud and an orthoimage was computed using the DSM. The first routine was processed without adding the GCP information. After the first orthoimage was generated, the GCPs were additionally added and the model was reconstructed by adjusting the point clouds with the GCP information. Furthermore, DSM and the orthoimage was reconstructed based on the GCP-provided point clouds. This processing method makes placing manual GCPs throughout the software easier, especially when a large quantity of images is used.

3.2.2. Multispectral Data Collection and Processing

On 17 October 2018, the flight campaign was conducted at the compartment area in the plantation site, which is the same as the corresponding area for the RGB acquisition. Right after the RGB collection flight, another flight was conducted to collect multispectral images of the area. The SlantRange 3p (SlantRange, San Diego, CA, USA) sensor was embedded in the gimbal of the VTOL UAS and images were collected throughout the flight path. The path was the same as the RGB flight. However, the forward (side) overlap was set to 40% (40%), and the flying altitude as 150 m, corresponding to an approximate 6 cm GSD. The SlantRange 3p can collect four bands, namely green, red, red-edge and near infrared. The ambient illumination sensor is placed on top of the VTOL UAS during the shooting (Figure 6) and the information is later used for calibrating each image for solar illuminations. The metadata of each image was processed with the SlantView software, which calibrates each image for solar illumination and includes coordinate information. The preprocessed data were exported and further processed via Photoscan to generate the overall view of the study area. The parameter slightly changes from RGB processing, where the alignment is "Very High", the dense point cloud is "Ultra high," while the other process stays the same. The multispectral data again performed its first routine without GCP and the second process was added with GCP calibration.

Figure 6. Firefly Pro 6 fixed-wing vertical takeoff and landing (VTOL) unmanned aerial system (UAS) carrying the multispectral sensor and the illumination sensor.

3.3. Image Classification

The study area is a small section of the compartment area of the *Acacia* plantation. However, there is a variety of LULC types in the area, as not only acacias but also water bodies, bare soils and grassy shrubs can be observed. Using the RGB image and multispectral imagery, a conventional supervised image classification was performed to develop a categorical map that shows the distribution of the general LULC types of the site. The Multilayer-Perceptron (MLP) classification [31] was implemented to classify each LULC class, which is a classifier that can handle even nonlinear trends between variables. A total of four classes were classified: *Acacia* trees, bare soil, water bodies and grass/shrubs. For the validation of the classified map, an accuracy assessment was additionally performed. Using the developed RGB orthoimage and ground information for the study area, 600 evenly distributed sampled points per class were manually collected throughout the scene, and the error matrix was computed. The LULC map was further used to compute the fraction of LULC classes within each gridded location. Other authors indicate that UAS information is highly correlated with traditional ground-based hemispherical photography, which can be used as reference for ground-truth information [34]. Therefore, the UAS-based classification result will be further used likewise as ground-truthing.

3.4. Relationship Analysis of VIs and Fraction of LULC

3.4.1. Various Vegetation Indices

Three different VIs were computed from the multispectral bands collected using SlantRange. Namely, the Normalized Difference Vegetation Index (NDVI), Green NDVI (GNDVI) and Red-edge NDVI (ReNDVI) were used, where each index can be computed using the following formula:

$$\mathrm{NDVI} = \frac{\mathrm{IR} - \mathrm{R}}{\mathrm{IR} + \mathrm{R}}, \quad (1)$$

$$\mathrm{GNDVI} = \frac{\mathrm{IR} - \mathrm{G}}{\mathrm{IR} + \mathrm{G}}, \quad (2)$$

$$\mathrm{ReNDVI} = \frac{\mathrm{IR} - \mathrm{Re}}{\mathrm{IR} + \mathrm{Re}}, \quad (3)$$

where IR is the infrared band, R is the red band, G is the green band and Re is the red-edge band. The indices all have similar functions. However, they use slightly different characteristics of the electromagnetic spectra. The IR and R are used often and characterize the vegetation function where the leaves' chlorophyll absorbs the red spectra and reflects infrared spectra [35]. Using this information one can sense the vegetation activity (greenness) of the area. GNDVI utilizes the green band instead of the red band. The GNDVI is considered an alternative index to NDVI, as it has a wider dynamic range and a higher sensitivity to chlorophyll [36]. ReNDVI also uses the slightly longer wavelength from the red channel instead of R or G, which corresponds to vegetation status information, especially water stress [37]. These three indices were examined with the fractions of the LULC classes to understand the sensitivity of the VIs.

3.4.2. Various Grid Scaling for Relationship Analysis

Four differently scaled square gridded polygons were continuously constructed along the study area with 5, 10, 30 and 60 m grid sizes. The 5 m resolution grid corresponds to satellites such as Rapideye, while the 10 m resolution grid corresponds to Sentinel-2 and the 30 m resolution corresponds to the satellites from the LANDSAT series. The 60 m grid is an additional size used to examine the viewability when the resolution is upscaled to a coarser grid size, which is almost the maximum size that can be determined in the current study area for a relevant analysis. Within each grid, the percentage of LULC classes and the average values of each VI were extracted and examined for the relationship analysis.

3.4.3. Estimating FVC Using UAS Data and Comparison with Satellite-Based FVC

The sample dataset that was used for relationship analysis is divided into sets of training and validation data. For the 5, 10, 30 and 60 m grid data, the samples were randomly divided into 50% for training and the remaining 50% for validating the estimated UAS-derived FVC. The model evaluation is considered by computing the R^2 and the root mean square error (RMSE) and the relative RMSE for each grid scale model [38].

Furthermore, Sentinel-2 imagery was downloaded via the webpage and processed through the Sentinel Application Platform (SNAP) software ver. 6.0.5 to compute the satellite-based FVC. The imagery was observed on 14 September, 2018, just one month before the UAS observation. The biophysical processor module within the SNAP software is used for computing biophysical parameters. While the UAS-based FVC uses a simple empirical model, the SNAP FVC is considered via the neural network method [39]. The satellite-based FVC is later compared with the result from the UAS-based FVC.

4. Results

4.1. UAS Photogrammetry and GCP

4.1.1. PPK Processing for Coordinate Precision

Using the base station and rover GNSS log collected through a small GNSS unit, postprocessing of the GNSS signals was performed using the RTKLIB. A total of six points were processed. Most of the targets are showing fixed positioning of the GNSS equipment (<1 cm precision in XYZ direction), while some GCPs give slightly larger errors when using the float solution. However, even though not all observations give a clear fixed solution, the PPK method can nonetheless give much higher precision than just a single GNSS observation, even with an extremely lightweight, small unit GNSS. The geolocation (XYZ coordinates) computed from the PPK process was used for each GCP target location.

4.1.2. Orthoimagery from UAS Flights for RGB and Multispectral Data

From the flight using the VTOL UAS, totals of 567 and 1204 images were obtained for RGB and multispectral data, respectively. Figure 7 shows the same location of the mosaicked images for the RGB and multispectral data. Due to the irrelevant overlap from the multispectral sensor, there are some gaps in the southern part of the compartment site. However, the procedure succeeded overall in developing an orthorectified image. Gap areas are omitted from further analysis.

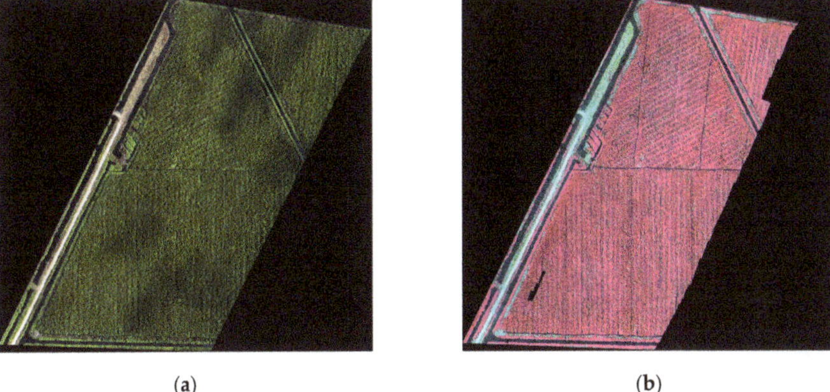

(a) (b)

Figure 7. Ortho imagery of the test site for (**a**) the true color composite (RGB) image from the digital camera and (**b**) the false color composite (RGB: NIR, Red and Green) imagery from the multispectral sensor. The resolution is reduced to 0.2 m from the original resolution for visual purposes.

4.2. LULC Map of the Study Area and Its Errors

Utilizing the RGB imagery and multispectral data, image classification was performed to develop a categorical map. A total of four classes were generated that consisted of the general landscape of the plantation site. Figure 8 shows the result of the mapping. For the validation of the map, an accuracy assessment was performed, and the error matrix is computed in Table 1. The result indicates the overall accuracy of the categorical map as 90.07%. The user accuracies (Producers accuracy) for *Acacia* forest, bare land, water bodies and grass/shrubs were 83% (96%), 96% (83%), 91% (95%) and 94% (89%), respectively.

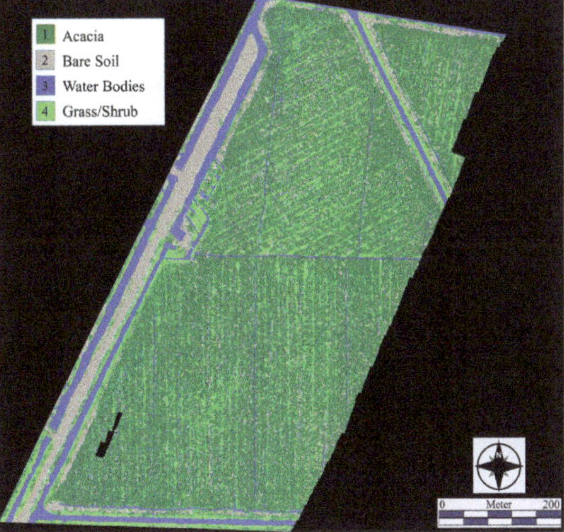

Figure 8. Land use/land cover (LULC) map of the compartment area, developed using RGB and multispectral data observed from the fixed-wing UAS using the multilayer-perceptron (MLP) method.

Table 1. Accuracy assessment and the error matrix of the generated LULC map. Error C is the error of commission, and Error O is the error of omission. The overall accuracy of the categorical map results is 90.7%.

		Classified					
		Acacia	**Bare Soil**	**Water Body**	**Grass/Shrub**	**Total**	**Error C**
Reference	*Acacia*	575	38	9	67	689	0.17
	Bare Soil	0	500	20	1	521	0.04
	Water Body	0	57	570	0	627	0.09
	Grass/Shrub	25	5	1	532	563	0.06
	Total	600	600	600	600	2400	
	Error O	0.04	0.17	0.05	0.11		0.093

4.3. Sensitivity of VI to the Fraction of Vegetation Cover (FVC)

Figure 9 shows the relationship analysis between the VIs and fractions of LULC covers for different scaling resolutions. Each VI is examined for three different LULC types: *Acacia*, grass/shrubs and non-vegetation (i.e., bare soil + water bodies). Each degree of correlation can be seen within the figure. In general, for the *Acacia* class, a clear exponential relationship occurs between the FVC and VIs. At the 5 and 10 m grid scales, ReNDVI shows the highest correlation, with GNDVI and NDVI showing lower R2 values. With the resolution becoming coarser, the sensitivity of the VIs to FVC shows a lesser error among all VIs, and for the 60 m grid, all VIs show a strong correlation ($R^2 > 0.9$). The grass/shrub class, which is classified as another vegetation type, does not show a clearer trend than the *Acacia* trees. The trend line for the grass/shrub type is shown with a second order polynomial for reference. The polynomial appears to not show many relevant relationships, but it does seem to show a combined trend of the area that has different land conditions. One example is a segment that shows a positive correlation between FVC and VIs, which is shown at areas such as on the west side of the study site, along with the track of the bare soils and water body that crosses from north to south direction. This area shows a simple positive correlation between vegetation and the VIs. Another segment is found within the compartment of the plantation area that shows a negative correlation between the FVC and VIs. In such areas, larger fractions of acacias are found when the grass/shrub class reduces, resulting in higher VIs. Other than the vegetative classes, non-vegetative areas were also examined for the sensitivity of VIs. Interestingly, at all grid scales, a clear negative correlation is observed between the fraction of non-vegetation cover (denoted hereafter as FnVC) and VIs. Similar to the *Acacia* class, the errors obviously reduce with coarser grids. However, instead of ReNDVI being more sensitive for acacias, the NDVI shows a higher correlation throughout different VIs and at different grid scales for FnVC.

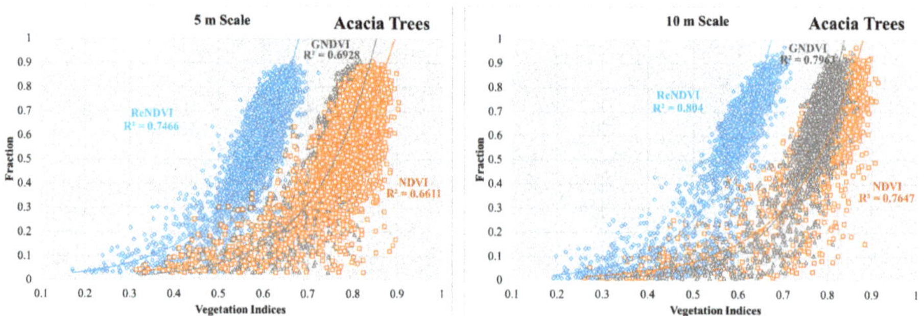

Figure 9. *Cont.*

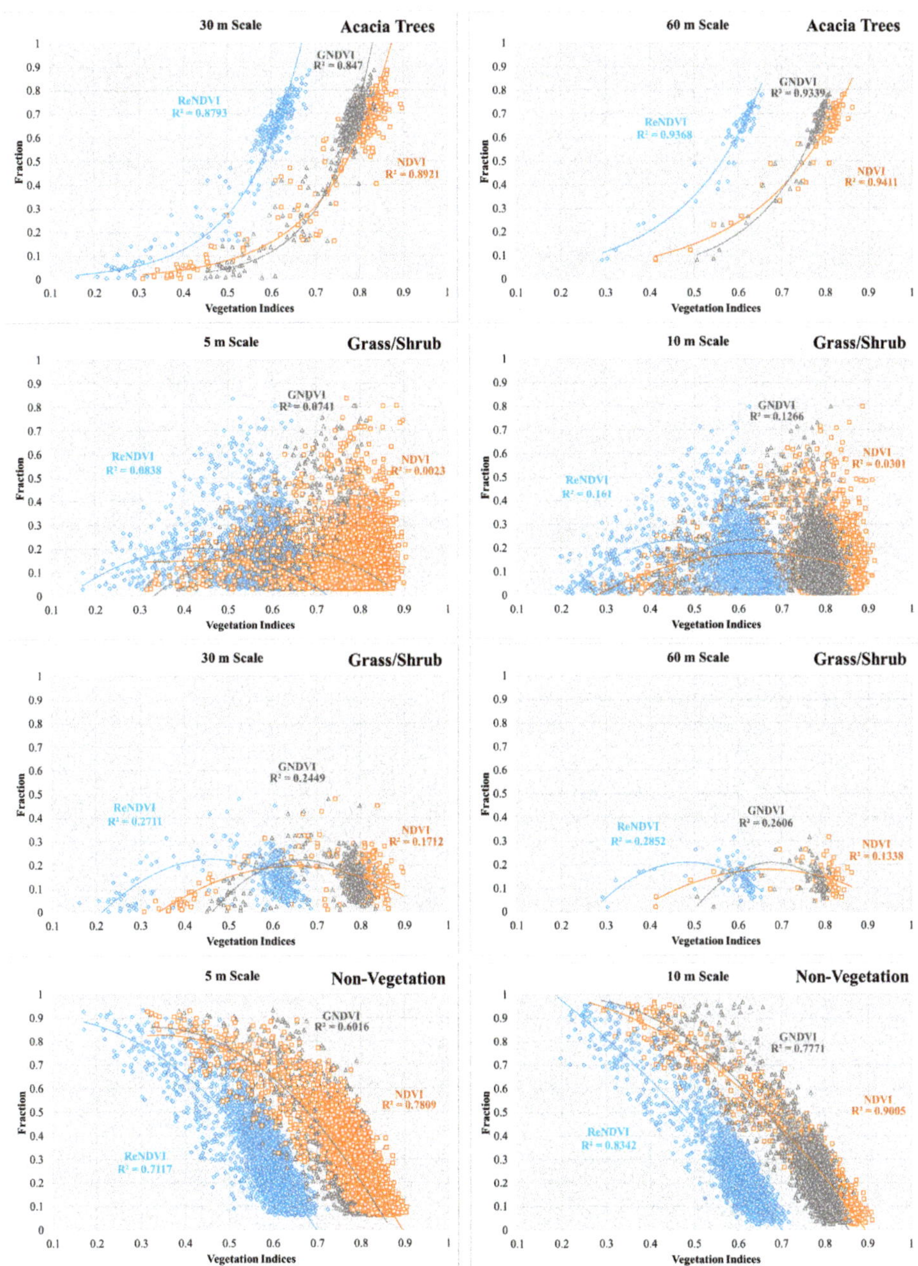

Figure 9. *Cont.*

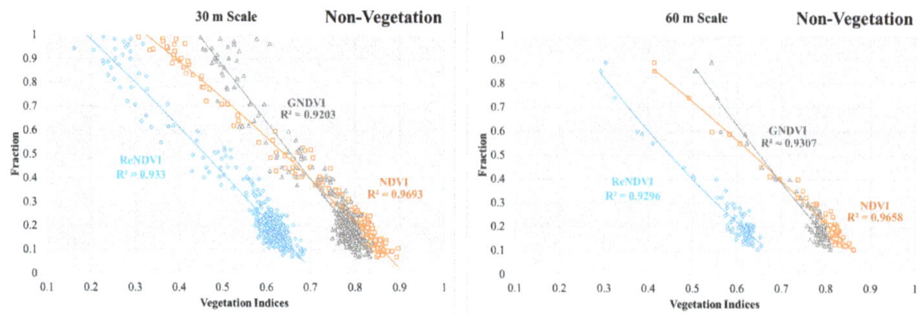

Figure 9. Relationship analysis between the different vegetation indices and fraction of LULC types: *Acacia*, grass/shrub and non-vegetation (bare soil + water) at different resolution scales.

4.4. FVC Estimation and Satellite-Derived FVC vs. UAS-Derived FVC

Figure 10 shows both the estimated FVC from the model training and the evaluation of the model for each grid scale together with the comparison of satellite-derived to UAS-derived FVC at the 10 m scale. The RMSE values of the model for 5, 10, 30 and 60 m were 0.107, 0.112, 0.085 and 0.05, respectively. The visual interpretation shows that the higher resolution image gives more detail of the area. However, the 30 m resolution data can characterize the overall trend of the site, as the bare soils on the west side can be seen clearly, and the difference between the north compartment with the lower FVC and the southern compartment with the higher FVC is acknowledged. The 60 m resolution can also characterize the area to some degree, however within the spatial extent of the study area, the information is much more aggregated compared to the other, finer resolutions.

The FVC computed from Sentinel-2 is shown in Figure 10e. The evaluation of the Sentinel-2 FVC is compared with the UAS FVC (Figure 10b). From the visual interpretation, the Sentinel-2 FVC shows a more smoothened result than the sharper UAS FVC. Since the Sentinel-2 FVC uses the neural network method for all different bands, the resampled information of the bands appears in the image (RGB is 10 m, however IR and SWIR are 20–30 m and the aerosol/water vapor bands are 60 m). The model validation shows overestimations for lower FVCs and underestimations for higher FVCs for the Sentinel-2 data. Two potential explanations can be made for the FVC difference. One is the error from the neural network approach, and the other is the temporal difference between the satellite and UAS. This subject will be discussed in the next section.

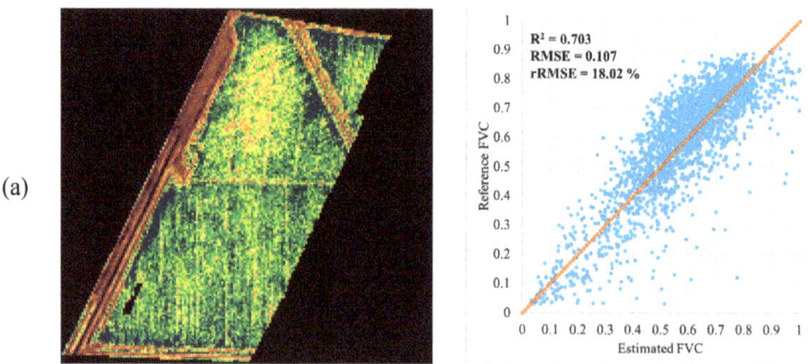

Figure 10. *Cont.*

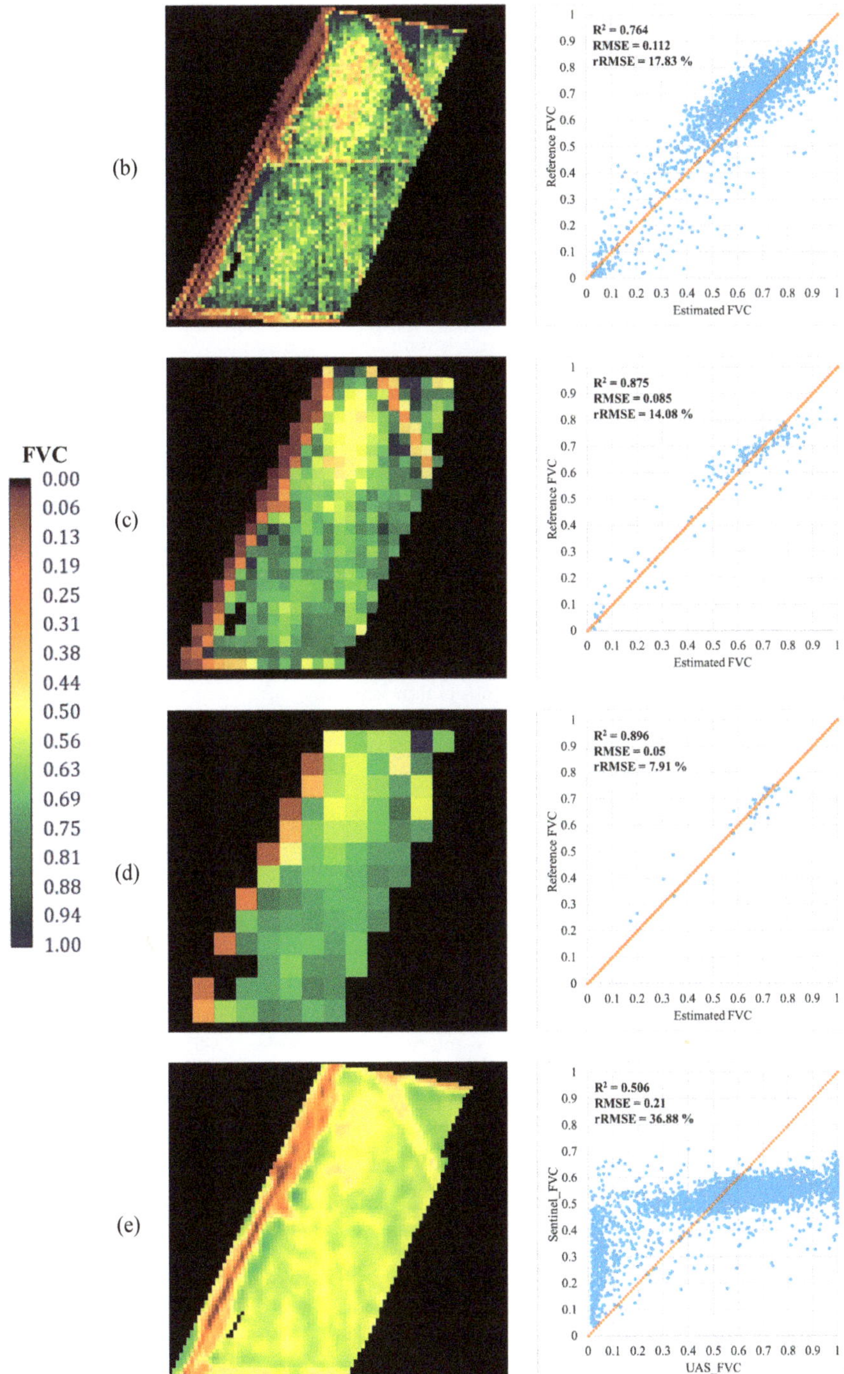

Figure 10. Estimated FVC from the UAS-based trained model for (**a**) 5 m (**b**) 10 m (**c**) 30 m and (**d**) 60 m. (**e**) The estimated FVC from Sentinel-2 data using the neural network method. The validation of the sentinel-based FVC is compared with (**b**).

5. Discussion

We will first discuss the trends of FVC to the sensitivity of VIs and the resolution. The resolution (grid scale) shows a straightforward trend, where a coarser resolution shows a stronger correlation. This also coincides with the findings of Riihimaki et al. (2019), which explains that when the finer data is aggregated to a larger unit, the variation in the data decreases, resulting in a higher correlation [29,40]. Starting from the 5 m up to the 60 m resolution, the R^2 improves for the explanatory power and the estimated FVC shows a much lower RMSE for coarser resolutions. This indicates that utilizing UAS data can improve the efficiency and quality of collected ground-truth information for validating coarser imagery FVCs. This approach may outperform expectations by using high-resolution Google Earth scenes [30] or direct observations in the field [28]; this dynamic remains in issues such as the temporal difference of scenes, coarse resolution images that make delineating fractional covers difficult and sampling scale/registration differences between field plots and imagery [29,30]. Since these issues can be controlled for UAS data, new possibilities in bridging to larger-scale earth observations may occur [41].

The sensitivity of VIs to FVC and FnVC was also clear. For the *Acacia* trees, an exponential increase of FVC occurred with increasing VIs, while for the FnVC, the decreasing trend of FVC with increasing VIs was seen throughout different grid scales. The grass/shrub class showed almost no trends when an overall trend is shown for the whole study area. However, their characteristics can be seen at two different segments, where the pure grass/shrub cover is dominant or if it is mixed within the *Acacia* plantation area. Many authors have indicated the issues caused by the mixing reflectance from vegetation or the background response to the spectral variability that is observed within the grid [42–44]. The background soil reflectance brings errors in observing correct vegetation signatures [42,44], and the mixture of vegetation types can lead to errors due to the spectral difference, especially for woody vegetation and green grass [43]. In our study, *Acacia* and grass/shrubs were mixed within the compartment area. When the FVC as separated clearly by species, it gave a strong correlation for woody vegetation between the VIs taken from UAS. As expressed by Xiao and Moody [43], the mixture of the vegetation types will result in uncertainties in the estimation. When both the *Acacia* and grass/shrub classes are aggregated, the R^2 of the exponential model was reduced from a minimum of 0.008 to a maximum of 0.086 depending on the grid scales (for ReNDVI). On the other hand, the correlation between the aggregated FVC (*Acacia* + Grass/Shrub) and VIs was the strongest for NDVI compared to other the VIs for all grid scales. Estimating single species fractions via satellite data also seems to show stronger correlations when using ReNDVI. However, when considering all vegetation types, NDVI is preferred, although the overall correlation would decrease if all vegetation types were considered. However, depending on different biomes this dynamic may also show trend changes [17]. Therefore, the relationship found here may be suitable mainly for IFP sites within Indonesia for the *Acacia* species. Notably, estimating the FnVC seems to also be possible, as it has a superior explanatory power compared to directly estimating the vegetation cover. The prospects of estimating the vegetation cover by inversely estimating FnVC look probable, although a large-scale analysis would need to be made for further confirmation.

The comparison between the satellite FVC and UAS FVC gave interesting results. The satellite FVC was based on the neural network approach, which generally tends to show a high accuracy [21,45]. However, from our observation, the lower boundary of the FVC was overestimated and the higher FVC was underestimated. The neural network approach needs to be treated with care during the training phase, as it can produce poor results if poor information is used during the training [46]. Therefore, the neural network approach might also show strong results in other areas, but this was not the case here. Another potential issue is the temporal difference between the satellite data and UAS. The observation period was valid for similar months. However, due to clouds, the data were unusable, which forced us to use only September's data. Although it is only a 1-month difference, the fast-growing acacias may show differences in the FVC result, hence, the validation between the UAS-FVC and satellite FVC could have had shown such a trend. This is the conventional issue of satellite data, where clouds and haze limit data collection on preferable observation dates. We would like to emphasize that this result does not conclude that the FVC estimation via regression is better than the neural network, but the conventional platforms could limit the desired information of the environment at various scenes. The UAS data can overcome this issue and may collect data at any occasion, time and resolution [31,47]. Thus, the UAS FVC may become a potential method for future FVC developments. The potential obstacles posed to future studies would be aviation law and engineering issues [47]. For example, the available flight time per UASs would limit the spatial extent, and the aviation law that limits the flight altitude would also limit the spatial extent. Even collecting land information at a 5 m resolution can provide sufficient green cover information, but expanding the observation area with a limited flight time would be constrained by the flying altitude, as it is mainly regulated by the law. Further investigation will be performed for more precise comparison between the satellite and UAS data in the future by coinciding both data sets.

6. Conclusions

This work demonstrated utilizing UAS to observe the RGB and multispectral data for analyzing the LULC of the plantation area, and further examined the relationship between the fraction of each LULC class with comparison to different VIs at multiple spatial resolutions. UAS observation was successful for RGB and multispectral data. An LULC map of the area was developed using the orthorectified RGB and multispectral data, and the fraction of each LULC type was extracted within different resolution grids. A comparison of the fractional vegetation cover (FVC) to NDVI, GNDVI and ReNDVI was performed. The result indicates the possibilities of computing the vegetation cover of terrestrial environments through UAS-derived data at IFP in Indonesia. The differences between the FVC and the grid scale and VIs were clear. ReNDVI showed a stronger correlation at a finer resolution for *Acacia* classes and NDVI was superior for estimating non-vegetation classes. Methods to delineate or monitor the land environment through UAS tended to show promising results and the possibility exists of expanding their potential applications to various fields and approaches. The emerging trend of UAS remote sensing is increasing globally. While airborne and spaceborne sensing are still developing, we hope that the new technology of UAS can integrate with the conventional sensing technology for new findings and developments in the near future. We believe our methods can also be useful for related restoration programs that are currently in progress in Indonesia. We will continue with the approach next to generalize the UAS result and modify the relationships using, for example, radiative transfer models.

Author Contributions: K.I. took the lead in designing, processing the research and writing the manuscript; T.K. arranged for the field surveys, research grants and revising the paper; S.S. and A.Y.S. contributed in field surveying, data collection, data process and revising/editing the manuscript; O.K. contributed in field surveys, research grants and revising of the paper; and all authors have substantially contributed to the work.

Funding: This research was funded by the Research Institute for Humanity and Nature (RIHN: a constituent member of NIHU) Project No.14200117 and by IDH The Sustainable Trade Initiative.

Acknowledgments: We thank PT. Wana Subur Lestari (WSL) and PT. Mayangkara Tanaman Industri (MTI) for all the support during the field work.

Conflicts of Interest: The authors declare no conflict of interest.

References

1. Yao, J.; He, X.Y.; Li, X.Y.; Chen, W.; Tao, D.L. Monitoring responses of forest to climate variations by MODIS NDVI: A case study of Hun River upstream, northeastern China. *Eur. J. Res.* **2012**, *131*, 705–716. [CrossRef]
2. Iizuka, K.; Tateishi, R. Estimation of CO_2 Sequestration by the Forests in Japan by Discriminating Precise Tree Age Category using Remote Sensing Techniques. *Remote Sens.* **2015**, *7*, 15082–15113. [CrossRef]
3. Masocha, M.; Murwira, A.; Magadza, C.H.D.; Hirji, R.; Dube, T. Remote sensing of surface water quality in relation to catchment condition in Zimbabwe. *Phys. Chem. Earth Parts A/B/C* **2017**, *100*, 13–18. [CrossRef]

4. Liu, Y.; Xue, Y.; MacDonald, G.; Cox, P.; Zhang, Z. Global vegetation variability and its response to elevated CO_2, global warming, and climate variability—a study using the offline SSiB4/TRIFFID model and satellite data. *Earth Syst. Dyn.* **2019**, *10*, 9–29. [CrossRef]
5. Gomes, L.; Simões, S.J.C.; Dalla Nora, E.L.; de Sousa-Neto, E.R.; Forti, M.C.; Ometto, J.P.H.B. Agricultural Expansion in the Brazilian Cerrado: Increased Soil and Nutrient Losses and Decreased Agricultural Productivity. *Land* **2019**, *8*, 12. [CrossRef]
6. Taubenböck, H.; Esch, T.; Felbier, A.; Wiesner, M.; Roth, A.; Dech, S. Monitoring urbanization in mega cities from space. *Remote Sens. Environ.* **2012**, *117*, 162–176. [CrossRef]
7. Satterthwaite, D.; McGranahan, G.; Tacoli, C. Urbanization and its implications for food and farming. *Philos. Trans. R. Soc. B.* **2010**, *365*, 2809–2820. [CrossRef] [PubMed]
8. Iizuka, K.; Johnson, B.A.; Onishi, A.; Magcale-Macandog, D.B.; Endo, I.; Bragais, M. Modeling Future Urban Sprawl and Landscape Change in the Laguna de Bay Area, Philippines. *Land* **2017**, *6*, 26. [CrossRef]
9. Climate Focus. Forests and Land Use in the Paris Agreement. The Paris Agreement Summary. Available online: http://www.climatefocus.com/publications/cop21-paris-2015-climate-focus-overall-summary-and-client-briefs (accessed on 22 December 2015).
10. Di Lallo, G.; Mundhenk, P.; Zamora López, S.E.; Marchetti, M.; Köhl, M. REDD+: Quick Assessment of Deforestation Risk Based on Available Data. *Forests* **2017**, *8*, 29. [CrossRef]
11. Ministry of Environment and Forestry (MoEF). *Decree of the Minister of Environment and Forestry of the Republic of Indonesia Number: No.129/MenLHK/Setjen/PKL.0/2/2017 Concerning Determination of Maps of National Peat Hydrological Units*; Ministry of Environment and Forestry: Jakarta, Indonesia, 2017.
12. Hooijer, A.; Silvius, M.; Wösten, H.; Page, S. *PEAT-CO_2: Assessment of CO_2 Emission from Drained Peatlands in SE Asia*; Delft Delft Hydraulics report Q3943, prepared in cooperation with Wetlands International and Alterra; Delf Hydraulics: Delft, The Netherland, 2006.
13. Department of Forestry. *Indonesian Forestry Statistics*; Department of Forestry of The Republic of Indonesia: Jakarta, Indonesia, 2007.
14. Haneda, N.F.; Retmadhona, I.Y.; Nandika, D.; Arinana, A. Biodiversity of subterranean termites on the Acacia crassicarpa plantation. *Biodivers. J. Biol. Divers.* **2017**, *18*, 1657–1662.
15. Atwood, E.C.; Englhart, S.; Lorenz, E.; Halle, W.; Wiedemann, W.; Siegert, F. Detection and Characterization of Low Temperature Peat Fires during the 2015 Fire Catastrophe in Indonesia Using a New High-Sensitivity Fire Monitoring Satellite Sensor (FireBird). *PLoS ONE* **2016**, *11*, e0159410. [CrossRef] [PubMed]
16. DiMiceli, C.M.; Carroll, M.L.; Sohlberg, R.A.; Huang, C.; Hansen, M.C.; Townshend, J.R.G. *Annual Global Automated MODIS Vegetation Continuous Fields (MOD44B) at 250 m Spatial Resolution for Data Years Beginning Day 65, 2000–2010*; Collection 5 Percent Tree Cover; University of Maryland: College Park, MD, USA, 2011.
17. Filipponi, F.; Valentini, E.; Nguyen Xuan, A.; Guerra, C.A.; Wolf, F.; Andrzejak, M.; Taramelli, A. Global MODIS Fraction of Green Vegetation Cover for Monitoring Abrupt and Gradual Vegetation Changes. *Remote Sens.* **2018**, *10*, 653. [CrossRef]
18. Dymond, J.R.; Stephens, P.R.; Newsome, P.F.; Wilde, R.H. Percentage vegetation cover of a degrading rangeland from SPOT. *Int. J. Remote Sens.* **1992**, *13*, 1999–2007. [CrossRef]
19. Kobayashi, T.; Tsend-Ayush, J.; Tateishi, R. A new global tree-cover percentage map using MODIS data. *Int. J. Remote Sens.* **2016**, *37*, 969–992. [CrossRef]
20. Hansen, M.C.; Defries, R.S.; Townshend, J.R.G.; Sohlberg, R.A.; Dimiceli, C.; Carroll, M.L. Towards an operational MODIS continuous field of percent tree cover algorithm: Examples using AVHRR and MODIS data. *Remote Sens. Environ.* **2002**, *83*, 303–319. [CrossRef]
21. Yang, L.; Jia, K.; Liang, S.; Liu, J.; Wang, X. Comparison of Four Machine Learning Methods for Generating the GLASS Fractional Vegetation Cover Product from MODIS Data. *Remote Sens.* **2016**, *8*, 682. [CrossRef]
22. Asner, G.P. Cloud cover in Landsat observations of the Brazilian Amazon. *Int. J. Remote Sens.* **2001**, *22*, 3855–3862. [CrossRef]
23. Turner, D.; Lucieer, A.; Watson, C. An Automated Technique for Generating Georectified Mosaics from Ultra-High Resolution Unmanned Aerial Vehicle (UAV) Imagery, Based on Structure from Motion (SfM) Point Clouds. *Remote Sens.* **2012**, *4*, 1392–1410. [CrossRef]
24. Guo, Q.; Wu, F.; Hu, T.; Chen, L.; Liu, J.; Zhao, X.; Gao, S.; Pang, S. Perspectives and prospects of unmanned aerial vehicle in remote sensing monitoring of biodiversity. *Biodivers. Sci.* **2016**, *24*, 1267–1278. [CrossRef]

25. Iizuka, K.; Yonehara, T.; Itoh, M.; Kosugi, Y. Estimating Tree Height and Diameter at Breast Height (DBH) from Digital Surface Models and Orthophotos Obtained with an Unmanned Aerial System for a Japanese Cypress (Chamaecyparis obtusa) Forest. *Remote Sens.* **2018**, *10*, 13. [CrossRef]
26. Csillik, O.; Cherbini, J.; Johnson, R.; Lyons, A.; Kelly, M. Identification of Citrus Trees from Unmanned Aerial Vehicle Imagery Using Convolutional Neural Networks. *Drones* **2018**, *2*, 39. [CrossRef]
27. Fang, S.; Tang, W.; Peng, Y.; Gong, Y.; Dai, C.; Chai, R.; Liu, K. Remote Estimation of Vegetation Fraction and Flower Fraction in Oilseed Rape with Unmanned Aerial Vehicle Data. *Remote Sens.* **2016**, *8*, 416. [CrossRef]
28. Chen, J.; Yi, S.; Qin, Y.; Wang, X. Improving estimates of fractional vegetation cover based on UAV in alpine grassland on the Qinghai–Tibetan Plateau. *Int. J. Remote Sens.* **2016**, *37*, 1922–1936. [CrossRef]
29. Riihimäkia, H.; Luotoa, M.; Heiskanena, J. Estimating fractional cover of tundra vegetation at multiple scales using unmanned aerial systems and optical satellite data. *Remote Sens. Environ.* **2019**, *224*, 119–132. [CrossRef]
30. Kobayashi, S.; Sanga-Ngoie, K. A comparative study of radiometric correction methods for optical remote sensing imagery: The IRC vs. other image-based C-correction methods. *Int. J. Remote Sens.* **2009**, *30*, 285–314. [CrossRef]
31. Iizuka, K.; Watanabe, K.; Kato, T.; Putri, N.A.; Silsigia, S.; Kameoka, T.; Kozan, O. Visualizing the Spatiotemporal Trends of Thermal Characteristics in a Peatland Plantation Forest in Indonesia: Pilot Test Using Unmanned Aerial Systems (UASs). *Remote Sens.* **2018**, *10*, 1345. [CrossRef]
32. Yoshimura, T.; Hasegawa, H. Comparing the precision and accuracy of GPS positioning in forested areas. *J. For. Res.* **2003**, *8*, 147–152. [CrossRef]
33. Takasu, T.; Yasuda, A. Development of the low-cost RTK-GPS receiver with an open source program package RTKLIB. In Proceedings of the International symposium on GPS/GNSS, Seogwipo-si Jungmun-dong, Korea, 4–6 November 2009.
34. Khokthon, W.; Zemp, D.C.; Irawan, B.; Sundawati, L.; Kreft, H.; Hölscher, D. Drone-Based Assessment of Canopy Cover for Analyzing Tree Mortality in an Oil Palm Agroforest. *Front. For. Glob. Chang.* **2019**, *2*, 12. [CrossRef]
35. Tucker, G.J. Red and photographic infrared linear combinations for monitoring vegetation. *Remote Sens. Environ.* **1979**, *8*, 127–150. [CrossRef]
36. Gitelson, A.A.; Kaufman, Y.J.; Merzlyak, M.N. Use of green channel in remote sensing of global vegetation from EOS-MODIS. *Remote Sens. Environ.* **1996**, *58*, 289–298. [CrossRef]
37. Eitel, J.U.H.; Vierling, L.A.; Litvak, M.E.; Long, D.S.; Schulthess, U.; Ager, A.A.; Krofcheck, D.J.; Stoscheck, L. Broadband, red-edge information from satellites improves early stress detection in a New Mexico conifer woodland. *Remote Sens. Environ.* **2011**, *115*, 3640–3646. [CrossRef]
38. Shen, X.; Cao, L.; Yang, B.; Xu, Z.; Wang, G. Estimation of Forest Structural Attributes Using Spectral Indices and Point Clouds from UAS-Based Multispectral and RGB Imageries. *Remote Sens.* **2019**, *11*, 800. [CrossRef]
39. Weiss, M.; Baret, F. S2ToolBox Level 2 Products: LAI, FAPAR, FCOVER. Available online: https://step.esa.int/docs/extra/ATBD_S2ToolBox_L2B_V1.1.pdf (accessed on 2 May 2016).
40. Dark, S.J.; Bram, D. The modifiable areal unit problem (MAUP) in physical geography. *Prog. Phys. Geogr.* **2007**, *31*, 471–479. [CrossRef]
41. Walker, D.A.; Daniëls, F.J.A.; Alsos, I.; Bhatt, U.S.; Breen, A.L.; Buchhorn, M.; Bültmann, H.; Druckenmiller, L.A.; Edwards, M.E.; Ehrich, D.; et al. Circumpolar Arctic vegetation: A hierarchic review and roadmap toward an internationally consistent approach to survey, archive and classify tundra plot data. *Environ. Res. Lett.* **2016**, *11*, 055005. [CrossRef]
42. Borel, C.C.; Gerstl, S.A.W. Nonlinear spectral mixing models for vegetative and soil surfaces. *Remote Sens. Environ.* **1994**, *47*, 403–416. [CrossRef]
43. Xiao, J.; Moody, A. A comparison of methods for estimating fractional green vegetation cover within a desert-to-upland transition zone in central New Mexico, USA. *Remote Sens. Environ.* **2005**, *98*, 237–250. [CrossRef]
44. Zhang, X.; Liao, C.; Li, J.; Sun, Q. Fractional vegetation cover estimation in arid and semi-arid environments using HJ-1 satellite hyperspectral data. *Int. J. Appl. Earth Obs. Geoinf.* **2013**, *21*, 506–512. [CrossRef]
45. Jia, K.; Liang, S.; Gu, X.; Baret, F.; Wei, X.; Wang, X.; Yao, Y.; Yang, L.; Li, Y. Fractional vegetation cover estimation algorithm for Chinese GF-1 wide field view data. *Remote Sens. Environ.* **2016**, *177*, 184–191. [CrossRef]

46. Verrelst, J.; Muñoz, J.; Alonso, L.; Delegido, J.; Rivera, J.P.; Camps-Valls, G.; Moreno, J. Machine learning regression algorithms for biophysical parameter retrieval: Opportunities for Sentinel-2 and-3. *Remote Sens. Environ.* **2012**, *118*, 127–139. [CrossRef]
47. Hogan, S.D.; Kelly, M.; Stark, B.; Chen, Y. Unmanned aerial systems for agriculture and natural resources. *Calif. Agric.* **2017**, *71*, 5–14. [CrossRef]

MDPI
St. Alban-Anlage 66
4052 Basel
Switzerland
Tel. +41 61 683 77 34
Fax +41 61 302 89 18
www.mdpi.com

Remote Sensing Editorial Office
E-mail: remotesensing@mdpi.com
www.mdpi.com/journal/remotesensing

www.ingramcontent.com/pod-product-compliance
Lightning Source LLC
LaVergne TN
LVHW071622170726
843515LV00009B/2459

* 9 7 8 3 0 3 6 5 2 1 9 2 3 *